Chemical Biology of Natural Products

Chemical Biology of Natural Products

Edited by

David J. Newman, Gordon M. Cragg,
and Paul G. Grothaus

CRC Press
Taylor & Francis Group
Boca Raton London New York

CRC Press is an imprint of the
Taylor & Francis Group, an **informa** business

CRC Press
Taylor & Francis Group
6000 Broken Sound Parkway NW, Suite 300
Boca Raton, FL 33487-2742

Printed on acid-free paper

International Standard Book Number-13: 978-1-4398-4193-8 (Hardback)

Library of Congress Cataloging-in-Publication Data

Names: Newman, David J., 1939- | Cragg, Gordon M. L. | Grothaus, Paul.
Title: Chemical biology of natural products/[edited by] David J. Newman, Gordon M. Cragg, and Paul Grothaus.
Description: Boca Raton : CRC Press, [2017] | Includes bibliographical references and index.
Identifiers: LCCN 2017011281 | ISBN 9781439841938 (hardback : alk. paper) | ISBN 9781315117089 (ebook)
Subjects: LCSH: Pharmacognosy. | Natural products.
Classification: LCC RS160 .C5227 2017 | DDC 615.3/21--dc23
LC record available at https://lccn.loc.gov/2017011281

Visit the Taylor & Francis Web site at
http://www.taylorandfrancis.com

and the CRC Press Web site at
http://www.crcpress.com

Contents

Preface

Chemical biology, biochemistry, and biological chemistry? So what, if anything, are the similarities and differences? One easy definition of the first and last terms would be that *Chemical biology = The biology of chemicals*, whereas *Biological chemistry = The chemistry of biology*, and *Biochemistry is the study of the chemistry of living systems*.

This definition dilemma is further illustrated by the fact that there are eminent universities across the United States where there are chemical biology programs in the College of Chemistry (Berkeley, for example*), but also biochemistry in the same university but housed in the College of Letters and Science†; and at the University of Pennsylvania, the Chemistry Department has a Biological Chemistry Resource Center‡ and a chemical biology postgraduate program,§ with a biochemistry program included within the Medical School.¶ We should add that one of the editors practiced as a *biological chemist* in the U.S. pharmaceutical industry 45 plus years ago, studying the effect of small synthetic molecules on oxygenation of hemoglobin. Today, he might well have been practicing chemical biology!

The Broad Institute, based in Cambridge, Massachusetts,** defines chemical biology as "the science of small molecules in the context of living systems to discover and to elucidate molecular pathways fundamental in cellular, developmental and disease biology."††

Furthermore, a survey based on *leading journals at the interface between chemistry and biology*, conducted by the American Chemical Society (ACS), of 4000 scientists working at the interface of chemistry and biology, indicated that natural products did not feature in the top eight disciplines (bioorganic chemistry, medicinal chemistry, molecular biology, enzymology, biophysics, biotechnology, cell biology, and structural biology) selected as being linked to chemical biology.‡‡ Interestingly, when natural products was used as the *lead* term, the survey selections indicated that they were linked in decreasing order to medicinal chemistry, bioorganic chemistry, chemical biology, plant science, and pharmacology, but not to microbiology.

Attempting to separate the three *disciplines* mentioned in the first paragraph is probably an exercise in frustration. However, what intrigues us about the Broad

---

* http://chemistry.berkeley.edu/ugrad/degrees/chembio.

† http://mcb.berkeley.edu/undergrad.

‡ https://www.chem.upenn.edu/content/penn-chemistry-biological-chemistry-resource-center.

§ https://www.chem.upenn.edu/content/graduate.

¶ http://www.med.upenn.edu/biocbiop/.

** https://www.broadinstitute.org/chembio-therapeutics.

†† http://www.broadinstitute.org/scientific-community/science/programs/csoft/chemical-biology/chemical-biology-program.

‡‡ http://pubs.acs.org/bio/.

Institute definition and the results of the ACS survey discussed above is the apparently widespread lack of appreciation of the role played by natural products in the area of chemical biology. Even more puzzling is the apparent failure to link natural products to microbiology. We therefore decided that a volume highlighting the role of natural products in *chemical biology* would be an enlightening undertaking and that such a volume should include examples of all three methods of interrogating Mother Nature in individual chapters.

The 15 chapters in this book range over the gamut of the definitions alluded to above and serve to emphasize the dominant role played by microbes in the production of bioactive metabolites. On the chemical biology front, they include the chemical biology of cyanobacteria, combinatorial biosynthesis, including synthetic biology, target identification from natural product inspired structures, and syntheses devised around active natural product structures. Moving to secondary metabolites that may be used in the future to *probe biological systems* or are themselves the products of complex interactions, there are discussions covering materials from insect–microbe symbioses, compounds from plant–endophytic microbes and rhizosphere interactions, and the coculture of microbes to induce production of fungal metabolites. Also covered are secondary metabolites from extremophilic sources, including toxic lakes and deep-sea sediments and vent organisms, and a chapter covering genomic mining of microbes to find novel bioactive natural products. Finally, moving closer to biological chemistry and/or biochemistry, there are significant discussions of neurotoxins from venomous *Conus* species and the somewhat similar active cyclic sulfide-bridged peptides from plants and animals.

There are many other very interesting topics at the interface of chemistry and biology, in particular if one looks at the burgeoning reports related to the actual sources of secondary metabolites in marine-related organisms, and perhaps in some cases, in plants as well. What has now become quite evident is that the majority of bioactive natural products described from the *Porifera* (sponges), and almost certainly in other marine phyla as well, are almost certainly produced by as yet uncultured microbes, whose secrets are now being revealed by the combination of genomic analyses of single microbial cells coupled to very sophisticated physico-chemical techniques.

As an example, the story of the pederin–mycalamide–onnamide locus is one that even a few years ago would have been science fiction but is now recognized as being correct. This story, in an abbreviated form, leading from the *Paederus* beetle toxin, via a German entomologist's suggestion that a microbe was involved, through the work done by Piel and his collaborators on an as yet uncultured *Entotheonella* species, was recently reported by one of the editors in an open-access paper.*

That the investigation of biological phenomena now requires a multidisciplinary approach, where chemists and biologists need to work together to uncover Mother

* Newman, D.J., The influence of Brazilian biodiversity on searching for human-use pharmaceuticals, *J. Braz. Chem. Soc.*, 2017, in press (http://jbcs.sbq.org.br/imagebank/pdf/160478RV_Biota.pdf).

Nature's secrets, has become evident today, and it is our hope that the examples in this book will further encourage scientists, be they chemical biologists, biochemists, biological chemists, or just *plain chemists and biologists*, to work together in order to further discover novel agents and their interplay, with the potential that some may lead to new treatments for human diseases.

Our sincere thanks to all who have participated in this project!

David J. Newman
Gordon M. Cragg
Paul G. Grothaus

Editors

David J. Newman retired from the position of chief of the Natural Products Branch (NPB) in the Developmental Therapeutics Program at the National Cancer Institute (NCI) in Frederick, Maryland, in early January 2015. Born in Grays, Essex, United Kingdom, in 1939, he received an MSc in synthetic organic chemistry from the University of Liverpool in 1963. Following time as a synthetic chemist at Ilford, Ltd., he joined the Agricultural Research Council's (ARC) Unit of Nitrogen Fixation at the Universities of London and Sussex, as a research assistant in metallo-organic chemistry, transferring to the microbial biochemistry group in early 1966 as a graduate student and being awarded a DPhil in 1968 for work on microbial electron transport proteins from *Desulfovibrio*. He moved to the United States in 1968 as a postdoc in the Biochemistry Department at the University of Georgia, working on protein sequencing of *Desulfovibrio* ferredoxins, and then in 1970, he joined Smith, Kline & French (SK&F) in Philadelphia as a biological chemist. At SK&F, most of his work was related to antibiotic discovery, and in 1985, when the antibiotic group was dissolved, he left SK&F. For the next six years, he worked in marine and microbial discovery programs at Air Products, SeaPharm, and Lederle, and then in 1991, he joined the NPB as a chemist responsible for marine and microbial collection programs. He was given the National Institutes of Health (NIH) Merit Award in 2003 for this work, and following Gordon M. Cragg's retirement from the position of chief of the National Products Branch of the National Cancer Institute (NPB/NCI), at the end of 2004, he was acting chief until appointed chief in late 2006. He is the author or coauthor of over 180 papers, reviews, and book chapters (and an editor, with Gordon M. Cragg and David Kingston, of *Anticancer Agents from Natural Products*) and holds 18 patents, mainly on microbial products. He is still associated with the NPB/NCI as a special volunteer and also has a small consulting business to *occupy his spare time!*

Gordon M. Cragg obtained his undergraduate training in chemistry at Rhodes University, South Africa, and his DPhil (organic chemistry) from Oxford University. After two years of postdoctoral research at the University of California, Los Angeles, he returned to South Africa to join the Council for Scientific and Industrial Research. In 1966, he joined the Chemistry Department at the University of South Africa, and he transferred to the University of Cape Town in 1972. In 1979, he returned to the United States to join the Cancer Research Institute at Arizona State University, working with Professor G.R. Pettit. In 1985, he moved to the National Cancer Institute (NCI), National Institutes of Health (NIH), in Bethesda,

Maryland, and was appointed chief of the NCI Natural Products Branch in 1989. He retired in December 2004 and is currently serving as an NIH special volunteer. His major interests lie in the discovery of novel natural product agents for the treatment of cancer and AIDS, with an emphasis on multidisciplinary and international collaboration. He has been awarded NIH merit awards for his contributions to the development of the anticancer drug Taxol (1991), leadership in establishing international collaborative research in biodiversity and natural products drug discovery (2004), contributions to developing and teaching NIH technology transfer courses (2004), and dedicated service to the NCI as a member of the PDQ Complementary and Alternative Medicine Editorial Board (2010). In 1998–1999, he was president of the American Society of Pharmacognosy and was elected to honorary membership in 2003 and named as a fellow in 2008. In 2006, he was given the William L. Brown Award for Plant Genetic Resources by Missouri Botanical Garden, which also named a recently discovered Madagascar plant in his honor, *Ludia craggiana*. He has established collaborations between the NCI and organizations in many countries, promoting drug discovery from their natural resources. He has authored or coauthored over 180 papers, reviews, and book chapters related to these interests.

Paul G. Grothaus earned a BSChem from Creighton University in 1977 and his PhD from Purdue University in 1983, where he completed the first enantiospecific total synthesis of a trichothecene mycotoxin, anguidine. His education was followed by a postdoctoral stint at the University of Washington, investigating the synthesis of germacrolides. In 1984, he joined the Natural Products Group in the Plant Sciences Division of Monsanto Agricultural Company, where he investigated the synthesis and structure–activity relationships of agriculturally useful natural products. In 1988, he became the head of chemistry at Hawaii Biotech, Inc., in Aiea, Hawaii, where he worked on drug discovery based on both terrestrial and marine natural product leads. In 2002, he joined the Medicinal Chemistry Department of Celera Genomics, Inc., in South San Francisco, California, and rose to become an associate director of medicinal chemistry in 2005. Research at Celera focused on the development of protease and kinase inhibitors and activity-based probes for chemical proteomics studies.

At Celera, Dr. Grothaus led the chemistry group that designed and synthesized irreversible and reversible inhibitors of Bruton's tyrosine kinase (BTK), culminating in the discovery of ibrutinib (CRA-32765/PCI-32765), a first-in-class, oral, once-daily therapy that inhibits BTK, a key protein in the B-cell receptor signaling complex. Following further preclinical and clinical development by Pharmacyclics, Inc., Ibrutinib received three Oncology Breakthrough Therapy Designations by the Food and Drug Association and was approved for the treatment of chronic lymphocytic leukemia/small lymphocytic lymphoma in 2014, mantle cell lymphoma in 2013, and Waldenstrom's macroglobulinemia in 2015.

In 2007, he joined the Natural Products Branch of the National Cancer Institute in Frederick, Maryland, where he coordinates biomass collections, biological screening of extracts, and collaborations with external natural product researchers. Dr. Grothaus is the author or coauthor of 24 papers, reviews, and book chapters and holds 5 patents.

Contributors

Emily R. Abraham
School of Chemistry
University of St. Andrews
St. Andrews, United Kingdom

Karl-Heinz Altmann
Department of Chemistry and Applied Biosciences
Institute of Pharmaceutical Sciences
ETH Zürich
Zürich, Switzerland

Elena Ancheeva
Institute of Pharmaceutical Biology and Biotechnology
Heinrich-Heine University
Düsseldorf, Germany

Eric H. Andrianasolo
Department of Marine and Coastal Sciences
Center for Deep-Sea Ecology and Biotechnology
Rutgers, The State University of New Jersey
New Brunswick, New Jersey

Christopher S. Bailey
School of Chemistry
University of St. Andrews
St. Andrews, United Kingdom

Richard H. Baltz
CognoGen Biotechnology Consulting
Sarasota, Florida

Tobias Brütsch
Department of Chemistry and Applied Biosciences
Institute of Pharmaceutical Sciences
ETH Zürich
Zürich, Switzerland

Andrés Mauricio Caraballo-Rodríguez
Laboratory of Microbial Chemistry
Department of Pharmaceutical Sciences
School of Pharmaceutical Sciences of Ribeirão Preto
University of Sao Paulo
Ribeirao Preto, Brazil

David J. Craik
Institute for Molecular Bioscience
The University of Queensland
Brisbane, Queensland, Australia

Ivana Crnovčić
Department of Chemistry
The Scripps Research Institute
Jupiter, Florida

Georgios Daletos
Institute of Pharmaceutical Biology and Biotechnology
Heinrich-Heine University
Düsseldorf, Germany

Maria C.F. de Oliveira
Laboratory of Biotechnology and Organic Synthesis
Post-graduate Program in Chemistry
Federal University of Ceará
Fortaleza, Brazil

Simon J. de Veer
Institute for Molecular Bioscience
The University of Queensland
Brisbane, Queensland, Australia

Weaam Ebrahim
Institute of Pharmaceutical Biology and Biotechnology
Heinrich-Heine University
Düsseldorf, Germany

and

Faculty of Pharmacy
Department of Pharmacognosy
Mansoura University
Mansoura, Egypt

Mona El-Neketi
Faculty of Pharmacy
Department of Pharmacognosy
Mansoura University
Mansoura, Egypt

David A. Gallegos
Department of Pharmaceutical Sciences
College of Pharmacy
Oregon State University
Corvallis, Oregon

William H. Gerwick
Center for Marine Biotechnology and Biomedicine
Scripps Institution of Oceanography
University of California, San Diego
La Jolla, California

Simon Glauser
Department of Chemistry and Applied Biosciences
Institute of Pharmaceutical Sciences
ETH Zürich
Zürich, Switzerland

Rebecca J.M. Goss
School of Chemistry
University of St. Andrews
St. Andrews, United Kingdom

A.A. Leslie Gunatilaka
Natural Products Center
School of Natural Resources and the Environment
College of Agriculture and Life Sciences
The University of Arizona
Tucson, Arizona

Lena Keller
Center for Marine Biotechnology and Biomedicine
Scripps Institution of Oceanography
University of California, San Diego
La Jolla, California

Parijat Kusari
Department of Biochemical and Chemical Engineering
Technical University of Dortmund
Dortmund, Germany

Souvik Kusari
Department of Chemistry and Chemical Biology
Institute of Environmental Research
Technical University of Dortmund
Dortmund, Germany

Luca Laraia
Department of Chemical Biology
Max Planck Institute of Molecular Physiology
Dortmund, Germany

Tiago Leão
Center for Marine Biotechnology and Biomedicine
Scripps Institution of Oceanography
University of California, San Diego
La Jolla, California

Wenhan Lin
State Key Laboratory of Natural and Biomimetic Drugs
Peking University
Beijing, People's Republic of China

Richard A. Lutz
Department of Marine and Coastal Sciences
Center for Deep-Sea Ecology and Biotechnology
Rutgers, The State University of New Jersey
New Brunswick, New Jersey

Jair Mafezoli
Laboratory of Applied Phytochemistry
Post-graduate Program in Chemistry
Federal University of Ceará
Fortaleza, Brazil

Kerry L. McPhail
Department of Pharmaceutical Sciences
College of Pharmacy
Oregon State University
Corvallis, Oregon

Baldomero M. Olivera
Department of Biology
University of Utah
Salt Lake City, Utah

Humberto Enrique Ortega-Domínguez
Laboratory of Microbial Chemistry
Department of Pharmaceutical Sciences
School of Pharmaceutical Sciences of Ribeirão Preto
University of Sao Paulo
Ribeirao Preto, Brazil

Rita de Cássia Pessotti
Laboratory of Microbial Chemistry
Department of Pharmaceutical Sciences
School of Pharmaceutical Sciences of Ribeirão Preto
University of Sao Paulo
Ribeirao Preto, Brazil

Peter Proksch
Institute of Pharmaceutical Biology and Biotechnology
Heinrich-Heine University
Düsseldorf, Germany

Mônica Tallarico Pupo
Laboratory of Microbial Chemistry
Department of Pharmaceutical Sciences
School of Pharmaceutical Sciences of Ribeirão Preto
University of Sao Paulo
Ribeirao Preto, Brazil

Shrinivasan Raghuraman
Department of Biology
University of Utah
Salt Lake City, Utah

Jeffrey D. Rudolf
Department of Chemistry
The Scripps Research Institute
Jupiter, Florida

Helena Safavi-Hemami
Department of Biology
University of Utah
Salt Lake City, Utah

Ben Shen
Department of Chemistry
Department of Molecular Medicine, and Natural Products Library Initiative
The Scripps Research Institute
Jupiter, Florida

Michael Spiteller
Department of Chemistry and Chemical Biology
Institute of Environmental Research
Technical University of Dortmund
Dortmund, Germany

Andrea Stierle
Department of Biomedical and Pharmaceutical Sciences
The University of Montana–Missoula
Missoula, Montana

Don Stierle
Department of Biomedical and Pharmaceutical Sciences
The University of Montana–Missoula
Missoula, Montana

Gary Strobel
Department of Plant Sciences
Montana State University
Bozeman, Montana

Russell W. Teichert
Department of Biology
University of Utah
Salt Lake City, Utah

Herbert Waldmann
Department of Chemical Biology
Max Planck Institute of Molecular Physiology
Dortmund, Germany

1 Microbial Genome Mining for Natural Product Drug Discovery

Richard H. Baltz

CONTENTS

1.1 INTRODUCTION

1.1.1 Importance of Natural Products and the Issue of Rediscovery

Since the commercial development of penicillin and streptomycin in the 1940s and 1950s, natural products (NPs) have played important, ever-expanding roles in human medicine, animal health, and plant crop protection.[1–4] In the early years of NP discovery, it was relatively easy to discover new and novel antibiotics that were produced abundantly by many soil microorganisms, primarily actinomycetes.[3,5–7] After the most commonly produced antibiotics were discovered, it became progressively more difficult to discover less-abundant antibiotics, and the previously discovered antibiotics became a nuisance because of continuous rediscovery.[5–7] Thus, as time progressed from the 1950s through the 1980s, the problem of *dereplication* of known compounds became evermore problematic, and the discovery process based on low-throughput fermentation analysis became much less productive.[3,6,8] In addition, NP discovery did not fit well with the new paradigm of the 1990s of high-throughput screening of combichem and other chemical libraries against multiple *in vitro* targets generated by microbial and human genomics projects.[3]

1.1.2 Genome Mining Enhances Discovery and Circumvents Rediscovery

The first two complete genomic sequences of *Streptomyces coelicolor* and *Streptomyces avermitilis* revealed that each organism encoded many more secondary metabolite gene clusters (SMGCs) than were anticipated from their known secondary metabolomes.[9,10] Many of the cryptic or silent gene clusters from these and other streptomycetes have been investigated,[11–14] and the observation of multiple cryptic SMGCs has been generalized to many additional actinomycetes and to other cosmopolitan bacteria with large genomes.[15] From these observations, the concept of microbial genome mining emerged as a new paradigm for drug discovery.[3,16–20] There are a number of advantages to genome mining discussed in this chapter, not the least of which is that it solves the confounding issues of rediscovery and dereplication of known secondary metabolites (SMs).

1.1.3 Scope of the Chapter

Microbial genome mining can be directed at members of the three domains of life: *Bacteria*, *Archaea*, and *Eukarya*. Because of space limitations and interests of the author, this chapter is focused on *Bacteria* with some discussion on *Archaea*. The fungi from the domain *Eukarya* are important sources of NPs, and interested readers are directed to recent publications relevant to fungal genome mining.[21–25]

1.2 THE NEW CENTRAL DOGMA OF GENOME MINING

1.2.1 What Is Genome Mining and Why Is It Important?

When we think of chemical biology, we often think in terms of the central dogma of DNA to RNA to protein, or DNA replication, transcription, and translation leading to enzyme function and cell biology. In the past, NPs were discovered in the absence of knowledge of the extended central dogma of DNA to RNA to protein (enzymes) to NPs.[16] Microorganisms were isolated, fermentations were carried out under different conditions, fermentation broths were screened for antibacterial and other activities, active chemical agents were purified, and structures determined. Often, novel chemical scaffolds with important activities were further embellished by chemical modifications, semisynthesis, mutasynthesis, and in some cases, combinatorial biosynthesis.[3] Now with inexpensive microbial genome sequencing, and growing knowledge of SM biosynthetic enzyme functions with linkages to the responsible genes located in SMGCs,[26] it is possible to predict new and novel SMGCs and possible structures bioinformatically,[27–30] and to predict or correlate associated SMGCs from metabolomics/proteomics data.[31–41]

The following is the new central dogma for genome mining: DNA sequence predicts new and novel chemical structures (genotype to chemotype); and new and novel chemical structures predict associated DNA sequences (chemotype to genotype).[16,41] In the forward direction, DNA that predicts new or novel chemistry exerts enormous leverage for discovery because it eliminates the confounding issues of rediscovery and dereplication that have stifled the productivity for NP discovery in recent years. Although bioinformatics predictions from genome sequences do not need inputs from the RNA and protein parts of the extended central dogma (transcription and translation), bringing genome mining into a productive discovery mode requires efficient transcription and translation of cryptic SMGCs by using a variety of approaches.[42]

1.2.2 What Are the Drivers for Successful Genome Mining?

Successful Pharmaceutical or Biotechnology Company research programs start with a vision and mission; then address these by developing strategies, tactics, and research plans; and follow by executing the plans. The vision and mission for NP discovery remain the same as 50 years ago: The vision is that NPs are important sources for drug discovery and the mission is to discover and develop NPs for human medicine, animal health, and plant crop protection. What has changed dramatically in the last few years is the strategy and tactics that have engaged genomics and metabolomics approaches to discovery. What is urgently needed to ensure success is refined strategies and tactics based upon relevant new information (e.g., which microbes are the most gifted for SM biosynthesis), and the development of appropriate research plans that can be executed in time frames consistent with robust drug discovery.

To address the overall vision of revitalized NP discovery, I have organized the various approaches and disciplines from the following standpoints: (i) what are the most critical strategic issues and approaches (primary drivers), and (ii) what are the key tactical approaches that directly support or enable the process to succeed (secondary drivers).

1.2.2.1 Primary Drivers

1.2.2.1.1 Microbes Gifted for Natural Product Biosynthesis

At the front end of the genome mining process is the genomic sequence information from microorganisms. It is well documented that not all microorganisms are created equal when it comes to NP production.[15,18,19,43–50] Culturable bacteria and archaea have genomes ranging from <2 to >13 Mb, and the ones with the largest genomes tend to encode large numbers of SMGCs.[15] Across the culturable bacterial phyla, the ability to encode multiple SMGCs is concentrated within the Actinobacteria, Proteobacteria, Cyanobacteria, and Firmacutes.[15,47–50] However, even among these major culturable phyla, many species have small genomes, and are not suitable choices for robust NP discovery.[15] A recent program sponsored by the Department of Energy—Genomic Encyclopedia of *Bacteria* and *Archaea* (GEBA-1)—has generated 1003 high-quality genome sequences of a wide range of bacterial- and archaeal-type strains. The 974 bacterial species include many Proteobacteria, Firmacutes, Bacteroidetes, Actinobacteria, and species from 17 other phyla.[51] The distribution

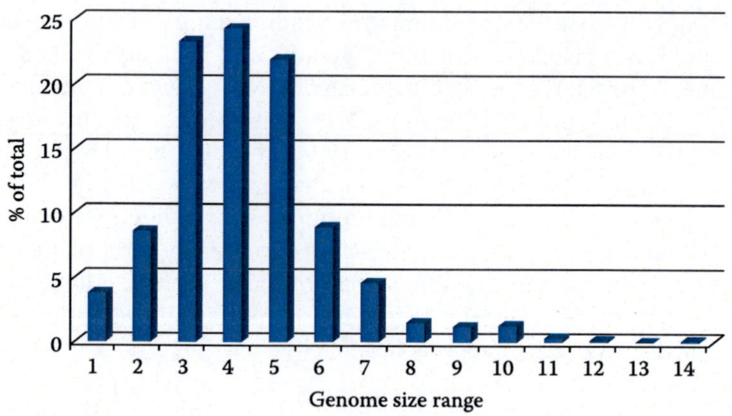

FIGURE 1.1 Distribution of genome sizes in bacterial-type strains sequenced in the GEBA-1 project.[51] Horizontal axis, 1 = 0–1 Mb; 2 = 1–2 Mb; and so forth.

of bacterial genome sizes is summarized in Figure 1.1. Less than 2% of the bacterial genomes are >8.0 Mb, and >80% have genomes ranging from <1.0 to 5.0 Mb. None of the 29 archaeal-type strains has genomes >6 Mb (not shown). In contrast, most bacteria gifted for SM production have genomes >8.0 Mb[15] (see below).

The vast majority of microbes (>99%)[52,53] are unculturable on typical microbiological media, and these have been the source of much speculation and conjecture about their potential for SM production.[54–58] In the following sections, I discuss the SM coding capacity of culturable and uncultured bacteria and archaea from a variety of environmental sources. Accurate information on microbial SM coding capacity is critical to inform future strategic and tactical approaches to revitalize NP discovery.

1.2.2.1.1.1 Actinomycetes Microbes with very large genomes tend to be cosmopolitan, and encode diverse primary metabolic functions that facilitate growth and reproduction on a wide range of nutrient sources.[59] Many of these also encode SMs, including antibacterial and antifungal agents among others. The most *gifted* microorganisms for SM production based on antiSMASH 3.0 analysis[30] are actinomycetes with genomes ranging from 8.0 to 12.7 Mb.[15] Table 1.1 shows the SM coding capacity for 18 actinomycetes with genomes >8.0 Mb that have been sequenced to completion. On average, they devote 1.64 Mb (16% of their coding capacity) to encode 35 SMs. The most gifted strains, *Kutzneria albida*, *Streptomyces bingchenggensis*, and *Streptomyces rapamycinicus*, devote 2.5–3.09 Mb (>20% of coding capacity) to encode 48–53 SMs. The 18 strains also encode an average of 17 SMs that employ PKS, NRPS, or mixed NRPS/PKS biosynthetic mechanisms. The three most gifted strains encode 24–30 of these pathways. In addition, the 18 actinomycetes encode an average of 4.8 MbtH homologs and 3.6 phosphopantetheinyl transferases (PPTases). MbtH homologs participate as chaperones in adenylation reactions carried out by NRPS multienzymes, and PPTases are required to convert apo-ACPs and apo-PCPs

TABLE 1.1
SM Coding Capacity in Select Strains of Cultured Bacteria and Archaea[a]

Microorganism	Size (Mb)	MbtH Genes	PPTase Genes	NRPS/PKS Clusters[b]	Total Clusters	Total SMs (Mb)	% of Genome
Actinomycetes with large genomes							
Streptomyces leeuwenhoekii DSM 42122	8.12	3	3	16	35	1.57	19.9
Saccharopolyspora erythraea NRRL 2338	8.21	3	3	16	36	1.53	18.6
Actinosynnema mirum DSM 43827^T	8.25	7	2	18	30	1.59	19.2
Streptomyces collinus Tü 365	8.27	4	6	14	31	1.37	16.6
Streptomyces ambofaciens ATCC 23877	8.39	2	4	10	27	1.11	13.2
Streptomyces griseus NBRC 13350	8.55	8	3	17	40	1.71	20.0
Streptomyces coelicolor A3(2)	8.67	2	3	11	27	0.93	10.7
Actinoplanes missouriensis ATCC 14538^T	8.77	2	2	6	11	0.46	5.3
Amycolatopsis orientalis HCCB10007	8.95	4	4	15	32	1.45	16.2
Streptomyces avermitilis MA-4680	9.03	4	7	17	37	1.51	16.7
Saccharothrix espanaensis DSM 44229^T	9.36	6	1	15	35	1.66	17.7
Kutzneria albida DSM 43870^T	9.88	9	3	24	48	2.50	25.3
Streptomyces hygroscopicus 5008	10.15	5	5	17	40	1.47	14.5
Amycolatopsis mediterranei U32	10.24	6	3	17	30	1.35	13.2
Streptosporangium roseum DSM 43021^T	10.34	8	4	15	26	1.18	11.4
Streptomyces violaceusniger Tu4113	10.66	3	4	22	43	2.33	21.9
Streptomyces bingchenggensis BCW-1	11.94	4	5	30	53	2.60	21.7
Streptomyces rapamycinicus NRRL 5491	12.70	7	3	28	52	3.09	24.3
Average	9.47	4.8	3.6	17.1	35.2	1.64	16.0
Bacteroidetes/Chlorbi							
Prevotella denticola F089	2.94	0	1	0	2	0.09	3.1
Elizabethkingia anophelis NUHP1	4.37	0	1	0	5	0.10	2.4
Bacteroides vulgaris ATCC 8462	5.16	0	1	0	1	0.05	0.9

(Continued)

TABLE 1.1 (*Continued*)
SM Coding Capacity in Select Strains of Cultured Bacteria and Archaea[a]

Microorganism	Size (Mb)	MbtH Genes	PPTase Genes	NRPS/PKS Clusters[b]	Total Clusters	Total SMs (Mb)	% of Genome
Bacteroides fragilis YCH46	5.28	0	1	0	1	0.05	0.8
Bacteroides ovatus ATCC 8483	6.47	0	1	0	1	0.05	0.7
Cyanobacteria							
Prochlorococcus marinus SS120	1.75	0	1	0	4	0.07	4.0
Pleurocapsa sp. PCC7319	4.99	0	2	5	11	0.33	6.7
Microcystis aeruginosa NIES-843	5.84	0	1	5	9	0.40	6.8
Cyanothece sp. PCC7424	6.55	0	1	3	10	0.28	4.2
Calothrix sp. PCC6303	6.96	0	2	2	12	0.41	5.9
Anabaena variabilis	7.11	0	2	6	14	0.59	8.3
Firmicutes—Bacilli							
Lactobacillus rhamnosus GG	3.01	0	1	0	0	0.00	0.0
Lactobacillus casei BD-II	3.07	0	1	0	1	0.01	0.3
Lactobacillus plantarum WCFS1	3.31	0	3	1	2	0.08	2.4
Bacillus subtilis 168	4.22	1	2	5	17	0.65	15.4
Bacillus thuringiensis St. 97-27	5.24	1	3	1	5	0.23	4.4
Paenibacillus polymyxa SC2	5.73	0	3	7	12	0.68	11.9
Paenibacillus mucilaginosus 3016	8.74	0	3	10	16	0.94	10.8
Firmicutes—Clostridia							
Ruminiclostridium therocellum DSM 1313	3.56	0	2	2	3	0.11	3.1
Clostridium kluyveri DSM 555	3.96	0	3	3	3	0.19	4.8
Clostridium pasteurianum BC1	4.99	0	2	1	2	0.07	1.4
Desulfitobacterium hafniense Y51	5.73	0	2	1	3	0.07	1.2
Clostridium beijerinckii NCIMB 8052	6.00	0	2	1	1	0.07	1.2

(*Continued*)

TABLE 1.1 (*Continued*)
SM Coding Capacity in Select Strains of Cultured Bacteria and Archaea[a]

Microorganism	Size (Mb)	MbtH Genes	PPTase Genes	NRPS/PKS Clusters[b]	Total Clusters	Total SMs (Mb)	% of Genome
Proteobacteria—Alpha							
Wolbachia endosymbiont	1.30	0	1	0	0	0.00	0.0
Acetobacter pasteurianus IFO 3283-01	3.34	0	1	0	3	0.06	1.8
Caulobacter crescentus CB15	4.01	0	1	0	4	0.09	2.1
Rhizobium leguninosarum WSM1689	6.90	1	3	1	5	0.13	1.9
Agrobacteriumm tumifaciens K84	7.27	0	3	3	8	0.24	3.3
Proteobacteria—Beta							
Neisseria meningitidis MC58	2.27	0	1	0	5	0.13	5.7
Limnohabitans sp. 103DPR2	2.95	0	1	0	2	0.04	1.4
Comamonas testosteroni CNB-2	5.37	0	1	0	2	0.06	1.1
Burkholderia thailandensis E264	6.72	1	5	8	19	0.92	13.7
Burkholderia pseudomallei 1710b	7.31	3	3	11	27	1.03	14.1
Burkholderia gladioli BSR3	9.05	5	3	9	19	0.84	9.3
Proteobacteria—Gamma							
Acinobacter johnsonii XBB1	4.08	0	1	0	3	0.07	1.7
Alteromonas mediterranea DE	4.48	0	1	0	1	0.01	0.2
Escherichia coli W	5.01	1	3	1	1	0.05	1.0
Aeromonas salmonicida A449	5.04	0	3	2	4	0.16	3.2
Azotobacter vinelandii DJ	5.37	2	2	4	8	0.30	5.6
Photorhabdus luminescens TT01	5.69	2	3	12	21	1.01	17.8
Pseudomonas syringae B728a	6.09	1	2	7	12	0.74	12.2
Pseudomonas fluorescens F113	6.85	1	2	3	9	0.35	5.1

(Continued)

TABLE 1.1 (*Continued*)
SM Coding Capacity in Select Strains of Cultured Bacteria and Archaea[a]

Microorganism	Size (Mb)	MbtH Genes	PPTase Genes	NRPS/PKS Clusters[b]	Total Clusters	Total SMs (Mb)	% of Genome
Proteobacteria—Delta							
Desulfovibrio desulfuricans ATCC 27774	2.87	0	3	1	1	0.06	2.1
Desulfovibrio alaskensis G20	3.73	0	2	1	1	0.06	1.6
Geobacter metallireducens GS-15	4.01	0	1	0	4	0.11	2.7
Myxococcus fulvus HW-1	9.00	1	3	13	25	1.13	12.6
Myxococcus xanthus DK 1622	9.14	1	3	15	23	1.09	11.9
Sorangium cellulosum so ce56	13.03	1	3	10	32	1.21	9.3
Proteobacteria—Epsilon							
Campylobacter jejuni ATCC 700819	1.64	0	1	0	0	0.00	0.0
Helicobacter pylori 26695	1.67	0	1	0	0	0.00	0.0
Nautilia profundicola AmH	1.68	0	1	0	0	0.00	0.0
Sulfurimonas autotrophica OK10	2.15	0	1	0	0	0.00	0.0
Arcobacter butzleri RM4018	2.34	0	1	0	0	0.00	0.0
Sulfurospirrilum barnesii SES-3	2.51	0	2	1	2	0.96	3.8
Archaea							
Methanococcus jannaschii	1.66	0	0	0	0	0.00	0.0
Pyrococcus sp. NA2	1.86	0	0	0	1	0.01	0.5
Methanobacterium sp. MB1	2.03	0	1	1	2	0.09	4.4
Methanobrevibacter sp. YE315	2.27	0	0	4	4	0.23	10.0
Vulcanisaeta distributa DSM.14429	2.37	0	0	0	1	0.01	0.5
Sulfolobus islandicus	2.78	0	0	0	1	0.01	0.4
Methanobrevbacter ruminantium M1	2.94	0	1	2	2	0.10	3.4
Methanocella paludicola SANAE	2.96	0	1	1	1	0.05	1.5
Methanosarcina barkeri Fusaro	4.84	0	0	0	0	0.00	0.0
Methylobacterium extorquens AM1	5.51	0	2	2	7	0.21	3.5

[a] Table modified from Baltz.[15]
[b] NRPS, PKS, or mixed NRPS/PKS pathways.

to holo-enzymes required for polyketide and nonribosomal peptide assembly by PKS and NRPS multienzymes.

There are many actinobacteria, including actinomycetes, with genomes ranging from <3.0 to 8.0 Mb.[51] Figure 1.2a shows the SM coding capacity of a sample of completed actinomycete genomes ranging in size from 3.6 to 12.7 Mb.[15] All encode SMs, but none of the strains with genomes <8.0 Mb devotes >1.0 Mb to encode SMGCs. The regression line drawn through the most productive strains may define an upper limit to coding capacity for SMs, 0.31 Mb SM coding capacity per Mb

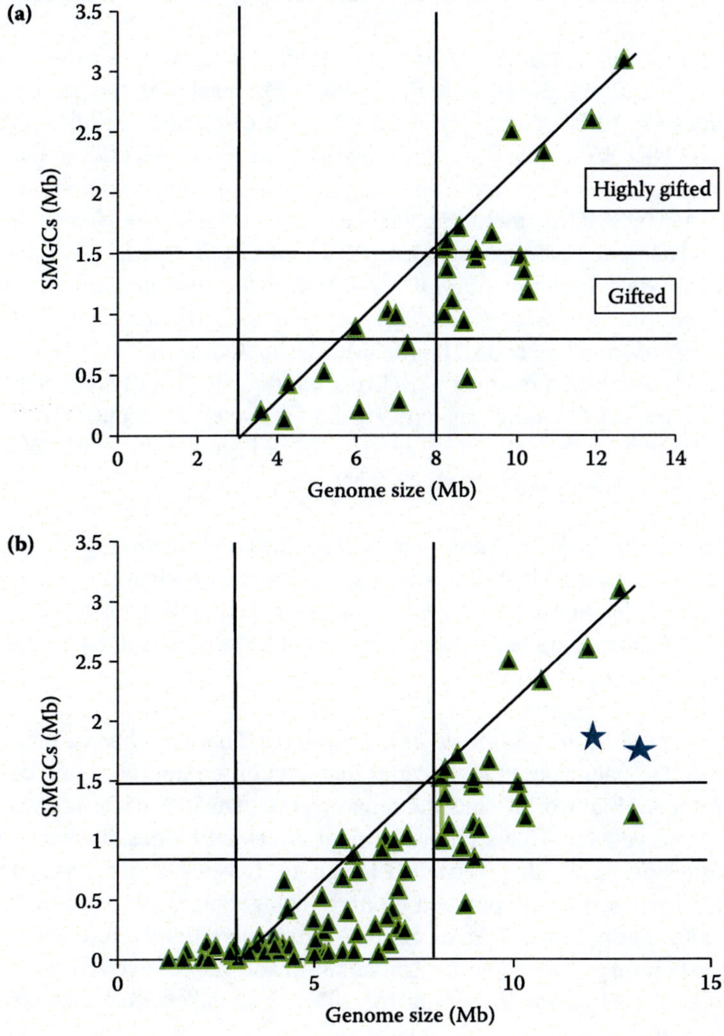

FIGURE 1.2 SM coding capacity versus genome size in select cultured bacteria. **(a)** Actinomycetes. **(b)** Actinomycetes and other bacteria. Data from Baltz[15] except that two additional myxobacterial genomes, *Archangium gephyra* DSM 2261 (12.5 Mb) and *Chondromyces crocatus* (11.3 Mb), are added as stars.

above the X intercept of 3.0 Mb, below which no significant SMGC coding capacity is predicted. This is consistent with data from an earlier study.[46] Thus, the relationship of SM coding capacity (Y) is related to genome size (X) by the following equation: $Y = 0.31(X-3)$. From this simple relationship, it is clear that actinomycetes with large genomes are the most likely to encode large numbers of SMs, including many of PKS, NRPS, and mixed NRPS/PKS origin, the hallmarks of the most commercially successful SMs historically.[3] It is noteworthy that the medium size of 325 draft genomes of unspeciated *Streptomyces* strains at the time of this writing was 8.2 Mb (http://www.ncbi.nlm.nih.gov/genome), suggesting that many more gifted *Streptomyces* sp. are already available for further analysis and genome mining.

1.2.2.1.1.2 Other Culturable Bacteria Table 1.1 also shows the SM coding capacity for several culturable bacterial phyla.[15] The bacterial and archaeal strains with genomes <6.0 Mb generally devote <0.1 Mb of coding capacity to SMGCs, including 0–3 NRPS/PKS clusters. Exceptions are strains of *Bacillus*, *Paenibacillus*, *Cyanobacteria*, *Azotobacter*, and *Photorhabdus*. Among the 56 strains surveyed, 12 devote >0.5 Mb to SMGCs, including 7 that encode >1.0 Mb. The 10 most gifted bacterial taxa—*Paenibacillus mucilaginosa*, *Burkholderia thailandensis*, *Burkholderia pseudomallei*, *Burkholderia gladioli*, *Photorhabdus luminescens*, *Myxococcus fulvus*, *Myxococcus xanthus*, *Chondromyces crocatus*, *Archangium gephyra*, and *Sorangium cellulosum*—encode 16–43 SMGCs, including 8–17 NRPS/PKS clusters. Seven of these have genomes >8.0 Mb, including all five myxobacterial strains (Table 1.1). The DNA coding capacity of the 56 bacterial species versus genome size is combined with the data on actinomycetes in Figure 1.2b. The SMGC coding capacity is 0 to very low in genomes <3.0 Mb.

1.2.2.1.1.3 Culturable Archaea In stark contrast to the actinomycetes with large genomes and other gifted bacterial species, members of domain *Archaea* generally have small genomes, mostly <3.0 Mb,[51] and are nearly devoid of relevant SMGCs (Table 1.1).[15] There are no examples of important SMs produced by microbes from the domain *Archaea*.[3]

1.2.2.1.1.4 Uncultured Bacteria and Archaea There has been a long-standing conjecture dating back about two decades that uncultured microbes, which make up >99% of the microbial world[52] and encompass more than 50% of the known bacterial phyla,[53] may serve as inexhaustible sources of novel antibiotics and other SMs.[54–58] Several companies, including TerraGen Diversity, Diversa, Ariad Pharmaceuticals, and Cubist Pharmaceuticals, pursued cloning metagenomic DNA in the late 1990s to early 2000s, but no novel NPs of clinical utility were discovered. More recently, the Brady laboratory at Rockefeller University made impressive advances on technical aspects of metagenomic mining for NP pathways.[60–66] One interesting result potentially useful for drug development was the discovery of novel glycopeptide antibiotics related to vancomycin.[67,68] However, this discovery is not relevant to cloning SMGCs from unculturable microbes, because many glycopeptide antibiotics (e.g., vancomycin, balhimycin, teicoplanin, dalbavancin, A47934) are produced by a wide range of readily culturable actinomycete genera. Importantly, additional new glycopeptide antibiotic-producing actinomycetes can be enriched and readily cultured from soil on media containing high levels of vancomycin. This stems from the observation that culturable actinomycetes that produce glycopeptides also express glycopeptide antibiotic resistance as part of the glycopeptide biosynthetic gene cluster.[69,70] The original metagenomic concept was to clone novel NPs from previously uncultured soil taxa.[54]

In spite of the lack of compelling evidence to support the conjecture that unculturable microbes will serve as robust sources of novel NPs for drug discovery, the notion that the unculturable majority has great potential remains pervasive.[27,41,71,72] It is now possible to test the validity of the conjecture with scientific evidence. With advances in metagenomic assembly methods and single-cell DNA sequencing, it is possible to obtain complete or nearly complete genome sequences from uncultured microbes.[73–83] This provides an opportunity to directly address the following question: Do uncultured microorganisms (i.e., so far unculturable) encode SMGCs that might be exploited in robust drug discovery programs? There is a growing body of evidence that many uncultured microbes have relatively small genomes. A sampling of 201 uncultured bacteria and archaea from diverse habitats covering 20 uncultivated lineages revealed genomes ranging from 0.6 to 3.1 Mb.[83] Switchgrass-adherent uncultured microbes from cow rumen yielded genomes of bacterial taxa ranging from 1.67 to 3.08 Mb.[79] Table 1.2 shows an antiSMASH 3.0 analysis of complete genome sequences of uncultured bacteria and archaea from diverse habitats. The 14 bacteria have genomes ranging from 0.60 to 1.57 Mb. Only two encode small SMGCs. Of the three archaea, one encodes two terpenes and a small polyketide. The combined SM coding capacity of the 17 microbes is 0.15 Mb, <10% of the coding capacity of an average actinomycete with a genome >8.0 Mb, and none of the SMs appears to be drug like. This snapshot of uncultured bacteria and archaea does not support the conjecture that the uncultured majority will provide an *inexhaustible* source for novel antibiotics and other SMs. They generally have genomes <3.0 Mb that encode minimal primary biosynthetic capacities to survive in specialized niches, often growing in microbial communities in mutualistic or synergistic relationships.[75,81] They do not have the luxury of devoting substantial coding capacity for drug-like NPs.[3]

1.2.2.1.1.5 Historical Record Katz and Baltz[3] surveyed 100 of the most important NPs from bacteria and fungi. Of these, actinomycetes produced 83 (68 by streptomycetes and 15 by other actinomycetes), fungi produced 12, other bacteria produced 5, and archaea produced none. Table 1.3 shows an expanded list of 100 important SMs produced by 93 actinomycetes. Of these, 79 are produced by streptomycetes, and 63 are of NRPS, PKS, or mixed NRPS-PKS biosynthetic origins. Of the 93 strains, only 49 have been sequenced (24 finished and 25 drafts). Notably, 38 of the 49 sequenced strains have genomes >8.0 Mb. This implies that actinomycetes that are known to produce important SMs are enriched for large genomes, a prerequisite for gifted status.[15]

From the forgoing analysis, it is clear that culturable bacteria, primarily actinomycetes, encompass genera and species that are truly gifted for the production of NPs. The tactical question of how to obtain large numbers of gifted microbes for genome mining is considered in Section 1.3.

TABLE 1.2
SM Coding Capacity of Uncultured Microorganisms[a]

Microorganism[b]	Habitat/Source	Size (Mb)	Nrps/PKS Clusters	Total Clusters	Total SMs (Mb)	Reference
Bacteria						
Ca. Kazan bacterium (CP011216.1)	Groundwater	0.60	0	0	0	[75]
Ca. Campbellbacteria bacterium (CP011215.1)	Groundwater	0.75	0	0	0	[75]
Ca. Woesebacteria bacterium (CP011214.1)	Groundwater	0.82	0	0	0	[75]
Ca. Berkelbacteria bacterium (CP011213.1)	Groundwater	0.92	0	0	0	[75]
Ca. Saccaribacteria bacterium (CP011211.1)	Groundwater	1.04	0	2	0.04	[75]
Ca. Beckwithbacteria bacterium (CP011210.1)	Groundwater	1.05	0	0	0	[75]
Ca. Wolfbacteria bacterium (CP011209.1)	Groundwater	0.98	0	0	0	[75]
Ca. Saccarimonas aalborgensis (CP005957.1)	Activated sludge	1.01	0	0	0	[73]
Ca. SR1 bacterium (NC_023002.1)	Aquifer sediment	1.18	0	0	0	[81]
Ca. WWE3 bacterium (CP006914.1)	Aquifer sediment	0.88	0	0	0	[81]
Ca. Saccaribacteria bacterium (NC_023004.1)	Aquifer sediment	0.85	0	0	0	[81]
SAR86 cluster bacterium (LUQZ00000000.1)	Red Sea	1.57	0	0	0	[76]
SAR324 bacterium lautmerah 10 (LNZD02000000.1)	Red Sea	3.50	1	1	0.02	[78]
Ca. Peribacter riflensis (CP013062)	Groundwater	1.25	0	0	0	[74]
Average		1.17	0.07	0.21	0.004	This report
Archaea						
Euryarchaeota MG-II (CM001443.1)	Puget sound (marine)	2.06	1	3	0.09	[80]
Ca. Nanosalina sp. J07AB43 (AEIY00000000.1)	Hypersaline lake	1.23	0	0	0	[82]
Ca. Nanosalinarum sp. J07AB56 (AEIX00000000.1)	Hypersaline lake	1.22	0	0	0	[82]
Average		1.50	0.33	1.0	0.03	This report

[a] Table modified from Baltz.[15] SM coding capacity determined by antiSMASH 3.0.[30]
[b] *Ca.*, candidate phylum. Accession numbers in ().

TABLE 1.3
One Hundred Important NPs Produced by Actinomycetes

Secondary Metabolite	Biosynthetic Origin	Major Activity/Use	Producing Actinomycete	Genome Size[a]
Acarbose	Glycoside	Antidiabetic	*Actinoplanes* sp. SE50/110	9.24 (C)
Actinomycin	NRPS	Antitumor	*Streptomyces anulatus* ATCC 11523	8.76
Actinorhodin	PKS II	RT	*Streptomyces coelicolor* A3(2)	9.05 (C)
Adriamycin	PKS II	Antitumor	*Streptomyces peucetius* NRRL WC-3868	9.53
Albomycin	Peptidyl-nucleoside	Antibacterial	*Streptomyces* sp. ATCC 700974	—
Amphomycin	NRPS	Antibacterial	*Streptomyces canus* FIM-0916	—
Amphotericin B	PKS I	Antifungal	*Streptomyces nodosus* ATCC 14899	7.71 (C)
Antimycin	NRPS-PKS I	Piscicide; RT	*Streptomyces albus* J1074	6.84 (C)
Apramycin	Aminoglycoside	Antibacterial; RT	*Streptoalloteichus tenebrarius* ATCC 17920	—
Ascomycin	NRPS-PKS I	Immunomodulator	*Streptomyces hygroscopicus* ATCC 14891	—
Avermectin	PKS I	Anthelmintic	*Streptomyces avermitilis* MA-4680	9.03 (C)
Avilamycin	Glycoside	Antibacterial	*Streptomyces viridochromogenes* Tü57	9.7
Bialphos	NRPS	Herbicide	*Streptomyces viridochromogenes* DSM 40736	8.65
Bleomycin	NRPS-PKS I	Antitumor	*Streptomyces verticillus* ATCC 15003	—
Calcimycin	PKS I	RT	*Streptomyces chartruensis* NRRL 3882	9.30
Calicheamicin	PKS I	Antitumor	*Micromonospora echinospora* NRRL 15839	—
Candicidin	PKS I	Antifungal	*Streptomyces* sp. FR-008	7.26 (C)
Capreomycin	NRPS	Antitubercular	*Saccharothrix mutabilis* ATCC 23892	—
Carbomycin	PKS I	Antibacterial	*Streptomyces halstedii* NRRL ISP-5068	7.74
Cephamycin	NRPS	Antibacterial	*Streptomyces clavuligerus* ATCC 27064	8.56
Chloramphenicol	Shikimate	Antibacterial	*Streptomyces venezuelae* ATCC 10712	8.23 (C)
Chlorobiocin	Aminocoumarine	Antibacterial	*Streptomyces roseochromogenes* DS 12.976	9.66
Chloroeremomycin	NRPS	Antibacterial	*Amycolatopsis orientalis* A82846	—

(Continued)

TABLE 1.3 (*Continued*)
One Hundred Important NPs Produced by Actinomycetes

Secondary Metabolite	Biosynthetic Origin	Major Activity/Use	Producing Actinomycete	Genome Size[a]
Clavulanic acid	Other	Antibacterial	*Streptomyces clavuligerus* ATCC 27064	8.56 (C)
Coumermycin	Aminocoumarin	Antibacterial	*Streptomyces rishiriensis* DSM 40489	—
D-cycloserine	NRPS	Antitubercular	*Streptomyces lavendulae* ATCC 11924	—
Daptomycin	NRPS	Antibacterial	*Streptomyces roseosporus* NRRL 158998	7.82
Dalbavancin	NRPS	Antibacterial	*Nonomuraea* sp. ATCC 39737	—
Daunorubicin	PKS II	Antitumor	*Streptomyces peucetius* NRRL WC-3868	9.53
Erythromycin	PKS I	Antibacterial	*Saccharopolyspora erythraea* NRRL 2338	8.08 (C)
Filipin	PKS I	Antifungal	*Streptomyces filipinensis* ATCC 23905	—
Fosphomycin	Phosphone	Antibacterial	*Streptomyces wedmorensis* NRRL 3426	9.38
Friulimicin	NRPS	Antibacterial	*Actinoplanes friuliensis* DSM 7358	9.38 (C)
GE2270	RiPP	Antibacterial	*Planobispora rosea* ATCC 53733	—
Geldanamycin	PKS I	Antitumor	*Streptomyces hygroscopicus* 17997	—
Gentamicin	Aminoglycoside	Antibacterial	*Micromonospora purpurea* ATCC 15835	—
Hygromycin B	Aminoglycoside	Antibacterial; RT	*Streptomyces hygroscopicus* NRRL 2387	—
Josamycin	PKS I	Antibacterial	*Streptomyces narbonensis* ATCC 19790	—
Kanamycin	Aminoglycoside	Antibacterial	*Streptomyces kanamyceticus* NRRL B3525	9.78
Kinamycin F	PKS II	Antitumor	*Streptomyces murayamaensis*	—
Kirromycin	NRPS-PKS I	Antibacterial	*Streptomyces collinus* Tü 365	8.27 (C)
Lactacystin	NRPS-PKS I	Antitumor	*Streptomyces lactacystinicus* DSM 43136	—
Lasalocid	PKS I	Coccidiostat	*Streptomyces lasaliensis* ATCC 35851	—
Lincomycin	Other	Antibacterial	*Streptomyces lincolnensis* ATCC 25466	—
Lipiaramycin	PKS I	Antibacterial	*Dactylosporangium aurantiacum* NRRL B8018	11.45
Lipstatin	Fatty acyl-lactone	Antiobesity	*Streptomyces toxitricini* NRRL 15443	—

(*Continued*)

TABLE 1.3 (*Continued*)
One Hundred Important NPs Produced by Actinomycetes

Secondary Metabolite	Biosynthetic Origin	Major Activity/Use	Producing Actinomycete	Genome Size[a]
Mannopeptimycin	NRPS	Antibacterial	*Streptomyces hygroscopicus* NRRL 30439	—
Micinamycin	PKS I	Antibacterial	*Micromonospora griseorubida* A11725	—
Milbemycin	PKS I	Antiparasidic	*Streptomyces hygroscopicus* SANK 60286	—
Mithramycin	PKS II	Antitumor	*Streptomyces argillaceus* ATCC 12956	—
Mitomycin C	Quinone	Antitumor	*Streptomyces lavendulae* NRRL 2564	—
Moenomycin	Phosphoglycolipid	Antibacterial	*Streptomyces ghanaensis* ATCC 14672	8.51
Monensin	PKS I	Coccidiostat	*Streptomyces cinnamonensis* ATCC 15413	—
Narasin	PKS I	Coccidiostat	*Streptomyces aureofaciens* NRRL 5758	—
Natamycin	PKS I	Antifungal	*Streptomyces natalensis* ATCC 27448	8.65
Neomycin	Aminoglycoside	Antibacterial	*Streptomyces fradiae* NCIMB 8233	—
Netropsin	NRPS	Antitumor	*Streptomyces ambofaciens* ATCC 23877	8.39 (C)
Nikkomycin	Peptidyl-nucleoside	Antifungal	*Streptomyces tendae* Tü901	—
Nocardicin A	Beta-lactam	Antibacterial	*Nocardia uniformis* ATCC 21806	8.77
Nosiheptide	RiPP	Antibacterial	*Streptomyces actuosus* ATCC 25421	—
Nystatin	PKS I	Antifungal	*Streptomyces noursei* ATCC 11455	—
Oligomycin	PKS I	RT	*Streptomyces avermitilis* MA-4680	9.03 (C)
Oxytetracycline	PKS II	Antibacterial	*Streptomyces rimosus* (multiple strains)	9.49
Pladienolide B	PKS I	Antitumor	*Streptomyces platensis* Mer-11107	—
Platencin	Diterpine	Antibacterial	*Streptomyces platensis* BS12029	—
Platensimycin	Diterpine	Antibacterial	*Streptomyces platensis* BS12029	—
Polyoxin (D)	Nucleoside	Antifungal	*Streptomyces cacaoi* AS4.1602	—
Pristinamycin IA, IIA	NRPS, NRPS-PKS I	Antibacterial	*Streptomyces pristinaespiralis* ATCC 25486	8.53 (C)
Puromycin	Nucleoside	Antibacterial; RT	*Streptomyces alboniger* NRRL B-1832	7.55

(*Continued*)

TABLE 1.3 (*Continued*)
One Hundred Important NPs Produced by Actinomycetes

Secondary Metabolite	Biosynthetic Origin	Major Activity/Use	Producing Actinomycete	Genome Size[a]
Ramoplanin	NRPS	Antibacterial	*Actinoplanes* sp. ATCC 33076	—
Rapamycin	NRPS-PKS I	Immunomodulator	*Streptomyces rapamycinicus* NRRL 5491	12.70 (C)
Rebeccamycin	Alkaloid	Antitumor	*Lechevalieria aerocolonegenes* ATCC 39243	10.64
Rifamycin	PKS I	Antibacterial	*Amycolatopsis mediterranei* U32	10.24 (C)
Ristocetin	NRPS	Antibacterial	*Amycolatopsis lurida* NRRL 2430	8.99 (C)
Salinomycin	PKS I	Coccidiostat	*Streptomyces albus* DSM 41389	8.38 (C)
Salinosporamide A	NRPS-PKS I	Antitumor	*Salinispora tropica* CNB-440	5.18 (C)
Sinefungin	Nucleoside	Antifungal; RT	*Streptomyces griseolus* NRRL B-2925	—
Spectinomycin	Aminoglycoside	Antibacterial	*Streptomyces spectabilis* UC 2294	—
Spinosad	PKS I	Insecticidal	*Saccharopolyspora spinosa* NRRL 18395	8.58
Spiramycin	PKS I	Antibacterial	*Streptomyces ambofaciens* ATCC 23877	8.27 (C)
Staurosporin	Alkaloid	Antitumor	*Streptomyces staurosporeus*	—
Streptomycin	Aminoglycoside	Antitubercular	*Streptomyces griseus* NBRC 13350	8.55 (C)
Streptothricin	Aminoglycoside	Antibacterial	*Streptomyces lavendulae* NRRL B-2774	8.47
Streptozotocin	Glucosamine-nitrosourea	Antitumor	*Streptomyces acromogenes* NRRL 3125	—
Tacrolimus	NRPS-PKS I	Immunomodulator	*Streptomyces tsukubaensis* NRRL 18488	7.67
Tallysomycin	NRPS-PKS I	Antitumor	*Streptoalloteichus hindustanus* ATCC 31158	—
Tautomycetin	PKS I	Antitumor	*Streptomyces* sp. CK4412	—
Teicoplanin	NRPS	Antibacterial	*Actinoplanes teichomyceticus* ATCC 31121	—
Tetracenomycin	PKS II	Antitumor	*Streptomyces glaucescens* DSM 40922	7.45 (C)
Tetracycline	PKS II	Antibacterial	*Streptomyces rimosus* (multiple)	8.2
Thienamycin	Other	Antibacterial	*Streptomyces cattleya* NRRL 8057	8.09 (C)
Thiostrepton	RiPP	Antibacterial; RT	*Streptomyces azureus* ATCC 14921	8.79

(*Continued*)

TABLE 1.3 (*Continued*)
One Hundred Important NPs Produced by Actinomycetes

Secondary Metabolite	Biosynthetic Origin	Major Activity/Use	Producing Actinomycete	Genome Size[a]
Tobramycin	Aminoglycoside	Antibacterial	*Streptoalloteicus tenebrarius* ATCC 17920	—
Tunicamycin	Nucleoside	Antibacterial	*Streptomyces chartreusis* NRRL 12338	8.89
Tylosin	PKS I	Antibacterial	*Streptomyces fradiae* ATCC 19609	7.67
Undecylprodigiosin	Other	RT	*Streptomyces coelicolor* A3(2)	9.05 (C)
Validamycin	Glycoside	Antifungal	*Streptomyces hygroscopicus jinggangensis* 5008	10.38 (C)
Vancomycin	NRPS	Antibacterial	*Amycolatopsis orientalis* HCCB 10007	8.95 (C)
Viomycin	NRPS	Antibacterial	*Streptomyces* sp. 11861	—
Virginiamycin	NRPS, NRPS-PKS I	Antibacterial	*Streptomyces virginiae* NRRL-ISP 5094	8.32

Abbreviations: PKS I, type I polyketide synthase; PKS II, type II PKS; NRPS, nonribosomal peptide synthetase; NRPS-PKS, mixed pathway; RiPP, ribosomally synthesized and posttranslationally modified peptide; RT, research tool.

[a] (C), complete sequence; —, not sequenced. All others have draft sequences.

1.2.2.1.2 DNA Sequencing Technology

Two key drivers for successful genome mining are inexpensive DNA sequencing, which can be directed at 10^5–10^7 microbial genomes, and accurate DNA assembly of giant-type I PKS and NRPS modular multidomain genes that often have internal regions of high sequence similarity. Since the early 2000s, the cost of DNA sequencing has dropped dramatically. For instance, the cost of sequencing a human genome has dropped from $100,000,000 in 2001 to $1,000 in 2014, or ~$10^5$-fold.[84] During this time frame, the quality, speed, and nucleotide read lengths have improved with the development of next-generation sequencing (NGS) technologies.[84] The first second-generation sequencing (SGS) technology, commercialized in 2005, was 454 pyrosequencing (Roche) followed by Solexa (now Illumina) technology in 2006.[84] Over the past decade, the sequencing methodologies of choice for actinomycete genomes have moved from Sanger sequencing to 454 to Illumina, or combinations of methods.[85,86] Illumina has become the primary method in recent years because of its low cost. It is estimated that the current cost of Sanger sequencing is $500/Mb, whereas Illumina costs $0.04/Mb.[87] Illumina sequencing is an inexpensive method to survey many microbes that encode multiple SMGCs, but it cannot by itself generate complete genome sequences with accurately assembled type I PKS and NRPS gene clusters. Its short reads are not sufficient to assemble highly repetitive DNA sequences such as those found in multimodular type I PKS and NRPS genes.

In 2011, the first third-generation sequencing (TGS) method was commercialized by Pacific Biosciences, called single-molecule real time (SMRT) or simply PacBio.[87] Even though PacBio has difficulty sequencing homopolymeric strings of three or more Cs or Gs correctly, resulting in frameshifts, it is very effective at correctly sequencing large gene clusters containing type I PKS or NRPS genes.[85,86] Illumina, on the other hand, does not have issues obtaining correct sequences of homopolymeric stretches. The Bibb group obtained an accurate full-length 8.2 Mb genomic sequence of *Streptomyces leeuwenhoekii* by using a combination of Illumina and PacBio sequencing, and identified 35 SMGCs.[85,86] This approach should accelerate the accumulation of complete and accurate genome sequences of actinomycetes and other microbes with large genomes encoding SMGCs containing type I PKS and NRPS genes, which in turn will streamline the process of genome mining.

1.2.2.1.3 Methods to Activate Expression of Cryptic SM Pathways

There are three general approaches to activate silent or cryptic SMGCs: fermentation optimization, genetic manipulation, and a combination of both. Other approaches, adding elicitors to the medium[88,89] and coculture,[90] may be useful in specific instances, but do not yet lend themselves to robust, large-scale drug discovery processes.

1.2.2.1.3.1 Fermentation Optimization For fermentation optimization, strains are grown in multiple media under different conditions (e.g., variations in time, temperature, agitation speed, fermentation vessel configuration).[7,91] The first step in fermentation optimization can be miniaturized, thus facilitating multiple parallel fermentations of many microbes in multiple media. It can be highly beneficial if genome mining provides multiple strains that encode the same SM. This approach has been exemplified by Metcalf and colleagues for the production of phosphonic acid–derived SMs.[18,19] Pathways that appear to be cryptic in one strain may be fully capable of expression in another. In addition, identifying multiple strains that encode the same product can be beneficial for further exploration of pathway intermediates and shunt products generated by genetic manipulation. A good example is the production of the novel antibacterial fatty acid synthase inhibitors platensin (Ptn) and platensimycin (Ptm). The original *Streptomyces platensis* strain isolated by Merck[92] produced low titers of both compounds. Shen and colleagues demonstrated that titers could be increased ~100-fold by deleting a pathway-associated negative regulatory gene *ptmR1* (or *ptnR1*),[93] but the recombinant could not be further engineered for lack of sporulation.[94,95] They developed a real-time polymerase chain reaction (PCR) method to identify six dual Ptm/Ptn–producing strains of *S. platensis* from a collection of 1911 strains, then sequenced several strains to confirm the presence of the Ptm/Ptn dual gene cluster.[94] One strain produced elevated Ptm/Ptn titers after deletion of *ptmR1*, but also expressed ample sporulation. They generated a derivative deleted for *ptmO4*, which encodes a long-chain acyl-CoA dehydrogenase, and the double mutant produced 14 Ptm/Ptn congeners, 10 of which were new. It is noteworthy that the original discovery of Ptm/Ptn required screening 250,000 extracts from 83,000 strains,[92] whereas the targeted real-time PCR method identified 6 Ptm/Ptn producers from 1911 strains, thus improving the discovery frequency by ~250-fold. This method can be used to first prioritize strains (prior to sequencing) that may produce congeners of important NPs by choosing appropriate sets of PCR primers, then sequencing the candidate strains to identify new pathways related to known pathways for further studies on expression.

1.2.2.1.3.2 Genetic Manipulations Genetic approaches generally applicable to express cryptic or poorly expressed SMGCs in actinomycetes include the following: (i) chemical or transposon mutagenesis; (ii) manipulation of transcription by disruption of negative regulators, overexpression of positive regulators, or refactoring all promoters for biosynthetic genes; (iii) manipulation of transcription by selection of spontaneous mutations in *rpoB*; (iv) manipulation of translation by mutations on *rpsL* or other genes involved in ribosome function; (v) inactivation of competing SMGC pathways; and (vi) heterologous expression in genetically amenable laboratory or commercial manufacturing strains. Many examples of these and related approaches in actinomycetes have been reviewed recently,[42] and interested readers are also directed to other reviews on this subject.[96–103] In addition to the example discussed above on disruption of negative regulatory genes in the Ptm/Ptn pathways, I summarize a few other examples in Section 1.2.3.

1.2.2.2 Secondary Drivers

In the past, information on chemical structures of SMs produced by microorganisms was published without knowledge of biosynthetic mechanisms or of the genes encoding their biosynthesis. Indeed, many complex polyketides such as erythromycin, spiramycin, tylosin, and rifampicin, and peptides such as penicillin, cephalosporins, and vancomycin were launched commercially before type I PKS and NRPS multienzymes were discovered. Over time, biosynthetic mechanisms have been deduced and biosynthetic gene clusters identified for some key SMs.

The degree of biochemical validation of specific enzyme functions encoded by SMGCs varied considerably, and data were spread over many different journals, not readily accessed for genome-mining efforts. In order to build a robust foundation for genome mining, it is imperative to learn from and exploit what is already known about the linkages between genes and gene clusters, biochemical mechanisms, and chemical structures, and to be able to use that information to assist in annotating new and novel SMGCs.

1.2.2.2.1 Bioinformatics

Many bioinformatics tools have been developed to help link SMGCs to chemical structures and vice versa, and many of the uses and merits of these have been discussed and reviewed extensively.[27–29,36,72,104–108] From a drug discovery perspective, there are certain bioinformatics approaches that merit highlighting—standard antiSMASH 3.0, MIBiG, IMG-ABC, and PRISM—as particularly important to support a productive discovery process. Other bioinformatics tools are pointed out in genome-mining examples in Section 1.2.3.

1.2.2.2.1.1 antiSMASH *Anti*biotics and *s*econdary *m*etabolite *a*nalysis *sh*ell (antiSMASH) 3.0[30] has become an indispensable resource for the identification of SMGCs (also referred to as biosynthetic gene clusters (BGCs)[30] encoded by finished microbial genomes. antiSMASH 3.0 is a powerful tool to identify the following: (i) the number of BGCs encoded per microbe, (ii) the biosynthetic types of BGCs, (iii) the relationships of BGCs to known BGCs encoding identical or related structures, (iv) the total coding capacity of the microbe devoted to BGCs, and (v) the potential novelty of BGCs. It also predicts some aspects of chemical structure from BGC content. antiSMASH 3.0 is less useful for the evaluation of draft genomes because misassembled large NRPS and type I PKS clusters result in SMGCs becoming fragmented and distributed over two or more different contigs. antiSMASH 3.0 was useful in identifying the most gifted microbes for genome mining as well as those to be avoided (Tables 1.1 and 1.2).[15]

antiSMASH 3.0 has the option of enabling the ClusterFinder algorithm[45] to detect *putative gene clusters*. The use of this algorithm appears to overestimate the number of validated SMGCs. For instance, the original annotation of the *S. coelicolor* genome predicted 22 SMGCs,[9] and standard antiSMASH 3.0 predicts 27 BGCs,[15] which is in reasonable agreement considering that antiSMASH 3.0 identifies BGCs such as RiPPs (ribosomally synthesized and posttranslationally modified peptides) that could not be identified in 2002. Furthermore, a number of laboratories have been expressing the cryptic pathways from *S. coelicolor* for well over a decade, and 17 SMGCs have been verified to encode SMs.[12] On the other hand, antiSMASH 3.0 analysis with ClusterFinder enabled predicts 97 BGCs in *S. coelicolor*. This number is 3.6-fold higher than the 27 predicted by standard antiSMASH 3.0, and the 70 additional BGCs are annotated as CF_putative (62), CF_saccharide (5), and CF_fatty acid (3). ClusterFinder has been characterized as a *low-confidence/high-novelty algorithm*.[27] From the extensive *S. coelicolor* expression work, there is no evidence that any of the 70 hypothetical BGCs identified by ClusterFinder are in fact SMGCs encoding molecules suitable for drug discovery. Until there is convincing evidence that most of the hypothetical BGCs identified by ClusterFinder encode SMs with drug-like qualities, it is prudent to use the well-validated standard antiSMASH 3.0 to evaluate microbes for potentially drug-like SMGCs, as has been done in the analyses presented in Tables 1.1 and 1.2, and in Figure 1.2.

1.2.2.2.1.2 PRISM *Pr*ediction *i*nformatics for *s*econdary *m*etabolomes (PRISM) was developed recently to improve the predictability of SM structures from gene sequences.[28] It has been developed with multiple hidden Markov models to improve the predictions of type II PKS structures, starter units for NRPS, and type I PKS multienzymes, and locations of a variety of tailoring reactions (e.g., glycosylations, hydroxylations, methylations, halogenations). PRISM also predicts antibiotic resistance genes, and may be useful for antibiotic discovery.[106] PRISM showed improved structure predictions over antiSMASH 3.0 in a variety of bacterial genomes, particularly *Streptomyces* and other actinomycetes, and improved dereplication of known compounds.[28] It remains to be seen if PRISM will become as widely used as antiSMASH 3.0.

1.2.2.2.1.3 MIBiG Recent initiatives are drawing from many of these bioinformatics tools to address robust genome mining. To address the harnessing of published information on linkages between BGCs (SMGCs), enzyme function, and chemical structures, a consortium of scientific leaders have come to consensus on the *mini*mal *i*nformation on *bi*osynthetic *g*ene clusters (MIBiG) that should be captured in a comprehensive public database.[109] The MIBiG standard covers general parameters applicable to all types of SMs, as well as those that apply only to specific classes, most notably to type I PKSs and NRPSs, the most important SM classes historically for drug discovery (Table 1.3).[3,42] The initial database contained 1170 BGCs chosen from the literature, and each was annotated with a minimal number of parameters, including genomic locus, chemical structure, biosynthetic class, and literature citations. The consortium of 81 academic and industrial groups annotated 405 of these clusters using more extensive parameters. Future efforts will focus on the other 765 BGCs (SMGCs) with minimal annotation. The system is set up for individual investigators to submit new information on existing and new BGCs. The MIBiG dataset has already been integrated into antiSMASH, and will be integrated into IMG-ABC (see below).

1.2.2.2.1.4 IMG-ABC IMG-ABC is an *a*tlas of *b*iosynthetic gene *c*lusters within the *i*ntegrated *m*icrobial *g*enomes (IMG) system of the Joint Genome Institute of the Department of Energy.[47] The mission is to couple the power of computational searching of big genomic datasets with the discovery of small molecules. They plan to integrate data from the MIBiG initiative to expand the number of linkages between known SMs and SMGCs (which they call BCs). The system currently has ~1,000 known structures, and over 900,000 predicted BCs not linked to chemical structures. However, the numbers of BCs predicted to be encoded by *Bacteria* and *Archaea* appear to be overestimated largely because of the use of ClusterFinder.[47] The overestimation of BCs may also be associated with the fact that most large microbial genomes remain unfinished, as witnessed by the current strategy of the

GEBA-1 project.[51] As such the predominant NRPS and type I PKS gene clusters are often not fully assembled, and genes from the same cluster can be found on different contigs. As mentioned earlier, this is a particular problem with Illumina sequencing. For instance, they cite 441,881 BCs encoded by 23,423 *Bacteria* (average of 18.9 per strain). This average is highly unlikely because the 23,423 strains include only ~25% actinobacteria, of which only a subset are actinomycetes with large genomes. The inclusion of putative BCs from ClusterFinder is of little or no current value for drug discovery, and it helps perpetuate the notion that there exists a vast untapped potential in microorganisms with small genomes that in fact have very limited potential (Tables 1.1 and 1.2). Also, in the future, it will be important to obtain finished genomes, particularly for actinomycetes and Proteobacteria with large genomes, for IMG-ABC to fulfill the mission of linking SMs and BCs to aid in drug discovery.

1.2.2.2.2 Metabolomics

Recent advances in mass spectrometry (MS),[38,110] MS/MS networking,[37,39,111] and nuclear magnetic resonance (NMR) spectroscopy[41] have aided dramatically in the discovery of new and novel SMs and dereplication of known compounds. They are critical for the success of microbial genome mining, and linking chemotype with genotype. MS and MS/MS analyses are particularly useful because they require orders of magnitude less material than required for NMR analyses.[110] This is particularly important for studies of SMGCs that begin as cryptic or silent pathways identified by genome sequencing. Recently, a Global Natural Products Social Molecular Networking (GNPS) initiative has been formed to establish an infrastructure to share information on MS/MS with the goal of bringing the analysis of chemotype up to the standard of genotype afforded by BLAST analyses.[112] Interested readers are directed to other key articles on metabolomics/proteomics approaches to dereplicate known compounds, to identify compounds related to known compounds, and to identify novel SMs.[32–36,40]

1.2.2.2.3 Enzyme Structural Biology and Function

There have been important recent advances on the basic understanding of the biochemistry and structural biology of NP biosynthesis.[3] Importantly, many advances have been made on the understanding of protein-protein interactions critical for NRPS and type I PKS mega-enzyme function.[113–120] Continued studies of the enzymology of NP biosynthesis will be critical to accelerate the establishment of connections between genotype and chemotype to aid in NP discovery, and to apply the newfound information from genome mining to combinatorial biosynthesis.[114,120,121]

1.2.3 Strategies and Tactics for Genome Mining

1.2.3.1 Mining Individual Microbes

There are several examples of genome mining from individual finished *Streptomyces* genomes.[9–14,72,122] These examples are important because they have helped establish heterologous expression hosts and other molecular genetic tools to activate cryptic SMGCs. Although the scale of discovery of new and novel SMs from single microbes is not adequate for industrial drug discovery, these ongoing studies are important to validate approaches that can be applied at a much larger scale. I discuss some examples below.

1.2.3.1.1 Streptomyces avermitilis

S. avermitilis, the producer of avermectin, was known to also produce oligomycin and filipin prior to genome sequencing.[123] Genome sequencing unveiled an additional 35 SMGCs,[10] 13 of which have been subsequently identified.[14] The genomic localization of the major SMGCs has allowed the Ōmura/Ikeda group to delete many of them, including avermectin, oligomycin, filipin, and terpene BGCs, and to minimize the genome size from 9.0 to 7.3 Mb.[14,123,124] The genome-minimized strains are good hosts for heterologous expression of SMGCs. In the initial studies, they expressed 26 diverse SMGCs either from native promoters or after manipulating transcription (e.g., by inserting a strong constitutive promoter in front of a multicistron).[14,123] A subsequent study used a strong constitutive *rpsJ* promoter to drive the expression of 29 terpene synthase genes from actinomycetes; 60 structures were identified, 13 of which were novel.[125] These genome-minimized hosts are particularly useful for expressing cryptic pathways because novel mass ions can be readily identified with minimal SM background.

1.2.3.1.2 Streptomyces coelicolor

S. coelicolor is a model streptomycete that has been the source of extensive academic studies on the genetics and regulation of antibiotic biosynthesis beginning with the seminal work on genetic recombination by David Hopwood in the 1960s, and was the first streptomycete developed for gene cloning. Prior to genome sequencing, it was known that the *S. coelicolor* chromosome encoded actinorhodin (Act), undecylprodigiosin (Red), and calcium-dependent antibiotic (CDA), and a plasmid encoded methylenomycin.[9] Genome sequencing revealed an additional 18 SMGCs, 13 of which have been identified in subsequent studies.[12] One of the cryptic pathways, the type I PKS coelimycin P1 encoded by the *cpk* BGC,[12] was revealed by deleting the *scbR2* gene, which encodes a gamma-butyrolactone binding protein that represses the expression of the *cpk* gene cluster.[126] This example helped validate the strategy of disrupting negative regulatory genes as a general approach to activate cryptic pathways in homologous hosts.[42,97] *S. coelicolor* has been one of the standard hosts for heterologous expression of SMGCs.[42,96,97] Recently, improved *S. coelicolor* expression hosts, strain M1154 and related strains,[127,128] have been developed by deleting the Act, Red, CDA, and Cpk BGCs, and by selecting for *rpsL* (K88E) and *rpoB* (S433L) mutations that enhance expression of many SMGCs. *S. coelicolor* M1154 and two related strains have been used to express at least 18 heterologous SMGCs encompassing many different classes,[128] thus establishing them as important hosts for genome mining. For example, *S. coelicolor* M1146 was used to express the BGC of taromycin A, a cryptic halogenated lipopeptide antibiotic related to daptomycin that is encoded by the marine actinomycete, *Saccharomonospora* sp. CQN-490. Taromycin A was produced in *S. coelicolor* only after deleting the *tar20* gene, which encodes a LuxR family transcriptional repressor.[129]

1.2.3.1.3 Streptomyces ambofaciens

S. ambofaciens ATCC 23877 is the parent strain of the commercial producers of the 16-membered macrolide antibiotic spiramycin, and has been the subject of many studies on genetic instability in the terminal arms of the linear chromosome.[130] The complete genomic sequences of *S. ambofaciens* ATCC 23877[11,131] and *S. ambofaciens* DSM 40697[132] have facilitated the mapping of loci involved in genome plasticity.

S. ambofaciens ATCC 23877 was known to produce spiramycin and netropsin (congocidine) prior to genome sequencing, which revealed 23 additional cryptic SMGCs, of which 10 have been expressed recently.[11] Two SMs were identified by manipulating pathway regulation. The giant 51-membered polyketide antibacterial stambomycin serves as an important example of activating a cryptic pathway by constitutively expressing SamR0469, a large ATP-binding LuxR (LAL) positive regulator.[11,133] Likewise, deletion of the negative regulator *alpW* facilitated the identification of two kinomycin gene clusters located in the long terminal repeats at the ends of the linear chromosome.[11,134]

Three decades ago, it was demonstrated that *S. ambofaciens* ATCC 15154 was a good host for protoplast transformation with plasmid DNA,[135] but ~50% of colonies from regenerated protoplasts lost the ability to produce spiramycin. A stable derivative that produced high levels of spiramycin was generated by several rounds of *N*-methyl-*N′*-nitro-*N*-nitrosoguanidine mutagenesis and protoplast regeneration.[136] A derivative of the stable high producer (BES2074) defective in spiramycin and netropsin production was generated for cloning purposes,[137] and was used more recently for heterologous expression of the lipopeptide antibiotic gene cluster of A54145 at ~400 mg/L.[138] *S. ambofaciens* ATCC 23877 has an Sfp-like phosphopantetheinyl transferase (PPTase) that can activate acyl carrier proteins (ACPs) and peptidyl carrier proteins (PCPs) to convert apo-enzymes into functional holo-enzymes.[139] Expression of this broad-specificity PPTase may distinguish *S. ambofaciens* as an advantageous host for the heterologous expression of SMGCs employing PKS and NRPS biosynthetic mechanisms.[42]

1.2.3.1.4 Streptomyces albus *J1074*

Streptomyces albus J1074 is a nonrestricting expression host that has been used for the heterologous expression of many SMGCs.[42,96,97] The complete genome sequence has revealed that *S. albus* J1074 contains a naturally minimized genome of 6.84 Mb that encodes 22 cryptic SMGCs.[140] Five of the cryptic SMGCs have been activated by inserting the strong constitutive *ermE*p* promoter in front of positive regulatory or other genes, or by disrupting a negative regulatory gene, thus further validating these approaches to activate cryptic SMGCs.[122] *S. albus* differs from other streptomycete expression hosts in that it harbors two very active *attB* sites for the insertion of plasmids containing the øC31 attachment/integration (*attP*/*int*) system, so SMGCs cloned in such vectors naturally generate gene cluster duplications which generally translate into higher product yields.[141,142] Other fascinating aspects of *S. albus* molecular biology and ecology have been reviewed.[42]

1.2.3.1.5 *Other Heterologous Expression Hosts*

There are a number of other expression hosts, mostly derived from commercial production strains, that can be considered for the expression of specific types of SMGCs. These have been reviewed elsewhere.[42,96,97]

1.2.3.2 Mining Multiple Microbes

For robust drug discovery, it is critical to be able to apply genome mining to very large sets of microbial genomes. There are a number of different strategies for mining multiple microbes directed at known chemistry, chemical targets, and novel chemistry. They have one element in common—molecular beacons—or specific genes or combinations of genes that can be used to seek out SMGCs of interest from large populations of cells, pooled DNA, or genome sequences. I discuss several examples of use of these strategies in the following sections.

1.2.3.2.1 Clinically Validated Chemistry

A somewhat conservative, highly predictable starting point for efficient mining of new SMGCs from much larger sets of microbes is to search for SMGCs related to known SMGCs that encode chemical scaffolds already validated with commercial products or promising clinical candidates.

1.2.3.2.1.1 Phosphonates Metcalf and colleagues were interested in discovering novel SMGCs encoding phosphonates[143] which have validated commercial products: phosphomycin, an antibacterial for human medicine; and bialophos, a herbicide for plant crop protection.[144] They first established that the *pepM* gene, encoding phosphoenolpyruvate phosphomutase that catalyzes the formation of phosphonopyruvate, is a good molecular beacon for phosphonate biosynthetic gene clusters.[145] They extracted DNA from over 10,000 actinomycetes from the Northern Regional Research Laboratory (NRRL) culture collection, and identified 403 strains that gave positive PCR signals for *pepM*.[143] They obtained draft sequences, and confirmed the presence of *pepM* in 278 strains, mostly *Streptomyces* species (87%). They explored additional genes in the vicinity of the *pepM* genes, and identified 64 discrete gene cluster types, only 9 of which could be assigned to known phosphonate clusters. Their efforts unveiled 19 new phosphonic acid NPs, and the collection of strains and genome sequences now available on the National Center for Biotechnology Information (NCBI) website provide valuable resources for future genome mining.

1.2.3.2.1.2 Enediynes Shen and colleagues were interested in the discovery of novel enediynes.[146,147] The enediyne calicheamicin has been clinically validated as a potent warhead coupled to different monoclonal antibodies to target double-strand DNA in specific tumor cells.[148,149] The Shen group identified a cluster of five genes conserved in 9- and 10-membered enediyne gene clusters as molecular beacons to search for related pathways in the NCBI and the Joint Genome Institute (JGI) genome databases by BLASTp, then coupled the hits with *g*enome *n*eighborhood *n*etwork (GNN) bioinformatics analysis.[150] GNN was developed as a bioinformatics tool to predict enzymatic functions on a large scale based upon the proximity of genes with uncharacterized functions to those with known functions from multiple genomes. This type of analysis can identify clusters of genes encoding identical, related, or highly diverged SMs. Shen and colleagues identified 87 potential enediyne gene clusters from 78 bacterial strains by analyzing 40 open reading frames upstream and downstream of the conserved iterative PKS gene.

It is noteworthy that 68 of the strains were actinomycetes, consistent with their general high capacity to encode multiple SMs (Table 1.1). About 2% of the actinobacteria surveyed encoded enediynes, compared to 0.035% of all other bacteria surveyed. It should be emphasized that many actinobacteria do not have large genomes. Data from the GEBA-1 genome sequencing project[51] indicate that >50% have genomes <4.0 Mb, and only 11% have genomes >8.0 Mb. When Warp Drive Bio surveyed newly sequenced actinomycete genomes, they noted that ~25% of the strains encoded enediyne gene clusters.[151] Both studies identified many novel enediyne gene clusters, demonstrating a robust genomics strategy to discover new enediyne warheads for drug development. Data from both groups further emphasize the special status of actinomycetes with large genomes as strategic starting points for genome mining and drug discovery.

1.2.3.2.1.3 Rapamycin/FK506 Scientists at Warp Drive Bio have taken an aggressive approach to genome mining by first developing a massive database of genome sequences from actinomycetes, then searching those sequences for SMGCs related to known SMGCs of clinical relevance.[151] For example, they sequenced ~135,000 genomes in pools, then identified pools containing molecular beacons for pathways related to the immune modulators rapamycin and FK506. After deconvolution, they obtained 157 finished genome sequences of candidate strains by using a combination of Illumina and PacBio sequencing. The finished genomes encoded SMCGs ranging from 10 to 56 SMGCs per strain as determined by standard antiSMASH 3.0, with a mean of 35 per genome. They identified a number of rapamycin and FK506 producers as well as ~15 novel pathways related to rapamycin or FK506. They expressed many of the pathways in heterologous hosts, and enhanced expression levels by promoter engineering and overexpression of LAL-positive regulatory genes. They screened the novel molecules using a *s*mall *m*olecule-*a*ssisted *r*eceptor *t*argeting (SMART) approach that exploits the properties of rapamycin and FK506 as facilitators of protein-protein interactions (e.g., rapamycin facilitates interaction between FKBP12 and mTor).[152] One novel compound (WDB-002)[153] facilitates a protein-protein interaction between FKBP12 and a flat coiled-coil in CEP250, a protein involved in centrosome function. The flat coiled-coil binding site does not contain a traditional pocket for drug binding, and therefore would be considered *undruggable* by conventional wisdom, yet the binding affinity was sub-nanomolar. This striking observation opens the possibility of using the SMART platform to target other important proteins involved in human diseases that have been considered to be not druggable by medicinal chemistry. Importantly, multiple new rapamycin and FK506 gene clusters were discovered, along with multiple copies of some of the novel analog clusters. These serve as valuable starting materials for computational learning about important protein-protein interactions in the type I PKS mega-enzymes that can be exploited in combinatorial biosynthesis to generate many more related SMART molecules (Verdine and Gray, unpublished).

The large Warp Drive Bio collection of genomic data from actinomycetes can also be mined for other molecules with validated or novel chemistry, or directed at specific targets as discussed below. It is noteworthy that the number of SMGCs in the Warp

Drive collection of fully sequenced actinomycete genomes increased with genome size, consistent with other finished actinomycete genomes in NCBI (Figure 1.2).[15] Also, the number of SMGCs observed in the fully sequenced Warp Drive collection indicates that >60% are from gifted or highly gifted actinomycetes.[15]

1.2.3.2.2 Target-Directed Genome Mining

Target-directed genome mining[154] can be used to discover novel antibiotics in the absence of prior knowledge of chemical structures, based on the strategy of targeting validated or nearly validated drug targets. The strategy is based on the considerable literature on antibiotics, establishing that genes for pathway regulation, host resistance, and compound transport are nearly always clustered with the biosynthetic pathway genes. Furthermore, two of the major mechanisms for host resistance are enzymatic modification of antibacterial targets and expression of redundant refractory targets, as exemplified by resistances to the following: (i) macrolide antibiotics such as erythromycin and tylosin (ribosome modification); (ii) aminoglycosides such as kanamycin, gentamycin, and tobramycin (ribosome modification); (iii) the ansamycin rifamycin (RNA polymerase refractory target); (iv) the aminocoumarins novobiocin and chlorobiocin (DNA gyrase refractory target); (v) the glycopeptide vancomycin (peptidoglycan refractory target); (vi) the elfamycins kirrothricin and efrotomycin (EF-Tu refractory target); (vii) cerulenin and platensin (fatty acid synthase refractory target); (viii) salinosporamide A and cinnabaramide (refractory proteasome target); and (ix) griselimycin (DNA replication refractory target).[155–159] The resistance mechanisms employing refractory targets have the common feature that the antibiotic-producing strains encode two copies of the target: one susceptible to the antibiotic as part of the core primary metabolic functions, and one resistant encoded in the antibiotic gene cluster, which serves as the molecular beacon. Tang et al.[154] have exploited this property to search for SMGCs encoding antibiotics directed at fatty acid biosynthesis targets. They carried out a bioinformatics search of 86 *Salinispora* genomes for duplications of housekeeping genes involved in fatty acid metabolism. They identified an *S. pacifica* strain that encoded two copies of FabB/F involved in fatty acid biosynthesis. One of the genes associated with an SMGC encoded a protein with high sequence similarity to PtmP3 and PtnP3 FabB/F homologs associated with resistance to platensimycin and platensin, respectively, both of which target FabB/F.[158] They cloned the previously uncharacterized gene cluster harboring the FabB/F homolog and expressed the pathway in *S. coelicolor* M1152.[127,128] The recombinant produced thiolactomycins. Further BLAST searches identified a novel SMGC in *Streptomyces afganiensis* that harbored two genes encoding FabB/F homologs. This pathway was cloned and expressed in *S. coelicolor* M1152, and the recombinant produced several new thiolactomycin analogs. This target-directed approach may have general utility for the discovery of new and novel antibacterial agents by genome mining.

1.2.3.2.3 Other Molecular Beacons

1.2.3.2.3.1 MbtH Homologs There are other approaches to identify gifted microbes in general or microbes that produce novel SMs that inhibit specific validated drug targets. Many NRPS pathways include *mbtH* homologs that encode nonenzymatic chaperones, typically of 65–80 amino acids, which enhance certain adenylation reactions.[43] MbtH homologs that are associated with very similar SM pathways show high (orthologous) sequence similarities, whereas those from unrelated pathways are highly divergent (paralogous). Conserved 60 amino acid segments from 24 MbtH homologs from actinomycetes were concatenated and used as a multiprobe to query microbial genomes by BLASTp analysis.[44] The sequence homologies of the individual 24 MbtH segments can be read out as a numerical code to identify NRPS pathways related to known pathways, or to identify potentially novel pathways.[44] For instance, the code for the structurally related glycopeptide antibiotics vancomycin, balhimycin, and dalbamycin (333,333,332, 333,322,222,223,322), encoded by *Amycolatopsis orientalis*, *Amycolatopsis balhimycini*, and *Nonomuraea* sp. ATTC39727, respectively, differs substantially from the consensus code for the structurally related antitumor antibiotics bleomycin, tallysomycin, and zorbamycin (222,222,223,222,222,002,112,223), encoded by *Streptomyces verticillus*, *Streptoalloteichus hindustanus*, and *Streptomyces flavoviridis*, respectively. Inspection of the MbtH homolog sequences from the latter three actinomycetes indicates that each encoded an MbtH homolog of 187–195 amino acids, over twice the size of typical MbtH homologs.[43,44] The *N*-terminal segment of these proteins is homologous to other MbtH homologs, whereas the C-terminal region has no motifs related to other proteins in GenBank. Therefore, these unique proteins represent novel beacons specifically for SM pathways related to bleomycin, tallysomycin, and zorbamycin. Bleomycin is a validated antitumor agent that is used to treat Hodgkin lymphoma and testicular germ-cell tumors, but it has significant toxic liabilities.[160] Genome mining might provide a means to identify many natural homologs of bleomycin to screen for improved efficacy and lower toxicity. BLASTp analysis with the bleomycin MbtH homolog ORF13 identified three additional actinomycete strains—*Streptomyces mobaraensis*, *Mycobacterium abscessus*, and *Actinosynnema mirum*—that encode ORF13 homologs ranging from 54% to 99% amino acid sequence identities. The bleomycin, tallysomycin, and zorbamycin gene clusters contain conserved NRPS genes (BlmX, TlmX, and ZmbX) that encode a di-module (CAT-CAT); BLASTp analysis of *S. mobaraensis*, *M. abscessus*, and *A. mirum* with BlmX identified homologs ranging from 53% to 99% sequence identities. Therefore, these three actinomycetes are likely to encode SMs in the bleomycin family. This single-protein MbtH beacon is highly specific, unlike the BlmX NRPS homologs which identify massive numbers of somewhat related NRPSs from BLASTp searches of nonredundant sequences in NCBI, and can be used to search for additional members of the bleomycin family among large sets of public and private genome databases.

The MbtH multiprobe can be used to identify gifted microbes and novel pathways by BLASTp analysis of pooled DNA from cultured microbes or from metagenomic samples for the discovery of novel NRPS pathways, and to help prioritize strains for complete genome sequencing. The small size of MbtH homologs makes the multiprobe particularly suitable for surveying draft genomes of actinomycetes and other bacteria to identify gifted strains for further analysis.[44]

1.2.3.2.3.2 Halogenases Many potent NPs have halogen residues that are important for *in vivo* activities. Over 4000 halogenated NPs have been described,[161] including, vancomycin, teicoplanin, chlorotetracycline, chloramphenicol, chlorobiocin, avilamycin, complestatin, and calicheamicin. The gene clusters for these diverse molecules contain genes encoding $FADH_2$-dependent halogenases which can be used as beacons to search for SMGCs related to known SMGCs or for unrelated SMGCs, some fraction of which will encode novel structures. Hornung et al.[161] made a degenerate probe for PCR amplification of genes from a random set of 550 actinomycetes. Of these, 103 gave positive signals. Of these strains, 35 were further analyzed by sequencing the adjacent genes, and 2 fell into a glycopeptide clade that also included vancomycin, balhimycin, chloroeremomycin, and teicoplanin, and 4 in a related glycopeptide clade that included complestatin and A47934. They carried out fermentations and supernatants were analyzed by HPLC-ESI-MS/MS, focusing on an *m/z* range for glycopeptides, and looked for the presence of isotope patterns characteristic of halogens. Several were confirmed to produce glycopeptides. They also characterized a novel halogenated type II polyketide with potent antibacterial activities against gram-positive pathogens and *Escherichia coli*, by cosmid cloning and expression in *S. albus* J1074. It is noteworthy that over 1% of the random actinomycetes chosen encode glycopeptide antibiotics, and that the halogenase beacon guided them to focus on mass ions that might have been overlooked by traditional analyses. Although they did not sequence the producing organisms in 2007, with the substantial reductions in sequencing costs in the meantime, this approach can now be used to identify microbes that produce halogenated SMs as a way to prioritize strains for sequencing.

1.2.3.2.3.3 NRPS and Type I PKS Clusters The majority of important secondary metabolites produced by microorganisms employ NRPS, PKS-I, or mixed NRPS/PKS-I biosynthetic mechanisms (Table 1.3).[3] One way to define gifted microbes is to count the number of these gene clusters.[15,162] This can be done effectively by surveying finished genomes with antiSMASH 3.0. However, these pathways are not effectively assembled in unfinished genomes,[162] which comprise >90% of *Streptomyces* and other actinomycete genomes (Table 1.4). Indeed, NRPS, PKS-I, and NRPS/PKS-I gene clusters are fragmented, misassembled in different ways, and overestimated in total numbers by two- to threefold by antiSMASH 3.0.[162] To more accurately estimate the numbers of these large SMGCs, a small beacon that could be adapted for counting the clusters was required. Most NRPS/PKS-I mega-enzymes contain single thioesterase (TE) domains preceded by a peptidyl-carrier protein (PCP) or acyl-carrier protein (ACP) in terminal modules to release the completed linear or cyclic structures. These PCP-TE and ACP-TE di-domains are relatively small, and their coding sequences have very low probability of misassembly. Five PCP-TEs and five ACP-TEs were concatenated to generate individual penta-probes to survey actinomycete genomes.[162] These multi-probes picked up ~70% of NRPS, type-I PKS, and mixed NRPS/type I PKS gene clusters identified by antiSMASH 3.0 analysis of finished actinomycete genomes, and were suitable to identify gifted actinomycetes from among strains with draft genome sequences, in spite of poor assemblies of the NRPS/type I PKS clusters as evidenced by antiSMASH 3.0 analyses.

TABLE 1.4
Genome Assemblies of Actinobacteria and Proteobacteria

Taxon	Genome Assemblies	% of Total	Finished Assemblies	% Finished
Actinobacteria	7,863	100	605	7.7
Mycobacterium sp.	4,527	57.6	153	3.4
M. tuberculosis	3,638	46.3	48	1.3
Streptomyces sp.	752	9.6	51	6.8
Nocardia sp.	103	1.3	5	4.9
Amycolatopsis sp.	35	0.4	9	25.7
Micromonosposa sp.	21	0.3	2	9.5
Actinoplanes sp.	12	0.2	4	33.3
Saccharopolyspora sp.	7	0.1	1	14.3
Streptosporangium sp.	3	0.04	1	33.3
Dactylosporangium sp.	2	0.02	0	0.00
Proteobacteria	33,627	100	2847	8.5
Escherichia coli	4,706	14.0	192	4.1
Pseudomonas sp.	2,625	7.8	169	6.4
Pseudomonas aeruginosa	1,657	4.9	58	3.5
Burkholderia sp.	1,210	3.6	126	10.4
B. pseudomallei	407	1.2	50	12.3
B. cepacia	92	0.3	5	5.4
B. mallei	46	0.1	15	32.6
Myxobacteria (total)	41	0.1	19	46.3
Sorangium sp.	10	0.03	2	20.0
Myxococcus sp.	7	0.02	5	71.4
Photorhabdus sp.	19	0.06	2	10.5

Source: Data generated by author from information available on the NCBI website at http://www.ncbi.nlm.nih.gov/assembly/organism/.

1.3 FUTURE DIRECTIONS

1.3.1 Expanded Sampling for Gifted Microbes

Janus, the Roman god of beginnings and transitions, had two faces, and could look to the future while looking to the past. We are in a transition to a new way of discovering NPs by microbial genome mining. We are also in an unprecedented position to look not only to the past for wisdom, but also to the future to predict with clarity the best path for successful NP drug discovery in the coming years.[15] When we look to the past, it is well established that the most productive microbes for NP discovery and commercialization have been the actinomycetes, most notably species of the genus *Streptomyces*.[3,15,163,164] In addition to their successful track record, their genomes contain a multitude of cryptic SMGCs that are fertile sources for NP discovery.[9–19] Inexpensive genome sequencing gives us a *molecular crystal ball* that lets us gaze into the future.[15] One way to gaze is to query finished genomes for the numbers and novelty of SMGCs bioinformatically by antiSMASH 3.0[30] and/or PRISM.[28] When microbes are queried in this way, three things become apparent: (i) most culturable bacteria and archaea are poor sources for SMGCs encoding drug-like NPs; (ii) uncultured microbes (>99% of all microbes) generally have small genomes containing little or no SMGCs encoding drug-like NPs; (iii) some bacterial species within the Proteobacteria, Cyanobacteria, and Firmacutes are good sources for NP production, and these generally have large genomes; and (iv) the best sources, including some truly gifted species, are found within the *Streptomyces* sp. and other actinomycetes with genomes >8.0 Mb.[15] Indeed, many gifted actinomycetes devote more coding capacity to NP biosynthesis than many other cultured and uncultured microbes devote to all functions.

Only a very small sampling of the surface of the globe for SM producers has occurred during the last six decades.[5] It is well documented that different soils and different climates favor the establishment of different actinomycete taxa,[5,7,165] and a wealth of culturable microbes with large genomes encoding multiple SMs await discovery. It is now possible to survey soil samples for new and novel SM genes directly by metagenomic sampling[61–64] before initiating the isolation of microbes. Coupling metagenomic sampling with new strain isolations and genome sequencing presents an unprecedented opportunity to reboot a previously very successful industry that went into decline in the 1980s and 1990s for lack of relevant new technologies. The new technologies are now available.

1.3.2 Expanded Efforts to Obtain Finished Genomes for Gifted Microbes

For microbial genome mining to have a substantial impact on drug discovery, it will be important to continue accumulating complete genome sequences from the most gifted microbes that encode large numbers of new and novel SMGCs encoding drug-like molecules. Historical records indicate that large, complex molecules encoded by NRPS, PKS, and mixed NRPS/PKS biosynthetic mechanisms will continue to be rich sources, and finished genomes are needed to accurately assemble NRPS and type I PKS gene clusters. The current acceptance of permanent draft sequences is not adequate to identify truly novel SMGCs from these classes.

There are large discrepancies between the numbers of publicly available genome sequences from pathogenic bacteria and the bacteria that are the major contributors to current and future antibiotic discovery. Table 1.4 shows examples from the Actinobacteria and Proteobacteria, two bacterial taxa that contain both pathogens and gifted producers of antibiotics and other SMs. On August 16, 2016, 71,863 genome sequences from the domain *Bacteria*, including 5,525 completed (or finished) assemblies were available on the NCBI website. Of the total, 33,627 and 7,863 genome sequences were from Proteobacteria and Actinobacteria, respectively. The Actinobacteria sequences were dominated by 4527 genome sequences from *Mycobacterium* sp., including 3638 from the human pathogen, *Mycobacterium tuberculosis*. In contrast, 752 sequences were from antibiotic-producing *Streptomyces* sp., of which only 51 have been completed. Other genera of Actinobacteria that produce important SMs (*Amycolatopsis*, *Micromonospora*, *Actinoplanes*, *Saccharopolyspora*, *Streptosporangium*, and *Dactylosporangium*) account for only 80 genome sequences, 17 of which have been completed.

The genome sequences from Proteobacteria are also dominated by pathogens (e.g., *E. coli* and *Pseudomonas aeruginosa*; Table 1.4). In contrast, the Myxobacteria and *Photorhabdus* sp., which have members gifted for SM production (Table 1.1), account for only 60 sequences (<0.2% of the total). Furthermore, a single species, *E. coli*, has 4706 sequences publicly available, or more than sixfold the number from all *Streptomyces* strains, which include >500 recognized species. Sequencing more *E. coli* strains will have minimal impact on drug discovery, whereas a concerted effort to *catch up* by sequencing multiple *Streptomyces* and other actinomycete strains could have a substantial impact on revitalizing antibiotic drug discovery.

In this regard, it would be worthwhile to identify a large set of actinomycetes that produce known antibiotics or other important SMs for finished genome sequencing as a reference set to help link genotype with chemotype. The 764 unfinished genomes from *Streptomyces*, *Amycolatopsis*, *Micromonospora*, *Actinoplanes*, *Saccharopolyspora*, *Streptosporangium*, and *Dactylosporangium*, along with other actinomycetes not included in Table 1.4, would be a good starting point. Warp Drive Bio has obtained complete sequences of ~150 actinomycete genomes by using a combination of Illumina and PacBio sequencing technologies,[151] so a worldwide effort to sequence 1000–2000 actinomycete genomes to completion for public availability is not a tall order, and would make a highly significant contribution to the future of NP discovery for drug development. These could also include a significant sampling of actinomycete symbionts of arthropods such as fungus-farming ants, fungus-growing termites, and southern pine beetles,[166] and actinomycetes from other specialized niches identified by metagenomic sampling.

There have been efforts to sample the SM coding capacity within the phylum Cyanobacteria,[38,167,168] and *Burkholderia* sp.[169] It would seem worthwhile to begin concerted efforts to obtain complete genome sequences from larger sets of gifted microbes identified among the Proteobacteria (e.g., Myxobacteria, nonpathogenic pseudomonads, *Photorhabdus* sp., including symbionts of nematodes).[170] It would also be prudent to de-emphasize drug discovery efforts on culturable and uncultured microbes with small genomes, and to dispense with the notion that unculturable microbes will be robust sources of new antibiotics. Genome sequencing data do not support this conjecture (Table 1.2).[15]

1.3.3 Leveraging Genomics for Combinatorial Biosynthesis

Combinatorial biosynthesis has been in the process of development for over three dec ades,[3,114,115,121,171–175] and much has been learned in recent years on the fundamental mechanistic interactions critical for NRPS and type I PKS function directly relevant to successful combinatorial biosynthesis.[114–120] Microbial genome mining is adding an enormous repertoire of new genes, domains, and modules (parts and devices) that can be used in combinatorial biosynthesis. Perhaps more importantly, a natural outcome of obtaining finished-quality genome sequences, and a side product of the dereplication process, is the accumulation of multiple complete copies of important SMGCs. This offers and unprecedented opportunity to apply computational learning to identify amino acid sequences in NRPS and type I PKS multienzymes critical for protein-protein interactions and maintenance of high catalytic function. Coupling microbial genome mining with combinatorial biosynthesis holds great promise for the future of NP discovery.

REFERENCES

1. Butler, M. S.; Robertson, A. A.; Cooper, M. A. Natural product and natural product derived drugs in clinical trials. *Nat. Prod. Rep.* **2014**, 31, 1612–1661.
2. Demain, A. L. Importance of microbial natural products and the need to revitalize their discovery. *J. Ind. Microbiol. Biotechnol.* **2014**, 41, 185–201.
3. Katz, L.; Baltz, R. H. Natural product discovery: Past, present, and future. *J. Ind. Microbiol. Biotechnol.* **2016**, 43, 155–176.
4. Newman, D. J.; Cragg, G. M. Natural products as sources of new drugs from 1981 to 2014. *J. Nat. Prod.* **2016**, 79, 629–661.
5. Baltz, R. H. Antibiotic discovery from actinomycetes: Will a renaissance follow the decline and fall? *SIM News* **2005**, 55, 186–196.
6. Baltz, R. H. Marcel Faber Roundtable: Is our antibiotic pipeline unproductive because of starvation, constipation or lack of inspiration? *J. Ind. Microbiol. Biotechnol.* **2006**, 33, 507–513.
7. Genilloud, O.; González, I.; Salazar, O.; Martín, J.; Tormo, J. R.; Vincente, F. Current approaches to exploit actinomycetes as a source of novel natural products. *J. Ind. Microbiol. Biotechnol.* **2011**, 38, 375–389.
8. Baltz, R. H. Renaissance in antibacterial discovery from actinomycetes. *Curr. Opin. Pharmacol.* **2008**, 8, 557–563.
9. Bentley, S. D.; Chater, K. F.; Cerdeño-Tárraga, A. M.; Challis, G. L.; Thomson, N. R.; James, K. D.; Harris, D. E. et al. Complete genome sequence of the model actinomycete *Streptomyces coelicolor* A3(2). *Nature* **2002**, 417, 141–147.
10. Ikeda, H.; Ishikawas, J.; Hanamoto, A.; Shinose, M.; Kikuchi, H.; Shiba, T.; Sakaki, Y.; Hattori, M.; Ōmura, S.. Complete genome sequence of and comparative analysis of the industrial microorganism *Streptomyces avermitilis*. *Nat. Biotechnol.* **2003**, 21, 526–531
11. Aigle, B.; Lautru, S.; Spiteller, D.; Dickschat, J. S.; Challis, G. L.; Leblond, P.; Pernodet, J. L. Genome mining of *Streptomyces ambofaciens*. *J. Ind. Microbiol. Biotechnol.* **2014**, 41, 251–264.
12. Challis, G. L. Exploitation of the *Streptomyces coelicolor* A3(2) genome sequence for discovery of new natural products and biosynthetic pathways. *J. Ind. Microbiol. Biotechnol.* **2014**, 41, 219–232.
13. Iftime, D.; Kulik, A.; Härtner, T.; Rohrer, S.; Niedermeyer, T. H.; Stegmann, E.; Weber, T.; Wohlleben, W. Identification and activation of novel biosynthetic gene clusters by genome mining in the kirromycin producer Tü 365. *J. Ind. Microbiol. Biotechnol.* **2016**, 43, 277–291.
14. Ikeda, H.; Shin-ya, K.; Ōmura, S. Genome mining of the *Streptomyces avermitilis* genome and development of genome-minimized hosts for heterologous expression of biosynthetic gene clusters. *J. Ind. Microbiol. Biotechnol.* **2014**, 41, 233–250.
15. Baltz, R. H. Gifted microbes for genome mining and natural product discovery. *J. Ind. Microbiol. Biotechnol.* **2017**, 44, 573–588.
16. Bachmann, B. O.; Van Lanen, S. G.; Baltz, R. H. Microbial genome mining for accelerated natural products discovery: Is a renaissance in the making? *J. Ind. Microbiol. Biotechnol.* **2014**, 41, 175–184.
17. Corre, C.; Challis, G. L. New natural product chemistry discovered by genome mining. *Nat. Prod. Rep.* **2009**, 26, 977–986.
18. Doroghazi, J. R.; Metcalf, W. W. Comparative genomics of actinomycetes with a focus on natural product biosynthetic genes. *BMC Genomics* **2013**, 14, 611.
19. Doroghazi, J. R.; Albright, J. C.; Goering, A. W.; Ju, K. S.; Haines, R. R.; Tchalukov, K. A.; Labeda, D. P.; Kelleher, N. L.; Metcalf, W. W. A roadmap for natural product discovery based on large-scale genomics and metabolomics. *Nat. Chem. Biol.* **2014**, 10, 963–968.

20. Zerikly, M.; Challis, G. L. Strategies for the discovery of new natural products by genome mining. *ChemBioChem* **2009**, 10, 625–633.
21. Brakhage, A. A. Regulation of fungal secondary metabolism. *Nat. Rev. Microbiol.* **2013**, 11, 21–32.
22. Li, Y. F.; Tsai, K. J.; Harvey, C. J.; Li, J. J.; Ary, B. E.; Berlew, E. E.; Boehman, B. L. et al. Comprehensive curation and analysis of fungal biosynthetic gene clusters of published natural products. *Fungal Genet. Biol.* **2016**, 89, 18–28.
23. Van der Lee, T. A.; Medema, M. H. Computational strategies for genome-based natural product discovery and engineering in fungi. *Fungal Genet. Biol.* **2016**, 89, 29–36.
24. Wiemann, P.; Keller, N. P. Strategies for mining fungal natural products. *J. Ind. Microbiol. Biotechnol.* **2014**, 41, 301–313.
25. Yaegashi, J.; Oakley, B. R.; Wang, C. C. Recent advances in genome mining of secondary metabolite biosynthetic gene clusters and the development of heterologous expression systems in *Aspergillus nidulans*. *J. Ind. Microbiol. Biotechnol.* **2014**, 41, 433–442.
26. Walsh, C. T.; Fischbach, M. A. Natural products version 2.0: Connecting genes to molecules. *J. Am. Chem. Soc.* **2010**, 132, 2469–2493.
27. Medema, M. H.; Fischbach, M. A. Computational approaches to natural product discovery. *Nat. Chem. Biol.* **2015**, 11, 639–648.
28. Skinnider, M. A.; Dejong, C. A.; Rees, P. N.; Johnston, C. W.; Li, H.; Webster, A. L.; Wyatt, M. A.; Magarvey, N. A. Genomes to natural products PRediction Informatics for Secondary Metabolomes (PRISM). *Nucleic Acids Res.* **2015**, 43, 9645–9662.
29. Tietz, J. I.; Mitchell, D. A. Using genomics for natural product structure elucidation. *Curr. Top. Med. Chem.* **2016**, 16, 1645–1694.
30. Weber, T.; Blin, K.; Duddela, S.; Krug, D.; Kim, H. U.; Bruccoleri, R.; Lee, S. Y. et al. antiSMASH 3.0: A comprehensive resource for the genome mining of biosynthetic gene clusters. *Nucleic Acids Res.* **2015**, 43, W237–W243.
31. Albright, J. C.; Goering, A. W.; Doroghazi, J. R.; Metcalf, W. W.; Kelleher, N. L. Strain-specific proteogenomics accelerates the discovery of natural products via their biosynthetic pathways. *J. Ind. Microbiol. Biotechnol.* **2014**, 41, 451–459.
32. da Silva, R. R.; Dorrestein, P. C.; Quinn, R. A. Illuminating the dark matter in metabolomics. *Proc. Natl. Acad. Sci. USA* **2015**, 112, 12549–12550.
33. Derewacz, D. K.; Covington, B. C.; McLean, J. A.; Bachmann, B. O. Mapping microbial response metabolomes for induced natural product discovery. *ACS Chem. Biol.* **2015**, 10, 1998–2006.
34. Goering, A. W.; McClure, R. A.; Doroghazi, J. R.; Albright, J. C.; Haverland, N. A.; Zhang, Y.; Ju, K. S.; Thomson, R. J.; Metcalf, W. W.; Kelleher, N. L. Metabologenomics: Correlation of microbial gene clusters with metabolites drives discovery of a nonribosomal peptide with an unusual amino acid monomer. *ACS Cent. Sci.* **2016**, 2, 99–108.
35. Gubbens, J.; Zhu, H.; Girard, G.; Song, L.; Florea, B. I.; Aston, P.; Ichinose, K. et al. Natural product proteomining, a quantitative proteomics platform, allows rapid discovery of biosynthetic gene clusters for different classes of natural products. *Chem. Biol.* **2014**, 21, 707–718.
36. Johnston, C. W.; Connaty, A. D.; Skinnider, M. A.; Li, Y.; Grunwald, A.; Wyatt, M. A.; Kerr, R. G.; Magarvey, N. A. Informatic search strategies to discover analogues and variants of natural product archetypes. *J. Ind. Microbiol. Biotechnol.* **2016**, 43, 293–298.
37. Liu, W. T.; Lamsa, A.; Wong, W. R.; Boudreau, P. D.; Kersten, R.; Peng, Y.; Moree, W. J. et al. MS/MS-based networking and peptidgenomics guided genome mining revealed the stenothricin gene cluster in *Streptomyces roseosporus*. *J. Antibiot.* **2014**, 67, 99–104.
38. Moss, N. A.; Bertin, M. J.; Kleigrewe, K.; Leao, T. F.; Gerwick, L.; Gerwick, W. H. Integrating mass spectrometry and genomics for cyanobacterial metabolite discovery. *J. Ind. Microbiol. Biotechnol.* **2016**, 43, 313–324.

39. Yang, J. Y.; Sanchez, L. M.; Rath, C. M.; Liu, X.; Boudreau, P. D.; Bruns, N.; Glukhov, E. et al. Molecular networking as a dereplication strategy. *J. Nat. Prod.* **2013**, 76, 1686–1699.
40. Yang, L.; Ibrahim, A.; Johnston, C. W.; Skinnider, M. A.; Ma, B.; Magarvey, N. A. Exploration of nonribosomal peptide families with an automated informatics search algorithm. *Chem. Biol.* **2015**, 22, 1259–1269.
41. Wu, C.; Choi, Y. H.; van Wezel, G. P. Metabolic profiling as a tool for prioritizing antimicrobial compounds. *J. Ind. Microbiol. Biotechnol.* **2016**, 43, 299–312.
42. Baltz, R. H. Genetic manipulation of secondary metabolite biosynthesis for improved production in *Streptomyces* and other actinomycetes. *J. Ind. Microbiol. Biotechnol.* **2016**, 43, 343–370.
43. Baltz, R. H. Function of MbtH homologs in nonribosomal peptide biosynthesis and applications in secondary metabolite discovery. *J. Ind. Microbiol. Biotechnol.* **2011**, 38, 1747–1760.
44. Baltz, R. H. MbtH homology codes to identify gifted microbes for genome mining. *J. Ind. Microbiol. Biotechnol.* **2014**, 41, 357–369.
45. Cimermancic, P.; Medema, M. H.; Claesen, J.; Kurita, K.; Wieland Brown, L. C.; Mavrommatis, K.; Pati, A. et al. Insights into secondary metabolism from a global analysis of prokaryotic biosynthetic gene clusters. *Cell* **2014**, 158, 412–421.
46. Donadio, S.; Monciardini, P.; Socio, M. Polyketide synthases and nonribosomal peptide synthases: The emerging view from bacterial genomics. *Nat. Prod. Rep.* **2007**, 24, 1073–1109.
47. Hadjithomas, M.; Chen, I. M.; Chu, K.; Ratner, A.; Palaniappan, K.; Szeto, E.; Huang, J. et al. IMG-ABC: A knowledge base to fuel discovery of biosynthetic gene clusters and novel secondary metabolites. *MBio* **2015**, 6, e00932-15.
48. Wang, H.; Fewer, D. P.; Holm, L.; Rouhianen, L.; Sivonen, K. Atlas of nonribosomal peptide and polyketide biosynthetic pathways reveals common occurrence of nonmodular enzymes. *Proc. Natl. Acad. Sci. USA* **2014**, 111, 9259–9264.
49. Wang, H.; Sivonen, K.; Fewer, D. P. Genome insights into the distribution, genetic diversity and evolution of polyketide synthases and nonribosomal peptide synthetases. *Curr. Opin. Genet. Dev.* **2015**, 35, 79–85.
50. Zhu, F.; Qin, C.; Tao, L.; Liu, X.; Shi, Z.; Ma, X.; Jia, J. et al. Clustered patterns of species origins of nature-derived drugs and clues for future bioprospecting. *Proc. Natl. Acad. Sci. USA* **2011**, 31, 12943–12948.
51. Mukherjee, S.; Seshadri, R.; Varghese, N. J.; Edoe-Fadrosh, E. A.; Meier-Kolthoff, J. P.; Göker, M.; Coates, C.; et al. 1,003 reference genomes of bacterial and archaeal isolates expand coverage of the tree of life. *Nat. Biotechnol.* **2017**, 35, 676–683.
52. Amann, R. I.; Ludwig, W.; Schleifer, K. H. Phylogenetic identification and in situ detection of individual microbial cells without cultivation. *Microbiol. Rev.* **1995**, 59, 143–169.
53. Pace, N. R. Mapping the tree of life: Progress and prospects. *Microbiol. Mol. Biol. Rev.* **2009**, 73, 565–576.
54. Handelsman, J.; Rondon, M. R.; Brady, S. F.; Clardy, J.; Goodman, R. M. Molecular biological access to the chemistry of unknown soil microbes: A new frontier for natural products. *Chem. Biol.* **1998**, 5, R245–R249.
55. Rondon, M. R.; August, P. R.; Bettermann, A. D.; Brady, S. F.; Grossman, T. H.; Liles, M. R.; Loiacono, K. A. et al. Cloning the soil metagenome: A strategy for accessing the genetic and functional diversity of uncultured microorganisms. *Appl. Environ. Microbiol.* **2000**, 66, 2541–2547.
56. Courtois, S.; Cappellano, C. M.; Ball, M.; Francou, F. X.; Normand, P.; Helynck, G.; Martinez, A. et al. Recombinant environmental libraries provide access to microbial diversity for drug discovery from natural products. *Appl. Environ. Microbiol.* **2003**, 69, 49–55.

57. Daniel, R. The soil metagenome: A rich resource for the discovery of novel natural products. *Curr. Opin. Biotechnol.* **2004**, 15, 199–204.
58. Pettit, R. K. Soil DNA libraries for anticancer drug discovery. *Cancer Chemother. Pharmacol.* **2004**, 54, 1–6.
59. Konstantinidis, K. T.; Tiedje, J. M. Trends between gene content and genome size in prokaryoteic species with larger genomes. *Proc. Natl. Acad. Sci. USA* **2004**, 101, 3160–3165.
60. Charlop-Powers, Z.; Milshteyn, A.; Brady, S. F. Metagenomic small molecule discovery methods. *Curr. Opin. Microbiol.* **2014**, 19, 70–75.
61. Charlop-Powers, Z.; Owen, J. G.; Reddy, B. V.; Ternei, M. A.; Brady, S. F. Chemical-biogeographic survey of secondary metabolism in soil. *Proc. Natl. Acad. Sci. USA* **2014**, 111, 3757–3762.
62. Charlop-Powers, Z.; Owen, J. G.; Reddy, B. V.; Ternei, M. A.; Guimaraes, D. O.; de Frias, U. A.; Pupo, M. T.; Seepe, P.; Feng, Z.; Brady, S. F. Global biogeographic sampling of bacterial secondary metabolism. *eLife* **2015**, 4, e05048.
63. Katz, M.; Hover, B. M.; Brady, S. F. Culture-independent discovery of natural products from soil metagenomes. *J. Ind. Microbiol. Biotechnol.* **2016**, 43, 129–141.
64. Milshteyn, A.; Schneider, J. S.; Brady, S. F. Mining the metabiome: Identifying novel natural products from microbial communities. *Chem. Biol.* **2014**, 21, 1211–1223.
65. Owen, J. G.; Charlop-Powers, Z.; Smith, A. G.; Teernei, M. A.; Calle, P. Y.; Reddy, B. V.; Montiel, D.; Brady, S. F. Multiplexed metagenome mining using short DNA sequence tags facilitates targeted discovery of epoxyketone protease inhibitors. *Proc. Natl. Acad. Sci. USA* **2015**, 112, 4221–4226.
66. Reddy, B. V.; Milshtyne, A.; Charlop-Powers, Z.; Brady, S. F. eSNaPD: A versatile, web-based bioinformatics platform for surveying and mining natural product biosynthetic diversity from metagenomes. *Chem. Biol.* **2014**, 21, 1023–1033.
67. Banek, J. J.; Brady, S. F. Cloning and characterization of new glycopeptide gene clusters found in an environmental DNA megalibrary. *Proc. Natl. Acad. Sci. USA* **2008**, 105, 17273–17277.
68. Banek, J. J.; Craig, J. W.; Calle, P. Y.; Brady, S. F. Tailoring enzyme-rich environmental DNA clones: A source of enzymes for generating libraries of unnatural natural products. *J. Am. Chem. Soc.* **2010**, 132, 15661–15670.
69. Thaker, M. N.; Wang, W.; Spanogionnopoulos, P.; Waglechner, N.; King, A. M.; Medina, R.; Wright, G. D. Identifying producers of antibacterial compounds by screening for antibiotic resistance. *Nat. Biotechnol.* **2013**, 31, 922–927.
70. Thaker, M. N.; Waglechner, N.; Wright, G. D. Antibiotic resistance-mediated isolation of scaffold-specific natural product producers. *Nat. Protoc.* **2014**, 9, 1469–1479.
71. Kolter, R.; van Wezel, G. P. Goodbye to brute force in antibiotic discovery? *Nat. Microbiol.* **2016**, 1, 1–2.
72. Ziemert, N.; Alanjary, M.; Weber, T. The evolution of genome mining in microbes: A review. *Nat. Prod. Rep.* **2016**, 33, 988–1005
73. Albertsen, M.; Hugenholtz, P.; Skarsheweski, A.; Nielsen, K. L.; Tyson, G. W.; Nielsen, P. H. Genome sequences of rare, uncultured bacteria obtained by differential coverage binning of multiple metagenomes. *Nat. Biotechnol.* **2013**, 31, 533–538.
74. Anantharaman, K.; Brown, C. T.; Burstein, D.; Castelle, C. J.; Probst, A. J.; Thomas, B. C.; Williams, K. H.; Banfield, J. F. Analysis of five complete genome sequences for members of the class *Peribacteria* in the recently recognized *Peregrinibacteria* bacterial phylum. *Peer J.* **2016**, 4, e1607.
75. Brown, C. T.; Hug, L. A.; Thomas, B. C.; Sharon, I.; Castelle, C. J.; Singh, A.; Wilkins, M. J.; Wrighton, K. C.; Williams, K. H.; Banfield, J. F. Unusual biology across a group comprising more than 15% of domain *Bacteria*. *Nature* **2015**, 523, 208–211.

76. Dupont, C. L.; Rusch, D. B.; Yooseph, S.; Lombardo, M. J.; Richter, R. A.; Valas, R.; Novotny, M. et al. Genomic insights to SAR86, an abundant and uncultivated marine bacterial lineage. *ISME J.* **2012**, 6, 1186–1199.
77. Garza, D. R.; Dutilh, B. E. From cultured to uncultured genome sequences: Metagenomics and modeling microbial ecosystems. *Cell. Mol. Life Sci.* **2015**, 72, 4287–4308.
78. Haroon, M. F.; Thompson, L. R.; Stingl, U. Draft genome of uncultured SAR324 bacterium lautmerah10, binned from a Red Sea metagenome. *Genome Announc.* **2016**, 4, e01711-15.
79. Hess, M.; Sczyrba, A.; Egan, R.; Kim, T. W.; Chokhawala, H.; Schroth, G.; Luo, S. et al. Metagenomic discovery of biomass-degrading genes and genomes from cow rumen. *Science* **2011**, 331, 463–467.
80. Iverson, V.; Morris, R. M.; Frazar, C. D.; Berthiaume, C. T.; Morales, R. L.; Armbrust, E. V. Untangling genomes from metagenomes: Revealing an uncultured class of marine *Eurarchaeota*. *Science* **2012**, 335, 587–590.
81. Kantor, R. S.; Wrighton, K. C.; Handley, K. M.; Sharon, I.; Hug, L. A.; Castelle, C. J.; Thomas, B. C.; Banfield, J. F. Small genomes and sparse metabolisms of sediment-associated bacteria from candidate phyla. *MBio* **2013**, 4, e00708-13.
82. Narasingarao, P.; Podell, S.; Ugalde, J. A.; Brochier, C.; Emerson, J. B.; Brocks, J. J.; Heidelberg, K. B.; Banfield, J. F.; Allen, E. E. *De novo* metagenomic assembly reveals abundant novel major lineage of *Archaea* in hypersaline microbial communities. *ISME J.* **2012**, 6, 81–93.
83. Rinke, C.; Schwientek, P.; Sczyrba, A.; Ivanova, N. N.; Anderson, I. J.; Cheng, J. F.; Darling, A. et al. Insights into the phylogeny and coding potential of microbial dark matter. *Nature* **2013**, 499, 431–437.
84. van Dijk, E. L.; Auger, H.; Jaszczyszyn, Y.; Thermes, C. Ten years of next-generation sequencing technology. *Trends Genet.* **2014**, 30, 418–426.
85. Gomez-Escribano, J. P.; Castro, J. F.; Razmilic, V.; Chandra, G.; Andrews, B.; Bibb, M. J. The *Streptomyces leeuwenhoekii* genome: De novo sequencing and assembly in single contigs of the chromosome, circular plasmid pSLE1 and linear plasmid pSLE2. *BMC Genomics* **2015**, 16, 485.
86. Gomez-Escribano, J. P.; Alt, S.; Bibb, M. J. Next generation sequencing of actinobacteria for the discovery of novel natural products. *Mar. Drugs* **2016**, 14, 78.
87. Rhoads, A.; Au, K. F. PacBio sequencing and its applications. *Genomics Proteomics Bioinformatics* **2015**, 13, 278–289.
88. Yoon, V.; Nodwell, J. R. Activating secondary metabolism with stress and chemicals. *J. Ind. Microbiol. Biotechnol.* **2014**, 41, 415–424.
89. Zhu, H.; Sandiford, S. K.; van Wezel, G. P. Triggers and cues that activate antibiotic production by actinomycetes. *J. Ind. Microbiol. Biotechnol.* **2014**, 41, 371–386.
90. Traxler, M. F.; Kolter, R. Natural products in soil microbe interactions and evolution. *Nat. Prod. Rep.* **2015**, 32, 956–970.
91. Bode, H. B.; Bethe, B.; Hofs, R.; Zeeck, A. Big effects from small changes: Possible ways to explore nature's chemical diversity. *ChemBioChem* **2002**, 3, 619–627.
92. Wang, J.; Soisson, S. M.; Young, K.; Shoop, W.; Kodali, S.; Galgoci, A.; Painter, R. et al. Platensimycin is a selective FabF inhibitor with potent antibiotic properties. *Nature* **2006**, 441, 358–361.
93. Smanski, M. J.; Peterson, R. M.; Rajski, S. R.; Shen, B. Engineered *Streptomyces platensis* strains that overproduce antibiotics platensimycin and platencin. *Antimicrob. Agents Chemother.* **2009**, 53, 1299–1304.
94. Hindra; Huang, T.; Yang, D.; Rudolf, J. D.; Xie, P.; Xie, G.; Teng, Q. et al. Strain prioritization for natural product discovery by a high-throughput real-time PCR method. *J. Nat. Prod.* **2014**, 77, 2296–2303.

95. Rudolf, J. D.; Dong, L. B.; Huang, T.; Shen, B. A genetically amenable platensimycin- and platencin-overproducer as a platform for biosynthetic explorations: A showcase of PtmO4, a long-chain acyl-CoA dehydrogenase. *Mol. BioSyst.* **2015**, 11, 2717–2726.
96. Baltz, R. H. *Streptomyces* and *Saccharopolyspora* hosts for heterologous expression of secondary metabolite gene clusters. *J. Ind. Microbiol. Biotechnol.* **2010**, 37, 759–772.
97. Baltz, R. H. Strain improvement in actinomycetes in the postgenomic era. *J. Ind. Microbiol. Biotechnol.* **2011**, 38, 657–666.
98. Ochi, K.; Hosaka, T. New strategies for drug discovery: Activation of silent or weakly expressed microbial gene clusters. *Appl. Microbiol. Biotechnol.* **2013**, 97, 87–98.
99. Ochi, K.; Tanaka, Y.; Tojo, S. Activating the expression of bacterial cryptic genes by *rpoB* mutations in RNA polymerase or by rare earth elements. *J. Ind. Microbiol. Biotechnol.* **2014**, 41, 403–414.
100. Rutledge, P. J.; Challis, G. L. Discovery of microbial natural products by activation of silent biosynthetic gene clusters. *Nat. Rev. Microbiol.* **2015**, 13, 509–523.
101. Weber, T.; Charusanti, P.; Musiol-Kroll, E. M.; Jiang, X.; Tong, Y.; Hu, K.; Lee, S. Y. Metabolic engineering of antibiotic factories: New tools for antibiotic production in actinomycetes. *Trends Biotechnol.* **2015**, 33, 15–26.
102. Kim, H. U.; Charusanti, P.; Lee, S. Y.; Weber, T. Metabolic engineering with systems biology tools to optimize production of prokaryotic secondary metabolites. *Nat. Prod. Rep.* **2016**, 33, 933–941.
103. Zarins-Tutt, J. S.; Barberi, T. T.; Gao, H.; Mearns-Spragg, A.; Zhang, L.; Newman, D. J.; Goss, R. J. M. Prospecting for new bacterial metabolites: A glossary of approaches for inducing, activating and upregulating the biosynthesis of bacterial *cryptic* or *silent* natural products. *Nat. Prod. Rep.* **2016**, 33, 54–72.
104. Bachmann, B. O.; Ravel, J. Chapter 8. Methods for *in silico* prediction of microbial polyketide and nonribosomal peptide biosynthetic pathways from DNA sequence data. *Methods Enzymol.* **2009**, 458, 181–217.
105. Boddy, C. N. Bioinformatics tools for genome mining of polyketide and non-ribosomal peptides. *J. Ind. Microbiol. Biotechnol.* **2014**, 41, 443–450.
106. Johnston, C. W.; Skinnider, M. A.; Wyatt, M. A.; Li, X.; Ranieri, M. R.; Yang, L.; Zechel, D. L.; Ma, B.; Magarvey, N. A. An automated Genomes-to-Natural Products (GNP) platform for the discovery of modular natural products. *Nat. Commun.* **2015**, 6, 8421.
107. Johnston, C. W.; Skinnider, M. A.; Dejong, C. A.; Rees, P. N.; Chen, G. M.; Walker, C. G.; French, S. et al. Assembly and clustering of natural antibiotics guides target identification. *Nat. Chem. Biol.* **2016**, 43, 9645–9662.
108. Weber, T.; Kim, H. U. The secondary metabolite bioinformatics portal: Computational tools to facilitate synthetic biology of secondary metabolite production. *Synth. Syst. Biotechnol.* **2016**, 1, 69–79.
109. Medema, M. H.; Kottmann, R.; Yilmaz, P.; Cummings, M.; Biggins, J. B.; Blin, K.; de Bruijn, I. et al. Minimal information about a biosynthetic gene cluster. *Nat. Chem. Biol.* **2015**, 11, 625–631.
110. Henke, M.; Kelleher, N. L. Modern mass spectrometry for synthetic biology and structure-based discovery of natural products. *Nat. Prod. Rep.* **2016**, 33, 942–950.
111. Nguyen, D. D.; Wu, C. H.; Moree, W. J.; Lamsa, A.; Medema, M. H.; Zhao, X.; Gavilan, R. G. et al. MS/MS networking guided analysis of molecule and gene cluster families. *Proc. Natl. Acad. Sci. USA* **2013**, 110, E2611–E2620.
112. Wang, M.; Carver, J. J.; Phelan, V. V.; Sanchez, L. M.; Garg, N.; Peng, Y.; Nguyen, D. D. et al. Sharing and community curation of mass spectrometry data with global natural products social molecular networking. *Nat. Biotechnol.* **2016**, 34, 828–837.
113. Marahiel, M. A structural model for multimodular NRPS assembly lines. *Nat. Prod. Rep.* **2016**, 33, 136–140.

114. Baltz, R. H. Combinatorial biosynthesis of cyclic lipopeptide antibiotics: A model for synthetic biology to accelerate the evolution of secondary metabolite biosynthetic pathways. *ACS Synth. Biol.* **2014**, 3, 748–759.
115. Ladner, C. C.; Williams, G. J. Harnessing natural product assembly lines: Structure, promiscuity, and engineering. *J. Ind. Microbiol. Biotechnol.* **2016**, 43, 371–387.
116. Dutta, S.; Whicher, J. R.; Hansen, D. A.; Hale, W. A.; Chemler, J. A.; Congdon, G. R.; Narayan, A. R. et al. Structure of a modular polyketide synthase. *Nature* **2014**, 510, 512–517.
117. Whicher, J. R.; Dutta, S.; Hansen, D. A.; Hale, W. A.; Chemler, J. A.; Dosey, A. M.; Narayan, A. R. et al. Structural rearrangements of a polyketide synthase module during its catalytic cycle. *Nature* **2014**, 510, 560–564.
118. Weissman, K. J. Uncovering the structures of modular polyketide synthases. *Nat. Prod. Rep.* **2015**, 32, 436–453.
119. Weissman, K. The structural biology of biosynthetic megaenzymes. *Nat. Chem. Biol.* **2015**, 11, 660–670.
120. Robbins, T.; Liu, Y. C.; Cane, D. E.; Khosla, C. Structure and mechanism of assembly line polyketide synthases. *Curr. Opin. Struct. Biol.* **2016**, 41, 10–18.
121. Rudolf, J. D.; Crnovcic, I.; Shen, B. The role of combinatorial biosynthesis in natural products discovery. In *Chemical Biology of Natural Products*; Newman, D.; Cragg, G.; Grothaus, P., Eds.; CRC Press: Boca Raton, FL, **2017**; pp. 87–125.
122. Olano, C.; Garcia, I.; González, A.; Rodriguez, M.; Rozas, D.; Rubio, J.; Sánchez-Hidalgo, M.; Braña, A. F.; Méndez, C.; Salas, J. A. Activation of and identification of five clusters for secondary metabolites in *Streptomyces albus* J1074. *Microb. Biotechnol.* **2014**, 7, 242–256.
123. Komatsu, M.; Uchiyama, T.; Omura, S.; Kazuo, S.; Cane, D. E.; Ikeda, H. Genome-minimized *Streptomyces* host for the heterologous expression of secondary metabolism. *Proc. Natl. Acad. Sci. USA* **2010**, 107, 2646–2651.
124. Komatsu, M.; Komatsu, K.; Koiwai, H.; Yamada, Y.; Kozone, I.; Izumikawa, M.; Hashimoto, J. et al. Engineered *Streptomyces avermitilis* host for heterologous expression of biosynthetic gene cluster for secondary metabolites. *ACS Synth. Biol.* **2013**, 2, 384–396.
125. Yamada, Y.; Kuzuyama, T.; Komatsu, M.; Shinya, L.; Ōmura, S.; Cane, D. E.; Ikeda, H. Terpene synthases are widely distributed in bacteria. *Proc. Natl. Acad. Sci. USA* **2015**, 21, 679–688.
126. Gottelt, M.; Kol, S.; Gomez-Escribano, J. P.; Bibb, M.; Takano, E. Deletion of a regulatory gene within the *cpk* gene cluster reveals novel antibacterial activity in *Streptomyces coelicolor* A3(2). *Microbiology* **2010**, 156, 2343–2353.
127. Gomez-Escribano, J. P.; Bibb, M. J. Engineering *Streptomyces coelicolor* for heterologous expression of secondary metabolite gene clusters. *Microb. Biotechnol.* **2011**, 4, 207–215.
128. Gomez-Escribano, J. P.; Bibb, M. J. Heterologous expression of natural product biosynthetic gene clusters in *Streptomyces coelicolor*: From genome mining to manipulation of biosynthetic pathways. *J. Ind. Microbiol. Biotechnol.* **2014**, 41, 425–431.
129. Yamanaka, K.; Reynolds, K. A.; Kersten, R. D.; Ryan, K. S.; Gonzalez, D. J.; Nizet, V.; Dorrestein, P. C.; Moore, B. S. Direct cloning and refactoring of a silent lipopeptide biosynthetic gene cluster yields the antibiotic taromycin A. *Proc. Natl. Acad. Sci. USA* **2014**, 111, 1957–1962.
130. Choulet, F.; Gallois, A.; Aigle, B.; Mangenot, S.; Gerbaud, C.; Truong, C.; Francou, F. X. et al. Intraspecific variability of the terminal inverted repeats of the linear chromosome of *Streptomyces ambofaciens*. *J. Bacteriol.* **2006**, 188, 6599–6610.
131. Thibessard, A.; Haas, D.; Gerbaud, C.; Aigle, B.; Lautra, S.; Pernodet, J. L.; Leblond, P. Complete genome sequence of *Streptomyces ambofaciens* ATCC 23877, the spiramycin producer. *J. Biotechnol.* **2015**, 214, 117–118.

132. Thibessard, A.; Leblond, P. Complete genome sequence of *Streptomyces ambofaciens* DSM 40697, a paradigm for genome plasticity studies. *Genome Announc.* **2016**, 4, e00470.
133. Laureti, L.; Song, L.; Corre, C.; Leblond, P.; Challis, G. L.; Aigle, B. Identification of a bioactive 51-membered macrolide complex by activation of a silent polyketide synthase in *Streptomyces ambofaciens. Proc. Natl. Acad. Sci. USA* **2011**, 108, 6258–6263
134. Bunet, R.; Song, L.; Vaz Mende, M.; Corre, C.; Hôtel, L.; Rouhier, N.; Framboisier, X.; Leblond, P.; Challis, G. L.; Aigle, B. Characterization and manipulation of the pathway-specific late regulator AlpW reveals *Streptomyces ambofaciens* as a new producer of kinamycins. *J. Bacteriol.* **2011**, 193, 1142–1153.
135. Baltz, R. H.; Matsushima, P. Efficient plasmid transformation of *Streptomyces ambofaciens* and *Streptomyces fradiae* protoplasts. *J. Bacteriol.* **1985**, 163, 180–185.
136. Ford, L. M.; Eaton, T. E.; Godfrey, O. W. Selection of *Streptomyces ambofaciens* mutants that produce large amounts of spiramycin and determination of optimal conditions for spiramycin production. *Appl. Environ. Microbiol.* **1990**, 56, 3511–3514.
137. Richardson, M. A.; Kuhstoss, S.; Huber, M. L.; Ford, L.; Godfrey, O.; Turner, J. R.; Rao, R. N. Cloning of spiramycin biosynthetic genes and their use in constructing *Streptomyces ambofaciens* mutants defective in spiramycin biosynthesis. *J. Bacteriol.* **1990**, 172, 3790–3798.
138. Alexander, D. A.; Rock, J.; He, X.; Miao, V.; Brian, P.; Baltz, R. H. Development of a genetic system for lipopeptide combinatorial biosynthesis in *Streptomyces fradiae* and heterologous expression of the A54145 biosynthetic gene cluster. *Appl. Environ. Microbiol.* **2010**, 76, 6877–6887.
139. Bunet, R.; Riclea, R.; Laureti, L.; Hôtel, L.; Paris, C.; Girardet, J. M.; Spiteller, D.; Dickschat, J. S.; Leblond, P.; Aigle, B. A single Sfp-type phosphopantetheinyl transferase plays a major role in the biosynthesis of PKS and NRPS derived metabolites in *Streptomyces ambofaciens* ATCC 23877. *PLoS One* **2014**, 9, e87607.
140. Zaburannyi, N.; Rabyk, M.; Ostach, B.; Federenko, V.; Luzhetskyy, A. Insights into naturally minimized *Streptomyces albus* J1074 genome. *BMC Genomics* **2014**, 15, 97.
141. Bilyk, B.; Luzhetskyy, A. Unusual site-specific integration into the highly active pseudo-*attB* of the *Streptomyces albus* J1074 genome. *Appl. Microbiol. Biotechnol.* **2014**, 98, 5096–5104.
142. Manderscheid, N.; Bilyk, B.; Busche, T.; Kalinowski, J.; Paululat, T.; Bechthold, A.; Petzke, L.; Luzhetskyy, A. An influence of the copy number of biosynthetic gene clusters on the production level of antibiotics in a heterologous host. *J. Biotechnol.* **2016**, 232, 110–117.
143. Ju, K. S.; Gao, J.; Doroghazi, J. R.; Wang, K. K.; Thibodeaux, C. J.; Li, S.; Metzger, E. et al. Discovery of phosphonic acid natural products by mining the genomes of 10,000 actinomycetes. *Proc. Natl. Acad. Sci. USA* **2015**, 112, 12175–12180.
144. Ju, K. S.; Doroghazi, J. R.; Metcalf, W. W. Genomics-enabled discovery of phosphonate natural products and their biosynthetic pathways. *J. Ind. Microbiol. Biotechnol.* **2014**, 41, 345–356.
145. Yu, X.; Doraghazi, J. R.; Janga, S. C.; Zhang, J. K.; Circello, B.; Griffin, B. M.; Labeda, D. P.; Metcalf, W. W. Diversity and abundance of phosphonate biosynthetic genes in nature. *Proc. Natl. Acad. Sci. USA* **2013**, 110, 20759–20764.
146. Shen, B.; Hindra; Yan, X.; Huang, T.; Ge, H.; Yang, D.; Teng, Q.; Rudolf, J. D.; Lohman, J. R. Enediynes: Exploration of microbial genomics to discover new anticancer drug leads. *Bioorg. Med. Chem. Lett.* **2015**, 25, 9–15.
147. Rudolf, J. D.; Yan, X.; Shen, B. Genome neighborhood network reveals insights into enediyne biosynthesis and faciliates prediction and prioritization for discovery. *J. Ind. Microbiol. Biotechnol.* **2016**, 43, 261–276.

148. O'Hear, C.; Rubnitz, J. E. Recent research and future prospects for gemtuzumab ozogamicin: Could it make a comeback? *Expert. Rev. Hematol.* **2014**, 7, 427–429.
149. Shor, B.; Gerber, H. P.; Sapra, P. Preclinical and clinical development of inotuzumab-ozogamicin in hematological malignancies. *Mol. Immunol.* **2015**, 67, 107–116.
150. Zhao, S.; Sakai, A.; Zhang, X.; Vetting, M. W.; Kumar, R.; Hillerich, B.; San Francisco, B. et al. Prediction and characterization of enzymatic activities guided by sequence similarity and genome neighborhood networks. *eLife* **2014**, 3, e03275.
151. Bowman, B. R. Rapid engineering of secondary metabolite gene clusters in the genomic era. *Society for Industrial Microbiology and Biotechnology Annual Meeting*, Philadelphia, PA, August, **2015** (http://www.warpdrivebio.com/news.php).
152. Yoo, Y. J.; Kim, H.; Park, S. R.; Yoon, Y. J. An overview of rapamycin: From discovery to future perspectives. *J. Ind. Microbiol. Biotechnol.* **2017**, 44, 537–553.
153. Verdine, G. SMART™ drugs: Engineering nature's solution to the undruggable target. *American Association for Cancer Research Annual Meeting*, New Orleans, LA, April, **2016** (http://www.warpdrivebio.com/news.php).
154. Tang, X.; Li, J.; Millán-Aguiñaga, N.; Zhang, J. J.; O'Neill, E. C.; Ugalde, J. A.; Jensen, P. R.; Mantovani, S. M.; Moore, B. S. Identification of thiotetronic acid antibiotic biosynthetic pathways by target-directed genome mining. *ACS Chem. Biol.* **2015**, 10, 2841–2849.
155. Cundliffe, E.; Demain, A. L. Avoidance of suicide in antibiotic-producing microbes. *J. Ind. Microbiol. Biotechnol.* **2010**, 37, 643–672.
156. Kale, A. J.; McGlinchley, R. P.; Lechner, A.; Moore, B. S. Bacterial self-resistance to the natural proteasome inhibitor salinosporamide A. *ACS Chem. Biol.* **2011**, 6, 1257–1264.
157. Kling, A.; Lukat, P.; Almeida, D. V.; Bauer, A.; Fontaine, E.; Sordello, S.; Zaburannyi, N. et al. Targeting DnaN for tuberculosis therapy using novel griselimycins. *Science* **2015**, 348, 1106–1112.
158. Peterson, R. M.; Huang, T.; Rudolf, J. D.; Smanski, M. J.; Shen, B. Mechanisms of self-resistance in the platensimycin- and platensin-producing *Streptomyces platensis* MA7327 and MA7339 strains. *Chem. Biol.* **2014**, 21, 389–397.
159. Rashid, S.; Huo, L.; Herrmann, J.; Stadler, M.; Köpcke, B.; Bitzer, J.; Müller, R. Mining the cinnabaramide biosynthetic pathway to generate novel proteasome inhibitors. *ChemBioChem* **2011**, 12, 922–931.
160. Froudarakis, M.; Hatzimichael, E.; Kyriazopoulou, L.; Lagos, K.; Pappas, P.; Tzakos, A. G.; Karavasilis, V.; Daliani, D.; Papandreou, C.; Briasoulis, E. Revisiting bleomycin from pathophysiology to safe clinical use. *Crit. Rev. Oncol. Hematol.* **2013**, 87, 90–100.
161. Hornung, A.; Bertazzo, M.; Dziarnowski, K.; Schneider, K.; Welzel, K.; Wohlert, S. E.; Holzenkämpfer, M. et al. A genomic screening approach to the structure-guided identification of drug candidates from natural sources. *ChemBioChem* **2007**, 8, 757–766.
162. Baltz, R. H. Molecular beacons to identify gifted microbes for genome mining. *J. Antibiot.* **2017**, 70, 639–646.
163. Bérdy, J. Bioactive microbial metabolites. *J. Antibiot.* **2005**, 58, 1–26.
164. Barka, E. A.; Vatas, P.; Sanchez, L.; Gaveau-Vaillant, N.; Jacquard, C.; Klenk, H. P.; Clement, C.; Ouhdouch, Y.; van Wezel, G. P. Taxonomy, physiology, and natural products of *Actinobacteria*. *Microbiol. Mol. Biol. Rev.* **2015**, 80, 1–43.
165. Smanski, M. J.; Schlatter, D. C.; Kinkel, L. L. Leveraging ecological theory to guide natural product discovery. *J. Ind. Microbiol. Biotechnol.* **2016**, 43, 115–128.
166. Challinor, V. L.; Bode, H. B. Bioactive natural products from novel microbial sources. *Ann. N. Y. Acad. Sci.* **2015**, 1354, 82–97.
167. Calteau, A.; Fewer, D. P.; Lafiti, A.; Coursin, T.; Laurent, T.; Jokela, J.; Kerfeld, C. A.; Sivonen, K.; Piel, J.; Gugger, M. Phylum-wide comparative genomics unravel the diversity of secondary metabolism in Cyanobacteria. *BMC Genomics* **2014**, 15, 977.

168. Shih, P. M.; Wu, D.; Latifi, A.; Axen, S. D.; Fewer, D. P.; Talla, E.; Calteau, A. et al. Improving the coverage of the cyanobacterial phylum using diversity-driven genome sequencing. *Proc. Natl. Acad. Sci. USA* **2013**, 110, 1053–1058.
169. Liu, X.; Cheng, Y. Q. Genome-guided discovery of diverse natural products from *Burkholderia* sp. *J. Ind. Microbiol. Biotechnol.* **2014**, 41, 275–284.
170. Ramadhar, T. R.; Beemelmanns, C.; Currie, C. R.; Clardy, J. Bacterial symbionts in agricultural systems provide a strategic source for antibiotic discovery. *J. Antibiot.* **2014**, 67, 53–58.
171. Baltz, R. H. Genetics and biochemistry of tylosin production: A model for genetic engineering in antibiotic-producing *Streptomyces*. *Basic Life Sci.* **1982**, 19, 431–444.
172. Baltz, R. H. Combinatorial biosynthesis of novel antibiotics and other secondary metabolites. *SIM News* **2006**, 56, 148–160.
173. Baltz, R. H. Molecular engineering approaches to peptide, polyketide and other antibiotics. *Nat. Biotechnol.* **2006**, 24, 1533–1540.
174. Wong, F. T.; Khosla, C. Combinatorial biosynthesis of polyketides: A perspective. *Curr. Opin. Chem. Biol.* **2012**, 16, 117–123.
175. Kim, E.; Moore, B. S.; Yoon, Y. J. Reinvigorating natural product combinatorial biosynthesis with synthetic biology. *Nat. Chem. Biol.* **2015**, 11, 649–659.

2 Chemical Biology of Marine Cyanobacteria

Lena Keller, Tiago Leão, and William H. Gerwick

CONTENTS

2.1 INTRODUCTION

Marine Cyanobacteria have especially captured the interest and fascination of many marine chemical biologists in that they produce a dizzying array of natural products (NPs) with many unique functional groups and atom arrangements.[1] Moreover, they have powerful biological properties and work through distinct and oftentimes unique pharmacological mechanisms, making them important tool compounds as well as drug leads. Indeed, one marine cyanobacterial metabolite, dolastatin 10, that has powerful antitubulin activity,[2] inspired the development of a related synthetic material, MMAE, which serves as the *warhead* of a Food and Drug Administration (FDA)-approved antibody-drug conjugate (ADC) known as brentuximab vedotin.[3]

Under suitable environmental conditions, filamentous marine Cyanobacteria such as members of the genus *Moorea* will bloom in a given region, and can be found in great abundance, making their collection by SCUBA or snorkeling quite easily accomplished.[4] Such bloom materials have served as the source biomass for hit compound discovery and early-stage preclinical development of numerous agents. Unfortunately, while some of these organisms can be grown in the laboratory and often do produce their metabolites under these conditions, they grow very slowly with doubling times measured in days or weeks. Thus, and in similarity to so many other classes of organisms, access to a reliable supply of compound has become an impediment to the development of marine cyanobacterial lead compounds.[5] Additionally, due to the filamentous form of these organisms with their cells encased in a thick, nearly impervious sheath, genetic manipulation of these Cyanobacteria has been very difficult, and this has prevented genetic and biochemical investigations of their metabolite biosynthesis. For this multitude of reasons, there is great interest to accomplish the heterologous expression of their NP biosynthetic pathways, and this serves as one focus of the present chapter.

From a related perspective, an ever-increasing number of genomes has been sequenced from marine Cyanobacteria, and, like other classes of microorganisms, is revealing that they have a much greater capacity for NP biosynthesis than is currently recognized from traditional NP isolation efforts. These observations have in turn spawned a growing interest in, and capacity to, more thoroughly explore the secondary metabolic capacities of these organisms through varied culture conditions, elicitors, and genome-driven compound discovery. As such, we chronicle in this chapter efforts to date to utilize these newer genome-inspired methodologies.

Finally, marine life forms have been remarkable for providing a wide variety of useful proteinaceous materials. These include such diverse products as thermophilic restriction endonucleases and fluorescent proteins, representing underpinning technological developments that have enabled much of the molecular biology revolution. From marine Cyanobacteria, valuable proteins have been obtained with applications in biomedical research and clinical diagnostics, food processing, as well as bioplastic production.

Several excellent reviews of the pharmacology[6] and biosynthesis[7–9] of marine cyanobacterial metabolites have recently appeared, and we direct the interested reader to those for an in-depth coverage of these topics.

2.2 HETEROLOGOUS EXPRESSION OF CYANOBACTERIAL NATURAL PRODUCT ENZYMES AND PATHWAYS

A major long-term goal of NP biosynthetic investigations is the functional heterologous expression of the entire pathway as well as its engineering to produce analog molecules. This is motivated by a desire to produce the substance via a highly controlled fermentation process so that a reliable supply is guaranteed. An additional motivation is to harness the power of the biosynthetic system to make analogs at relatively lower cost, thereby examining structure-activity relationships. Finally, the heterologous expression of cryptic biosynthetic gene clusters (BGCs) represents a newly emerging discovery approach for novel NPs.

A number of problems have emerged in efforts aimed at the cloning and heterologous expression of marine cyanobacterial NP pathways, some of which are common to most polyketide synthase (PKS) and nonribosomal peptide synthetase (NRPS) pathways, such as their relatively large size and their requirement for cofactors including phosphopantetheinyl (PPant). Other emergent problems include the low GC content of marine cyanobacterial genomes (typically around 45%), unusual codon usage, and nonaxenic nature of source materials (typically marine Cyanobacteria are grown in mixed culture with an apparently obligate surface microbiome). Additionally, some marine Cyanobacteria, especially the chemically rich genus *Moorea* (previously described as marine *Lyngbya*), have a thick sheath that resists breakage and thus inhibits the yield of intact DNA. In some cases, exonucleases are reported from Cyanobacteria that further reduce the yield of isolated DNA. Finally, isolated DNA from marine Cyanobacteria is frequently contaminated with polysaccharides that render the nucleic acids unresponsive to subsequent manipulation such as polymerase chain reaction (PCR).

A primary need for heterologous expression of marine cyanobacterial NP pathways is to meet *supply needs*. The long-standing and well-described *supply* problem of NPs has impeded the development of numerous promising lead molecules.[5] A variety of approaches to meet supply requirements have been developed, such as total synthesis, partial synthesis, aquaculture or fermentation of the producing species, and wild harvest. The latter, however, is especially problematic in that the ecology and natural history of most marine organisms is fragmentary, and there exists a significant possibility to damage the natural population as well as associated or dependent species. Total synthesis can provide a highly reliable supply, but for more complex agents, such as Eribulin, the large number of required chemical steps adds significantly to the cost of the agent.[5] Semisynthesis from a fermentation product is the current route to the European-approved agent Trabectidin, although an aquacultural method was tried as an intermediate approach. However, low yield and sporadic pond failures rendered the latter approach nonviable.[5] Hence, for reliability of production and cost efficiency, production via fermentation is highly desired. Unfortunately, marine Cyanobacteria are rather difficult to reliably culture, and even in culture, they grow extremely slowly with doubling times measured in days to weeks.[10] This is especially true for agents at an intermediate stage in development wherein it is unknown if the compound will become a true clinical candidate, and thus uncertainty exists as to how much investment in developing a production method is reasonable.

An additional need for heterologous expression focuses on individual enzymes rather than entire pathways. The focus here is on enzymes of unusual specificity or reactivity, and thus the motivation is to understand the structural determinants that confer these properties. In some cases, the nature of the transformation that is catalyzed has potential engineering or economic value, providing an additional impetus. The sections that follow first consider efforts to heterologously express individual proteins involved in NP biosynthesis, and then consider the few cases in which entire NP pathways have been cloned and heterologously expressed from marine Cyanobacteria.

2.2.1 Expression of Biosynthetic Enzymes

2.2.1.1 Adenylation Domains

The jamaicamides A–C (Figure 2.1) were reported as modestly active neurotoxins from cultures of the marine cyanobacterium *Moorea producens* (originally reported as *Lyngbya majuscula*), collected from Hector's Bay, Jamaica. The structures are remarkable for the presence of several unique functional groups, including a vinyl chloride and acetylenic bromide, as well as a dissection into a number of integrated amino acid and polyketide subunits. To explore this fascinating molecule's origin, the fundamental biosynthetic subunits were mapped from a series of isotope-labeled feeding experiments. Ensuing, the BGC was cloned and sequenced using distinctive genetic signatures for its locations, namely the β-ketosynthase (KS) and β-hydroxyl-β-methylglutaryl-CoA synthase (HMGCoA) domains.[11] A precise colinearity between genes in the 58 kb BGC and their utilization in jamaicamide assembly was observed. To obtain partial validation that this cluster truly encoded for jamaicamide biosynthesis, the adenylation domain of the final NRPS domain was cloned and overexpressed as a polyhistidine His6-tagged protein in *Escherichia coli*. The purified protein was then evaluated for its ability to activate various amino acids. The residue activated to the highest degree was alanine, in full agreement with the structure and proposed BGC predictions.

A second class of metabolites from this Jamaican strain was named hectochlorin, and possessed a distinctive cyclic structure with two thiazole rings and a gem-dichloro functionality on the penultimate carbon of an alkyl chain (Figure 2.1).[12] Its gene cluster was similarly cloned into a fosmid and sequenced to yield a 42 kb BGC candidate.[13] Substantiation that this colinear arrangement of biosynthetic genes coded for hectochlorin biosynthesis was obtained from subcloning both adenylation domains from HctE and demonstrating their preferred activation of 2-hydroxyisovaleric acid and cysteine, respectively.

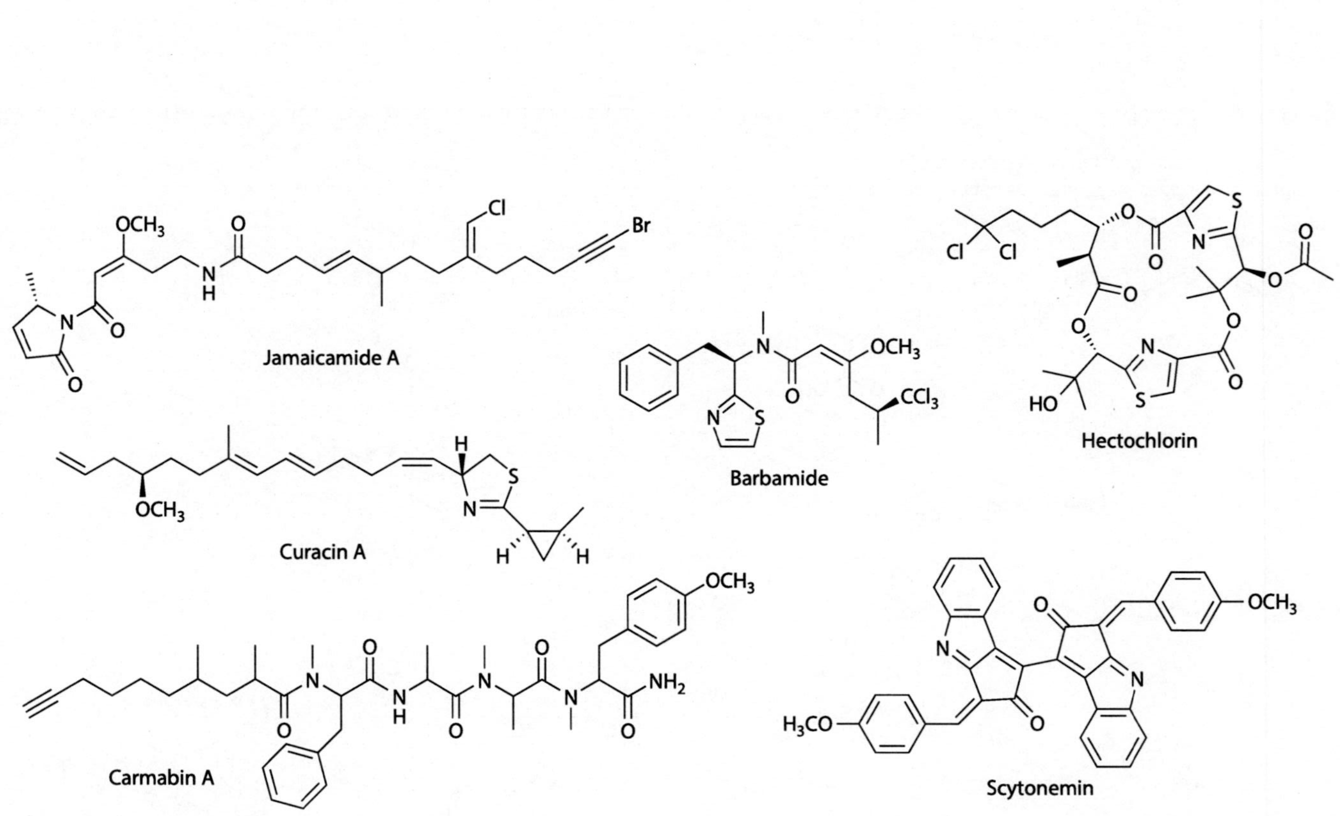

FIGURE 2.1 Structures of cyanobacterial natural products involved in the heterologous expression of individual biosynthetic enzymes. (*Continued*)

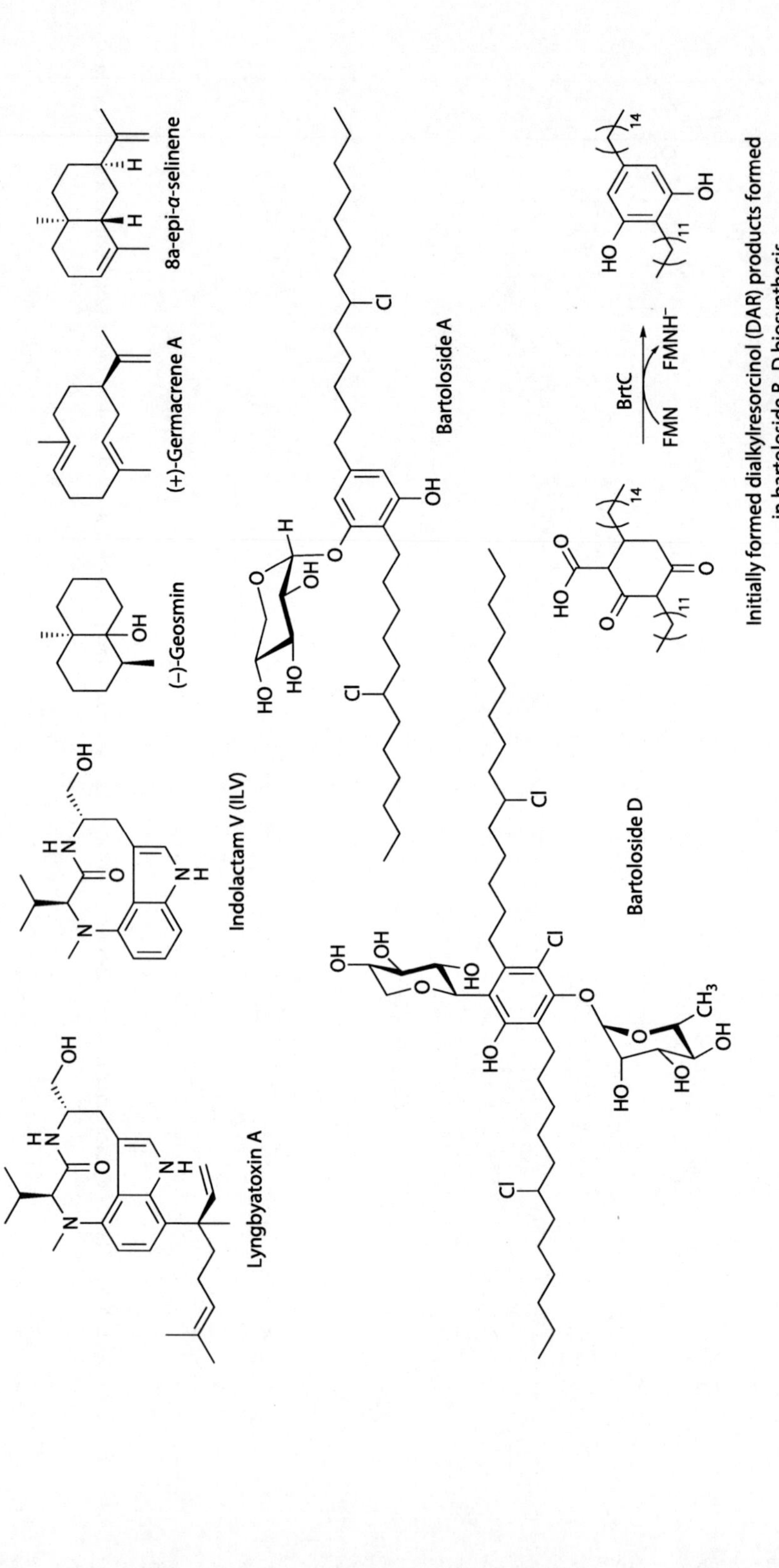

FIGURE 2.1 (*Continued*) Structures of cyanobacterial natural products involved in the heterologous expression of individual biosynthetic enzymes.

Another strain of *M. producens*, this one from Curaçao, was found to contain a rich complement of unique secondary metabolites, such as the curacin A, carmabin A, and barbamide (Figure 2.1). Similar to the approach taken above with jamaicamide and hectochlorin, substantiation of the identity of the barbamide BGC was obtained from work with *in vitro* expressed proteins, in this case the adenylation domain incorporating the leucine-derived fragment.[14] Cloning and overexpression of this adenylation domain of BarE from the barbamide gene cluster gave a protein that selectively activated trichloromethyl-2-oxopentanoic acid, ultimately deriving from L-leucine.

2.2.1.2 Barbamide Halogenase

The barbamide halogenase catalyzes the conversion of (2*S*)-leucine into (2*S*,4*S*)-5,5,5-trichloroleucine, and this forms the initial substrate for the trimodular NRPS, which assembles the molluscicidal barbamide molecule (Figure 2.1). This was initially studied in the wild-type organism, *M. producens* (originally *L. majuscula*), through various feeding experiments, and revealed the radical nature of the chlorination reaction with (2*S*)-leucine.[15] Cloning and sequencing of the barbamide pathway further revealed two candidate halogenase genes, *barB1* and *barB2*, and a proposal was advanced based on bioinformatics that these enzymes were nonheme iron-containing 2-oxoglutarate-dependent radical halogenases involving a high-energy $Fe^{IV}{=}O$ species.[14] Using N-His6-tagged synthetic genes for several of the barbamide pathway genes (*barA*, *barB1*, *barB2*, and *barD*) and radioactive substrates, the system was carefully dissected to reveal that BarD was able to load (2*S*)-leucine onto BarA, and (in the presence of 2-oxoglutarate, O_2, and Cl^-) was able to be transformed into (2*S*,4*S*)-5,5,5-trichloroleucine by the combined action of BarB1 and BarB2.[16] Moreover, it was found that BarB1 catalyzed the initial production of (2*S*,4*S*)-5,5-dichloroleucine, whereas BarB2 could further chlorinate this latter substrate into (2*S*,4*S*)-5,5,5-trichloroleucine.[17] The barbamide halogenases were found to be quite O_2 sensitive and their purification was thus achieved by working in an anaerobic glovebox.

2.2.1.3 Curacin A Biosynthetic Enzymes

Curacin A (Figure 2.1) was discovered as a major secondary metabolite of the Curaçao marine cyanobacterium *Moorea producens* (previously *L. majuscula*), and has powerful cancer cell toxicity due to its ability to block tubulin polymerization through interaction at the colchicine drug-binding site.[18] Its biosynthesis was first explored through isotope-labeled precursor-feeding studies, followed by determination of their specificity of incorporation by nuclear magnetic resonance (NMR).[19] This helped focus the isolation and identification of a 63.7 kb BGC from a cosmid clone library, and revealed a remarkable number of unique biosynthetic steps involved in its formation. Subsequently, a number of the enzymes catalyzing these transformations have been cloned and heterologously expressed in efforts to study their structures by X-ray crystallography and their functioning through *in vitro* experiments.

The initiation of the curacin A biosynthetic sequence is encoded by an unusual set of three enzymes, annotated as the adaptor region (AR), GCN5-related *N*-acetyltransferase (GNAT), and an acyl carrier protein (ACP), that are encoded by the N-terminal section of the *curA* gene. Overexpression of this soluble construct and derived fragments followed by biochemical assays, X-ray crystallography of the GNAT section with and without malonyl-CoA, mutagenesis studies, and modeling have provided insights into a new mode of pathway initiation.[20] The GNAT possesses two opposing tunnels that meet deep within the protein, one of which accepts malonyl CoA and the other the phosphopantetheinyl arm of the ACP. The GNAT first catalyzes decarboxylation of the substrate, and then *S*-acetyl transfer from acetyl-CoA to the ACP. By bioinformatics, this previously uncharacterized mechanism for PKS initiation is utilized in a number of other significant NPs, including rhizoxin, myxovirescin A, and onnamide A.

Another unique feature of curacin A is a cassette of genes now known to be involved in the β-branching of a polyketide, in this case to form the distinctive cyclopropyl ring. One feature of this cassette is an enoyl-CoA hydratase, which was cloned and overexpressed to demonstrate a decarboxylase activity toward 3-methylglutaconyl-ACP, forming 3-methyl-crotonyl-ACP.[21] X-ray crystallography of this protein revealed an unusual trimeric structure, and identified an active site with a crotonase fold. In combination with site-directed mutagenesis, it was proposed that a His and Lys residue are responsible for the decarboxylation and proton donation at C-4, respectively, and that Ala and Gly residues help stabilize the oxyanion intermediate.

A detailed study of the β-branching reactions, catalyzed by very similar cassettes of genes in the curacin A and jamaicamide A biosynthetic pathways, made extensive use of heterologously expressed proteins from each.[22] In the former case, a cyclopropyl ring is formed, whereas in the latter a pendant vinyl chloride functionality is created. The heterologously expressed proteins were combined with biochemical assays and detection by Fourier transform ion cyclotron resonance mass spectrometry (FTICR-MS) and infrared multiphoton dissociation (IRMPD) methods, giving a profound insight into their divergent mechanisms of formation. Chlorination of an ACP-tethered β-hydroxy-β-methylglutaryl moiety in curacin A, or its equivalent with a longer acyl chain in jamaicamide A, is introduced at the γ-position by a nonheme Fe(II) α-ketoglutarate-dependent halogenase. This is followed by dehydration in each case to form an α,β-unsaturated-γ-chloro-glutaryl-CoA derivative. Divergence occurs between the pathways through the next enzyme in the pathway, an enoyl CoA hydratase. In the Cur case, this creates an α,β-enoyl thioester, whereas in the Jam pathway, a β,γ-enoyl thioester is formed; in the latter case, this terminates the modification sequence as the next enzyme in the pathway, an enoyl reductase, is inactive to this β,γ-unsaturated substrate. In the Cur case, the enoyl reductase was shown to be a cyclopropanase through delivery of a hydride to the β-position, causing rearrangement of the α,β-olefin to displace the γ-chloro atom and form a cyclopropyl ring. X-ray crystallography of several of these intriguing proteins has occurred (e.g., the CurA halogenase, ECH_2, and cyclopropanase, where the latter can be compared with structures of another enoyl reductase from both the curacin and jamaicamide pathways).

2.2.1.4 Scytonemin Biosynthetic Enzymes

Scytonemin (Figure 2.1) is a greenish brown pigment produced by a wide variety of terrestrial and marine Cyanobacteria in response to stress conditions, especially those caused by high ultraviolet (UV) irradiance. While first reported in the mid-1800s[23] and given this name by Nägeli and Schwendener,[24] it was not until nearly 150 years later that its careful adaptive function and chemical characterization were accomplished.[25,26] Subsequently, a transposon mutagenesis approach was used in *Nostoc punctiforme* to locate and then sequence its BGC,[27] and additional transcriptomic investigations widened the size of the relevant biosynthetic gene cluster.[28] This set the stage for the heterologous expression of specific genes from the cluster in order to investigate their biochemical function. Of particular interest were the enzymes involved in forming the fundamental carbon framework and carbocyclization reactions.

The initial steps in scytonemin biosynthesis were predicted by bioinformatic analysis of genes in the scytonemin BGC, and this led to cloning and overexpression of two pivotal enzymes as C-His6-tagged fusion proteins.[29] The first was a proposed tryptophan dehydrogenase, which was demonstrated to convert this amino acid to its corresponding α-keto acid. As this enzyme was very sluggish in the corresponding conversion of tyrosine to its α-keto acid, it was predicted that a putative prephenate dehydrogenase in the gene cluster is involved in the conversion of prephenate into this α-keto acid. Another encoded enzyme in the pathway was annotated as a thiamin diphosphate-dependent acetolactate synthase, and treatment of an equimolar mixture of indole-3 pyruvic and *p*-hydroxyphenyl pyruvic acids in the presence of $MgCl_2$ and thiamin diphosphate at pH 7.5 resulted in a single product by high-performance liquid chromatography (HPLC). However, this product was unstable and rapidly underwent decarboxylation to two regioisomeric α-ketols. Treatment of the initially formed reaction product resulted in two epimeric diols in which the carboxylate was still present, confirming the nature of the initially formed β-keto acid product arising from a highly selective acyloin coupling of these two α-keto acid precursors. Subsequently, a third enzyme from the scytonemin gene cluster, ScyC, was overexpressed as N- and C-His6-tagged fusion proteins, and was found to catalyze the cyclization of the intermediate β-keto acid into a cyclopentenone product that is conceptually one monomeric half of scytonemin.[30] Additional experimentation identified a likely mechanism for this transformation, in which C-2 of the indole ring intramolecularly adds to the side chain carbonyl, followed by dehydration and decarboxylation to form the cyclopentenone ring.

2.2.1.5 Lyngbyatoxin Biosynthetic Enzymes

Field-collected *M. producens* (previously *L. majuscula*) from Kahala Beach, Oahu, Hawaii, was the source of DNA from which the 11.3 kb lyngbyatoxin BGC was obtained and sequenced.[31] It was composed of four open reading frames (ORFs) *ltxABCD*, the first of which was a bimodular NRPS that encoded for the assembly of the dipeptide Val-Trp in which the Trp carboxyl group was reduced to a primary alcohol upon offloading.[32] The second ORF encoded for a cytochrome P450 that was believed to be responsible for cyclization of the dipeptide *N*-methyl-L-valyl-L-tryptophanol into (−)-indolactam V (ILV; Figure 2.1), while the third ORF was unrelated by sequence to any known protein, and was proposed to catalyze the reverse geranylation of ILV. Overexpression in *E. coli* and purification of the His6-tagged putative geranyl transferase LtxC, followed by incubation with ILV, geranyl pyrophosphate, and 2 mM $MgCl_2$ led to a time-dependent production of lyngbyatoxin A (Figure 2.1). The function of the fourth ORF, *ltxD*, has not yet been demonstrated, although it was speculated that it may be involved in forming the further oxidized species, lyngbyatoxin B and C. A subsequent report investigated the production of LtxB and LtxC in a *Streptomyces coelicolor* host.[33] While it was hoped that the entire Ltx gene cluster would be functionally expressed, it appears that transcriptional termination occurs for the large NRPS gene *ltxA* yielding a truncated and nonfunctional product. However, heterologous expression of the LtxC geranyl transferase in *S. coelicolor* was successful, and it once again catalyzed the production of lyngbyatoxin A when provided with ILV, geranyl pyrophosphate, and 2 mM $MgCl_2$.

The function of LtxB was explored through heterologous expression of the gene as a His6-tagged protein in *E. coli*.[34] The overexpressed protein showed many of the characteristics of a cytochrome P450 enzyme, and its catalytic ability was demonstrated by the conversion of the acyclic dipeptide derivative *N*-methyl-L-valyl-L-tryptophanol to ILV. A number of valine analogs of this dipeptide were also evaluated for reactivity with LtxB, and were generally found tolerated so as to produce a variety of analogs. The ensuing enzyme in the pathway, LtxC geranyl transferase, was able to accept these analog precursors and produce geranylated derivatives of lyngbyatoxin A. Thus, the functions of these components of the lyngbyatoxin BGC were unequivocally demonstrated through *in vitro* protein expression experiments. As described below in the whole pathway expression section, lyngbyatoxin A has been produced through heterologous expression of the entire pathway, both in *E. coli* and in the freshwater cyanobacterium *Anabaena* sp. PCC 7120.

2.2.1.6 *Nostoc punctiforme* Terpene Synthases and Cytochrome P450 (CYP)

Freshwater Cyanobacteria are known to contribute musty smells to water and soil, and this is due to their production of terpenes such as geosmin (Figure 2.1) and 2-methyl-isoborneol. Genome sequences of two such Cyanobacteria, a *Nostoc* sp. PCC7120 and *N. punctiforme* PCC73102, showed that both contained genes for terpene synthases.[35] From the former strain, a terpene synthase gene *NS1* was found in close proximity to genes encoding for a CYP as well as a hybrid two-component signaling protein. In the latter strain, a similar synthase/CYP/two-component gene cluster NP1 was found in addition to a fusion-type sesquiterpene synthase NP2 gene, which was similar to a fusion protein that has been shown to be responsible for geosmin biosynthesis in *Streptomyces* spp. Cloning and expression studies of these three, using the pUCmodRBS expression vector in *E. coli*, revealed that the *Nostoc* sp. synthase NS1 produced germacrene A (Figure 2.1) (detected largely as the thermal rearrangement product β-elemene), whereas the *N. punctiforme* NP1 gene cluster produced 8a-epi-α-selinene (Figure 2.1). Moreover, when NS1 was coexpressed *in vivo* in *E. coli* with its adjacent CYP as well as a ferredoxin/ferredoxin reductase gene from *Nostoc* sp., a new product of *m*/*z* 220 was produced. Unfortunately, due to the low titer of its production, it could not be rigorously identified, but by mass it was suggested to be an oxygenated germacrene derivative. Expression of the fusion-type terpene synthase NP2 in *E. coli* using the pUCmodRBS expression vector led to production of (*E*,*E*)-germacradienol as well as smaller amounts of germacrene D and β-elemene. This was interesting in that the product of the second catalytic domain, geosmin, was not produced in the *E. coli* construct, but is apparently the sole product synthesized in the native situation by *N. punctiforme*.

2.2.1.7 Bartoloside Biosynthetic Enzymes

Four unique chlorinated metabolites, including bartoloside A and D (Figure 2.1), were obtained from a beach in Portugal isolated from two unrelated Cyanobacteria: *Nodosilinea* sp. LEGE 06102 and *Synechocystis salina* LEGE 06155.[36] These four metabolites possessed in common a highly substituted dialkylresorcinol (DAR) core with a variety of phenol, *O*-glycoside, *C*-glycoside, and chlorine atom substituents. The alkyl groups were challenging to define in several regards, including their position, overall carbon chain length, and secondary chloro-substituent locations, due to signal degeneracy and their highly substituted aromatic ring core. Hence, a combination of 2D NMR methods, principally 1D TOCSYs with variable delay times, as well as the HSQC-TOCSY and HMBC experiments, and key heterologous expression investigations, were used to establish the lengths and relative positions of the carbon chains. The heterologous expression experiments were based on TBLASTN identification of the likely BGC from genome sequencing of *S. salina*. Informatic analysis of the gene cluster suggested two genes, *brtC* and *brtD*, encoded for proteins involved in the biosynthesis of the DAR moiety. Indeed, when BrtD was overexpressed in *E. coli* BL21 cells as the N-His6-tagged protein, isolated, and provided with 2-octadecenoyl-ACP thioester and 3-oxo-hexadecanoyl-ACP thioester, a couple of conceptual DAR intermediates were formed. When BrtC was added to this reaction as its FMN holoenzyme, or when a *brtC/brtD* construct was expressed *in vivo* in *E. coli*, a DAR product was formed with C-12 and C-15 alkyl substituents positioned *para* to one another (Figure 2.1). These experiments greatly helped to reduce the number of structural possibilities for the bartolosides as well as the relative positions of alkyl groups on the aromatic core, and thus aided in the resolution of the carbon skeleton. Location of chloro-substituents on the alkyl chains and other substituents on the aromatic ring was then achieved through careful application of the 2D NMR dataset.

2.2.2 Heterologous Expression of Entire Biosynthetic Pathways

2.2.2.1 4-*O*-Demethylbarbamide

The first hybrid NRPS-PKS pathway to be heterologously expressed from a marine cyanobacterium was that encoding for barbamide (Figure 2.1), a unique trichloromethyl group containing NP reported from a Curaçao collection of *M. producens*.[37] The host for this experiment was *Streptomyces venezuelae*, and the biosynthetic genes *barA-barK* were cloned into the replicative *E. coli-Streptomyces* shuttle vector DHS 2001 under control by the *pikAI* promoter. Cultivation of this transformant on solid medium for 6 days, followed by extraction and LC-MS/MS analysis, revealed that a new compound was produced which contained three chlorine atoms, but was 14 mass units lighter than barbamide. Interestingly, a careful analysis of the wild-type organism showed that this same barbamide analog was a minor NP, and its preparative isolation and spectroscopic structure elucidation from cultured *M. producens* identified it as 4-*O*-demethylbarbamide (Figure 2.2). Why the engineered system only produced this *O*-demethyl product, whereas the wild-type organism produces barbamide as the major component, is unknown at the present time, although it was speculated that the *O*-methyltransferase transcript might be unstable, or that a gene

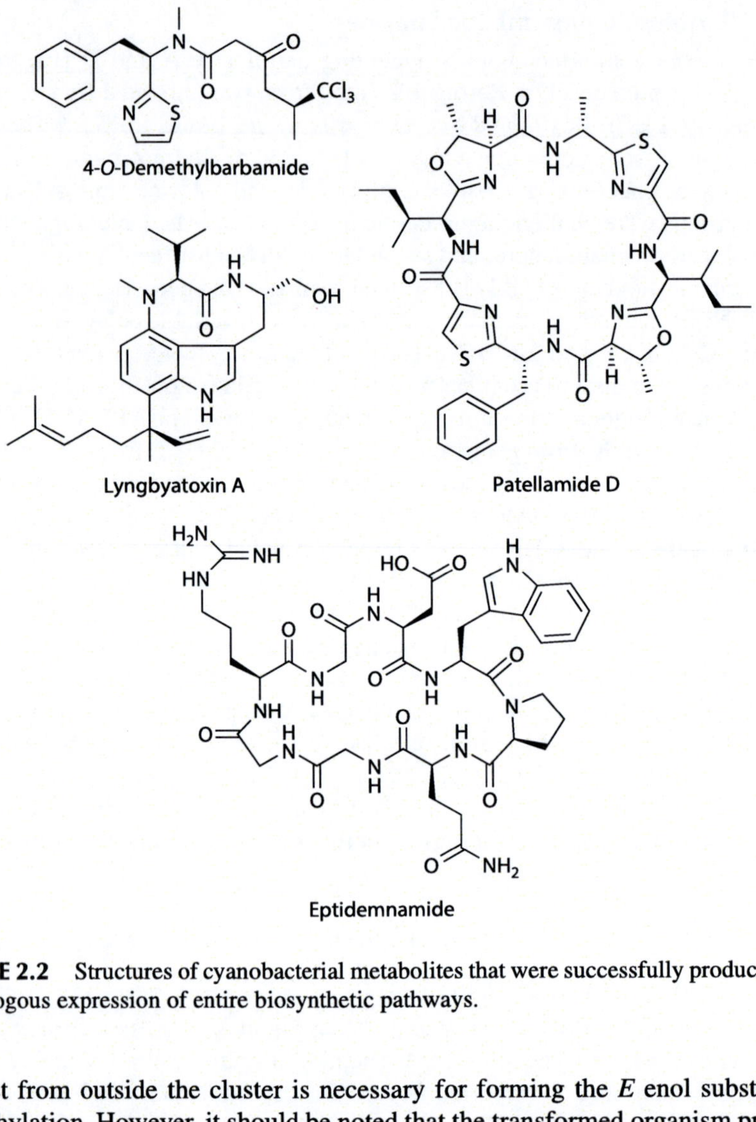

FIGURE 2.2 Structures of cyanobacterial metabolites that were successfully produced using heterologous expression of entire biosynthetic pathways.

product from outside the cluster is necessary for forming the *E* enol substrate for O-methylation. However, it should be noted that the transformed organism produced 4-*O*-demethylbarbamide in extremely low yield (less than 1 μg/L), such that other explanations for this lack of O-methylation are conceivable.

2.2.2.2 Lyngbyatoxin Production in *Escherichia coli*

As noted above, lyngbyatoxin (Figure 2.2) is comprised of amino acids (tryptophan and valine), terpene (geranyl), and SAM methyl group components, and these are concisely assembled as a result of the enzyme products of an 11.3 kb gene cluster. The cluster, present on a fosmid, was heterologously expressed in an *E. coli* strain that carries a myxobacterial PPTase with a broad substrate tolerance.[38] A tetO tetracycline-inducible promoter was inserted upstream of the biosynthetic cluster, and the modified fosmid was then introduced into *E. coli* using electroporation with maintenance under multiple antibiotic selection. At 30°C, no products were observed; however, at the lower temperature of 18°C, substantial levels of lyngbyatoxin (25.6 mg/L), as well as the biosynthetic intermediates *N*-methylvalyl-tryptophanol (not quantitated) and indolactam-V (Figure 2.1) (150 mg/L), were formed. This favorable result is likely the result of improved folding of the large NRPS protein. A number of potential mechanisms and strategies for further enhancing production levels were inspired from these results, but not experimentally examined.

2.2.2.3 Lyngbyatoxin Production in *Anabaena* sp. PCC7120

Because the lyngbyatoxin BGC is relatively small (11.3 kb) and its core biosynthesis has been well characterized,[31] it was selected for exploratory heterologous expression in the cyanobacterial host *Anabaena* sp. PCC 7120.[39] This freshwater cyanobacterium has attractive features for heterologous expression in that it has been extensively used as a model system in which to explore nitrogen fixation and cellular differentiation, and is genetically tractable with established protocols for DNA insertion and manipulation.[40] The minimal *ltx* gene cassette for lyngbyatoxin A biosynthesis, *ltx A-C,* was cloned into a replicative plasmid suitable for selection and replication in *Anabaena* PCC 7120. Triparental conjugation was successful in introducing the plasmid into this strain, and following growth on solid media for 28 days, an extract was analyzed by various HPLC, mass spectrometry (MS), and ^{1}H NMR methods and shown to contain the expected product in small yield. A transformation-associated recombination (TAR) cloning approach was also explored, and *ltx A–C* were captured and introduced into *Anabaena* 7120 again by conjugation; production levels were equivalent to those achieved by traditional plasmid cloning, and under certain nitrogen supplementation conditions, equal to levels produced in the wild-type organism *M. producens*. A variety of different promoters were explored for their impact on lyngbyatoxin expression, and this led in one case to a 13-fold increase in functional expression. The effect of nitrogen sources and concentrations was also explored for their impact on lyngbyatoxin expression.

2.2.2.4 Expression of Ribosomally Encoded Peptides

Two papers published in 2005 reported on the biosynthesis of the patellamide cyclic peptides (e.g., patellamide D; Figure 2.2) as obtained from the Indo-Pacific tunicate *L. patella*. These cyclic peptides are distinctive for their alternating composition of aliphatic amino acids and heterocyclized threonine, serine, and cysteine residues, and their attendant cytotoxic and other biological properties. Their production was believed to be due to a symbiotic cyanobacterium *Prochloron* sp., which is found in large abundance within specialized channel structures present in the tunicate. The Jaspars group produced a shotgun library of *Prochloron* DNA in *E. coli*, and arrayed the clones in some fourteen 96-well plates.[41] DNA hybridization techniques using NRPS genes were unsuccessful in localizing the BGC; however, chemical screening of the transformants was successful in identifying several clusters that produced one or another of the patellamide-type NPs. This was accomplished by forming mixed cultures of *E. coli* transformants (e.g., 14 plate cultures, then 8 rows and 12 column cultures from positive plates), and analyzing for patellamide production using an optimized liquid chromatography-mass spectrometry (LCMS) procedure. Several strains were located with the capacity to produce low levels of patellamide-type products, generally less than 1 mg/L. Failure to associate these producing transformants with NRPS genes suggested that a new pathway was responsible for their production, and this was indeed the case as shown by the Schmidt laboratory.

The latter group took a genome sequencing approach toward locating the patellamide BGC.[42] DNA harvested from *Prochloron* sp. that had been isolated from *L. patella* was sequenced to produce a draft genome comprised of 734 scaffolds. BLAST analysis revealed a single NRPS not matching the expected architecture for patellamide biosynthesis, and therefore a ribosomally synthesized and post-translationally modified peptide (RIPP) pathway was hypothesized. Hence, all possible octapeptide sequences that could result in patellamide production (e.g., the beginning and end of the molecule were unknown) were searched in the draft genome, and resulted in a single biosynthetic locus that had the appropriate amino acid sequence for both patellamides A (Figure 2.4) and C. This precursor peptide sequence was surrounded by six other genes that had reasonable similarity to other lantibiotic and microcin biosynthetic machinery, and thus it was reasoned that by a combination of proteases, oxidases, and heterocyclases, the patellamides arose from a ribosomal pathway. This was firmly demonstrated by cloning this seven-gene cassette into *E. coli* and demonstrating patellamide A biosynthesis, albeit at quite low levels (~20 μg/mL).

This same study identified that five of these seven genes (*patA-G*, except *patB* and *patC*) were essential for the *in vivo* biosynthesis of patellamide A. In order to investigate the abundance and variability of patellamide-like NPs, the Schmidt group conducted genome sequencing of an additional 46 *Prochloron* symbionts from diverse hosts containing diverse NPs. They proposed the presence of two conserved recognition sites in the precursor peptide (*patE* homologs) responsible for recruiting the enzymes that modify the precursor peptide, indicating that the actual NP coding region could be variable. This hypothesis implied that the ascidian symbionts could yield a natural combinatorial peptide library by varying this small region with the conserved biosynthetic background. Using this information, the authors engineered and heterologously expressed a novel peptidic NP, eptidemnamide (Figure 2.2).[43] Moreover, they identified 29 *patE* gene variants and concluded that ascidians harbored a mix of cyanobacterial symbionts that collectively encoded for libraries of these cyclic and heterocyclized peptides (later designated as cyanobactins). Further NP and biosynthetic investigations identified a global assembly line for this type of NP.[44]

Characterization of the macrocyclase domain (PatG, the first subtilisin-like macrocyclase to be characterized) and its mechanism allowed rational engineering of this enzyme, and it is now a promising tool for organic synthesis and biotechnological applications.[45] The biosynthesis and biotechnological applications of the cyanobactin biosynthetic system, including substrate flexibility, enhancing turnover rates of key enzymes (e.g., the macrocyclase PatG), cyclization of peptides without leader sequences, and general utility of other enzymes in the system (e.g., PatD, the heterocyclase), has recently been comprehensively reviewed.[46]

2.3 GENETIC DIVERSITY OF SECONDARY METABOLISM IN CYANOBACTERIA

Despite progress in finding new NPs during the twenty-first century, large pharmaceutical companies have reduced investments in NP drug discovery for several reasons, including rediscovery of known compounds or the discovery of less unique chemical structures. However, the genomics revolution has revealed that the number and chemical diversity of compounds isolated from microbes using classical NP isolation methods are only a small percentage of its potential. Therefore, new approaches have emerged to complement classic methods and promise to yield the discovery of many novel and bioactive compounds, termed *bottom-up approaches* (Figure 2.3).[47] They consist of evaluating the drug discovery potential of a given microbe by studying its BGCs, thus providing initial directions in the search for novel metabolites. This approach is founded on observations that most NPs are formed by many enzymes working together in an assembly line–like configuration. These enzymes derive from biosynthetic genes, which are conveniently grouped together into gene clusters in microbes. Other genes are also commonly found clustered together with the biosynthetic genes, such as regulatory, mobility (allow the transfer of portions or even complete BGCs between genomes of two microbes), and resistance (mutated copies of the NP target, which is not sensitive to the effect of the NP anymore, thereby allowing survival of the producer).

The toolbox of bottom-up approaches can be divided into three broad categories: bioinformatic tools, native host expression, and heterologous host expression. Bioinformatic tools can range from software programs that simply identify prokaryotic biosynthetic genes (antiSMASH, ClusterFinder) to those performing very complex tasks such as correlating a gene cluster to its final products (Pep2Path, RiPPquest, NRPquest, GARLIC).[48] Native host expression can include the genomisotopic approach (consisting of the use of isotope-labeled substrates to identify the NP even when present in very low abundance), induction of the targeted secondary metabolite, and engineering of BGCs in the native host (overexpression of promoters or deletion of repressors) for the activation of a *silent* gene cluster.[49]

Bottom-up approaches have contributed many new techniques to the NP discovery toolbox; however, the promises of the genomic revolution have not yet yielded many new and bioactive NPs (especially from Cyanobacteria). This is due to the existence of biological, chemical, and computational challenges; the latter were recently reviewed by Medema and Fischbach.[50] The first computational challenge arises from the difficulty of obtaining complete genomes from environmental or nonaxenic cultures (e.g., filamentous Cyanobacteria with associated heterotrophs). This occurs because the abundance of shared repeated regions poses considerable problems for the current assembly algorithms. A second core computational challenge is the lack of decent and readily available methods by which to group BGCs into families. This would facilitate dereplication at a genomic level as well as highlight unique gene clusters that could be further investigated. Previous efforts to perform gene cluster networking have been extremely laborious and inefficient, and some did not utilize publicly available BGCs. Medema and Fischbach[50] emphasize that an efficient networking approach would be a powerful tool to analyze BGCs, and the characterization of gene cluster families would facilitate the comparison

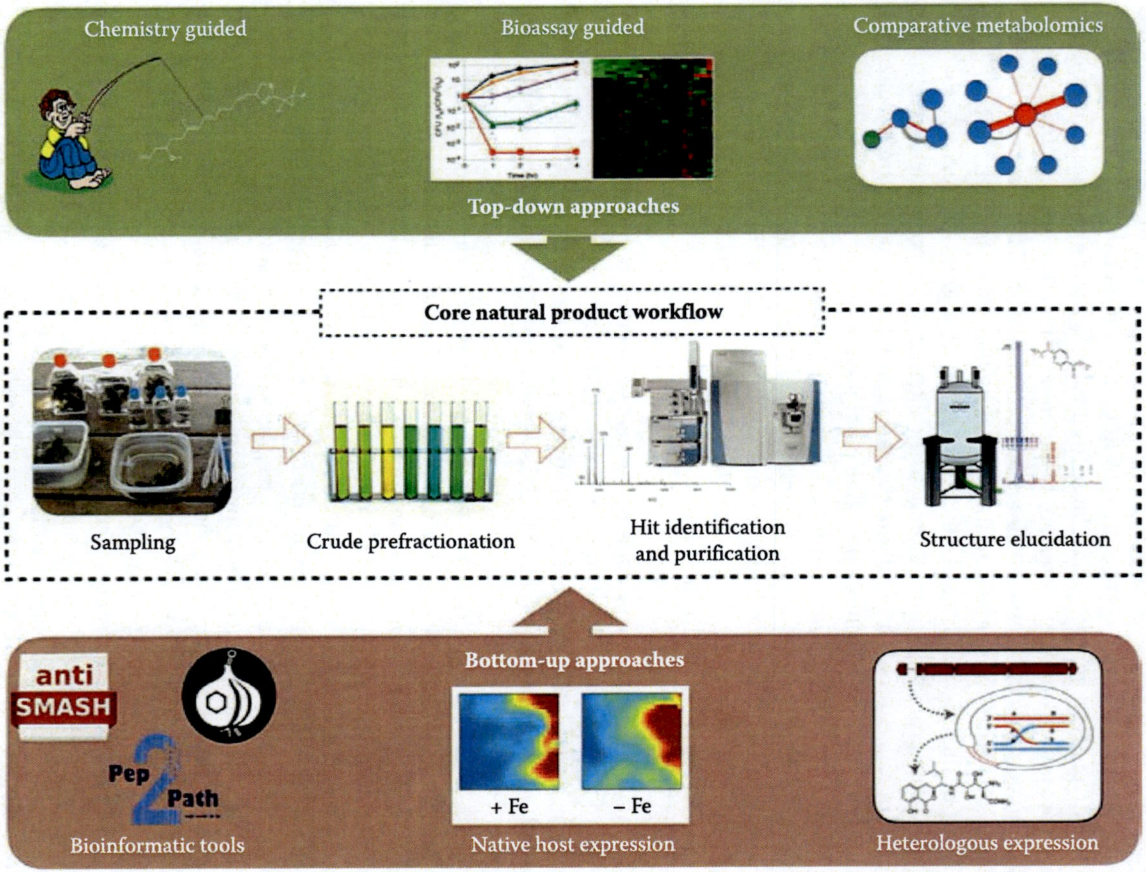

FIGURE 2.3 Bottom-up versus top-down approach for natural products discovery.

between genomic data (genotype) and molecular data (phenotype). This is especially true for Cyanobacteria where genetically related strains can have quite different metabolic profiles. If one could reliably connect BGCs with their encoded metabolites, a process known as *pattern-based genome mining*,[51,52] this would greatly accelerate the targeted isolation of new NPs. Finally, given these emerging bottom-up approaches for assessing and investigating BGCs, there is a need for a better and more universal BGC nomenclature which would allow for the rapid and accurate communication of gene cluster identity.

Cyanobacteria are especially interesting for bottom-up approaches because it was recently identified from genome studies that their potential for NP discovery is much larger than appreciated from traditional isolation efforts.[7,53,54] In this section, we provide an overview of the NPs gene cluster diversity and the potential for the discovery of novel NPs from Cyanobacteria. We also identify the main compound families found in Cyanobacteria (and their respective gene cluster families), as well as review the use of bottom-up approaches for the discovery of cyanobacterial secondary metabolites to date.

2.3.1 Genetic Diversity and Distribution of Cyanobacterial Natural Product Pathways

Cyanobacterial NP pathways are interesting for their presence of unique biosynthetic genes, some of which are responsible for unprecedented chemical transformations, as recently reviewed by Kleigrewe et al. (2016).[55] However, knowledge of cyanobacterial genomes needs improvement, both in terms of quantity and quality. Recent sequencing efforts, both large[56] and small,[57] are allowing for a better assessment of the genetic basis for NPs biosynthesis in Cyanobacteria.

The following provides some background on the current situation for those cyanobacterial genomes that are available in public databases, as well as a correlation to

TABLE 2.1
Summary of Publicly Available Cyanobacterial Genomes in the JGI/IMG Database

Subsection	# Genera	# Strains
Unknown	N/A	14
Unclassified	2	2
Melainabacteria	4	6
I. Gloeobacterales	1	2
II. Synechococcales	14	181
III. Spirulinales	1	2
IV. Chroococcales	7	54
V. Pleurocapsales	4	5
VI. Oscillatoriales	11	55
VII. Chroococcidiopsidales	3	5
VIII. Nostocales	22	67
Total	69	393

their NPs when known. Table 2.1 illustrates the number of cyanobacterial genomes currently available at the IMG/JGI database, organized according to phylogenetic subsections. Since the time of inclusion of Cyanobacteria under the bacteriological code in 1978, their taxonomic classification has been under considerable discussion and restructuring. Originally, Cyanobacteria were classified into five subsections, mainly categorized based on morphological attributes.[58] The recent expansion of genomic information for Cyanobacteria has led to a further revision.[59] This new classification, comprised of eight subsections, relies mainly on phylogenetic analysis of several conserved protein sequences (phylogenomic analysis) in combination with key morphological traits, such as thylakoid patterns, the formation of filaments, and the presence of heterocysts. Despite this recent and quite thorough examination, the authors noted that further changes are likely as more data become available. For example, the two organisms *Halothece* sp. and *Rubidibacter lacune* do not fit into any of the current subsections, and therefore remain unclassified.

The subsections can be divided into three categories according to the number of genomes sequenced in each: overrepresented, medially represented, and underrepresented. The *Synechococcales* is the only overrepresented subsection, and many of the sequenced strains are significantly similar to one another. The overrepresentation of this subsection is partially due to the average size of its genome, which tends to be the smallest among Cyanobacteria (excluding Melainabacteria) and therefore facilitates sequencing and assembly. Also, Cyanobacteria of this subsection have considerable importance in biogeochemical cycles of the ocean. *Synechococcus* and *Prochlorococcus* are among the major sources of biological carbon and nitrogen in oligotrophic environments such as the open ocean.[60] However, there are relatively few novel NPs deriving from Cyanobacteria of this subsection, and therefore they are of less interest from a drug discovery perspective. Their genomes tend to harbor only few RiPP and terpene NP pathways, with the exception of two *Prochlorococcus marinus* strains (MIT 9303 and MIT 9313), which are remarkably rich in RiPP pathways. Despite the small number of novel secondary metabolites from this subsection, there have been a couple of highly significant discoveries, such as patellamide A (Figure 2.4) and prochlorosins,[61] and this is further discussed in subsequent sections.

Subsections VI and VIII comprise the most prolific Cyanobacteria in terms of numbers of secondary metabolites. The largest number of sequenced genomes appear in the medially represented range, whereas the number of NPs discovered from *Oscillatoriales* (VI) and *Nostocales* (VIII) is the largest among this phylum. In 2015, it was estimated that these two subsections were responsible for more than 65% of all cyanobacterial NPs discovered to date,[7] with the greatest emphasis on the freshwater genera *Nostoc* (VIII) and marine genera *Lyngbya* (VI, reclassified in 2012 as *Moorea*).[62] The majority of compounds observed in these subsections derive from PKSs, NRPSs, hybrid PKS/NRPS, and indole-alkaloids. In the same review,[7] the third most prolific subsection was the *Chroococcales* (IV, another medially represented subsection), with great emphasis on the freshwater genus *Microcystis*, which is a producer of a number of toxins. All of the remaining subsections (I, III, V, and VII) are underrepresented in terms of genomes sequenced, and this parallels the number of reported NPs; there are few to no studies reporting novel NPs from these subsections to date.

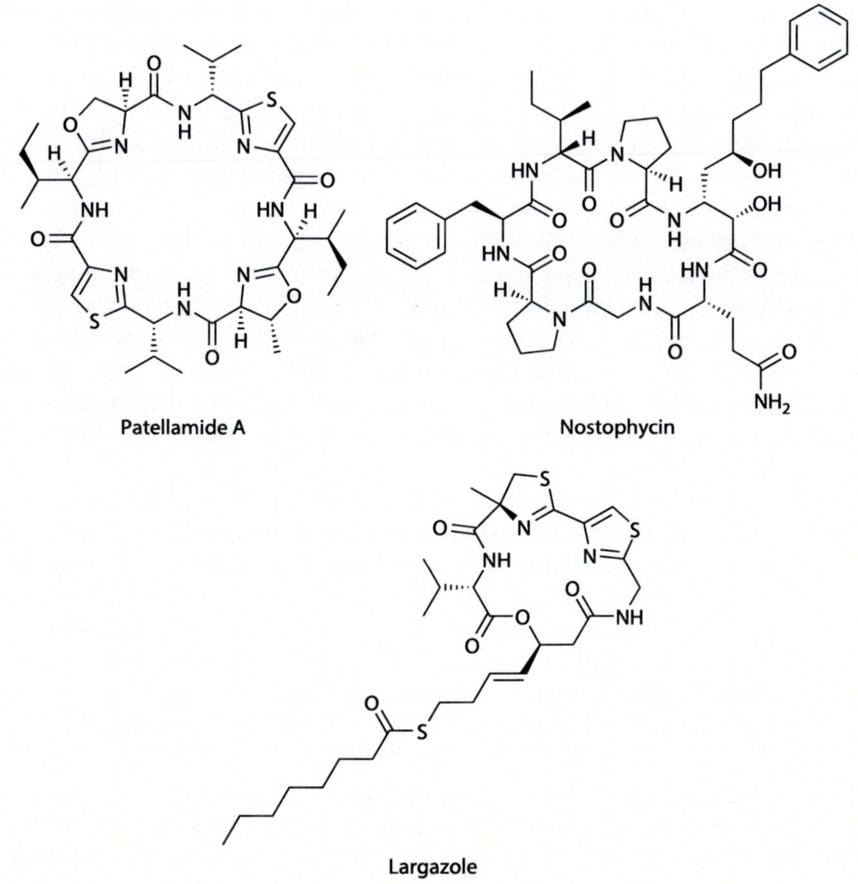

FIGURE 2.4 Structures of diverse natural products isolated from Cyanobacteria.

2.3.1.1 Polyketides and Nonribosomal Peptides

Cyanobacteria produce structurally diverse metabolites that possess many interesting chemical features such as unusual methylations, halogenations, and oxidations. Examples include curacin A (Figure 2.1),[63] a compound that contains a long unsaturated fatty acid chain that is attached to a thiazolene ring adjacent to a cyclopropyl ring. It shows exquisite activity (low nM to high pM IC_{50}s) against several cancer cell lines due to its antitubulin effects. Largazole[64] (Figure 2.4) is a novel cyclodepsipeptide with potent antiproliferative bioactivity that preferentially targets cancer cells. Nostophycin (Figure 2.4)[65] is a cyanotoxin that contains a *β*-amino acid residue. All of these examples are hybrid metabolites deriving from a mixture of PKS and NRPS pathways, a biosynthetic type which is especially abundant in Cyanobacteria.

In 2014, Calteau and collaborators[53] performed a phylum-wide analysis of 425 NRPS and PKS pathways from 126 cyanobacterial genomes, and grouped the homologous pathways (found in more than one genome) into cluster families. Since NRPS/PKS hybrids are one of the major types of biosynthetic pathways in Cyanobacteria, this broad analysis gave an interesting perspective on NP evolution and biosynthesis in this phylum. For example, a known NP could be assigned as deriving from only 20% of the NRPS/PKS gene clusters in this phylum, and were mainly fatty acids, proteasome inhibitors, UV protectants, and toxins. This finding highlights the fact that genome-driven drug discovery could be a valuable tool in Cyanobacteria, and that there are many NPs left to be discovered from these organisms. Most pathways without an assigned known NP, described as *cryptic* pathways, were not found to be broadly distributed among the different strains. About one-third of the cryptic gene clusters were shared among less than five genomes, therefore constituting small cluster families. The remaining large majority of cryptic gene clusters were present in the genomes of a single strain and, hence, did not belong to any cluster family (termed *orphan* gene cluster). The orphan gene clusters had a lower mean size compared to other cryptic gene clusters (i.e., those that were somewhat more broadly distributed), highlighting the great diversity of small and unique NRPS/PKSs in the Cyanobacteria.

In terms of distribution, this phylum-wide analysis[53] concluded that most of the pathways with known NPs were constrained to a small taxonomic range. An exception was the cluster encoding for the biosynthesis of polyunsaturated fatty acids (PUFAs) and α-olefins (the olefin synthase [OLS] pathway).[66] This trend also applied for cryptic pathways. The only exception in this latter group was a short PKS cluster (highly conserved in its gene content and order) that was distributed throughout numerous freshwater picocyanobacteria that have highly syntenic genomes. The most remarkable trend was that close to half of all PKS/NRPS pathways were found to represent orphan gene clusters.

Observation that only a small number of PKS/NRPS pathways are widely dispersed among different cyanobacterial genera led to the hypothesis that pathway mobility is rare in these organisms. While difficult to investigate, features such as GC content, dinucleotide signature, presence of genes encoding for mobile elements (transposases, phages, or integrases), and number of clusters found in plasmids (common vectors for gene transfer) can provide some insights into this topic. Calteau and collaborators found that about 30% of the NRPS/PKS gene clusters in Cyanobacteria had mobility features and/or were present in plasmids, and therefore suggested the possibility of horizontal gene transfer. However, most of these mobile elements were characterized as highly degraded, probably representing pseudogenes (remainder of a gene that was once functional). Moreover, phylogenetic analysis of the NRPS condensation domains and PKS ketoacyl synthase domains indicated that the evolution of these clusters involved more complex mechanisms than horizontal gene transfer, such as gene duplications, punctual mutations, and domain recombinations (deletion/substitution).

From the small taxonomic range of NRPS/PKS gene clusters, the limited evidence of mobility traces, and the indications of other complex evolutionary mechanisms, the authors concluded that horizontal gene transfer is not a major mechanism for NP diversification in Cyanobacteria. Rather, diversification of vertically transferred gene clusters can be achieved by duplication/deletion/addition of genes and changes in domain specificity. In support of this latter hypothesis, there are highly syntenic pathways encoding for hectochlorin- and lyngbyabellin-type NPs in three different strains of the genus *Moorea*.[8] In this gene cluster family, two classes of compounds are produced with highly similar backbones. Both contain a chlorinated lipid moiety followed by four or five amino acids, the last of which condenses its carboxyl group with a hydroxy group from the lipid moiety, generating a lactone ring. A main difference is that the hectochlorin pathway is shorter by one NRPS module (one amino acid unit), which probably occurred by partial deletion of the *hctF* gene. A second major difference between these compounds is the amino acid specificity of the adenylation domains, indicating that punctual mutations probably altered the domain specificity. Lastly, a transposase gene is present within the hectochlorin BGC, a major driver of pathway diversification in Cyanobacteria. Other examples of this diversification hypothesis have recently been reviewed.[55]

Another interesting aspect of the distribution of NRPS/PKS pathways in the Cyanobacteria is the higher abundance of gene clusters in late branches of the phylogenetic tree. Out of 75 strains, 65 contain at least one NRPS/PKS gene cluster in the late branches, with an average of 6.2 gene clusters per genome. This contrasts with organisms deriving from early branches in which only 24 of 51 strains contain NRPS/PKS gene clusters, and these averaged merely 2 clusters per genome.

2.3.1.2 Ribosomally Synthesized and Post-Translationally Modified Peptides

Apart from NRPS and PKS pathways, another abundant NP class in Cyanobacteria are the RiPPs. This class of pathway encodes a short precursor peptide that is ribosomally synthesized. The precursor peptide plus leader sequence then undergoes a number of modifications, such as heterocyclization, methylation, oxidation, prenylation, and then macrocyclization to generate a final mature peptide. Among the 11 different classes of RiPP pathways, the most predominant in cyanobacterial genomes are the cyanobactins, lathionines, and microviridines. Given the abundance of lanthionine-containing peptide (lantipeptide) pathways in cyanobacterial genomes, it is surprising that only one group of cyanobacterial lantipeptides has been reported to date, the prochlorosins.[61] Due to the predictability of the precursors and their post-translational modifications, RiPPs are one of the most promising classes of NPs to be pursued via genome-guided efforts. It is estimated that about one-third of all cyanobactins have been isolated as a result of predictions drawn from their precursor peptides and possible modifications.[67] Examples include trichamide, the aeruginosamides, and viridisamide (Figure 2.6), and these will be discussed further in the section on genome-guided isolation in Cyanobacteria. The distribution of cyanobactins is quite ubiquitous among this phylum, even within genomes containing fewer NP pathways (e.g., the many Cyanobacteria in early phylogenetic branches). Cyanobactins are more abundant in the genera *Oscillatoria*, *Arthrospira*, and *Microcystis*, but are scarce to nonexistent in the genera *Prochlorococcus* and *Synechococcus*.[68] In addition, to the best of our knowledge, all cyanobactins reported to date are exclusively from Cyanobacteria, or from marine invertebrates containing cyanobacterial associates. In terms of size, cyanobactins tend to be comprised of 6–20 amino acid residues, and cyclic cyanobactins are commonly generated by head-to-tail macrocyclization. In this aspect, they differ from cyclic bacteriocins, which tend to be large head-to-tail peptides comprising 35–70 amino acid residues.[67] In fact, many bacteriocin BGCs are predicted in cyanobacterial genomes due to the presence of the C39 gene, which encodes for a peptidase. However, no true bacteriocin has been reported from Cyanobacteria to date. The final class of RiPPs in Cyanobacteria are the microviridins; these are a small family of cyclic N-acetylated trideca- or tetradeca-peptides containing lactam and lactone rings. This NP family has generated considerable interest due to their potent serine protease inhibitory activity. Interestingly, many cryptic microviridin gene clusters are predicted in cyanobacterial genomes, and are widely distributed, indicating that there is much chemical diversity left to be discovered in this chemical class.

2.3.1.3 Indole-Alkaloids and Mycosporine-Like Amino Acids

Indole-alkaloids are another biosynthetically interesting NP class commonly produced by Cyanobacteria. Members of the order *Stigonematales* (subsection VIII) are especially well known to produce a group of structurally complex indole alkaloids including hapalindole A, welwitindolinones A isonitrile, and 12-*epi*-fisherindole G (Figure 2.5).[69] In addition to the indole moiety which is produced by tryptophan biosynthetic genes (found within a hapalindole-like alkaloid gene cluster), their biosynthesis involves the terpene pathway with genes encoding for the production of isopentenyl pyrophosphate (IPP) and dimethylallyl pyrophosphate (DMAPP). Diversification of these structurally related indole alkaloids occurs as a result of several chlorination, oxidation, cyclization, prenylation, and methylation reactions. A novel nonheme iron-dependent halogenase (WelO5) was characterized in the biosynthesis of welwitindolinone.[70] This tailoring enzyme is capable of catalyzing monochlorination of substrates free of any peptidyl or acyl carrier protein. Genome mining of 11 *Stigonematales* strains indicated that additional hapalindole-like alkaloid BGCs are present and could be the subject of future investigations.

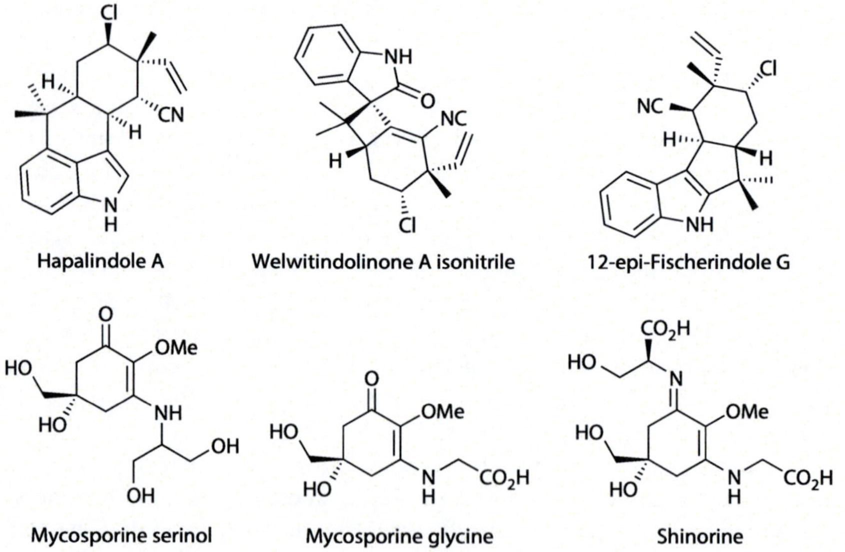

FIGURE 2.5 Structures of indole-alkaloids and mycosporine-like amino acids from Cyanobacteria.

The discovery of WelO5 emphasizes the importance of genome mining not only in the search of novel bioactive compounds but also for helping to expand our toolbox of useful enzymes and their corresponding genes.

Scytonemin is another interesting indole-alkaloid that acts as a UV-protecting metabolite (Figure 2.1). It results from a complex biosynthetic pathway that involves several transformations of L-tryptophan (see Section 2.2.1). Another interesting class of UV-protecting metabolites are the mycosporines and mycosporine-like amino acids (MAAs) (Figure 2.5), a superfamily of UV-absorbing compounds (in the range of 310–360 nm) considered as sunscreen pigments. This superfamily can be divided into two large groups according to the number of amino acid residues linked to the cyclohexenone core. A single amino acid residue constitutes the mycosporine family, such as in mycosporine serinol and mycosporine glycine, whereas a double substitution defines the MAA family, such as shinorine (Figure 2.5).[71] One of the most interesting aspects of the biosynthesis of MAA is that, despite the conservation of three core biosynthetic genes (an *O*-methyltransferase, a 3-dehydroquinate synthase, and a C-N-ligase), two quite different mechanisms are utilized for the attachment of the serine residue in the final product, for example, shinorine. In the strain *Anabaena variabilis* ATCC 29413, an NRPS gene is responsible for this condensation, while in *N. punctiforme* ATCC 29133 an ATP-grasp ligase is the catalyst for this final biosynthetic step. In general, these UV-absorbing NPs are widely spread among cyanobacterial phylogeny, and they can be upregulated in their production through manipulation of UV light or other stress conditions.[25,72]

2.3.2 Genome-Driven Drug Discovery in Cyanobacteria

Genome-driven NP investigations with Cyanobacteria provide two important observations: (1) cyanobacterial BGCs have many novel enzymatic features, confirming what has been previously observed by classical NP discovery; and (2) the NPs discovered prior to the genomic revolution represent a very small percentage of the potential for secondary metabolite production in these organisms. Nevertheless, at the present time, genome-driven drug discovery has not significantly enhanced the chemical space in Cyanobacteria. Indeed, most of the currently characterized cyanobacterial NPs have been discovered using top-down approaches.[7,54]

One example of a genome-driven NP discovery resulted in the characterization of columbamides A (Figure 2.6) and B from a *M. bouillonii* culture.[73] The columbamide BGC was located as a result of scanning the genome of this cultured cyanobacterium for the presence of a sensor histidine kinase gene that is believed to constitutively regulate NP biosynthesis in related Cyanobacteria. Indeed, this search led to the discovery of one that was highly homologous to two other sensor histidine kinases from the curacin and jamaicamide BGCs. Because jamaicamides and curacins are major compounds from cultured biomass, the authors hypothesized that this third gene cluster would also be significantly expressed under standard culture conditions. Structure predictions were informatically developed based on the biosynthetic genes, and this was matched to compounds of the same approximate size and heteroatom and halogen atom content as detected in the MS^2-based molecular network. The MS signature was then used to direct their isolation, and their structures were determined

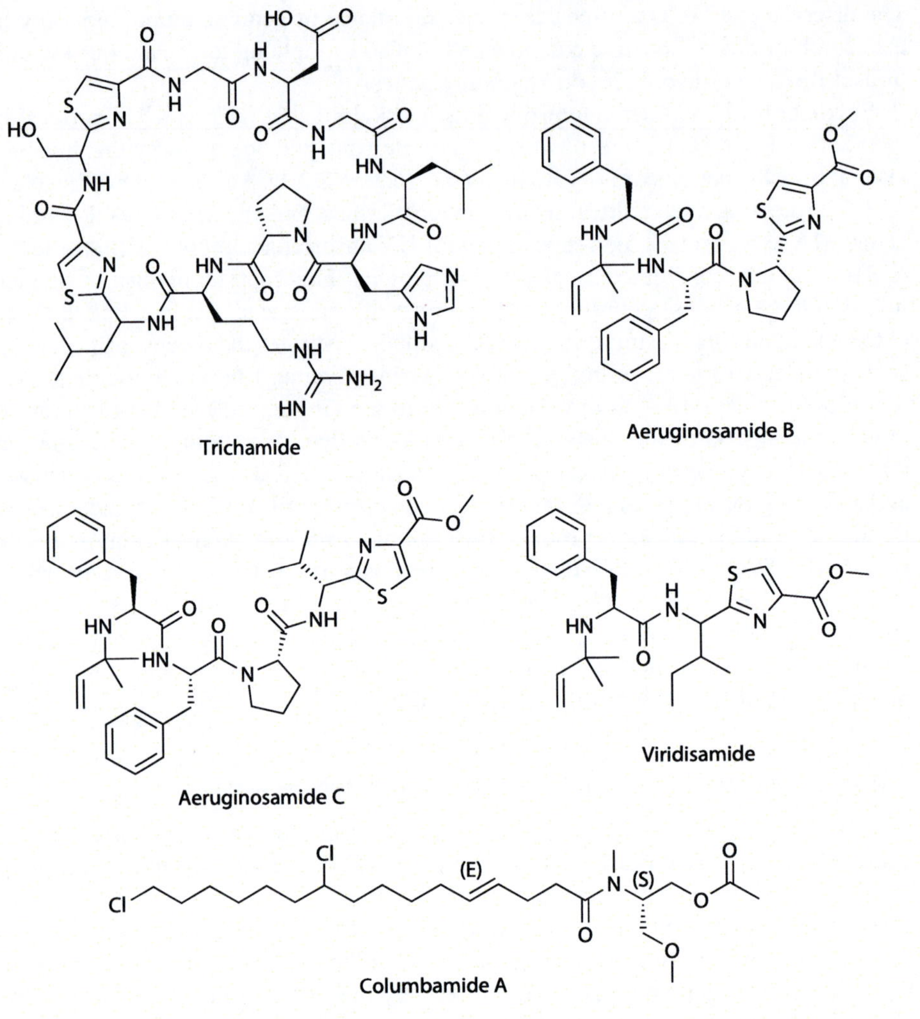

FIGURE 2.6 Structures of cyanobacterial natural products identified by genome-driven drug discovery approaches.

by NMR and other spectroscopic methods. The columbamides are unique chlorinated acyl amides with relatively potent cannabimimetic activities.

A second example of genome-driven NP discovery comes from the cyanobactins (see Section 2.3.1). Approximately one-third of all known cyanobactins have been discovered through genome mining. Major examples of this discovery path include trichamide, the prochlorosins, the aeruginosamides, and viridisamide (Figure 2.6). The discovery of trichamide was one of the earliest genome-guided NP discoveries, occurring even before the development of antiSMASH in 2011.[74] The patellamide pathway (see Section 2.2.2), the first discovered cyanobactin pathway, was used to mine for similar gene clusters in the genome of the toxic bloom-forming cyanobacterium *Trichodesmium*. One small cluster of 12.5 kb was identified in *T. erythraeum*

IMS101, and six genes were homologous to the patellamide pathway. The predicted precursor peptide in this gene cluster contained sequences for cleavage sites very similar to the precursor of patellamide, allowing for a relatively high-quality structure prediction for the NP. These predictions were confirmed via LCMS/MS in which mass differences in the fragments were linked to predicted amino acid residues, yielding the discovery of this circular post-translationally modified peptide named trichamide. This discovery clearly illustrated the effective alignment that is possible between predictions from genome mining and mass fragmentation experiments.

Analogously, Li et al.[61] found a number of *lanA*-like genes but only one *lanM*-like homolog in the genomes of several marine *Prochlorococcus* and *Synechococcus* strains. LanM is a synthetase that dehydrates and cyclizes a precursor LanA peptide to produce a mature lantipeptide. Most lantipeptide-producing bacteria encode a LanM synthetase that acts on only a single LanA precursor peptide. However, *in vitro* experiments demonstrated that this LanM has low substrate specificity, and suggested that these bacteria may take advantage of this substrate promiscuity to produce NP libraries. This hypothesis led to the discovery of the prochlorosins.

The discovery of the first linear cyanobactins, the aeruginosamides and viridisamide (Figure 2.6), was possible via a large genome-mining effort of 126 cyanobacterial genomes, and led to the identification of 31 cyanobactin BGCs.[75] Cyanobactins were present in 24% of these genomes, including many different cyanobacterial subsections. One gene encoding for an unusual 70 kDa bimodular protein was of special interest, and led to the examination of three cyanobactin BGCs from *Leptolyngbya* sp. PCC 7376, *Microcystis aeruginosa* PCC 9432, and *Oscillatoria nigro-viridis* PCC 7112. Except in *Leptolyngbya* sp. PCC 7376, the precursor peptides contained conserved leader and cleavage sites, allowing accurate predictions of partial structures of the final NP. These predictions directed subsequent LCMS-guided isolation efforts. The structures of the three newly discovered linear prenylated cyanobactins (viridisamide, aeruginosamides B and C) were elucidated by NMR and Marfey's analysis. This study reinforced the concept that genome mining can facilitate isolation and structure elucidation efforts in Cyanobacteria, and also showed that it can direct those efforts toward the discovery of NPs with novel chemical features.

2.4 BIOTECHNOLOGICAL APPLICATIONS OF CYANOBACTERIAL PROTEINS

Cyanobacteria produce an array of useful proteins that have found applications in diverse areas. As photoautotrophic prokaryotes they do not require organic substrates for energy, but rather directly utilize carbon dioxide and sunlight to produce useful products. The protein content of some species makes up a large fraction of the biomass of actively growing Cyanobacteria. A prominent example are the filamentous Cyanobacteria commonly known as Spirulina that are used as an important source of protein. The most common and widely available Spirulina, *Arthrospira platensis*, contains over 50% of protein in the dry biomass[76] and is employed as the major commercial source of phycocyanin. *A. platensis* has gained considerable popularity in the human health food industry and has been applied as a protein and vitamin supplement as well as a functional food due to its richness in various nutritional components.[77]

2.4.1 Phycobiliproteins

Phycobiliproteins are water-soluble proteins found in Cyanobacteria (blue-green algae), Rhodophyta (red algae), and a class of biflagellate unicellular eukaryotic algae (cryptomonads). These proteins can appear blue-green, purple, red, orange, or yellow in color.[78] The pigmentation results from an interplay of chlorophyll α and β-carotene as well as the predominating phycobiliprotein(s). These intensely colored proteins are photosynthetic accessory pigments that act as light-harvesting protein-pigment complexes during photosynthesis. They capture light energy that is otherwise not accessible to the chlorophylls, and enable Cyanobacteria to perform efficient photosynthesis over a broad region (450–650 nm) of the solar spectrum.[79] The most prominent representatives are phycoerythrin (PE), phycocyanin (PC), and allophycocyanin (APC). The phycobiliproteins owe their intense color to the covalently attached linear tetrapyrrole (bilin) prosthetic groups that act as chromophores. They are organized in large multiprotein particles called phycobilisomes that provide 30%–50% of the total light-harvesting capacity of the producing organism, and are found attached to the outer surface of the thylakoids of cyanobacterial cells.[80,81]

Whereas PE shows a bright red color in daylight, PC and APC cover the blue color range exhibiting maximum absorbance (A_{max}) at 565, 620, and 650 nm, respectively. The first crystal structure of C-phycoerythrin from the marine cyanobacterium *Phormidium* sp. A09DM revealed small changes compared to PE isolated from red algae, and explain the observed changes in the fluorescence characteristics of this solar energy–harvesting protein. This structural analysis suggested that the microenvironment of the protein dictates the conformation of the bound chromophores (Figure 2.7).[82]

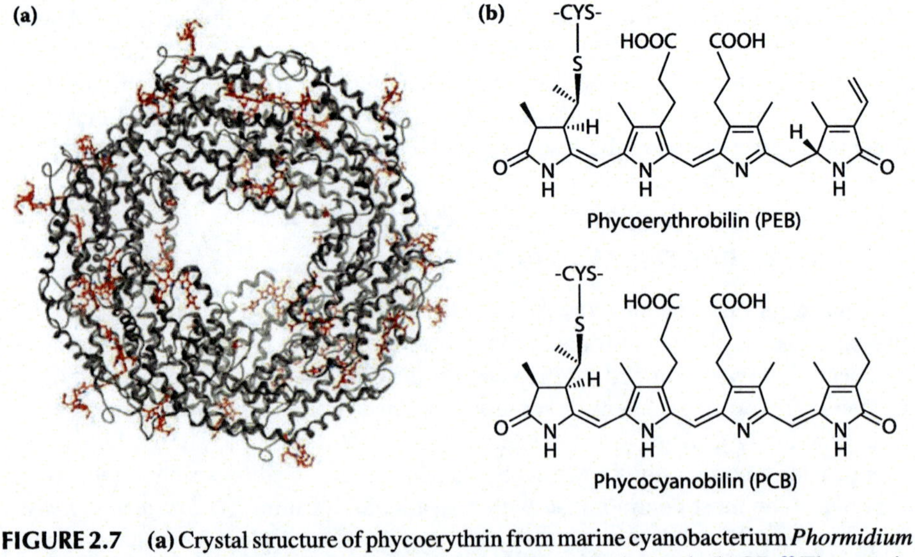

FIGURE 2.7 (a) Crystal structure of phycoerythrin from marine cyanobacterium *Phormidium* sp. A09DM; figure created using data obtained from PDB accession code 5AQD.[82] The protein (grey) microenvironment dictates the conformation of bound PEB chromophores (red). **(b)** Chromophores of phycoerythrin (PEB), phycocyanin, and allophycocyanin (PCB).

Phycobiliproteins possess a wide spectrum of biotechnological applications for biomedical research and clinical diagnostics, in the food industry as well as in pharmaceuticals.[83] A patent analysis of Sekar and Chandramohan[84] identified 297 existing patents on phycobiliproteins divided into 216 patents on fluorescence-based applications of phycobiliproteins; 30 patents on applications including therapeutic, cosmetic, and other applications; and 51 patents on phycobiliprotein production, extraction, and purification processes.

2.4.1.1 Phycobiliproteins as Fluorescent Agents

Besides their intense color, phycobiliproteins are also bright fluorescent pigments. Phycobiliprotein conjugates were introduced as a novel class of fluorescent reagents in 1982,[85] and have found applications in flow cytometry, fluorescent immunoassays, and fluorescence microscopy for diagnostics and biomedical research.[86] The amino and carboxyl groups found in the apoprotein chains of phycobiliproteins provide the possibility to synthesize conjugates with molecules possessing biological specificity, for example, immunoglobulins, protein A, biotin, and avidin. These conjugates represent valuable reagents for two-color fluorescence analysis of single cells using fluorescence-activated cell sorting (FACS).[85,87] Moreover, they can be used as internal protein markers for electrophoretic techniques, such as monitoring protein blotting and focusing protein samples during isoelectric focusing.[88]

2.4.1.2 Phycobiliproteins as Natural Protein Dyes

There is an increasing demand for natural colorants as safe food colors that can replace synthetic colorants, which have potential toxicity and carcinogenicity.[89,90] Despite this, there are difficulties to overcome, including the relatively low stability of natural dyes during processing and storage. The bright color of the phycobiliproteins, their water solubility, and their nontoxic/noncarcinogenic nature allows their use as a natural protein dye in various food and cosmetic applications. Mainly the blue color of PC is used as a colorant in chewing gums, soft drinks, and dairy products, whereas the cosmetic industry also utilizes the red color of PE for applications such as lipstick and eyeliners.[91] They are also used for coloring many other food products such as fermented milk products, ice creams, desserts, sweet cake decoration, and milk shakes.[84]

2.4.1.3 Pharmaceutical Potential of Phycobiliproteins

The potential therapeutic applications of the phycobiliproteins cover a broad area, and a variety of pharmacological activities have been reported including anticancer, anti-inflammatory, neuroprotective, and hepatoprotective activity.[84] PC has been investigated for its antioxidant or free radical scavenging potential using various *in vitro* and *in vivo* experimental systems,[92] and has been reported to inhibit cell proliferation and induce apoptosis in cancer cell lines.[93,94] Various studies underpin the use of phycobiliproteins as antioxidants in the treatment of clinical conditions related to oxidative stress such as atherosclerosis, arthritis, diabetes, and cancer.[95,96] It was shown that the covalently linked chromophore, phycocyanobilin, is involved in the antioxidant and radical scavenging activity of PC.[97]

2.4.2 Cyanophycin

Cyanophycin (Figure 2.8), also referred to as cyanophycin grana protein (CGP), is a polymer that was discovered in Cyanobacteria by Borzi about 120 years ago. It is a nonribosomally synthesized protein-like polymer, which consists of equal amounts of aspartic acid and arginine that are arranged as a polyaspartate backbone, with arginine moieties linked to the *β*-carboxyl group of each aspartate by its *α*-amino group. With five nitrogen atoms in every building block, CGP serves as a storage molecule for carbon, nitrogen, and energy that is accumulated in insoluble cyanophycin granules in the cytoplasm (Figure 2.8).[98] To provide nitrogen and carbon to the cell, cyanophycin is broken down into arginine and aspartic acid by cyanophycinase. Subsequently, the amino acids can be broken down by arginine decarboxylases and agmantinase and/or arginase.[99] Cyanophycin is a biocompatible and completely biodegradable polymer that makes it an ideal candidate for many applications in the fields of biomedicine, agro-chemistry, agriculture, personal care, and pharmacy.[100] In order to reduce pollution, energy consumption, and global warming, it can be chemically converted into a polymer with reduced arginine content that can be used as an alternative to nonbiodegradable synthetic polyacrylate.[101] This poly-aspartic acid backbone polymer can also serve as an environment-friendly water softener.

Berg et al. investigated the biosynthesis of this reserve polymer.[102] It is synthesized without participation of ribosomes by a single enzyme, the cyanophycin synthetase CphA, which exhibits similarities to both D-alanyl-D-alanine synthetase and Mur-ligases.[103] The enzyme has the unusual property of catalyzing the incorporation of two different amino acids into the growing molecule. CphA incorporates Asp and

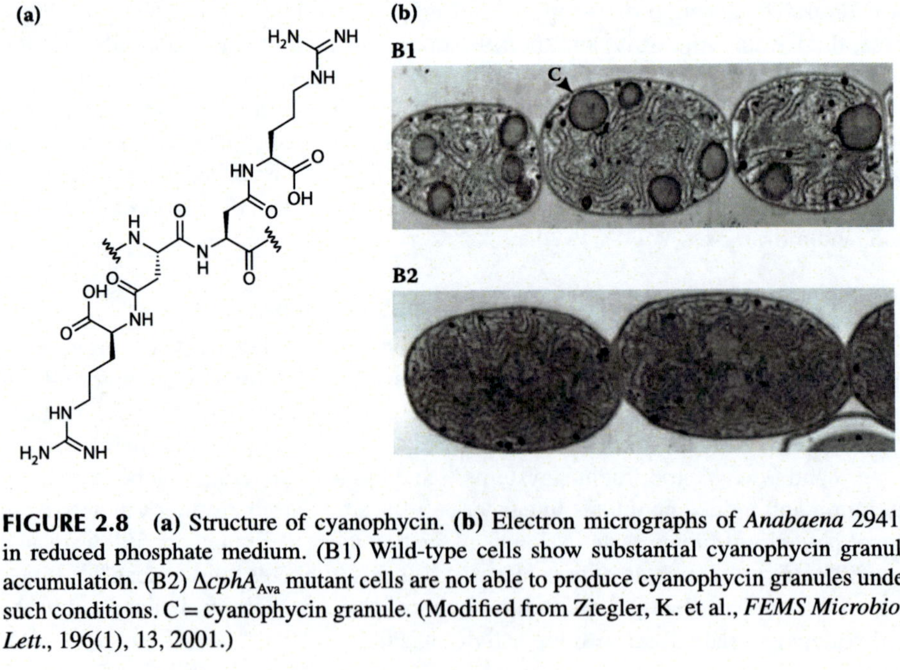

FIGURE 2.8 (a) Structure of cyanophycin. (b) Electron micrographs of *Anabaena* 29413 in reduced phosphate medium. (B1) Wild-type cells show substantial cyanophycin granule accumulation. (B2) $\Delta cphA_{Ava}$ mutant cells are not able to produce cyanophycin granules under such conditions. C = cyanophycin granule. (Modified from Ziegler, K. et al., *FEMS Microbiol. Lett.*, 196(1), 13, 2001.)

Arg and thus catalyzes an elongation reaction via an ATP-dependent stepwise mechanism. It catalyzes the formation of normal peptide bonds that build up the aspartic acid backbone as well as the formation of iso-peptide bonds to link the α-amino groups of arginine residues to the β-carboxy groups of the polyaspartate backbone. ATP is used to activate the amino acids by phosphorylation of the carboxyl groups as acylphosphates.[104] The suggested mechanism consists of two subsequent elongation steps that utilize two different ATP-binding sites. In the first ATP-activated elongation step, the activated carboxyl group of the Asp backbone is coupled to the amino group of the following Asp. The product is then transferred to the second active site where the β-carboxyl group of Asp is activated and coupled to the α-amino group of Arg. The next cycle of elongation requires back-transfer of the product to the first active site.[102,105] Interruption of the gene *cphA* in *A. variabilis* ATCC 29413 by insertional mutagenesis demonstrated that the mutant lacked cyanophycin granules in the cells but was still capable of nitrogen fixation (Figure 2.8).[106]

It has been shown that using chloramphenicol to inhibit protein biosynthesis results in an increase in the production of this nonribosomally produced polymer.[107] However, Cyanobacteria are unfavorable for the large-scale production of cyanophycin because of their relatively low polymer content and slow growth rate. Since the polymerization reaction is catalyzed by a single enzyme CphA, several recombinant strains of *E. coli*, *Ralstonia eutropha*, *Pseudomonas putida*, and *Acinetobacter baylyi* have been used for the heterologous production of CGP. CGP produced from recombinant strains exhibits a lower molecular mass range of 25–30 kDa than cyanobacterial CGP where the molecular mass of the polymer ranges from 25 to 100 kDa.[108] The polymer isolated from recombinant strains also contains the essential amino acid L-lysine, which partially replaces arginine.[104,109]

There are numerous nutritional and therapeutic applications of the nonessential amino acid L-aspartate, the conditionally essential amino acid L-arginine, and the essential amino acid L-lysine. Hydrolyzed protein diets that contain di- and tripeptides are used as nitrogen sources for the recovery of malnourished patients. CGP delivered as β-dipeptides has been suggested to be a highly bioavailable form for the administration of its constituting amino acids in nutritional therapy.[110]

2.4.3 Cyanovirin-N, an Anti-HIV Cyanobacterial Lectin

Cyanovirin-N (CV-N) is a virucidal lectin protein isolated from the cyanobacterium *Nostoc ellipsosporum*. Lectins are carbohydrate-binding proteins that are present in a variety of plant, fungal, and cyanobacterial species. With a size of 11 kDa, the polypeptide comprises 101 amino acid residues and was first identified in 1997 based on its potent inhibition of HIV cytopathicity in a phenotypic screen.[111] The protein structure was solved by NMR spectroscopy and later confirmed by crystallography. It contains two pseudosymmetric domains that are formed by strand exchange between two sequence repeats. The protein is a monomer in solution and exhibits a domain-swapped dimer in crystal form (Figure 2.9).[112,113]

The recombinant production of cyanovirin-N in *E. coli* provided an initial basis for further development of the protein.[114] By utilizing a chaperon-fused expression system, problems such as a low yield, aggregation, and abnormal modifications could

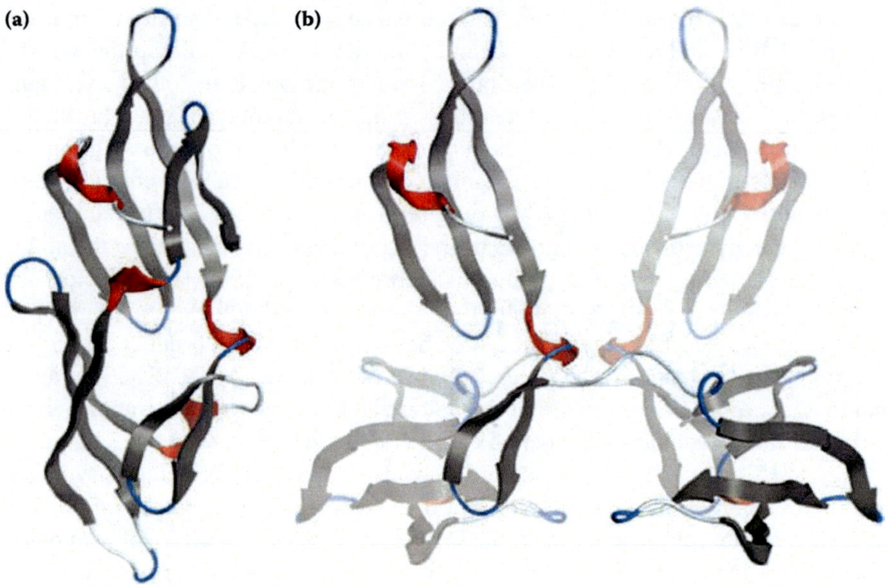

FIGURE 2.9 **(a)** NMR solution structure of the monomeric CV-N in solution, figure created using data obtained from PDB accession code 2EZM.[112] **(b)** Crystal structure of CV-N showing the domain-swapped dimer, figure created using data obtained from PDB accession code 3EZM.[113]

later be eradicated.[115] The fusion gene consisting of *cvn*, a small ubiquitin-related modifier (SUMO), and a His6-tag led to the expression of the SUMO-fused CVN in the cytoplasm of *E. coli* in a folded and soluble form, allowing the system to be used for large-scale production of this antiviral candidate. Other anti-HIV proteins isolated from Cyanobacteria include scytovirin and Cyt-CVNH.[116,117] These anti-HIV proteins have no sequence or structural homology with known proteins, and their physiological or ecological function in Cyanobacteria remains unknown.

2.4.3.1 Mode of Action of Cyanovirin-N

The virucide CV-N interacts directly with HIV virions to inhibit viral transmission by aborting the initial infection process at nanomolar concentrations. The potent HIV-inactivating protein binds with high affinity and specificity to the HIV surface envelope protein gp120, and therefore blocks the interaction of gp120 to cell-associated CD4.[118,119] This binding of CV-N to the glycosylated HIV envelope protein is carbohydrate dependent, with a specifically high binding affinity to mannose-rich oligosaccharides such as mannose-8 and mannose-9 moieties. The relatively low abundance of these oligosaccharides in mammalian glycoproteins compared to their higher abundance in certain viral glycoproteins such as gp120 is the basis for the potential utility of this agent as an anti-HIV microbicide.[120] Out of the 24 N-linked oligosaccharides found on the gp120 surface, 11 are high mannose-type or hybrid-type oligosaccharide structures.[121]

Other viruses are covered with N-glycosylated sites on their envelope proteins, and inhibition studies with members of the *Nidovirales* order showed that CV-N exhibited antiviral activity against several coronaviruses as well as the mouse hepatitis virus.[122] Another study that screened CV-N against a broad range of respiratory and enteric viruses, as well as flaviviruses and herpesviruses, found that CV-N showed highly potent antiviral activity against almost all strains of influenza A and B viruses together with a moderate activity against some herpes and hepatitis viruses. The influenza virus hemagglutinin was identified as a target for CV-N.[123] The interaction of CV-N with hemagglutinin was shown to involve oligosaccharides by binding to specific high-mannose oligosaccharides (oligomannose-8 and -9) at glycosylation sites on the viral hemagglutinin HA1 subunit.[124] CV-N also shows antiviral activity against ebola virus, feline immunodeficiency virus, human herpesvirus 6, and measles virus.[119,125]

As a potent fusion inhibitor, CV-N was tested for potential use as a microbicide that would be the first line of defense in preventing the sexual transmission of HIV. It was shown to be effective as a microbicide gel formulation for intravaginal delivery, and also as live microbicide that employs a natural vaginal strain of *Lactobacillus jensenii* engineered to deliver CV-N.[126–129] Another interesting application of CVN is its potential use as a targeting reagent. Productively infected, HIV-producing host cells express gp120 at the cell surface. Fused to *Pseudomonas* exotoxin (PE38), CV-N can confer gp120 binding activity and therefore enhance selectivity for killing HIV-infected cells.[130]

2.4.4 Cyanobacteria as Green Cell Factories to Produce Biocatalytic Enzymes

Cyanobacteria are the perfect candidates to investigate for production of biocatalysts suitable for industrial applications because they only require sunlight, water, nutrient salts, carbon dioxide, and N_2. Biocatalysts function under very mild reaction conditions compared to synthetic reactions and are therefore considered sustainable catalysts.[131] Another advantage is their high selectivity, which allows for multicomponent reactions and shortened reaction routes. These benefits have led to numerous applications for the environment-friendly synthesis of fine chemicals and pharmaceuticals.[132,133]

2.4.4.1 Bioconversion of Hydrocortisone and Other Steroids

Microbial transformations of natural and synthetic steroids provide an important method for obtaining new steroid derivatives as well as producing certain steroids for the pharmaceutical industry. Steroid compounds are among the most widely marketed pharmaceutical products with a wide range of medical applications, including those based on their immunoprotective, immunoregulatory, and anti-inflammatory functions, as anticancer agents in humans, as well as allosteric modulators of neuronal receptors.[134,135] Steroids show a very high specificity related to small chemical modifications, and in many cases hydroxylated derivatives show much higher biological activity.[136] Considering the complexity of steroid molecules, the use of biocatalysts is favorable because of the desired high regio- and stereo-specificity of the necessary reactions. Furthermore, the use of microorganisms provides an efficient alternative to chemical synthesis, which usually results in low yields. Whole cell systems are preferred over enzymes as biocatalysts because there is no need for enzyme isolation, purification, and stabilization.

In 1986, a systematic study of transformations of steroids by algae included several cyanobacterial strains, and showed that most of the tested cultures were capable of effecting transformation.[137] The steroid AD (androst-4-ene-3,17-dione; Figure 2.10a), one of the most useful substrates for production of steroid derivatives, was used to test the capabilities of 22 algal cultures. These experiments demonstrated the presence of primarily 17-ketoreductases, as well as 6β-hydroxylases, 14α-hydroxylases, and Δ^1-dehydrogenases, together with several unidentified enzymes. Since then, countless strains have been investigated for their ability to convert easily accessible steroids into pharmaceuticals. *Nostoc muscorum* was shown to be able to successfully transform AD and ADD (androst-1,4-diene-3,17-dione) to their 17-hydroxy derivatives, testosterone, and 1-dehydrotestosterone (Figure 2.10a).[138] *Nostoc* sp. were also shown to transform hydrocortisone, an important steroid compound, to 11β-hydroxyandrost-4-ene-3,17-dione, 11β,17β-dihydroxyandrost-4-ene-3-one, and 11β,17α,20β,21-tetrahydroxy-pregn-4-ene-3-one (Figure 2.10b).[139,140] Hydrocortisone conversion was also shown for *Fischerella ambigua*,[141] *Chorococcus disperses*,[142] and *Synechococcus nidulans*.[143]

Furthermore, cyanobacterial electron carrier proteins can complement other plant and bacterial cytochrome P450s by providing the reducing equivalents necessary for the hydroxylation process, and therefore allowing for the design of hybrid systems.[144,145]

2.4.4.2 Production of Other Commercially Important Enzymes

Cyanobacteria produce an array of additional enzymes with potential commercial applications. Nine filamentous cyanobacterial isolates were shown to produce amylases, proteases, as well as phosphatases.[146] Amylases, enzymes that catalyze the hydrolysis of polysaccharides into sugars, are important industrial enzymes. Possible applications include brewing and alcohol production, textile, biofuels, and other industries. Increased attention has been paid to enzymes from mesophilic and psychrophilic organisms (optimal growth at moderate and low temperatures) in the last few years because they have the capacity to manage the decrease in their reaction rates triggered by low temperatures. A novel α-amylase with interesting properties for industrial applications has been functionally characterized from the mesophilic cyanobacterium *Nostoc* sp. PCC 7119.[147] The *Nostoc* Amy1 protein has its pH optimum at a higher pH than most α-amylases, and is active until pH 10.5. Its high turnover number exceeds that of other enzymes with industrial uses, and could therefore be interesting for various industrial processes. Interestingly, it was also shown that the mesophilic Cyanobacteria, *Phormidium* spp., are able to produce highly thermophilic restriction endonucleases that have an optimum activity at 65°C–80°C, temperatures that are much higher than the lethal temperature of the host organisms.[148]

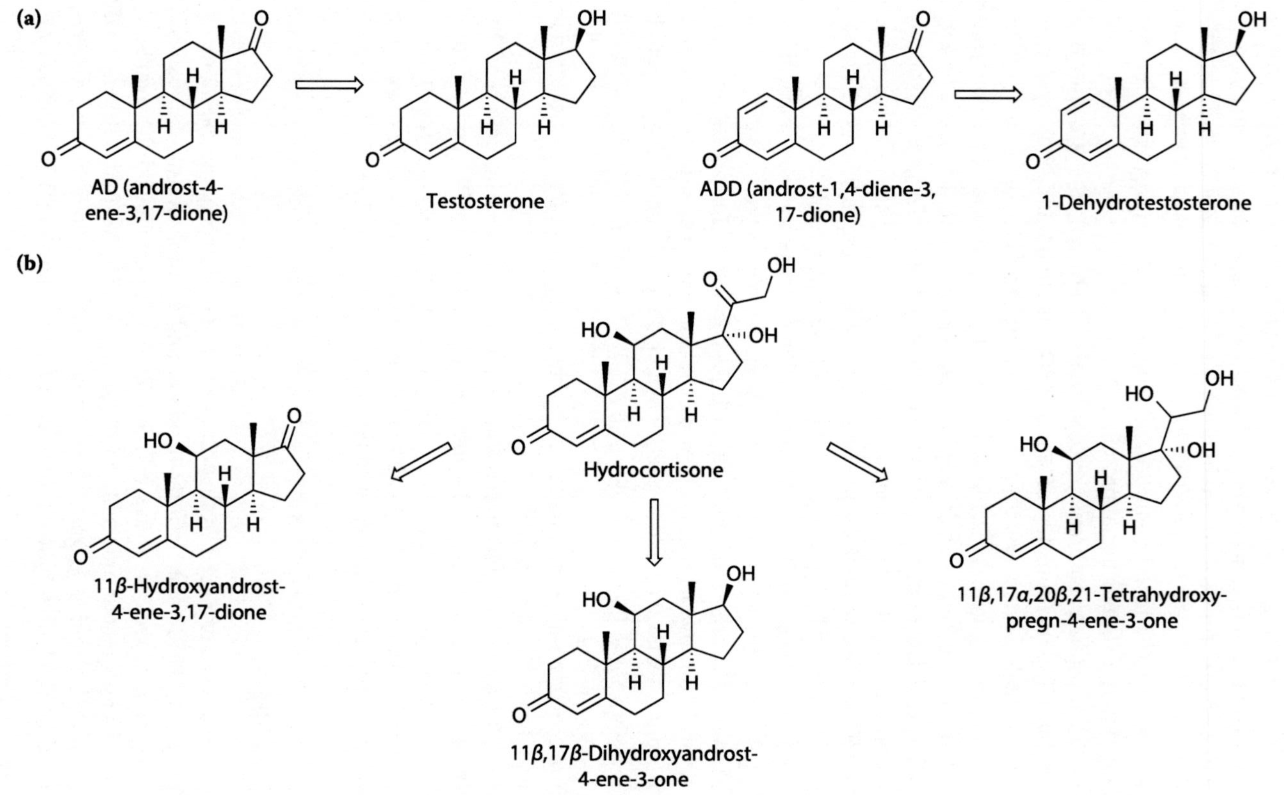

FIGURE 2.10 **(a)** The structures of steroids AD (androst-4-ene-3,17-dione) and ADD (androst-1,4-diene-3,17-dione) together with the cyanobacterial biotransformed products testosterone and 1-dehydrotestosterone. **(b)** Structures of hydrocortisone together with the biotransformed products 11β-hydroxyandrost-4-ene-3,17-dione, 11β,17β-dihydroxyandrost-4-ene-3-one, and 11β,17α,20β,21-tetra-hydroxy-pregn-4-ene-3-one.

2.4.4.3 Attempts to Increase the Production Efficacy in Cyanobacteria

Utilizing Cyanobacteria for the large-scale synthesis of proteins for biotechnological applications is challenging, and few genetic engineering approaches have proven successful to improve their efficiency. Most efforts have focused on the genetic engineering of Cyanobacteria for biofuel production, a topic that is not covered in this book chapter and was recently reviewed by Savakis and Hellingwerf.[149] However, improving the photosynthetic efficiency of Cyanobacteria is a key step for the increased production of cyanobacterial proteins.

Ribulose-1,5-bisphosphate carboxylase/oxygenase (Rubisco), the enzyme that catalyzes the initial CO_2–fixation step of photosynthesis, limits the photosynthesis rate due to its slow catalytic rate, low affinity for atmospheric CO_2, and use of O_2 as an alternative substrate.[150] Overexpression of Rubisco has led to increased isobutyraldehyde and isobutanol production in the engineered cyanobacterium *Synechococcus elongatus* PCC7942.[151] The integration of an additional set of *rbcLS* genes responsible for Rubisco production into the chromosome of *S. elongatus* PCC7942 increased photosynthetic efficiencies about 1.4-fold, and resulted in a roughly twofold higher production rate of isobutyraldehyde in the genetically engineered strain.

As mentioned in Section 2.4.1, Cyanobacteria have evolved a very specialized light-harvesting system, the phycobilisome, which allows photosynthesis even under low light conditions. Nevertheless, under high light conditions for production purposes, this system is disadvantageous. Cells on the surface layer experience a rate of photon absorption that exceeds the rate by which they can be utilized through photosynthesis, and excess light that could otherwise be used by cells lower in the water column is lost through thermal dissipation. It was shown that PC deletion in the cyanobacterium *Synechococcus* sp. led to a 1.5 times increased biomass accumulation under simulated bright sunlight.[152]

Furthermore, it was shown that photosynthetic efficiency can be enhanced by increasing NADPH consumption.[153] By introducing extra NADPH consumption capability into *Synechocystis* sp. PCC 6803 to improve the coupling of light and dark reactions, the growth rate was doubled and the engineered cyanobacterium was able to utilize more light and thus produce more biomass.

2.5 CONCLUSION

Three main themes were explored in this chapter on the biotechnological value of Cyanobacteria. First, we detailed efforts to clone and heterologously express NP biosynthetic enzymes as well as entire NP BGCs. We then reviewed the available literature on the use of genetic information to guide the discovery and isolation of novel bioactive NPs from these organisms, a pursuit called *bottom-up* drug discovery. Finally, we described other biotechnological products deriving from marine cyanobacteria, namely their unique enzymes and proteins that have useful applications in medicine, foods, and industrial processes. From these discussions, it is clear that these very ancient organisms have exceptional metabolic capacities to make adaptive molecules, both large and small. For example, in response to light of various qualities they produce small molecules that are protective against photo-damage (e.g., microsporines, scytonemin) as well as large molecules that are especially adapted to harvest energy from a broader range of wavelengths (phycobilins).

Cyanobacteria have proven useful to generate various products for biotechnological applications, not least because of their ability to transform solar energy and carbon dioxide directly into industrial products. They show a great potential considering their high versatility to synthesize an array of valuable compounds, such as proteins, small molecules, toxins, and other metabolic products. However, the low yields obtained from Cyanobacteria so far suggests that efforts should be focused on the production of high-value products, such as biocatalyic enzymes or fluorescent dyes for biomedical research, more than on the mass production of high-volume compounds.

But for reasons unknown, their capacity to produce structurally unique secondary metabolites is truly astounding. As discussed in the chapter, this exceptional capacity is not uniformly distributed in all cyanobacteria, but largely concentrated in two of the eight taxonomic subsections, and thus species within these two are all the more exceptional. Recent unpublished genome-sequencing studies from the authors' laboratory indicate that some filamentous marine cyanobacteria have the genetic capacity to produce over 40 different classes of metabolites, and that the genes for this ability occupy nearly 20% of the genome. These studies reveal that as rich as cyanobacteria have shown themselves to be in *top-down* NP studies, there are new frontiers left to explore with these organisms, and *bottom-up* approaches will certainly expand and be highly productive in the future.

REFERENCES

1. Burja, A. M.; Banaigs, B.; Abou-Mansour, E.; Grant Burgess, J.; Wright, P. C. Marine cyanobacteria—A prolific source of natural products. *Tetrahedron* **2001**, *57* (46), 9347–9377.
2. Pettit, G. R.; Kamano, Y.; Herald, C. L.; Tuinman, A. A.; Boettner, F. E.; Kizu, H.; Schmidt, J. M.; Baczynskyj, L.; Tomer, K. B.; Bontems, R. J. The isolation and structure of a remarkable marine animal antineoplastic constituent: Dolastatin 10. *J. Am. Chem. Soc.* **1987**, *109* (22), 6883–6885.
3. Katz, J.; Janik, J. E.; Younes, A. Brentuximab vedotin (SGN-35). *Clin. Cancer Res.* **2011**, *17* (20), 6428–6436.
4. Tan, L. T. Filamentous tropical marine cyanobacteria: A rich source of natural products for anticancer drug discovery. *J. Appl. Phycol.* **2010**, *22* (5), 659–676.
5. Newman, D. J. Developing natural product drugs: Supply problems and how they have been overcome. *Pharmacol. Ther.* **2016**, *162*, 1–9.
6. Salvador-Reyes, L. A.; Luesch, H. Biological targets and mechanisms of action of natural products from marine cyanobacteria. *Nat. Prod. Rep.* **2015**, *32* (3), 478–503.
7. Dittmann, E.; Gugger, M.; Sivonen, K.; Fewer, D. P. Natural product biosynthetic diversity and comparative genomics of the cyanobacteria. *Trends Microbiol.* **2015**, *23* (10), 642–652.
8. Moss, N. A.; Bertin, M. J.; Kleigrewe, K.; Leao, T. F.; Gerwick, L.; Gerwick, W. H. Integrating mass spectrometry and genomics for cyanobacterial metabolite discovery. *J. Ind. Microbiol. Biotechnol.* **2015**, *43* (2), 1–12.
9. Pattanaik, B.; Lindberg, P. Terpenoids and their biosynthesis in cyanobacteria. *Life* **2015**, *5* (1), 269–293.

10. Rossi, J. V.; Roberts, M. A.; Yoo, H.-D.; Gerwick, W. H. Pilot scale culture of the marine cyanobacterium *Lyngbya majuscula* for its pharmaceutically-useful natural metabolite curacin A. *J. Appl. Phycol.* **1997**, *9* (3), 195–204.
11. Edwards, D. J.; Marquez, B. L.; Nogle, L. M.; McPhail, K.; Goeger, D. E.; Roberts, M. A.; Gerwick, W. H. Structure and biosynthesis of the jamaicamides, new mixed polyketide-peptide neurotoxins from the marine cyanobacterium *Lyngbya majuscula. Chem. Biol.* **2004**, *11* (6), 817–833.
12. Marquez, B. L.; Watts, K. S.; Yokochi, A.; Roberts, M. A.; Verdier-Pinard, P.; Jimenez, J. I.; Hamel, E.; Scheuer, P. J.; Gerwick, W. H. Structure and absolute stereochemistry of hectochlorin, a potent stimulator of actin assembly. *J. Nat. Prod.* **2002**, *65* (6), 866–871.
13. Ramaswamy, A. V.; Sorrels, C. M.; Gerwick, W. H. Cloning and biochemical characterization of the hectochlorin biosynthetic gene cluster from the marine cyanobacterium *Lyngbya majuscula. J. Nat. Prod.* **2007**, *70* (12), 1977–1986.
14. Chang, Z.; Flatt, P.; Gerwick, W. H.; Nguyen, V.-A.; Willis, C. L.; Sherman, D. H. The barbamide biosynthetic gene cluster: A novel marine cyanobacterial system of mixed polyketide synthase (PKS)-non-ribosomal peptide synthetase (NRPS) origin involving an unusual trichloroleucyl starter unit. *Gene* **2002**, *296* (1–2), 235–247.
15. Sitachitta, N.; Rossi, J.; Roberts, M. A.; Gerwick, W. H.; Fletcher, M. D.; Willis, C. L. Biosynthesis of the marine cyanobacterial metabolite barbamide. 1. Origin of the trichloromethyl group. *J. Am. Chem. Soc.* **1998**, *120* (28), 7131–7132.
16. Galonić, D. P.; Vaillancourt, F. H.; Walsh, C. T. Halogenation of unactivated carbon centers in natural product biosynthesis: Trichlorination of leucine during barbamide biosynthesis. *J. Am. Chem. Soc.* **2006**, *128* (12), 3900–3901.
17. Flatt, P. M.; O'Connell, S. J.; McPhail, K. L.; Zeller, G.; Willis, C. L.; Sherman, D. H.; Gerwick, W. H. Characterization of the initial enzymatic steps of barbamide biosynthesis. *J. Nat. Prod.* **2006**, *69* (6), 938–944.
18. Blokhin, A. V.; Yoo, H. D.; Geralds, R. S.; Nagle, D. G.; Gerwick, W. H.; Hamel, E. Characterization of the interaction of the marine cyanobacterial natural product curacin A with the colchicine site of tubulin and initial structure-activity studies with analogues. *Mol. Pharmacol.* **1995**, *48* (3), 523–531.
19. Chang, Z.; Sitachitta, N.; Rossi, J. V.; Roberts, M. A.; Flatt, P. M.; Jia, J.; Sherman, D. H.; Gerwick, W. H. Biosynthetic pathway and gene cluster analysis of curacin A, an antitubulin natural product from the tropical marine cyanobacterium *Lyngbya majuscula. J. Nat. Prod.* **2004**, *67* (8), 1356–1367.
20. Gu, L.; Geders, T. W.; Wang, B.; Gerwick, W. H.; Håkansson, K.; Smith, J. L.; Sherman, D. H. GNAT-like strategy for polyketide chain initiation. *Science* **2007**, *318* (5852), 970–974.
21. Geders, T. W.; Gu, L.; Mowers, J. C.; Liu, H.; Gerwick, W. H.; Håkansson, K.; Sherman, D. H.; Smith, J. L. Crystal structure of the ECH2 catalytic domain of CurF from *Lyngbya majuscula*. Insights into a decarboxylase involved in polyketide chain beta-branching. *J. Biol. Chem.* **2007**, *282* (49), 35954–35963.
22. Gu, L.; Wang, B.; Kulkarni, A.; Geders, T. W.; Grindberg, R. V.; Gerwick, L.; Håkansson, K. et al. Metamorphic enzyme assembly in polyketide diversification. *Nature* **2009**, *459* (7247), 731–735.
23. Nägeli, C. Gattungen einzelliger Algen, physiologisch und systematisch bearbeitet. *Neue Denkschr. Allg. Schweiz. Ges. Naturw.* **1849**, *10*, 1–139.
24. Nägeli, C.; Schwendener, S. *Das Mikroskop: Theorie Und Anwendung Desselben*, 2nd edn.; W. Engelmann: Leipzig, Germany, **1877**.
25. Garcia-Pichel, F.; Castenholz, R. W. Characterization and biological implications of scytonemin, a cyanobacterial sheath pigment. *J. Phycol.* **1991**, *27* (3), 395–409.
26. Proteau, P. J.; Gerwick, W. H.; Garcia-Pichel, F.; Castenholz, R. The structure of scytonemin, an ultraviolet sunscreen pigment from the sheaths of cyanobacteria. *Experientia* **1993**, *49* (9), 825–829.

27. Soule, T.; Stout, V.; Swingley, W. D.; Meeks, J. C.; Garcia-Pichel, F. Molecular genetics and genomic analysis of scytonemin biosynthesis in *Nostoc punctiforme* ATCC 29133. *J. Bacteriol.* **2007**, *189* (12), 4465–4472.
28. Sorrels, C. M.; Proteau, P. J.; Gerwick, W. H. Organization, evolution, and expression analysis of the biosynthetic gene cluster for scytonemin, a cyanobacterial UV-absorbing pigment. *Appl. Environ. Microbiol.* **2009**, *75* (14), 4861–4869.
29. Balskus, E. P.; Walsh, C. T. Investigating the initial steps in the biosynthesis of cyanobacterial sunscreen scytonemin. *J. Am. Chem. Soc.* **2008**, *130* (46), 15260–15261.
30. Balskus, E. P.; Walsh, C. T. An enzymatic cyclopentyl[b]indole formation involved in scytonemin biosynthesis. *J. Am. Chem. Soc.* **2009**, *131* (41), 14648–14649.
31. Edwards, D. J.; Gerwick, W. H. Lyngbyatoxin biosynthesis: Sequence of biosynthetic gene cluster and identification of a novel aromatic prenyltransferase. *J. Am. Chem. Soc.* **2004**, *126* (37), 11432–11433.
32. Read, J. A.; Walsh, C. T. The lyngbyatoxin biosynthetic assembly line: Chain release by four-electron reduction of a dipeptidyl thioester to the corresponding alcohol. *J. Am. Chem. Soc.* **2007**, *129* (51), 15762–15763.
33. Jones, A. C.; Ottilie, S.; Eustáquio, A. S.; Edwards, D. J.; Gerwick, L.; Moore, B. S.; Gerwick, W. H. Evaluation of *Streptomyces coelicolor* A3(2) as a heterologous expression host for the cyanobacterial protein kinase C activator lyngbyatoxin A. *FEBS J.* **2012**, *279* (7), 1243–1251.
34. Huynh, M. U.; Elston, M. C.; Hernandez, N. M.; Ball, D. B.; Kajiyama, S.; Irie, K.; Gerwick, W. H.; Edwards, D. J. Enzymatic production of (–)-indolactam V by LtxB, a cytochrome P450 monooxygenase. *J. Nat. Prod.* **2010**, *73* (1), 71–74.
35. Agger, S. A.; Lopez-Gallego, F.; Hoye, T. R.; Schmidt-Dannert, C. Identification of sesquiterpene synthases from *Nostoc punctiforme* PCC 73102 and *Nostoc* sp. strain PCC 7120. *J. Bacteriol.* **2008**, *190* (18), 6084–6096.
36. Leão, P. N.; Nakamura, H.; Costa, M.; Pereira, A. R.; Martins, R.; Vasconcelos, V.; Gerwick, W. H.; Balskus, E. P. Biosynthesis-assisted structural elucidation of the bartolosides, chlorinated aromatic glycolipids from cyanobacteria. *Angew. Chem. Int. Ed. Engl.* **2015**, *54* (38), 11063–11067.
37. Kim, E. J.; Lee, J. H.; Choi, H.; Pereira, A. R.; Ban, Y. H.; Yoo, Y. J.; Kim, E. et al. Heterologous production of 4-*O*-demethylbarbamide, a marine cyanobacterial natural product. *Org. Lett.* **2012**, *14* (23), 5824–5827.
38. Ongley, S. E.; Bian, X.; Zhang, Y.; Chau, R.; Gerwick, W. H.; Müller, R.; Neilan, B. A. High-titer heterologous production in *E. coli* of lyngbyatoxin, a protein kinase C activator from an uncultured marine cyanobacterium. *ACS Chem. Biol.* **2013**, *8* (9), 1888–1893.
39. Videau, P.; Wells, K. N.; Singh, A. J.; Gerwick, W. H.; Philmus, B. Assessment of *Anabaena* sp. Strain PCC 7120 as a heterologous expression host for cyanobacterial natural products: Production of lyngbyatoxin A. *ACS Synth. Biol.* **2016**, *5* (9), 978–988.
40. Koksharova, O.; Wolk, C. Genetic tools for cyanobacteria. *Appl. Microbiol. Biotechnol.* **2002**, *58* (2), 123–137.
41. Long, P. F.; Dunlap, W. C.; Battershill, C. N.; Jaspars, M. Shotgun cloning and heterologous expression of the patellamide gene cluster as a strategy to achieving sustained metabolite production. *ChemBioChem* **2005**, *6* (10), 1760–1765.
42. Schmidt, E. W.; Nelson, J. T.; Rasko, D. A.; Sudek, S.; Eisen, J. A.; Haygood, M. G.; Ravel, J. Patellamide A and C biosynthesis by a microcin-like pathway in *Prochloron didemni*, the cyanobacterial symbiont of *Lissoclinum patella*. *Proc. Natl. Acad. Sci. USA* **2005**, *102* (20), 7315–7320.
43. Donia, M. S.; Hathaway, B. J.; Sudek, S.; Haygood, M. G.; Rosovitz, M. J.; Ravel, J.; Schmidt, E. W. Natural combinatorial peptide libraries in cyanobacterial symbionts of marine ascidians. *Nat. Chem. Biol.* **2006**, *2* (12), 729–735.

44. Donia, M. S.; Ravel, J.; Schmidt, E. W. A global assembly line for cyanobactins. *Nat. Chem. Biol.* **2008**, *4* (6), 341–343.
45. Koehnke, J.; Bent, A.; Houssen, W. E.; Zollman, D.; Morawitz, F.; Shirran, S.; Vendome, J. et al. The mechanism of patellamide macrocyclization revealed by the characterization of the PatG macrocyclase domain. *Nat. Struct. Mol. Biol.* **2012**, *19* (8), 767–772.
46. Martins, J.; Vasconcelos, V. Cyanobactins from cyanobacteria: Current genetic and chemical state of knowledge. *Mar. Drugs* **2015**, *13* (11), 6910–6946.
47. Luo, Y.; Cobb, R. E.; Zhao, H. Recent advances in natural product discovery. *Curr. Opin. Biotechnol.* **2014**, *30*, 230–237.
48. Ziemert, N.; Alanjary, M.; Weber, T. The evolution of genome mining in microbes—A review. *Nat. Prod. Rep.* **2016**, *33*, 988–1005.
49. Rutledge, P. J.; Challis, G. L. Discovery of microbial natural products by activation of silent biosynthetic gene clusters. *Nat. Rev. Microbiol.* **2015**, *13* (8), 509–523.
50. Medema, M. H.; Fischbach, M. A. Computational approaches to natural product discovery. *Nat. Chem. Biol.* **2015**, *11* (9), 639–648.
51. van der Lee, T. A. J.; Medema, M. H. Computational strategies for genome-based natural product discovery and engineering in fungi. *Fungal Genet. Biol.* **2016**, *89* (2015), 29–36.
52. Duncan, K. R.; Crüsemann, M.; Lechner, A.; Sarkar, A.; Li, J.; Ziemert, N.; Wang, M. et al. Molecular networking and pattern-based genome mining improves discovery of biosynthetic gene clusters and their products from *Salinispora* species. *Chem. Biol.* **2015**, *22* (4), 460–471.
53. Calteau, A.; Fewer, D. P.; Latifi, A.; Coursin, T.; Laurent, T.; Jokela, J.; Kerfeld, C. A.; Sivonen, K.; Piel, J.; Gugger, M. Phylum-wide comparative genomics unravel the diversity of secondary metabolism in cyanobacteria. *BMC Genomics* **2014**, *15* (1), 977.
54. Micallef, M. L.; D'Agostino, P. M.; Al-Sinawi, B.; Neilan, B. A.; Moffitt, M. C. Exploring cyanobacterial genomes for natural product biosynthesis pathways. *Mar. Genomics* **2014**, *21*, 1–12.
55. Kleigrewe, K.; Gerwick, L.; Sherman, D. H.; Gerwick, W. H. Unique marine derived cyanobacterial biosynthetic genes for chemical diversity. *Nat. Prod. Rep.* **2016**, *33*, 348–364.
56. Shih, P. M.; Wu, D.; Latifi, A.; Axen, S. D.; Fewer, D. P.; Talla, E.; Calteau, A. et al. Improving the coverage of the cyanobacterial phylum using diversity-driven genome sequencing. *Proc. Natl. Acad. Sci. USA* **2013**, *110* (3), 1053–1058.
57. Micallef, M. L.; D'Agostino, P. M.; Sharma, D.; Viswanathan, R.; Moffitt, M. C. Genome mining for natural product biosynthetic gene clusters in the subsection V cyanobacteria. *BMC Genomics* **2015**, *16*, 669.
58. Stanier, R. Y.; Sistrom, W. R.; Hansen, T. A.; Whitton, B. A.; Castenholz, R. W.; Pfennig, N.; Gorlenko, V. N. et al. Proposal to place the nomenclature of the cyanobacteria (blue-green algae) under the rules of the international code of nomenclature of bacteria. *Int. J. Syst. Bacteriol.* **1978**, *28* (2), 335–336.
59. Komarek, J.; Kastovsky, J.; Mares, J.; Johansen, J. R. Taxonomic classification of cyanoprokaryotes (cyanobacterial genera) 2014, using a polyphasic approach. *Preslia* **2014**, *86* (4), 295–335.
60. Flores, E.; López-lozano, A.; Herrero, A. Nitrogen fixation in the oxygenic (cyanobacteria): The fight against oxygen. *Biol. Nitrogen Fixat.* **2015**, *2*, 879–889.
61. Li, B.; Sher, D.; Kelly, L.; Shi, Y.; Huang, K.; Knerr, P. J.; Joewono, I.; Rusch, D.; Chisholm, S. W.; van der Donk, W. A. Catalytic promiscuity in the biosynthesis of cyclic peptide secondary metabolites in planktonic marine cyanobacteria. *Proc. Natl. Acad. Sci. USA* **2010**, *107* (23), 10430–10435.
62. Engene, N.; Rottacker, E. C.; Kaštovský, J.; Byrum, T.; Choi, H.; Ellisman, M. H.; Komárek, J.; Gerwick, W. H. *Moorea producens* gen. nov., sp. nov. and *Moorea Bouillonii* comb. nov., tropical marine cyanobacteria rich in bioactive secondary metabolites. *Int. J. Syst. Evol. Microbiol.* **2012**, *62* (Pt 5), 1171–1178.

63. Gerwick, W. H.; Proteau, P. J.; Nagle, D. G.; Hamel, E.; Blokhin, A.; Slate, D. L. Structure of curacin A, a novel antimitotic, antiproliferative and brine shrimp toxic natural product from the marine cyanobacterium *Lyngbya majuscula*. *J. Org. Chem.* **1994**, *59* (6), 1243–1245.
64. Taori, K.; Paul, V. J.; Luesch, H. Structure and activity of largazole, a potent antiproliferative agent from the floridian marine cyanobacterium *Symploca* sp. *J. Am. Chem. Soc.* **2008**, *130* (6), 1806–1807.
65. Fujii, K.; Sivonen, K.; Kashiwagi, T.; Hirayama, K.; Harada, K. I. Nostophycin, a novel cyclic peptide from the toxic cyanobacterium *Nostoc* sp. 152. *J. Org. Chem.* **1999**, *64* (16), 5777–5782.
66. Coates, R. C.; Podell, S.; Korobeynikov, A.; Lapidus, A.; Pevzner, P.; Sherman, D. H.; Allen, E. E.; Gerwick, L.; Gerwick, W. H. Characterization of cyanobacterial hydrocarbon composition and distribution of biosynthetic pathways. *PLoS One* **2014**, *9* (1), e85140.
67. Arnison, P. G.; Bibb, M. J.; Bierbaum, G.; Bowers, A. A.; Bugni, T. S.; Bulaj, G.; Camarero, J. A. et al. Ribosomally synthesized and post-translationally modified peptide natural products: Overview and recommendations for a universal nomenclature. *Nat. Prod. Rep.* **2013**, *30* (1), 108–160.
68. Leikoski, N.; Liu, L.; Jokela, J.; Wahlsten, M.; Gugger, M.; Calteau, A.; Permi, P.; Kerfeld, C. A.; Sivonen, K.; Fewer, D. P. Genome mining expands the chemical diversity of the cyanobactin family to include highly modified linear peptides. *Chem. Biol.* **2013**, *20* (8), 1033–1043.
69. Hillwig, M. L.; Fuhrman, H. A.; Ittiamornkul, K.; Sevco, T. J.; Kwak, D. H.; Liu, X. Identification and characterization of a welwitindolinone alkaloid biosynthetic gene cluster in the stigonematalean cyanobacterium *Hapalosiphon welwitschii*. *ChemBioChem* **2014**, *15* (5), 665–669.
70. Hillwig, M. L.; Liu, X. A new family of iron-dependent halogenases acts on freestanding substrates. *Nat. Chem. Biol.* **2014**, *10* (11), 921–923.
71. Balskus, E. P.; Walsh, C. T. The genetic and molecular basis for sunscreen biosynthesis in cyanobacteria. *Science* **2010**, *329* (5999), 1653–1656.
72. Kehr, J.-C.; Gatte Picchi, D.; Dittmann, E. Natural product biosyntheses in cyanobacteria: A treasure trove of unique enzymes. *Beilstein J. Org. Chem.* **2011**, *7*, 1622–1635.
73. Kleigrewe, K.; Almaliti, J.; Tian, I. Y.; Kinnel, R. B.; Korobeynikov, A.; Monroe, E. A.; Duggan, B. M. et al. Combining mass spectrometric metabolic profiling with genomic analysis: A powerful approach for discovering natural products from cyanobacteria. *J. Nat. Prod.* **2015**, *78*, 1671–1682.
74. Sudek, S.; Haygood, M. G.; Youssef, D. T. A.; Schmidt, E. W. Structure of trichamide, a cyclic peptide from the bloom-forming cyanobacterium *Trichodesmium erythraeum*, predicted from the genome sequence. *Appl. Environ. Microbiol.* **2006**, *72* (6), 4382–4387.
75. Leikoski, N.; Fewer, D. P.; Jokela, J.; Alakoski, P.; Wahlsten, M.; Sivonen, K. Analysis of an inactive cyanobactin biosynthetic gene cluster leads to discovery of new natural products from strains of the genus microcystis. *PLoS One* **2012**, *7* (8), e43002.
76. González López, C. V.; García, M. del Carmen, C.; Fernández, F. G. A.; Bustos, C. S.; Chisti, Y.; Sevilla, J. M. F. Protein measurements of microalgal and cyanobacterial biomass. *Bioresour. Technol.* **2010**, *101* (19), 7587–7591.
77. Cohen, Z. The chemicals of spirulina. In *Spirulina platensis (Arthrospira): Physiology, Cell-Biology and Biotechnology*; Vonshak, A., Ed.; Taylor & Francis Ltd: London, U.K., **1997**; pp. 175–204.
78. Grossman, A. R.; Schaefer, M. R.; Chiang, G. G.; Collier, J. L. The phycobilisome, a light-harvesting complex responsive to environmental conditions. *Microbiol. Rev.* **1993**, *57* (3), 725–749.

79. Glazer, A. N.; Clark, J. H. Phycobilisomes. *Biophys. J.* **1986**, *49* (1), 115–116.
80. Glazer, A. N. Phycobilisomes: Structure and dynamics. *Annu. Rev. Microbiol.* **1982**, *36* (1), 173–198.
81. Glazer, A. N. Light guides: Directional energy transfer in a photosynthetic antenna. *J. Biol. Chem.* **1989**, *264* (1), 1–4.
82. Kumar, V.; Sonani, R. R.; Sharma, M.; Gupta, G. D.; Madamwar, D. Crystal structure analysis of C-phycoerythrin from marine cyanobacterium *Phormidium* sp. A09DM. *Photosynth. Res.* **2016**, *129* (1), 17–28.
83. Pandey, V. D.; Pandey, A.; Sharma, V. Review article biotechnological applications of cyanobacterial phycobiliproteins. *Int. J. Curr. Microbiol. Appl. Sci.* **2013**, *2* (9), 89–97.
84. Sekar, S.; Chandramohan, M. Phycobiliproteins as a commodity: Trends in applied research, patents and commercialization. *J. Appl. Phycol.* **2008**, *20* (2), 113–136.
85. Oi, V. T.; Glazer, A. N.; Stryer, L. Fluorescent phycobiliprotein conjugates for analyses of cells and molecules. *J. Cell Biol.* **1982**, *93* (3), 981–986.
86. Sonani, R. R. Recent advances in production, purification and applications of phycobiliproteins. *World J. Biol. Chem.* **2016**, *7* (1), 100.
87. Glazer, A. N. Phycobiliproteins—A family of valuable, widely used fluorophores. *J. Appl. Phycol.* **1994**, *6* (2), 105–112.
88. Araoz, R.; Lebert, M.; Haeder, D. P. Electrophoretic applications of phycobiliproteins. *Electrophoresis* **1998**, *19* (2), 215–219.
89. Sasaki, Y. F.; Kawaguchi, S.; Kamaya, A.; Ohshita, M.; Kabasawa, K.; Iwama, K.; Taniguchi, K.; Tsuda, S. The comet assay with 8 mouse organs: Results with 39 currently used food additives. *Mutat. Res.* **2002**, *519* (1–2), 103–119.
90. Ben Mansour, H.; Corroler, D.; Barillier, D.; Ghedira, K.; Chekir, L.; Mosrati, R. Evaluation of genotoxicity and pro-oxidant effect of the azo dyes: Acids yellow 17, violet 7 and orange 52, and of their degradation products by *Pseudomonas putida* Mt-2. *Food Chem. Toxicol.* **2007**, *45* (9), 1670–1677.
91. Spolaore, P.; Joannis-Cassan, C.; Duran, E.; Isambert, A. Commercial applications of microalgae. *J. Biosci. Bioeng.* **2006**, *101* (2), 87–96.
92. Bermejo, P.; Piñero, E.; Villar, Á. M. Iron-chelating ability and antioxidant properties of phycocyanin isolated from a protean extract of *Spirulina platensis*. *Food Chem.* **2008**, *110* (2), 436–445.
93. Liu, Y.; Xu, L.; Cheng, N.; Lin, L.; Zhang, C. Inhibitory effect of phycocyanin from *Spirulina platensis* on the growth of human leukemia K562 cells. *J. Appl. Phycol.* **2000**, *12* (2), 125–130.
94. Roy, K. R.; Arunasree, K. M.; Reddy, N. P.; Dheeraj, B.; Reddy, G. V.; Reddanna, P. Alteration of mitochondrial membrane potential by *Spirulina platensis* C-phycocyanin induces apoptosis in the doxorubicinresistant human hepatocellular-carcinoma cell line HepG2. *Biotechnol. Appl. Biochem.* **2007**, *47* (Pt 3), 159–167.
95. Benedetti, S.; Benvenuti, F.; Scoglio, S.; Canestrari, F. Oxygen radical absorbance capacity of phycocyanin and phycocyanobilin from the food supplement aphanizomenon flos-aquae. *J. Med. Food* **2010**, *13* (1), 223–227.
96. Bhat, V. B.; Madyastha, K. M. Scavenging of peroxynitrite by phycocyanin and phycocyanobilin from *Spirulina platensis*: Protection against oxidative damage to DNA. *Biochem. Biophys. Res. Commun.* **2001**, *285* (2), 262–266.
97. Bhat, V. B.; Madyastha, K. M. C-Phycocyanin: A potent peroxyl radical scavenger *in vivo* and *in vitro*. *Biochem. Biophys. Res. Commun.* **2000**, *275* (1), 20–25.
98. Li, H.; Sherman, D. M.; Bao, S.; Sherman, L. A. Pattern of cyanophycin accumulation in nitrogen-fixing and non-nitrogen-fixing cyanobacteria. *Arch. Microbiol.* **2001**, *176* (1–2), 9–18.

99. Jones, A. C.; Monroe, E. A.; Podell, S.; Hess, W. R.; Klages, S.; Esquenazi, E.; Niessen, S. et al. Genomic insights into the physiology and ecology of the marine filamentous cyanobacterium *Lyngbya majuscula*. *Proc. Natl. Acad. Sci. USA* **2011**, *108* (21), 8815–8820.
100. Obst, M.; Steinbüchel, A. Microbial degradation of poly(amino acid)s. *Biomacromolecules* **2004**, *5* (4), 1166–1176.
101. Schwamborn, M. Chemical synthesis of polyaspartates: A biodegradable alternative to currently used polycarboxylate homo- and copolymers. *Polym. Degrad. Stab.* **1998**, *59* (1–3), 39–45.
102. Berg, H.; Ziegler, K.; Piotukh, K.; Baier, K.; Lockau, W.; Volkmer-Engert, R. Biosynthesis of the cyanobacterial reserve polymer multi-L-arginyl-poly-L-aspartic acid (cyanophycin). *Eur. J. Biochem.* **2000**, *267* (17), 5561–5570.
103. Krehenbrink, M.; Steinbüchel, A. Partial purification and characterization of a non-cyanobacterial cyanophycin synthetase from *Acinetobacter calcoaceticus* strain ADP1 with regard to substrate specificity, substrate affinity and binding to cyanophycin. *Microbiology* **2004**, *150* (8), 2599–2608.
104. Ziegler, K.; Diener, A.; Herpin, C.; Richter, R.; Deutzmann, R.; Lockau, W. Molecular characterization of cyanophycin synthetase, the enzyme catalyzing the biosynthesis of the cyanobacterial reserve material multi-L-arginyl-poly-L-aspartate (cyanophycin). *Eur. J. Biochem.* **1998**, *254* (1), 154–159.
105. Simon, R. D. The biosynthesis of multi-L-arginyl-poly(L-aspartic acid) in the filamentous cyanobacterium *Anabaena cylindrica*. *Biochim. Biophys. Acta Enzymol.* **1976**, *422* (2), 407–418.
106. Ziegler, K.; Stephan, D. P.; Pistorius, E. K.; Ruppel, H. G.; Lockau, W. A mutant of the cyanobacterium *Anabaena variabilis* ATCC 29413 lacking cyanophycin synthetase: Growth properties and ultrastructural aspects. *FEMS Microbiol. Lett.* **2001**, *196* (1), 13–18.
107. Simon, R. D. The effect of chloramphenicol on the production of cyanophycin granule polypeptide in the blue green alga *Anabaena cylindrica*. *Arch. Mikrobiol.* **1973**, *92* (2), 115–122.
108. Sallam, A.; Kast, A.; Przybilla, S.; Meiswinkel, T.; Steinbüchel, A. Biotechnological process for production of beta-dipeptides from cyanophycin on a technical scale and its optimization. *Appl. Environ. Microbiol.* **2009**, *75* (1), 29–38.
109. Aboulmagd, E.; Voss, I.; Oppermann-Sanio, F. B.; Steinbüchel, A. Heterologous expression of cyanophycin synthetase and cyanophycin synthesis in the industrial relevant bacteria *Corynebacterium glutamicum* and *Ralstonia eutropha* and in *Pseudomonas putida*. *Biomacromolecules* **2001**, *2* (4), 1338–1342.
110. Sallam, A.; Steinbüchel, A. Dipeptides in nutrition and therapy: Cyanophycin-derived dipeptides as natural alternatives and their biotechnological production. *Appl. Microbiol. Biotechnol.* **2010**, *87* (3), 815–828.
111. Boyd, M. R.; Gustafson, K. R.; McMahon, J. B.; Shoemaker, R. H.; O'Keefe, B. R.; Mori, T.; Gulakowski, R. J. et al. Discovery of cyanovirin-N, a novel human immunodeficiency virus-inactivating protein that binds viral surface envelope glycoprotein gp120: Potential applications to microbicide development. *Antimicrob. Agents Chemother.* **1997**, *41* (7), 1521–1530.
112. Bewley, C. A.; Gustafson, K. R.; Boyd, M. R.; Covell, D. G.; Bax, A.; Clore, G. M.; Gronenborn, A. M. Solution structure of cyanovirin-N, a potent HIV-inactivating protein. *Nat. Struct. Biol.* **1998**, *5* (7), 571–578.
113. Yang, F.; Bewley, C. A.; Louis, J. M.; Gustafson, K. R.; Boyd, M. R.; Gronenborn, A. M.; Clore, G. M.; Wlodawer, A. Crystal structure of cyanovirin-N, a potent HIV-inactivating protein, shows unexpected domain swapping. *J. Mol. Biol.* **1999**, *288*, 403–412.

114. Mori, T.; Gustafson, K. R.; Pannell, L. K.; Shoemaker, R. H.; Wu, L.; McMahon, J. B.; Boyd, M. R. Recombinant production of cyanovirin-N, a potent human immunodeficiency virus-inactivating protein derived from a cultured cyanobacterium. *Protein Expr. Purif.* **1998**, *12* (2), 151–158.
115. Gao, X.; Chen, W.; Guo, C.; Qian, C.; Liu, G.; Ge, F.; Huang, Y.; Kitazato, K.; Wang, Y.; Xiong, S. Soluble cytoplasmic expression, rapid purification, and characterization of cyanovirin-N as a His-SUMO fusion. *Appl. Microbiol. Biotechnol.* **2010**, *85* (4), 1051–1060.
116. Bokesch, H. R.; O'Keefe, B. R.; McKee, T. C.; Pannell, L. K.; Patterson, G. M. L.; Gardella, R. S.; Sowder, R. C. et al. A potent novel anti-HIV protein from the cultured cyanobacterium *Scytonema varium*. *Biochemistry* **2003**, *42* (9), 2578–2584.
117. Matei, E.; Basu, R.; Furey, W.; Shi, J.; Calnan, C.; Aiken, C.; Gronenborn, A. M. Structure and glycan binding of a new cyanovirin-N homolog. *J. Biol. Chem.*. **2016**, *291* (36), 18967–18976.
118. Esser, M. T.; Mori, T.; Mondor, I.; Sattentau, Q. J.; Dey, B.; Berger, E. A.; Boyd, M. R.; Lifson, J. D. Cyanovirin-N binds to gp120 to interfere with CD4-dependent human immunodeficiency virus type 1 virion binding, fusion, and infectivity but does not affect the CD4 binding site on gp120 or soluble CD4-induced conformational changes in gp120. *J. Virol.* **1999**, *73* (5), 4360–4371.
119. Dey, B.; Lerner, D. L.; Lusso, P.; Boyd, M. R.; Elder, J. H.; Berger, E. A. Multiple antiviral activities of cyanovirin-N: Blocking of human immunodeficiency virus type 1 gp120 interaction with CD4 and coreceptor and inhibition of diverse enveloped viruses. *J. Virol.* **2000**, *74* (10), 4562–4569.
120. Botos, I.; O'Keefe, B. R.; Shenoy, S. R.; Cartner, L. K.; Ratner, D. M.; Seeberger, P. H.; Boyd, M. R.; Wlodawer, A. Structures of the complexes of a potent anti-HIV protein cyanovirin-N and high mannose oligosaccharides. *J. Biol. Chem.* **2002**, *277* (37), 34336–34342.
121. Leonard, C. K.; Spellman, M. W.; Riddle, L.; Harris, R. J.; Thomas, J. N.; Gregory, T. J. Assignment of intrachain disulfide bonds and characterization of potential glycosylation sites of the type 1 recombinant human immunodeficiency virus envelope glycoprotein (gp120) expressed in chinese hamster ovary cells. *J. Biol. Chem.* **1990**, *265* (18), 10373–10382.
122. van der Meer, F. J. U. M.; de Haan, C. A. M.; Schuurman, N. M. P.; Haijema, B. J.; Peumans, W. J.; Van Damme, E. J. M.; Delputte, P. L.; Balzarini, J.; Egberink, H. F. Antiviral activity of carbohydrate-binding agents against nidovirales in cell culture. *Antivir. Res.* **2007**, *76* (1), 21–29.
123. O'Keefe, B. R.; Smee, D. F.; Turpin, J. A.; Saucedo, C. J.; Gustafson, K. R.; Mori, T.; Blakeslee, D.; Buckheit, R.; Boyd, M. R. Potent anti-influenza activity of cyanovirin-N and interactions with viral hemagglutinin. *Antimicrob. Agents Chemother.* **2003**, *47* (8), 2518–2525.
124. Smee, D. F.; Bailey, K. W.; Wong, M.-H.; O'Keefe, B. R.; Gustafson, K. R.; Mishin, V. P.; Gubareva, L. V. Treatment of influenza A (H1N1) virus infections in mice and ferrets with cyanovirin-N. *Antivir. Res.* **2008**, *80* (3), 266–271.
125. Barrientos, L. G.; O'Keefe, B. R.; Bray, M.; Sanchez, A.; Gronenborn, A. M.; Boyd, M. R. Cyanovirin-N binds to the viral surface glycoprotein, GP1,2 and inhibits infectivity of Ebola virus. *Antivir. Res.* **2003**, *58* (1), 47–56.
126. Liu, X.; Lagenaur, L. A.; Simpson, D. A.; Essenmacher, K. P.; Frazier-Parker, C. L.; Liu, Y.; Tsai, D. et al. Engineered vaginal *Lactobacillus* strain for mucosal delivery of the human immunodeficiency virus inhibitor cyanovirin-N. *Antimicrob. Agents Chemother.* **2006**, *50* (10), 3250–3259.
127. Tsai, C.-C.; Emau, P.; Jiang, Y.; Agy, M. B.; Shattock, R. J.; Schmidt, A.; Morton, W. R.; Gustafson, K. R.; Boyd, M. R. Cyanovirin-N inhibits AIDS virus infections in vaginal transmission models. *AIDS Res. Hum. Retrovir.* **2004**, *20* (1), 11–18.

128. Brichacek, B.; Lagenaur, L. A.; Lee, P. P.; Venzon, D.; Hamer, D. H. *In vivo* evaluation of safety and toxicity of a *Lactobacillus jensenii* producing modified cyanovirin-N in a rhesus macaque vaginal challenge model. *PLoS One* **2013**, *8* (11).
129. Tsai, C.-C.; Emau, P.; Jiang, Y.; Tian, B.; Morton, W. R.; Gustafson, K. R.; Boyd, M. R. Cyanovirin-N gel as a topical microbicide prevents rectal transmission of shiv89.6P in macaques. *AIDS Res. Hum. Retrovir.* **2003**, *19* (7), 535–541.
130. Mori, T.; Shoemaker, R. H.; McMahon, J. B.; Gulakowski, R. J.; Gustafson, K. R.; Boyd, M. R. Construction and enhanced cytotoxicity of a [cyanovirin-N]-[pseudomonas exotoxin] conjugate against human immunodeficiency virus-infected cells. *Biochem. Biophys. Res. Commun.* **1997**, *239* (3), 884–888.
131. Wohlgemuth, R. Biocatalysis-key to sustainable industrial chemistry. *Curr. Opin. Biotechnol.* **2010**, *21* (6), 713–724.
132. Sanchez, S.; Demain, A. L. Enzymes and bioconversions of industrial, pharmaceutical, and biotechnological significance. *Org. Process. Res. Dev.* **2011**, *15* (1), 224–230.
133. Kourist, R.; Guterl, J. K.; Miyamoto, K.; Sieber, V. Enzymatic decarboxylation—An emerging reaction for chemicals production from renewable resources. *ChemCatChem* **2014**, *6* (3), 689–701.
134. Lambert, J. J.; Cooper, M. A.; Simmons, R. D. J.; Weir, C. J.; Belelli, D. Neurosteroids: Endogenous allosteric modulators of GABAA receptors. *Psychoneuroendocrinology* **2009**, *34* (Suppl. 1), 48–58.
135. Camacho Arroyo, I.; Morales-Montor, J. Non-reproductive effects of sex steroids: Their immunoregulatory role. *Curr. Top. Med. Chem.* **2011**, *11* (13), 1661–1662.
136. Naidoo, B. K.; Witty, T. R.; Remers, W. A.; Besch, H. R. Cardiotonic steroids I: Importance of 14β-hydroxy group in digitoxigenin. *J. Pharm. Sci.* **1974**, *63* (9), 1391–1394.
137. Abul-Hajj, Y. J.; Qian, X. D. Transformation of steroids by algae. *J. Nat. Prod.* **1986**, *49* (2), 244–248.
138. Faramarzi, M. A.; Tabatabaei Yazdi, M.; Ghostinroudi, H.; Amini, M.; Ghasemi, Y.; Jahandar, H.; Arabi, H. *Nostoc muscorum*: A regioselective biocatalyst for 17-carbonyl reduction of androst-4-en-3,17-dione and androst-1,4-dien-3,17-dione. *Ann. Microbiol.* **2006**, *56* (3), 253–256.
139. Yazdi, M. T.; Arabi, H.; Faramarzi, M. A.; Ghasemi, Y.; Amini, M.; Shokravi, S.; Mohseni, F. A. Biotransformation of hydrocortisone by a natural isolate of *Nostoc muscorum*. *Phytochemistry* **2004**, *65* (15), 2205–2209.
140. Ghasemi, Y.; Yazdi, M. T.; Dehshahri, A.; Niknahad, H.; Shokravi, S.; Amini, M.; Ghasemian, A.; Faramarzi, M. A. Algal transformation of hydrocortisone by the cyanobacterium *Nostoc ellipsosporum*. *Chem. Nat. Compd.* **2006**, *42* (6), 702–705.
141. Yazdi, M. T.; Ghasemi, Y.; Ghasemian, A.; Shokravi, S.; Niknahad, H.; Amini, M.; Dehshahri, A.; Faramarzi, M. A. Bioconversion of hydrocortisone by cyanobacterium *Fischerella ambigua* PTCC 1635. *World J. Microbiol. Biotechnol.* **2005**, *21* (6–7), 811–814.
142. Ghasemi, Y.; Faramarzi, M. A.; Arjmand-Inalou, M.; Mohagheghzadeh, A.; Shokravi, S.; Morowvat, M. H. Side-chain cleavage and C-20 ketone reduction of hydrocortisone by a natural isolate of *Chorococcus dispersus*. *Ann. Microbiol.* **2007**, *57* (4), 577–581.
143. Rasoul-Amini, S.; Ghasemi, Y.; Morowvat, M. H.; Ghoshoon, M. B.; Raee, M. J.; Mosavi-Azam, S. B.; Montazeri-Najafabady, N. et al. Characterization of hydrocortisone bioconversion and 16S RNA gene in *Synechococcus nidulans* cultures. *Appl. Biochem. Microbiol.* **2010**, *46* (2), 191–197.
144. Goñi, G.; Zöllner, A.; Lisurek, M.; Velázquez-Campoy, A.; Pinto, S.; Gómez-Moreno, C.; Hannemann, F.; Bernhardt, R.; Medina, M. Cyanobacterial electron carrier proteins as electron donors to CYP106A2 from *Bacillus megaterium* ATCC 13368. *Biochim. Biophys. Acta Proteins Proteomics* **2009**, *1794* (11), 1635–1642.

145. Bernhardt, R.; Urlacher, V. B. Cytochromes P450 as promising catalysts for biotechnological application: Chances and limitations. *Appl. Microbiol. Biotechnol.* **2014**, *98* (14), 6185–6203.
146. Padmapriya, V.; Anand, N. Evaluation of some industrially important enzymes in filamentous cyanobacteria. *ARPN J. Agric. Biol. Sci.* **2010**, *5* (5), 86–97.
147. Reyes-Sosa, F. M.; Molina-Heredia, F. P.; De la Rosa, M. A. A novel alpha-amylase from the cyanobacterium *Nostoc* sp. PCC 7119. *Appl. Microbiol. Biotechnol.* **2010**, *86* (1), 131–141.
148. Piechula, S.; Waleron, K.; Åwiatek, W.; Biedrzycka, I.; Podhajska, A. J. Mesophilic cyanobacteria producing thermophilic restriction endonucleases. *FEMS Microbiol. Lett.* **2001**, *198* (2), 135–140.
149. Savakis, P.; Hellingwerf, K. J. Engineering cyanobacteria for direct biofuel production from CO_2. *Curr. Opin. Biotechnol.* **2015**, *33*, 8–14.
150. Spreitzer, R. J.; Salvucci, M. E. Rubisco: Structure, regulatory interactions, and possibilities for a better enzyme. *Annu. Rev. Plant Biol.* **2002**, *53* (1), 449–475.
151. Atsumi, S.; Higashide, W.; Liao, J. C. Direct photosynthetic recycling of carbon dioxide to isobutyraldehyde. *Nat. Biotechnol.* **2009**, *27* (12), 1177–1180.
152. Kirst, H.; Formighieri, C.; Melis, A. Maximizing photosynthetic efficiency and culture productivity in cyanobacteria upon minimizing the phycobilisome light-harvesting antenna size. *Biochim. Biophys. Acta* **2014**, *1837* (10), 1653–1664.
153. Zhou, J.; Zhang, F.; Meng, H.; Zhang, Y.; Li, Y. Introducing extra NADPH consumption ability significantly increases the photosynthetic efficiency and biomass production of cyanobacteria. *Metab. Eng.* **2016**, *38*, 217–227.

3 The Role of Combinatorial Biosynthesis in Natural Products Discovery

Jeffrey D. Rudolf, Ivana Crnovčić, and Ben Shen

CONTENTS

3.1 INTRODUCTION

Natural products possess unrivaled chemical and structural diversity and, correspondingly, a wide range of biological activities (Figure 3.1).[1,2] They serve as excellent small-molecule probes for exploring fundamental biological processes; inspire novel chemistry, enzymology, and biology; and continue to be the best sources of drugs and drug leads.[2–5] Natural products are crafted from simple building blocks into complex and highly functionalized scaffolds. The structural diversity found in natural products is the result of Nature's intrinsic use of combinatorial biosynthesis.

Combinatorial biosynthesis is traditionally defined as the generation of natural product analogues through the use of genetic engineering of biosynthetic pathways. It is a mix-and-match system of metabolic pathways, genes, enzymes, and small molecules with the potential for unlimited structural outcomes. Since the report of *hybrid* antibiotics by Hopwood, Floss, and Omura in 1985[6]—a demonstration that gene transfer among producers of the benzoisochromanequinone antibiotics resulted

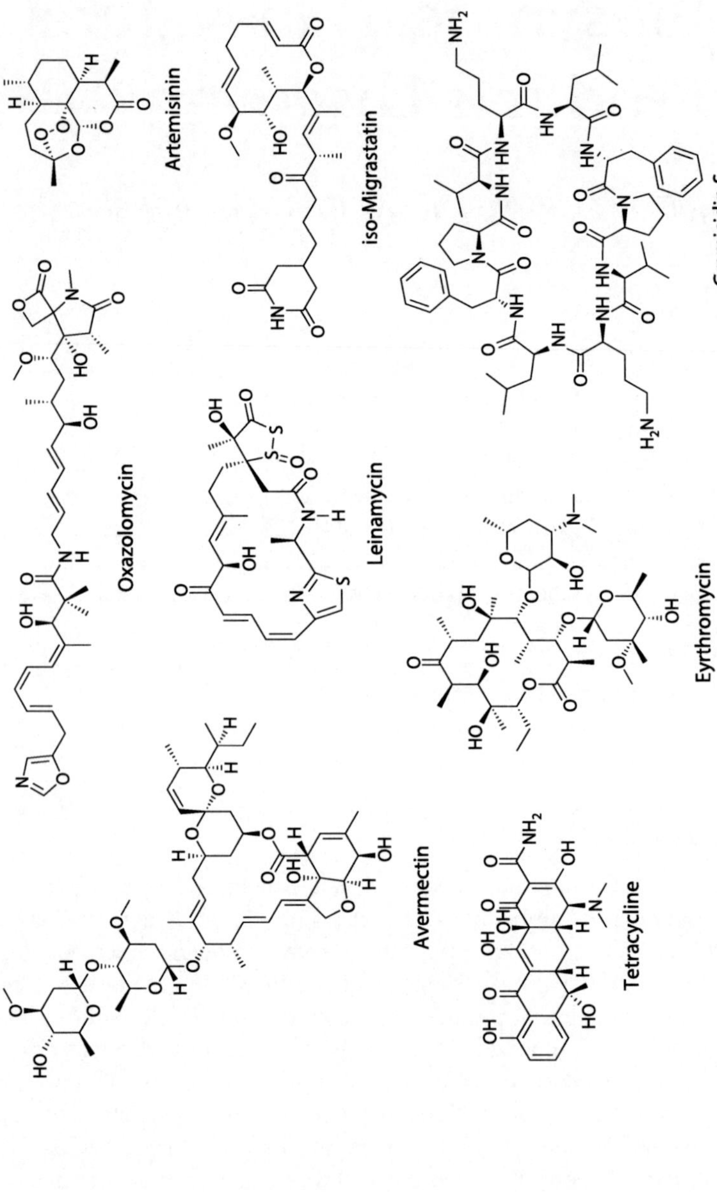

FIGURE 3.1 Structures of representative natural products, some of which are discussed in this chapter, highlighting their chemical and structural diversities. *(Continued)*

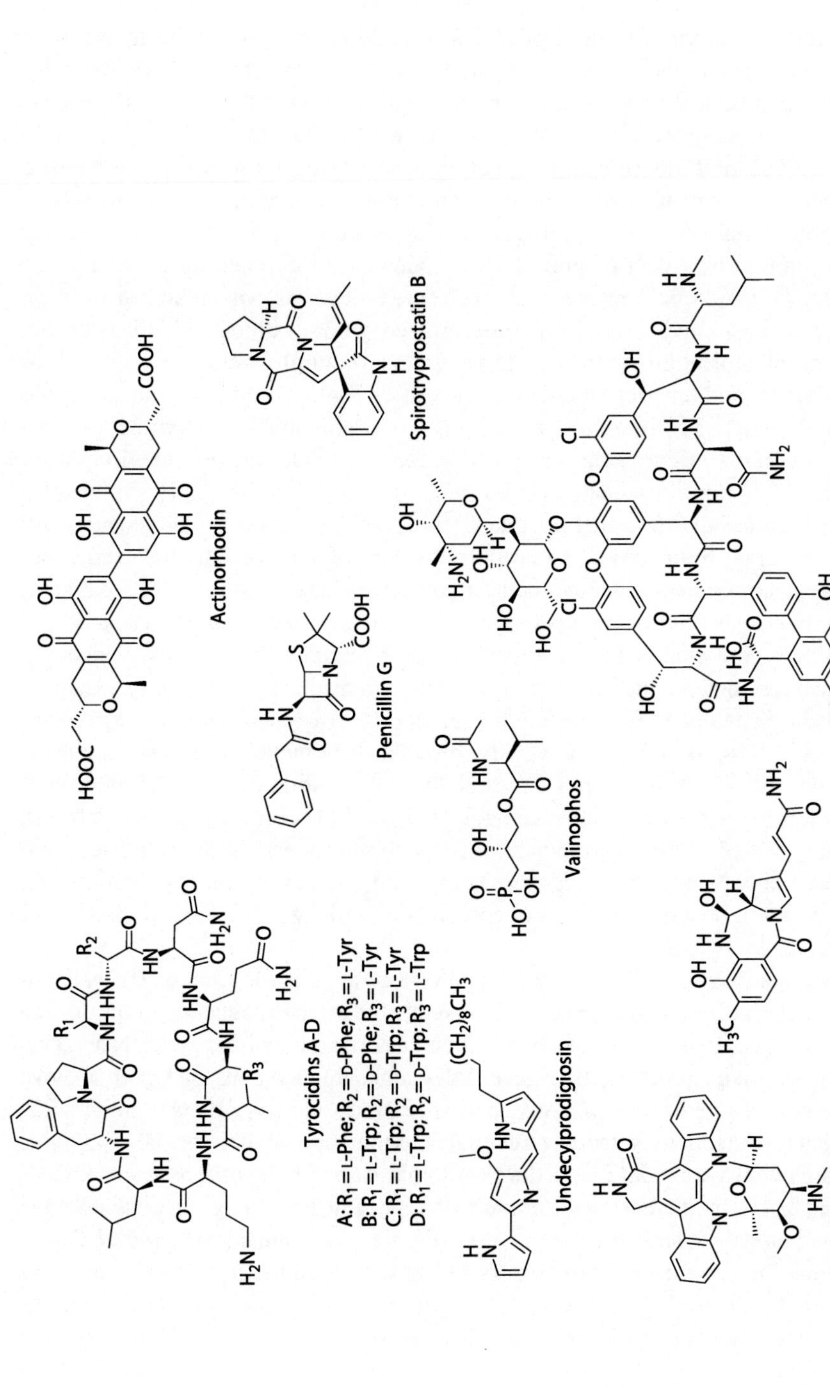

FIGURE 3.1 *(Continued)* Structures of representative natural products, some of which are discussed in this chapter, highlighting their chemical and structural diversities.

in production of *unnatural* natural products—combinatorial biosynthesis has been widely accepted as an innovative and feasible technique to create structural diversity among natural products.[7,8] The discovery that many polyketide and nonribosomal peptide natural products were biosynthesized by modular megasynthases (polyketide synthases [PKS] and nonribosomal peptide synthetases [NRPS])[9–15] galvanized both the academic and pharmaceutical communities to develop PKS- and NRPS-based combinatorial biosynthesis programs. Key examples, including the extensive work done on erythromycin (Figure 3.1),[16,17] showcase the potential of manipulating modular biosynthetic enzymes, but widespread success was achieved only on a case-by-case basis and often with compromised product titers.[18–20] However, the idea that combinatorial biosynthesis can generate structural diversity, optimize biological activities, and unravel biochemical novelties continues to permeate scientific thought and, thus, the literature. Accordingly, combinatorial biosynthetic strategies are now used in every natural product family including, but not limited to, polyketides,[18,21–25] nonribosomal peptides,[24,26] terpenes,[27,28] ribosomally synthesized and post-translationally modified peptides (RiPPs),[29,30] alkaloids,[31–33] glycosides,[33,34] and hybrid natural products (i.e., natural products of mixed biosynthetic origin).[35,36]

All existing combinatorial biosynthetic strategies are based on the collective knowledge of genetics, evolution, biosynthetic machineries, enzymology, and structural biology. Simply, proteins of known functions can be mapped back to their encoding genes, which can then be manipulated to generate organisms that produce the designer natural product analogues (Figure 3.2a). This *knowledge-based* approach to combinatorial biosynthesis is limited, obviously, by what information (i.e., knowledge) is gleaned from the above disciplines and how that information can be applied to construct the designer pathways. Understandably, in regard to combinatorial biosynthesis, the dominant source of this knowledge is what natural products have been discovered in Nature and how Nature produces these compounds. Nature has used evolution over billions of years to become an expert in combinatorial biosynthesis, and we have only begun to tap its knowledge.

Typically, combinatorial biosynthesis requires four aspects for success: (i) availability of producing strains and/or gene clusters responsible for the production of the desired natural product, (ii) genetic and biochemical characterizations of the gene clusters and encoded biosynthetic proteins, (iii) genetically amenable systems in which in vivo manipulation in native or heterologous hosts is feasible, and (iv) production titers that are sufficient for detection, isolation, structural determination, and biological evaluation.

There are two ways to continue to develop combinatorial biosynthesis: either knowledge-based, that is, by increasing the amount of applicable knowledge by detailed study of key genes, enzymes, and protein structures (Figure 3.2a); or *discovery-based*, that is, by leveraging the preexisting combinatorial biosynthesis found in Nature to discover novel natural products (Figure 3.2b). Using genomic data, natural product structures, scaffolds, or functional groups can be targeted by using a key gene or DNA sequence as a probe. There are two possible outcomes from a positive hit. The hit strain, containing the targeted DNA, has an identical or highly homologous genotype, and produces an identical natural product (i.e., rediscovery). Alternatively, the hit strain contains the DNA probe, but it is in a different genetic environment and produces a novel natural product or a novel variant of a known scaffold. Discovery-based combinatorial

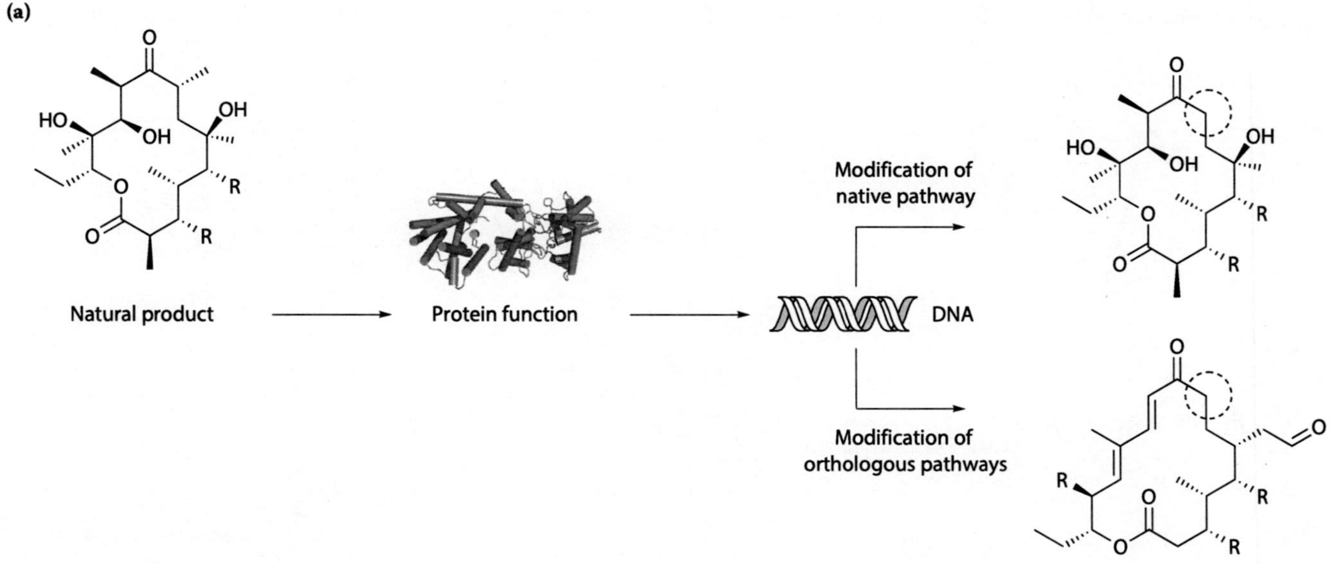

FIGURE 3.2 Schematic representation of **(a)** knowledge-based and **(b)** discovery-based combinatorial biosynthesis. **(a)** Proteins, with known functions that are involved in the production of natural products, are mapped back to their encoding genes, which are then manipulated to generate organisms that produce the designer natural product analogues. Knowledge-based combinatorial biosynthesis typically modifies one structural feature (dotted circles) while leaving the rest of the skeleton intact. *(Continued)*

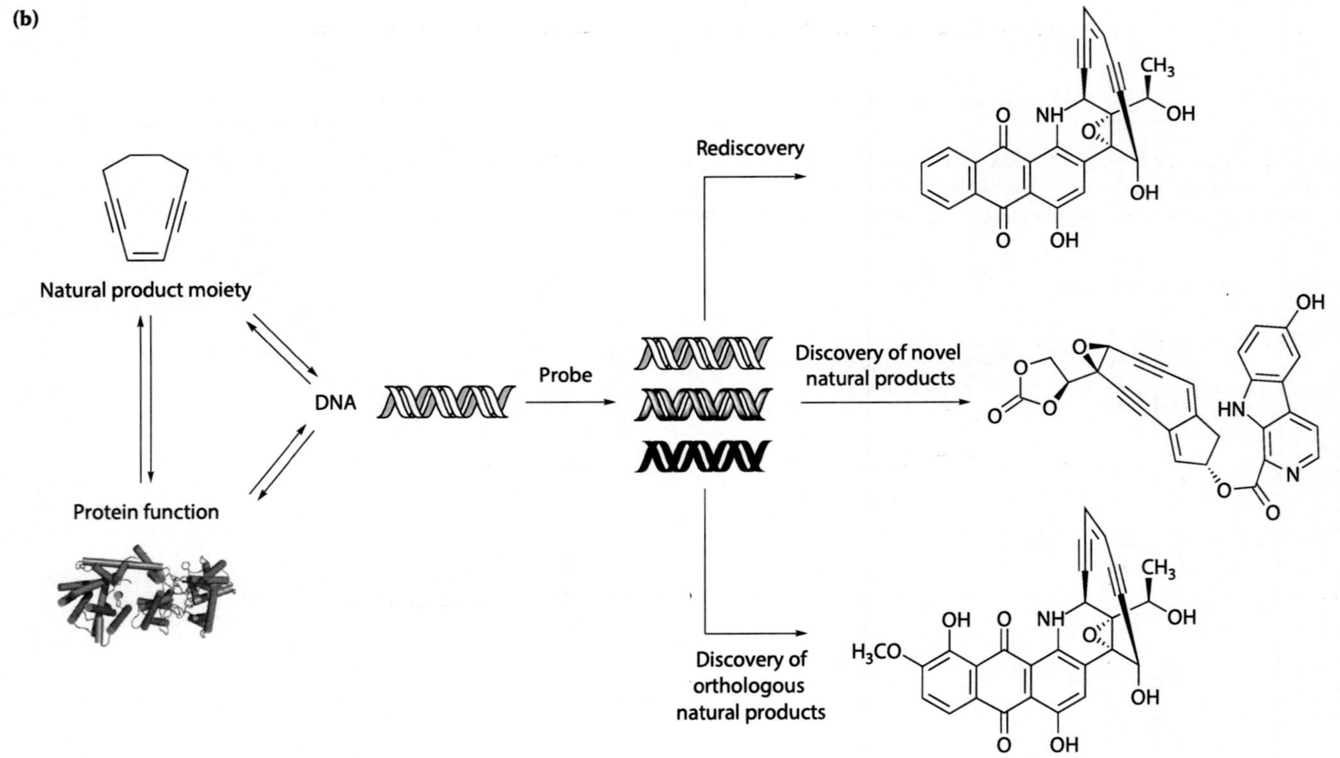

FIGURE 3.2 (*Continued*) Schematic representation of **(a)** knowledge-based and **(b)** discovery-based combinatorial biosynthesis. **(b)** Strains and genomic DNA libraries are screened for key genes or DNA sequences that are implicated in the construction of a desired natural product scaffold or moiety. The hit strains produce either an identical, orthologous, or completely novel natural product. Discovery-based combinatorial biosynthesis searches for one structural feature (e.g., enediyne core) while allowing the rest of the skeleton to change.

biosynthesis not only provides new structures, it simultaneously contributes to the knowledge needed for knowledge-based combinatorial biosynthesis.

This chapter is not intended to comprehensively review the field of combinatorial biosynthesis in regard to natural products. Instead, it is meant to highlight Nature as an inspiration for current practices in combinatorial biosynthesis and the role of combinatorial biosynthesis in future natural product discovery programs. For a more comprehensive exploration of natural products and combinatorial biosynthesis, readers are directed to excellent review articles on natural products in drug discovery,[2,3,5,37] combinatorial biosynthesis,[3,19,24,38–40] natural products in the genomic era,[41–44] and synthetic biology,[40,45] as well as the cited references throughout the chapter.

3.2 NATURE, THE ULTIMATE COMBINATORIAL BIOSYNTHETIC CHEMIST

Nature left hints early in the history of natural product discovery and biosynthesis, revealing itself as a combinatorial biosynthetic chemist. For example, tyrocidine A and gramicidin S (Figure 3.1), two cyclodecapeptides isolated from *Bacillus brevis* in the mid twentieth century, are composed of several amino acids in two distinct ways.[46–49] In fact, there are four different forms of tyrocidines (A–D) with differing amino acid compositions at three distinct positions (Figure 3.1). Nature's disposition for combinatorial biosynthesis was further supported by early cloning and sequencing experiments of the actinorhodin (*act*) and granaticin (*gra*) biosynthetic gene clusters.[50–60] Homologous genes for a minimal PKS, a ketoreductase, and aromatase were found in the same order in the *act* and *gra* clusters, but the remaining genetic makeup of the clusters was varied. The rise of the genomic era,[41,61] and seemingly countless data sets of microbial gene clusters, cemented Nature's status as the ultimate combinatorial biosynthetic chemist.

3.2.1 Platensimycin and Platencin

Platensimycin (PTM) and platencin (PTN) are a new class of promising antibiotic and antidiabetic drug leads.[62–66] They are effective against a wide range of gram-positive bacteria including methicillin-resistant *Staphylococcus aureus* (MRSA), vancomycin-resistant enterococci (VRE), and *Mycobacterium tuberculosis*.[62,63,67] PTM and PTN potently inhibit bacterial fatty acid synthase (FASII) by binding to the acyl-enzyme intermediate of the condensation reactions.[62,63] PTM selectively inhibits the elongation-condensing enzyme FabF/B, and PTN dually inhibits both FabF and the initiation-condensing enzyme FabH. PTM was also shown to be a potent and highly selective inhibitor of mammalian fatty acid synthase.[64]

PTM and PTN were discovered by the systematic screening of 250,000 natural product extracts, from over 80,000 bacterial strains, using a target-based whole-cell antisense *fabF* RNA assay.[68] *Streptomyces platensis* MA7327, collected in South Africa, and *S. platensis* MA7339, collected in Mallorca Spain, were identified as the producing strains of PTM and PTN, respectively.[62,63] PTM and PTN are constructed using two distinct moieties linked by an amide bond: a diterpene-derived aliphatic cage, or ketolide, and a 3-amino-2,4-dihydroxybenzoic acid (ADHBA, Figure 3.3a).[69,70]

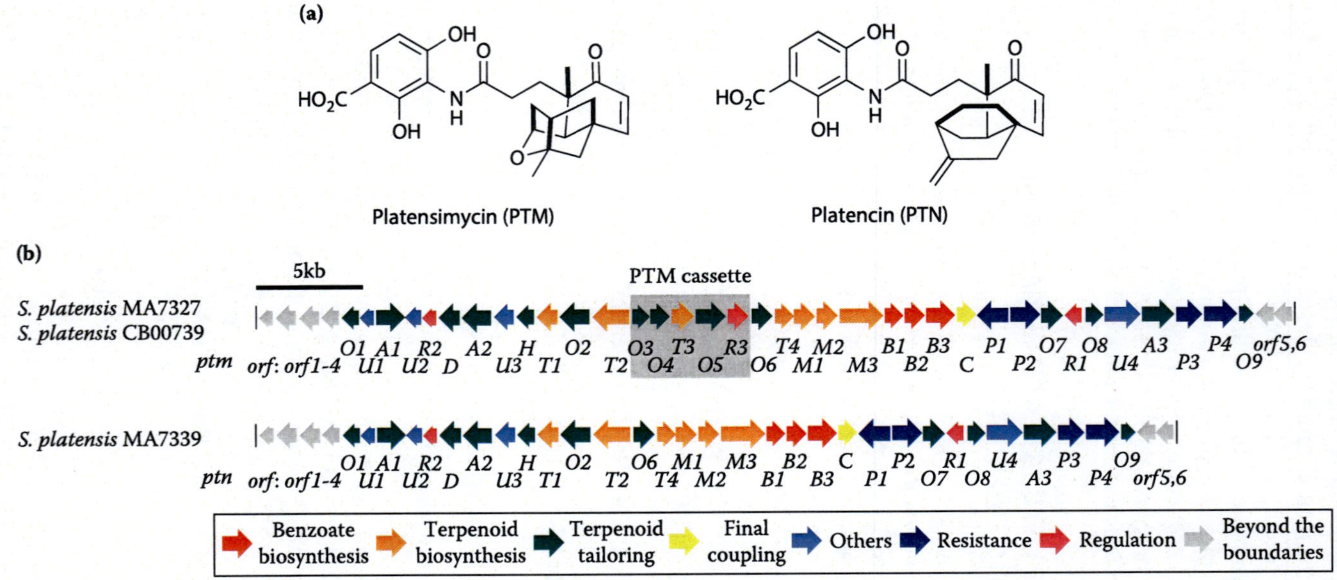

FIGURE 3.3 PTM and PTN structures, biosynthetic gene clusters, and a unified pathway for PTM and PTN biosynthesis. **(a)** Structures of PTM and PTN featuring diterpene-derived ketolide moieties linked to ADHBA. **(b)** Genetic organization of the *ptm* and *ptn* gene clusters from *S. platensis* MA7327, MA7339, and CB00739. Genes are color-coded based on their predicted functions. The PTM cassette that is found in the *ptm* gene cluster and not in the *ptn* gene cluster is shaded in gray.

(Continued)

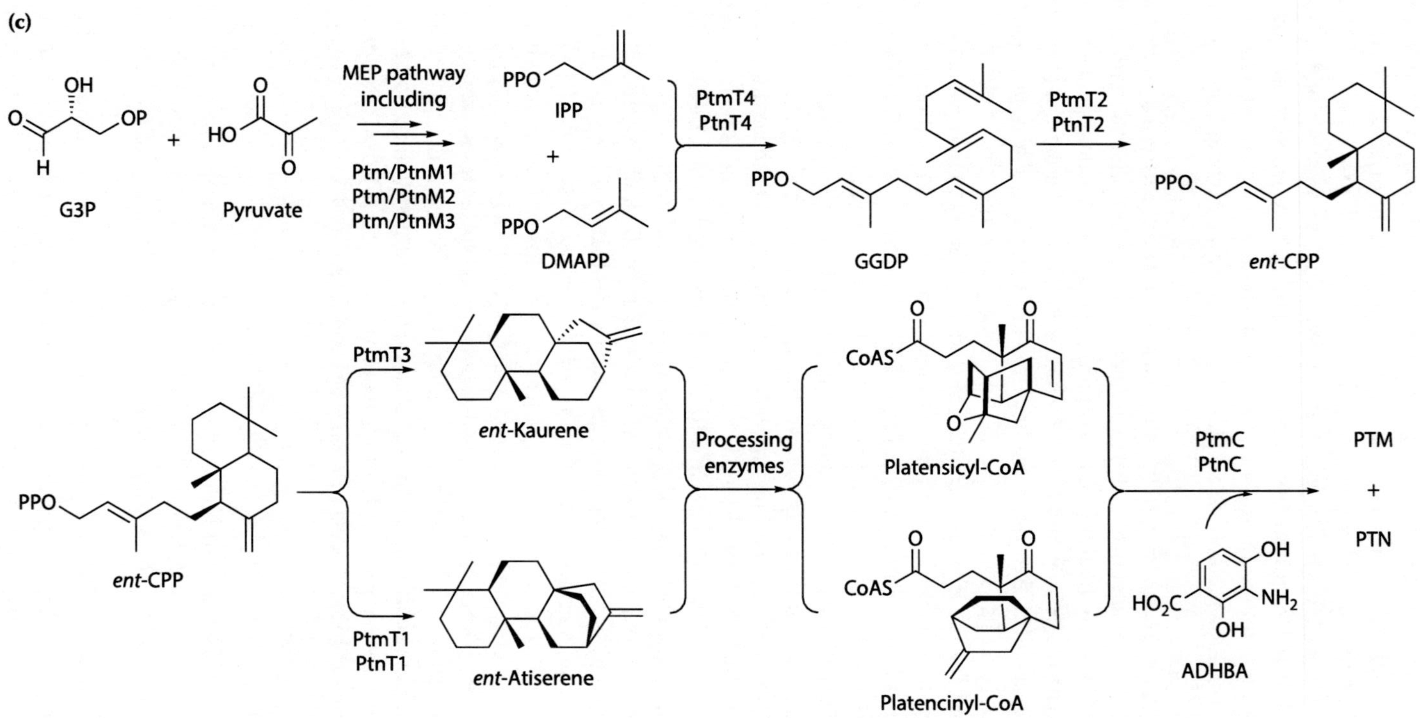

FIGURE 3.3 (*Continued*) PTM and PTN structures, biosynthetic gene clusters, and a unified pathway for PTM and PTN biosynthesis. **(c)** *ent*-CPP, the biosynthetic result of enzymes in the MEP pathway, PtmT4, and PtmT2, is the final common intermediate in PTM and PTN biosynthesis. *ent*-CPP is then converted into the *ent*-kaurene and *ent*-atiserene scaffolds, which are then processed by one set of enzymes into platensicyl- and platencinyl-CoA. These penultimate intermediates are then coupled with ADHBA to yield PTM and PTN.

The differences in the structures of PTM and PTN, and thus in their biological selectivities, are found solely in their diterpene-derived moieties. Based on their structures, the biogenesis of PTM and PTN was initially proposed to originate from the *ent*-kaurane and *ent*-atiserene skeletons, respectively.[70] *S. platensis* MA7327 was later identified as a dual PTM-PTN producer.[71] It was unclear whether one set of genes was responsible for the production of both PTM and PTN or whether two separate sets of genes were required for PTM and PTN biosynthesis. In fact, *ent*-atiserene had been seen as a minor metabolite in plant by enzymes catalyzing *ent*-kaurene biosynthesis.[72]

Upon the cloning and sequencing of the *ptm* and *ptn* gene clusters from *S. platensis* MA7327 and MA7339, respectively, it was immediately evident that the *ptm* gene cluster (from the PTM-PTN dual producing strain) possessed additional genes compared to the *ptn* gene cluster (Figure 3.3b).[73] The 5.4 kb DNA fragment, named the PTM cassette, consisted of five open reading frames (ORFs) and was positioned in the middle of the *ptm* gene cluster. Several of those genes encoded proteins that could be envisaged for the divergence of PTM and PTN biosynthesis including a type I diterpene synthase (PtmT3) and a cytochrome P450 (PtmO5).[74] Another diterpene synthase, PtmT1, was found outside of the PTM cassette, as well as in the *ptn* gene cluster (Figure 3.3b). It was proposed that after cyclization of geranylgeranyl diphosphate (GGPP) into *ent*-copalyl diphosphate (CPP) by PtmT2,[75] *ent*-CPP is diverted into *ent*-kaurene and *ent*-atiserene scaffolds for PTM and PTN, respectively (Figure 3.3c).[73] Both scaffolds are then processed into the appropriate ketolide acids by enzymes specific for PTM (e.g., PtmO5[74]) or enzymes with enough substrate flexibility to accommodate both scaffolds. In a final step, the ketolides are then coupled with ADHBA by a promiscuous amide synthase (Figure 3.3c).[73,76]

The *ptm* and *ptn* gene clusters show snapshots of natural product gene cluster evolution in action and Nature's ability to perform combinatorial biosynthesis via promiscuous processing enzymes. Two 1 kb repeats on either side of the PTM cassette are likely suspects for either deletion, to form a *ptn* cluster from a *ptm* cluster, or insertion, to form a *ptm* cluster from a *ptn* cluster, of the PTM cassette by homologous recombination.[73] The PTM-PTN dual-producing strain utilizes two distinct diterpene synthases to construct preliminary diterpene scaffolds in the second step of biosynthesis.[73] These scaffolds are then processed in an impressive combinatorial biosynthetic manner by enzymes (of which there are at least 10) that are flexible enough to complete the biosynthesis of both PTM and PTN.

3.2.2 Tiancimycin, Uncialamycin, and Dynemycin

The enediyne family of antitumor antibiotics comprises structurally complex natural products with obvious combinatorial biosynthetic origins. Enediynes all share a unique molecular architecture of an unsaturated carbocycle containing two acetylenic groups conjugated to a double bond or incipient double bond (Figure 3.4).[77–80] Electronic rearrangement, via a Bergman or Myers-Saito cyclization, of the enediyne core produces a transient benzenoid diradical, which, when positioned within the minor groove of DNA, abstracts hydrogen atoms from the deoxyribose backbone of duplex DNA (Figure 3.5).[81,82] The DNA radicals can then cause interstrand

(a)

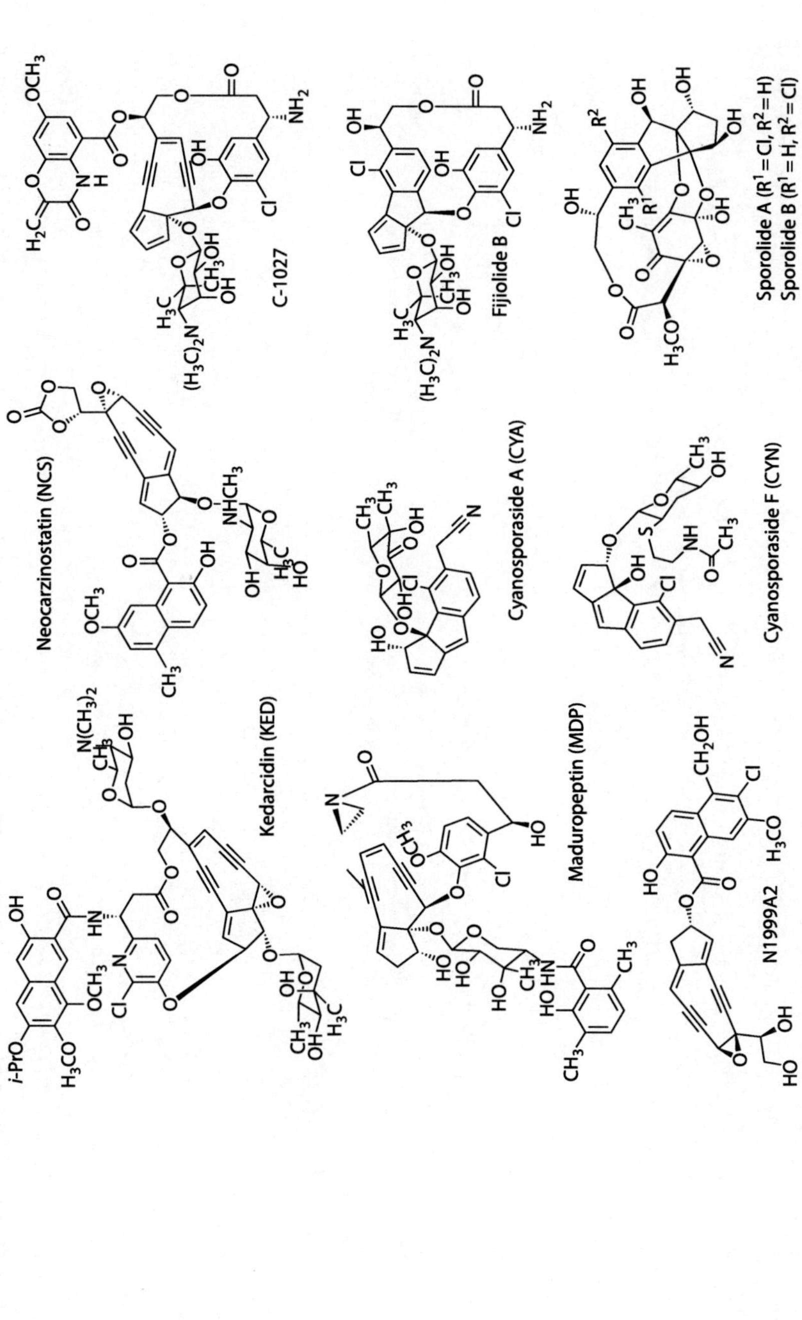

FIGURE 3.4 Structures of the 12 enediynes. **(a)** Five 9-membered enediynes, with four additional members isolated in their cycloaromatized form.

(Continued)

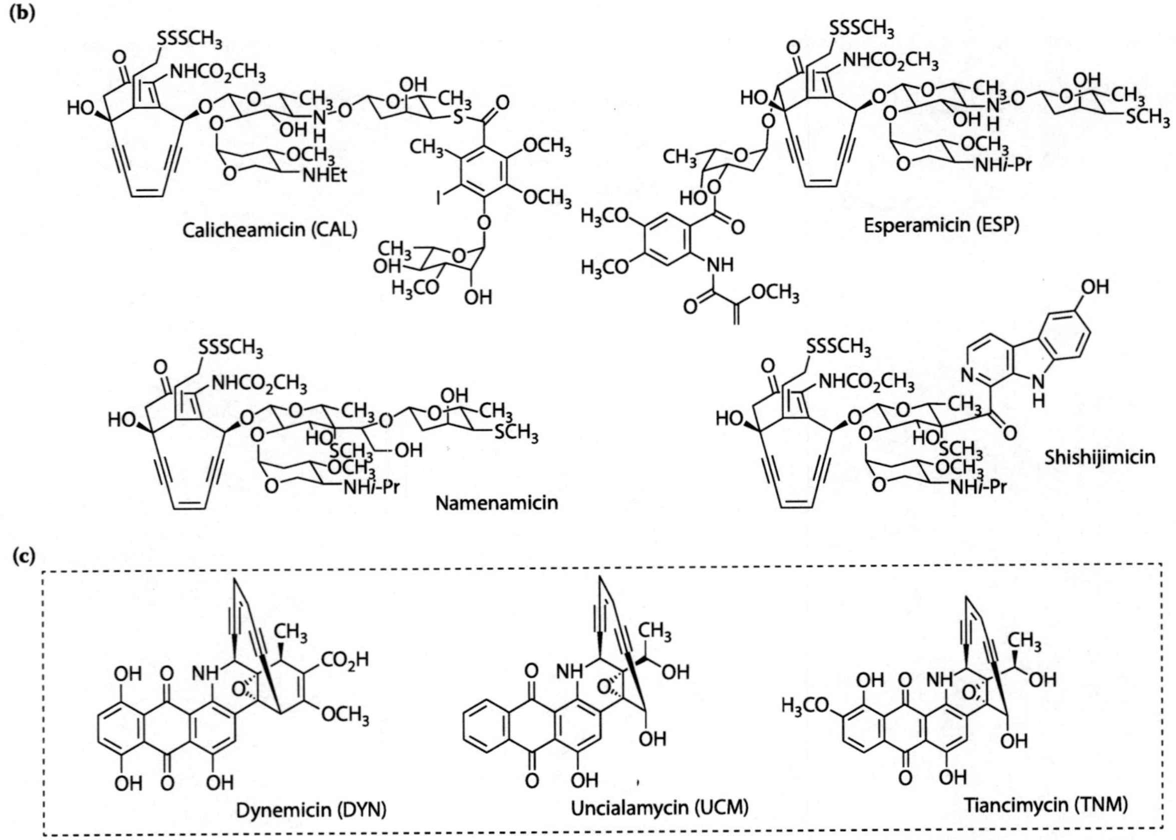

FIGURE 3.4 (*Continued*) Structures of the 12 enediynes. **(b)** Seven 10-membered enediynes with the three anthraquinone-containing 10-membered enediynes highlighted in the dotted rectangle **(c)**.

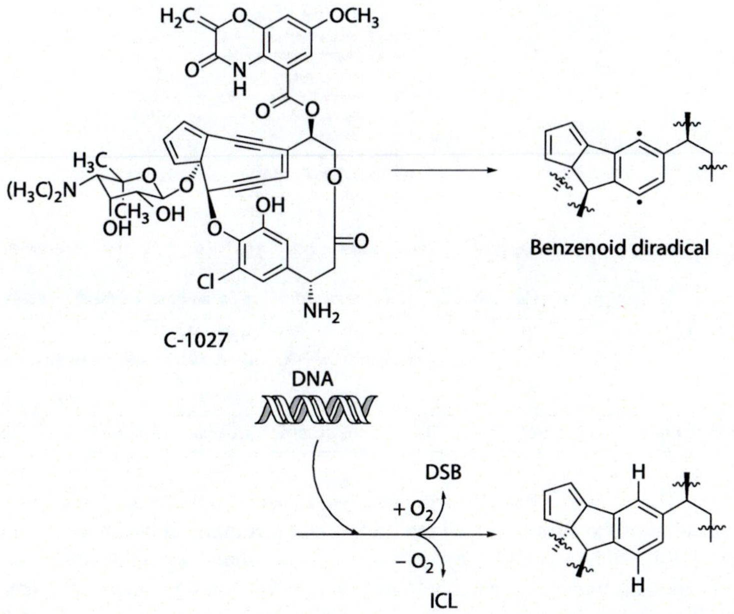

FIGURE 3.5 Enediyne mode of action. Upon electronic rearrangement of the enediyne core via a Bergman or Myers-Saito cyclization, a transient diradical species is formed, which upon interaction with DNA causes DNA damage. The DNA radicals can then cause interstrand cross-links (ICLs) or react with molecular oxygen, resulting in DNA double-strand breaks.

cross-links (ICLs) or react with molecular oxygen, resulting in DNA double-strand breaks (DSBs).[77,79,83–85] Due to their extreme cytotoxicity, enediynes have been translated into clinical drugs.

Enediynes are classified into two subcategories according to the size of their enediyne core scaffolds, 9- or 10-membered. There are only 12 structurally characterized members, five 9-membered with four additional members isolated in their cycloaromatized form and seven 10-membered (Figure 3.4). The 9-membered enediyne subfamily is comprised of neocarzinostatin (NCS) from *Streptomyces carzinostaticus*,[86] C-1027 from *Streptomyces globisporus*,[87,88] kedarcidin (KED) from *Streptoalloteichus* sp. ATCC 53650,[89,90] maduropeptin (MDP) from *Actinomadura madurae*,[91] and N1999A2 from *Streptomyces* sp. AJ9493.[92] The proposed 9-membered sporolides (SPOs) from *Salinispora tropica* CNB-440,[93] cyanosporasides (CYAs) from *Salinispora pacifica* CNS-143,[94] cyanosporasides (CYNs) from *Streptomyces* sp. CNT-179,[95] and fijiolides from *Nocardiopsis* sp. CNS-653[96] were isolated in their cycloaromatized forms. The 10-membered subfamily consists of calicheamicin (CAL) from *Micromonospora echinospora*,[97] esperamicin (ESP) from *Actinomadura verrucosopora*,[98] dynemicin (DYN) from *Micromonospora chersina*,[99] uncialamycin (UCM) from *Streptomyces uncialis*,[100] tiancimycin (TNM) from *Streptomyces* sp. CB03234,[101] and namenamicin[102] and the shishijimicins[103] from two distinct marine ascidia.

DYN, UCM, and TNM are members of the anthraquinone subtype of enediynes; they are unique hybrid structures consisting of an anthraquinone moiety fused into

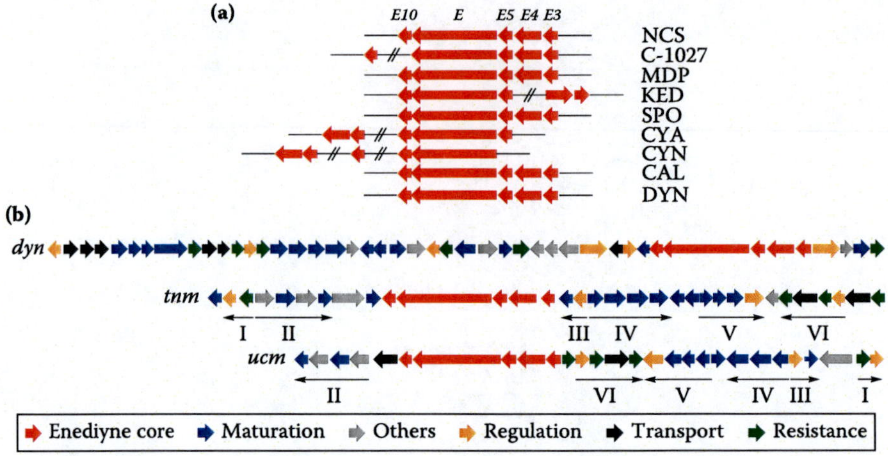

FIGURE 3.6 The enediyne PKS cassette and *dyn*, *tnm*, and *ucm* biosynthetic gene clusters. **(a)** Representation of the minimal enediyne PKS cassette comprised five genes common to all enediynes (*E3*, *E4*, *E5*, *E*, and *E10*). **(b)** Many of the genes found within the *dyn*, *tnm*, and *ucm* gene clusters are conserved; however, their genetic organizations are vastly different suggesting a convergent evolution. In the *tnm* and *ucm* gene clusters, in addition to the enediyne PKS cassette, there are six conserved operons (I–VI), whose organization and reading direction are different.

the 10-membered enediyne core (Figure 3.4c). Given their structural similarity, it would be reasonable to suggest that the gene clusters responsible for the production of these natural products would also be similar. In fact, the *dyn*,[104] *ucm*, and *tnm* gene clusters[101] showed little conservation in genetic organization beyond the enediyne PKS gene cassette, a set of five genes common to all enediynes and consisting of *E3*/*E4*/*E5*/*E*/*E10* (Figure 3.6a).[105–108] This trend is also obvious in the 12 known enediyne gene clusters, whose structures mirror this trend. The enediyne cassette is strictly conserved in each gene cluster, but the surrounding genes show various degrees of or no conservation. Therefore, the known enediynes possess conserved enediyne cores with high structural diversity in their peripheral moieties (Figure 3.4), a desired result of combinatorial biosynthesis.

At first glance, the gene clusters of DYN, UCM, and TNM would suggest high structural diversity among these three natural products (Figure 3.6b). Upon closer inspection (see GNN analysis in Section 3.4.2), many of the genes throughout the three clusters are conserved, but they reside in genetically different positions in relation to the enediyne PKS cassette (Figure 3.6b). In the *tnm* and *ucm* gene clusters (both from *Streptomyces* spp.), there are six conserved operons (I–VI) in addition to the enediyne PKS cassette. The organization and reading direction of operons I–VI are different and are not found in the *dyn* gene cluster. This genetic diversity surely suggests that the *dyn*, *ucm*, and *tnm* gene clusters converged and evolved to produce structurally similar, complex, natural products (Figure 3.4c). This intricate evolution highlights Nature's ability to use a mix-and-match system of genes and operons to construct designer scaffolds with varying functional groups.

3.3 KNOWLEDGE-BASED COMBINATORIAL BIOSYNTHESIS

The biosynthesis of natural products requires a complex combination of genetics, chemistry, enzymology, and structural biology. It is difficult, if not impossible, to design successful combinatorial biosynthesis experiments without considering what Nature has been able to do throughout evolution. Only by looking at how Nature forms simple and complex molecules and understanding the functions, possibilities, and limitations of each step involved in the biosynthesis can the scientific community be effective. Basic understanding of functions is typically established by either gene inactivation and subsequent isolation of biosynthetic intermediates and shunt pathway metabolites or in vitro enzymology studies that directly correlate protein with substrate and function. Studies of combinatorial biosynthesis in Nature have greatly contributed to a knowledge base that helps us understand what changes can be made in a laboratory environment. This has resulted in knowledge-based combinatorial biosynthesis methodologies including point mutations, gene deletions, gene additions, modifying and exchanging domains or modules of enzymes, mobilizing parts of or whole gene clusters, and precursor-directed feeding experiments (Figure 3.7).

Complementary to chemical synthesis, combinatorial biosynthesis provides an alternative to generating natural product structural diversity. Using knowledge-based combinatorial biosynthetic strategies, novel natural products have been generated resulting in compounds with altered biological activities, including increased bioactivities[109,110] or altered modes of action,[84,85,111] compared with the parent natural products. It is even conceivable that the unnatural natural products generated through combinatorial biosynthesis may later be found from natural sources, considering Nature is the ultimate combinatorial biosynthetic chemist and has been practicing combinatorial biosynthesis for millions of years.

3.3.1 Deschloro-C-1027 and Desmethyl-C-1027

Enediynes have limited use as clinical drugs mainly because of substantial toxicity; however, their exquisite potency and mechanism of action make them ideal payloads for anticancer antibody-drug conjugates (ADCs). C-1027 is one of the most potent enediyne members and is currently in clinical development as an ADC for hepatoma.[112] Under aerobic conditions, C-1027 almost exclusively induces DNA DSBs.[113] However, under cell-free anaerobic conditions, C-1027 can induce additional types of DNA damage, such as ICLs (Figure 3.5).[84,85]

First isolated in 1993 from *S. globisporus*,[114–116] C-1027 has been studied as a model for 9-membered enediyne biosynthesis and engineering.[79,107,108,117,118] The C-1027 structure can be divided into four discrete biochemical units: an enediyne core, a deoxyaminosugar, a benzoxazolinate, and a β-amino acid moiety (Figure 3.4a).[88] The enediyne core acts as the warhead for DNA damage and the other three moieties are responsible for binding the chromophore to DNA.[119] The 9-membered enediynes are unique in that they are produced as chromoproteins consisting of an apoprotein (e.g., CagA for C-1027) and the enediyne chromophore. While the chromophores are responsible for biological activities, the apoproteins act as stabilizers for maintaining the integrity of the unstable chromophore until

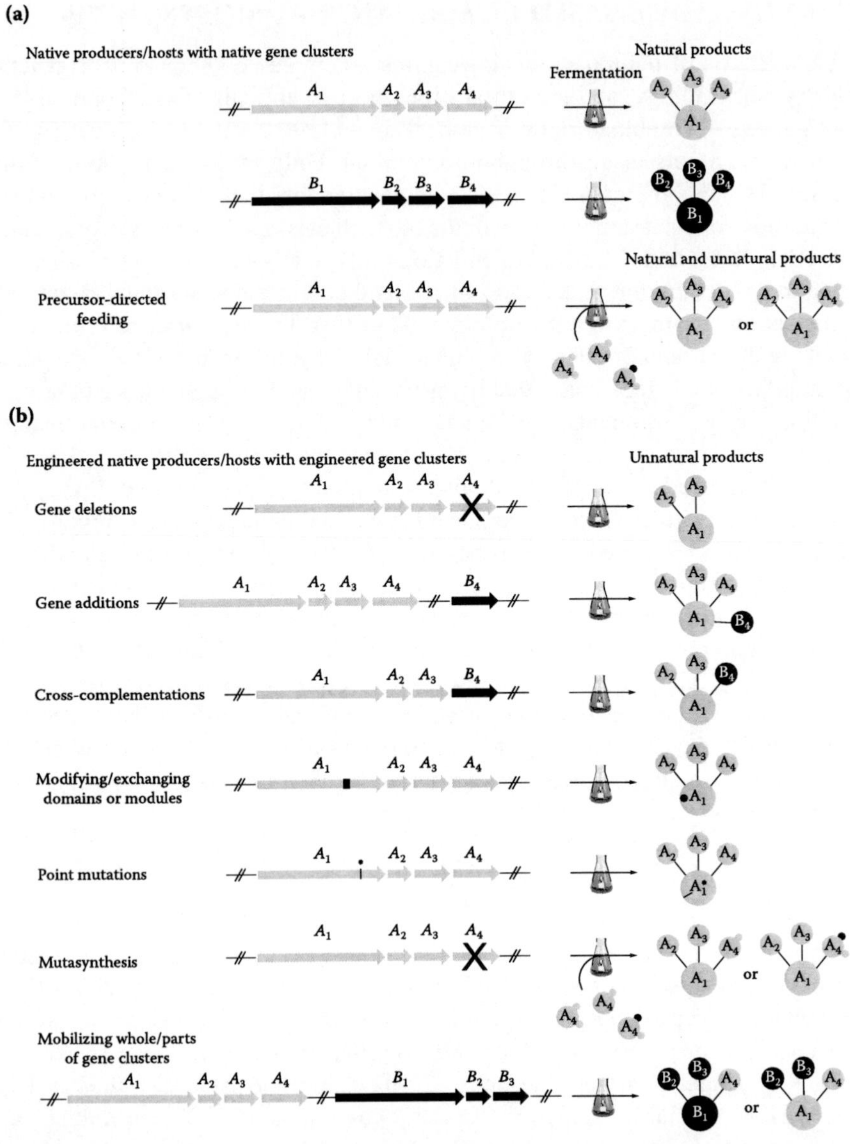

FIGURE 3.7 Knowledge-based combinatorial biosynthetic strategies for generating unnatural products. **(a)** Natural products are the result of fermentation of native producers or heterologous hosts carrying the native biosynthetic gene cluster. Precursor-directed feeding experiments in producers/hosts carrying native gene clusters can yield both natural and unnatural products. **(b)** Unnatural products are generated in engineered native producers or engineered heterologous hosts using a variety of methods including gene deletions, gene additions, cross-complementations, modifying or exchanging of domains or modules, point mutations, mutasynthesis, and mobilizing whole or parts of gene clusters. The unnatural products depicted for the precursor-directed feeding, mutasynthesis, and gene cluster mobilization experiments are selected representatives of potential products, that is, not all possible combinations are shown.

it interacts with target DNA.[120] Genetic manipulations of genes governing C-1027 biosynthesis offer a promising combinatorial biosynthesis approach for preparation of novel C-1027 analogues.

Incorporation of functional groups (e.g., oxygen, halogen atoms) into natural products plays an important role in increasing the diversity and biological activity of natural products. SgcC3 and SgcC were identified as a C-3 (C-20′ in C-1027) halogenase and C-5 (C-22′ in C-1027) hydroxylase, respectively, which catalyze modification of the *β*-tyrosine moiety coupled to the peptidyl carrier protein SgcC2. Gene inactivation of *sgcC3* or *sgcC* generated new C-1027 analogues, 20′-deschloro-C-1027 and 22′-deshydroxy-C-1027, respectively (Figure 3.8).[121,122] In vitro analysis of SgcC3 additionally revealed that chlorine can be substituted with bromine, enabling potential application for precursor-directed biosynthesis.[121] Deletion of another gene, *sgcD4*, which encodes an *O*-methyltransferase responsible for the methylation of the 7″-position of the benzoxazolinate moiety, also led to the production of a new C-1027 analogue, 7″-desmethyl-C-1027 (Figure 3.8).[85]

Single genetic modifications, like the deletion of *sgcC3*, *sgcC*, or *sgcD4*, led to simple modifications of the C-1027 chromophore that affect both the drug potency and the mode of action of the resultant chromoprotein complexes. All three novel analogues, 20′-deschloro-C-1027, 22′-deshydroxy-C-1027, and 7″-desmethyl-C-1027, were less active in inducing DSB than C-1027 (4-, 30-, and 50-fold, respectively).[84] In addition, 7″-desmethyl-C-1027 primarily induces ICLs in cells, whereas 22′-deshydroxy-C-1027-induced ICLs were not detectable in cell-free studies and 20′-deschloro-C-1027 significantly lost its ability to cause ICL DNA damage.[84] Modifications of the *β*-amino acid of the C-1027 chromophore can suppress ICLs, while substitutions on the benzoxazolinate result in a preference of anaerobic induction of ICLs. These results demonstrate that modification of peripheral moieties on the enediyne family of natural products may lead to the fine-tuning of these compounds as drugs. Generation of new C-1027 antitumor antibiotics can produce C-1027 analogues exclusively acting as agents inducing DSB, ICLs, or a combination of both.

3.3.2 Bleomycin

The bleomycins (BLMs) are a family of glycopeptide-polyketide antitumor antibiotics first isolated from *Streptomyces verticillus*.[123,124] BLMs induce sequence-specific single- and double-strand breaks in DNA via a metal- and oxygen-dependent mechanism.[125] Extensive study on the biosynthesis of BLMs revealed that they are derived from nine amino acids, one acetate, and two molecules of *S*-adenosyl methionine (AdoMet) and therefore represent an excellent model system for the biosynthesis of hybrid peptide-polyketide natural products.[77,126] The structures of BLMs can be divided into four different structural motifs, all contributing to their outstanding antitumor activity (Figure 3.9).[127] On the N-terminal region of BLMs, a pyrimidoblamic acid and the adjacent *β*-hydroxyhistidine are responsible for binding the divalent metal ion and, eventually, molecular oxygen. The DNA-binding domain is comprised of a bithiazole moiety and various C-terminal amines. The disaccharide moiety plays a crucial role in the activity, and it has been speculated that it is involved in cellular

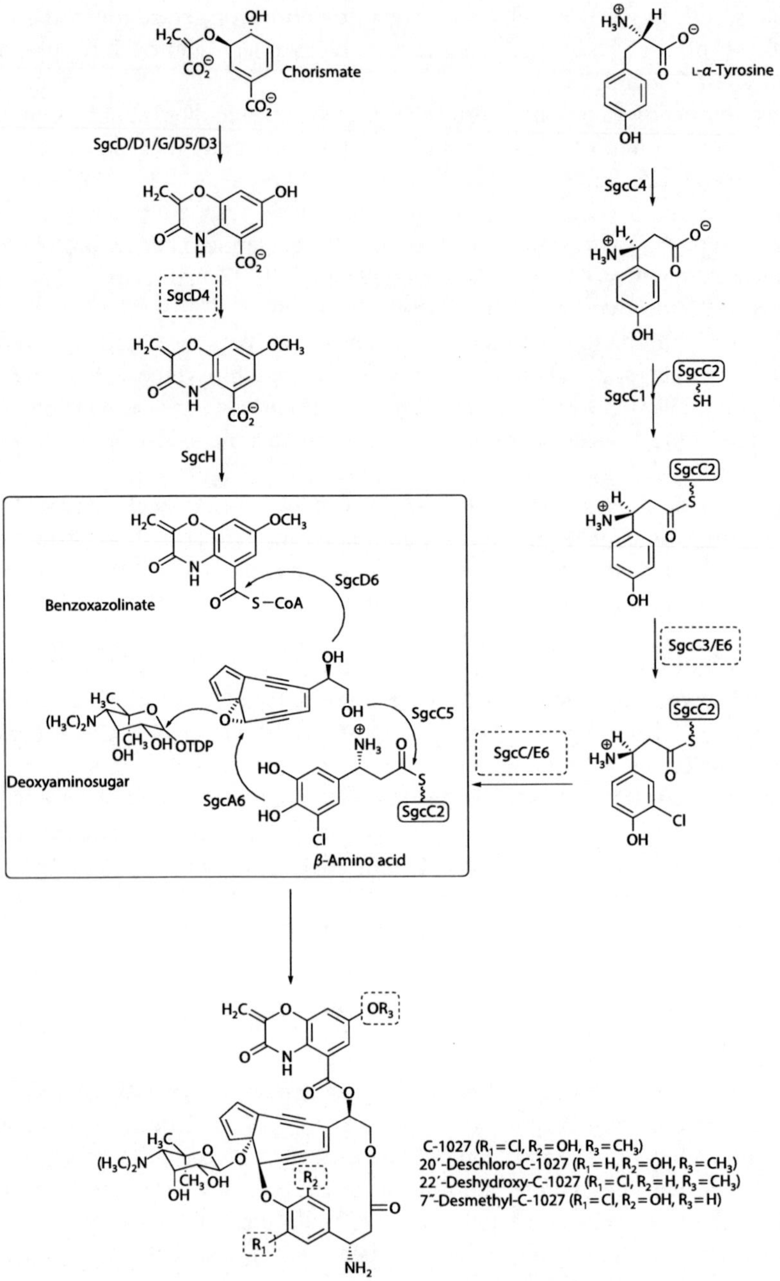

FIGURE 3.8 Engineered production of C-1027 analogues by gene inactivations. Deletion of genes *sgcC3*, *sgcC*, and *sgcD4* encoding enzymes that modify the peripheral moieties of C-1027 resulted in the production of 20′-deschloro-C-1027, 22′-deshydroxy-C-1027, and 7″-desmethyl-C-1027, respectively. The SgcC3, SgcC, and SgcD4 enzymes and the corresponding modifications to the peripheral moieties on the final natural product are highlighted with dotted boxes.

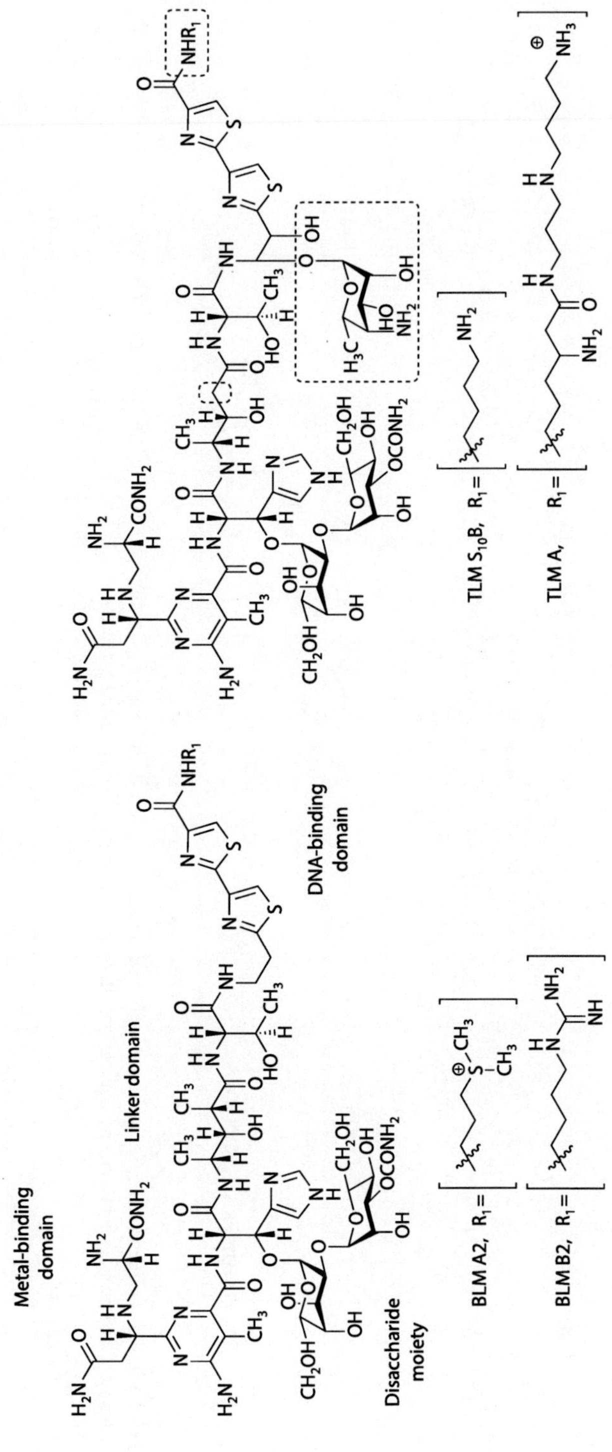

FIGURE 3.9 Structures of selected members of the BLM family of natural products. The four structural domains of BLMs are labeled. Structural differences between the natural products are highlighted with dotted boxes. *(Continued)*

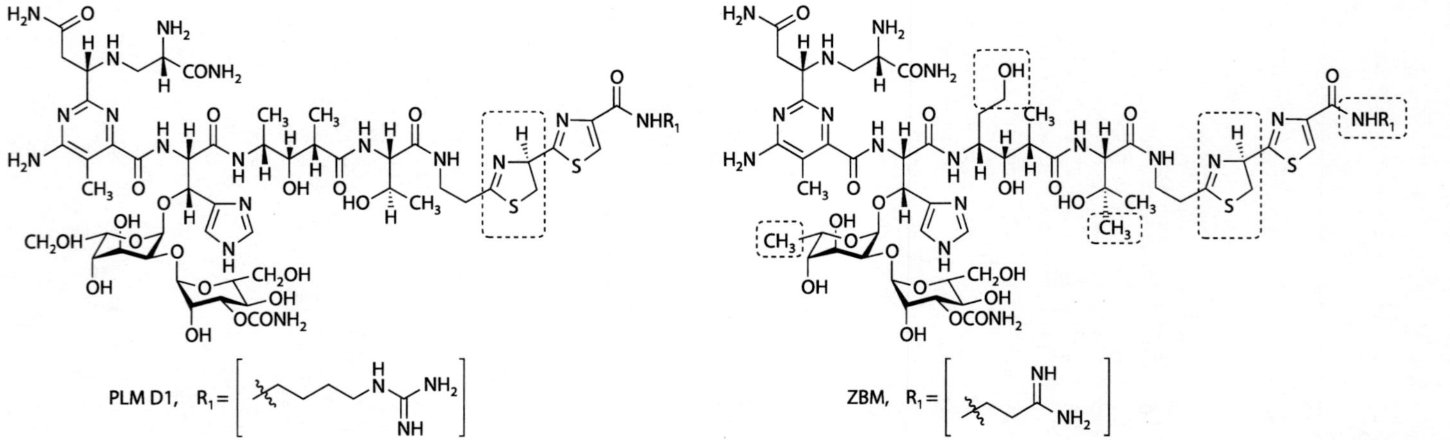

FIGURE 3.9 *(Continued)* Structures of selected members of the BLM family of natural products. The four structural domains of BLMs are labeled. Structural differences between the natural products are highlighted with dotted boxes.

uptake and metal ion coordination.[128] The metal- and DNA-binding domains are held together by a linker domain consisting of a pentanoic acid and an adjacent threonine.

BLMs are clinically relevant drugs and are currently used in palliative chemotherapy treatments against squamous cell carcinoma, non-Hodgkin's lymphoma, Hodgkin's disease, and testicular cancer.[77,125] The exact mechanism of action of BLMs as a cytotoxic glycopeptide antibiotic remains elusive.[125] Evidence indicates that besides their ability to inhibit DNA, RNA, and protein synthesis, BLMs are able to cause cell cycle arrest in G2 and mitosis.[129] However, early development of drug resistance and dose-dependent pulmonary fibrosis—one of the serious complications during cancer therapy—restricts its broader application.[77,130] There is an urgent need to discover or engineer novel antitumor compounds that can overcome the limitations of BLMs.

Four well-known members of the BLM family of glycopeptides—the BLMs, tallysomycins (TLMs), phleomycin (PLM), and zorbamycin (ZBM)—highlight their natural combinatorial biosynthetic disposition. While the metal-binding domain of each of these natural products is strictly conserved, three major structural variations differentiate each member (Figure 3.9): (i) the bithiazole moieties of the BLMs and TLMs are replaced with thiazolinyl-thiazole moieties in PLM and ZBM; (ii) ZBM has a unique C-terminal amine compared to the BLMs, TLMs, and PLMs, which share many of the same C-terminal amines; and (iii) ZBM also has a unique 6-deoxy-L-gulose-containing disaccharide while the others consist of an L-gulose-containing disaccharide. In addition, TLM has a 4-amino-4,6-dideoxy-L-talose moiety in its linker region. Besides the naturally produced members of this family, including the clinically used BLMs, ZBM, and TLM, more than 100 different BLM family analogues have been prepared by synthetic routes or direct biosynthesis.[124,127] While these analogues were essential for the elucidation of the fundamental roles of the individual domains of the BLMs, none of these analogues showed improved potency compared to the natural BLMs.[109] A very recent study, however, showed that deglyco-BLM retained antitumor activity but did not induce pulmonary toxicity or fibrosis, raising optimism that a safer BLM analogue may soon be therapeutically available.[131]

Because of their structural complexity, one practical approach is to engineer novel BLMs using combinatorial biosynthesis in the native producers or heterologous hosts. Bioinformatics analysis of the BLM and ZBM gene clusters revealed putative genes encoding proteins for the biosynthesis of their sugar moieties.[126,132,133] In comparison to the BLM-producing strain, which was recalcitrant to genetic manipulation, a successful gene deletion in the ZBM-producing strain of *zbmL*, a gene encoding for a GDP-mannose-4,6-dehydratase, led to the complete abolishment of ZBM.[133] The Δ*zbmL* mutant strain could be complemented with the *zbmL* gene, restoring ZBM production.[109,133] In true combinatorial biosynthetic fashion, introduction of orthologous sugar biosynthetic genes from the *blm* gene cluster (*blmGF*, *blmGEF*, or *blmG*) led to the production of a new metabolite 6′-hydroxy-ZBM (Figure 3.10).[109]

Application of combinatorial biosynthesis to generate structurally diverse metabolites can also take advantage of larger orthologous parts of or whole gene clusters. In the case of the Δ*zbmL* mutant strain, cross complementation with the whole BLM gene cluster not only led to the production of 6′-hydroxy-ZBM, but two new compounds were also produced and isolated, BLM Z and 6′-deoxy-BLM Z (Figure 3.10).[109] BLM Z is a hybrid BLM structure with the ZBM C-terminal amine;

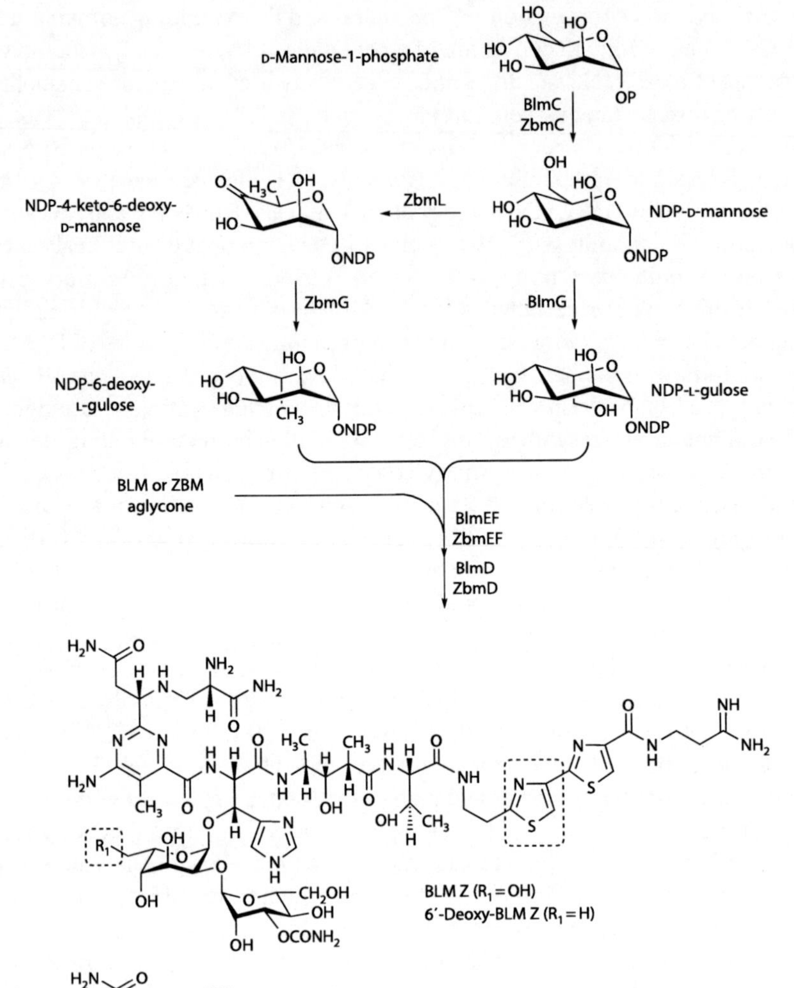

FIGURE 3.10 Engineered production of BLM and ZBM analogues by manipulating their sugar biosynthetic pathways. Cross-complementation of orthologous sugar biosynthetic genes from the *blm* gene clusters (*blmGF*, *blmGEF*, or *blmG*) leads to production of 6′-hydroxy-ZBM. Mobilizing the entire *blm* gene cluster into the native ZBM-producing strain yielded 6′-hydroxy-ZBM, BLM Z, and 6′-deoxy-BLM Z. 6′-Deoxy-BLM Z is currently the most potent BLM analogue. Structural differences between the natural products and engineered products are highlighted with dotted boxes.

6′-deoxy-BLM Z is effectively BLM with the ZBM C-terminal amine and the ZBM disaccharide. In an effort to abolish the native ZBM biosynthesis in *S. flavoviridis* and generate a cleaner background in the host, the PKS machinery for ZBM production (Δ*zbmVIII*) was deleted and expression of the *blm* gene cluster yielded sole production of BLM Z and 6′-deoxy-BLM Z.[109] The designer 6′-deoxy-BLM Z exhibited outstanding DNA cleavage activity in vitro; 6′-deoxy-BLM Z had an EC_{50} one order of magnitude smaller than BLM.[109] Therefore, 6′-deoxy-BLM Z represents the most potent BLM known to date. Inspired by the recent finding that the BLM aglycon may be an effective and less toxic alternative to BLM as a next-generation therapy, engineered production of the BLM aglycon can easily be envisaged.

3.4 COMBINATORIAL BIOSYNTHESIS BY DISCOVERY

In addition to the requirements of a basic understanding of the genes and enzymes involved in natural product biosynthesis and the availability of a genetically amenable and producing host, knowledge-based combinatorial biosynthesis is also limited to the diversification of known scaffolds, typically making only minor modifications. Discovery-based combinatorial biosynthesis offers a variation on generating structural diversity in natural products. By leveraging the preexisting chemical space developed by Nature, new natural products, of known or novel scaffolds, will be discovered. Throughout evolution, Nature has sampled innumerable combinations of genes, proteins, and biosynthetic pathways that have been optimized for production and a selected biological activity (although not necessarily the biological activity that the scientific community look for or that is ultimately utilized in pharmaceutical applications). Why not exploit what Nature has already developed?

In its simplest form, knowledge-based combinatorial biosynthesis modifies one targeted position while keeping the majority of the skeleton intact. In contrast, discovery-based combinatorial biosynthesis searches for one key structural feature while allowing the rest of the natural product scaffold to change. Each natural product can be distilled down to a specific scaffold, functional group, or chemical transformation that can be mapped back to a gene or DNA sequence. If the probe DNA can be targeted and found, the strain/genome/gene cluster that contains the targeted DNA has the potential to produce a natural product containing the desired structural moiety.

The arrival of the genomic era, and the sequencing of actinomycetes in particular, revealed that discovery-based combinatorial biosynthesis is a plausible way to find structural diversity. It is estimated that in Actinobacteria, the traditional workhorses of natural product discovery, we are missing ~90% of their biosynthetic potential.[41] With the exponential growth of genomic data, new and innovative methods are needed to process these data, generate hypotheses, and translate this untapped reservoir into new natural products.

3.4.1 Strain Prioritization

Traditional microbial natural product discovery programs employ a brute force approach of fermenting each strain in various media and analyzing their crude extracts (Figure 3.11b). This approach, while successful, is not time or cost effective

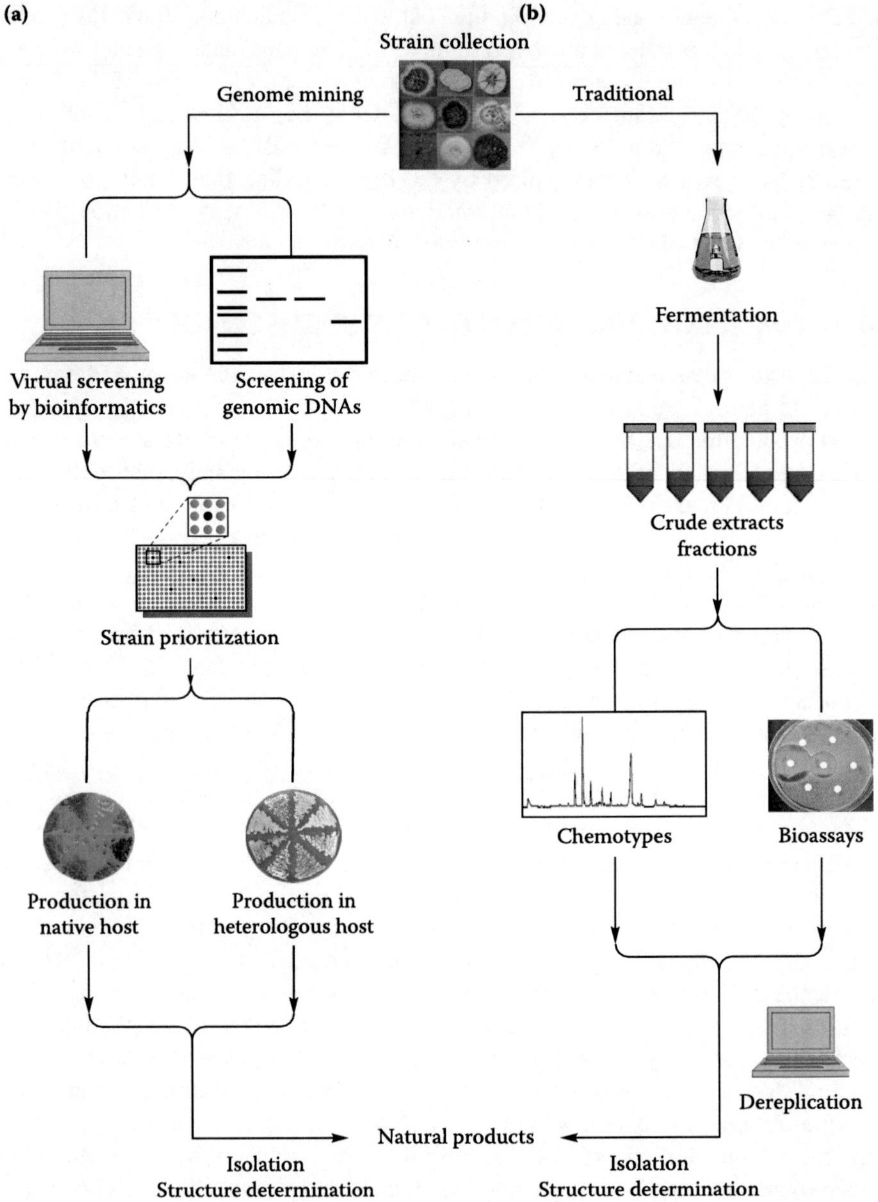

FIGURE 3.11 Strategies for discovering natural products from microorganisms. **(a)** Genome mining (postgenomic) programs utilize genome screening, strain prioritization, and production in native or heterologous hosts for production and isolation of targeted natural products. **(b)** Traditional (pregenomic) programs employing brute force fermentations, bioassays or chemotypes, and isolations. *(Continued)*

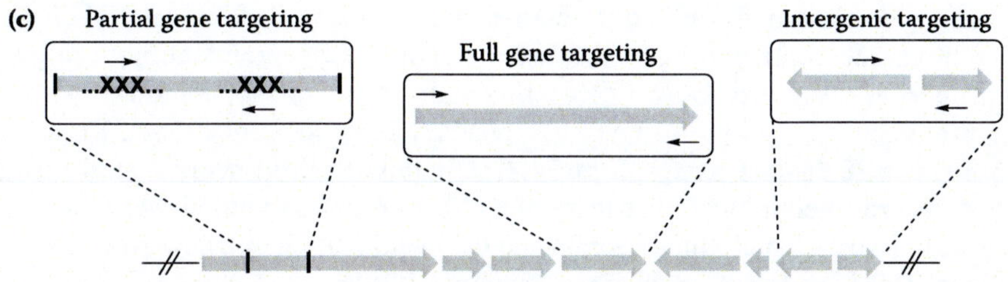

FIGURE 3.11 ***(Continued)*** Strategies for discovering natural products from microorganisms. **(c)** Schematic strategy of virtual or PCR-based genome screening. Depending on the situation and conserved nature of the gene or gene clusters, an ideal probe may target a partial gene, full gene, or an intergenic region. The design of specific or degenerate primers (small arrows), usually of conserved motifs (...XXX...), is essential for specificity of the probe.

and has severe limitations; many strains are unculturable, taxonomically or phenotypically identical, genetically unamenable, with low and unstable titers. In addition, brute force fermentation is a random lottery that can easily result in the rediscovery of known compounds. Genome mining has become a preferred alternative to the traditional approach as natural products of desired structure or predicted function can be targeted (Figure 3.11a).

PCR-based strain prioritization has emerged as a feasible and high-throughput practice of genome mining to target specific natural product gene clusters and reduce the number of strains chosen for fermentation. For discovery-based combinatorial biosynthetic applications, an ideal probe may target a partial or whole gene that encodes a protein that catalyzes the formation of the characteristic core scaffold (Figure 3.11c). The gene clusters found to contain the target would then be more likely to generate different products due to biosynthetic divergence after the targeted protein acts. This idea has been demonstrated for several systems,[105,134–137] but one recent example epitomizes this strategy. The targeting of a key gene in the biosynthesis of phosphonate natural products resulted in the identification of 278 strains, classified into 64 distinct groups, and isolation of 11 previously unknown phosphonic acid natural products from a collection of 10,000 actinomycetes.[138] As all but two characterized phosphonic acid natural product biosynthetic pathways begin with phosphoenol pyruvate (PEP) mutase (encoded by *pepM*), targeting *pepM* using degenerate primers allowed an unbiased search for phosphonate-containing natural products.

In some situations, particularly those involving PKS and NRPS, targeting a DNA sequence that is conserved in many different genes or gene clusters can result in the identification of extraneous hits. Enediynes are a unique subfamily of PKS-derived natural products. If the regions of *pksE*, a gene encoding a type I PKS responsible for enediyne biosynthesis, are targeted, other non-*pksE* type I PKS genes could appear as positive hits. In this case, genetic organization of typical enediyne gene clusters was taken advantage of to differentiate *pksE*s from other type I PKS genes (Figure 3.6a). The *E5-E-E10* gene organization allowed degenerate primers to be constructed that target the terminal and intergenic regions between two genes (Figure 3.11c), in this case, *E5-E* or *E-E10*. In a virtual survey of >25,000 bacterial genomes, 87 potential enediyne gene clusters from 78 different bacteria strains (an additional 180 clusters representing 120 *Salinispora* genomes were dereplicated) were identified using virtual primers as described above.[139] Experimentally, this strategy, combined with the high-throughput nature of real-time PCR, resulted in the identification of 81 strains, classified into 28 distinct groups, from 3400 actinomycetes.[101] Enediyne strain prioritization also resulted in the rediscovery of C-1027 producers with higher titers compared with the original producing strain and the identification of a new 10-membered enediyne, TNM (as discussed above in Section 3.2.2).[101]

Although not usually desired, rediscovery of known natural products is valuable when alternative natural product producers with improved characteristics (e.g., genetic amenability, higher titers) are needed. The wild-type dual producer of PTM and PTN, *S. platensis* MA7327, was genetically amenable.[71,73] This was evident after a negative transcriptional regulator in *S. platensis* MA7327, *ptmR1*, was deleted to improve the production of both PTM and PTN.[71] The resultant strain, *S. platensis* SB12001, showed a dramatic increase (>100×) in production of PTM and PTN, but was recalcitrant to additional genetic modifications. High-throughput strain prioritization of 1911 actinomycetes for PTM/PTN discovery, by targeting the required diterpene synthase genes, resulted in the discovery of six additional dual PTM-PTN producers, which were genetically and phenotypically distinct from *S. platensis* MA7327.[137] Deletion of *ptmR1* in one of the new strains, *S. platensis* CB00739, reestablished the high titers of SB12001 and retained the genetic amenability of the wild-type strain.[140] This new, genetically amenable dual PTM-PTN–overproducing strain has now been showcased as a reliable platform for novel natural product discovery and biosynthetic and bioengineering applications of PTM and PTN.[76,140]

3.4.2 Bioinformatics

After the genomes for *Streptomyces coelicolor*, the actinorhodin-producing model actinomycete, and *Streptomyces avermitilis*, the avermectin-producing industrial strain, were sequenced, it was clear that the actinomycetes had much more biosynthetic potential than they were letting on.[41,141–144] To highlight this notion, the *S. avermitilis* genome possesses at least 38 secondary metabolite gene clusters, more than half of which encode biosynthetic machineries for unknown natural products.[141,145] It was, and continues to be, a thrilling realization that there appears to be unlimited potential of microorganisms to produce novel natural products. However, with the vast influx of genomic data now publicly available, it is vital to develop and utilize new methodologies to analyze and translate the massive amount of bioinformatics data into working hypotheses.

To address these needs, many programs and tools have been developed with varying degrees of success for the natural product community. Commonly used programs include BLAST,[146] HMMER,[147] FramePlot,[148] 2ndFind (biosyn.nih.go.jp/2ndfind), antiSMASH,[149,150] MultiGeneBlast,[151] NaPDoS,[152] ClusterFinder,[153] NP.searcher,[154] PKS/NRPS Analysis,[155] NRPSpredictor,[156] and Mauve.[157] Online databases such as ClusterMine360[158] and the MIBiG repository[159] have been constructed and are constantly updated based on the predicted and experimentally determined relationships between genes and natural products. For a more comprehensive summary and a perspective of computational approaches to natural product discovery, readers are directed to a recent perspective.[44]

Sequence similarity and genome neighborhood networks (GNNs) are emerging as powerful tools for analyzing relationships among protein sequences and proteins encoded by gene clusters.[160,161] Using a database of protein sequences and *E* values from a complete set of protein-protein sequence alignments, sequence similarity networks (SSNs) display protein sequence homologies in a holistic manner (Figure 3.12a).[162] Adjusting the stringency (i.e., *E* value) of the SSN reorganizes the observed protein families; as stringency increases, protein families separate from one another. At relevant *E* value thresholds, inferences can be made for individual proteins and protein families to predict sequence-function relationships. Utilization of SSNs have resulted in accurate predictions of protein function for the polyprenyl transferase family,[163] exploration of the sequence-function space in the mannonate dehydratase subgroup of the enolase superfamily,[164] and the revelation

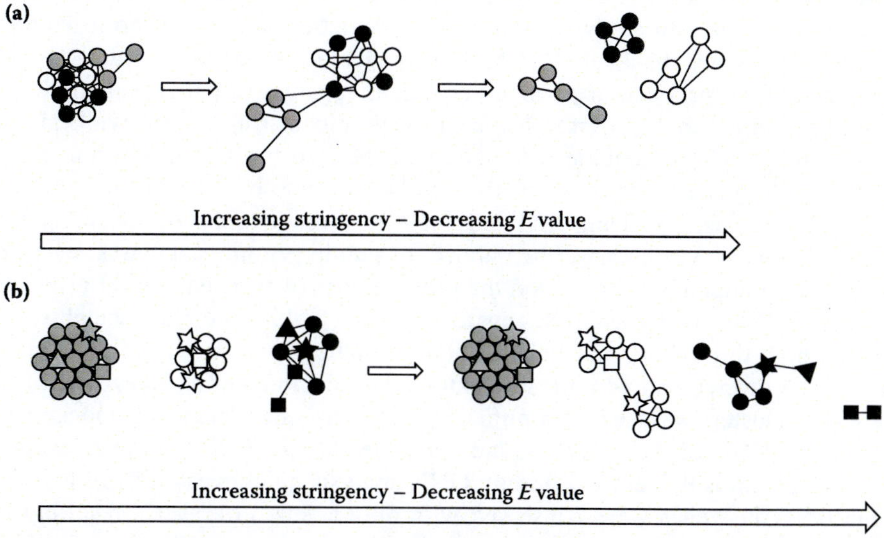

FIGURE 3.12 Sequence similarity and genome neighborhood networks as tools for analyzing relationships among proteins and proteins within gene clusters. Nodes (circles) represent proteins and edges (connecting lines) between nodes represent *E* values. **(a)** SSNs display relationships among proteins (i.e., *E* values). In this example, white, gray, and black nodes represent subclasses of a superfamily of enzymes. At increasing levels of stringency, highly homologous proteins remain clustered while low-homology relationships begin to disappear. **(b)** GNNs display relationships among proteins that are found within a genomic neighborhood. In this example, white, gray, and black nodes represent different families of enzymes. Star, triangle, and square nodes represent three different genome neighborhoods (e.g., gene clusters). Some gene clusters encode representatives of all protein families and some do not, highlighting how GNNs are a versatile tool for analyzing large sets of natural product genomic data.

that ketosynthase (KS) domains of acyltransferase (AT)-less (trans-AT) PKSs cluster together based not only on sequence but also on substrate preference.[165] Structural comparisons, based on X-ray data, of representative KS domains also revealed that structural clustering is similar to sequence clustering, suggesting that SSNs, and therefore sequence and/or function, foreshadow structural conservation or divergence.[165]

GNNs are a recent variation of SSNs. They are constructed in a similar manner, but GNNs are not limited to one (super)family of proteins, as SSNs typically are. By definition, a GNN is a collection of proteins that are encoded by genes contained within a specified genetic region, or neighborhood (Figure 3.12b). By observing the genetic context, especially in bacteria and archaea where metabolic pathways are commonly found in gene clusters, clues can be found regarding both protein and pathway function. Application of GNNs was first shown to facilitate prediction of functions for the functionally diverse proline racemase superfamily (PRS).[166] The genetic context, along with the protein sequence, from a corresponding SSN, was used to successfully predict ~85% of proline racemase proteins in the database and reveal new members of the PRS family.

Application of GNNs to large and complex natural product gene clusters revealed their value in analysis, annotation, prediction, and strain prioritization for natural product discovery.[139] The unprecedented molecular architecture of enediyne natural products begs many questions regarding their fascinating biosynthesis. While much progress has been made elucidating the biosynthesis of the various peripheral moieties of enediynes, the size (up to 100 kb) and complexity of the gene clusters, as well as the abundance of functionally uncharacterized proteins, has hampered biosynthetic investigations of the enediyne core. A virtual survey, as described above, of public genome databases revealed enediynes are much more common than the number of isolated natural products suggests.[78,139] An enediyne GNN, constructed from 87 biosynthetic gene clusters, allowed high-throughput and holistic bioinformatics analysis of the gene clusters.[139] The diverse and complicated nature (i.e., combinatorial biosynthetic) of these gene clusters was easily simplified by GNN analysis. This was demonstrated by a simple GNN of the *dyn*, *ucm*, and *tnm* gene clusters.[101] While the gene cluster organization is evidently different, GNN analysis quickly established that the majority of the genes were conserved with only a few differences, resulting in only minor structural differences between natural products. GNN analysis of 87 enediyne gene clusters also revealed a simple prediction scheme to determine 9- versus 10-membered enediyne gene clusters.[139] With only nine complete gene clusters (four 9-membered, three cycloaromatized, and two 10-membered) available at the time, it was impossible to differentiate between 9- and 10-membered enediyne gene clusters using bioinformatics. With the total number of potential enediyne gene clusters increased by 10-fold, the GNN was able to decrease the background genetic noise. At a low *E* value threshold (10^{-6}), the functionally uncharacterized E2 and E3 proteins cluster together; at an *E* value of 10^{-75}, E2 and E3 separated into discrete subfamilies, revealing that all known 9-membered, but not 10-membered, enediyne gene clusters possessed the *E2* genes. One large protein family was left untouched after using E2 as an indicator of 9-membered enediynes, the CalR3 family of putative regulators. Using these two indicators of 9- versus 10-membered enediynes, that is, E2 for 9-membered and CalR3 for 10-membered, ~80% of the putative gene clusters could be predicted. While these indicators have yet to be experimentally confirmed, this bioinformatics technique facilitates a fresh way to analyze, predict, and prioritize natural product gene clusters for future discovery efforts.

Global natural product similarity networks have also been constructed.[153,167] Using an adapted distance method for multidomain proteins,[168] over 10,000 *high-confidence* gene clusters from 1,154 prokaryotic genomes were assembled into a *global map* of biosynthesis for novel natural product discovery.[153] This global network revealed 947 candidate biosynthetic gene cluster classes, of which 655 are distinct from known families. Characterization of one prominent gene cluster family resulted in the production of aryl polyenes.[153] This is another example of Nature's combinatorial biosynthetic character, three widely divergent subfamilies producing remarkably similar aryl polyene products. Sequence networking is most powerful when incredibly large numbers of sequences are used, and given the ever-increasing number of genes and proteins available in public databases, SSNs, GNNs, and other networking variations are poised to become a mainstay in natural product biosynthetic research.

3.5 SUMMARY, CHALLENGES, AND FUTURE PERSPECTIVES

Natural product discovery has had a profound impact on modern human health and has inspired both biology and drug therapy. This was exemplified by the 2015 Nobel Prize in Physiology or Medicine awarded to William C. Campbell and Satoshi Omura, and Youyou Tu for their discoveries of avermectins and artemisinin (Figure 3.1), respectively. These natural product therapies revolutionized the treatments of the parasitic diseases river blindness and elephantiasis (avermectins) and malaria (artemisinin).[169] With countless human afflictions still left unsolved and the ever-increasing threat of multidrug-resistant pathogens, the scientific communities are faced with a continuing question of how to address these needs. The success of natural products, not only as pharmaceutical options in the clinic but also as small-molecule probes for investigating basic principles of biology, chemistry, and enzymology, endorses their continued study.

Combinatorial biosynthesis has fundamentally changed the landscape of natural products research by integrating an aspect of rational design. Just as Nature continues to provide inspiration for synthetic chemistry and synthetic biology,[170] it equally inspires combinatorial biosynthesis of natural products. Knowledge-based combinatorial biosynthesis relies on the study of biological processes, namely how Nature uses simple building blocks to form structurally complex natural products with diverse biological activities. Using genetic modifications of natural product biosynthetic gene clusters, combinatorial biosynthesis is very successful at rational and targeted construction of unnatural natural products. It is, however, typically limited to making only minor changes on the scaffold. It is difficult to predict how structural modifications of a small molecule, which has been naturally optimized over billions of years, will impact its interactions with the targets that direct biological processes.

While minor modifications normally do not result in changes in modes of action, and may result in decreased biological activities and production titers, they can help to fine-tune a natural product for designer properties.

Discovery-based combinatorial biosynthesis can supplement the knowledge-based approach by discovering novel natural products, biosynthetic gene clusters, and enzyme functions. The power of combinatorial biosynthesis by discovery is the ability to probe Nature for a desired structural motif while allowing the rest of the natural product scaffold to change. Depending on the targeted structural motif, this approach may result in the discovery of completely distinct natural products (e.g., terpenes) or natural products with similar core scaffolds but varying modifications or peripheral moieties (e.g., enediynes). Both options, as opposed to knowledge-based combinatorial biosynthesis, will likely be the result of catalytically efficient biosynthetic pathways, as they have evolved to produce those specific natural products. While there is a seemingly endless supply of new natural products waiting to be found, discovery-based combinatorial biosynthesis still requires preliminary study of the genes and enzymes responsible for the biosynthesis of the targeted structural motif. There is also a lingering question of how to discover natural products with structures or modifications that are still unknown. Large-scale, holistic bioinformatics, such as SSNs and GNNs, may help answer this question. However, the ultimate goal of precisely predicting the structures of natural products solely based on the genes encoding their biosynthetic machineries remains a distant fantasy.

The promising biosynthetic potential of microbes and the vast, untapped ecological biodiversity in Nature (as a relatively small number of organisms from only a small portion of the biosphere have been studied) create great optimism that the discovery of natural products is still in its infancy. The future of natural product discovery will benefit from the incorporation of a combinatorial biosynthetic mindset. Technological advances in DNA sequencing, genome mining, and bioinformatics will surely lead to increases in genomics-based natural products discovery. Methodology advances in genetic manipulations in amenable microbes will lead to the continued construction of strains that produce novel natural products. The enthusiasm for DNA synthesis and synthetic biology will complement traditional combinatorial biosynthesis by facilitating the construction and engineering of natural product gene clusters for heterologous expression in industrial or model hosts. Harnessing the power of both knowledge-based and discovery-based combinatorial biosynthesis will undoubtedly prove to be vital in the quest for novel natural products.

ACKNOWLEDGMENTS

We would like to thank Drs. Guohui Pan and Xiaohui Yan for preliminary discussions regarding this chapter. Studies on natural product biosynthesis, engineering, and drug discovery in the Shen laboratory are currently supported in part by National Institutes of Health grants CA078747, CA106150, GM114353, and GM115575. JDR is supported by the Arnold and Mabel Beckman Foundation and IC is supported by the DFG (Deutsche Forschungsgemeinschaft).

REFERENCES

1. Dictionary of Natural Products. dnp.chemnetbase.com (accessed June 1, 2016).
2. Newman, D. J.; Cragg, G. M. Natural products as sources of new drugs from 1981 to 2014. *J. Nat. Prod.* **2016**, *79*, 629–661.
3. Katz, L.; Baltz, R. H. Natural product discovery: Past, present, and future. *J. Ind. Microbiol. Biotechnol.* **2016**, *43*, 155–176.
4. Schreiber, S. L. Small molecules: The missing link in the central dogma. *Nat. Chem. Biol.* **2005**, *1*, 64–66.
5. Ji, H.-F.; Li, X.-J.; Zhang, H.-Y. Natural products and drug discovery. Can thousands of years of ancient medical knowledge lead us to new and powerful drug combinations in the fight against cancer and dementia? *EMBO Rep.* **2009**, *10*, 194–200.
6. Hopwood, D. A.; Malpartida, F.; Kieser, H. M.; Ikeda, H.; Duncan, J.; Fujii, I.; Rudd, B. A. M.; Floss, H. G.; Omura, S. Production of 'hybrid' antibiotics by genetic engineering. *Nature* **1985**, *314*, 642–644.
7. Wilkinson, B.; Moss, S. J. Biosynthetic engineering of natural products for lead optimization and development. *Curr. Opin. Drug Discov. Devel.* **2005**, *8*, 748–756.
8. Ortholand, J.-Y.; Ganesan, A. Natural products and combinatorial chemistry: Back to the future. *Curr. Opin. Chem. Biol.* **2004**, *8*, 271–280.
9. Donadio, S.; Staver, M. J.; McAlpine, J. B.; Swanson, S. J.; Katz, L. Modular organization of genes required for complex polyketide biosynthesis. *Science* **1991**, *252*, 675–679.
10. Donadio, S.; Katz, L. Organization of the enzymic domains in the multifunctional polyketide synthase involved in erythromycin formation in *Saccharopolyspora erythraea*. *Gene* **1992**, *111*, 51–60.
11. Marahiel, M. A. Multidomain enzymes involved in peptide synthesis. *FEBS Lett.* **1992**, *307*, 40–43.
12. Marahiel, M. A.; Stachelhaus, T.; Mootz, H. D. Modular peptide synthetases involved in nonribosomal peptide synthesis. *Chem. Rev.* **1997**, *97*, 2651–2673.
13. Marahiel, M. A. A structural model for multimodular NRPS assembly lines. *Nat. Prod. Rep.* **2016**, *33*, 136–140.
14. Stachelhaus, T.; Marahiel, M. A. Modular structure of genes encoding multifunctional peptide synthetases required for non-ribosomal peptide synthesis. *FEMS Microbiol. Lett.* **1995**, *125*, 3–14.
15. Shen, B. Polyketide biosynthesis beyond the type I, II and III polyketide synthase paradigms. *Curr. Opin. Chem. Biol.* **2003**, *7*, 285–295.
16. Staunton, J. Combinatorial biosynthesis of erythromycin and complex polyketides. *Curr. Opin. Chem. Biol.* **1998**, *2*, 339–345.
17. Walsh, C. T. Combinatorial biosynthesis of antibiotics: Challenges and opportunities. *ChemBioChem* **2002**, *3*, 125–134.
18. Weissman, K. J.; Leadlay, P. F. Combinatorial biosynthesis of reduced polyketides. *Nat. Rev. Microbiol.* **2005**, *3*, 925–936.
19. Floss, H. G. Combinatorial biosynthesis-potential and problems. *J. Biotechnol.* **2006**, *124*, 242–257.
20. Wong, F. T.; Khosla, C. Combinatorial biosynthesis of polyketides—A perspective. *Curr. Opin. Chem. Biol.* **2012**, *16*, 117–123.
21. Staunton, J.; Weissman, K. J. Polyketide biosynthesis: A millennium review. *Nat. Prod. Rep.* **2001**, *18*, 380–416.
22. Floss, H. G. Antibiotic biosynthesis: From natural to unnatural compounds. *J. Ind. Microbiol. Biotechnol.* **2001**, *27*, 183–194.
23. Weissman, K. J. Genetic engineering of modular PKSs: From combinatorial biosynthesis to synthetic biology. *Nat. Prod. Rep.* **2016**, *33*, 203–230.

24. Walsh, C. T. Polyketide and nonribosomal peptide antibiotics: Modularity and versatility. *Science* **2004**, *303*, 1805–1810.
25. Helfrich, E. J. N.; Piel, J. Biosynthesis of polyketides by trans-AT polyketide synthases. *Nat. Prod. Rep.* **2016**, *33*, 231–316.
26. Winn, M.; Fyans, J. K.; Zhuo, Y.; Micklefield, J. Recent advances in engineering nonribosomal peptide assembly lines. *Nat. Prod. Rep.* **2016**, *33*, 317–347.
27. Jia, M.; Potter, K. C.; Peters, R. J. Extreme promiscuity of a bacterial and a plant diterpene synthase enables combinatorial biosynthesis. *Metab. Eng.* **2016**, *37*, 24–34.
28. Andersen-Ranberg, J.; Kongstad, K. T.; Nielsen, M. T.; Jensen, N. B.; Pateraki, I.; Bach, S. S.; Hamberger, B. et al. Expanding the landscape of diterpene structural diversity through stereochemically controlled combinatorial biosynthesis. *Angew. Chem. Int. Ed.* **2016**, *55*, 2142–2146.
29. Arnison, P. G.; Bibb, M. J.; Bierbaum, G.; Bowers, A. A.; Bugni, T. S.; Bulaj, G.; Camarero, J. A. et al. Ribosomally synthesized and post-translationally modified peptide natural products: Overview and recommendations for a universal nomenclature. *Nat. Prod. Rep.* **2013**, *30*, 108–160.
30. Sardar, D.; Schmidt, E. W. Combinatorial biosynthesis of RiPPs: Docking with marine life. *Curr. Opin. Chem. Biol.* **2016**, *31*, 15–21.
31. Sanchez, C.; Zhu, L.; Brana, A. F.; Salas, A. P.; Rohr, J.; Mendez, C.; Salas, J. A. Combinatorial biosynthesis of antitumor indolocarbazole compounds. *Proc. Natl. Acad. Sci. USA* **2005**, *102*, 461–466.
32. Mendez, C.; Moris, F.; Salas, J. A. Biosynthesis of indolocarbazole alkaloids and generation of novel derivatives by combinatorial biosynthesis. *RSC Drug Discov. Ser.* **2012**, *25*, 99–115.
33. Luzhetskyy, A.; Bechthold, A. It works: Combinatorial biosynthesis for generating novel glycosylated compounds. *Mol. Microbiol.* **2005**, *58*, 3–5.
34. Yang, J.; Hoffmeister, D.; Liu, L.; Fu, X.; Thorson, J. S. Natural product glycorandomization. *Bioorg. Med. Chem.* **2004**, *12*, 1577–1584.
35. Heide, L.; Gust, B.; Anderle, C.; Li, S. M. Combinatorial biosynthesis, metabolic engineering and mutasynthesis for the generation of new aminocoumarin antibiotics. *Curr. Top. Med. Chem.* **2008**, *8*, 667–679.
36. Baltz, R. H. Combinatorial glycosylation of glycopeptide antibiotics. *Chem. Biol.* **2002**, *9*, 1268–1270.
37. Li, J. W. H.; Vederas, J. C. Drug discovery and natural products end of an era or an endless frontier? *Science* **2009**, *325*, 161–165.
38. Sun, H.; Liu, Z.; Ang Ee, L.; Zhao, H. Recent advances in combinatorial biosynthesis for drug discovery. *Drug Des. Dev. Ther.* **2015**, *9*, 823–833.
39. Weber, T.; Charusanti, P.; Musiol-Kroll, E. M.; Jiang, X.; Tong, Y.; Kim, H. U.; Lee, S. Y. Metabolic engineering of antibiotic factories: New tools for antibiotic production in actinomycetes. *Trends Biotechnol.* **2015**, *33*, 15–26.
40. Kim, E.; Moore, B. S.; Yoon, Y. J. Reinvigorating natural product combinatorial biosynthesis with synthetic biology. *Nat. Chem. Biol.* **2015**, *11*, 649–659.
41. Nett, M.; Ikeda, H.; Moore, B. S. Genomic basis for natural product biosynthetic diversity in the actinomycetes. *Nat. Prod. Rep.* **2009**, *26*, 1362–1384.
42. Winter, J. M.; Behnken, S.; Hertweck, C. Genomics-inspired discovery of natural products. *Curr. Opin. Chem. Biol.* **2011**, *15*, 22–31.
43. Letzel, A.-C.; Pidot, S. J.; Hertweck, C. A genomic approach to the cryptic secondary metabolome of the anaerobic world. *Nat. Prod. Rep.* **2013**, *30*, 392–428.
44. Medema, M. H.; Fischbach, M. A. Computational approaches to natural product discovery. *Nat. Chem. Biol.* **2015**, *11*, 639–648.
45. Smanski, M. J.; Zhou, H.; Claesen, J.; Shen, B.; Fischbach, M. A.; Voigt, C. A. Synthetic biology to access and expand nature's chemical diversity. *Nat. Rev. Microbiol.* **2016**, *14*, 135–149.

46. Van Epps, H. L. Rene Dubos: Unearthing antibiotics. *J. Exp. Med.* **2006**, *203*, 259.
47. Hunter, F. E., Jr.; Schwartz, L. S. Tyrocidines and gramicidin S (J_1, J_2). *Antibiotics* **1967**, *1*, 636–641.
48. Hotchkiss, R. D. Gramicidin, tyrocidine and tyrothricin. *Adv. Enzymol. Relat. Subj. Biochem.* **1944**, *4*, 153–199.
49. Lipmann, F. Attempts to map a process evolution of peptide biosynthesis. *Science* **1971**, *173*, 875–884.
50. Fernandez-Moreno, M. A.; Martinez, E.; Caballero, J. L.; Ichinose, K.; Hopwood, D. A.; Malpartida, F. DNA sequence and functions of the *act*VI region of the actinorhodin biosynthetic gene cluster of *Streptomyces coelicolor* A3(2). *J. Biol. Chem.* **1994**, *269*, 24854–24863.
51. Fernandez-Moreno, M. A.; Martinez, E.; Boto, L.; Hopwood, D. A.; Malpartida, F. Nucleotide sequence and deduced functions of a set of cotranscribed genes of *Streptomyces coelicolor* A3(2) including the polyketide synthase for the antibiotic actinorhodin. *J. Biol. Chem.* **1992**, *267*, 19278–19290.
52. Fernandez-Moreno, M. A.; Caballero, J. L.; Hopwood, D. A.; Malpartida, F. The *act* cluster contains regulatory and antibiotic export genes, direct targets for translational control by the *bldA* tRNA gene of *Streptomyces*. *Cell* **1991**, *66*, 769–780.
53. Sherman, D. H.; Malpartida, F.; Bibb, M. J.; Kieser, H. M.; Bibb, M. J.; Hopwood, D. A. Structure and deduced function of the granaticin-producing polyketide synthase gene cluster of *Streptomyces violaceoruber* Tu22. *EMBO J.* **1989**, *8*, 2717–2725.
54. Hallam, S. E.; Malpartida, F.; Hopwood, D. A. Nucleotide sequence, transcription and deduced function of a gene involved in polyketide antibiotic synthesis in *Streptomyces coelicolor*. *Gene* **1988**, *74*, 305–320.
55. Malpartida, F.; Hopwood, D. A. Physical and genetic characterization of the gene cluster for the antibiotic actinorhodin in *Streptomyces coelicolor* A3(2). *Mol. Gen. Genet.* **1986**, *205*, 66–73.
56. Malpartida, F.; Hopwood, D. A. Molecular cloning of the whole biosynthetic pathway of a *Streptomyces* antibiotic and its expression in a heterologous host. *Nature* **1984**, *309*, 462–464.
57. Okamoto, S.; Taguchi, T.; Ochi, K.; Ichinose, K. Biosynthesis of actinorhodin and related antibiotics: Discovery of alternative routes for quinone formation encoded in the *act* gene cluster. *Chem. Biol.* **2009**, *16*, 226–236.
58. Ichinose, K.; Taguchi, T.; Ebizuka, Y.; Hopwood, D. A. Biosynthetic gene clusters of benzoisochromanequinone antibiotics in *Streptomyces* spp. Identification of genes involved in post-PKS tailoring steps. *Actinomycetologica* **1998**, *12*, 99–109.
59. Ichinose, K.; Bedford, D. J.; Tornus, D.; Bechthold, A.; Bibb, M. J.; Revill, W. P.; Floss, H. G.; Hopwood, D. A. The granaticin biosynthetic gene cluster of *Streptomyces violaceoruber* Tu22: Sequence analysis and expression in a heterologous host. *Chem. Biol.* **1998**, *5*, 647–659.
60. Bechthold, A.; Sohng, J. K.; Smith, T. M.; Chu, X.; Floss, H. G. Identification of *Streptomyces violaceoruber* Tu22 genes involved in the biosynthesis of granaticin. *Mol. Gen. Genet.* **1995**, *248*, 610–620.
61. Cox, R. J.; Piel, J.; Moore, B. S.; Weissman, K. J., eds. Themed issue on genomics. *Nat. Prod. Rep.* **2009**, *26*, 1353–1508.
62. Wang, J.; Soisson, S. M.; Young, K.; Shoop, W.; Kodali, S.; Galgoci, A.; Painter, R. et al. Platensimycin is a selective FabF inhibitor with potent antibiotic properties. *Nature* **2006**, *441*, 358–361.
63. Wang, J.; Kodali, S.; Lee, S. H.; Galgoci, A.; Painter, R.; Dorso, K.; Racine, F. et al. Discovery of platencin, a dual FabF and FabH inhibitor with in vivo antibiotic properties. *Proc. Natl. Acad. Sci. USA* **2007**, *104*, 7612–7616.
64. Wu, M.; Singh, S. B.; Wang, J.; Chung, C. C.; Salituro, G.; Karanam, B. V.; Lee, S. H. et al. Antidiabetic and antisteatotic effects of the selective fatty acid synthase (FAS) inhibitor platensimycin in mouse models of diabetes. *Proc. Natl. Acad. Sci. USA* **2011**, *108*, 5378–5383.

65. Martens, E.; Demain, A. L. Platensimycin and platencin: Promising antibiotics for future application in human medicine. *J. Antibiot.* **2011**, *64*, 705–710.
66. Rudolf, J. D.; Dong, L.-B.; Shen, B. Platensimycin and platencin: Inspirations for chemistry, biology, enzymology, and medicine. *Biochem. Pharmacol.* **2016**, *133*, 139–151.
67. Brown, A. K.; Taylor, R. C.; Bhatt, A.; Futterer, K.; Besra, G. S. Platensimycin activity against mycobacterial β-ketoacyl-ACP synthases. *PLoS One* **2009**, *4*, e6306.
68. Young, K.; Jayasuriya, H.; Ondeyka, J. G.; Herath, K.; Zhang, C.; Kodali, S. et al. Discovery of FabH/FabF inhibitors from natural products. *Antimicrob. Agents Chemother.* **2006**, *50*, 519–526.
69. Singh, S. B.; Jayasuriya, H.; Ondeyka, J. G.; Herath, K. B.; Zhang, C.; Zink, D. L.; Tsou, N. N. et al. Isolation, structure, and absolute stereochemistry of platensimycin, a broad spectrum antibiotic discovered using an antisense differential sensitivity strategy. *J. Am. Chem. Soc.* **2006**, *128*, 11916–11920.
70. Jayasuriya, H.; Herath, K. B.; Zhang, C.; Zink, D. L.; Basilio, A.; Genilloud, O.; Diez, M. T. et al. Isolation and structure of platencin: A FabH and FabF dual inhibitor with potent broad-spectrum antibiotic activity. *Angew. Chem. Int. Ed.* **2007**, *46*, 4684–4688.
71. Smanski, M. J.; Peterson, R. M.; Rajski, S. R.; Shen, B. Engineered *Streptomyces platensis* strains that overproduce antibiotics platensimycin and platencin. *Antimicrob. Agents Chemother.* **2009**, *53*, 1299–1304.
72. Xu, M.; Wilderman, P. R.; Morrone, D.; Xu, J.; Roy, A.; Margis-Pinheiro, M.; Upadhyaya, N. M.; Coates, R. M.; Peters, R. J. Functional characterization of the rice kaurene synthase-like gene family. *Phytochemistry* **2007**, *68*, 312–326.
73. Smanski, M. J.; Yu, Z.; Casper, J.; Lin, S.; Peterson, R. M.; Chen, Y.; Wendt-Pienkowski, E.; Rajski, S. R.; Shen, B. Dedicated *ent*-kaurene and *ent*-atiserene synthases for platensimycin and platencin biosynthesis. *Proc. Natl. Acad. Sci. USA* **2011**, *108*, 13498–13503.
74. Rudolf, J. D.; Dong, L.-B.; Manoogian, K.; Shen, B. Biosynthetic origin of the ether ring in platensimycin. *J. Am. Chem. Soc.* **2016**, *138*, 16711–16721.
75. Rudolf, J. D.; Dong, L.-B.; Cao, H.; Hatzos-Skintges, C.; Osipiuk, J.; Endres, M.; Chang, C.-Y. et al. Structure of the *ent*-copalyl diphosphate synthase PtmT2 from *Streptomyces platensis* CB00739, a bacterial type II diterpene synthase. *J. Am. Chem. Soc.* **2016**, *138*, 10905–10915.
76. Dong, L.-B.; Rudolf, J. D.; Shen, B. A mutasynthetic library of platensimycin and platencin analogues. *Org. Lett.* **2016**, *18*, 4606–4609.
77. Galm, U.; Hager, M. H.; Van Lanen, S. G.; Ju, J.; Thorson, J. S.; Shen, B. Antitumor antibiotics: Bleomycin, enediynes, and mitomycin. *Chem. Rev.* **2005**, *105*, 739–758.
78. Shen, B.; Hindra; Yan, X.; Huang, T.; Ge, H.; Yang, D.; Teng, Q.; Rudolf, J. D.; Lohman, J. R. Enediynes: Exploration of microbial genomics to discover new anticancer drug leads. *Bioorg. Med. Chem. Lett.* **2015**, *25*, 9–15.
79. Van Lanen, S. G.; Shen, B. Biosynthesis of enediyne antitumor antibiotics. *Curr. Top. Med. Chem.* **2008**, *8*, 448–459.
80. Liang, Z.-X. Complexity and simplicity in the biosynthesis of enediyne natural products. *Nat. Prod. Rep.* **2010**, *27*, 499–528.
81. Dedon, P. C.; Goldberg, I. H. Sequence-specific double-strand breakage of DNA by neocarzinostatin involves different chemical mechanisms within a staggered cleavage site. *J. Biol. Chem.* **1990**, *265*, 14713–14716.
82. Xu, Y.-J.; Xi, Z.; Zhen, Y.-S.; Goldberg, I. H. A single binding mode of activated enediyne C1027 generates two types of double-strand DNA lesions: Deuterium isotope-induced shuttling between adjacent nucleotide target sites. *Biochemistry* **1995**, *34*, 12451–12460.

83. Beerman, T. A.; Gawron, L. S.; Shin, S.; Shen, B.; McHugh, M. M. C-1027, a radiomimetic enediyne anticancer drug, preferentially targets hypoxic cells. *Cancer Res.* **2009**, *69*, 593–598.
84. Kennedy, D. R.; Ju, J.; Shen, B.; Beerman, T. A. Designer enediynes generate DNA breaks, interstrand cross-links, or both, with concomitant changes in the regulation of DNA damage responses. *Proc. Natl. Acad. Sci. USA* **2007**, *104*, 17632–17637.
85. Kennedy, D. R.; Gawron, L. S.; Ju, J.; Liu, W.; Shen, B.; Beerman, T. A. Single chemical modifications of the C-1027 enediyne core, a radiomimetic antitumor drug, affect both drug potency and the role of ataxia-telangiectasia mutated in cellular responses to DNA double-strand breaks. *Cancer Res.* **2007**, *67*, 773–781.
86. Edo, K.; Mizugaki, M.; Koide, Y.; Seto, H.; Furihata, K.; Otake, N.; Ishida, N. The structure of neocarzinostatin chromophore possessing a novel bicyclo[7.3.0]dodecadiyne system. *Tetrahedron Lett.* **1985**, *26*, 331–334.
87. Otani, T.; Yasuhara, T.; Minami, Y.; Shimazu, T.; Zhang, R.; Xie, M. Purification and primary structure of C-1027-AG, a selective antagonist of antitumor antibiotic C-1027, from *Streptomyces globisporus*. *Agric. Biol. Chem.* **1991**, *55*, 407–417.
88. Otani, T.; Yoshida, K.-I.; Sasaki, T.; Minami, Y. C-1027 enediyne chromophore: Presence of another active form and its chemical structure. *J. Antibiot.* **1999**, *52*, 415–421.
89. Ren, F.; Hogan, P. C.; Anderson, A. J.; Myers, A. G. Kedarcidin chromophore: Synthesis of its proposed structure and evidence for a stereochemical revision. *J. Am. Chem. Soc.* **2007**, *129*, 5381–5383.
90. Lohman, J. R.; Huang, S.-X.; Horsman, G. P.; Dilfer, P. E.; Huang, T.; Chen, Y.; Wendt-Pienkowski, E.; Shen, B. Cloning and sequencing of the kedarcidin biosynthetic gene cluster from *Streptoalloteichus* sp. ATCC 53650 revealing new insights into biosynthesis of the enediyne family of antitumor antibiotics. *Mol. Biosyst.* **2013**, *9*, 478–491.
91. Komano, K.; Shimamura, S.; Norizuki, Y.; Zhao, D.; Kabuto, C.; Sato, I.; Hirama, M. Total synthesis and structure revision of the (−)-maduropeptin chromophore. *J. Am. Chem. Soc.* **2009**, *131*, 12072–12073.
92. Kobayashi, S.; Ashizawa, S.; Takahashi, Y.; Sugiura, Y.; Nagaoka, M.; Lear, M. J.; Hirama, M. The first total synthesis of N1999-A2: Absolute stereochemistry and stereochemical implications into DNA cleavage. *J. Am. Chem. Soc.* **2001**, *123*, 11294–11295.
93. Buchanan, G. O.; Williams, P. G.; Feling, R. H.; Kauffman, C. A.; Jensen, P. R.; Fenical, W. Sporolides A and B: Structurally unprecedented halogenated macrolides from the marine actinomycete *Salinispora tropica*. *Org. Lett.* **2005**, *7*, 2731–2734.
94. Oh, D.-C.; Williams, P. G.; Kauffman, C. A.; Jensen, P. R.; Fenical, W. Cyanosporasides A and B, chloro- and cyano-cyclopenta[a]indene glycosides from the marine actinomycete "*Salinispora pacifica*". *Org. Lett.* **2006**, *8*, 1021–1024.
95. Lane, A. L.; Nam, S.-J.; Fukuda, T.; Yamanaka, K.; Kauffman, C. A.; Jensen, P. R.; Fenical, W.; Moore, B. S. Structures and comparative characterization of biosynthetic gene clusters for cyanosporasides, enediyne-derived natural products from marine actinomycetes. *J. Am. Chem. Soc.* **2013**, *135*, 4171–4174.
96. Nam, S.-J.; Gaudencio, S. P.; Kauffman, C. A.; Jensen, P. R.; Kondratyuk, T. P.; Marler, L. E.; Pezzuto, J. M.; Fenical, W. Fijiolides A and B, inhibitors of TNF-α-induced NFκB activation, from a marine-derived sediment bacterium of the genus *Nocardiopsis*. *J. Nat. Prod.* **2010**, *73*, 1080–1086.
97. Lee, M. D.; Dunne, T. S.; Chang, C. C.; Ellestad, G. A.; Siegel, M. M.; Morton, G. O.; McGahren, W. J.; Borders, D. B. Calichemicins, a novel family of antitumor antibiotics. 2. Chemistry and structure of calichemicin γ1I. *J. Am. Chem. Soc.*. **1987**, *109*, 3466–3468.
98. Golik, J.; Dubay, G.; Groenewold, G.; Kawaguchi, H.; Konishi, M.; Krishnan, B.; Ohkuma, H.; Saitoh, K.; Doyle, T. W. Esperamicins, a novel class of potent antitumor antibiotics. 3. Structures of esperamicins A1, A2, and A1b. *J. Am. Chem. Soc.* **1987**, *109*, 3462–3464.

99. Myers, A. G.; Fraley, M. E.; Tom, N. J.; Cohen, S. B.; Madar, D. J. Synthesis of (+)-dynemicin A and analogs of wide structural variability: Establishment of the absolute configuration of natural dynemicin A. *Chem. Biol.* **1995**, *2*, 33–43.
100. Davies, J.; Wang, H.; Taylor, T.; Warabi, K.; Huang, X.-H.; Andersen, R. J. Uncialamycin, a new enediyne antibiotic. *Org. Lett.* **2005**, *7*, 5233–5236.
101. Yan, X.; Ge, H.-M.; Huang, T.; Hindra; Yang, D.; Teng, Q.; Crnovcic, I. et al. **2016**, *MBio*, *7*, e2104–e2116.
102. McDonald, L. A.; Capson, T. L.; Krishnamurthy, G.; Ding, W.-D.; Ellestad, G. A.; Bernan, V. S.; Maiese, W. M. et al. Namenamicin, a new enediyne antitumor antibiotic from the marine ascidian *Polysyncraton lithostrotum*. *J. Am. Chem. Soc.* **1996**, *118*, 10898–10899.
103. Oku, N.; Matsunaga, S.; Fusetani, N. Shishijimicins A-C, novel enediyne antitumor antibiotics from the ascidian *Didemnum proliferum*. *J. Am. Chem. Soc.* **2003**, *125*, 2044–2045.
104. Gao, Q.; Thorson, J. S. The biosynthetic genes encoding for the production of the dynemicin enediyne core in *Micromonospora chersina* ATCC53710. *FEMS Microbiol. Lett.* **2008**, *282*, 105–114.
105. Liu, W.; Ahlert, J.; Gao, Q.; Wendt-Pienkowski, E.; Shen, B.; Thorson, J. S. Rapid PCR amplification of minimal enediyne polyketide synthase cassettes leads to a predictive familial classification model. *Proc. Natl. Acad. Sci. USA* **2003**, *100*, 11959–11963.
106. Zazopoulos, E.; Huang, K.; Staffa, A.; Liu, W.; Bachmann, B. O.; Nonaka, K.; Ahlert, J.; Thorson, J. S.; Shen, B.; Farnet, C. M. A genomics-guided approach for discovering and expressing cryptic metabolic pathways. *Nat. Biotechnol.* **2003**, *21*, 187–190.
107. Horsman, G. P.; Chen, Y.; Thorson, J. S.; Shen, B. Polyketide synthase chemistry does not direct biosynthetic divergence between 9- and 10-membered enediynes. *Proc. Natl. Acad. Sci. USA* **2010**, *107*, 11331–11335.
108. Zhang, J.; Van Lanen, S. G.; Ju, J.; Liu, W.; Dorrestein, P. C.; Li, W.; Kelleher, N. L.; Shen, B. A phosphopantetheinylating polyketide synthase producing a linear polyene to initiate enediyne antitumor antibiotic biosynthesis. *Proc. Natl. Acad. Sci. USA* **2008**, *105*, 1460–1465.
109. Huang, S.-X.; Feng, Z.; Wang, L.; Galm, U.; Wendt-Pienkowski, E.; Yang, D.; Tao, M.; Coughlin, J. M.; Duan, Y.; Shen, B. A designer bleomycin with significantly improved DNA cleavage activity. *J. Am. Chem. Soc.* **2012**, *134*, 13501–13509.
110. Ju, J.; Rajski, S. R.; Lim, S.-K.; Seo, J.-W.; Peters, N. R.; Hoffmann, F. M.; Shen, B. Evaluation of new migrastatin and dorrigocin congeners unveils cell migration inhibitors with dramatically improved potency. *Bioorg. Med. Chem. Lett.* **2008**, *18*, 5951–5954.
111. Huang, S.-X.; Yun, B.-S.; Ma, M.; Basu, H. S.; Church, D. R.; Ingenhorst, G.; Huang, Y. et al. Leinamycin E1 acting as an anticancer prodrug activated by reactive oxygen species. *Proc. Natl. Acad. Sci. USA* **2015**, *112*, 8278–8283.
112. Brukner, I. C-1027 Taiho Pharmaceutical Co Ltd. *Curr. Opin. Oncol., Endocr. Metab. Invest. Drugs* **2000**, *2*, 344–352.
113. Povirk, L. F. DNA damage and mutagenesis by radiomimetic DNA-cleaving agents: Bleomycin, neocarzinostatin and other enediynes. *Mutat. Res., Fundam. Mol. Mech. Mutagen.* **1996**, *355*, 71–89.
114. Minami, Y.; Yoshida, K.; Azuma, R.; Saeki, M.; Otani, T. Structure of an aromatization product of C-1027 chromophore. *Tetrahedron Lett.* **1993**, *34*, 2633–2636.
115. Yoshida, K.-I.; Minami, Y.; Azuma, R.; Saeki, M.; Otani, T. Structure and cycloaromatization of a novel enediyne, C-1027 chromophore. *Tetrahedron Lett.* **1993**, *34*, 2637–2640.
116. Iida, K.-I.; Fukuda, S.; Tanaka, T.; Hirama, M. Absolute configuration of C-1027 chromophore. *Tetrahedron Lett.* **1996**, *37*, 4997–5000.
117. Liu, W.; Christenson, S. D.; Standage, S.; Shen, B. Biosynthesis of the enediyne antitumor antibiotic C-1027. *Science* **2002**, *297*, 1170–1173.

118. Van Lanen, S. G.; Lin, S.; Shen, B. Biosynthesis of the enediyne antitumor antibiotic C-1027 involves a new branching point in chorismate metabolism. *Proc. Natl. Acad. Sci. USA* **2008**, *105*, 494–499.
119. Yu, L.; Mah, S.; Otani, T.; Dedon, P. The benzoxazolinate of C-1027 confers intercalative DNA binding. *J. Am. Chem. Soc.* **1995**, *117*, 8877–8878.
120. Tanaka, T.; Fukuda-Ishisaka, S.; Hirama, M.; Otani, T. Solution structures of C-1027 apoprotein and its complex with the aromatized chromophore. *J. Mol. Biol.* **2001**, *309*, 267–283.
121. Lin, S.; Van Lanen, S. G.; Shen, B. Regiospecific chlorination of (*S*)-β-tyrosyl-*S*-carrier protein catalyzed by SgcC3 in the biosynthesis of the enediyne antitumor antibiotic C-1027. *J. Am. Chem. Soc.* **2007**, *129*, 12432–12438.
122. Lin, S.; Van Lanen, S. G.; Shen, B. Characterization of the two-component, FAD-dependent monooxygenase SgcC that requires carrier protein-tethered substrates for the biosynthesis of the enediyne antitumor antibiotic C-1027. *J. Am. Chem. Soc.* **2008**, *130*, 6616–6623.
123. Umezawa, H.; Maeda, K.; Takeuchi, T.; Okami, Y. New antibiotics, bleomycin A and B. *J. Antibiot. Ser. A* **1966**, *19*, 200–209.
124. Hecht, S. M. Bleomycin group antitumor agents. In *Anticancer Agents from Natural Products*, 2nd edn., Cragg, G. M.; Kingston, D. G. I.; Newman, D. J., eds. CRC Press: Boca Raton, FL, **2012**, pp. 451–478.
125. Chen, J.; Stubbe, J. Bleomycins: Towards better therapeutics. *Nat. Rev. Cancer* **2005**, *5*, 102–112.
126. Galm, U.; Wendt-Pienkowski, E.; Wang, L.; Huang, S.-X.; Unsin, C.; Tao, M.; Coughlin, J. M.; Shen, B. Comparative analysis of the biosynthetic gene clusters and pathways for three structurally related antitumor antibiotics: Bleomycin, tallysomycin, and zorbamycin. *J. Nat. Prod.* **2011**, *74*, 526–536.
127. Boger, D. L.; Cai, H. Review of bleomycin: Synthetic and mechanistic studies. *Angew. Chem. Int. Ed.* **1999**, *38*, 448–476.
128. Schroeder, B. R.; Ghare, M. I.; Bhattacharya, C.; Paul, R.; Yu, Z.; Zaleski, P. A.; Bozeman, T. C.; Rishel, M. J.; Hecht, S. M. The disaccharide moiety of bleomycin facilitates uptake by cancer cells. *J. Am. Chem. Soc.* **2014**, *136*, 13641–13656.
129. Cloos, J.; Temmink, O.; Ceelen, M.; Snel, M. H. J.; Leemans, C. R.; Braakhuis, B. J. M. Involvement of cell cycle control in bleomycin-induced mutagen sensitivity. *Environ. Mol. Mutagen.* **2002**, *40*, 79–84.
130. Reinert, T.; Baldotto, C. S. d. R.; Nunes, F. A. P.; Scheliga, A. A. d. S. Bleomycin-induced lung injury. *J. Cancer Res.* **2013**, *2013*, 1–10.
131. Burgy, O.; Wettstein, G.; Bellaye Pierre, S.; Goirand, F.; Decologne, N.; Racoeur, C.; Bettaieb, A. et al. Deglycosylated bleomycin has the antitumor activity of bleomycin without pulmonary toxicity. *Sci. Transl. Med.* **2016**, *8*, 326ra20.
132. Du, L.; Sanchez, C.; Chen, M.; Edwards, D. J.; Shen, B. The biosynthetic gene cluster for the antitumor drug bleomycin from *Streptomyces verticillus* ATCC15003 supporting functional interactions between nonribosomal peptide synthetases and a polyketide synthase. *Chem. Biol.* **2000**, *7*, 623–642.
133. Galm, U.; Wendt-Pienkowski, E.; Wang, L.; George Nicholas, P.; Oh, T.-J.; Yi, F.; Tao, M.; Coughlin Jane, M.; Shen, B. The biosynthetic gene cluster of zorbamycin, a member of the bleomycin family of antitumor antibiotics, from *Streptomyces flavoviridis* ATCC 21892. *Mol. Biosyst.* **2009**, *5*, 77–90.
134. Borchert, S.; Patil, S. S.; Marahiel, M. A. Identification of putative multifunctional peptide synthetase genes using highly conserved oligonucleotide sequences derived from known synthetases. *FEMS Microbiol. Lett.* **1992**, *92*, 175–180.
135. Xie, P.; Ma, M.; Rateb, M. E.; Shaaban, K. A.; Yu, Z.; Huang, S.-X.; Zhao, L.-X. et al. Biosynthetic potential-based strain prioritization for natural product discovery: A showcase for diterpenoid-producing actinomycetes. *J. Nat. Prod.* **2014**, *77*, 377–387.

136. Waldman, A. J.; Pechersky, Y.; Wang, P.; Wang, J. X.; Balskus, E. P. The cremeomycin biosynthetic gene cluster encodes a pathway for diazo formation. *ChemBioChem* **2015**, *16*, 2172–2175.
137. Hindra; Huang, T.; Yang, D.; Rudolf, J. D.; Xie, P.; Xie, G.; Teng, Q. et al. Strain prioritization for natural product discovery by a high-throughput real-time PCR method. *J. Nat. Prod.* **2014**, *77*, 2296–2303.
138. Ju, K.-S.; Gao, J.; Doroghazi, J. R.; Wang, K.-K. A.; Thibodeaux, C. J.; Li, S.; Metzger, E. et al. Discovery of phosphonic acid natural products by mining the genomes of 10,000 actinomycetes. *Proc. Natl. Acad. Sci. USA* **2015**, *112*, 12175–12180.
139. Rudolf, J. D.; Yan, X.; Shen, B.; Shen, B.; Shen, B. Genome neighborhood network reveals insights into enediyne biosynthesis and facilitates prediction and prioritization for discovery. *J. Ind. Microbiol. Biotechnol.* **2016**, *43*, 261–276.
140. Rudolf, J. D.; Dong, L.-B.; Huang, T.; Shen, B. A genetically amenable platensimycin- and platencin-overproducer as a platform for biosynthetic explorations: A showcase of PtmO4, a long-chain acyl-CoA dehydrogenase. *Mol. Biosyst.* **2015**, *11*, 2717–2726.
141. Omura, S.; Ikeda, H.; Ishikawa, J.; Hanamoto, A.; Takahashi, C.; Shinose, M.; Takahashi, Y. et al. Genome sequence of an industrial microorganism *Streptomyces avermitilis*: Deducing the ability of producing secondary metabolites. *Proc. Natl. Acad. Sci. USA* **2001**, *98*, 12215–12220.
142. Bentley, S. D.; Chater, K. F.; Cerdeno-Tarraga, A. M.; Challis, G. L.; Thomson, N. R.; James, K. D.; Harris, D. E. et al. Complete genome sequence of the model actinomycete *Streptomyces coelicolor* A3(2). *Nature* **2002**, *417*, 141–147.
143. Ikeda, H.; Ishikawa, J.; Hanamoto, A.; Shinose, M.; Kikuchi, H.; Shiba, T.; Sakaki, Y.; Hattori, M.; Omura, S. Complete genome sequence and comparative analysis of the industrial microorganism *Streptomyces avermitilis*. *Nat. Biotechnol.* **2003**, *21*, 526–531.
144. Weber, T.; Welzel, K.; Pelzer, S.; Vente, A.; Wohlleben, W. Exploiting the genetic potential of polyketide-producing streptomycetes. *J. Biotechnol.* **2003**, *106*, 221–232.
145. Ikeda, H.; Shin-ya, K.; Omura, S. Genome mining of the *Streptomyces avermitilis* genome and development of genome-minimized hosts for heterologous expression of biosynthetic gene clusters. *J. Ind. Microbiol. Biotechnol.* **2014**, *41*, 233–250.
146. Altschul, S. F.; Gish, W.; Miller, W.; Myers, E. W.; Lipman, D. J. Basic local alignment search tool. *J. Mol. Biol.* **1990**, *215*, 403–410.
147. Finn, R. D.; Clements, J.; Eddy Sean, R. HMMER web server: Interactive sequence similarity searching. *Nucleic Acids Res.* **2011**, *39*, W29–W37.
148. Ishikawa, J.; Hotta, K. Frameplot: A new implementation of the frame analysis for predicting protein-coding regions in bacterial DNA with a high G+C content. *FEMS Microbiol. Lett.* **1999**, *174*, 251–253.
149. Medema, M. H.; Blin, K.; Cimermancic, P.; de Jager, V.; Zakrzewski, P.; Fischbach, M. A.; Weber, T.; Takano, E.; Breitling, R. Antismash: Rapid identification, annotation and analysis of secondary metabolite biosynthesis gene clusters in bacterial and fungal genome sequences. *Nucleic Acids Res.* **2011**, *39*, W339–W346.
150. Blin, K.; Medema Marnix, H.; Kazempour, D.; Fischbach Michael, A.; Breitling, R.; Takano, E.; Weber, T. antiSMASH 2.0—A versatile platform for genome mining of secondary metabolite producers. *Nucleic Acids Res.* **2013**, *41*, W204–W212.
151. Medema, M. H.; Takano, E.; Breitling, R. Detecting sequence homology at the gene cluster level with multigeneblast. *Mol. Biol. Evol.* **2013**, *30*, 1218–1223.
152. Ziemert, N.; Podell, S.; Penn, K.; Badger, J. H.; Allen, E.; Jensen, P. R. The natural product domain seeker NaPDoS: A phylogeny based bioinformatic tool to classify secondary metabolite gene diversity. *PLoS One* **2012**, *7*, e34064.
153. Cimermancic, P.; Medema, M. H.; Claesen, J.; Kurita, K.; Wieland Brown, L. C.; Mavrommatis, K.; Pati, A. et al. Insights into secondary metabolism from a global analysis of prokaryotic biosynthetic gene clusters. *Cell* **2014**, *158*, 412–421.

154. Li, M. H. T.; Ung, P. M. U.; Zajkowski, J.; Garneau-Tsodikova, S.; Sherman David, H. Automated genome mining for natural products. *BMC Bioinformatics* **2009**, *10*, 185.
155. Bachmann, B. O.; Ravel, J. Methods for in silico prediction of microbial polyketide and nonribosomal peptide biosynthetic pathways from DNA sequence data. *Methods Enzymol.* **2009**, *458*, 181–217.
156. Roettig, M.; Medema, M. H.; Blin, K.; Weber, T.; Rausch, C.; Kohlbacher, O. NRPSpredictor2—A web server for predicting NRPS adenylation domain specificity. *Nucleic Acids Res.* **2011**, *39*, W362–W367.
157. Darling, A. C. E.; Mau, B.; Blattner, F. R.; Perna, N. T. Mauve: Multiple alignment of conserved genomic sequence with rearrangements. *Genome Res.* **2004**, *14*, 1394–1403.
158. Conway, K. R.; Boddy, C. N. Clustermine360: A database of microbial PKS/NRPS biosynthesis. *Nucleic Acids Res.* **2013**, *41*, D402–D407.
159. Medema, M. H.; Kottmann, R.; Yilmaz, P.; Cummings, M.; Biggins, J. B.; Blin, K.; de Bruijn, I. et al. Minimum information about a biosynthetic gene cluster. *Nat. Chem. Biol.* **2015**, *11*, 625–631.
160. Brown, S. D.; Babbitt, P. C. New insights about enzyme evolution from large scale studies of sequence and structure relationships. *J. Biol. Chem.* **2014**, *289*, 30221–30228.
161. Leuthaeuser, J. B.; Knutson, S. T.; Kumar, K.; Babbitt, P. C.; Fetrow, J. S. Comparison of topological clustering within protein networks using edge metrics that evaluate full sequence, full structure, and active site microenvironment similarity. *Protein Sci.* **2015**, *24*, 1423–1439.
162. Atkinson Holly, J.; Morris John, H.; Ferrin Thomas, E.; Babbitt Patricia, C. Using sequence similarity networks for visualization of relationships across diverse protein superfamilies. *PLoS One* **2009**, *4*, e4345.
163. Wallrapp, F. H.; Pan, J.-J.; Ramamoorthy, G.; Almonacid, D. E.; Hillerich, B. S.; Seidel, R.; Patskovsky, Y. et al. Prediction of function for the polyprenyl transferase subgroup in the isoprenoid synthase superfamily. *Proc. Natl. Acad. Sci. USA* **2013**, *110*, E1196–E1202.
164. Wichelecki, D. J.; Balthazor, B. M.; Chau, A. C.; Vetting, M. W.; Fedorov, A. A.; Fedorov, E. V.; Lukk, T. et al. Discovery of function in the enolase superfamily: D-Mannonate and D-gluconate dehydratases in the D-mannonate dehydratase subgroup. *Biochemistry* **2014**, *53*, 2722–2731.
165. Lohman, J. R.; Ma, M.; Osipiuk, J.; Nocek, B.; Kim, Y.; Chang, C.; Cuff, M. et al. Structural and evolutionary relationships of "AT-less" type I polyketide synthase ketosynthases. *Proc. Natl. Acad. Sci. USA* **2015**, *112*, 12693–12698.
166. Zhao, S.; Jacobson, M. P.; Sakai, A.; Zhang, X.; Kumar, R.; San Francisco, B.; Solbiati, J. et al. Prediction and characterization of enzymatic activities guided by sequence similarity and genome neighborhood networks. *eLife* **2014**, *3*, e03275.
167. Donia, M. S.; Cimermancic, P.; Schulze, C. J.; Wieland Brown, L. C.; Martin, J.; Mitreva, M.; Clardy, J.; Linington, R. G.; Fischbach, M. A. A systematic analysis of biosynthetic gene clusters in the human microbiome reveals a common family of antibiotics. *Cell* **2014**, *158*, 1402–1414.
168. Lin, K.; Zhu, L.; Zhang, D.-Y. An initial strategy for comparing proteins at the domain architecture level. *Bioinformatics* **2006**, *22*, 2081–2086.
169. Shen, B. A new golden age of natural products drug discovery. *Cell* **2015**, *163*, 1297–1300.
170. Keasling, J. D.; Mendoza, A.; Baran, P. S. Synthesis a constructive debate. *Nature* **2012**, *492*, 188–189.

4 Generation of New-to-Nature Natural Products through Synthesis and Biosynthesis

Blending Synthetic Biology with Synthetic Chemistry

Christopher S. Bailey, Emily R. Abraham, and Rebecca J.M. Goss

CONTENTS

4.1 INTRODUCTION

Historically, natural products have proven to be a matchless starting point for drug discovery. More than 70% of antimicrobials and more than 60% of chemotherapeutics entering clinical trials over the past three decades have been based on natural products. As they are biosynthesized by biomolecules, natural products are inherently predisposed to interact with biomolecules, and their interactions with a myriad of biological targets should not be surprising. Natural products may be considered privileged compounds, and now it would seem that industrial interest in these compounds may be returning. Though natural products are an excellent starting point for drug discovery, their physicochemical properties are often far from ideal, and in order to tune these there is significant need for analogue development. Oftentimes, in order to fully understand the natural product's molecular mode of action there is a need for analogue generation. Tethered and tagged analogues may be used for target identification and structure-activity relationship (SAR) studies with modified variants of the natural product can be used to pinpoint functionality within a molecule that is essential to its activity. The misperception that natural products cannot be subjected to such a style of medchem analysis arises from the perceived intractability of a series of analogues; however, advances in biosynthetic manipulation, enzymology, selective chemical modification, and the ability to strategically blend synthetic chemistry with synthetic biology enable expeditious access to previously intractable series of analogues.

The traditional method of analogue generation was through total synthesis, or semisynthesis, where the parent molecule is extracted, purified, and subjected to selective chemical modification. The need to employ total synthesis to access analogues of compounds from unculturable organisms or unclonable pathways remains; however, in recent years, even faster access to libraries of synthetically intractable analogues has been permitted by combining (or hyphenating) synthetic and biosynthetic approaches. With genomic data becoming increasingly available and with the development of new and fast methods for deciphering and manipulating encoded genes, it is likely that in the next few years we will see significant increase in the combined and complementary use of chemical and biological tools in order to enable expeditious generation of libraries of natural product analogues.

This chapter will take the reader through recent applications of synthesis and the variations of synthetic biology and of synthetic chemistry blended with synthetic biology.

A wide variety of means of flexibly blending synthesis and biosynthesis to enable expeditious access to natural product analogues are available; in order to better reflect on these hyphenated approaches, we utilize the simple classification as first defined by Kirschning.[1] We use the two abbreviations *Chem* and *Bio*; the former refers to a chemical synthesis or partial chemical synthesis while the latter describes a biosynthesis or biological step. We reflect on the application of total synthesis (abbreviated as *Chem*) in generating natural products and their analogues, semisynthesis (abbreviated as *Bio-Chem*) to reflect the natural biosynthesis of the compound followed by its selective synthetic modification, precursor-directed biosynthesis (PDB) (*Chem-Bio*), mutasynthesis (*Bio-Chem-Bio*), mutasynthesis enabling postbiosynthetic functionalization (*Bio-Chem-Bio-Chem*), combinatorial biosynthesis (*Bio-Bio*), and the new approach of GenoChemetics: gene expression enabling synthetic diversification (*Bio-Bio-Chem*).

We describe and reflect upon the uses of each blend in turn as tools to access medicinally relevant compounds. Our focus rests predominantly, though not exclusively, on the generation of antibiotics.

It must be noted that strategies to access libraries of natural product analogues using synthesis, engineered biosynthesis, or combining biosynthesis and chemical synthesis are labor intensive; therefore, before embarking on such an endeavor, one needs to consider three key issues:

1. The natural product chosen should merit the effort. It should either have excellent biological activity in a pharmaceutically relevant field or, alternatively, perhaps serve as a tool for chemical biology.
2. If choosing to blend synthesis and biosynthesis, the target natural product should be so structurally complex that a combined chemical and biological approach is clearly superior to chemical total synthesis for the rapid preparation of new derivatives.
3. If applying a biosynthetic step, the biological tool, for example, a fermentation or an enzyme, must yield sufficient amounts of the desired metabolite, be it either a final product or an intermediate for further functionalization, to be useful for downstream biological study.

4.1.1 Bioactive Natural Products and Their Roles in Nature and Medicine

For millennia, natural products have been utilized by humans for medicinal purposes, with some of the earliest examples documented being that of the ancient Egyptians who would use honey to aid wound healing and opium-soaked cloths during surgical procedures.[2,3] Today, we are far more aware of the chemistry and biology that surround their use. For example, the antimicrobial activity of honey has been shown to be the result of peroxide activity[4] and, to this day, medicinal-grade honey is utilized as a surgical dressing. Another such example is the natural product morphine **1**, which has been isolated and characterized from the opium poppy *Papaver somniferum* **1** and codeine are extensively used in modern medicine to control pain. During the first half of the twentieth century, natural products from microbes were brought to the forefront of modern medicine. Sir Alexander Fleming demonstrated that the mold *Penicillum notatum* was able to produce compounds with antibiotic properties.[5] In 1945, the structure of the key bioactive compound was determined by Dorothy Hodgkin, and this structure revealed a central beta-lactam motif, the warhead of penicillin **2**.[6]

Such early discoveries inspired the search for further medicinally useful natural products from both plants and microbes and resulted in the discovery of numerous compounds with a diverse array of chemical architectures and biological activities. Examples include antimicrobial compounds such as erythromycin **3**, daptomycin **4**, and vancomycin **5**, the cholesterol-lowering lovastatin **6**, the antiparisitic avermectin **7**, and from plants, the anticancer drug Taxol **8** and the antimalarial artemisinin **9** (Figure 4.1). The importance of natural products to medicine and society, both historically and today, was acknowledged through the 2015 Nobel Prize in Physiology or Medicine being awarded for the discovery of **6** and **8**.

4.1.2 Antibiotic Crisis and the Need for New Medicines

Antibiotics are chemical agents that are capable of inhibiting or perturbing bacterial growth. In order to exert a toxic effect, the chemical agent must prevent or interfere with cellular processes that are essential to the survival and proliferation of the bacterium. For this to be of medicinal use there must be a high level of selectivity, exploiting differences between the biochemistry of the bacterium and the host such that the bacterium is targeted but the animal host is not. There are six main targets: cell wall biosynthesis, protein synthesis, cell membrane integrity, RNA synthesis, folate synthesis, and DNA synthesis and integrity; these are summarized in Figure 4.2.

Antibiotic resistance is by no means a new phenomenon; indeed, Fleming, in his 1945 lecture, warned of the danger of resistance resulting from the inappropriate usage of antibiotics, by exposing bacteria to sufficient compound to train them to be resistant.[8] Antibiotic resistance has now been observed for all clinically used classes of antibiotics,[9] and multidrug-resistant (MDR) strains exist including strains that are resistant to all clinically available drugs: extreme drug resistance (EDR). The problem is exacerbated by the lack of introduction, into the clinic, of new antibiotics with novel structures and new modes of action. This means that few alternatives exist for the treatment of resistant strains. The rise in antibiotic resistance has many causes; however, poor antibiotic stewardship has no doubt exacerbated the problem.[10]

Bacteria acquire resistance to antibiotics through a number of different pathways; these include proactive responses such as mutations in preexisting genes to render antibiotic targets resistant (e.g., resistance to rifampicin through mutations in the RNA polymerase)[11–14]; horizontal gene transfer (HGT), where resistance genes are transferred between environmental species through transfection,

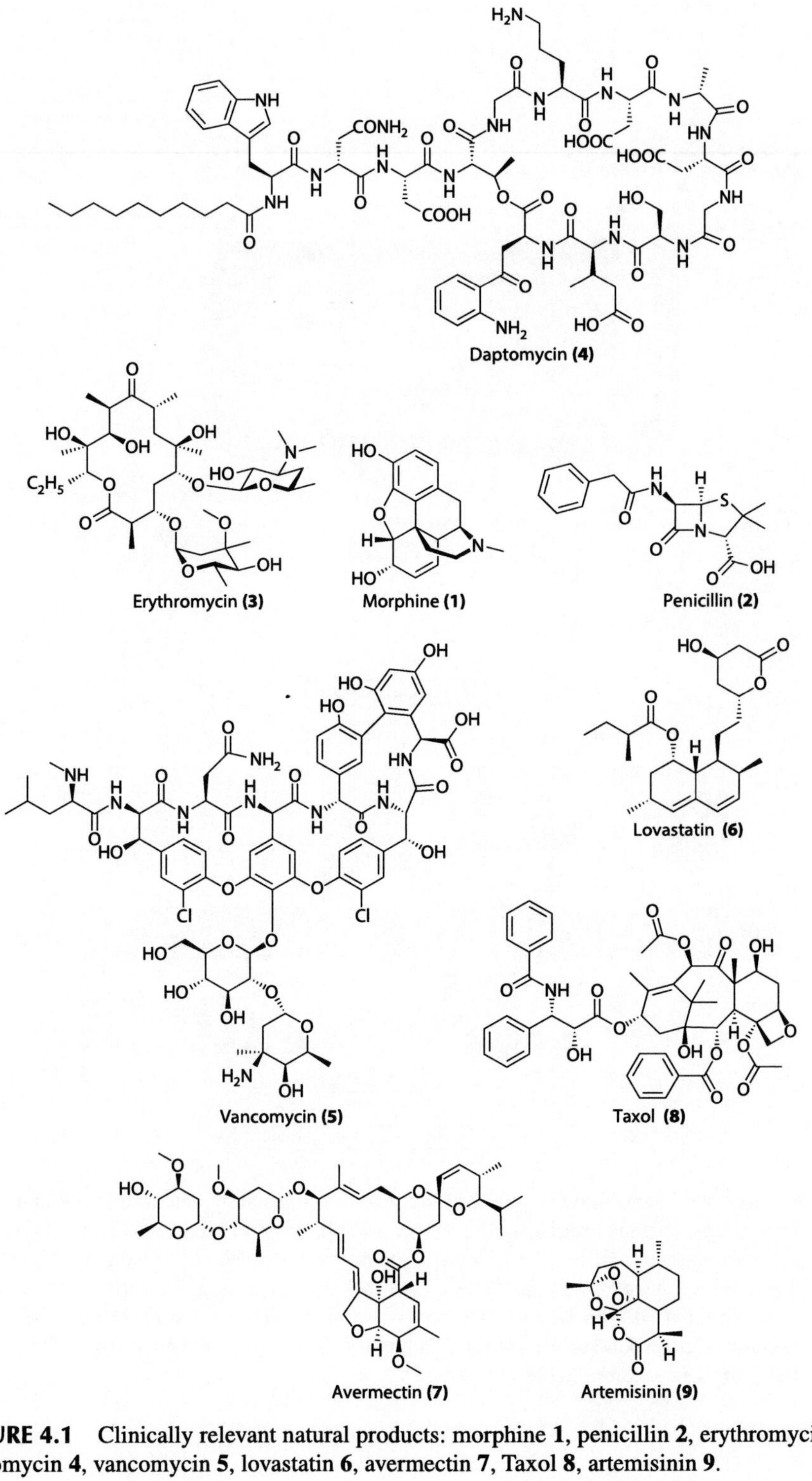

FIGURE 4.1 Clinically relevant natural products: morphine **1**, penicillin **2**, erythromycin **3**, daptomycin **4**, vancomycin **5**, lovastatin **6**, avermectin **7**, Taxol **8**, artemisinin **9**.

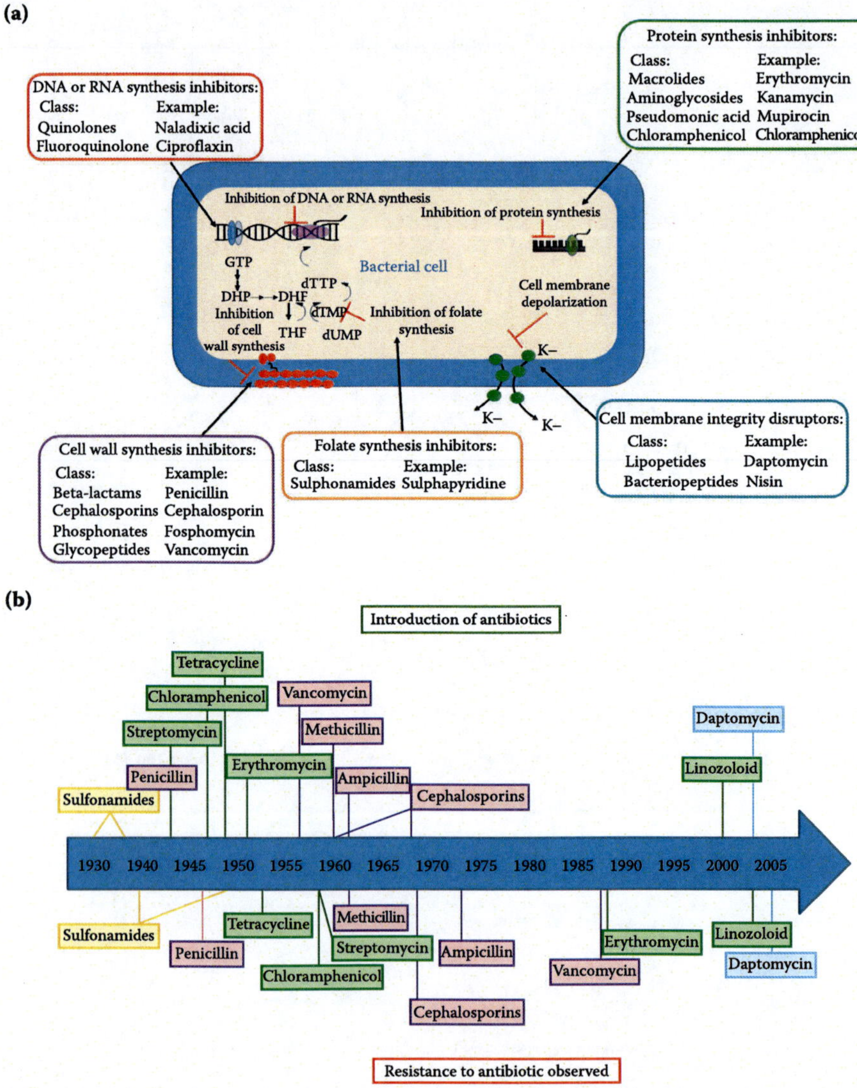

FIGURE 4.2 Main antibiotic targets with clinical examples alongside a timeline highlighting when resistance to the clinical example was first observed. **(a)** The cartoon illustrates the main antibiotic targets and highlights classes of drugs that affect each bacterial target and clinical examples that belong to each class. **(b)** Timeline illustrating the introduction of antibiotics, (top) compared with when resistance to the antibiotic was first observed. The colors of the antibiotic correspond to the antibiotic's target shown in **(a)**. (Adapted from Clatworthy, A.E. et al., *Nat. Chem. Biol.*, 3(9), 541, 2007.[7])

conjugation, or transformation; and physiological and behavioral changes such as biofilm formation in which the bacterial cells are protected by an extracellular polysaccharide matrix and metabolic dormancy enables persistor cells to escape inhibition.[15]

In the vast majority of cases, the antibiotics are natural products or derivatives of natural products and the organisms generating them have an endemic resistance system built in so that their own survival is not compromised through the production of the antibiotic compound. As such, these resistance genes are already present in nature and can be transmitted between bacterial species. These genes include detoxifying enzymes, such as the beta-lactamases, which break down beta-lactam antibiotics through the hydrolysis of the beta-lactam ring, and genes responsible for the rapid export or efflux of the compound.

Due to this widespread resistance, there is an urgent need for new antibiotics with novel structures that are not recognized by or susceptible to existing resistance mechanisms and preferably that inhibit clinically unexploited targets.

4.1.3 Natural Products and the Clinic: Toward New Antibiotics in the Battle against Resistance

In the past three decades, more than 70% of antibiotics entering clinical trials have been based on compounds generated by organisms, in particular microbes.[16] Strikingly, investment in the exploration of streptomyces, perhaps the most chemically characterized microbial genus to date, has resulted in the discovery of over two-thirds of all known antibiotics.

While the direct use of a natural product only currently accounts for 6% of all clinically used medicines today, a further 27% and 17% result from derivatization of a natural product or the inclusion of a pharmacophore from a natural product respectively. Therefore, a staggering 50% of medicines are based on natural products (Figure 4.3)[17] with further synthetic compounds designed to inhibit targets that have been revealed through the use of natural products.[17] Only a small proportion (0.5%) of antibiotic-producing microbes are currently cultivable in the lab, although innovative culturing methods have enabled success in antibiotic discovery with the recent identification of teixobactin from a previously uncultured microbe.[4] The marine environment represents a rich source of biologically active compounds, with marine organisms accounting for approximately half of the earth's biodiversity. Though the marine microbiome has been little explored for compound discovery, evidence indicates searches will be fruitful. In July 2014, a total of seven marine-derived compounds had already received EU/FDA approval, with a further 25 in phase I–III clinical trials. Studies thus far have focused predominantly on collection and examination of marine invertebrates (such as sponges and tunicates) rather than microbes, leaving this a largely unexplored resource. Marine microbes therefore represent an important discovery opportunity; in many of these systems, the mutualistic microbes are suspected and, in some cases, shown to be the originators of the isolated bioactive compounds.

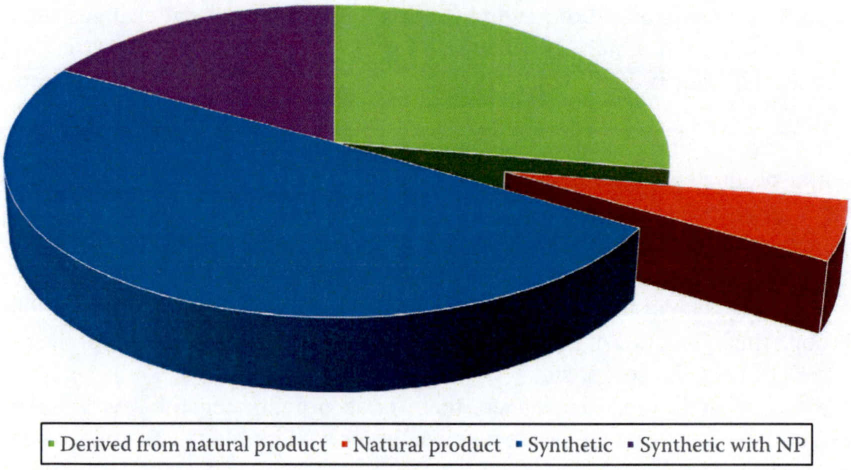

FIGURE 4.3 Percentage and origin of all clinically used drugs. (Adapted from Newman, D.J. and Cragg, G.M., *J. Nat. Prod.*, 75(3), 311, 2012.)

As described above, natural products have evolved for utility within their producing organisms; therefore, it is essential to access analogues of natural products, not only to determine the target of interaction and the way in which the natural product interrelates with it at the molecular level, but also to generate materials with improved activities and oral bioavailability.

The remainder of this chapter looks particularly at approaches that may be utilized to enable access to a series of analogues of natural products, with particular focus on natural products with antibiotic activities.

4.2 APPROACHES TO GENERATING NATURAL PRODUCT ANALOGUES: BLENDING SYNTHETIC BIOLOGY WITH SYNTHETIC CHEMISTRY

Natural products are made by biomolecules and therefore naturally predisposed to interact with biomolecules; thus, as discussed above, they make an excellent starting point for drug discovery. The ability to access a series of natural product analogues is crucial in order to gain an understanding of the structure activity relationship of a natural product, investigate its precise interactions with its target at the molecular level, and develop compounds with improved bioactivity as well as other desirable qualities such as good solubility, low toxicity, appropriate stability, and oral bioavailability. Due to the structural and chemical complexity of natural products they have been perceived to be *unmedchemable* and challenging to make analogues of; this viewpoint comes from focusing on the more traditional approaches to analogue generation, including semisynthesis and total synthesis; however, opportunities are being developed that utilize synthetic biology and synthetic chemistry combining the best of both worlds, in order to more readily generate designer analogues of natural products.

Today, approaches to natural product analogue generation include semisynthesis (in which a natural product is directly chemically modified), total synthesis (chemically generating the compound *de novo* from simple readily abundant starting materials), and a variety of ways in which biosynthesis is harnessed.

4.2.1 Synthetic Approaches (Chem)

Total synthesis, the art of building complex natural products from simple precursor building blocks (Scheme 4.1), is the oldest and most established method of generating natural product analogues. It offers an unparalleled flexibility with respect to modification of natural product scaffolds by the substitution, though often not chemically trivial, of any one or a combination of synthetic building blocks for analogues.

While total synthesis is highly flexible, the complexity of many natural products often renders their total synthesis a substantial challenge to chemists. This frequently results in lengthy multistep synthesis with multiple purification, protection and deprotection steps needed throughout. Further complexity is introduced if an asymmetric route to the product is needed; if this is not possible, the FDA requires separation and characterization of each compound in a racemic mixture to be studied separately.[18] Such a challenge inspires creativity, and chemists devise ingenious routes to access complex scaffolds.

SCHEME 4.1 Assembly of precursors enables the construction of analogues with diversity at almost any point on the given structure.

One such example is the total synthesis of the potent antibiotic and anticancer agent—marinomycin A **10**. To date there have been three successful syntheses. The first was developed by K. C. Nicolaou in 2003 using a Suzuki cyclization strategy, which enabled the formation of the core cyclic structure. The use of an intramolecular cyclization allowed them to avoid a potentially tricky dimerization.[19] A second insightful approach recognized the importance of reactivity switching in the salicylate moiety and its key role in the control of a stepwise dimerization and cyclization of the molecule. This switching could be cleverly potentiated by hydrogen bonding modification, and this approach enabled access to the molecule via a triply convergent synthesis, with a longest 18-step linear sequence, in 3.5% overall yield by Evans et al. (Scheme 4.2).[20] The third synthesis, reported 2 years later in 2014 with a 24-step longest linear sequence, gave an overall yield of 4% and utilized a direct dimerization methodology, achieving dimerization and cyclization in one step. This was achieved through the activation of the hydroxy ester using a dioxane protecting group, building on observations made by Evans et al. in their total synthesis.[21] These three syntheses provide asymmetric access to this challenging polyene natural product and build upon nearly 200 years of developments in organic chemistry,[22] combining modern Stille and Suzuki cross-couplings with protecting group chemistry and use of Grignard reagents.

Total synthesis can be incredibly challenging. If the final, or one of the final steps, of a long synthetic route poor yielding or not possible, it can lead to a significant amount of lost time and resource. The real breakthrough in the synthesis of marinomycin A **10** is appreciating and controlling the reactivity of the salicylate in the dimerization step to afford the full, cyclic molecule.

The total synthesis of the antibiotic moenomycin A **11** showcases the flexible nature of chemical synthesis for access to a series of natural product analogues. As moenomycin A **11** is the only known active site inhibitor of peptidoglycan glycosyltransferases,[23] it is an attractive starting point for developing new antibiotics as there is less chance of resistant strains existing. A series of stereoselective sulfoxide glycosylation reactions gave the pentaglycoside moiety, which was then coupled to the moenocinyl phosphoglycerate unit to produce moenomycin A **11**. Due to the modular nature of the synthesis, stepwise glycosylation reactions employed by coupling to the phosphoglycerate unit may be substituted for different sugars or phosphoglycerates, enabling diverse series of analogues to be readily accessed. This approach was taken by Adachi et al. to create a derivative of moenomycin A **11** with different phosphoglycerates (see Reference 23 for relevant references).

It had been observed by Welzel et al. that only the disaccharide of moenomycin A **11** was necessary for activity rather than the whole pentasaccharide, simplifying the synthesis of analogues to a series of analogous disaccharides.[24] This enabled a solid-phase synthesis to be developed by Silva et al. Having previously developed a route for effective stereo-controlled solid-phase disaccharide synthesis,[25] four separate disaccharides were synthesized that could be selectively functionalized at three separate positions. This enabled the efficient synthesis of 1,300 disaccharide analogues of moenomycin A **11**. It was found that 6 of the compounds synthesized IC_{50}s for inhibiting cell wall biosynthesis of <15 μg/mL (Figure 4.4).[26] The vast nature of this library also allowed for SARs to be built up to direct future analogue synthesis.

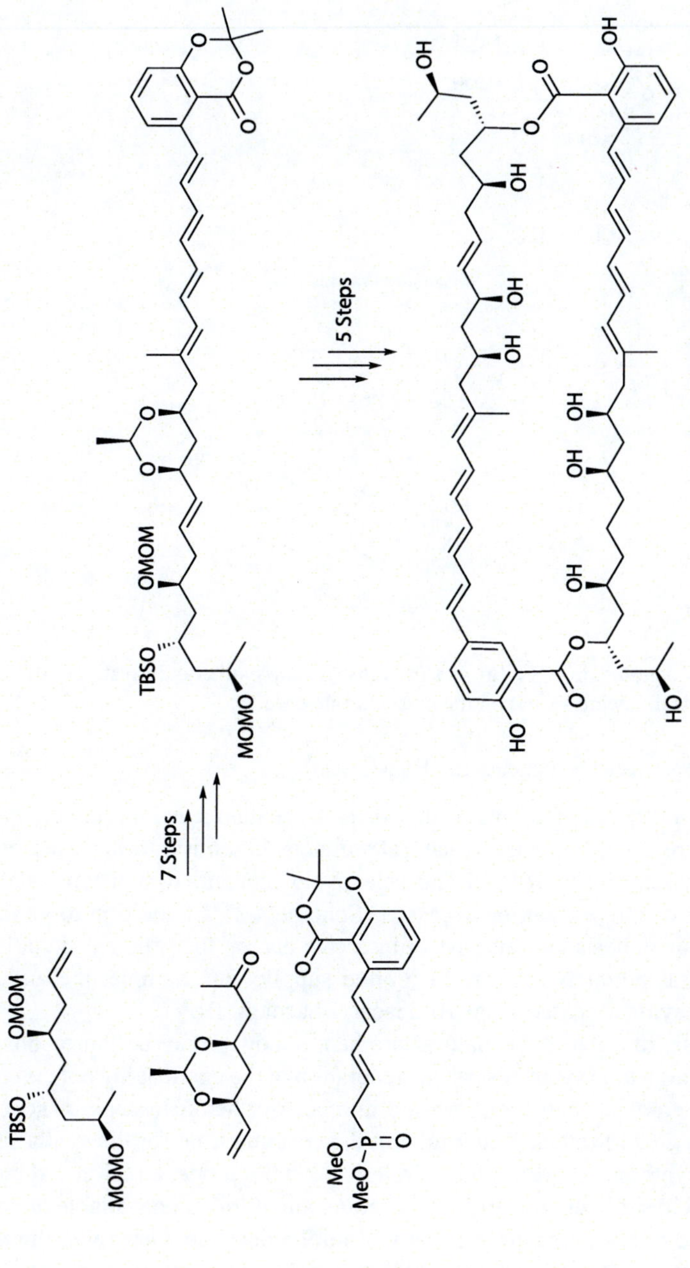

Marinomycin A (10)

SCHEME 4.2 Total synthesis of marinomycin A **10** utilizing a salicylate to switch and control reactivity during dimerization of the molecule.[20]

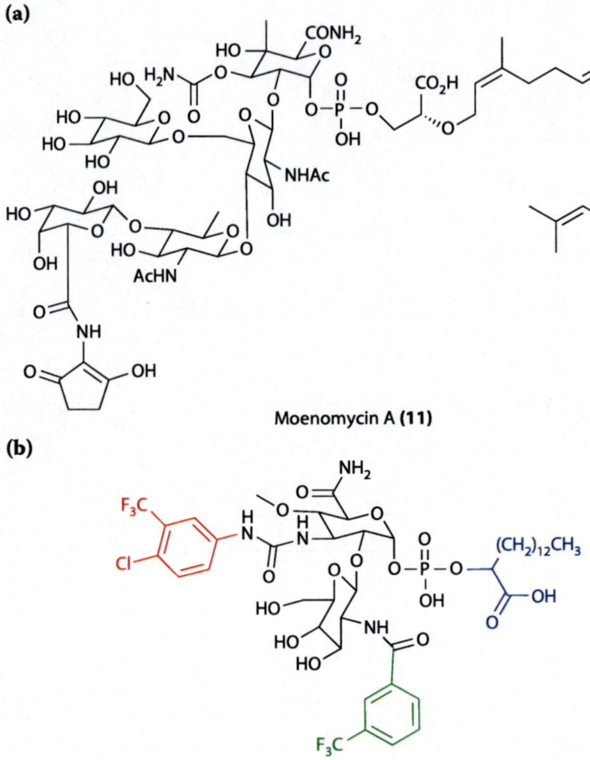

FIGURE 4.4 Moenomycin A **11** (**a**) and the most potent synthetic derivatives (**b**) showing the three positions that were varied on the disaccharide core.[26]

4.2.2 Semisynthetic Approaches (Bio-Chem)

One of the major difficulties of total synthesis is the challenge in building up the complex three-dimensional core of many natural products. In semisynthesis, a pre-existing natural product scaffold is utilized and selectively diversified, and thus analogues of a compound of interest can be accessed (Scheme 4.3). In addition to generating analogues, semisynthesis can also be utilized for access to sufficient quantities of important natural products that are in limited supply; for example, the anticancer drug Taxol **8** may be accessed from 10-deacetyl-baccatin III.[27]

Semisynthesis of a particular natural product analogue may be more convenient than access through total synthesis, as the generation of the challenging core is enabled by the natural producer and therefore the number of chemical steps in generating the analogue is greatly reduced, minimizing the cost and time for the synthesis. The nature of semisynthesis, however, does limit its potential scope; an advanced building block is needed in sufficient quantity from nature, and if this is unavailable or prohibitively expensive or rare, a semisynthesis will not be possible. Generally, the natural product must have chemically orthogonal functional groups that can enable the selective diversification required; selective modification of a reactive and often unstable natural product that possesses a series of functional groups is often not a trivial task.

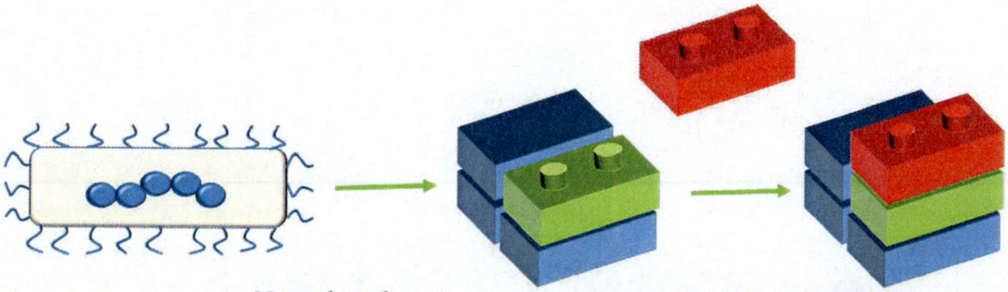

SCHEME 4.3 A natural product is produced, which can then be chemically modified to generate either a new-to-nature analogue of the natural product or to access a structurally related natural product of greater interest.

Despite the difficulties outlined above, if an appropriate advanced building block can be readily sourced, semisynthesis can very rapidly provide access to a whole host of analogues. This is demonstrated well in the case of tetracycline **12a**. First reported in 1948,[28] tetracycline **12a** is a broad-spectrum antibiotic used to treat a variety of infections from *Propionibacterium acnes*, one of the causes of acne, to *Yersinia pestis*, the bacterium responsible for causing the plague.[29] Semisynthetic analogues of **12a** were developed to try to improve the potency and efficacy of this natural product. Notable among these examples are doxycycline **12b**, minocycline **12c**,[30] and, more recently, the glycylcycline **12d** derivatives such as tigecycline **12e**, approved for use as an IV broad-spectrum antibiotic in 2005.[31] The ability of the core of tetracycline **12a** to undergo relatively simple chemical modifications, such as catalytic hydrogenations and dehydrogenations, led Blackwood to employ a disjunctive approach to make simplified tetracycline analogues. This approach yielded doxycycline **12b**, a broad-spectrum antibiotic and antimalarial agent generated by a single palladium-catalyzed hydrogenolysis (Scheme 4.4).[32] The adaptability of the tetracycline core has allowed concerns about the pharmacokinetics and resistance to this class of antibiotic to be addressed.

Miller and coworkers have begun to revolutionize semisynthesis using libraries of peptide-based catalysts to enable previously impossible selective modifications of series of natural products. One such example is their diversification of the glycopeptide teicoplanin **13**, an antibiotic utilized for the treatment of bacterial infections. Miller's work involved screening a series of synthetic peptides to promote selectivity of chemical modifications; two different members of the library enabled the selective bromination of each biaryl system separately using the peptide catalyst and *N*-bromophthalimide (NBP) (Figure 4.5).[33] Creatively they are also able to effect the site-selective bromination of vancomycin **5**, a broad-spectrum antibiotic of last resort. Vancomycin **5** is well known to derive its activity by being able to bind to DAla-DAla motifs within the bacterial cell wall; by including this motif within the tripeptide catalysts and taking advantage of the natural binding of vancomycin, the group was able to achieve selective bromination again using NBP. Bromination would then allow for downstream cross-coupling chemistries to be used to generate further derivatives of vancomycin **5**.[33]

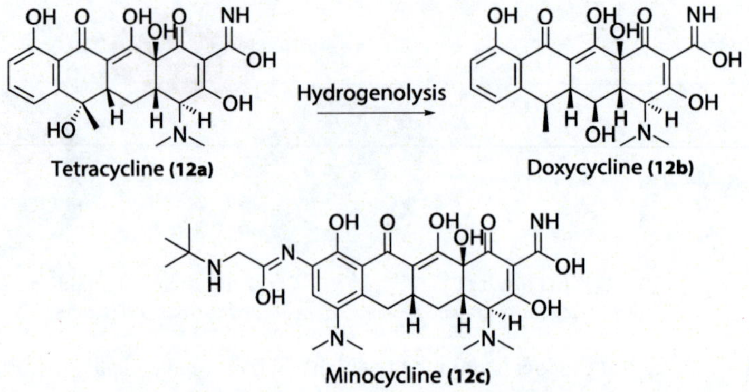

SCHEME 4.4 Hydrogenolysis of tetracycline **12a** to give doxycycline **12b**, shown alongside the third-generation tetracycline class antibiotic minocycline **12c**.

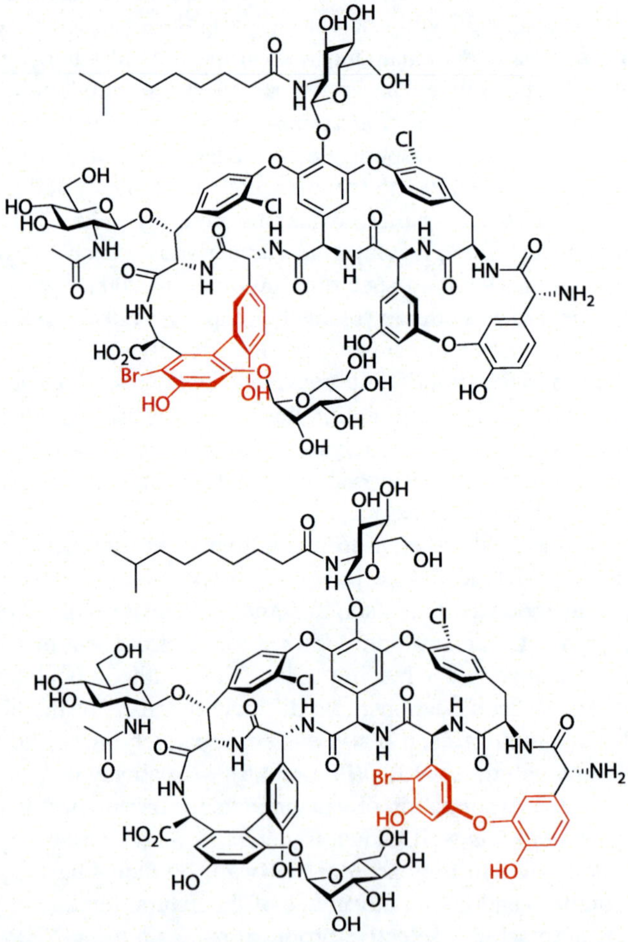

FIGURE 4.5 Brominated derivatives of teicoplanin **13** generated by addition of NBP and a peptide additive.[34]

4.2.3 Precursor-Directed Biosynthesis (Chem-Bio)

Many of the difficulties associated with semisynthesis can be attributed to the challenges of selectively modifying complex natural products, and the presence of naturally occurring, chemically orthogonal, and readily functionalized motifs within the molecule. One way of circumventing such problems is to use other mechanisms of introducing different functionality into molecules; one such method is PDB. This approach involves synthesizing unnatural substrates that can potentially be incorporated as the building blocks from which the complex structure of a new-to-nature natural product is assembled by the native biosynthetic pathways. The alternative biosynthetic building blocks may simply be *fed* to the unmodified natural product–producing organism at the point at which it starts the synthesis of the natural products. While representing a powerful means of harnessing a biosynthetic pathway in order to enable rapid access to designer natural product analogues, the generation of analogues using the PDB approach is completely dependent on the substrate flexibility of the enzymes involved in the biosynthetic pathway (Scheme 4.5).

Many biosynthetic pathways are highly selective for their natural substrates; in the case of polyketides, selectivity is governed both by the loading module and by enzymes within the downstream modules that mediate further chain elongation. This selectivity can result in the synthetic substrate analogue not being used by the biosynthetic pathway. This usually limits synthetic precursors to close analogues of the natural precursors.

In the case of rapamycin **14** biosynthesis, in which modified analogues of the cyclohexanoate starter acid were administered to cultures of the producing organism, *Streptomyces hygroscopicus*, it became apparent that the biosynthetic enzymes would only accept cyclohexanoic acids with a hydroxyl group at the 3-, 4-, or 5-position.[35] The fact that a pyranose analogue was also utilized as a substrate perhaps indicates the need for a hydrogen bond acceptor at this position of the ring. There are, however, numerous examples of very flexible biosynthetic enzymes; one striking example is the loading module from the avermectin **7** polyketide synthase (PKS), which is known to load over 40 carboxylic acid starter units (Figure 4.6).[36,37]

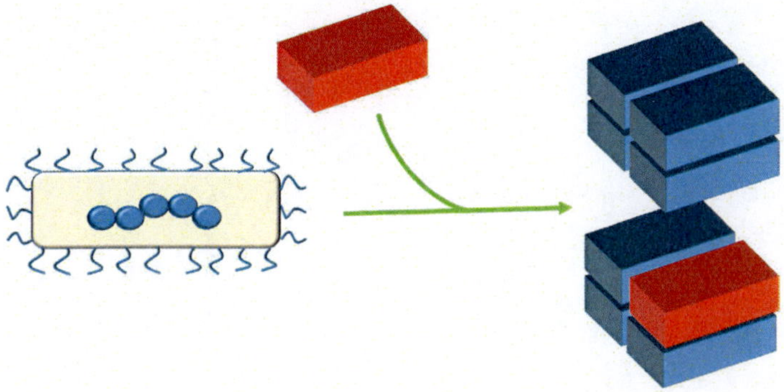

SCHEME 4.5 Feeding of precursor analogues to be incorporated into the biosynthetic pathway produces new analogues of the natural product. These are produced alongside the natural compound, and a mixture results.

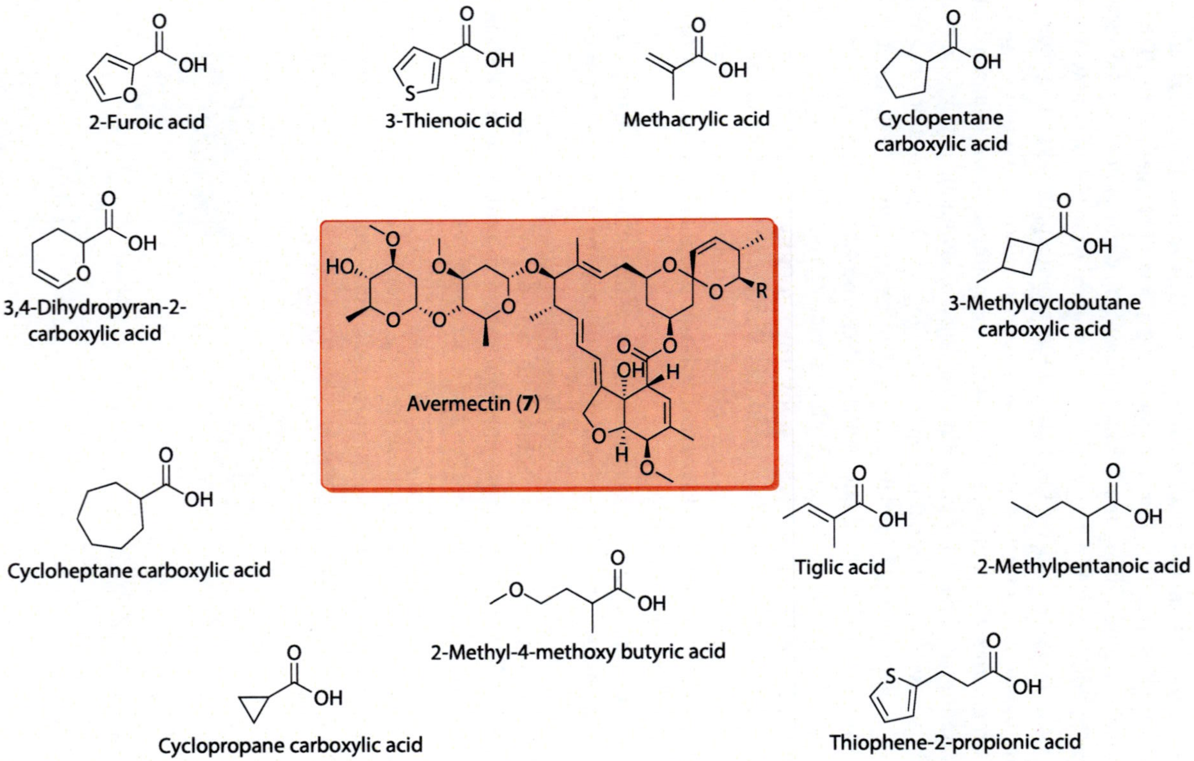

FIGURE 4.6 Avermectin **7** shown alongside a selection of carboxylic acids the PKS loading module for avermectin **7** is known to take up.[37]

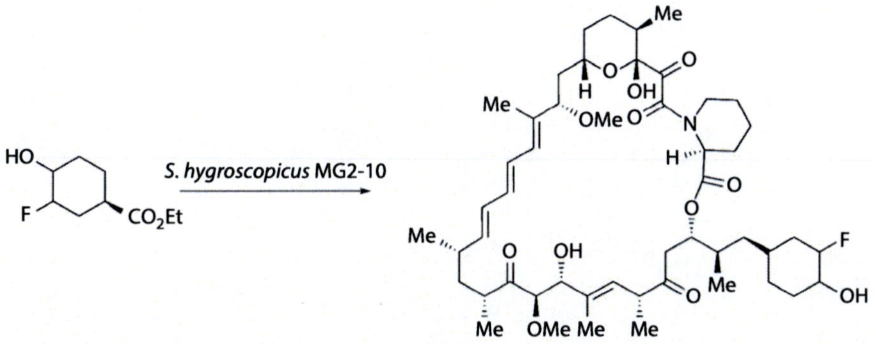

SCHEME 4.6 Feeding *S. hygroscopicus* with fluorohydrin-modified benzoic acid to give fluorinated analogues of rapamycin **14**.[38]

Though the substrate analogues are small and considerably less complex than the full natural product, they may still present a challenge for chemical synthesis. For example, once the substrate requirement for a hydroxyl group within the cyclohexane ring was revealed in the biosynthesis, six cyclohexanoic acids with a fluorohydrin modification were synthesized for use in the generation of fluorinated analogues of rapamycin **14**, where each involved a demanding six-step synthesis (Scheme 4.6).[38]

A further limitation of the PDB approach to natural product analogue generation is that the synthetic substrates are in competition with the natural substrates, for which the enzyme usually has a preference, and the organism continues to make the parent natural product in addition to any analogues. It can be a challenge to then separate the modified natural products from the chemically and physically similar natural product.

In spite of the challenges, PDB provides a very useful approach by which to expeditiously access libraries of complex new-to-nature natural products. One such example is given by Xu et al., who used PDB to build a library of the mycotoxin beauvericin **15a**. An extensive array of precursors was used consisting of 30 analogues of D-2-hydroxyisovalerate and L-phenylalanine, which were fed to the producer of beauvericin **15a**, *Beauveria bassiana*. It was reasoned that the presence of each precursor would lead to three products by replacing one, two, or all three of the corresponding amino acid or hydroxycarboxylic acid residues. Six of the analogues generated beauvericin G_1–G_6 (**15b–15g**) (Figure 4.7). These were isolated and tested against a metastatic prostate cancer cell line, and the results were compared to the activity of natural beauvericin **15a** to enable an understanding of SARs. It was shown that the stepwise replacement of the natural D-Hiv moieties with D-Hbu decreased the activity of the compound, indicating the importance of a branched side chain at these positions for activity.[39]

Kreutzer et al. demonstrated that it is possible to use an *in vitro* assay to determine which substrates are likely to be incorporated by a biosynthetic pathway. They set out to produce a series of phenol ring–modified analogues of micacocidin **16a**, a natural compound with activity against the bacteria *Mycoplasma pneumoniae*.[40] The substrate flexibility of the gatekeeper domain from the biosynthetic pathway of micacocidin **16a**, MicC-FAAL, was probed *in vitro* using an ATP-[^{32}P]-pyrophosphate

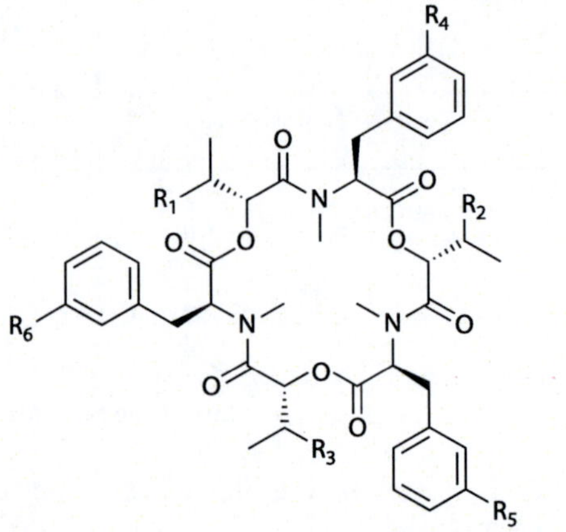

Compound		R_1	R_2	R_3	R_4	R_5	R_6
Beauvericin	**15a**	Me	Me	Me	H	H	H
Beauvericin G_1	**15b**	H	Me	Me	H	H	H
Beauvericin G_2	**15c**	H	H	Me	H	H	H
Beauvericin G_3	**15d**	H	H	H	H	H	H
Beauvericin G_4	**15e**	Me	Me	Me	F	H	H
Beauvericin G_5	**15f**	Me	Me	Me	F	F	H
Beauvericin G_6	**15g**	Me	Me	Me	F	F	F

FIGURE 4.7 Derivatives of beauvericin **15a** formed by PDB.

exchange assay.[41] A total of 29 compounds were tested in this assay including substrates with varying degrees of desaturation. The assay enabled them to determine a turnover rate for each compound, which was compared to the turnover of the natural substrate to work out a relative activation for each substrate. Out of the tested compounds, the seven with the best activation relative to the natural substrate were taken forward for feeding experiments. These seven precursors were fed to 50 mL cultures of the producing fungus and incubated for 3 days. The cultures were then analyzed by high-performance liquid chromatography (HPLC), revealing that six out of the seven precursors had been incorporated to make new derivatives of micacocidin **16a**, micacocidin P1–P6 (**16b–16g**). Large-scale cultures of between 3 and 15 L followed by HPLC purification led to isolation of the new derivatives.

An earlier report showed that micacocidin **16a** may act as a siderophore for gallium, increasing cellular uptake. The researchers formed the corresponding gallium-micacocidin derivative chelates after purification. The chelated derivatives were tested against *M. pneumoniae*, with increased activity seen for four of the compounds with the most active compound, micacocidin P1 **16b**, having almost double the activity of the natural compound (Figure 4.8).[40,41]

Pacidamycin **17** is a uridyl peptide antibiotic active against MraY, a previously unexploited target involved in cell wall biosynthesis.[42] It was found that the pacidamycin biosynthetic pathway was remarkably flexible at both the N and C termini. The natural suite of pacidamycins contains a phenyl-alanine at the N terminus and either a

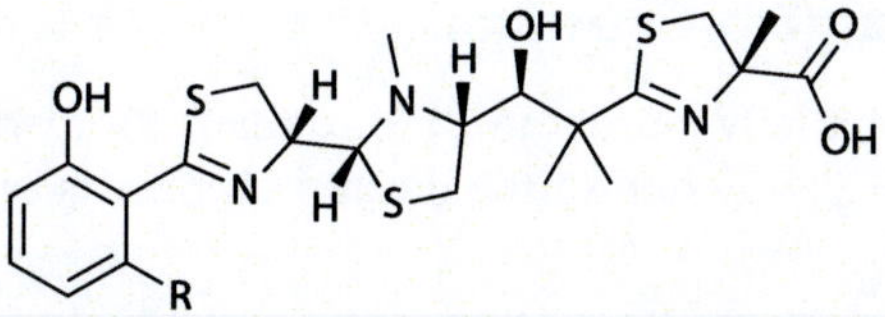

Compound		R	MIC_{50} (μM)
Micacocidin	**16a**	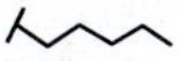	6.9
Micacocidin P1	**16b**		3.1
Micacocidin P2	**16c**		26.3
Micacocidin P3	**16d**		5.9
Micacocidin P4	**16e**		Not detected
Micacocidin P5	**16f**		4.2
Micacocidin P6	**16g**	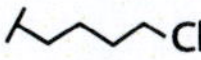	3.5

FIGURE 4.8 Structures of micacocidin **16a** and its derivatives with the MIC_{50} values of their Ga^{3+} complexes.[40]

tryptophan, phenyl alanine, or *meta*-tyrosine at the C terminus. By feeding a variety of halogenated halo-phenyl alanines or halo-tryptophans, a large suite of new analogues of pacidamycin **17** could be generated, including several examples of the precursors being incorporated at both the N and C termini simultaneously. This approach also addressed one of the drawbacks of PDB; by using a cell-free lysate containing the enzyme tryptophan synthase it was possible to cheaply and easily produce substantial quantities of halo-tryptophans from the corresponding halo-indoles (Scheme 4.7).[43,44]

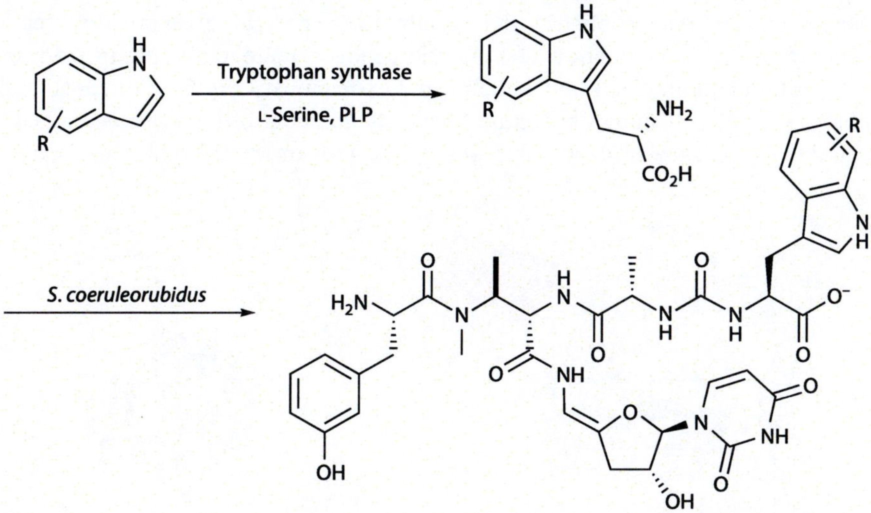

SCHEME 4.7 Production of halogenated analogues of tryptophan through feeding of halo-tryptophans for PDB to give halogenated analogues of pacidamycin **17**. The halo-tryptophans are efficiently produced using a cell-free lysate containing tryptophan synthase.

4.2.4 Mutasynthesis (Bio-Chem-Bio)

Mutasynthesis is conceptually similar to PDB; a distinction with this more sophisticated approach is that production of the desired product is optimized by either of the following:

1. Engineering of the biosynthetic gene cluster that makes the natural product, making it more substrate flexible.
2. Disrupting the enzymes that make the natural precursor so that only synthetic precursors are available for incorporation into the natural product by the biosynthetic pathway (Scheme 4.8). This latter approach leads to production of only the desired modified natural product, without the parent compound, and simplifies the purification of the desired analogue.

Mutasynthesis enabling expansion of substrate scope is probably most clearly demonstrated for PKS and nonribosomal peptide synthetase biosynthetic pathways. Leadlay's elegant replacement of the loading module for erythromycin **3** biosynthesis with the substrate-flexible loading module from the biosynthesis of avermectin **7** enabled a broad series of analogues to be generated.[45] This engineered strain could be utilized to enable access to a new fluorinated analogue of erythromycin **3**.[46] Additionally, when triketide analogues, containing a hydroxyl on C5, of the product of the DEBS module 1 were fed to the host strain, analogues of erythromycin **3** were produced, which had undergone a ring expansion. The ring expansion was a result of the macrolactonization occurring on the new hydroxyl on the synthetic precursor.[47] This approach successfully brought about large structural changes with only modest changes to the starting unit.

Mutasynthesis that abolishes the supply of a natural biosynthetic precursor enables the sole generation of a natural product analogue resulting from the administration and uptake of any suitable and exogenously supplied surrogate biosynthetic precursor. Though substrate flexibility remains relatively constrained it is not always an issue in developing analogues with improved biological activity

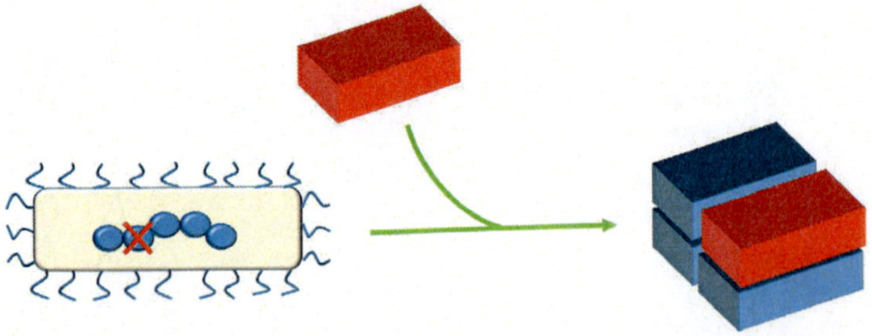

SCHEME 4.8 Feeding of biosynthetic precursor analogues into a modified biosynthetic pathway, blocked in the production of the natural substrate, enables selective production of the desired product.

as minor structural changes can cause substantial changes in biological properties. This was the case with the development of the fluorogeldanamycin compounds, analogues of the heat shock protein 90 (Hsp90) inhibitor geldanamycin **18**, a series of compounds that showed potential as anticancer agents.[48] The natural producer of geldanamycin **18**, *Streptomyces pretiosum*, was engineered in order to disrupt the synthesis of the PKS starter unit, 3-amino-5-hydroxybenzoic acid (AHBA).[49] Feeding four different mono-fluorinated and one tri-fluorinated 3-amino-fluoro benzoic acid derivatives to the mutant strain resulted in the production of six new compounds. While five of these compounds were the direct result of the incorporation of the new precursors, an additional analogue resulted from the hydroxylation of one of the substrates *in vivo*. The compounds were isolated and tested for antiproliferative activity against a series of cancer cell lines. It was observed that the hydroxylated species had significantly lower activity than any of the other derivatives. It was also noted that the other fluorinated derivatives had analogous activity to geldanamycin **18**. While this strategy didn't result in any substantial increases in activity, mutasynthesis allowed for a simple route enabling the removal of the quinone group from the natural compound, a group often avoided by medicinal chemists due to hepatotoxic effects (Figure 4.9).[50]

These minor alterations through mutasynthesis are an attractive proposition for a medicinal chemist. Fluorination can increase metabolic stability and facilitate the passage of a drug through membranes.[51]

A further example of mutasynthesis, enabling the construction of libraries of a complex natural product, in the absence of the parent compound, is the generation of 28 novel derivatives of clorobiocin **19a** (Figure 4.12). The natural compound novobiocin **19b** (Figure 4.12) is an antibiotic and acts by inhibiting the β subunit of bacterial DNA gyrase.[52] However, toxicity and efficacy issues caused much of the exploration of novobiocin **19b**, along with other bacterial DNA gyrase inhibitors, as a potential drug to be discontinued.[53,54] However, to probe into whether it would be possible to improve the properties of novobiocin **19b**, a mutant of the natural producer was produced that blocked the dimethyl transferase gene *cloQ*. This resulted in the loss of production of the precursor 3-dimethylallyl-4-hydroxybenzoyl (DMAHB).

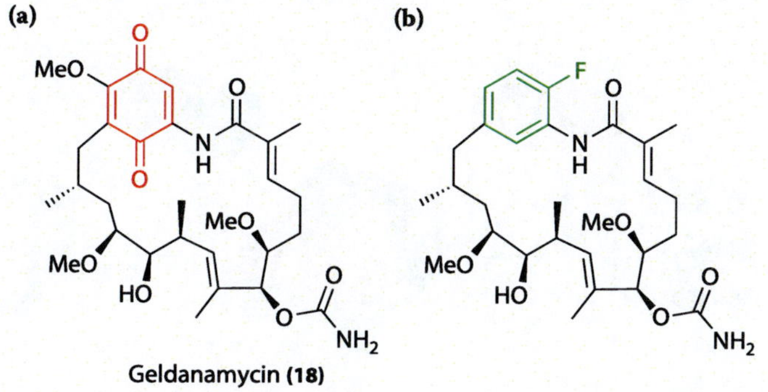

FIGURE 4.9 Geldanamycin **18** (**a**) and one new-to-nature 19-fluorogeldanamycin analogue (**b**). The potentially hepatotoxic quinone (red) is replaced by a safer fluorobenzene (green).

Feeding a range of DMAHB derivatives led to a series of clorobiocin **19a** derivatives. These derivatives were tested in an *Escherichia coli* DNA gyrase assay. None of the analogues showed improved activity, leading to the conclusion that natural novobiocin **19b** was a highly evolved structure optimized for DNA gyrase inhibition, and that it might not be possible to improve upon it.

4.2.5 Mutasynthesis Enabling Postbiosynthetic Modification (Bio-Chem-Bio-Chem)

An extension to the mutasynthesis approach is to utilize functionalized biosynthetic precursor analogues, which allow for downstream synthetic modification (Scheme 4.9). By placing a chemically orthogonal and reactive functional group into a precursor analogue, this approach can be utilized to overcome the Achilles' heel of PDB and mutasynthesis—that of relatively limited substrate scope. In principle, if precursors containing chemically functionalizable handles are tolerated by the biosynthetic pathway, any number of modifications can then be made using synthetic chemistry post biosynthesis to build upon the functionalized precursor that has been incorporated into the natural product.

Chemically reactive and orthogonal functional groups enabling selective functionalization of the natural product, without the need to employ protecting group strategies, are particularly desirable. Bioorthogonal chemistries are defined as "chemical reactions that neither interact with nor interfere with a biological system."[55] Even more desirable is the incorporation of a motif into a natural product which is not only chemically orthogonal but also enables bioorthogonal chemistry. Particular focus has been placed by the group of Bertozzi and others in the development of the copper-catalyzed azide-alkyne cycloaddition (CuAAC reaction) for the *in vivo* modification of biomolecules such as proteins and nucleic acids.[56]

Due to the utility of chemo- and bio-orthogonal reactions, both in biology and in synthetic chemistry, precursors containing the appropriate functionality are often readily available from commercial sources; however, this is not always the case, meaning the challenges of synthesizing bespoke precursor analogues can persist. As with PDB and mutasynthesis, the approach is limited by substrate scope flexibility. These limitations aside, mutasynthesis enabling postbiosynthetic modification brings

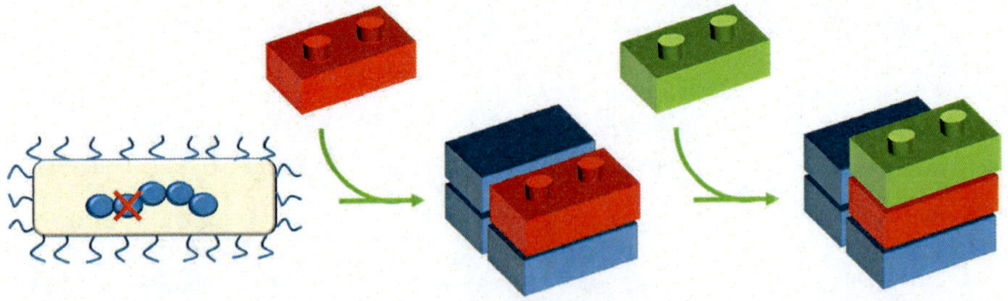

SCHEME 4.9 Mutasynthesis enabling postbiosynthetic modification. The feeding of precursors containing functionalizable handles allows for downstream chemo- or bioorthogonal chemistries to be carried out.

together many of the advantages of chemical synthesis and biosynthesis, allowing the chemist expeditious access to complex natural product structures that are modified to enable ready and site-selective modification post biosynthesis. This gives the chemist the freedom to explore wider regions of chemical space than is usually available through PDB and standard mutasynthesis.

A useful example of this approach is the generation of a series of analogues of ansamitocin **20** by Kirschning and coworkers. A mutant of the ansamitocin **20** producer *Actinosynnema pretiosum* (HGF073) was developed so it would be incapable of synthesizing the PKS starter unit 3-amino-5-hydroxybenzoic acid. This mutant was then fed a series of alkynylated amino-benzoic acids, which were successfully incorporated to form several alkynylated forms of the natural product. The alkyne within the modified polyketide was then diversified using the CuACC reaction to give a series of analogues which were taken forward for testing against several human cancer cell lines.[57]

In an analogous manner, brominated derivatives of ansamitocin **20** were produced by feeding 3-amino-5-hydroxybenzoic acid to the engineered strain. Using Stille cross-coupling methodology, the group utilized the reactive bromide handle to join ansamitocin **20** to folic acid via a short linker containing a disulfide (Figure 4.10). The folic acid was intended to selectively deliver the drug to folate receptors, overexpressed in many cancer cell lines. It was postulated that reduction of the disulfide bond in the anoxic environment of a tumor would release the active ansamitocin **20** analogue.[58–60] It was observed that while the conjugate was inactive against noncancerous cells, the alkynylated derivative of ansamitocin **20** still maintained strong antiproliferative activity (IC_{50} < 10 nM) against a series of cancer cell lines.[57]

In contrast to the use of starter units, alkynylated synthetic chain terminators have been creatively utilized. These β-keto acid substrates have been employed to act in place of malonate extender units in an engineered PKS pathway.[61] Incorporation of the β-keto acid leads to the disassociation of the intermediate from the enzyme, thereby forcing the formation of new polyketide structures (Scheme 4.10). This approach may partially address the problem of substrate scope, as by incorporating the synthetic unit at the end of biosynthesis there are fewer requirements for it to be tolerated by numerous downstream enzymes.

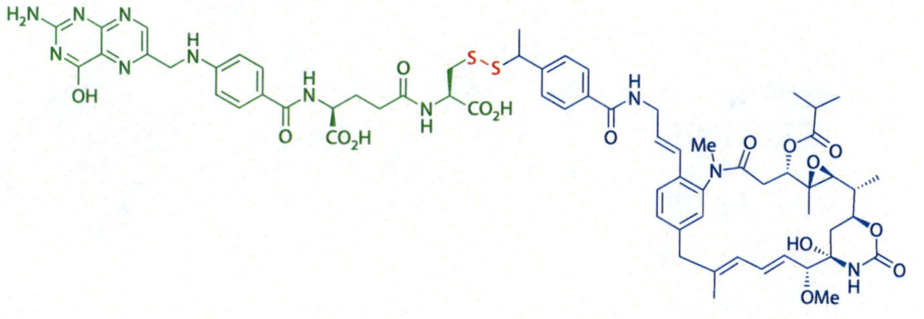

FIGURE 4.10 Folate linked to a derivative of ansamitocin **20**. Folate (green) enables delivery to folate receptors, overexpressed in many cancer cell lines, and the disulfide bridge (red) can be cleaved, releasing the ansamitocin **20** derivative warhead (blue).[57]

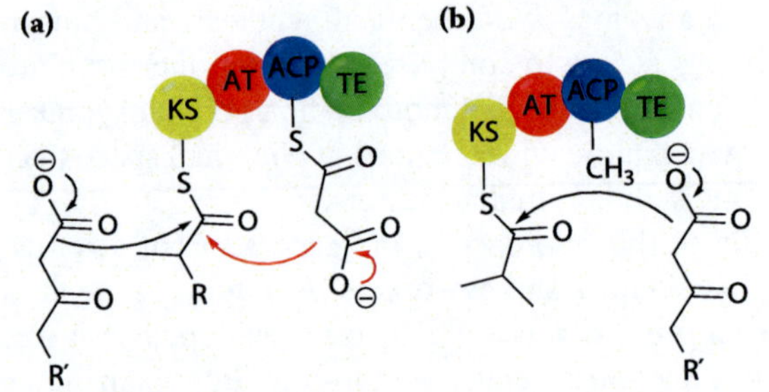

SCHEME 4.10 Terminator unit competing with malonyl CoA in the natural producer **(a)** and with no competition with the engineered strain **(b)**.[61]

This approach was used to generate a series of derivatives of lasalocid A **21** (Figure 4.12) using an engineered *Streptomyces lasaliensis*. A series of point mutations on the active serine for 4′-phosphopantetheine attachment on ACP12 and ACP5 converting it to alanine deactivated these enzymes to malonate uptake. The mutant still produced intermediates of lasalocid A **21** bound to the biosynthetic enzymes but the biosynthesis could not progress to completion. A series of alkyne and azide-containing terminator units were then fed to the *S. lasaliensis* mutant giving a series of alkynylated and azide-containing polyketide compounds. Using the CuACC reaction, a range of different groups were successfully added onto the lasalocid A derivatives including alkyl chains, protected acids, sugars, and phosphates.[61]

4.2.6 Combinatorial Biosynthesis (Bio-Bio)

Combinatorial biosynthesis, often referred to now as the synthetic biological modification of biosynthetic gene clusters, is the genetic manipulation of biosynthetic enzymes within or into a pathway. It may involve either the assembly of genes encoding enzymes from different organisms to create entirely new designer biosynthetic pathways or the switching around of enzymatic domains of multifunctional proteins to modify the natural product produced by the organism (Scheme 4.11). As with the previous approaches described, combinatorial biosynthesis can be

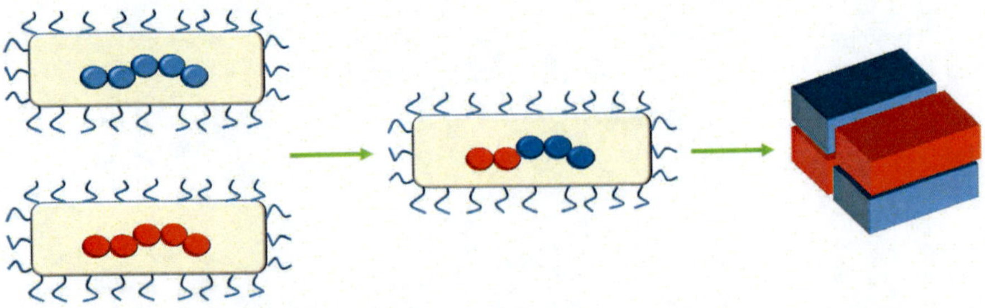

SCHEME 4.11 Combinatorial biosynthesis genetic engineering whereby genes from the same or different organisms are combined in a new-to-nature fashion to produce novel compounds.

utilized to create natural product analogues for determining SARs and interactions of the natural product and its biological target.

A combinatorial biosynthesis approach has been utilized to create many new-to-nature natural products, including a small subgroup of novel halogenated metabolites on which we will now focus. The incorporation of a halogen into a molecule is often desirable as it can alter the bioactivity and bioavailability of a natural product. Physical properties of the molecule may also be altered. For instance, the stability of the compound may be affected, as well as the compound's electronics and lipophilicity.[62] Importantly, halogenating a molecule can also change its bioactivity. For example, the chlorinated cytotoxic anticancer compound salinosporamide A **22a**[63] is over 500 times more active than salinosporamide B **22b**, its deschloro analogue (Figure 4.12).[64]

An early example of the use of combinatorial biosynthesis to create halogenated compounds was demonstrated by Sánchez et al.[65] This work investigated two closely related indolocarbazole alkaloids, rebeccamycin **23** and staurosporine **24** (Figure 4.11), which display antitumor properties.

The study aimed at generating indolocarbazole derivatives in order to help with efforts to treat diseases including cancer and neurodegenerative disorders. The biosynthetic pathway of rebeccamycin **23** was dissected, and different combinations of the genes of rebeccamycin **23** were coexpressed into the heterologous host *Streptomyces albus*. When two halogenase genes from different microorganisms, *pyrH* and *thal*, were incorporated into the biosynthetic gene cluster, over 30 new chlorinated derivative compounds were created. Another early study targeted the biosynthetic pathway of an aminocoumarin antibiotic. The authors substituted the 8′-methyltransferase from the biosynthetic gene cluster of novobiocin **19b** with the 8′-halogenase from a close structurally related aminocoumarin antibiotic, clorobiocin **19a**, and subsequently produced 8′-chlorinated derivatives of novobiocin **19b** (Figure 4.12).[66] Similarly, new 5-chloropyrrole analogues of clorobiocin **19a** were also generated after the halogenase *hrmQ* gene was cloned from the horaomycin biosynthetic pathway into the producer of clorobiocin **19a**.[67]

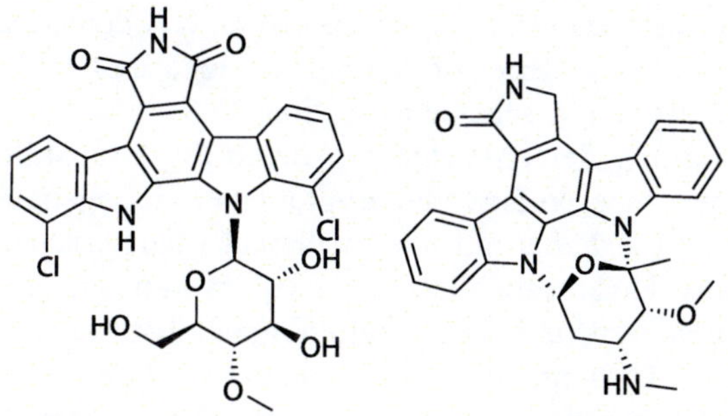

FIGURE 4.11 The structures of closely related rebeccamycin **23** and staurosporine **24**, two indolocarbazole alkaloids, which exhibit antitumor effects.

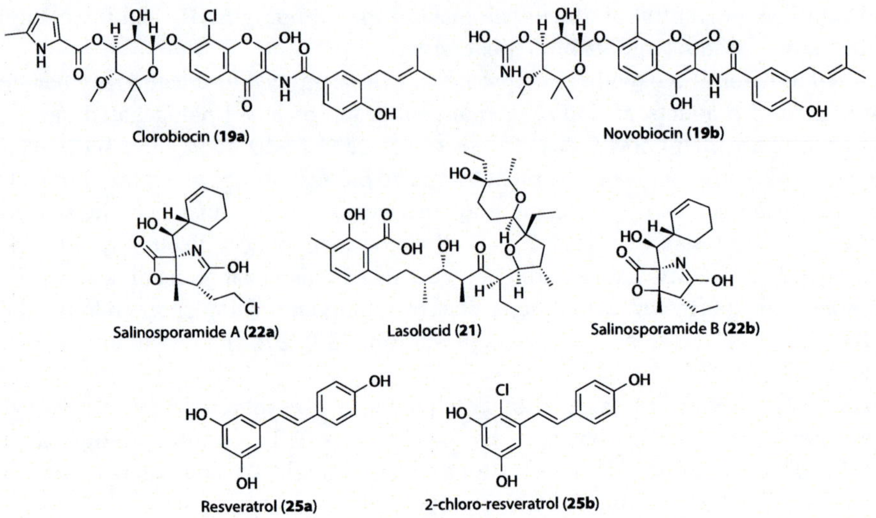

FIGURE 4.12 The structures of several natural products and some of their derivatives: clorobiocin **19a**, novobiocin **19b**, lasolocid **21**, salinosporamide A **22a**, salinosporamide B **22b**, resveratrol **25a**, and 2-chloro-resveratrol.

Tryptophan halogenases have been demonstrated to be highly useful in combinatorial biosynthesis approaches due to the fact that many bioactive peptidic natural products and alkaloids are derived from tryptophan. The work of Goss and O'Connor showed that halogenases can be applied completely out of context on structurally unrelated metabolites. The two groups introduced tryptophan halogenase to act in concert with biosynthetic pathways, making uridyl peptide antibiotics and alkaloids respectively. The work by the Goss group introduced tryptophan halogenases to complement the biosynthetic pathway to pacidamycin **17** in order to create corresponding chlorinated analogues (Scheme 4.7).[44] Switching on the expression of the halogenase gene such that it acted in concert with the pacidamycins biosynthetic gene cluster enabled the *in vivo* generation of 7-chlorotryptophan, which was subsequently incorporated into the pacidamycin biosynthetic pathway to afford chloropacidamycin **17b**. These chloropacidamycin analogues can be functionalized using chemical cross-coupling to generate a series of alternate analogues of pacidamycin **17**. This selective functionalization is discussed in Section 4.2.7.

The O'Connor group demonstrated the flexibility and utility of tryptophan halogenases as they incorporated the tryptophan 5-halogenase *Pry*H and the tryptophan 7-halogenase *Reb*H into the medicinal Madagascar periwinkle plant, *Catharanthus roseus*. This resulted in chlorinated tryptophan, which was then used as a substrate in monoterpene indole alkaloid metabolism to generate chlorinated alkaloid analogues.[68] Biosynthetic work in plants is often difficult as secondary metabolite biosynthesis genes are often found dispersed throughout the genome rather than clustered together, as in bacteria. Using combinatorial biosynthesis to insert halogenases into *C. roseus* was therefore deemed an attractive method to produce alkaloid analogues, as the majority of the monoterpene indole alkaloid biosynthesis genes are unknown and so heterologous expression of the monoterpene indole alkaloid was not possible. However, a bottleneck in this study was that while transforming *C. roseus* with *Reb*H resulted in the successful production of the chlorotryptophan intermediate, the chlorinated intermediate accumulated in the plant tissues as it is not the optimal substrate for the next biosynthetic enzyme, tryptophan decarboxylase.[68] This is undesirable as it means that the accumulated product is not altered further into the more valuable final halogenated alkaloid product. In order to eliminate this bottleneck, *Reb*H was successfully reengineered so that it preferentially installs chlorine onto tryptamine, which is not accepted by the native halogenase.[69]

Recently, the Zhan group inserted the fungal flavin-dependent halogenase *Rdc*2 into the resveratrol biosynthetic pathway[70] in order to generate a new chlorinated derivative of resveratrol **25a**, called 2-chloro-resveratrol **25b** (Figure 4.12). This chlorinated analogue has been shown to have greater antioxidant and antimicrobial activity compared to resveratrol **25a**.[71] Resveratrol **25a** is a stilbenoid, which has a wide range of bioactive properties, including antibacterial and antifungal activity, as well as anti-inflammatory and neuroprotective effects. While this molecule has been previously synthesized,[71] the potential of combinatorial biosynthesis is clearly demonstrated in this study. Moreover, the chlorination of molecules by chemical synthesis is often not ideal as there is often a lack of selection and the generation of undesirable side products.[71] These halogenase combinatorial biosynthesis examples illustrate that new natural product analogues can be generated using this approach. However, in order to utilize the full potential of combinatorial biosynthesis, more research is needed into biosynthetic enzymes; we require a greater understanding of enzyme interactions, the structures of multifunctional enzymes, and how we can manipulate substrate specificity.[72]

4.2.7 GenoChemetics: Gene Expression Enabling Synthetic Diversification (Bio-Bio-Chem)

A very new approach to producing natural product analogues is GenoChemetics. This approach has been described as "gene expression enabling synthetic diversification"[73] as it makes use of molecular biology to install a chemically reactive and orthogonal handle, which may be further functionalized using mild synthetic chemistry methodologies.

In this approach a foreign gene is introduced into the natural product–producing organism. The foreign gene will work in concert with the biosynthetic gene cluster to install a chemically orthogonal handle in the natural product, which then allows for the natural product to be selectively functionalized (Scheme 4.12).[74] The foreign gene encodes an enzyme that will work within the native biosynthetic gene cluster to produce a novel modified natural compound, sometimes in addition to and alongside the natural compound. The introduction of the chemically orthogonal handle into the natural product is important as it allows for selective modification of the new unnatural product, and therefore new analogues of the natural compound can be easily generated. The handles should enable simple and effective selective chemistry, such as cross-coupling or click chemistry reactions. It is most

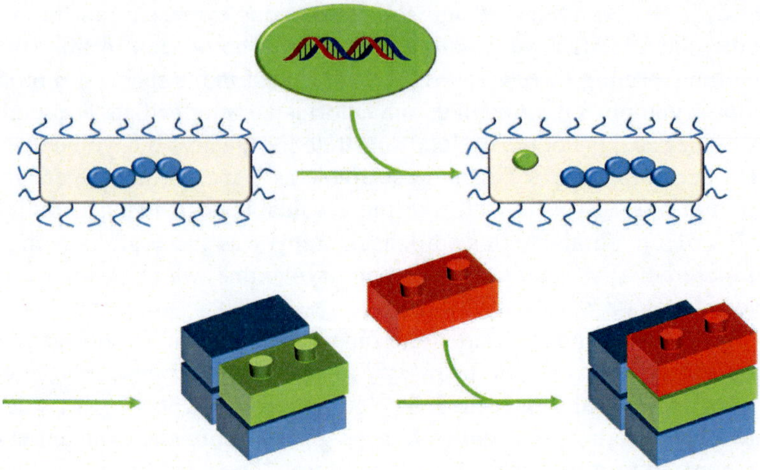

SCHEME 4.12 The GenoChemetics approach to natural product analogue generation. An unrelated foreign gene is introduced into the natural product–producing organism. The foreign gene works within the biosynthetic gene cluster to install a chemically orthogonal handle.

beneficial if this selective derivatization can be performed on the natural product without needing to purify the compound from the crude extracts and without protecting group chemistry.

The Goss group has investigated the use of halogenases in a GenoChemetics approach and has developed methods of selectively modifying the resulting halogenated unnatural compounds. The first example of a GenoChemetics approach, demonstrated by the Goss group, successfully created analogues of Pacidamycin **17** (Scheme 4.7). Pacidamycin **17** was a robust test bed for this method of natural product analogue generation, as it is a challenging system to work with. Challenges in the system included the fact that the pacidamycin producer, *Streptomyces coeruleorubidus*, had been previously untransformed, and that pacidamycin is a thermally unstable compound, which has poor solubility in all solvents apart from water.

The *prnA* gene, a tryptophan halogenase responsible for chlorinating tryptophan at the 7-position, was introduced into the genome of *S. coeruleorubidus*, resulting in the production of chloropacidamycin **17b**, an unnatural halogenated analogue of pacidamycin **17** (Scheme 4.13).[73] The chlorine handle was subsequently functionalized by palladium-catalyzed Suzuki-Miyaura cross-coupling, which had been developed by Buchwald et al.[75] The functionalization of the chlorine handle was so effective, protecting group chemistry was not needed and the compound could even be modified within the crude culture extract without need for prior purification.[73]

Another example of this powerful GenoChemetics approach has been demonstrated by the O'Connor group, who used this approach for the first time in a eukaryotic system. The O'Connor group aimed to exploit monoterpene indole alkaloid biosynthesis within *C. roseus*, a medicinal plant. In this instance, a halide

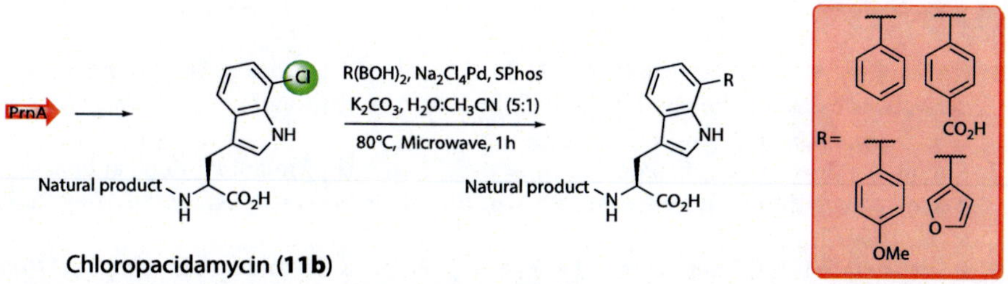

Chloropacidamycin (**11b**)

SCHEME 4.13 The gene *prn*A was introduced into the biosynthetic pathway of pacidamycin **17**, resulting in chloropacidamycin **17b**. The chlorine handle was then used as a chemical handle to selectively functionalize the compound by Suzuki-Miyaura cross-coupling.

was introduced as a chemical handle to allow for chemical functionalization of the indole moiety of the monoterpene indole alkaloids. Chemical functionalization was again achieved using Suzuki-Miyaura cross-coupling[76] to generate monoterpene indole alkaloid analogs, illustrating the utility of this postbiosynthetic modification technique.

4.2.8 Future Directions in Natural Product Analogue Generation

It is an exciting time for natural product research; advances in sequencing technologies and tools for molecular biological manipulation and synthesis are enabling significant acceleration in this field. From the large numbers of microbial sequences that have now been acquired it is apparent from the sequence data that, even from well-known laboratory-culturable microbes, the microbes can produce many more compounds than have been previously observed.[77,78]

Natural products remain an unprecedented source of leads for drug discovery[17] and, as a wealth of new compounds is unlocked from cryptic pathways and as yet unculturable microbes, they will perhaps develop increasing importance in the future. New understanding of biosynthetic pathways and how the enzymes communicate is being revealed, an impressive recent advancement being the publication of a Cryo-EM structure of a modular PKS.[79] This greater understanding as to how such complex systems function goes hand in hand with advancements in harnessing these biosynthetic machines, and the further marrying together of synthetic chemistry and synthetic biology in order to generate new-to-nature designer metabolites. Other significant recent advancements include demonstrating that not just loading modules but midpathway modules of these multifunctional enzymes may be intercepted to enable access to more greatly structurally diversified compounds.[57,80–82] This moves mankind steps closer to the goal of being able to dial into any bioactive, designer new-to-nature natural product.

ACKNOWLEDGMENTS

The authors are grateful to the European Research Council under the European Union's Seventh Framework Programme (FP7/2007–2013/ERC Grant agreement no. 614779) and EMBRIC (Grant no. 654008)) for generous financial support.

REFERENCES

1. Kirschning, A.; Taft, F.; Knobloch, T. Total synthesis approaches to natural product derivatives based on the combination of chemical synthesis and metabolic engineering. *Org. Biomol. Chem.* **2007**, *5* (20), 3245.
2. Simon, A.; Traynor, K.; Santos, K.; Blaser, G.; Bode, U.; Molan, P. Medical honey for wound care-still the "latest resort"? *Evid. Based Complement. Alternat. Med.* **2009**, *6* (2), 165–173.
3. Brownstein, M. J. A brief history of opiates, opioid peptides, and opioid receptors. *Proc. Natl. Acad. Sci. USA* **1993**, *90* (12), 5391–5393.
4. Irish, J.; Blair, S.; Carter, D. A. The antibacterial activity of honey derived from Australian flora. *PLoS One* **2011**, *6* (3), e18229.
5. Fleming, A. Classics in infectious diseases: On the antibacterial action of cultures of a *Penicillium*, with special reference to their use in the isolation of *B. influenzae* by Alexander Fleming. *Br. J. Exp. Pathol.* **1929**, *10*, 226–236.
6. Robinson, F. A. The chemistry of penicillin. *J. Pharm. Pharmacol.* **1949**, *1* (1), 634–635.
7. Clatworthy, A. E.; Pierson, E.; Hung, D. T. Targeting virulence: A new paradigm for antimicrobial therapy. *Nat. Chem. Biol.* **2007**, *3* (9), 541–548.
8. Fleming, S. A. Nobel Lecture, December 11, 1945. In: *Nobel Lecture: Physiology or Medicine, 1942–1962*, Elsevier Publishing Company, Amsterdam, the Netherlands, **1945**, pp. 83–93.
9. Liu, Y.-Y. et al. Emergence of plasmid-mediated colistin resistance mechanism MCR-1 in animals and human beings in China: A microbiological and molecular biological study. *Lancet Infect. Dis.* **2015**, *16* (2), 161–168.
10. Goossens, H.; Ferech, M.; Vander Stichele, R.; Elseviers, M. Outpatient antibiotic use in europe and association with resistance: A cross-national database study. *Lancet* **2005**, *365* (9459), 579–587.
11. Srivastava, A. et al. New target for inhibition of bacterial RNA polymerase: "Switch region". *Curr. Opin. Microbiol.* **2011**, *14* (5), 532–543.
12. Ho, M. X.; Hudson, B. P.; Das, K.; Arnold, E.; Ebright, R. H. Structures of RNA polymerase-antibiotic complexes. *Curr. Opin. Struct. Biol.* **2009**, *19* (6), 715–723.
13. Floss, H. G.; Yu, T.-W. Rifamycin-mode of action, resistance, and biosynthesis. *Chem. Rev.* **2005**, *105* (2), 621–632.
14. Mariani, R.; Maffioli, S. I. Bacterial RNA polymerase inhibitors: An organized overview of their structure, derivatives, biological activity and current clinical development status. *Curr. Med. Chem.* **2009**, *16* (4), 430–454.
15. Cantón, R. Antibiotic resistance genes from the environment: A perspective through newly identified antibiotic resistance mechanisms in the clinical setting. *Clin. Microbiol. Infect.* **2009**, *15* (Suppl. 1), 20–25.
16. Van Lanen, S. G.; Shen, B. Microbial genomics for the improvement of natural product discovery. *Curr. Opin. Microbiol.* **2006**, *9* (3), 252–260.
17. Newman, D. J.; Cragg, G. M. Natural products as sources of new drugs over the 30 years from 1981 to 2010. *J. Nat. Prod.* **2012**, *75* (3), 311–335.
18. Nguyen, L. A.; He, H.; Pham-Huy, C. Chiral drugs: An overview. *Int. J. Biomed. Sci.* **2006**, *2* (2), 85–100.
19. Nicolaou, K. C.; Nold, A. L.; Milburn, R. R.; Schindler, C. S. Total synthesis of marinomycins A-C. *Angew. Chem. Int. Ed.* **2006**, *45* (39), 6527–6532.
20. Evans, P. A.; Huang, M.-H.; Lawler, M. J.; Maroto, S. Total synthesis of marinomycin A using salicylate as a molecular switch to mediate dimerization. *Nat. Chem.* **2012**, *4*, 680–684.
21. Nishimaru, T.; Kondo, M.; Takeshita, K.; Takahashi, K.; Ishihara, J.; Hatakeyama, S. Total synthesis of marinomycin a based on a direct dimerization strategy. *Angew. Chem. Int. Ed.* **2014**, *53* (32), 8459–8462.

22. Wöhler, F. Ueber Künstliche Bildung Des Harnstoffs. *Ann. Phys.* **1828**, *88* (2), 253–256.
23. Gampe, C. M.; Tsukamoto, H.; Doud, E. H.; Walker, S.; Kahne, D. Tuning the moenomycin pharmacophore to enable discovery of bacterial cell wall synthesis inhibitors. *J. Am. Chem. Soc.* **2013**, *135* (10), 3776–3779.
24. Welzer, P. et al. Moenomycin A: Minimum structural requirements for biological activity. *Tetrahedron* **1987**, *43* (3), 585–598.
25. Silva, D. J.; Wang, H.; Allanson, N. M.; Jain, R. K.; Sofia, M. J. Stereospecific solution- and solid-phase glycosylations. Synthesis of β-linked saccharides and construction of disaccharide libraries using phenylsulfenyl 2-deoxy-2-trifluoroacetamido glycopyranosides as glycosyl donors. *J. Org. Chem.* **1999**, *64* (16), 5926–5929.
26. Sofia, M. J. et al. Discovery of novel disaccharide antibacterial agents using a combinatorial library approach. *J. Med. Chem.* **1999**, *42* (17), 3193–3198.
27. Gennari, C.; Vulpetti, A.; Donghi, M.; Mongelli, N.; Vanotti, E. Semisynthesis of taxol: A highly enantio- and diastereoselective synthesis. *Angew. Chem. Int. Ed.* **1996**, *35* (15), 1723–1725.
28. Nelson, M. L.; Levy, S. B. The history of the tetracyclines. *Ann. N. Y. Acad. Sci.* **2011**, *1241* (1), 17–32.
29. Boulanger, L. L.; Ettestad, P.; Fogarty, J. D.; Dennis, D. T.; Romig, D.; Mertz, G. Gentamicin and tetracyclines for the treatment of human plague: Review of 75 cases in New Mexico, 1985–1999. *Clin. Infect. Dis.* **2004**, *38* (5), 663–669.
30. Chopra, I.; Roberts, M. Tetracycline antibiotics: Mode of action, applications, molecular biology, and epidemiology of bacterial resistance. *Microbiol. Mol. Biol. Rev.* **2001**, *65* (2), 232–260.
31. Rose, W. E.; Rybak, M. J. Tigecycline: First of a new class of antimicrobial agents. *Pharmacotherapy* **2006**, *26* (8), 1099–1110.
32. Stephens, C. et al. 6-Deoxytetracyclines. IV. Preparation, C-6 stereochemistry, and reactions. *J. Am. Chem. Soc.* **1963**, *85* (1961), 2643–2652.
33. Pathak, T. P.; Miller, S. J. Site-selective bromination of vancomycin. *J. Am. Chem. Soc.* **2012**, *134* (14), 6120–6123.
34. Pathak, T. P.; Miller, S. J. Chemical tailoring of teicoplanin with site-selective reactions. *J. Am. Chem. Soc.* **2013**, *135* (22), 8415–8422.
35. Goss, R. J. M.; Lanceron, S. E.; Wise, N. J.; Moss, S. J. Generating rapamycin analogues by directed biosynthesis: Starter acid substrate specificity of mono-substituted cyclohexane carboxylic acids. *Org. Biomol. Chem.* **2006**, *4* (22), 4071.
36. Wang, F.; Wang, Y.; Ji, J.; Zhou, Z.; Yu, J.; Zhu, H.; Su, Z.; Zhang, L.; Zheng, J. Structural and functional analysis of the loading acyltransferase from avermectin modular polyketide synthase. *ACS Chem. Biol.* **2015**, *10* (4), 1017–1025.
37. Dutton, C. J.; Gibson, S. P.; Goudie, A. C.; Holdom, K. S.; Pacey, M. S.; Ruddock, J. C.; Bu'Lock, J. D.; Richards, M. K. Novel avermectins produced by mutational biosynthesis. *J. Antibiot. (Tokyo)* **1991**, *44* (3), 357–365.
38. Goss, R. J. M.; Lanceron, S.; Roy, A. D.; Sprague, S.; Nur-e-Alam, M.; Hughes, D. L.; Wilkinson, B.; Moss, S. J. An expeditious route to fluorinated rapamycin analogues by utilising mutasynthesis. *ChemBioChem* **2010**, *11* (5), 698–702.
39. Xu, Y.; Zhan, J.; Wijeratne, E. M. K.; Burns, A. M.; Gunatilaka, A. A.; Molnár, I. Cytotoxic and antihaptotactic beauvericin analogues from precursor-directed biosynthesis with the insect pathogen *Beauveria bassiana* ATCC 7159. *J. Nat. Prod.* **2007**, *70* (9), 1467–1471.
40. Kreutzer, M. F.; Kage, H.; Herrmann, J.; Pauly, J.; Hermenau, R.; Müller, R.; Hoffmeister, D.; Nett, M. Precursor-directed biosynthesis of micacocidin derivatives with activity against *Mycoplasma pneumoniae*. *Org. Biomol. Chem.* **2014**, *12* (1), 113–118.

41. Kreutzer, M. F.; Kage, H.; Gebhardt, P.; Wackler, B.; Saluz, H. P.; Hoffmeister, D.; Nett, M. Biosynthesis of a complex yersiniabactin-like natural product via the mic locus in phytopathogen *Ralstonia solanacearum*. *Appl. Environ. Microbiol.* **2011**, *77* (17), 6117–6124.
42. Winn, M.; Goss, R. J. M.; Kimura, K.; Bugg, T. D. H. Antimicrobial nucleoside antibiotics targeting cell wall assembly: Recent advances in structure–function studies and nucleoside biosynthesis. *Nat. Prod. Rep.* **2010**, *27* (2), 279–304.
43. Ragab, A. E.; Grüschow, S.; Rackham, E. J.; Goss, R. J. M. New pacidamycins biosynthetically: Probing N- and C-terminal substrate specificity. *Org. Biomol. Chem.* **2010**, *8* (14), 3128–3129.
44. Grüchow, S.; Rackham, E. J.; Elkins, B.; Newill, P. L. A.; Hill, L. M.; Goss, R. J. M. New pacidamycin antibiotics through precursor-directed biosynthesis. *ChemBioChem* **2009**, *10* (2), 355–360.
45. Marsden, A. F.; Wilkinson, B.; Cortés, J.; Dunster, N. J.; Staunton, J.; Leadlay, P. F. Engineering broader specificity into an antibiotic-producing polyketide synthase. *Science* **1998**, *279* (5348), 199–202.
46. Goss, R. J. M.; Hong, H. A novel fluorinated erythromycin antibiotic. *Chem. Commun. (Camb.).* **2005**, 31, 3983–3985.
47. Cane, D. E.; Kudo, F.; Kinoshita, K.; Khosla, C. Precursor-directed biosynthesis: Biochemical basis of the remarkable selectivity of the erythromycin polyketide synthase toward unsaturated triketides. *Chem. Biol.* **2002**, *9* (1), 131–142.
48. Whitesell, L.; Mimnaugh, E. G.; De Costa, B.; Myers, C. E.; Neckers, L. M. Inhibition of heat shock protein HSP90-pp60v-Src heteroprotein complex formation by benzoquinone ansamycins: Essential role for stress proteins in oncogenic transformation. *Proc. Natl. Acad. Sci. USA* **1994**, *91* (18), 8324–8328.
49. Eichner, S.; Floss, H. G.; Sasse, F.; Kirschning, A. New, highly active nonbenzoquinone geldanamycin derivatives by using mutasynthesis. *ChemBioChem* **2009**, *10* (11), 1801–1805.
50. Hermane, J. et al. New, non-quinone fluorogeldanamycin derivatives strongly inhibit Hsp90. *ChemBioChem* **2015**, *16* (2), 302–311.
51. Kirk, K. L. Editorial for special issue of JFC on biomedicinal chemistry. *J. Fluor. Chem.* **2008**, *129* (9), 730.
52. Lewis, R. J.; Tsai, F. T. F.; Wigley, D. B. Molecular mechanisms of drug inhibition of DNA gyrase. *Bioessays* **1996**, *18* (8), 661–671.
53. Perronne, C. M.; Malinverni, R.; Glauser, M. P. Treatment of *Staphylococcus aureus* endocarditis in rats with coumermycin A1 and ciprofloxacin, alone or in combination. *Antimicrob. Agents Chemother.* **1987**, *31* (4), 539–543.
54. Kirby, W. M.; Hudson, D. G.; Noyes, W. D. Clinical and laboratory studies of novobiocin, a new antibiotic. *AMA Arch. Intern. Med.* **1956**, *98* (1), 1–7.
55. Sletten, E. M.; Bertozzi, C. R. Bioorthogonal reactions. *Acc. Chem. Res.* **2011**, *44* (9), 666–676.
56. Prescher, J. A.; Bertozzi, C. R. Chemistry in living systems. *Nat. Chem. Biol.* **2005**, *1* (1), 13–21.
57. Harmrolfs, K.; Mancuso, L.; Drung, B.; Sasse, F.; Kirschning, A. Preparation of new alkyne-modified ansamitocins by mutasynthesis. *Beilstein J. Org. Chem.* **2014**, *10*, 535–543.
58. Taft, F.; Brünjes, M.; Floss, H. G.; Czempinski, N.; Grond, S.; Sasse, F.; Kirschning, A. Highly active ansamitocin derivatives: Mutasynthesis using an AHBA-blocked mutant. *ChemBioChem* **2008**, *9* (7), 1057–1060.
59. Salazar, M. D. A.; Ratnam, M. The folate receptor: What does it promise in tissue-targeted therapeutics? *Cancer Metastasis Rev.* **2007**, *26* (1), 141–152.
60. Low, P. S.; Antony, A. C. Folate receptor-targeted drugs for cancer and inflammatory diseases. *Adv. Drug Deliv. Rev.* **2004**, *56* (8), 1055–1058.

61. Riva, E.; Wilkening, I.; Gazzola, S.; Li, W. M. A.; Smith, L.; Leadlay, P. F.; Tosin, M. Chemical probes for the functionalization of polyketide intermediates. *Angew. Chem. Int. Ed.* **2014**, *53* (44), 11944–11949.
62. Grüschow, S.; Smith, D. R. M.; Gkotsi, D. S.; Goss, R. J. M. Halogenases. In: *Biocatalysis in Organic Synthesis*, Vol. 3, Faber, K., Fessner, W.-D., Turner, N., Eds., Thieme Medical Publishers, Stuttgart, Germany, **2015**, 313–352.
63. Feling, R. H.; Buchanan, G. O.; Mincer, T. J.; Kauffman, C. A.; Jensen, P. R.; Fenical, W. Salinosporamide A: A highly cytotoxic proteasome inhibitor from a novel microbial source, a marine bacterium of the new genus *Salinospora*. *Angew. Chem. Int. Ed.* **2003**, *42* (3), 355–357.
64. Eustáquio, A. S.; Pojer, F.; Noel, J. P.; Moore, B. S. Discovery and characterization of a marine bacterial SAM-dependent chlorinase. *Nat. Chem. Biol.* **2008**, *4* (1), 69–74.
65. Sánchez, C.; Zhu, L.; Braña, A. F.; Salas, A. P.; Rohr, J.; Méndez, C.; Salas, J. A. Combinatorial biosynthesis of antitumor indolocarbazole compounds. *Proc. Natl. Acad. Sci. USA* **2005**, *102* (2), 461–466.
66. Eustáquio, A. S.; Gust, B.; Li, S.-M.; Pelzer, S.; Wohlleben, W.; Chater, K. F.; Heide, L. Production of 8′-halogenated and 8′-unsubstituted novobiocin derivatives in genetically engineered *Streptomyces coelicolor* strains. *Chem. Biol.* **2004**, *11* (11), 1561–1572.
67. Heide, L.; Westrich, L.; Anderle, C.; Gust, B.; Kammerer, B.; Piel, J. Use of a halogenase of hormaomycin biosynthesis for formation of new clorobiocin analogues with 5-chloropyrrole moieties. *ChemBioChem* **2008**, *9* (12), 1992–1999.
68. Runguphan, W.; Qu, X.; O'Connor, S. E. Integrating carbon-halogen bond formation into medicinal plant metabolism. *Nature* **2010**, *468* (7322), 461–464.
69. Glenn, W. S.; Nims, E.; O'Connor, S. E. Reengineering a tryptophan halogenase to preferentially chlorinate a direct alkaloid precursor. *J. Am. Chem. Soc.* **2011**, *133* (48), 19346–19349.
70. Wang, S.; Zhang, S.; Xiao, A.; Rasmussen, M.; Skidmore, C.; Zhan, J. Metabolic engineering of *Escherichia coli* for the biosynthesis of various phenylpropanoid derivatives. *Metab. Eng.* **2015**, *29*, 153–159.
71. Li, X.-Z.; Wei, X.; Zhang, C.-J.; Jin, X.-L.; Tang, J.-J.; Fan, G.-J.; Zhou, B. Hypohalous acid-mediated halogenation of resveratrol and its role in antioxidant and antimicrobial activities. *Food Chem.* **2012**, *135* (3), 1239–1244.
72. Mahoney, K. P. P.; Smith, D. R. M.; Bogosyan, E. J. A.; Goss, R. J. M. Access to high value natural and unnatural products through hyphenating chemical synthesis and biosynthesis. *Synthesis (Germany)*. **2014**, *46*, 2122–2132.
73. Roy, A. D.; Grüschow, S.; Cairns, N.; Goss, R. J. M. Gene expression enabling synthetic diversification of natural products: Chemogenetic generation of pacidamycin analogs. *J. Am. Chem. Soc.* **2010**, *132* (35), 12243–12245.
74. Goss, R. J. M.; Shankar, S.; Fayad, A. A. The generation of "unnatural" products: Synthetic biology meets synthetic chemistry. *Nat. Prod. Rep.* **2012**, *29* (8), 870–889.
75. Barder, T. E.; Walker, S. D.; Martinelli, J. R.; Buchwald, S. L. Catalysts for suzuki-miyaura coupling processes: Scope and studies of the effect of ligand structure. *J. Am. Chem. Soc.* **2005**, *127* (13), 4685–4696.
76. Runguphan, W.; O'Connor, S. E. Diversification of monoterpene indole alkaloid analogs through cross-coupling. *Org. Lett.* **2013**, *15* (11), 2850–2853.
77. Rutledge, P. J.; Challis, G. L. Discovery of microbial natural products by activation of silent biosynthetic gene clusters. *Nat. Rev. Microbiol.* **2015**, *13* (8), 509–523.
78. Zarins-Tutt, J. S.; Barberi, T. T.; Gao, H.; Mearns-Spragg, A.; Zhang, L.; Newman, D. J.; Goss, R. J. M. Prospecting for new bacterial metabolites: A glossary of approaches for inducing, activating and upregulating the biosynthesis of bacterial cryptic or silent natural products. *Nat. Prod. Rep.* **2016**, *33*, 54–72.

79. Whicher, J. R. et al. Structural rearrangements of a polyketide synthase module during its catalytic cycle. *Nature* **2014**, *510* (7506), 560–564.
80. Ye, Z.; Musiol, E. M.; Weber, T.; Williams, G. J. Reprogramming acyl carrier protein interactions of an acyl-CoA promiscuous trans-acyltransferase. *Chem. Biol.* **2014**, *21* (5), 636–646.
81. Koryakina, I.; McArthur, J.; Randall, S.; Draelos, M. M.; Musiol, E. M.; Muddiman, D. C.; Weber, T.; Williams, G. J. Poly specific trans-acyltransferase machinery revealed via engineered acyl-Coa synthetases. *ACS Chem. Biol.* **2013**, *8* (1), 200–208.
82. Parascandolo, J. S.; Havemann, J.; Potter, H. K.; Huang, F.; Riva, E.; Connolly, J.; Wilkening, I.; Song, L.; Leadlay, P. F.; Tosin, M. Insights into 6-methylsalicylic acid bio-assembly by using chemical probes. *Angew. Chem. Int. Ed.* **2016**, *55* (10), 3463–3467.

5 Terrestrial Microbial Natural Products Discovery Guided by Symbiotic Interactions and Revealed by Advanced Analytical Methods

Rita de Cássia Pessotti, Andrés Mauricio Caraballo-Rodríguez, Humberto Enrique Ortega-Domínguez, and Mônica Tallarico Pupo

CONTENTS

5.1 INTRODUCTION

Natural products continue to play a significant role in the discovery and development of new drugs,[1] and this is probably one of the great motivations of the continuous research in the field. Natural products from different sources have impacted almost all therapeutic areas; however, the most impressive contribution can be exemplified by the antibacterial and anticancer drugs. Just considering natural products and their semisynthetic derivatives, 73% of 112 molecules approved (by the Food and Drug Administration [FDA] or their equivalents in other countries) as antibacterial agents from 1981 to 2014 were derived from microorganisms, while 40% of 136 small molecules approved as anticancer drugs in the same period were derived from natural products and their semisynthetic derivatives, and of these, microorganisms have contributed to 10% of the total number of approved anticancer drugs.[1] The success of naturally occurring small molecules in drug discovery is likely related to their evolutionary history, by which they have been selected to fulfill physiological functions for the producing organisms.[2] However, the knowledge of these physiological and/or biological roles for the producers is still incomplete.[3]

Microorganisms coexist in nature in complex interactions with other microorganisms and macro-organisms; therefore, secondary metabolites might play a role in the communication and mediation of such biotic interactions. Using an ecological approach in natural product drug discovery could improve the search efficiency to isolate new therapeutic agents instead of using conventional strategies.[4]

In fact, most advances by mankind are inspired through direct observation of nature. According to Darwin's theory of evolution, all living organisms are products of millions of years of evolution under constant pressure of natural selection,[5] which makes learning from nature a wise choice. Interestingly, although Darwin did not include microorganisms in his theory, they are the best evidence of natural selection for his theory of evolution.[6]

A famous observation that led to the discovery of a product that changed our history was reported back in the late 1920s. By incidentally observing an interaction between a bacterium and a fungus, Alexander Fleming reported that it is possible to use microorganisms to heal bacterial infections. He observed that a bacterial strain could not grow close to a contaminant fungus,[7] later identified as *Penicillium notatum*. After many years of work on this observation, researchers found that the inhibition zone caused by this fungus was due to a secreted antibiotic compound named penicillin. The medicinal use of penicillin is a milestone in public health (the *miracle drug* of the twentieth century)[8] and has boosted the study of microorganisms from a pharmacological perspective.[3] After the discovery of this promising class of bioactive natural products, scientists all over the world put their efforts into studying microorganisms focusing on the discovery of antimicrobial agents. Since then different classes of compounds have been discovered: aminoglycosides, cephalosporins, tetracyclines, macrolides, and glycopeptides among others.[9] At this point, it is important to highlight that *antibiotic activity* may represent only a partial description of the biological function of a molecule in nature; an antibiotic may be better described as "a therapeutic agent produced by a pharmaceutical company."[10]

These microbial compounds (natural products) are products of their specialized metabolism, often called secondary metabolism. This comprises not only antimicrobials, but also antitumor agents, pesticides, and immunosuppressants among others,[11] including isolated compounds with no biological activity reported as yet. Regulation of this secondary metabolism is not completely understood, and it is not constitutively expressed as the primary metabolism.[10] Some factors that can trigger the biosynthesis of some compounds are the onset of microbial stationary phase, nutrient depletion, and/or some inducer compounds like mycolic acid and butyrolactones.[11]

The role of these compounds for their producers is not fully understood.[3] They are not required for microbial growth.[10,12] If they are not essential but are still kept in microbial genome that is under a constant natural selection pressure, it leads to the inference that natural products somehow increase the fitness of their producer in specific situations.[13] Genome sequencing of different microorganisms has shown a vast number of biosynthetic gene clusters (BGCs) that are not expressed under laboratory conditions.[10,14] Studying the natural role of secondary metabolites would help us in understanding their specific regulation. This knowledge can further help in awakening these silent gene clusters and potentially get new useful compounds for therapeutics.[15] The research field that brings a biological perspective to the study of natural products, in trying to describe their role in nature, is called chemical ecology.[16]

In their natural niche, microorganisms are in constant interaction with abiotic and biotic factors of the environment. One well-known and widespread interaction is symbiosis. The concept of symbiosis has been used broadly with diverging meanings.[17] In this review, we adopt de Bary's classical concept: "close, long-term associations between different organisms."[18] One way microorganisms communicate with their partners in a symbiotic relationship is through secreted compounds—natural products.[19,20] In a positive symbiosis, partners benefit from this association by increasing each other's fitness. In this sense, symbiotic interactions are a rich source of bioactive compounds that were selected along the course of evolution. Therefore, natural selection is a valuable tool for natural products research aiming to discover potential candidates for therapeutic agents.

A classic example of symbiosis that has been studied as a source of natural products is the plant-microorganisms system. More recently, the insect-microorganisms system also became an important part of the natural products research field. The following sections describe these systems and how their study has been contributing to bioactive natural products research.

5.2 CHEMICAL ECOLOGY KNOWLEDGE FOR SELECTING MICROBIAL STRAINS AS SOURCES OF BIOACTIVE NATURAL PRODUCTS

5.2.1 Plant-Microorganisms System

Microorganisms that live inside plant tissues and cells without harming their host are known as endophytes. It has been suggested that the plant host provides nutrients and protection for the endophytes, and the latter provides bioactive natural products that can, for instance, promote growth and protection against pathogens and herbivores.[21]

A study performed by Mousa and collaborators[22] exemplifies how endophytes can provide protection to their host plant. It is reported that wild maize is less susceptible to pests than nonwild genotypes, and it was hypothesized that endophytes may play a role in this observed difference.[22] Endophytes were isolated from diverse maize genotypes, with 46% of the strains isolated from wild maize and 54% from nonwild maize genotypes. Endophytes were tested for their ability to control the gibberella ear rot (GER) disease that is caused by the phytopathogen *Fusarium graminearum*. Four strains were active, with three that were isolated from wild maize genotypes being the most active ones. Out of these four strains, three were identified as *Paenibacillus polymyxa*, which is known to be a good plant colonizer and a conserved endophyte of maize. Chemical and molecular investigation showed the production of the nonribosomal peptides fusaricidins A, C, and D by these endophytic strains, which had already been described as anti-*Fusarium* compounds. Therefore, these results corroborate the hypothesis that wild maize genotypes are less susceptible to GER due to specific endophytes that may defend the plant by producing antifungal compounds.[22]

Compounds that protect the plant host against pathogens are of special interest for therapeutics, since these pathogens include bacteria and fungi, and, therefore, antimicrobial compounds. Endophytes have been extensively studied over the last two decades and have indeed proved to be a good source of bioactive compounds. Several bioactive compounds were isolated from endophytic fungi with diverse structures, such as the antibacterial epicolactone (**1**) and dothideomycetide A (**2**),[23–25] the antifungal pestalotether A (**3**) and microsphaerophthalide A (**4**),[26,27] and the anticancer annulosquamulin (**5**) and MBJ-0011 (**6**),[28,29] among many other compounds (Figure 5.1).

5.2.1.1 Taxol: The Blockbuster Anticancer Drug

A classic and outstanding example of a bioactive compound also produced by plant endophytes is the diterpenoid paclitaxel (**7**) (Figure 5.2), which is sold as Taxol, an important anticancer drug with a unique mode of action. The history of paclitaxel has been evolving in an interesting way over the last decades. It started as a bioactive compound found in random National Cancer Institute (NCI) screening for anticancer compounds. It was initially isolated from a plant (*Taxus brevifolia*), and was later developed as an important anticancer drug. The remarkable activity of this compound and the problem of getting enough quantities of it from *T. brevifolia* bark stimulated the research to find more reasonable sources of this compound.

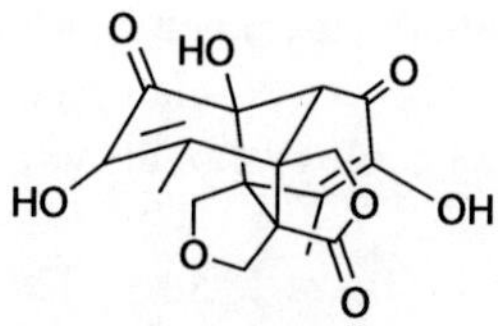

Epicolactone (**1**)

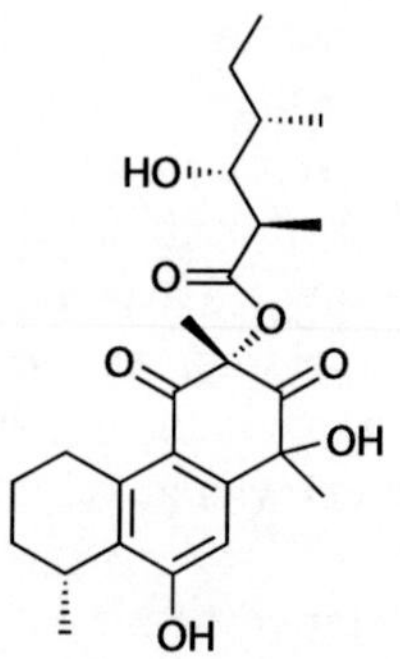

Dothideomycetide A (**2**)

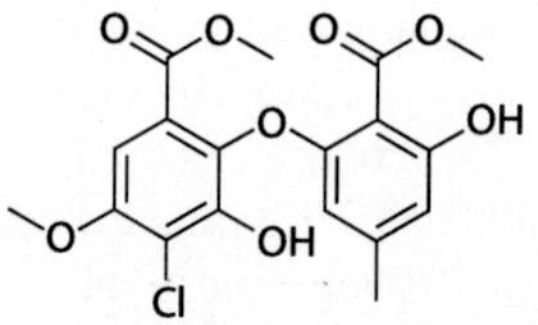

Pestalotether A (**3**)

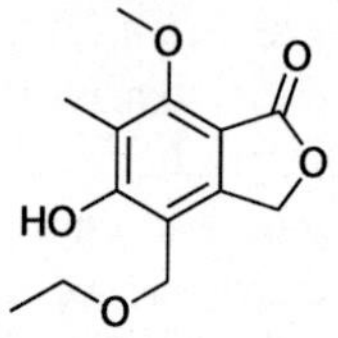

Microsphaerophthalide A (**4**)

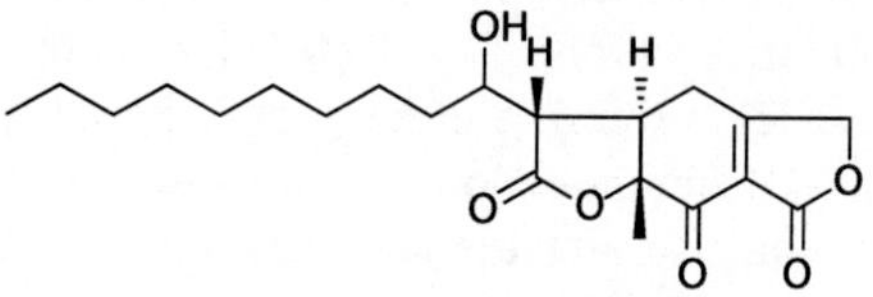

Annulosquamulin (**5**)

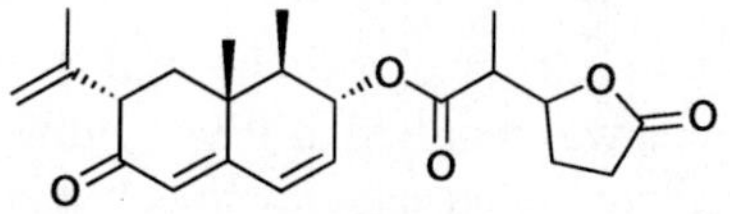

MBJ-0011 (**6**)

FIGURE 5.1 Bioactive compounds isolated from endophytes.

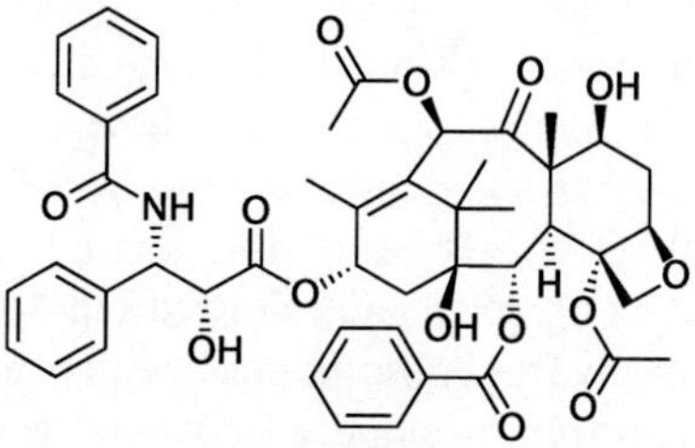

Paclitaxel (**7**)

FIGURE 5.2 Paclitaxel, generic name for the anticancer drug Taxol.

Initially, other trees from *Taxus* genera were investigated, and many alternatives were tried, such as synthesis, semisynthesis, and plant tissue culture. (For a more detailed review on this trajectory, see Cragg and collaborators.[30]) Total synthesis was first published in 1994, but it is still too expensive and complex, and presents low yields and some toxic side products, making it not viable for the industry.[31] Semisynthesis has been successful and is used to commercially produce Taxol. Two precursors can be used—baccatin III and 10-deacetylbaccatin III—that are found in *Taxux* spp. needles, which have the advantage of being a renewable source (the tree doesn't need to be cut to collect them), but it still depends on old *Taxus* spp. individuals.[30,31] Plant-cell tissue culture is also currently used to produce Taxol commercially. Some advantages of this approach are that it is a renewable and reliable source of this compound.[31] Interestingly, a plant hormone—methyl-jasmonate—promotes increased yields of paclitaxel in plant-cell cultures.[32] It is even more interesting to note that jasmonates play a role in the regulation of defense-related genes in plants,[33] corroborating the idea that knowledge of chemical ecology can be helpful in bioactive natural products research.

In the 1990s, microorganisms started to play a role in Taxol history. The first report of an endophytic fungus of *T. brevifolia* that was able to produce paclitaxel dates back to 1993 (fungus: *Taxomyces andreanae*). It is now well reported in the literature that several compounds first isolated from plants are actually produced by their microbial symbionts (see Reference 34 for a review on this issue). Over the last two decades, several research groups have reported the production of paclitaxel by at least 40 different endophytic fungal genera.[35] Nevertheless, endophytic fungi cultures do not give a stable and substantial yield, and are therefore not currently interesting for the industry.[35]

5.2.1.2 Chemical Biology History behind the Anticancer Drug Taxol

The interesting history of paclitaxel goes beyond its activity and source. The fungicide activity of paclitaxel was reported in 1994[36]; however, it was not known whether this compound indeed acted as a fungicide in nature. A study performed by Soliman and collaborators[37] showed in an elegant way the biology behind the biosynthesis of this compound in nature.

Taxus spp. develops some branch-associated bark cracks (points where branches grow), which are a favorable place for pathogens to attack the plant. At this site, it becomes difficult to imagine how the plant would use paclitaxel as a fungicide to protect itself, since this compound inhibits cell division, which in turn stops branch growth. It was observed (in nature and also *in vitro*) that when the bark cracks, hyphae of the endophytic strain *Paraconiothyrium* SSM001 start to infiltrate and dominate it from the plant vascular system. This fungus was already known to be a paclitaxel producer. A series of experiments showed that this strain keeps paclitaxel stored intracellularly inside hydrophobic bodies, which prevents it from harming plant cells. Interestingly, this strain is induced by wood-decaying fungi (phytopathogens) to release these hydrophobic bodies at the endophyte-pathogen interface. It results in an extraordinary local extracellular fungicide barrier, protecting the plant host against invasion of fungal pathogens at these vulnerable points just in the presence of an attack, with the bonus of reducing the phytotoxic collateral effect of paclitaxel on the host.

Since it was also proved that paclitaxel inhibits the growth of a couple of wood-decaying fungi, the authors suggest that this endophyte plays an important role as a novel plant defense mechanism. The authors argue that while *Taxus* spp. produce paclitaxel as a systemic protection, endophytes act locally at sites, where it would be harmful for the plant to protect itself using this same compound (cell-division sites such as bud branches), releasing paclitaxel close to the pathogen and only when needed.[37] This is a neat and outstanding example that corroborates the hypothesis that endophytes increase the fitness of their plant host.

Taxol was discovered randomly. But the way its history evolved reinforces the idea of learning by observing nature. The discovery and study of paclitaxel shows the importance of doing a backward investigation by considering the chemical ecology of symbiosis on bioactive natural products research. Plants and microorganisms have been writing a history of symbiosis that is continually shaped by evolution through natural selection. Therefore, endophytes are a rich source of natural products, which plants can take advantage of, as in the example of paclitaxel.

5.2.2 Insect-Associated Microbe Systems

There have been many published examples of different symbiotic interactions between insects and microorganisms, but most of the better described ones are based on nutritional or defensive services provided by the symbionts to their hosts. In nutritional mutualisms, they can provide nutrients or digestive enzymes; and in defensive interactions, they can protect them against pathogens, parasites, parasitoids, or predators, through the production of antibiotics or toxins that may, therefore, be useful as therapeutics. These interactions help insects in their evolutionary process to survive in different environments.[38]

Several actinobacteria have been described that produce secondary metabolites that defend insects such as ants, beetles, and wasps against pathogenic microorganisms. These bacteria are suitable symbionts since they develop the capability to exploit energy sources available from the insect, produce spores for their transmission and survival under unfavorable conditions, and produce a wide variety of potent antimicrobial compounds.[39] The next topics will cover some examples of bioactive compounds produced by microbial symbionts isolated from insects.

5.2.2.1 Fungus-Growing Ants

One of the best known examples in mutualistic relationship between microorganisms and insects is the complex symbioses in a fungus-growing ant (tribe Attini) colony. The fungus-growing ants can be divided into five agricultural systems: lower agriculture; coral fungus agriculture; yeast agriculture; generalized higher agriculture; and leaf-cutter agriculture, which has evolved remarkably recently (≈8–12 million years ago) to become the dominant herbivores of the New World tropics. Leaf-cutter agriculture involves different species of two major genera, *Atta* and *Acromyrmex*, with the ability to cut and process fresh vegetation as a nutritional substrate for their fungal crop.[40] These ants culture a fungus (phylum Basidiomycota) that serve as a major food source, and have a symbiotic bacterium (phylum Actinobacteria) that produces secondary metabolites, which can suppress growth of the specialized pathogenic fungus of the genus *Escovopsis*.[41] Additionally, the presence of black yeast inside these colonies was revealed, which is a potential competitor that reduces the bacterial symbiont's growth.[42]

Some chemical and biological studies have been carried out on fungus-growing ant ecosystems. A neat example is the description of a new cyclic depsipeptide, dentigerumycin (**8**) (Figure 5.3), produced by the symbiotic *Pseudonocardia* strain, isolated from the exoskeleton of the coral fungus agriculture ant, *Apterostigma dentigerum*. This compound has a selective inhibition against the pathogenic fungus *Escovopsis* sp. Dentigerumycin also inhibits wild-type *Candida albicans*, *C. albicans* ATCC10231, and amphotericin-resistant *C. albicans* ATCC200955 with minimum inhibitory concentration (MIC) values of 1.1 μM.[43] More recently, new smaller analogues of dentigerumycin, named gerumycins A–C (**9–11**) (Figure 5.3), have been isolated from *Pseudonocardia* spp. strains associated with *Apterostigma* spp. and *Trachymyrmex cornetzi* ants. These compounds are at least three orders of magnitude less potent than dentigerumycin at suppressing *Escovopsis* growth. The chemical structure of gerumycins is similar to that of the cytotoxic hexadepsipeptides, piperazimycins, isolated from a marine *Streptomyces*.[41]

Additional studies were carried out on *Pseudonocardia* strains isolated from *A. dentigerum* ants from different colonies to better understand their highly specialized ecological niche. It was demonstrated that some of these actinobacteria can inhibit growth of other *Pseudonocardia* strains isolated from different colonies,

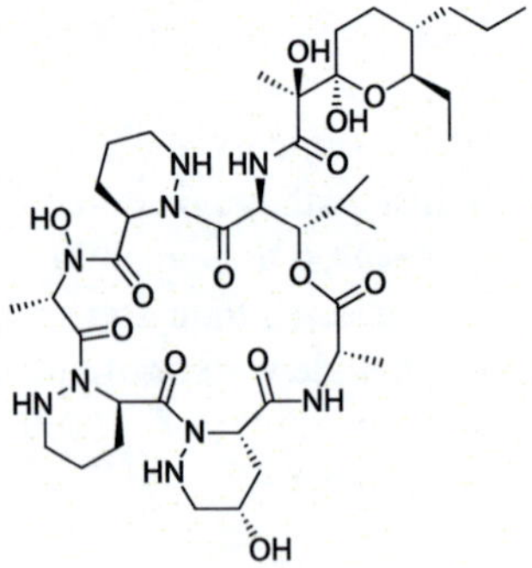

Dentigerumycin (**8**)

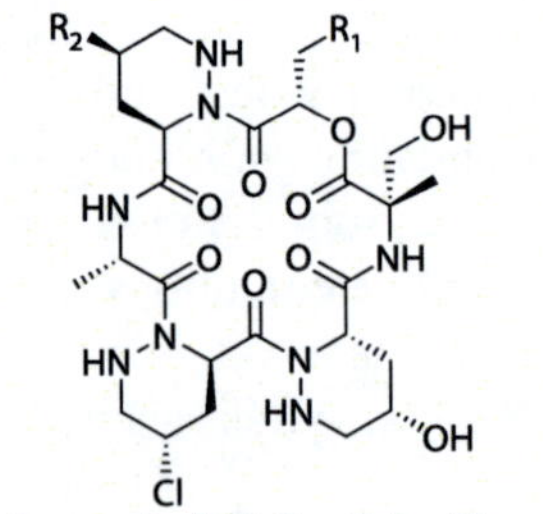

Gerumycin A (**9**): $R_1 = H$; $R_2 = H$
Gerumycin B (**10**): $R_1 = CH_3$; $R_2 = H$
Gerumycin C (**11**): $R_1 = H$; $R_2 = OH$

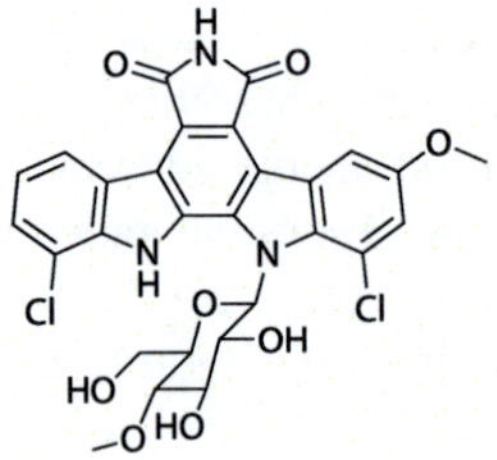

9-Methoxyrebeccamycin (**12**)

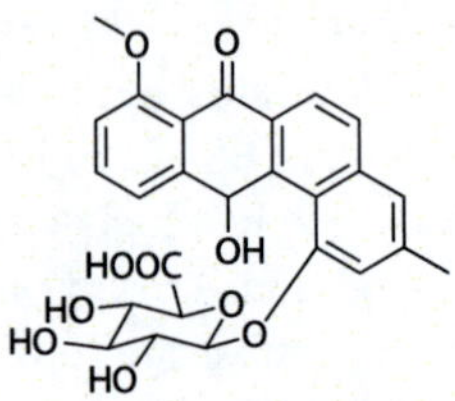

Pseudonocardone A (**13**)

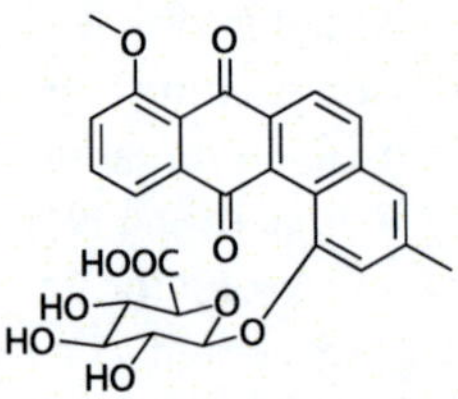

Pseudonocardone B (**14**)

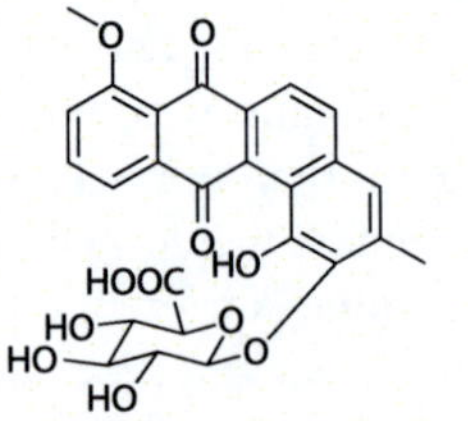

Pseudonocardone C (**15**)

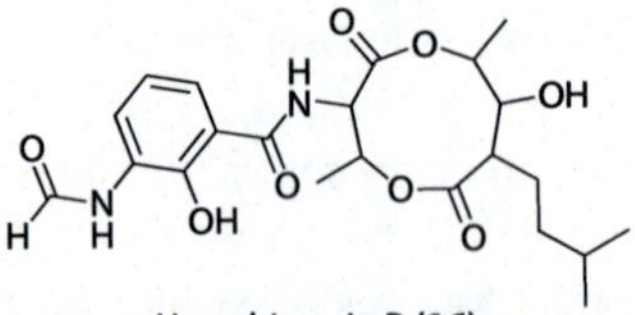

Urauchimycin B (**16**)

FIGURE 5.3 Compounds isolated from symbionts of fungus-growing ants.

showing a possible competition among strains for establishing the symbiotic relationship with an ant colony. Further studies lead to the discovery of a new indolocarbazole named 9-methoxyrebeccamycin (**12**) (Figure 5.3), which was isolated through activity-guided fractionation. It showed potent activity in an agar-based assay against a small panel of *Pseudonocardia*. This compound has several analogues that have been used in clinical trials for different cancer types.[44] It is a neat example of how an ecologically driven bioassay can lead to the isolation of new bioactive compounds.

From the same ant species *A. dentigerum* three new angucyclines, pseudonocardones A–C (**13–15**) (Figure 5.3), together with the known antibiotic 6-deoxy-8-*O*-methylrabelomycin and X-14881 E have been isolated from a *Pseudonocardia* strain. The new angucyclines did not show significant biological activity, while 6-deoxy-8-*O*-methylrabelomycin and X-14881 E showed activity against *Bacillus subtilis* 3610 and liver-stage *Plasmodium berghei*.[45]

Several known antibiotic compounds, candicidin D, actinomycin D, actinomycin X_2, valinomycin, and antimycin A_1–A_4 have also been identified from *Streptomyces* strains isolated from several leaf-cutter ant colonies of the genus *Acromyrmex*. These compounds showed high inhibition activity against *Escovopsis*. The presence of valinomycin on the integument of *Acromyrmex* workers and in the waste of some colonies was shown, while the actinomycins were only observed in the waste, supporting the importance of these compounds to keep colonies healthy against pathogenic microorganisms. Actinomycins also have the capability to affect the growth of soil bacteria, as well as other *Streptomyces* and *Pseudonocardia* symbionts. It was also observed that antimycins inhibit the mutualistic fungal garden, *Leucoagaricus gongylophorus*.[46,47] These examples suggest a complex chemical interaction among the different microbial strains associated with ants.

As part of a natural product drug discovery from symbionts of fungus-growing ants, the rare antimycins urauchimycins A and B were identified from *Streptomyces* sp. TD025 isolated from workers of the higher agricultural *Trachymyrmex* ants. These compounds were evaluated for antifungal activity against a panel of *Candida* spp. Urauchimycin B (**16**) (Figure 5.3) showed a high activity with MIC values equivalent to nystatin. Urauchimycins A and B were previously reported from a marine *Streptomyces* strain.[48]

The literature describes many examples of bioactive compounds produced by microorganisms associated with ant colonies that might be helping their host by regulating the growth of the pathogenic fungus *Escovopsis*, demonstrating how ants can benefit from this symbiosis. Some of these compounds, which were found on screening against a natural pathogen, exhibited good bioactivity against human pathogens and cancer cell lines. Chemical ecology studies describing this association between fungus-growing ants and microorganisms are leading to the discovery of many interesting compounds, demonstrating how ecological knowledge can help in finding new candidates for future drug development.

5.2.2.2 Fungus-Growing Termites

The fungus-growing termites (Macrotermitinae: single Termitidae subfamily) are other social insects with similar behavior to the fungus-growing ants. There are around 11 genera and 330 species reported. Termites have had important symbiotic

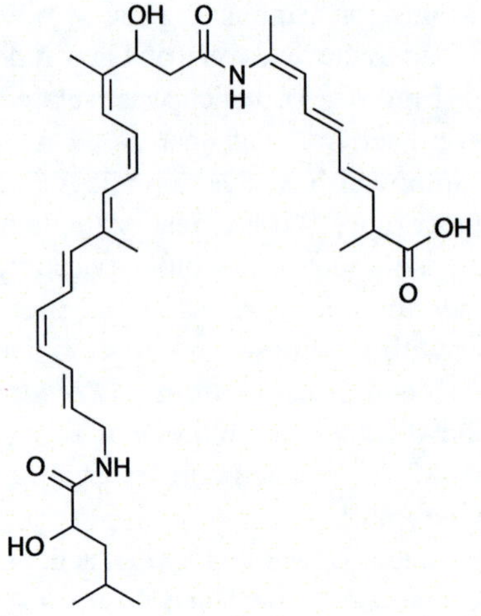

Bacillaene A (**17**)

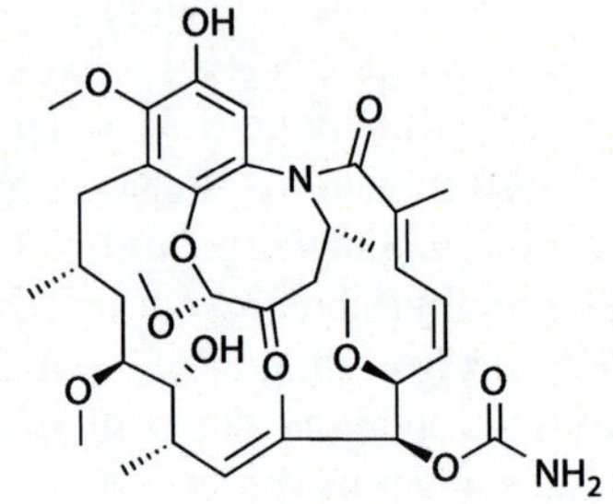

Natalamycin A (**18**)

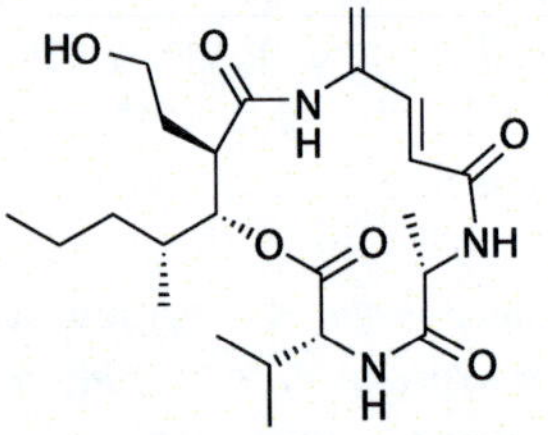

Microtermolide A (**19**)

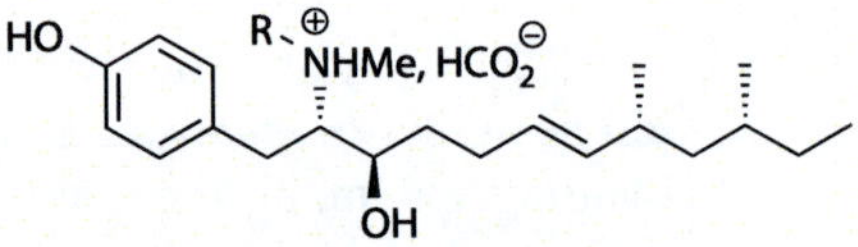

Tyroscherin (**20**): R = H
N-methyltyroscherin (**21**): R = Me

FIGURE 5.4 Compounds isolated from symbionts of fungus-growing termites.

relationships during their evolution with gut microorganisms such as protists, methanogenic archaea, and bacteria. However, the Macrotermitinae have only shown mutualistic symbiosis with a basidiomycete fungus of the genus *Termitomyces*, which helps them in degradation of plant material and as food source for the colony.[49] It was reported that the fungi *Pseudoxylaria* and *Trichoderma* can be potential antagonists against *Termitomyces*. So, fungus-growing termites need a mechanism of defense to maintain their cultivar fungus. Through chemical and biological analysis of this system, bacillaene A (**17**) (Figure 5.4) was identified as being produced by a *Bacillus* sp. strain isolated from colonies of *Macrotermes natalensis*. This compound showed a selective inhibition against the fungi *Pseudoxylaria* and *Trichoderma*. It also inhibited the fungi *Coriolopsis* sp., *Umbelopsis* sp., and *Fusarium* sp., all considered as other possible competitors isolated from the same colony.[50] In another study, a new fused bicyclic [6.4.0] ansa macrolide, named natalamycin A (**18**) (Figure 5.4), together with other analogues produced by a *Streptomyces* sp. symbiont from an *M. natalensis* Mn802 colony were described. These compounds were isolated by activity-guided fractionation against *Pseudoxylaria* X802.[51]

Further analyses have been carried out on symbionts of fungus-growing termites. Two new compounds, named 2-formylpyrrole-4-acrylamide and dihydrostreptazolin, were isolated from a *Streptomyces* strain residing in the gut of *Odontotermes formosanus*.[52] Two new polyketide synthase-nonribosomal peptide synthetase hybrid, microtermolides A (**19**) (Figure 5.4) and B, were isolated from a *Streptomyces* strain associated with *Microtermes* sp.[53] Two additional antifungal compounds, tyroscherin (**20**) and *N*-methyltyroscherin (**21**) (Figure 5.4), were discovered from the fungus *Pseudallescheria boydii*, isolated from *Nasutitermes* sp. Both compounds showed activity against *C. albicans* and *Trichophyton rubrum*.[54]

As exemplified, fungus-growing termites seem also to rely on antibiotics produced by symbionts to keep their colony healthy, and are therefore another interesting source of bioactive compounds that can be useful in therapeutics.

5.2.2.3 Beetles

In other complex insect symbiotic association studies, it was observed that the southern pine beetle, *Dendroctonus frontalis*, utilizes the novel polyene peroxide, mycangimycin (**22**) (Figure 5.5), produced by a mutualisitc *Streptomyces* sp. SPB74, to control the growth of the antagonistic fungus *Ophiostoma minus* and protect its larval food fungus *Entomocorticium* sp. Mycangimycin inhibits different *C. albicans* strains and has a potent activity against *Plasmodium falciparum*, comparable to clinical antimalarial drugs such as artemisinin, chloroquine, pyrimethamine, and mefloquine.[55] In another study, two new polycyclic tetramate macrolactams, named frontalamides A (**23**) and B (**24**) (Figure 5.5), were discovered from *Streptomyces* sp. SPB78 isolated from the same beetle. Frontalamides have a similar chemical structure to the antibiotics dihydromaltophilin, maltophilin, cylindramide, ikarugamycin, alteramide, and discodermide.[56]

Chemical and biological studies were performed on the symbionts of the dung beetle ecosystem. This insect uses feces as a reservoir for its eggs. Due to the vulnerability of the larvae against pathogenic microorganisms, they need mechanisms of defense. Therefore, it may be suggested that microbial symbionts play a role in this process. In fact, some bioactive natural products were isolated from this system, with interesting chemical structures. Two new phenylpyridines, coprismycins A (**25**) (Figure 5.5) and B, along with some known dipyridines; an unprecedented cyclobutane-bearing tricyclic lactam compound, named tripartilactam (**26**); a new dichlorinated indanone, named tripartin (**27**); and two new cyclic heptapeptides, coprisamides A (**28**) (Figure 5.5) and B, were isolated from *Streptomyces* sp. strains associated with the brood ball and gut of the dung beetle, *Copris tripartitus*.[57–60] The coprismycins A and B and the dipyridines displayed comparable neuroprotective effects against MPP+ (1-methyl-4-phenylpyrimidium)-induced neurotoxicity in neuroblastoma SH-SY5Y cells. Coprismycins A and B are the first phenylpyridines isolated from a microbial source. These classes of compounds have been reported from the plants *Nicotiana tabacum* and *Abelmoschus moschatus*.[57] Tripartilactam was reported with Na^+/K^+ ATPase inhibitor activity.[58] Tripartin was described with specific activity as an inhibitor of the histone H3 lysine 9 demethylase KDM4 in HeLa cells. Several histone demethylases (KDMs) are targets for the treatment of diseases such as leukemia, and breast and prostate cancer and inflammation.[59] Coprisamides showed significant activity for the induction of quinone reductase, a representative phase II detoxification enzyme that is considered to have significance

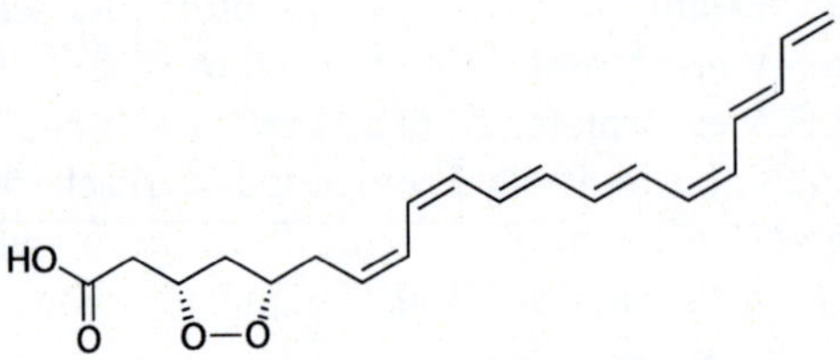

Mycangimycin (**22**)

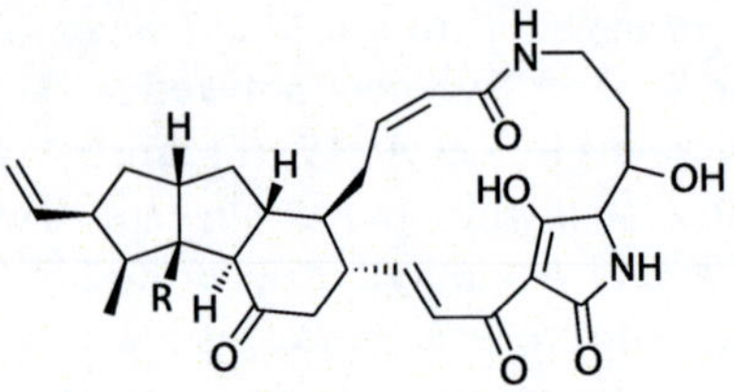

Frontalamide A (**23**): R=OH
Frontalamide B (**24**): R=H

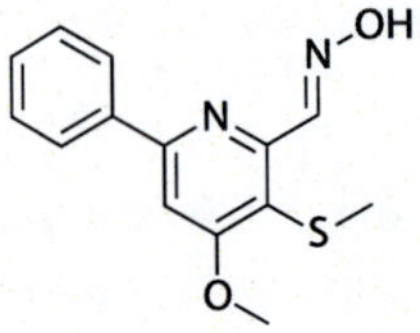

Coprismycin A (**25**)

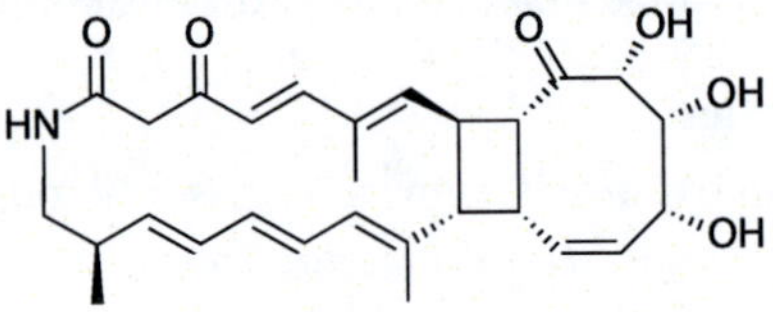

Tripartilactam (**26**)

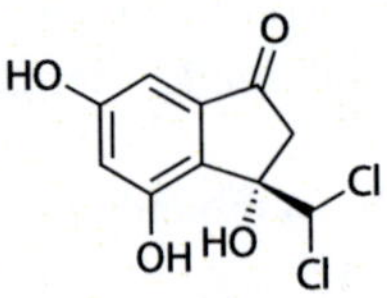

Tripartin (**27**)

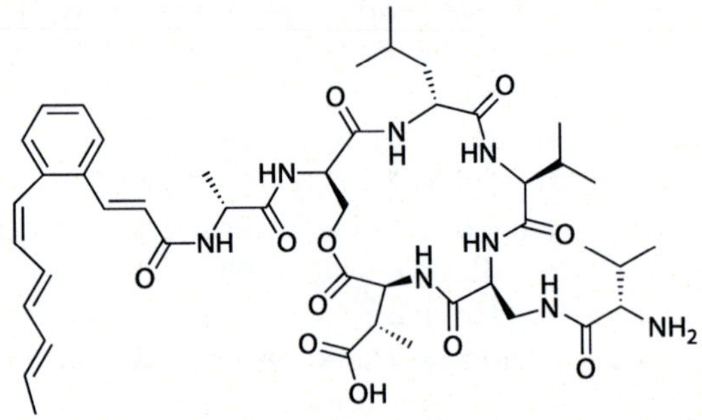

Coprisamide A (**28**)

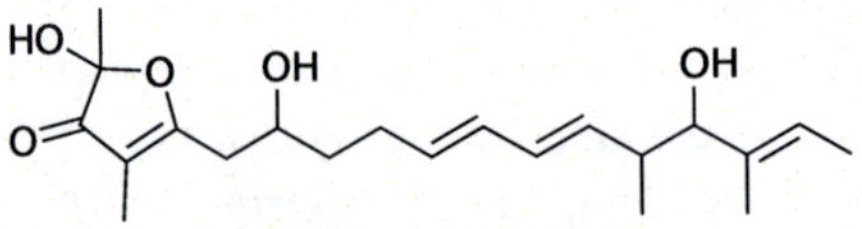

Actinofuranone C (**29**)

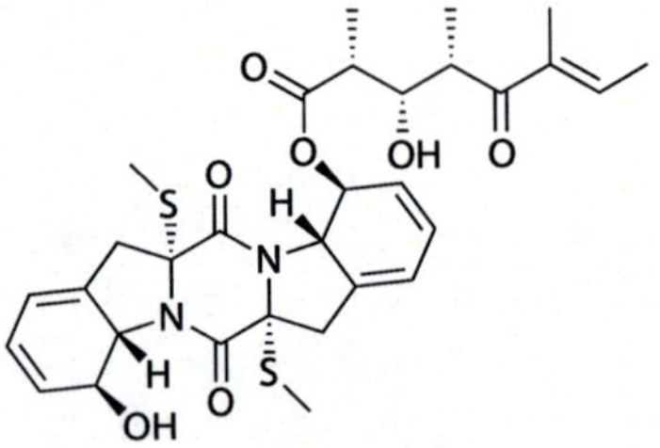

Boydine B (**30**)

FIGURE 5.5 Compounds isolated from symbionts of different beetle ecosystems.

in the prevention of cancer.[60] Studying different symbionts from the same ecosystem, a new 3-furanone-bearing polyketide, named actinofuranone C (**29**) (Figure 5.5), was described from an actinobacteria of genus *Amycolatopsis* associated with a female of the dung beetle *C. tripartitus*.[61] The ecological function of all these compounds in the dung beetle ecosystem remains unknown.

As part of an additional study on another beetle ecosystem, several compounds have been reported, including four new epipolythiodioxopiperazines, boydines

A–D, and two novel sesquiterpenes, boydines A and B, together with the known compounds bisdethiobis(methylthio)-deacetylaranotin, bisdethiodi(methylthio)-deacetylapoaranotin, AM6898 A, and ovalicin, produced by the fungus *P. boydii*, isolated from the gut of the larvae of the beetle *Holotrichia parallela*. Boydine B (**30**) (Figure 5.5) showed significant biological activity against clinically relevant bacterial strains *Bifidobacterium* sp., *Veillonella parvula*, *Anaerosterptococcus* sp., *Bacteroides vulgatus*, and *Peptostreptococcus* sp.[62] Therefore, the beetle-microorganisms system proved to be an important source of bioactive natural products that can be considered as potential candidates for drug discovery programs.

5.2.2.4 Wasps

Wasp-associated microbes are another valuable source of active natural products. For instance, streptochlorin (**31**), piericidin A_1 (**32**) (Figure 5.6), and seven other analogues were identified from cocoon extracts of the European beewolf, *Philanthus triangulum*. These antibiotics had not been detected in the cocoons when the *Streptomyces* symbionts of beewolf brood cells were removed, supporting the importance of the actinobacteria in the protection against pathogens.[63] A previously unreported 26-membered polyene macrocyclic lactam, sceliphrolactam (**33**) (Figure 5.6), was isolated from a *Streptomyces* symbiont of the mud dauber wasp, *Sceliphron caementarium*. This compound showed pronounced antifungal activity against amphotericin B-resistant *C. albicans*.[64] The chemical analyses of several *Streptomyces* strains from two species of solitary mud dauber wasps, *S. caementarium* and *Chalybion californicum*, revealed the production of a diverse collection of antibiotic compounds from six structural classes: antimycins, bafilomycins A_1 and B_1, daunomycin, mycangimycin, streptazolin and streptazon B, and sceliphrolactam.[65]

5.2.2.5 Bees

Symbiotic studies of social bees have not been properly explored as yet. It has been reported that the Brazilian social stingless bee *Scaptotrigona depilis* must cultivate

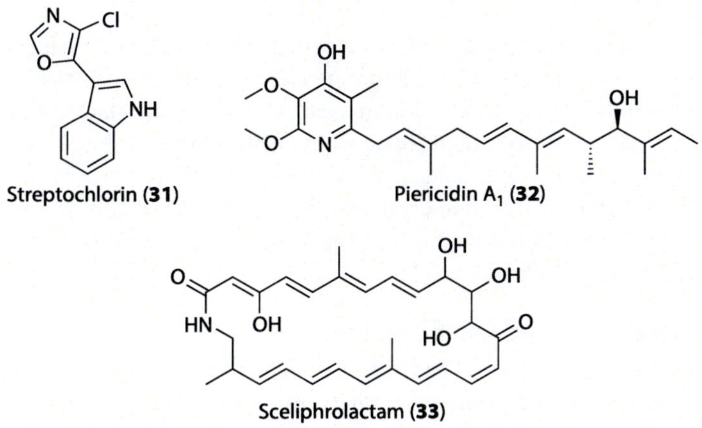

FIGURE 5.6 Compounds isolated and identified in wasp ecosystems.

a fungus to survive.[66] This fungus grows inside the brood cell as a source of food for the larvae. The mechanism of defense that may control the growth of pathogenic microorganisms in this system is unknown.[66] This finding opens an unprecedented opportunity for chemical ecology–guided studies in the search for natural products that might be involved in the mediation of microbial interactions and microbe-host interactions in *S. depilis* and other colonies of stingless bees. As observed in the previously described insect-microorganism systems, it is also expected that bioactive natural products isolated from bee-associated microorganisms might be useful to guide further pharmaceutical development.

5.3 CHEMICAL ECOLOGY KNOWLEDGE FOR SELECTING METHODOLOGICAL APPROACH

Laboratory culture conditions are far different from the niche where strains were originally collected, with modifications such as selected nutrients, shaking condition, constant temperature, and lack of competition among others. Secondary metabolism is not expressed constitutively[3] and may need some cues to be activated, which should be present in the natural environment in which the strains have evolved.[15] Therefore, several BGCs remain silent under classical culture approaches. These silent gene clusters may encode for useful compounds for therapeutics, which motivates the use of strategies seeking their activation.[67]

Competition for space and nutrients among microbes is possibly mediated by chemical compounds that are able to inhibit the growth of other strains as a strategy to eliminate the competitor, thus providing potential compounds for therapeutic use. Therefore, the presence of other strains in the culture medium in a coculture approach (as opposed to an axenic culture) can be a potential cue for triggering the expression of gene clusters related to secondary metabolism.[15,68] The following topics show how this approach, driven by chemical ecology knowledge, can help in isolating bioactive compounds.

5.3.1 Coculture Approach

This section presents only published articles that report induced or increased production of one or more identified compounds as a result of coculture conditions. Several other articles report induced/increased production of some compounds but their structures have not been identified; others also report inhibited production of some compounds in coculture. Literature reports many more cases of novel bioactive compounds that are induced in coculture, but here only cocultures among strains that share an ecological relationship are reported. For a comprehensive coculture approach review, see Bertrand et al.[68]

5.3.1.1 Coculture of Endophytic Strains

As discussed above, endophytes share a history of evolution in which both plant host and microorganisms benefit from each other. Inside the plant host, there are several microbial strains that coexist and are possibly sharing a history of evolution as well. Coculturing endophytic strains that were isolated from the same host is an approach that tries to mimic at least in part the natural niche of these strains and, therefore, it might include some needed cues for increasing the production of natural compounds or, more interestingly, triggering the expression of gene clusters that are silent under pure culture conditions.

Based on this ecologically driven approach, Soliman and Raizada[69] tested the influence of coculturing strains isolated from the same host on the production of paclitaxel (**7**). They cocultured cohabitating fungi from *Taxus* bark: *Alternaria alternata* (resident pathogen), *Phomopsis* sp. (endophyte), and the paclitaxel-producing endophyte *Paraconiothyrium* sp. SSM001. Paclitaxel production was increased by 7.8-fold.

In a coculture of the fungi *Alternaria tenuissima* and *Nigrospora spherica*, both isolated from the edible plant *Smallanthus sonchifolius* (yácon), the production of four polyketides by *A. tenuissima*—alterperylenol (**34**), altertoxin I (**35**), alternariol (**36**), and alternariol monomethyl ether (**37**)—was increased, and the antifungal stemphyperylenol (**38**) (Figure 5.7) was detected only in coculture.[70] Despite being cytotoxic, this latter compound was not harmful for the plant host when tested *in vitro*. This example sheds light on the regulation of endophytic community growth inside the plant host, which is important for keeping endophytes as they are not harmful for their host.

Two new quinolones were found in a coculture of two endophytes from a mangrove plant: marinamide (**39**) and its methyl ester derivative (**40**) (Figure 5.8).[71,72] Both exhibited cytotoxic activity. Another coculture, also of mangrove-derived endophytes, was performed by Zhu and collaborators in 2007, in which they observed the induction of two antimicrobial compounds: 6-methyl salicylic acid (**41**) and the diketopiperazine cyclo-Phe-Phe (**42**).[73]

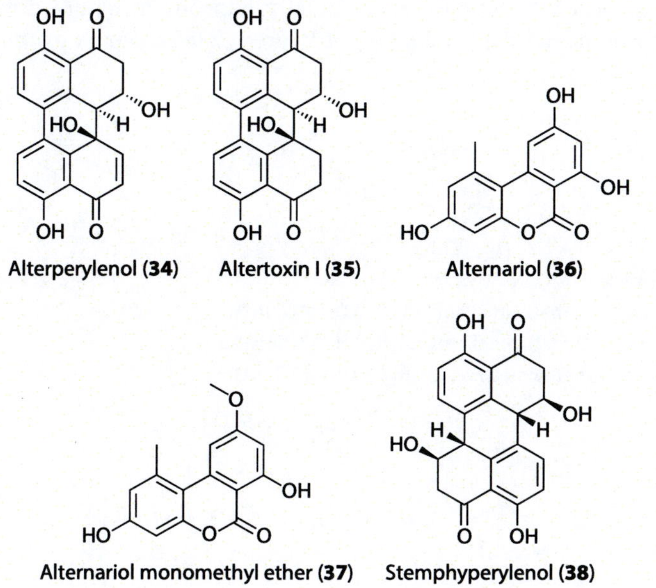

FIGURE 5.7 Polyketides isolated from the coculture of fungal endophytes isolated from *Smallanthus sonchifolius*.

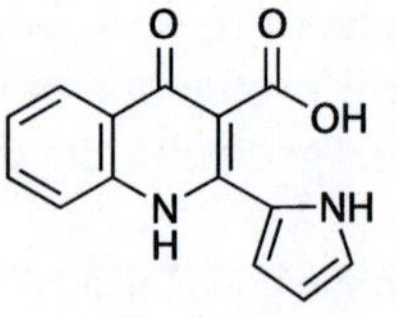

Marinamide (**39**)

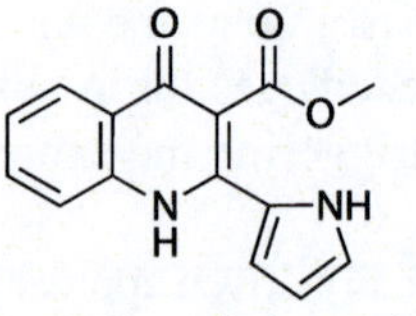

Marinamide methyl ester (**40**)

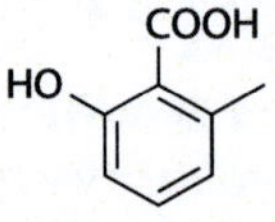

6-methyl salicylic acid (**41**)

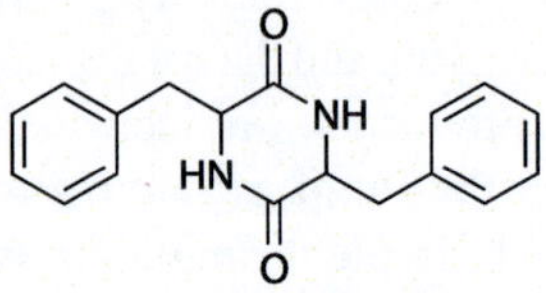

Cyclo-Phe-Phe (**42**)

FIGURE 5.8 Compounds isolated from cocultures of mangrove endophytes.

5.3.1.2 Coculture of Pathogens

Another meaningful way of taking advantage of the ecological knowledge of choosing strains for coculture is by observing interactions that involve pathogens. Pathogens can potentially coexist in the same host and compete for the same sources.

Three cocultures performed between pathogens that share a history of evolution reported great improvement on the yield of bioactive compounds. Glauser and collaborators[74] worked on a coculture of grape wine fungal pathogens, and the yield of *O*-methylmellein (**43**) (Figure 5.9) was increased by hundredfold, and, interestingly, it showed antifungal and phytotoxic activities, highlighting the ecological meaning of this harmful interaction that potentially happens in nature. Moreover, this same coculture induced the production of hydroxylated isomers of this compound

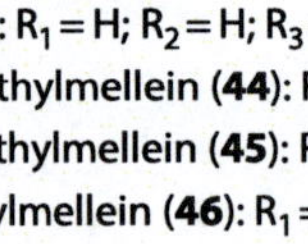

O-methylmellein (**43**): $R_1 = H$; $R_2 = H$; $R_3 = H$
4(*R*)-Hydroxy-8-*O*-methylmellein (**44**): $R_1 = OH$; $R_2 = H$; $R_3 = H$
4(*S*)-Hydroxy-8-*O*-methylmellein (**45**): $R_1 = H$; $R_2 = OH$; $R_3 = H$
5-Hydroxy-8-*O*-methylmellein (**46**): $R_1 = H$; $R_2 = H$; $R_3 = OH$

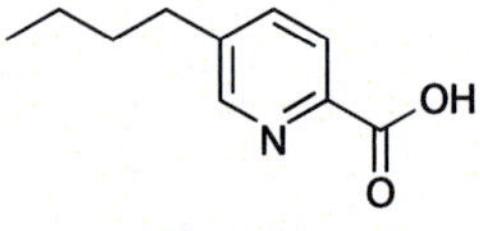

Fusaric acid (**47**)

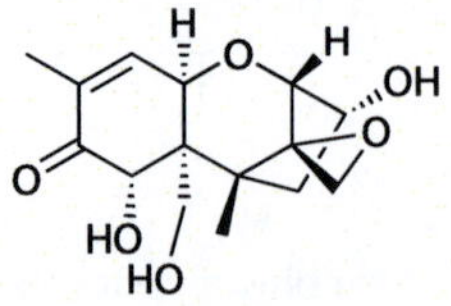

Deoxynivalenol (**48**)

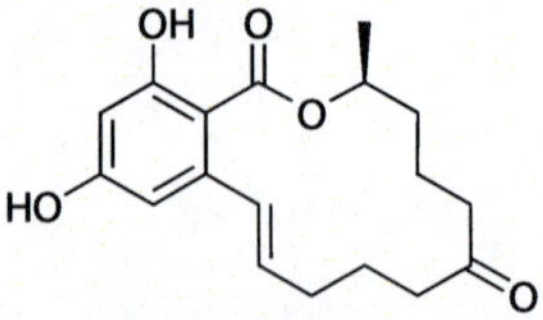

Zearalenone (**49**)

FIGURE 5.9 Compounds isolated from cocultures of pathogenic microbial strains.

(**44–46**) (Figure 5.9).[74] Coculture among maize-infecting fungi and wheat-infecting fungi also showed increased production of the antifungal compounds fusaric acid (**47**),[75] deoxynivalenol (**48**), and zearalenone (**49**).[76]

There is a lack in the literature for coculturing strains that share a history of evolution. Evolution and ecology of symbioses should be considered when choosing strains for coculturing. This can potentially increase the rate of success on finding novel bioactive compounds and/or increase the yield of bioactive compounds. Moreover, it will aggregate information on the chemical ecology field, improving our knowledge of the role of natural products in nature, which can in turn help in finding better strategies in the search for bioactive natural products.

5.4 ANALYTICAL TECHNIQUES FOR STUDYING MICROBIAL NATURAL PRODUCTS

There is certainly a need of different techniques for the early detection of natural products in order to avoid long and exhausting periods of time to rediscover products of microbial metabolism. Fortunately, these techniques are starting to spread among the scientific community in the field of natural products. As has been highlighted in a comprehensive review[68] already cited in this chapter, several analytical strategies may be applied for the study of microbial interactions mediated by natural products.

As microbial chemical diversity is huge, the use of different techniques that help in correct identification of chemical entities is necessary. Chromatographic separations generally include thin layer chromatography (TLC), gas chromatography (GC), and high-pressure liquid chromatography (HPLC and UHPLC), while the most popularly used detector is the ultraviolet/visible detector (UV/Vis). For additional chemical information, mass spectrometry (MS) allows high sensitivity for detecting lesser quantities of small molecules, plus structural information from molecular formula (high resolution MS) to fragmentation patterns (tandem MS). Besides that, MS data are also useful for metabolomic approaches. Finally, in order to determine or confirm chemical structures, nuclear magnetic resonance (NMR) approaches are undoubtedly required.

Taking into account the diversity of microbial chemical entities, it is logical to say that there is no single technique that allows detection of the whole metabolome from a single microorganism. Differences in functional groups allow the application of specific analytical techniques for detection of specific compounds. Some of them are easily detected by simple TLC analysis, but sometimes the limits of detection of some techniques are involved in the invisibility of most microbial metabolites. MS has solved some detection problems, enabling to go steps beyond and use large quantities of data for metabolomic approaches.[77] However, detection does not necessarily imply an easy and unchallenging way until purification and structural determination of a target compound. An overview of recent analytical approaches that can be applied for revealing the chemistry behind the microbial world will be provided in the following sections.

5.4.1 Classical Detectors

As particular chemical structures can be detected by specific detecting techniques, the analytical method should be correctly selected. Some microbial natural products are part of complex pigments where chomophore-based detection techniques can be easily applied, such as ultraviolet-visible spectroscopy (UV/Vis) for detection of the recently discovered polyphenolic polyketide clostrubin from the anaerobic bacterium *Clostridium beijerinckii*.[78] But for compounds that do not possess chromophores and are hence not color related, other detection systems have to be used, such as refractive index (RI) and evaporative light scattering detector (ELSD).

The principle of ELSD lies in the measurement of scattered light by dried particles in a sample. As ELSD can be coupled to HPLC systems, this universal detector is very useful for detecting all classes of compounds even without chromophores, and for natural products such as polysaccharides, lipids, saponins, or terpenes. Although ELSD shows lower sensitivity levels in comparison to, for instance, UV detectors, its use has been increasing significantly since 1985.[79] Since microbial interactions can lead to the production of even more diverse chemical entities, ELSD should be considered as good detecting devices for natural products analysis, in spite of not giving any chemical information.

5.4.2 MS-Related Methods

Besides just detecting chemical entities, it is necessary to obtain as much chemical information as possible, and MS has been applied successfully for this purpose. In addition to detecting ionized molecules, tandem MS allows one to obtain typical fragmentation patterns for a particular compound. From this basic method, several approaches have been developed in order to go beyond and identify chemical families, such as molecular networking.[80] Molecular networking consists of organizing chemical data by matching chemical families in a network due to similarities in fragmentation patterns with the help of a recently developed online and free platform named Global Natural Products Social Molecular Networking (GNPS, gnps.ucsd.edu).[81] Acquired data can be deposited onto this platform, which remains active for future search and comparison of tandem MS/MS data. Molecular networking combined with MALDI-TOF imaging was applied to study interactions of *Pseudomonas fluorescens* and *Naegleria americana*, allowing the authors to suggest that lipopeptide accumulation at the bacteria-protozoa interface may be a defense mechanism for at least two *P. fluorescens* strains against protozoa.[82]

Modern natural products discovery is associated with early identification of known compounds, so a dereplication strategy should be included into the workflow of studying microbial metabolites. In order to correctly identify microbial metabolites, obtaining enough information about known compounds is fundamental, and, as a consequence, the identification will be more reliable. For instance, dereplication of fungal terpenes by using monoisotopic mass is still challenging, since this chemical family shows high redundancy in composition, resulting in several candidates for one chemical formula when searching against natural products databases such as AntiBase.[83] For this reason, comprehensive approaches for dereplication of natural products, including information such as chemical structure, chromatographically validated monoisotopic ions from ESI (electrospray ionization), MS/MS fragmentation data, UV/Vis spectrum, UV absorption maxima, and refractive index, as well as comparison and matching against available databases[83,84] would be ideal in order to reveal novel chemistries from microorganisms—a very useful step to study microbial interactions.

Beyond dereplication, automated workflows for identification of peptides from complex samples have been conducted. An algorithm for detecting known and novel analogues of peptide natural products, named informatics Search of Natural Products (iSNAP), screens tandem MS data and matches peaks with real and *in silico* MS/MS fragments.[85,86] The analogue search algorithm was tested on crude extracts of *Bacillus parabrevis*, which produces one of the most diverse and well-annotated natural products: the tyrocidine family of cyclic decapeptides. Modified peptides produced by *P. polymyxa* were also investigated by iSNAP, revealing the largest natural product complex discovered from a single organism, resulting in more than 50 identified structures. One advantage of this approach is that there is no need for genomic sequence information, HRMS systems, or prior knowledge of the sample although the complementary chemical information of NMR still remains necessary for exact structural confirmation, as well as the dependence for the analyzed compounds to ionize under the specific settings of the MS instruments.[86]

Other approaches involve the correct matching of tandem mass spectra to gene clusters responsible for the biosynthesis of bioactive microbial peptides. The Natural Products Peptidogenomics (NPP)[87] uses high-throughput MS to connect peptide natural products to their BGCs, while the Antibiotics and Secondary Metabolite Analysis Shell (antiSMASH)[88] matches genomic data with BGCs. By using together NPP, iSNAP, and antiSMASH tools, a gap in automated-matching mass shift sequences to BGCs still remains. To solve this issue, the Pep2Path, which consists of a set of algorithms to facilitate identification of candidate BGCs corresponding to NRP- and RiPP-derived mass shift sequence tags detected by MS,[89] was recently developed. For a review on computational approaches for natural product discovery, see Reference 90.

Although tandem MS is now widely and necessarily used among the natural products scientific community, there are always more improvements that can be made. For instance, when automated methods are applied and more intense signals are the targets (data-dependent methods), the low-intensity signals are relegated at times to not being considered. When those low-abundance precursor ions are indeed the interesting signals to be targeted, different analytical methods need to be developed. In order to uncover this issue, improved MS/MS scan coverage for low-abundance precursor ions present in complex myxobacterial extracts was developed. The main goal of this method is to increase the coverage of MS/MS spectra of interesting features, corresponding to microbial small molecules, irrespective of their peak abundances.[91] As this method was developed and proved by testing monocultures from *Sorangium cellulosum*, it can be extended and applied to analyze, for instance, cocultures in order to capture the variation of chemical features when microbial interactions occur.

As analysis of microbial interactions can lead, besides *de novo* production of small molecules, to up- or downregulation of metabolite biosynthesis, so chemometric approaches need to be included as part of the workflow to unravel microbial metabolic exchange. To specifically study microbial interactions by using the coculture strategy, the Projected Orthogonal CHemical Encounter Monitoring (POCHEMON) has been recently introduced to the scientific community in this field.[92] A fungal coculture of *T. rubrum* and *Fusarium solani*, similar to those isolated from onychomycosis patients,[93] was used to demonstrate the usefulness of the POCHEMON method. This method was developed in order to answer how monocultures differ chemically, how cocultures chemically resemble monocultures, and what systematic information is specific to cocultures. Although POCHEMON was presented in conjunction with metabolomics, other high-dimensional biological data may be subjected to this method, including dynamic studies involving more than two species.[92]

Other interesting metabolomic tools that can be applied to study microbial metabolism involve the combination of Self-Organizing Map (SOM)[94] and Molecular Expression Dynamic Investigator (MEDI).[95] Basically, MEDI is used to organize LC-MS features from complex multidimensional data sets in order to cluster features with similar profiles. The SOM output results in heat maps that facilitate rapid visualization of the different metabolic features across samples. Difference maps from SOM capture all responding peaks, irrespective of their intensity, and organize them by intensity profiles across multiple stimuli, providing a visualization of the metabolites resulting from stimulus.[96]

5.4.3 Imaging-Related Techniques

MS imaging approaches have been useful to reflect a real mapping of the metabolic repertoire in the microbial world. From the first approaches of MS imaging, a lot of improvement has occurred since every technique had challenging issues. Nowadays it is possible to directly visualize the spatial distribution of metabolic exchange during microbial interactions, allowing the suggestion of different hypotheses for the biological role of the compounds in such interactions, which makes this approach very useful in the chemical ecology field.

For getting a real distribution map of microbial metabolites care has to be taken in order to not alter, or at least alter as little as possible, the sample to be analyzed. By applying MALDI-TOF IMS to microbial samples directly from the culture plates, a lot of effort is taken in order to minimize varying factors such as the culture media, time of culturing, mycelia removal, and many others.[97] MALDI imaging has been previously used during the investigation of the role of natural antibiotics of microbial symbionts of leaf-cutting ants. Using this technique, it was possible to monitor and reveal the distribution of valinomycin in the integument of *Acromyrmex echinatior* workers. The presence of this known microbial antibiotic at different positions on the ant's body supports its ecological role as a protective agent of individual workers, not just against microbial pathogens but also against parasites.[46]

Another option of an improved imaging-related technique is Replica-EXtraction-transfer Nanostructure-Initiator Mass Spectrometry (REX-NIMS). Basically, this approach consists of a liquid-liquid extraction, transference to a NIMS chip, and then analysis by MALDI-TOF.[98] This technique was successfully used to reveal microbial exchange between *Pseudomonas stutzeri* and *Shewanella oneidensis*, and it was even tested for detecting small molecules in the range of 100–300 Da. Although the published work was carried out on microbial cultures on agar and by using just one extraction solvent, this workflow can be extended to a variety of surfaces using several extraction solvents, taking into account the variety of systems containing microorganisms and the chemodiversity to be studied.

For future characterization of interactions of bacterial colonies, a spatial profiling by using nanospray desorption electrospray ionization (nano-DESI) MS has useful results. Chemical gradients of sucrose and glycosylglycerol generated by microbial communities on agar plates were demonstrated during cultivation of the cyanobacterium *Synechococcus* sp. More interestingly, this results in the combination of spatial distribution of metabolites with their temporal variation, showing, for instance, in the case of glucosylglycerol increments in diffusion through the culture media for older colonies, suggesting that some kind of mechanisms should be involved, such as active excretion or passive cell lysis.[99] Now that it is possible to track several metabolites and capture their spatial/temporal distribution, it would help in hypothesizing what could be the functions of those metabolites not just in single colonies but also in microbial communities, thus contributing important knowledge in the chemical ecology field.

Several challenges arise when analyzing fungi, such as imaging from irregular surfaces due to the presence of aerial hyphae and sometimes spores on the culture media. On this challenging issue, a prior transference of the sample to a hard surface, such as common packaging tape, followed by analysis using DESI-MS worked well and was named imprint-DESI-MS. This approach was applied in order to analyze the antagonistic interaction between the phytopathogen *Moniliophtora roreri* and the endophyte *Trichoderma harzianum*, two fungi cohabiting cacao plants (*Theobroma cacao* L.). Resulting from the analyses of this fungal interaction, four metabolites produced by *T. harzianum* were only observed during coculture.[100] Although additional antifungal bioassays are needed to confirm their activity against *M. roreri*, this is a good example of how imprint-DESI-MS can be applied to microbial interactions in a biocontrol context mediated by small molecules.

A recent technique that can give complementary information when studying microbial interactions is the fluorescence-based coculture screen. This screening is based on the well-known bacterial model *B. subtilis* taking advantage of the different cell types that can coexist within a single colony. This screening method was used to identify soil microorganisms that alter the physiology of *B. subtilis* through the secretion of small molecules. When molecules from an interacting colony induce phenotypic changes in another bacterium, in this case *B. subtilis*, it is possible to visualize the phenotypic response as fluorescent colonies as a consequence of metabolic exchange exposure.[101] This strategy is a valuable tool in the context of natural products mediating microbial interactions as they occur in nature, not just against *B. subtilis* but also extended to other microorganisms, including fungi.

5.4.4 Structural Elucidation

After detecting and revealing or discovering microbial natural products, a necessary step remains: structural elucidation or confirmation of a structural proposal. In order to accelerate the establishment of natural products containing peptidic backbones, an NMR strategy typically applied to proteins was recently developed and proved in uniformly labeled peptides carrying ^{13}C and ^{15}N. Triple-resonance 3D NMR, which observes correlations of three resonances (a proton, a carbon, and nitrogen), was successfully applied for structural elucidation of two novel antibiotic peptides, eudistamide A and B, produced by *Streptomyces* sp. previously isolated from ascidian specimens.[102] At this point, these kinds of experiments highlight the importance of continuous development in NMR experiments that allow remarkable reduction of structure determination times from months to hours.

As demonstrated above, several techniques have been developed, and are continuously being improved, with different purposes but all of them can be definitively applied to study the microbial world in several contexts, such as the ecological role of natural products in nature. As understanding microbial interactions implicates elucidation of the secreted chemical entities as well as the response mechanisms involved, it is necessary to combine different techniques to get a whole and trusty panorama of microbial interactions in nature.

5.5 FINAL REMARKS

Nature continues offering promising compounds with a wide variety of biological activities. The evolving field of research on microbial natural products now is focusing on understanding the role of small molecules in nature as a prior step for a more rational search for therapeutic compounds. By studying a natural symbiotic system, where it is possible to identify the involved organisms as well as the chemical entities they use to interact, it will be feasible to explain and even predict the presence of specific molecules as well as the role they play in that system. The symbioses described for plants and insects are interesting areas to explore and understand the functionality of microbial secondary metabolites in different ecosystems. Besides that, coculture can be used to mimic natural ecological interactions that can possibly occur inside these symbiotic niches where they have been evolving, which can potentially trigger or increase the expression of BGCs related to secondary metabolism. However, there are only few studies on coculturing strains isolated from these systems. By using the recent advances in detecting analytical techniques, bioinformatics tools, as well as platforms to combine scientific efforts, the encrypted natural products will be revealed faster during the years to come.

We therefore encourage the combined application of these research approaches: host microbial chemically mediated interactions, microbial coculture, and modern analytical tools. This strategy will not only provide new insights into the chemical ecology knowledge of symbiotic systems but also foster the rational discovery of new bioactive compounds that might fulfill therapeutic needs. However, focused and united efforts among academia, industry, and government are required for achieving success in translating basic science discoveries to a drug development pipeline.

REFERENCES

1. Newman, D. J.; Cragg, G. M. Natural products as sources of new drugs from 1981 to 2014. *J. Nat. Prod.* **2016**, 79, 629–661.
2. Clardy, J.; Walsh, C. Lessons from natural molecules. *Nature* **2004**, 432, 829–837.
3. Davies, J.; Ryan, K. S. Introducing the parvome: Bioactive compounds in the microbial world. *ACS Chem. Biol.* **2012**, 7, 252–259.

4. Choi, H.; Oh, D. C. Considerations of the chemical biology of microbial natural products provide an effective drug discovery strategy. *Arch. Pharm. Res.* **2015**, 38, 1591–1605.
5. Darwin, C.; Wallace, A. On the tendency of species to form varieties; and on the perpetuation of varieties and species by natural means of selection. *Zool. J. Linn. Soc. Lond.* **1858**, 3, 45–62.
6. Davies, J. Darwin and microbiomes. *EMBO Rep.* **2009**, 10, 805.
7. Fleming, A. On the antibacterial action of cultures of a *Penicillium*, with special reference to their use in the isolation of *B. influenzae*. *Br. J. Exp. Pathol.* **1929**, 10, 226–236.
8. Bennett, J. W.; Chung, K. T. Alexander Fleming and the discovery of penicillin. *Adv. Appl. Microbiol.* **2001**, 49, 163–184.
9. Clatworthy, A. E.; Pierson, E.; Hung, D. T. Targeting virulence: A new paradigm for antimicrobial therapy. *Nat. Chem. Biol.* **2007**, 3, 541–548.
10. Davies, J. Are antibiotics naturally antibiotics? *J. Ind. Microbiol. Biotechnol.* **2006**, 33, 496–499.
11. Demain, A. L. Induction of microbial secondary metabolism. *Int. Microbiol.* **1998**, 1, 259–264.
12. Berdy, J. Bioactive microbial metabolites: A personal view. *J. Antibiot.* **2005**, 58, 1–26.
13. Williams, D. H.; Stone, M. J.; Hauck, P. R.; Rahman, S. K. Why are secondary metabolites (natural products) biosynthesized? *J. Nat. Prod.* **1989**, 52, 1189–1208.
14. Nett, M.; Ikeda, H.; Moore, B. S. Genomic basis for natural product biosynthetic diversity in the actinomycetes. *Nat. Prod. Rep.* **2009**, 26, 1362–1384.
15. Zhu, H.; Sandiford, S. K.; van Wezel, G. P. Triggers and cues that activate antibiotic production by actinomycetes. *J. Ind. Microbiol. Biotechnol.* **2014**, 41, 371–386.
16. Brakhage, A. A. Regulation of fungal secondary metabolism. *Nat. Rev. Microbiol.* **2013**, 11, 21–32.
17. Martin, B. D.; Schwab, E. Current usage of symbiosis and associated terminology. *Int. J. Biol.* **2013**, 5, 32–45.
18. Wilkinson, D. M. At cross purposes: How do we cope with scientific terms that have two different definitions? *Nature* **2001**, 412, 485.
19. Goh, E. B.; Yim, G.; Tsui, W.; McClure, J.; Surette, M. G.; Davies, J. Transcriptional modulation of bacterial gene expression by subinhibitory concentrations of antibiotics. *Proc. Natl. Acad. Sci. USA* **2002**, 99, 17025–17030.
20. Scherlach, K.; Hertweck, C. Triggering cryptic natural product biosynthesis in microorganisms. *Org. Biomol. Chem.* **2009**, 7, 1753–1760.
21. Tan, R. X.; Zou, W. X. Endophytes: A rich source of functional metabolites. *Nat. Prod. Rep.* **2001**, 18, 448–459.
22. Mousa, W. K.; Shearer, C. R.; Limay-Rios, V.; Zhou, T.; Raizada, M. N. Bacterial endophytes from wild maize suppress *Fusarium graminearum* in modern maize and inhibit mycotoxin accumulation. *Front. Plant Sci.* **2015**, 6, 19.
23. Araujo, F. D. D.; Favaro, L. C. D.; Araujo, W. L.; de Oliveira, F. L.; Aparicio, R.; Marsaioli, A. J. Epicolactone: Natural product isolated from the sugarcane endophytic fungus *Epicoccum nigrum*. *Eur. J. Org. Chem.* **2012**, 2012, 5225–5230.
24. Senadeera, S. P. D.; Wiyakrutta, S.; Mahidol, C.; Ruchirawat, S.; Kittakoop, P. A novel tricyclic polyketide and its biosynthetic precursor azaphilone derivatives from the endophytic fungus *Dothideomycete* sp. *Org. Biomol. Chem.* **2012**, 10, 7220–7226.
25. Talontsi, F. M.; Dittrich, B.; Schuffler, A.; Sun, H.; Laatsch, H. Epicoccolides: Antimicrobial and antifungal polyketides from an endophytic fungus *Epicoccum* sp. associated with *Theobroma cacao*. *Eur. J. Org. Chem.* **2013**, 15, 3174–3180.
26. Klaiklay, S.; Rukachaisirikul, V.; Tadpetch, K.; Sukpondma, Y.; Phongpaichit, S.; Buatong, J.; Sakayaroj, J. Chlorinated chromone and diphenyl ether derivatives from the mangrove-derived fungus *Pestalotiopsis* sp. PSU-MA69. *Tetrahedron* **2012**, 68, 2299–2305.

27. Sommart, U.; Rukachaisirikul, V.; Tadpetch, K.; Sukpondma, Y.; Phongpaichit, S.; Hutadilok-Towatana, N.; Sakayaroj, J. Modiolin and phthalide derivatives from the endophytic fungus *Microsphaeropsis arundinis* PSU-G18. *Tetrahedron* **2012**, 68, 10005–10010.
28. Cheng, M. J.; Wu, M. D.; Yuan, G. F.; Chen, Y. L.; Su, Y. S.; Hsieh, M. T.; Chen, I. S. Secondary metabolites and cytotoxic activities from the endophytic fungus *Annulohypoxylon squamulosum*. *Phytochem. Lett.* **2012**, 5, 219–223.
29. Kawahara, T.; Itoh, M.; Izumikawa, M.; Sakata, N.; Tsuchida, T.; Shin-ya, K. Three eremophilane derivatives, MBJ-0011, MBJ-0012 and MBJ-0013, from an endophytic fungus *Apiognomonia* sp. f24023. *J. Antibiot.* **2013**, 66, 299–302.
30. Cragg, G. M. Paclitaxel (Taxol): A success story with valuable lessons for natural product drug discovery and development. *Med. Res. Rev.* **1998**, 18, 315–331.
31. Howat, S.; Park, B.; Oh, I. S.; Jin, Y. W.; Lee, E. K.; Loake, G. J. Paclitaxel: Biosynthesis, production and future prospects. *New Biotechnol.* **2014**, 31, 242–245.
32. Tabata, H. Paclitaxel production by plant-cell-culture technology. *Adv. Biochem. Eng. Biotechnol.* **2004**, 87, 1–23.
33. Thaler, J. S.; Owen, B.; Higgins, V. J. The role of the jasmonate response in plant susceptibility to diverse pathogens with a range of lifestyles. *Plant Physiol.* **2004**, 135, 530–538.
34. Newman, D. J.; Cragg, G. M. Endophytic and epiphytic microbes as "sources" of bioactive agents. *Front. Chem.* **2015**, 3, 1–13.
35. Kusari, S.; Singh, S.; Jayabaskaran, C. Rethinking production of taxol (paclitaxel) using endophyte biotechnology. *Trends Biotechnol.* **2014**, 32, 304–311.
36. Elmer, W. H.; Mattina, M. J. I.; Maceachern, G. J. Sensitivity of plant-pathogenic fungi to taxane extracts from ornamental yews. *Phytopathology* **1994**, 84, 1179–1185.
37. Soliman, S. S. M.; Greenwood, J. S.; Bombarely, A.; Mueller, L. A.; Tsao, R.; Mosser, D. D.; Raizada, M. N. An endophyte constructs fungicide-containing extracellular barriers for its host plant. *Curr. Biol.* **2015**, 25, 2570–2576.
38. Berasategui, A.; Shukla, S.; Salem, H.; Kaltenpoth, M. Potential applications of insect symbionts in biotechnology. *Appl. Microbiol. Biotechnol.* **2016**, 100, 1567–1577.
39. Kaltenpoth, M. Actinobacteria as mutualists: General healthcare for insects? *Trends Microbiol.* **2009**, 17, 529–535.
40. Schultz, T. R.; Brady, S. G. Major evolutionary transitions in ant agriculture. *Proc. Natl. Acad. Sci. USA* **2008**, 105, 5435–5440.
41. Sit, C. S.; Ruzzini, A. C.; Van Arnam, E. B.; Ramadhar, T. R.; Currie, C. R.; Clardy, J. Variable genetic architectures produce virtually identical molecules in bacterial symbionts of fungus-growing ants. *Proc. Natl. Acad. Sci. USA* **2015**, 112, 13150–13154.
42. Little, A. E. F.; Currie, C. R. Black yeast symbionts compromise the efficiency of antibiotic defenses in fungus-growing ants. *Ecology* **2008**, 89, 1216–1222.
43. Oh, D. C.; Poulsen, M.; Currie, C. R.; Clardy, J. Dentigerumycin: A bacterial mediator of an ant-fungus symbiosis. *Nat. Chem. Biol.* **2009**, 5, 391–393.
44. Van Arnam, E. B.; Ruzzini, A. C.; Sit, C. S.; Currie, C. R.; Clardy, J. A rebeccamycin analog provides plasmid-encoded niche defense. *J. Am. Chem. Soc.* **2015**, 137, 14272–14274.
45. Carr, G.; Derbyshire, E. R.; Caldera, E.; Currie, C. R.; Clardy, J. Antibiotic and antimalarial quinones from fungus-growing ant-associated *Pseudonocardia* sp. *J. Nat. Prod.* **2012**, 75, 1806–1809.
46. Schoenian, I.; Spiteller, M.; Ghaste, M.; Wirth, R.; Herz, H.; Spiteller, D. Chemical basis of the synergism and antagonism in microbial communities in the nests of leaf-cutting ants. *Proc. Natl. Acad. Sci. USA* **2011**, 108, 1955–1960.
47. Haeder, S.; Wirth, R.; Herz, H.; Spiteller, D. Candicidin-producing *Streptomyces* support leaf-cutting ants to protect their fungus garden against the pathogenic fungus *Escovopsis*. *Proc. Natl. Acad. Sci. USA* **2009**, 106, 4742–4746.

48. Mendes, T. D.; Borges, W. S.; Rodrigues, A.; Solomon, S. E.; Vieira, P. C.; Duarte, M. C. T.; Pagnocca, F. C. Anti-Candida properties of urauchimycins from actinobacteria associated with *Trachymyrmex* ants. *Biomed. Res. Int.* **2013**, 2013, 8.
49. Aanen, D. K.; Eggleton, P.; Rouland-Lefevre, C.; Guldberg-Froslev, T.; Rosendahl, S.; Boomsma, J. J. The evolution of fungus-growing termites and their mutualistic fungal symbionts. *Proc. Natl. Acad. Sci. USA* **2002**, 99, 14887–14892.
50. Um, S.; Fraimout, A.; Sapountzis, P.; Oh, D. C.; Poulsen, M. The fungus-growing termite macrotermes natalensis harbors bacillaene-producing *Bacillus* sp. that inhibit potentially antagonistic fungi. *Sci. Rep.* **2013**, 3, 7.
51. Kim, K. H.; Ramadhar, T. R.; Beemelmanns, C.; Cao, S. G.; Poulsen, M.; Currie, C. R.; Clardy, J. Natalamycin A, an ansamycin from a termite-associated *Streptomyces* sp. *Chem. Sci.* **2014**, 5, 4333–4338.
52. Bi, S. F.; Li, F.; Song, Y. C.; Tan, R. X.; Ge, H. M. New acrylamide and oxazolidin derivatives from a termite-associated *Streptomyces* sp. *Nat. Prod. Commun.* **2011**, 6, 353–355.
53. Carr, G.; Poulsen, M.; Klassen, J. L.; Hou, Y. P.; Wyche, T. P.; Bugni, T. S.; Currie, C. R.; Clardy, J. Microtermolides A and B from termite-associated *Streptomyces* sp. and structural revision of vinylamycin. *Org. Lett.* **2012**, 14, 2822–2825.
54. Nirma, C.; Eparvier, V.; Stien, D. Antifungal agents from *Pseudallescheria boydii* SNB-CN73 isolated from a *Nasutitermes* sp. termite. *J. Nat. Prod.* **2013**, 76, 988–991.
55. Oh, D. C.; Scott, J. J.; Currie, C. R.; Clardy, J. Mycangimycin, a polyene peroxide from a mutualist *Streptomyces* sp. *Org. Lett.* **2009**, 11, 633–636.
56. Blodgett, J. A. V.; Oh, D. C.; Cao, S. G.; Currie, C. R.; Kolter, R.; Clardy, J. Common biosynthetic origins for polycyclic tetramate macrolactams from phylogenetically diverse bacteria. *Proc. Natl. Acad. Sci. USA* **2010**, 107, 11692–11697.
57. Kim, S. H.; Ko, H.; Bang, H. S.; Park, S. H.; Kim, D. G.; Kwon, H. C.; Kim, S. Y.; Shin, J.; Oh, D. C. Coprismycins A and B, neuroprotective phenylpyridines from the dung beetle-associated bacterium, *Streptomyces* sp. *Bioorg. Med. Chem. Lett.* **2011**, 21, 5715–5718.
58. Park, S. H.; Moon, K.; Bang, H. S.; Kim, S. H.; Kim, D. G.; Oh, K. B.; Shin, J.; Oh, D. C. Tripartilactam, a cyclobutane-bearing tricyclic lactam from a *Streptomyces* sp. in a dung beetle's brood ball. *Org. Lett.* **2012**, 14, 1258–1261.
59. Kim, S. H.; Kwon, S. H.; Park, S. H.; Lee, J. K.; Bang, H. S.; Nam, S. J.; Kwon, H. C.; Shin, J.; Oh, D. C. Tripartin, a histone demethylase inhibitor from a bacterium associated with a dung beetle larva. *Org. Lett.* **2013**, 15, 1834–1837.
60. Um, S.; Park, S. H.; Kim, J.; Park, H. J.; Ko, K.; Bang, H. S.; Lee, S. K.; Shin, J.; Oh, D. C. Coprisamides A and B, new branched cyclic peptides from a gut bacterium of the dung beetle copris tripartitus. *Org. Lett.* **2015**, 17, 1272–1275.
61. Um, S.; Bang, H.-S.; Shin, J.; Oh, D.-C. Actinofuranone C, a new 3-furanone-bearing polyketide from a dung beetle-associated bacterium. *Nat. Prod. Sci.* **2013**, 19, 71–75.
62. Wu, Q.; Jiang, N.; Han, W. B.; Mei, Y. N.; Ge, H. M.; Guo, Z. K.; Weng, N. S.; Tan, R. X. Antibacterial epipolythiodioxopiperazine and unprecedented sesquiterpene from *Pseudallescheria boydii*, a beetle (coleoptera)-associated fungus. *Org. Biomol. Chem.* **2014**, 12, 9405–9412.
63. Kroiss, J.; Kaltenpoth, M.; Schneider, B.; Schwinger, M. G.; Hertweck, C.; Maddula, R. K.; Strohm, E.; Svatos, A. Symbiotic streptomycetes provide antibiotic combination prophylaxis for wasp offspring. *Nat. Chem. Biol.* **2010**, 6, 261–263.
64. Oh, D. C.; Poulsen, M.; Currie, C. R.; Clardy, J. Sceliphrolactam, a polyene macrocyclic lactam from a wasp-associated *Streptomyces* sp. *Org. Lett.* **2011**, 13, 752–755.
65. Poulsen, M.; Oh, D. C.; Clardy, J.; Currie, C. R. Chemical analyses of wasp-associated *Streptomyces* bacteria reveal a prolific potential for natural products discovery. *PLoS One* **2011**, 6, 8.

66. Menezes, C.; Vollet-Neto, A.; Marsaioli, A. J.; Zampieri, D.; Fontoura, I. C.; Luchessi, A. D.; Imperatriz-Fonseca, V. L. A Brazilian social bee must cultivate fungus to survive. *Curr. Biol.* **2015**, 25, 2851–2855.
67. Seyedsayamdost, M. R. High-throughput platform for the discovery of elicitors of silent bacterial gene clusters. *Proc. Natl. Acad. Sci. USA* **2014**, 111, 7266–7271.
68. Bertrand, S.; Bohni, N.; Schnee, S.; Schumpp, O.; Gindro, K.; Wolfender, J. L. Metabolite induction via microorganism co-culture: A potential way to enhance chemical diversity for drug discovery. *Biotechnol. Adv.* **2014**, 32, 1180–1204.
69. Soliman, S. S. M.; Raizada, M. N. Interactions between co-habitating fungi elicit synthesis of taxol from an endophytic fungus in host taxus plants. *Front. Microbiol.* **2013**, 4, 14.
70. Chagas, F. O.; Dias, L. G.; Pupo, M. T. A mixed culture of endophytic fungi increases production of antifungal polyketides. *J. Chem. Ecol.* **2013**, 39, 1335–1342.
71. Zhu, F.; Lin, Y. C. Marinamide, a novel alkaloid and its methyl ester produced by the application of mixed fermentation technique to two mangrove endophytic fungi from the South China Sea. *Chin. Sci. Bull.* **2006**, 51, 1426–1430.
72. Zhu, F.; Chen, G. Y.; Wu, J. S.; Pan, J. H. Structure revision and cytotoxic activity of marinamide and its methyl ester, novel alkaloids produced by co-cultures of two marine-derived mangrove endophytic fungi. *Nat. Prod. Res.* **2013**, 27, 1960–1964.
73. Zhu, F.; Lin, Y.-C.; Ding, J.-H.; Wang, X.-P.; Huang, L.-S. Secondary metabolites of two marine-derived mangrove endophytic fungi (strains nos. 1924 and 3983) by mixed fermentation. *Linchan Huaxue Yu Gongye* **2007**, 27, 8–10.
74. Glauser, G.; Gindro, K.; Fringeli, J.; De Joffrey, J.-P.; Rudaz, S.; Wolfender, J.-L. Differential analysis of mycoalexins in confrontation zones of grapevine fungal pathogens by ultrahigh pressure liquid chromatography/time-of-flight mass spectrometry and capillary nuclear magnetic resonance. *J. Agric. Food Chem.* **2009**, 57, 1127–1134.
75. Jonkers, W.; Estrada, A. E.; Lee, K.; Breakspear, A.; May, G.; Kistler, H. C. Metabolome and transcriptome of the interaction between *Ustilago maydis* and *Fusarium verticillioides in vitro*. *Appl. Environ. Microbiol.* **2012**, 78, 3656–3667.
76. Muller, M. E. H.; Steier, I.; Koppen, R.; Siegel, D.; Proske, M.; Korn, U.; Koch, M. Cocultivation of phytopathogenic *Fusarium* and *Alternaria* strains affects fungal growth and mycotoxin production. *J. Appl. Microbiol.* **2012**, 113, 874–887.
77. Garg, N.; Kapono, C. A.; Lim, Y. W.; Koyama, N.; Vermeij, M. J. A.; Conrad, D.; Rohwer, F.; Dorrestein, P. C. Mass spectral similarity for untargeted metabolomics data analysis of complex mixtures. *Int. J. Mass Spectrom.* **2015**, 377, 719–727.
78. Pidot, S.; Ishida, K.; Cyrulies, M.; Hertweck, C. Discovery of clostrubin, an exceptional polyphenolic polyketide antibiotic from a strictly anaerobic bacterium. *Angew. Chem. Int. Ed.* **2014**, 53, 7856–7859.
79. Ganzera, M.; Stuppner, H. Evaporative light scattering detection (ELSD) for the analysis of natural products. *Curr. Pharm. Anal.* **2005**, 1, 135–144.
80. Watrous, J.; Roach, P.; Alexandrov, T.; Heath, B. S.; Yang, J. Y.; Kersten, R. D.; van der Voort, M. et al. Mass spectral molecular networking of living microbial colonies. *Proc. Natl. Acad. Sci. USA* **2012**, 109, E1743–E1752.
81. Wang, M. X.; Carver, J. J.; Phelan, V. V.; Sanchez, L. M.; Garg, N.; Peng, Y.; Nguyen, D. D. et al. Sharing and community curation of mass spectrometry data with Global Natural Products Social Molecular Networking. *Nat. Biotechnol.* **2016**, 34, 828–837.
82. Song, C. X.; Mazzola, M.; Cheng, X.; Oetjen, J.; Alexandrov, T.; Dorrestein, P.; Watrous, J.; van der Voort, M.; Raaijmakers, J. M. Molecular and chemical dialogues in bacteria-protozoa interactions. *Sci. Rep.* **2015**, 5, 13.
83. Nielsen, K. F.; Mansson, M.; Rank, C.; Frisvad, J. C.; Larsen, T. O. Dereplication of microbial natural products by LC-DAD-TOFMS. *J. Nat. Prod.* **2011**, 74, 2338–2348.

84. Kildgaard, S.; Mansson, M.; Dosen, I.; Klitgaard, A.; Frisvad, J. C.; Larsen, T. O.; Nielsen, K. F. Accurate dereplication of bioactive secondary metabolites from marine-derived fungi by UHPLC-DAD-QTOFMS and a MS/HRMS library. *Mar. Drugs* **2014**, 12, 3681–3705.
85. Ibrahim, A.; Yang, L.; Johnston, C.; Liu, X. W.; Ma, B.; Magarvey, N. A. Dereplicating nonribosomal peptides using an informatic search algorithm for natural products (iSNAP) discovery. *Proc. Natl. Acad. Sci. USA* **2012**, 109, 19196–19201.
86. Yang, L.; Ibrahim, A.; Johnston, C. W.; Skinnider, M. A.; Ma, B.; Magarvey, N. A. Exploration of nonribosomal peptide families with an automated informatic search algorithm. *Chem. Biol.* **2015**, 22, 1259–1269.
87. Kersten, R. D.; Yang, Y. L.; Xu, Y. Q.; Cimermancic, P.; Nam, S. J.; Fenical, W.; Fischbach, M. A.; Moore, B. S.; Dorrestein, P. C. A mass spectrometry-guided genome mining approach for natural product peptidogenomics. *Nat. Chem. Biol.* **2011**, 7, 794–802.
88. Weber, T.; Blin, K.; Duddela, S.; Krug, D.; Kim, H. U.; Bruccoleri, R.; Lee, S. Y. et al. antiSMASH 3.0: A comprehensive resource for the genome mining of biosynthetic gene clusters. *Nucleic Acids Res.* **2015**, 43, W237–W243.
89. Medema, M. H.; Paalvast, Y.; Nguyen, D. D.; Melnik, A.; Dorrestein, P. C.; Takano, E.; Breitling, R. Pep2Path: Automated mass spectrometry-guided genome mining of peptidic natural products. *PLoS Comput. Biol.* **2014**, 10, 7.
90. Medema, M. H.; Fischbach, M. A. Computational approaches to natural product discovery. *Nat. Chem. Biol.* **2015**, 11, 639–648.
91. Hoffmann, T.; Krug, D.; Huttel, S.; Muller, R. Improving natural products identification through targeted LC-MS/MS in an untargeted secondary metabolomics workflow. *Anal. Chem.* **2014**, 86, 10780–10788.
92. Jansen, J. J.; Blanchet, L.; Buydens, L. M. C.; Bertrand, S.; Wolfender, J. L. Projected Orthogonalized CHemical Encounter MONitoring (POCHEMON) for microbial interactions in co-culture. *Metabolomics* **2015**, 11, 908–919.
93. Bertrand, S.; Schumpp, O.; Bohni, N.; Bujard, A.; Azzollini, A.; Monod, M.; Gindro, K.; Wolfender, J. L. Detection of metabolite induction in fungal co-cultures on solid media by high-throughput differential ultra-high pressure liquid chromatography-time-of-flight mass spectrometry fingerprinting. *J. Chromatogr. A* **2013**, 1292, 219–228.
94. Goodwin, C. R.; Covington, B. C.; Derewacz, D. K.; McNees, C. R.; Wikswo, J. P.; McLean, J. A.; Bachmann, B. O. Structuring microbial metabolic responses to multiplexed stimuli via self-organizing metabolomics maps. *Chem. Biol.* **2015**, 22, 661–670.
95. Goodwin, C. R.; Sherrod, S. D.; Marasco, C. C.; Bachmann, B. O.; Schramm-Sapyta, N.; Wikswo, J. P.; McLean, J. A. Phenotypic mapping of metabolic profiles using self-organizing maps of high-dimensional mass spectrometry data. *Anal. Chem.* **2014**, 86, 6563–6571.
96. Derewacz, D. K.; Covington, B. C.; McLean, J. A.; Bachmann, B. O. Mapping microbial response metabolomes for induced natural product discovery. *ACS Chem. Biol.* **2015**, 10, 1998–2006.
97. Yang, J. Y.; Phelan, V. V.; Simkovsky, R.; Watrous, J. D.; Trial, R. M.; Fleming, T. C.; Wenter, R. et al. Primer on agar-based microbial imaging mass spectrometry. *J. Bacteriol.* **2012**, 194, 6023–6028.
98. Louie, K. B.; Bowen, B. P.; Cheng, X. L.; Berleman, J. E.; Chakraborty, R.; Deutschbauer, A.; Arkin, A.; Northen, T. R. "Replica-extraction-transfer" nanostructure-initiator mass spectrometry imaging of acoustically printed bacteria. *Anal. Chem.* **2013**, 85, 10856–10862.
99. Lanekoff, I.; Geydebrekht, O.; Pinchuk, G. E.; Konopka, A. E.; Laskin, J. Spatially resolved analysis of glycolipids and metabolites in living *Synechococcus* sp. PCC 7002 using nanospray desorption electrospray ionization. *Analyst* **2013**, 138, 1971–1978.

100. Tata, A.; Perez, C.; Campos, M. L.; Bayfield, M. A.; Eberlin, M. N.; Ifa, D. R. Imprint desorption electrospray ionization mass spectrometry imaging for monitoring secondary metabolites production during antagonistic interaction of fungi. *Anal. Chem.* **2015**, 87, 12298–12304.
101. Shank, E. A. Using coculture to detect chemically mediated interspecies interactions. *J. Vis. Exp.* **2013**, 80, 50863.
102. Zhang, F.; Adnani, N.; Vazquez-Rivera, E.; Braun, D. R.; Tonelli, M.; Andes, D. R.; Bugni, T. S. Application of 3D NMR for structure determination of peptide natural products. *J. Org. Chem.* **2015**, 80, 8713–8719.

6 Natural Products from Endophytic Microbes

Historical Perspectives, Prospects, and Guidance

Gary Strobel

CONTENTS

6.1 INTRODUCTION

Endophyte biology has its origins with Darnel in 1904.[1] Since that time, this emerging field of biology has taken on a kaleidoscope of activity starting with the wonderment of a few earlier investigators in being able to isolate a plethora of microorganisms from plant tissues that seemed to bear no external evidence of the presence of any life form within them. This offered an interesting situation since plant pathology was an important science emerging in the nineteenth century, but whose major strides were contemporary with the development of the field of endophyte biology in the twentieth century.[2] Thus, it was soon obvious that plants could be hosts for microorganisms that were not necessarily pathogens. But in the early part of the twentieth century, to most people, microbes generally associated with plants were considered as pathogens or agents of destruction. Also, it appeared that endophytes seemed not to have any economic consequence and therefore the effort and monies committed to this field in the past were miniscule to nonexistent. As expected, departments and units of plant pathology were formed in virtually every country having an agricultural-based economy, but to date, there is yet to form a Department of Endophytics at any university, college, or institute anywhere in the world.

Since the early times, the definition of an endophyte has more or less remained the same—"a microorganism associated with living plant tissues that produces no apparent indication of its presence in the plant and seems not to cause harm to the host."[3] Endophytes have been isolated from virtually all plant organs (roots, stems, leaves, flowers, fruits, and seeds), and the most commonly observed endophytes are the fungi (usually Ascomycteous fungi), as well as bacteria, and this includes the filamentous bacteria, Actinomycetes. Periodically, a Phycomycete, a Basidiomycete, and a representative of the Fungi Imperfecti are also isolated. In the future, it may be the case that other life forms may be found that can be considered endophytes such as the mycoplasmas. There is some certainty that all plant forms, including those in the world's oceans, are hosts or are potential hosts for one or more endophytes.[4] And many of the lower plant forms such as mosses and liverworts are also hosts to endophytes.[5]

As with most fields, the initial work on endophytes began as an observational science with investigators who were keenly interested in isolating and identifying every possible endophyte from a given plant and then moving on to the next plant species.[3,6] Much of this early work was being done by the Petrinis and their group in Switzerland.[7] The efforts on the distribution of endophytes quickly led to studies concerning the relationship of endophytes to their host plants.[3] However, in spite of all of the reports on endophytes from Europe, Canada, and the United States, it appears that hundreds, if not thousands, of novel endophytic species still remain to the discovered, especially in those remote, unique, and untouched forests and fields of the world. The likelihood of finding novel taxonomic microbes is the greatest in the world's tropical and temperate rain forests.[8] Finally, with taxonomic novelty, it seems, the chances of finding chemical novelty greatly increases.[1]

6.2 ENDOPHYTIC MICROBES: PROSPECTS

6.2.1 ENDOPHYTE BIOLOGY AND THE *PLANT MICROBIOME*

The relationship between the endophyte and its host remains shrouded in many unknowns and uncertainties. For instance, as the endophyte enters the plant via natural openings or by direct penetration, it usually becomes localized interstitially between plant cells. In some cases, it seems that the endophyte carries out several cell divisions and then ceases growth. And in other cases, if the host plant is under some nutritional, environmental, or age-related stress, the endophyte can become pathogenic and symptoms of its presence occur as the microbe begins its growth cycle. Thus, the growth of the endophyte is subjected to the physiological state of its host and its environmental conditions. It appears that factors in the host keep the growth of the endophyte suppressed. Plant-associated compounds that may be responsible for inhibiting endophytes are the phenolic substances and, as these are released by leaching of dead plant materials, this allows for massive growth of the endophyte.[9] But specific details on one or more compounds with this antimicrobial activity are lacking.

Usually, endophytes are considered as symbiotic or mutualistic in their relationship to the host. If the endophyte is deemed symbiotic, then there is an implication that something must be contributed by the microbe to the relationship and vice versa. It is obvious that the main contribution by the plant is the nutritional factors, and the protection that it provides to the microbe. On the other hand, many authors have implied that various antibiotics, antioxidants, anti-insect substances, and other potentially bioactive products that may be beneficial to the plant are contributed by the microbe, and this has a direct bearing on the potential of endophytes in natural products chemistry.[1] Most interestingly, however, there are few well-conceived scientific studies that actually show the presence of such compounds in or associated with the microbe-plant interaction. This produces uncertainty for the role of any bioactive compound in the host/endophyte relationship. Most authors make assertions and conclusions based on what the endophyte produces in pure culture and then assume that comparable processes are occurring in nature.[1] There appears to be a great need for further work in this area.

Besides the putative protection offered by endophytes to their hosts, these organisms may carry out a plethora of other activities that may benefit the host, and these are as follows: (a) facilitation of nutrient uptake as per the mycorrhizal fungi; (b) enhancement of plant growth as seen with some plant-associated bacteria; (c) induction of heat tolerance to plants living under such stress; and (d) perhaps others, including increased salt tolerance and the neutralization of toxic compounds that would otherwise harm plants.[10] Thus, there is a huge potential of endophytes benefiting plants which will result in major strides being made in agriculture, forestry, and the improvement of the world's environment. Suddenly, the world of endophyte biology has taken on an entirely new scope. Interestingly, it has a counterpart in a modern and important area of human biology which is the human microbiome. What has been learned from this enormous effort is now having impacts on modern medicine and health care in general.[11]

The techniques used in solving some of the problems in the human microbiome project involve metagenomics, transcriptomics, and proteomics combined with improved and new methods in chemical instrumentation.[11] These same tools are available for the probing of the *plant microbiome*. However, the numbers of organisms involved are much more numerous and complex merely by the total number of plants on the planet and the number of organs in their structures. What is being realized is that most of the world's major crop plants may have had their original endophyte populations so modified that they no longer even closely resemble the original host microbiome.[10] The original endophytes of all the important plants that feed mankind are most likely to be found in or around the center of origin of the host plant. The loss of the endophyte population may have resulted in dramatic changes in disease defense, environmental adaptability, insect resistance, and, ultimately, yield. Obviously, there is a chemical relationship between the endophytic microbe and its host, suggesting major potential roles of natural products chemistry for the benefit of the world's agricultural enterprise via endophyte biology/chemistry.

6.2.2 Grass Endophytes: A Diversion

Interestingly, the first solid work on the natural products chemistry of any endophyte came via interest and observations made on endophytes of grasses.[12] Tall fescue and perennial ryegrass symbiota have long held associations with animal performance difficulties and toxicities.[13] *Acremonium* spp., and other endophytic associates of the fungal family Clavicipitaceae, produce ergot alkaloids resulting in animal toxicities.[14] These alkaloids consist of both the ergopeptine and clavine types. In addition, there are tremorgenic neurotoxins, known as the lolitrems from the perennial ryegrass symbiotum.[15] These compounds are considered responsible for the *staggers of sheep* syndrome.[16]

Neotyphodium coenophialum (Acremonium-like fungus) is an endophyte commonly associated with tall fescue, and the ability of this plant to tolerate drought and high temperatures has been related to the presence of this endophyte.[17] However, strangely enough, the agronomic diversity attributed to this endophyte comes with the price of its toxicity to livestock in places such as New Zealand, Australia, Italy, and the United States.[3,6,17] The chief toxic component involved in the toxicology of this endophyte is the alkaloid, ergovaline.[6] Other alkaloids are also known from this fungus including ergocornine, ergonine, ergotine, and ergotamine.[14] Ergovaline is an agonist for dopamine D2 receptors which initiates a plethora of pathologies in horse, cattle, and swine. Peramine is the only known pyrrolpyrazine alkaloid in the tall fescue-endophyte associations, and it is not active in mammalian bioassays and is primarily known as an insect-feeding deterrent.[18]

Interestingly, the efforts of numerous investigators on a number of continents on the ergot alkaloids of fescue resulted in hundreds of papers on such subjects as the endophyte, the alkaloids, the mammalian toxicities, and other aspects of the biology of this complex, starting in the early 1980s. Basically, in the minds of many people, the concept of an *endophyte* became synonymous with organisms having interactions with plants that result in a detrimental outcome. In fact, to this day, many still hold to this concept when hearing the term *endophyte*. The possibility that endophytes could hold something of a benefit to society as a whole seemed remote to nonexistent in 1980.

6.2.3 Taxol: A New Paradigm for Endophytes

Taxol, from *Taxus brevifolia* (the Pacific yew), was first described as a novel, highly functionalized terpenoid by Wani and Wall in 1971. The report languished in the chemical literature until Susan Horwitz and her group showed that the molecule had a novel mode of action. Taxol prevents the depolymerization of tublin molecules during the G2-M phase of mitosis, effectively blocking cell division, resulting in cell death.[19] Taxol was especially effective in certain breast and ovarian cancer cell lines, as well as in several animal xenograft models, and it was advanced to human clinical trials by the U.S. National Cancer Institute in the early 1980s. These developments have been reviewed by Wani and Horwitz.[20] As a result, this molecule stirred up immense interest in the pharmaceutical industry, and by the early 1990s taxol was on its way to becoming the world's first billion-dollar anticancer drug. This was in spite of the fact that the compound was found in diminishingly small quantities in the bark of the yew, its price tag was $6000 per gram, and there was no basic patent protection filed at the time of the original discovery.

Strobel's group in the early 1990s reasoned that an endophyte may exist in yew that also makes this compound. A systematic search was launched in the northern yew forests of Montana to find such an organism. To the surprise of all concerned, a novel endophytic fungus, *Taxomyces andreanae*, was described and all available evidence indicated that it was making extremely small amounts of taxol in culture.[21] Since that time, taxol has been found in cultures of over 20 endophytic fungi of yews and other plants as well as in some saprophytic fungal forms.[22,23] These organisms make between 1 and 800 μg of taxol per liter.[23] The original report on fungal taxol in *Science* (1993) effectively launched a global search for bioactive compounds from other endophytic microbes associated with a plethora of medicinal as well as nonmedicinal plants. Since that time, many reviews on this topic have appeared.[1,24–28]

6.2.4 Endophytes and Important Anticancer Compounds

The original work on fungal taxol inspired others to also seek for products normally associated with plants in their corresponding suite of endophytic fungi. That is, if one has a rare plant as a source of an invaluable natural product, is it possible to find and eventually isolate that same product from one or more of its associated endophytes? Camptothecin is a pentacyclic quinoline alkaloid [(*S*)-4-ethyl-4-hydroxy-1*H* pyrano[3′,4′:6,7]indolizino[1,2-*b*] quinoline-3,14-(4*H*,12*H*)-dione] which serves as the lead molecule for Topotecan and Irinotecan, which are Food and Drug Administration (FDA)-approved and heavily sought-after anticancer drugs. This compound has now been discovered in an endophytic fungus of the host plant *Campotheca acuminata*.[29] Campothecin binds to the topoisomerase I-DNA complex in cancer cells which prevents DNA replication, leading to cell death.[30] Other reports suggest that campothecin derivatives show some effectiveness in the treatment of acquired immunodeficiency syndrome (AIDS), as general antiviral agents, antimalarial agents, and antifungal agents. Until certain endophytic fungi were discovered the only known source of this compound was *C. acuminata* and *Nothapodytes nimmoniana*, which are Asiatic trees whose populations are being heavily reduced due to the demand for this drug.

The discovery of an endophytic fungus, *Entrophospora infrequens*, and other endophytes producing camptothecin has provided some hope for an alternative source of this drug.[30,31] In fact, other endophytes are now known from *C. acuminata* that also make camptothecin, including *Aspergillus* spp. and *Trichoderma* spp.[32] A strain of *Fusarium solani* is known from *Apodytes dimidiate* that also is a producer of this compound.[30] It appears that in all cases the yields are low and the production of the compound is under heavy regulation by the fungal genome. Also, in many cases, an organism, after multiple transfers away from its host material, will become attenuated in secondary product formation, and product formation can be restored with added factors from the host plant.[33] Nevertheless, the story of fungal camptothecin seems to be following the same path as that of fungal taxol. The promises are great for future breakthroughs that will lead to commercial-range microbial production of these valuable molecules, but much more work is needed at the molecular biological and biochemical levels in order to make major improvements in product yields.

Finally, both vincristine and vinblastine are alkaloids from the Madagascar periwinkle plant, *Catharanthus roseus*, which are used to treat various leukemias including childhood leukemia. An endophytic *Fusarium oxysporum* was isolated, characterized, and shown via a multitude of chemical methods to produce both these compounds in yields in the range of 70 μg/L.[34] And in another case, the promising anticancer prodrug podophyllotoxin found in *Podophyllum peltatum* has now been reported in two strains of the endophytic fungus *Phialocephala fortinii*.[35]

From these results it appears possible to find natural products normally associated with plants via endophytes associated with them that make the same or related products. Ultimately, this may be vital when the source plant of a critically important compound is threatened with extinction. Certainly, the tools for the genetic manipulation of microorganisms, to realize greater product yields, are further advanced and technically more refined than those known for higher organisms. Thus, the potential is great for the eventual realization of the goal to have endophytes domesticated for the production of valuable natural products.

6.3 DEALING WITH ENDOPHYTES

This particular section of the review is based on my long experience in dealing with thousands of endophytic microbes from all corners of the world, including polar regions, tropical and temperate rain forests, savannas, tropical dry forests, cloud forests, and desert areas. Endophytes have been recovered from plants growing in each of these locations. The logic of sampling plants in all of these regions is that plant specificity among endophytic microbe populations may exist, and the biochemistry and diversity of microbes will vary dependent upon the region one is sampling. Presently, only about 1%–2% of the plants in the world have been sampled for their endophytic microbial populations and, furthermore, less than a handful of these have ever been completely surveyed for the entire set of microbes occurring in all plant tissues and organs.

The logic associated with a search for bioactive natural products from endophytic microbes is that it is commonly believed that endophytes have the innate capability to make secondary products that might protect their respective host plant from subsequent attack and infection by microbial pathogens(s). Thus, it seems reasonable to isolate, identify, and screen endophytic microbes for such products. The most critical aspect of such work is the establishment of a set of bioassays to identify target organisms for further study.[1] Such assays will also be critical for the eventual isolation via bioassay-guided fractionation and purification of the product(s).[1] Most interestingly, one can presuppose a biological activity for a molecule but, in fact, another more important utility for it may exist. For instance, taxol is a known anticancer agent but, in fact, its role in its host plant is probably as an antifungal agent.[36] It acts against the phycomycetous fungi in the same manner that it kills human cancer cells via the inhibition of the depolymerization of tubulin molecules. One can presume that endophytes are making taxol inhibit and destroy microbes that may be pathogenic to the host plant, or it is acting as an anticompetitor compound. Thus, we commonly use

Pythium ultimum in our initial screening bioassays since it is as sensitive to taxol as are human breast cancer cell lines.

Central for work on endophytes is the realization that virtually any kind of microbe might arise during the isolation procedures being used.[1,26] It is critical to use both standard microbiological methods for identification of the endophyte and more modern methods including internal transcribed spacer (ITS) and 18S rDNA sequence methodologies. Sometimes the data sets will not be in agreement, and that will require further in-depth study. It is commonly the case that comprehensive morphological data should have precedence over limited sequence data. Endophytic microorganisms need to be deposited in national or university collections, and ITS and other sequence information needs to be deposited in GenBank. Thus, one should never rely strictly on molecular information to provide the absolute identity of an organism since the method is somewhat restrictive, and there are examples of morphology disparities with molecular data, especially when only ITS or 18S rDNA sequence information is obtained.[37] Likewise, it is not acceptable to simply isolate and place a numerical tag on a microbe without using proper identification methods. When this is done there will be no proper placement of the compound to its origins in nature or to its relationship with other microbes or life forms.

Sometimes an unexpected bioactive endophyte may arise whose identification is unknown. This has happened many times in my experience. One must be ready and able to assume the role of a taxonomist and establish the identity of the endophyte prior to publishing any report on its bioactive natural products. Furthermore, it is not common to combine all of the taxonomic information gleaned on a novel microbe with information on the chemistry of its natural product into one scientific report. One of our greatest surprises was the discovery of numerous plants harboring actinomycetes. Many of these were biologically active (antifungal and antibacterial activities).[38,39] None of those studied had 16S rDNA-ITS regions that were identical to any actinomycete deposited in GenBank.[38,40] Interestingly, during the wide search by the pharmaceutical industry in the 1940s–1980s, there was no effort to isolate and screen microbes (endophytes) from plants. Basically, it appears as though there is a treasure trove of biologically active endophytic actinomycetes yet to be isolated and studied.

6.4 EXAMPLES FROM THE AUTHOR'S LABORATORY OF BIOACTIVE NATURAL PRODUCTS

In the search for biologically active natural products from endophytes, one must be prepared to deal with any one of a plethora of various natural products, including compounds, which are, among many others, carbohydrate derivatives, peptides, benzene derivatives, terpenoids, alkaloids, organic acids, and volatiles of all types and descriptions. Isolation and identification of the biological active natural product(s) from the source endophyte is the ultimate goal. Using modern methods of separation science, combined with spectroscopic methods, one will have some sense of the identity of the compound. Ultimately, however, it is most desirable to acquire a crystal of the product or a chemical derivative for its application to x-ray crystallographic methods. Furthermore, spectroscopic data from all sources including x-ray, mass spectroscopy, nuclear magnetic resonance (NMR), ultraviolet (UV), and infrared (IR) all need to be in agreement with each other. Reliance on NMR data alone can be fraught with difficulty, since at least 30% of the chemical structures published with these data alone have eventually needed revision.[41]

Over the years, we have isolated and characterized many novel endophytic microorganisms and particularly many of their bioactive natural products. A few compounds are listed which demonstrate the bioassays that have been used as well as the techniques applied to acquire chemical structures. Each of the first three examples is a compound isolated from *Pestalotiopsis microspora*, but the organism had its origins in significantly different plants located in widely different places. This endophyte is commonly found in temperate and tropical rain forests, is rich in natural products, and has only infrequently been studied.

6.4.1 Pestacin

Pestacin was obtained from a culture of *P. microspora*, an endophytic fungus obtained from *Taxus wallichiana* sampled in the Himalayan foothills. It produces a new 1,3-dihydro isobenzofuran and exhibits antioxidant activity 11 times greater than the vitamin E derivative troxol, and moderate antifungal activities (Figure 6.1).[42] Isolation of pestacin was achieved by extraction of culture fluid with methylene chloride followed by silica gel chromatography. Its structure was established by x-ray diffraction and ^{13}C and ^{1}H NMR. The x-ray data demonstrated that pestacin occurs naturally as a racemic mixture. Mechanisms for antioxidant activity and postbiosynthetic racemization have been proposed. Isopestacin is also produced by this endophyte and it possesses similar bioactivities as pestacin (Figure 6.1).[43]

6.4.2 Ambuic Acid

Ambuic acid is a highly functionalized cyclohexenone, which was isolated from *P. microspora* from *Fagraea bodenii* found in the highlands of Papua New Guinea (Figure 6.2).[44] The compound possesses weak antifungal properties. Its absolute

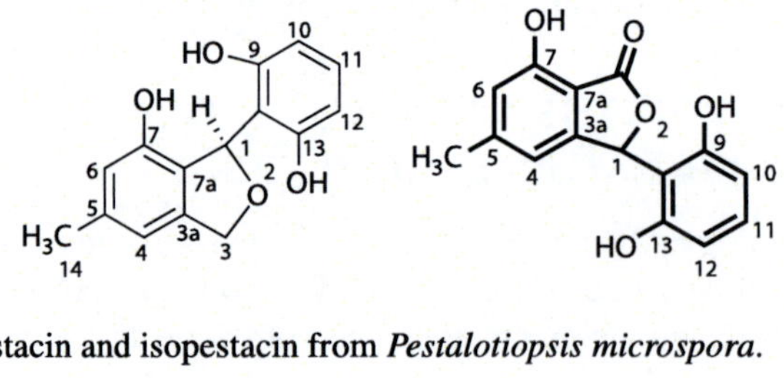

FIGURE 6.1 Pestacin and isopestacin from *Pestalotiopsis microspora*.

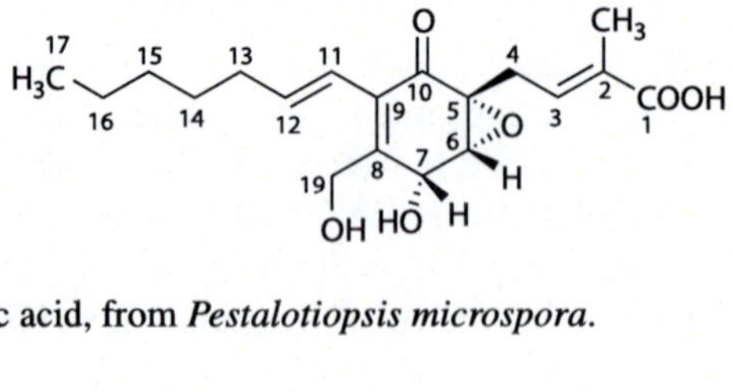

FIGURE 6.2 Ambuic acid, from *Pestalotiopsis microspora*.

structure was the first natural product to have its structure established by solid-state NMR methods, allowing a spatial assignment to the hydroxyl group on carbon 7 (Figure 6.2).[45,46] Quite surprisingly, after the initial work on the isolation and structural determination of ambuic acid, it was later learned that it is one of the best compounds known for its antiquorum sensing activity in gram-positive bacteria.[47] Ambuic acid inhibits the biosynthesis of the cyclic peptide quormones of *Staphylococcus aureus* and *Listeria innocua*. Ambuic acid is a lead compound in the search for antipathogenic drugs that target quorum sensing–mediated virulent expression of gram-positive bacteria. Once again, what initially was supposed to be a weak antifungal agent turned out to have a totally unsuspected biological activity as an inhibitor of quorum sensing in bacteria.

6.4.3 Torreyanic Acid

Torreyanic acid is a dimeric *quinone* first isolated from the endophyte *P. microspora* obtained from *Torreya taxifolia* in Northern Florida (Figure 6.3).[48] The compound was cytotoxic against 25 different human cancer cell lines with an average IC_{50} value of 9.4 μg/mL, ranging from 3.5 (G3-neuroendocrine carcinomas; NEC) to 45 (A549 human lung carcinoma) μg/mL. Torreyanic acid is 5–10 times more potent in cell lines sensitive to protein kinase C (PKC) agonists, such as 12-*O*-tetradecanoyl phorbol-13-acetate (TPA), and was shown to cause cell death via apoptosis. Torreyanic acid also promoted G1 arrest of G0 synchronized cells at 1–5 μg/mL levels, depending on the cell line. It has been proposed that the eukaryotic translation initiation factor EIF-4a is a potential biochemical target for the natural compound.

6.4.4 Colutellin A

Colutellin A is an immunosuppressive agent produced by *Colletotrichum dematium*, an endophytic fungus recovered from a *Pteromischum* sp. growing in a tropical forest in Costa Rica.[49] This fungus makes a novel peptide antimycotic, Colutellin A, with MICs of 3.6 μg/mL (48 h) against *Botrytis cinerea* and *Sclerotinia sclertiorum*. This peptide has a mass of 1127.70 and contains residues of Ile, Val, Ser, *N*-methyl-Val, and *β*-amino-isobutyric acid in nominal molar ratios

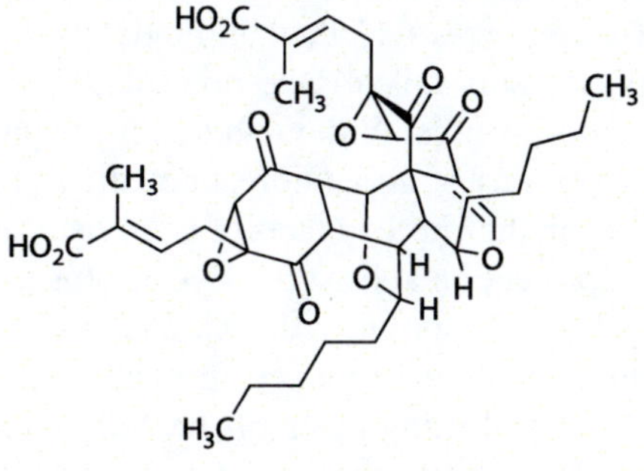

FIGURE 6.3 Torreyanic acid from *Pestalotiopsis microspora*.

of 3:2:1:1:1, respectively. Independent lines of evidence suggest that the peptide is cyclic, and sequences of Val-Ileu-Ser-Ileu as well as Ileu-Pro-Val have been deduced by MS/MS as well as Edman degradation methods. Colutellin A inhibited CD4 T cell activation of IL-2 production with an IC_{50} of 167.3 ± 0.38 nM, whereas cyclosporine A, in the same test, yielded a value of 61.8 nM. Since IL-2 production is inhibited by Colutellin A, at such a low concentration, this is an effective measure of the potential immunosuppressive activity of this compound. On the other hand, in repeated experiments, cyclosporin A at or above 8 μg/mL exhibited high levels of cytotoxicity on human peripheral blood mononuclear cells, whereas Colutellin A or DMSO alone, after 24 and 48 h of culture, exhibited no toxicity. Because of these properties Colutellin A has potential as a novel immunosuppressive drug.[49]

6.4.5 Fungal Volatiles

Fungal volatiles are made and exuded by numerous endophytic microbes. A novel endophytic fungus, *Muscodor albus*, was described from *Cinnamomum zelanicum* in a Honduran rain forest which produced a gas mixture with potent antibiotic properties.[50] The gas mixture contained organic acids, esters, ketones, naphthalene, and azulene-related compounds. A synthetic mixture of these compounds could also mimic the antimicrobial effects of the fungal gases.[51] Now, more than 12 other species of this fungus are known from various parts of the world, including Australia, China, Thailand, India, Peru, Bolivia, and Mexico. Each isolate makes a unique set of biological active volatiles.[52,53]

Other endophytes, such as *Ascocoryne sarcoides* and *Nodulisporium* sp., produce volatiles whose composition contains some hydrocarbons related or identical to those found in diesel fuel.[54–56] These discoveries have led to the proposition that endophytes may have contributed to the formation of crude oil in the first place.[9] Studies on the performance of these fungal-derived hydrocarbons are now under way at the U.S. Department of Energy, and other reviews have appeared.[54–60]

6.5 GUIDANCE AND PERSPECTIVE ON THE FUTURE OF ENDOPHYTE BIOLOGY AND NATURAL PRODUCTS CHEMISTRY

In the recent past, there has been an enormous worldwide effort on the isolation and characterization of endophytes and their biologically active natural products. However, in most cases, workers fail to take note of the possibility of filing and acquiring patents on their discoveries. Without this step, the potential for a novel useful product to be adopted by industry, agriculture, or medicine becomes greatly diminished to nonexistent. Filing in the United States of a provisional patent is inexpensive and will guarantee patent protection at least for 1 year. This time frame should allow the inventor the flexibility and opportunity to find one or more interested parties to license and begin development of the product. In most cases, the company partner will assume all of the legal and filing costs of the full patent if an agreement for licensing can be reached. The product may be the endophyte itself as is the case with *M. albus*, which is currently being developed as an agricultural biological control agent by MBI of Davis, California, with registration recently approved in November, 2016 by the U.S. Environmental Protection Agency. Or in the case of *M. crispans*, a patent on its gas mixture has been issued, and a form of the fungal gas formula is being developed as one or more products for the microbial decontamination of foods, feeds, plants, and industrial processes.[61] Antibiotics and other medicinals will require enormous sums for product development and it behooves the investigator to make appropriate company connections before the natural product search begins. On the other hand, there will be less expense if the product is being developed for agricultural or industrial purposes. These are important considerations if one ever expects to see practical utility for the discovery of a useful natural product. Obviously, patentability of an organism or a useful natural product requires novelty, utility, and a demonstration that the discovery was not obvious.

There are enormous opportunities of finding novel endophytes along with novel natural products that have utility. The likely places of discovery are in areas of high to extreme biodiversity along with areas having unique biological niches, for instance, endemic plants in specialized zones. Furthermore, geological consequences may have an important influence on the biology of an area, and that also needs consideration. This would be true of islands or island nations whose origins may have been continental millions of years ago. As such, plants and their microbial symbionts associated with these places would have had time to evolve in totally unique circumstances, in contrast to their continental counterparts. New Caledonia, which was once a part of Australia, nicely illustrates this point.

Finally, besides the scientific complexities in dealing with novel microbial endophytes, natural product isolation, compound identification, bioassay work, and patenting, other major issues may arise. Virtually, every country has a permitting process or processes. Australia seems to have the most complex system, with requirements for a national permit, a state permit, and a local park or reservation permit. Sharing agreements may also be a part of the process. In the United States, there is a requirement for a U.S. Department of Agriculture (USDA) permit to handle plant materials being brought to U.S. labs from foreign locations. This involves a lengthy form, lab inspection, and notification when samples are being carried into the United States from foreign lands. The USDA process is justified given the threat that is constantly being posed by the accidental introduction of pests and diseases from abroad. Finally, once permits are obtained, and one is heading into jungle settings, there should be concerns over the threats of malaria, dengue fever, a host of parasitic diseases, poisonous snakes, and a myriad of other creatures that view you either as a threat or as a morsel of food.

The world has an urgent need of new agents, methods, and processes to improve conditions in health, agriculture, and industry. Endophytes and their secondary products offer some hope in solving some of these problems. The immediate past has shown hope for the future. For instance, the discovery of endophytes making volatile antibiotics, the finding of plants hosting novel actinomycetes making bioactive compounds, and endophytes making a plethora of fuel-related hydrocarbons all provide examples of surprising discoveries that may help solve some of these problems.

On the larger scale, the search for interesting and novel endophytes has just begun. Only a tiny fraction of the world's plants has been sampled. The discovery of the basic plant biome of each of the world's top 20 food plants is largely unknown, and there is promise that work in this area will be fruitful. In addition, only a few of the known 300,000 plant species have ever had their roots, flowers, leaves, or fruits sampled for endophytes. Lower plant forms also host endophytes including those in the world's oceans. The possibilities are enormous and the challenges are exciting.

ACKNOWLEDGMENTS

The author acknowledges the support of the National Science Foundation (NSF) Emerging Frontiers in Research and Innovation (NSF-EFRI), Grant No. 0937613 to Dr. Brent Peyton and Gary Strobel of MSU, and a DoE grant to GAS to carry out some of the work referenced in this chapter.

REFERENCES

1. Strobel, G. A.; Daisy, B. Bioprospecting for microbial endophytes and their natural products. *Microbiol. Mol. Biol. Rev.* **2003**, 67, 491–502.
2. Walker, J. C. *Plant Pathology*. McGraw-Hill, New York, **1957**.
3. Bacon, C. W.; White, J. F. *Microbial Endophytes*. Marcel Dekker Inc, New York, **2000**.
4. Jones, E. B. G.; Stanley, S. J.; Pinruan, U. Marine endophytes sources of new chemical natural products: A review. *Bot. Mar.* **2008**, 51, 163–170.
5. Davis, C.; Franklin, J. B.; Shaw, A. J. Biogeographic and phylogenetic patterns in diversity of liverwort-associated endophytes. *Am. J. Bot.* **2003**, 90, 1661–1667.
6. Redlin, S. C.; Carris, L. M. *Endophytic Fungi in Grasses and Woody Plants*. APS Press, St Paul, MN, **1996**.
7. Petrini, A. E.; Petrini, O. Xylarious fungi as endophytes. *Sydowia* **1985**, 38, 216–234.
8. Smith, S. A.; Tank, D. C.; Boulanger, L. A.; Bascom-Slack, C. A.; Eisenman, K.; Kingery, D.; Babbs, B. et al. Bioactive endophytes warrant intensified exploration and conservation. *PLoS One* **2008**, 3, 3052–3059.
9. Strobel, G. A.; Booth, E.; Schaible, G.; Mends, M. T.; Sears, J.; Geary, B. The paleobiosphere: A novel device for the *in vivo* testing of hydrocarbon production-utilizing microorganisms. *Biotechnol. Lett.* **2013**, 35, 539–552.
10. Reid, A.; Greene, S. E. *How Microbes Can Help Feed the World*. American Society of Microbiology/American Academy of Microbiology, Washington, DC, **2013**. ofrf.org/sites/ofrf.org/files/FeedTheWorld_0.pdf.
11. Peterson, J.; Garges, S.; Giovanni, M.; McInnes, P.; Wang, L.; Schloss, J. A.; Bonazzi, V. et al. The NIH human microbiome project. *Genet. Res.* **2009**, 19, 2317–2323.
12. Morgan-Jones, G.; Gams, W. Notes on hyphomycetes, XLI. An endophyte of *Festuca arundinaceae* and the anamorph of *Epichloe typhina*, new taxa in one of two new sections of *Acremonium*. *Mycotaxon* **1982**, 15, 318–331.
13. Fribourg, H. A.; Hannaway, D. B.; West, C. P. *Tall Fescue Online Monograph*. Oregon State University, Corvallis, OR, **2009**. http://forages.oregonstate.edu/tallfescuemonograph.
14. Lyons, P. C.; Plattner, R. D.; Bacon, C. W. Occurrence of peptide and clavine ergot alkaloids in tall fescue grass. *Science* **1986**, 232, 487–489.
15. Siegel, M. R.; Latch, G. C. M.; Johnson, M. C. Fungal endophytes of grasses. *Annu. Rev. Phytopathol.* **1987**, 25, 293–315.
16. Siegel, M. R.; Latch, G. C. M.; Bush, L. P.; Fammin, N.; Rowen, D.; Tapper, B. A.; Bacon, C. W. Alkaloids and insecticidal activity of grasses infected with fungal endophytes. *J. Chem. Ecol.* **1991**, 16, 3301–3315.

17. Read, J. C.; Camp, B. J. The effect of the fungal endophytes *Acremonium coenophialum* in tall fescue on animal performance toxicity and stand maintenance. *Agron. J.* **1986**, 78, 848–850.
18. Siegel, M. R.; Bush, L. P. Toxin production in grass/endophyte associations. In *The Mycota*, Vol. V. Plant Relationships, Part A, Carrol, G. C.; Tudzynski, P., Eds. Springer-Verlag, Berlin, Germany, **1997**, pp. 185–207.
19. Schiff, P. B.; Horwitz, S. B. Taxol stabilizes microtubles in mouse fibroblast cells. *Proc. Natl. Acad. Sci. USA* **1980**, 77, 1561–1565.
20. Wani, M. C.; Horwitz, S. B. Nature as a remarkable chemist: A personal story of the discovery and development of Taxol. *Anticancer Drugs* **2014**, 25, 482–487.
21. Stierle, A.; Strobel, G.; Stierle, D. Taxol and taxane production by *Taxomyces andreanae*, an endophytic fungus of Pacific yew. *Science* **1993**, 260, 154–155.
22. Zhou, X.; Zhu, H.; Tang, K. A review: Recent advances and future prospects of taxol-producing endophytic fungi. *Appl. Biochem. Microbiol.* **2010**, 86, 1707–1717.
23. Gond, S. K.; Kharwar, R. N.; White, J. F. Will fungi be the new source of the blockbuster drug taxol? *Fungal Biol. Rev.* **2014**, 28, 77–84.
24. Tan, R. X.; Zau, W. X. Endophytes: A rich source of functional metabolites. *Nat. Prod. Rep.* **2001**, 18, 448–459.
25. Strobel, G. A. Rainforest endophytes and bioactive products. *Crit. Rev. Biotechnol.* **2002**, 22, 315–333.
26. Strobel, G. A.; Daisy, B. H.; Castillo, U.; Harper, J. Natural products from endophytic microorganisms. *J. Nat. Prod.* **2004**, 67 257–268.
27. Guo, B. Y.; Wang, Y.; Sun, X.; Tang, K. Bioactive natural products from endophytes: A review. *Appl. Biochem. Microbiol.* **2008**, 44, 136–144.
28. Verma, V. C.; Kharwar, R. N.; Strobel, G. A. Chemical and functional diversity of natural products from plant associated endophytic fungi. *Nat. Prod. Commun.* **2009**, 4, 1511–1532.
29. Kusari, S.; Zuehlke, S.; Spiteller, M. An endophytic fungus from *Camptotheca acuminata* that produces camptothecin and analogues. *J. Nat. Prod.* **2009**, 72, 2–7.
30. Shweta, S.; Zuehlke, S.; Ramesha, B. T.; Priti, V. M.; Kumar, P.; Ravikanth, G.; Spiteller, M.; Vasudeva, R.; Shaanker, R. Endophytic fungal strains of *Fusarium solani*, from *Apodytes dimidiata* E. Mey. ex Arn (Icacinaceae) produce camptothecin, 10-hydroxycamptothecin and 9-methoxycamptothecin. *Phytochemistry* **2010**, 71, 117–122.
31. Amna, T.; Puri, S. C.; Verma, V.; Sharma, J. P.; Khajuria, R. K.; Musarrat, J.; Spiteller, M.; Qazi, G. N. Bioreactor studies on the endophytic fungus *Entrophospora infrequens* for the production of an anticancer alkaloid camptothecin. *Can. J. Microbiol.* **2006**, 52, 189–196.
32. Pu, X.; Qu, S.; Chen, F.; Bao, J.; Shang, G.; Luo, Y. Camptothecin-producing endophytic fungus *Trichoderma atroviride* LY 357: Isolation, identification and fermentation conditions optimization for camptothecin production. *Appl. Microbiol. Biotechnol.* **2013**, 97, 9365–9375.
33. Li, J. Y.; Sidhu, R. S.; Ford, E.; Hess, W. M.; Strobel, G. A. The induction of taxol production in the endophytic fungus—*Periconia* sp. from *Torreya grandifolia*. *J. Ind. Microbiol.* **1998**, 20, 259–264.
34. Kumar, A.; Rajamohanan, P. R.; Absar, A. Isolation, purification and characterization of vinblastine and vincristine from endophytic fungus *Fusarium oxysporum* isolated from *Catharanthus roseus*. *PLoS One* **2013**, 8, e71805. DOI 10.1371/journal.pone.0071805.
35. Eyberger, A. L.; Dondapati, R.; Porter, J. R. Endophyte fungal isolates from *Podophyllum peltatum* produce podophyllotoxin. *J. Nat. Prod.* **2006**, 69, 1121–1124.
36. Young, D. H.; Michelotti, E. J.; Sivendell, C. S.; Krauss, N. E. Antifungal properties of taxol and various analogues. *Experientia* **1992**, 48, 882–885.

37. Xie, J.; Strobel, G. A.; Mends, M.; Hilmer, J.; Nigg, J.; Geary, B. *Collophora aceris*, a novel antimycotic producing endophyte associated with Douglas maple. *Microb. Ecol.* **2013**, 66, 784–795.
38. Castillo, U.; Strobel, G. A.; Ford, E. J.; Hess, W. M.; Porter, H.; Jensen, J. B.; Albert, H. et al. Munumbicins, wide spectrum antibiotics produced by *Streptomyces munumbi*, endophytic on *Kennedia nigriscans*. *Microbiology* **2002**, 148, 2675–2685.
39. Castillo, U.; Strobel, G. A.; Mullenberg, K.; Condron, M. M.; Teplow, D.; Folgiano, V.; Gallo, M. et al. Munumbicins E-4 and E-5: Novel broad spectrum antibiotics from *Streptomyces* NRRL 3052. *FEMS Lett.* **2006**, 255, 296–300.
40. Castillo, U. F.; Browne, L.; Strobel, G. A.; Hess, W. M.; Ezra, S.; Pacheco, G.; Ezra, D. Biologically active endophytic streptomycetes from *Nothofagus* spp. and other plants in Patagonia. *Microb. Ecol.* **2007**, 53, 12–19.
41. Harper, J. K. Natural products structural analysis enhancements. In *Encyclopedia of NMR*, Grant, D. M.; Harris, R. K., Eds. Wiley, Chichester, U.K., **2002**, Vol. 9, pp. 589–597.
42. Harper, J. K.; Ford, E. J.; Strobel, G. A.; Arif, A.; Grant, D.; Porco, J.; Tomer, D. P.; O'Neill, K. Pestacin: A 1,3-dihydro isobenzofuran from *Pestalotiopsis microspora* possessing antioxidant and antimycotic activities. *Tetrahedron* **2003**, 59, 2471–2476.
43. Strobel, G. A.; Ford, E.; Worapong, J.; Harper, J. K.; Arif, A. M.; Grant, D. M.; Fung, P.; Chau, R. M. Isopestacin, a unique isobenzofuranone from *Pestalotiopsis microspora* possessing antifungal and antioxidant properties. *Phytochemistry* **2002**, 60, 179–183.
44. Li, J. Y.; Harper, J. K.; Grant, D. M.; Tombe, B.; Bashyal, B.; Hess, W. M.; Strobel, G. A. Ambuic acid, a highly functionalized cyclohexenone with bioactivity from *Pestalotiopsis* spp. and *Monochaetia* sp. *Phytochemistry*. **2001**, 56, 463–468.
45. Harper, J. K.; Li, J. Y.; Grant, D. M.; Strobel, G. A. Characterization of stereochemistry and molecular conformation using solid-state NMR tensors. *J. Am. Chem. Soc.* **2001**, 123, 9837–9842.
46. Harper, J. K.; Barich, D. H.; Hu, J. Z.; Strobel, G. A.; Grant, D. M. Stereochemical analysis by solid-state NMR: Structural predictions in ambuic acid. *J. Org. Chem.* **2003**, 68, 4609–4614.
47. Nakayama, J.; Uemura, Y.; Nishiguchi, K.; Yoshimura, N.; Igarashi, Y.; Sonomoto, K. Ambuic acid inhibits the biosynthesis of cyclic peptide quormones in Gram-positive bacteria. *Antimicrob. Agents Chemother.* **2009**, 53, 580–586.
48. Lee, J. C.; Strobel, G. A.; Lobkovsky, E.; Clardy, J. Torreyanic acid: A selectively cytotoxic quinone dimer from the endophytic fungus *Pestalotiopsis microspora*. *J. Org. Chem.* **1996**, 61, 3232–3233.
49. Ren, Y.; Strobel, G. A.; Graff, J. C.; Jutila, M.; Park, S. G.; Gosh, S.; Teplow, D.; Condron, M.; Pang, E.; Hess, W. M. Colutellin A, an immunosuppressive peptide from *Colletotrichum dematium*. *Microbiology* **2008**, 154, 1973–1979.
50. Worapong, J.; Strobel, G. A.; Ford, E. J.; Li, J. Y.; Baird, G.; Hess, W. M. *Muscodor albus* gen. et sp. nov. an endophyte from *Cinnamomum zeylanicum*. *Mycotaxon* **2001**, 79, 67–79.
51. Strobel, G. A.; Dirksie, E.; Sears, J.; Markworth, C. Volatile antimicrobials from a novel endophytic fungus. *Microbiology* **2001**, 147, 2943–2950.
52. Strobel, G. A. *Muscodor albus*—The anatomy of an important biological discovery. *Microbiol. Today* **2012**, 39, 108–111.
53. Strobel, G. A. *Muscodor albus* and its biological promise. *Phytochem. Rev.* **2011**, 10, 165–172.
54. Strobel, G.; Knighton, B.; Kluck, K.; Ren, Y.; Livinghouse, T.; Griffen, M.; Spakowicz, D.; Sears, J. The production of myco-diesel hydrocarbons and their derivatives by the endophytic fungus *Gliocladium roseum*. *Microbiology* **2008**, 154, 3319–3328.

55. Griffin, M. A.; Spakowicz, D. J.; Gianoulis, T. A.; Strobel, S. A. Volatile organic compound production by organisms in the genus *Ascocoryne* and a re-evaluation of myco-diesel production by NRRL 50072. *Microbiology* **2010**, 156, 3814–3829.
56. Tomsheck, A.; Strobel, G. A.; Booth, E.; Geary, B.; Spakowicz, D.; Knighton, B.; Floerchinger, C.; Sears, J.; Liarzi, O.; Ezra, D. *Hypoxylon* sp. an endophyte of *Persea indica*, producing 1,8-cineole and other bioactive volatiles with fuel potential. *Microb. Ecol.* **2010**, 60, 903–914.
57. Gladden, A. M.; Taatjes, C. A.; Gao, C.; O'Bryan, G.; Powell, A. J.; Scheer, A. M.; Turner, K.; Wu, W.; Yu, E. T. Tailoring next-generation biofuels and their combustion in next-generation engines. *Sandia Rep.* **2013**, 2013, 10094.
58. Strobel, G. A. Methods of discovery and techniques to study endophytic fungi producing fuel-related hydrocarbons. *Nat. Prod. Rep.* **2014**, 39, 259–272.
59. Strobel, G. A. The story of mycodiesel. *Curr. Opin. Microbiol.* **2014**, 19, 52–58.
60. Strobel, G. A. The use of endophytic fungi for the conversion of agricultural wastes to hydrocarbons. *Biofuels* **2014**, 5, 447–455.
61. Gandhi, N. R.; Skebba, V. P.; Strobel, G. A. Antimicrobial composition and methods of use. US 8728462 B2, **2014**.

7 Novel Insights in Plant-Endophyte Interactions

Souvik Kusari, Parijat Kusari, and Michael Spiteller

CONTENTS

7.1 INTRODUCTION

Endophytic microorganisms (so-called endophytes) are inherent and decisive partners in the biological marketplace of microbiota associated with plants in any ecosystem. Endophytes are typically ubiquitous in all plant tissues, where they establish remarkably multifarious groups of complex communities that deliver vital eco-specific functional traits. The fundamental role of chemical and molecular crosstalk in plants and associated microbiota, including endophytes, is increasingly being unearthed in a plethora of studies. Thus far, it is clear that elucidating all factors affecting the nature, distribution, and dynamics of these crosstalk molecules and how they influence the communication between endophytes and associated organisms, such as the host plant, associated endophytes, and invading pathogens, is indispensable to gain fundamental insights into endophytic functional traits in distinct ecological niches. The biology and chemistry behind these associations and the driving forces or triggers involved should be elucidated on a case-by-case basis to get a holistic view of plant-endophyte mutualism in an ecosystem. Herein, we provide a conceptual overview of plant-endophyte interactions that have been elucidated by several recent studies on different facets of microbial relationship with plants, including relevant examples from our group.

7.2 ENDOPHYTES AT THE CENTER STAGE OF NATURE'S BIOLOGICAL MARKETPLACE

Plant-associated endophytes (Figure 7.1) live in a beneficial symbiosis with their hosts. Although the strategies coevolved by endophytes for living and functioning in different ecological niches are still open to future exploration, multitrophic physiological and biochemical interactions underlying their crosstalk are known to be crucial.[1–4] For proper functioning of ecosystem processes, a plethora of multiplexed and dynamic crosstalk occurs between endophytes (exemplified by endophytic fungi, endophytic bacteria, endophytic viruses, and other micro- and nano-organisms), concordant with their associated host plants, under various biotic and abiotic conditions (Figure 7.1).[5,6]

The intricate network of interactions occurring between plants and associated microbiomes might be further exemplified using the "biological market theory."[7,8] Notably, an analogy between the interactions and the concept of a "biological marketplace" could be drawn, in which numerous fair-trade microbial partners such as symbionts or mutualists (e.g., endophytes) engage in cooperative and social interactions with cheaters, antagonists, and pathogens.[6–8] Endophytic microorganisms are ubiquitous in plant tissues, not only in below ground root tissues but also in the above-ground foliar tissues (e.g., in the apoplastic compartments), where they

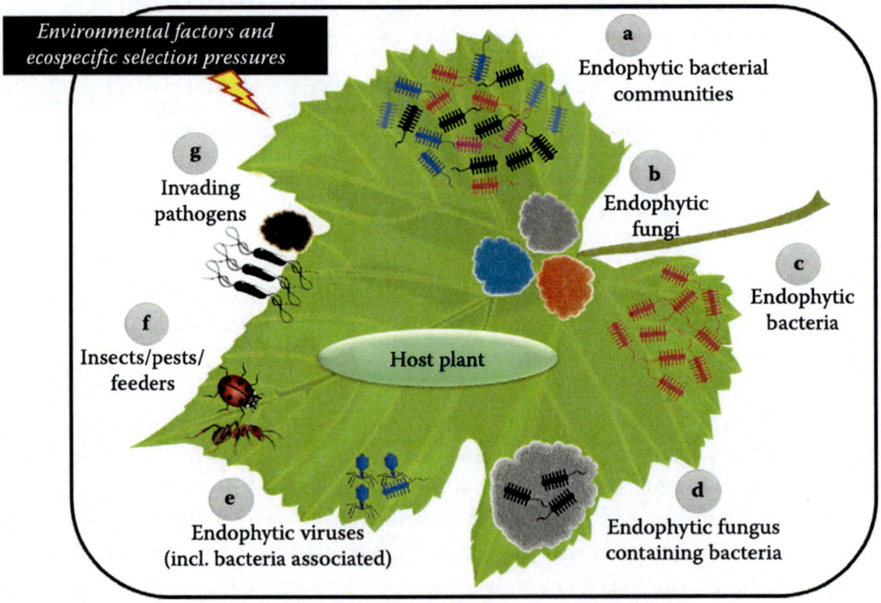

FIGURE 7.1 Schematic representation of the network of endophytic microbiota in plants comprising a plethora of macro- and microorganisms. **(a)** Endophytic bacterial communities. **(b)** Endophytic fungi. **(c)** Endophytic bacteria. **(d)** Endophytic fungus containing endosymbiotic bacteria. **(e)** Endophytic viruses. **(f)** Invading insects, pests, and feeders. **(g)** Invading pathogens.

establish multifarious communities that can be beneficial to host plant health. Plant-microbe (mainly plant-endophyte) interactions and endophyte-endophyte interactions (such as fungus-fungus, fungus-bacterium, bacterium-bacterium, and/or more complex endophytic community/microbiome interactions) play a major role in the production of communication molecules and secondary metabolites, including compounds mimetic to the host plants (Figure 7.2). In microbial communities within plants, the interplay between endophytes and plants, and the production of these metabolites, results in an assortment of functional traits of both agricultural and ecological importance, such as promoting defense by plants toward biotic and abiotic

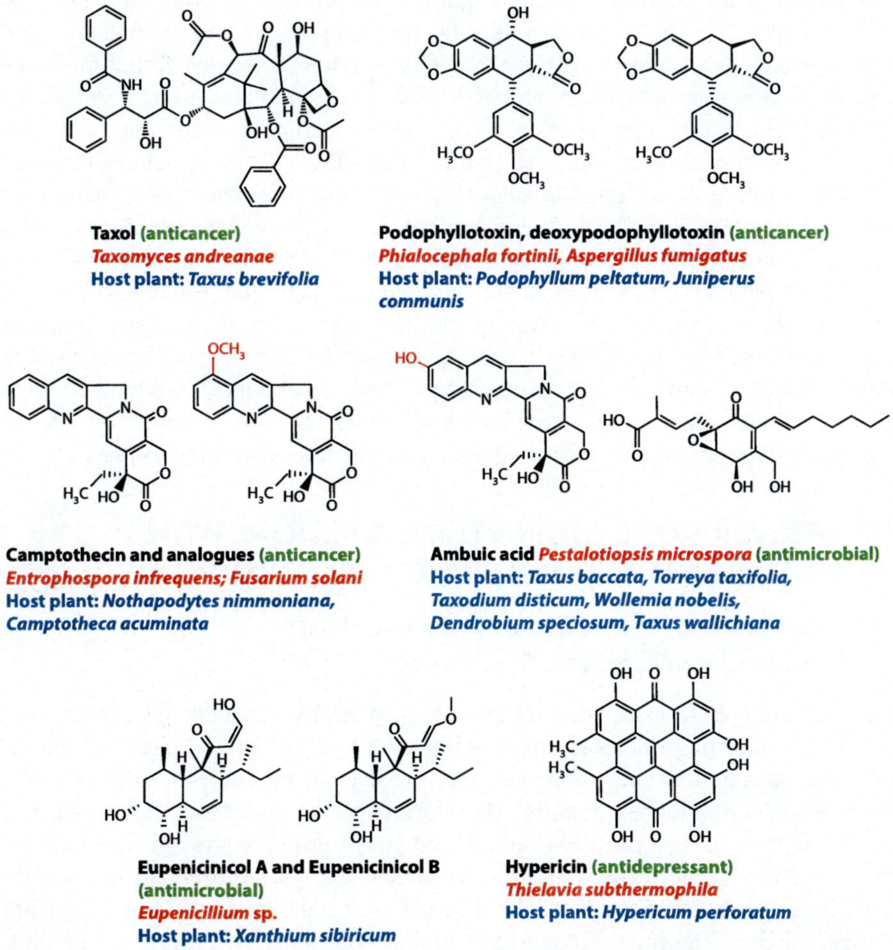

FIGURE 7.2 Representation of selected natural products produced by endophytes belonging to different genera harbored by diverse plant species in varied ecosystems, and their structures belonging to different chemical scaffolds with a variety of biological activities. This infinitesimal representation exemplifies the innumerable number of compounds isolated so far from endophytes and the virtually unlimited bioresource provided by endophytes colonizing plants around the world.

stress, and defense suppression or promotion by endophytes concomitant to their interaction with the host plant immune system. The secondary metabolites produced by endophytes also include bioactive natural products of medicinal and industrial value.[6,9–12] The consequence of plant-endophyte interactions is determined by a set of dynamic molecular and cellular interactions in which endophytes colonize plant tissues, interact, and evade or suppress host immune mechanisms.[1,13] Such interactions are in turn dependent not only on the metabolome of endophytes, but also on their secretome, which modulate host plant physiology and immunity, concomitant to the *in situ* environment within the plant.[14–16]

Plant-endophyte interactions in specific ecological niches generate *signals* such as biochemical and molecular triggers that activate cryptic biosynthetic pathways, precursors of compounds with eco-specific functions, epigenetic modulators, and quorum-sensing molecules, to name a few. These signal molecules further modulate the host plant immune network encompassing salicylate, indole-acetic acid and related molecules, reactive oxygen species, nitric oxide, γ-amino butyric acid, jasmonic acid, ethylene, cyanide, calcium, and extracellular ATP.[13,17–24] For example, the plant growth–promoting bacterium *Gluconacetobacter diazotrophicus*, which associates with sugarcane (*Saccharum* spp.) as an endophyte, has been shown to differentially produce proteins that modulate protein degradation and lipid metabolism pathways in the host plants.[25] Furthermore, the endophytic bacterium is able to produce elicitor molecules capable of activating the host plant defense against pathogenic *Xanthomonas albilineans*.[26] Thus far, these signals have demonstrated potential to function bidirectionally, in plant-endophyte as well as endophyte-endophyte cross talk and communication, because in several plants they have been shown to be engendered by microorganisms during plant colonization in specific tissues.

7.3 DYNAMICS OF ENDOPHYTE INTERACTIONS WITH HOST PLANTS AND ASSOCIATED ORGANISMS

7.3.1 Endophytic Fungal Interactions with Host Plants and Associated Endophytes

Any plant-endophyte association is preceded by a physical encounter between a plant and an invading microorganism, followed by several physical, chemical, and molecular barriers that must be overcome, resisted, or neutralized to successfully establish an endophytic association. Therefore, selected plants and their associated endophytic microbiome must have developed coevolutionary strategies to acclimatize to coexistence with each other. In particular, once a microorganism from the environment comes into contact with a target host plant and breaches the plant's physical barriers, the plant recognizes certain evolutionarily conserved microbial fingerprints known as microbe-associated molecular patterns (MAMPs) for mutualists, and pathogen-associated molecular patterns (PAMPs) for pathogens.[13,27] Such early signal recognition is achieved by the cell-surface-resident pattern recognition receptors (PRRs) and leads to activation of the plant's pattern-triggered immunity (PTI), one of the main innate immune responses activated by the invaded plant.[27–29] Notably, an important early signaling event in many pathogenic and mutualistic plant-microbial interactions is an increase of intracellular calcium levels in the cells of plants.[30] On one hand, in the case of pathogenic interactions, elevation of cytoplasmic calcium levels induces downstream defense responses due to activation of ion fluxes, an oxidative burst, phosphorylation events via mitogen-activated protein kinase (MAPK) signaling, and elicitation of phytohormone(s) signaling. On the other hand, endophytic interactions (mutualistic) involve sequential elevation of cytoplasmic and nuclear calcium levels.[31] Plant genomes typically contain a large number of calcium-binding proteins. For example, the model plant *Arabidopsis thaliana* has been shown to contain at least 100 calcium-binding and/or receptor genes.[32] Response to pathogens and beneficial endophytes depends on activation of specific members of these protein families. Nevertheless, pathogenic and mutualistic plant-microbial interactions may share some common signaling components such as MAPKs. Events downstream of MAPK generally involve elicitation of phytohormones. Growth promotion by endophytes involves production of phytohormones such as auxins and cytokinin, or of enzymes, such as 1-aminocyclopropane-1-carboxylate (ACC) deaminase, which lower ethylene levels. For instance, an endophytic strain, *Aspergillus fumigatus*, isolated from soybean roots, was shown to biosynthesize gibberellins when subjected to selected abiotic stress such as high salt.[33] This endophytic fungus was capable of reprograming the host plant in both lowering the production of endogenous abscisic acid and elevating free proline, salicylic acid, and jasmonic acid levels. In another study, it was demonstrated that the fungal endophyte, *Piriformospora indica*, eludes defense of *Arabidopsis* plants by founding a biotrophic association.[34] In order to achieve such an association, *P. indica* initially impedes *Arabidopsis* hormone levels while simultaneously secreting lectins and effectors to subdue plant defenses. Finally, after *P. indica* establishes in *Arabidopsis* as an endophyte, it selectively colonizes the dilapidated host cells and secretes digestive enzymes to degrade proteins and maintain its colonization.[34]

Several studies have further revealed that some microorganisms are able to evade the host plant defenses, either by regulating the expression of plant secondary metabolite (e.g., defense compounds) biosynthetic genes, or by resisting the action of the plant defense compounds altogether. For instance, we isolated an endophytic fungal strain (*Fusarium solani*) from the Chinese *Happy Tree*, *Camptotheca acuminata* Decaisne (Nyssaceae), which was able to biosynthesize the important anticancer compound camptothecin by a cross-species biosynthetic pathway contributed in part by the host plant.[35–37] Camptothecin is a potent antineoplastic agent that causes DNA damage by stabilizing a typically transient covalent complex between topoisomerase I enzyme and DNA.[35] This discovery, therefore, led us to question how the endophytic fungus ensures self-resistance before being incapacitated or even killed *in situ* in the plant by camptothecin. We discovered its survival strategy by examining the fungal topoisomerase I structure with emphasis on the camptothecin-binding and catalytic domains. Notably, typical amino acid residues, such as *Asn352*, *Glu356*, *Arg488*, *Gly503*, *Gly717*, and others, were identified that ensure fungal resistance against camptothecin.[37] This work revealed that some plant-endophyte interactions might not be just equilibrium between virulence and defense, but a much more complex, precisely controlled interaction that plausibly results from intrinsic evolutionary preadaptation or target-based coevolutionary adaptation.

Another interesting evolutionary trait of some microbes, particularly filamentous fungi, is evading host plant immune response by structural and chemical modifications of their own cell wall constituents, as well as by secretion of certain effector proteins that bypass the plant's PRRs-mediated initiation of PTI.[38] In order for fungi to fruitfully facilitate infection and establish either mutualism or pathogenicity, not only PTI but also the next line of the host plant's immune response called effector-triggered immunity (ETI) must be suppressed or bypassed.[39] It is compelling that the driving force in the convergent evolution of own cell wall component modulation by plant-associated fungi is the necessity to avoid triggering PTI and/or ETI.[38,39] For example, Veneault-Fourrey et al.[40] demonstrated that the ectomycorrhizal fungus, *Laccaria bicolor*, is capable of actively remodeling its own cell wall components, not only during establishment of mutualism with a host plant, but also for further *in planta* proliferation. In another recent study, it was shown that when *Colletotrichum graminicola* infects maize plants, it downregulates the synthesis of β-1,6-glucan in the biotrophic hyphae in order to circumvent triggering the host plant immune responses.[41]

In one of our recent studies on plant-endophyte cross talk, we further provided a proof-of-concept of how a fungus capable of producing bioactive compounds might avoid triggering the host plant defense during infection and colonization as an endophyte. We isolated an endophytic fungus, *Colletotrichum* sp. BS4, from the boxwood plant *Buxus sinica* (Buxaceae), which is used not only in traditional Chinese medicine (TCM) but also as a topiary shrubbery in People's Republic of China.[42] We employed the "one strain many compounds" (OSMAC) approach[43] to discover that the fungus is capable of producing the azaphilones, colletotrichones A–C, along with chermesinone B, under static solid-state fermentation conditions (Figure 7.3a).[42] Azaphilones belong to a group of polyketides biosynthesized typically by polyketide synthase (PKS) pathways.[44] Azaphilones are frequently produced by many species of fungi belonging to both Ascomycota and Basidiomycota, particularly the genera *Aspergillus*, *Chaetomium*, *Penicillium*, *Pestalotiopsis*, *Phomopsis*, *Talaromyces*, *Emericella*, *Monascus*, *Epicoccum*, and *Hypoxylon*.[44] For selected fungal genera, azaphilones even serve as important chemotaxonomic markers.[45–47] Interestingly, azaphilones display

(a)

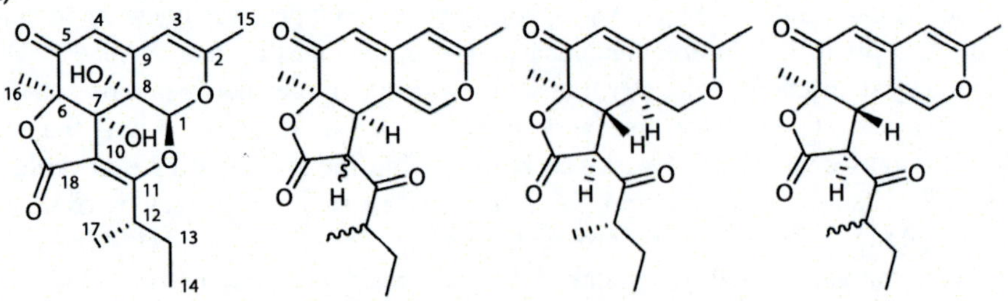

FIGURE 7.3 Spatial distribution of colletotrichone A and colletotrichone B and/or chermesinone B produced by endophytic *Colletotrichum* sp. BS4 on rice agar after 16 days. **(a)** Structures of the compounds. *(Continued)*

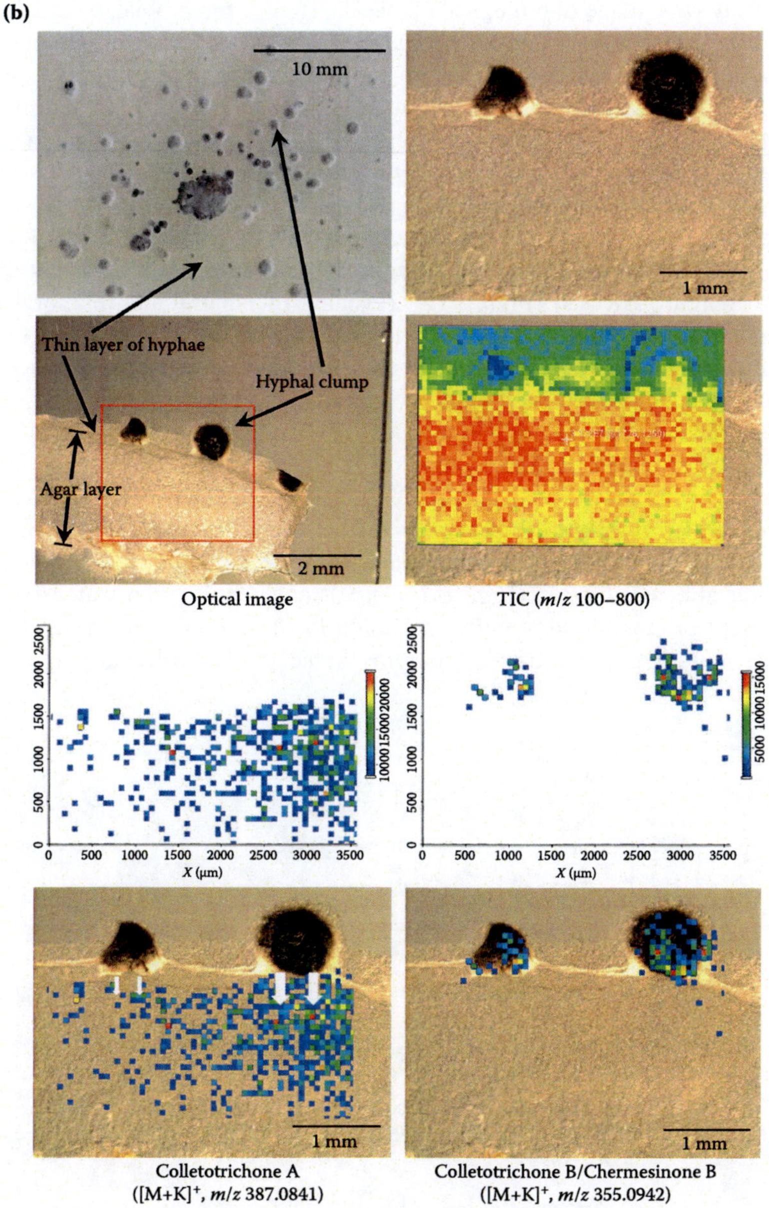

FIGURE 7.3 (*Continued*) Spatial distribution of colletotrichone A and colletotrichone B and/or chermesinone B produced by endophytic *Colletotrichum* sp. BS4 on rice agar after 16 days. **(b)** Optical image of the colony and the cross section of the agar layer on a glass slide, along with total ion current (TIC) of the scanned area (*m/z* 100–800), and spatial distribution of colletotrichone A ($[M+K]^+$, *m/z* 387.0841), colletotrichone B, and/or chermesinone B ($[M+K]^+$, *m/z* 355.0942). (With kind permission from Springer Science+Business Media: *Fungal Applications in Sustainable Environmental Biotechnology*, Unraveling the chemical interactions of fungal endophytes for exploitation as microbial factories, 2016, 353–370, Wang, W.-X., Kusari, S., and Spiteller, M.)

an ominously wide range of biological efficacies, from antimicrobial, cytotoxic, antiviral, and anti-inflammatory to anticancer activities.[44] Further, some azaphilones have been reported to demonstrate inhibition of gp120-CD4 binding in anti-HIV assays,[48] and Grb2-SH2 and MDM2-p53 interactions in anticancer assays.[49,50] In our investigation on the compounds produced by endophytic *Colletotrichum* sp., colletotrichone A showed significant antibacterial efficacy against environmental bacterial strains of *Escherichia coli* and *Bacillus subtilis*.[42]

Further detailed examination of the endophyte using microscopy and matrix-assisted laser desorption ionization imaging high-resolution mass spectrometry (MALDI-imaging-HRMS) (Figure 7.4) revealed the spatial and temporal localization of the compounds vis-à-vis their plausible ecological function(s).[12] On amended rice agar medium, *Colletotrichum* sp. BS4 displayed the characteristic morphological features of the genus,[51,52] notably the thick-walled, dark-colored hyphal clumps (called setae) (Figure 7.3b). MALDI-imaging-HRMS experiments clearly revealed in high spatial resolution that even though the inactive compounds, colletotrichone B and/or chermesinone B, were retained at the site of production at or around the vicinity of the hyphal clumps, highly bioactive colletotrichone A was effectively removed from the hyphal clumps by secreting it into agar after production (Figure 7.3b).[12] It has been demonstrated in earlier studies that *Colletotrichum* species, particularly those that associate with plants either as mutualists or pathogens, employ setae or the associated appressoria

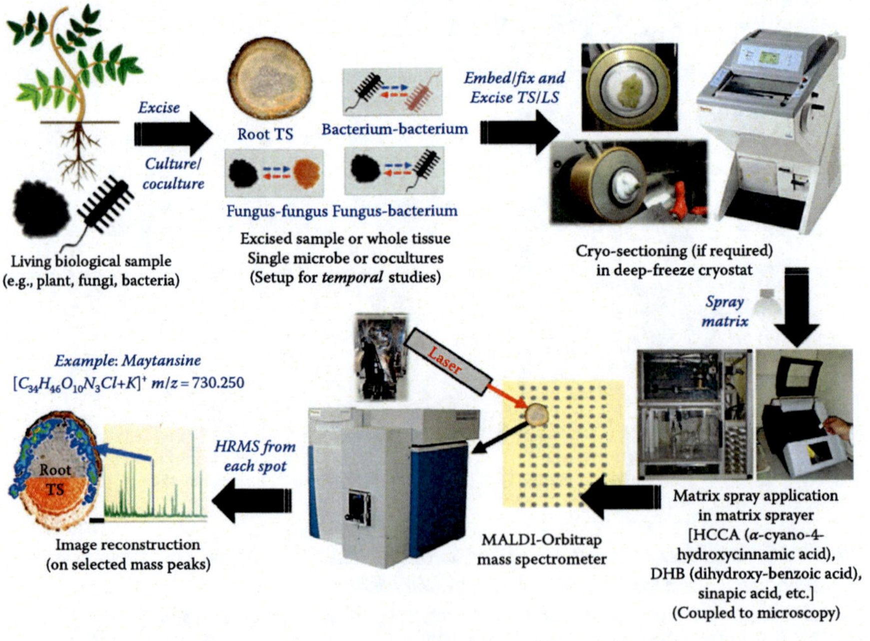

FIGURE 7.4 Schematic overview of matrix-assisted laser desorption ionization imaging high-resolution mass spectrometry (MALDI-imaging-HRMS), which can be used to visualize, in high spatial resolution, the distribution, localization, and dynamics of both target and nontarget molecules in plants and microorganisms with limited sample preparation.

for attachment to the host surface for permeation into the host tissue or apoplast compartments.[51] Concurrently, it was remarkable to note that endophytic *Colletotrichum* sp. BS4 effectively removed the bioactive compound (colletotrichone A) from the site of production (i.e., hyphal clump). The other inactive or less active compounds (colletotrichone B and/or chermesinone B) were not actively dispersed away from the hyphal clumps where they were produced. It is compelling that endophytic *Colletotrichum* sp. BS4 might have coevolved this selective functional trait of removing colletotrichone A from the site of production, which is also the conceivable site of its association with the host plant. Notably, this might allow the fungus to circumvent triggering of the plant immune signals and/or to thwart it from prying with the endophyte-mediated ecological balance in plant tissues.[53]

In every ecosystem, plant-associated microorganisms face innumerable biotic and abiotic selection pressures as they colonize the internal tissues and/or apoplastic compartments of host plants and establish themselves as endophytes.[14–16,24,54] The host plant's internal environment is a complex biological marketplace with a variety of endophytes that are coexisting (see Figure 7.1), which drives each microbial partner to evolve and adapt to coexisting organisms. One of the mechanisms evolved by endophytes is the so-called species-to-species communication, which serves not only as an essential strategy to persist in any given niche, but also to carry out important ecological functions. The concept of cell-to-cell communication and crosstalk between allied cells or systems is not new. A classic example is the phenomena of quorum sensing and quorum quenching (see Section 7.3.3) between bacterial cells for "self-communication" as well as "neighbor communication,"[55,56] which has been well explored in the last decades.[57–61] Neighbor communication is not only prevalent in microorganisms, but also in many other systems, with the same basic circuitry comprising of signal production and secretion by one partner and sensing of secreted signal(s), followed by utilization by an associated partner, up to a system-dependent threshold limit (Figure 7.5). Examples include growth regulation in human T-cells mediated by production and utilization of the cytokine interleukin-2 (IL-2),[62–65] and insulin-effectuated modulation of human pancreatic β-cells.[66,67] Recently, we studied the communication between two neighboring endophytes, namely the fungus *F. solani* coexisting with the bacterium *Achromobacter xylosoxidans* in the bulbs of the Chinese sacred lily (*Narcissus tazetta*) originating from Zhangzhou, Fujian province in the People's Republic of China. We demonstrated that the endophytic fungus is able to produce and secrete a plethora of unique hexacyclopeptides at physiological concentrations, which were sensed and selectively accumulated by the bacterium in a time-dependent manner.[68]

However, interactions between coexisting endophytes within the mesh of plant-associated microbiota are not always synergistic. The concept of "balanced antagonism" was proposed more than a decade back by Schulz et al.,[69] who hypothesized that the existence of an endophyte in a plant and the maintenance of colonization are an endless and dynamic balance of antagonisms between the endophyte and the host plant. Given that there is almost always an assortment of endophytes in any ecological niche in a plant, this concept has been expanded to

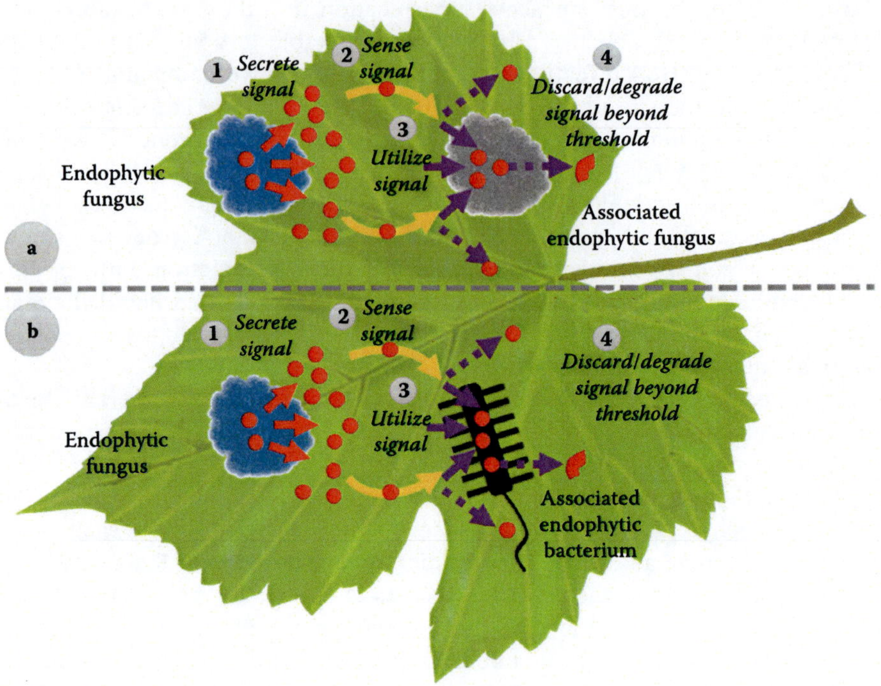

FIGURE 7.5 Schematic representation of microbial neighbor communication between endophytes comprising signal production and secretion by one microbial partner, sensing of secreted signal(s), followed by utilization by an associated microbial partner up to a system-dependent threshold limit. **(a)** Endophytic fungal-fungal communication. **(b)** Endophytic fungal-bacterial communication.

cover not only endophyte-plant antagonism, but also endophyte-endophyte and plant-endophyte-associated microbe tripartite interactions (Figure 7.6).[10] In fact, it was demonstrated in an interesting study on wild lima bean plants that an endophyte is effectively able to antagonize an invading pathogen once the endophyte has already colonized the host plant; however, if the pathogen colonized the host tissues before the endophyte, it not only fails to inhibit the pathogen but also facilitates the growth of the pathogen.[70] Recently, we explored the interactions of endophytes harbored in a Chinese medicinal plant, *Mahonia fortunei*, originating from Guangdong, People's Republic of China. It was revealed that an endophytic fungal strain isolated from the stem, identified as *Fusarium decemcellulare*, produced the allelochemical fusaristatin A, which could antagonize the fungal endophyte *Glomerella acutata* (see Figure 7.6a for a conceptual representation).[71] In another recent study, a tripartite antagonistic interaction between an endophytic fungus (*Acremonium strictum*), an endophytic bacterium (*Acinetobacter* sp.), and the host plant (*Atractylodes lancea*) was explored.[72] The paper demonstrated that an antagonistic interaction between the fungal and the bacterial endophyte (see Figure 7.6b for a conceptual representation) led to the hindrance in endophyte-mediated protection of the host plant.[72]

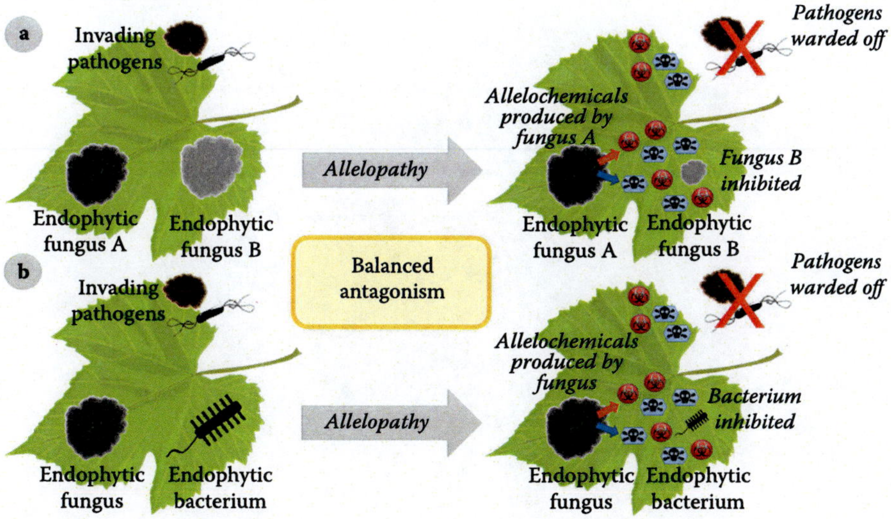

FIGURE 7.6 Schematic representation of balanced antagonism between coexisting endophytes. **(a)** Allelopathic effect of an endophytic fungus on a coexisting endophytic fungus. **(b)** Allelopathic effect of an endophytic fungus on a coexisting endophytic bacterium.

7.3.2 Endophytic Fungal Interactions with Endosymbiotic Bacteria

Recent studies on plant-microbe and microbe-microbe interactions are increasingly demonstrating the complexity of organismal association in Nature. For plant-associated microorganisms in particular, the interface of plants and endophytic communities are very complex ecological systems that go beyond fungus-fungus, bacterium-bacterium, and fungus-bacterium associations.[6,10,13] It has now been firmly established that many endophytic fungi harbor an endosymbiotic bacterium (Figure 7.1d) in order to survive and function in the complex mesh of microbial networks in selected plant ecosystems. Such fungal-bacterial endosymbiosis[73,74] can range from simple "bacterial mycophagy,"[75] encompassing necrotrophy, extracellular biotrophy, or intracellular biotrophy, to more complex evolutionarily primed "reciprocal coupling" of fungus and endosymbiotic bacterium for selected ecological functions.[76] Ecological functions can range from endosymbiont-induced phenotypic traits of the fungal host to fungal-endosymbiotic cross-species traits (Figure 7.7). A notable example is that of the cultivable endosymbiotic bacterium, *Burkholderia rhizoxinica*, which resides in the rice seedling blight fungus, *Rhizopus microspores*.[77,78] The endosymbiotic bacterium initially biosynthesizes the basic scaffold of the highly efficacious antimitotic toxin rhizoxin,[77] which is further tailored by a host fungal oxygenase enzyme provided by *R. microspores*, the fungal host, to abridge it to an extremely potent phytotoxin.[79] The scenario of endophytes containing uncultivable endosymbionts further adds to a higher level of complexity operational in natural ecological niches.[10] These hidden endosymbiotic endophytes and their corresponding functions are likely essential parts of the whole interaction network at the plant-microbe interface.

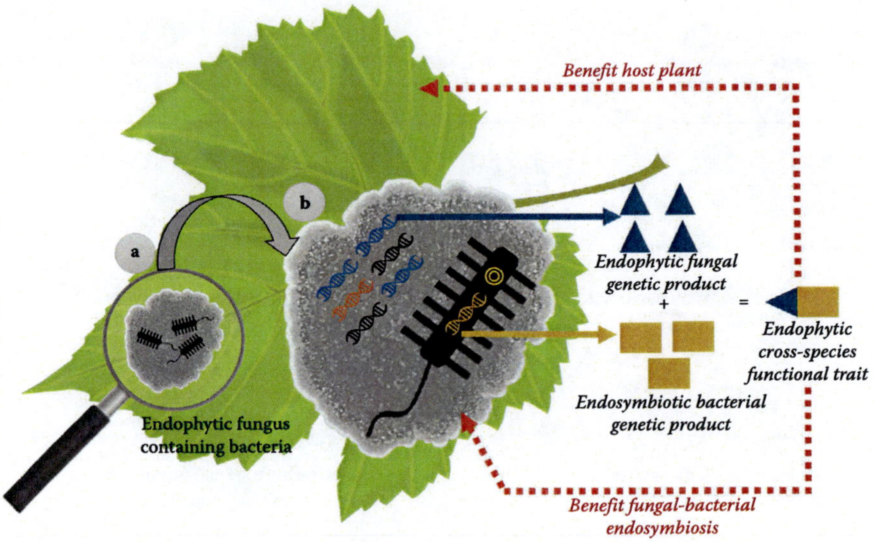

FIGURE 7.7 Schematic representation of endophytic fungal-bacterial endosymbiosis. **(a)** An endophytic fungus containing endosymbiotic bacteria. **(b)** Magnified illustration of endosymbiont-induced phenotypic trait of the fungal host and fungal-endosymbiotic cross-species functional trait.

In the course of our research on endophytes harbored in traditional medicinal plants, we recently started investigating the endophytic microflora of the well-known boxwood tree used in TCM, *B. sinica*. We could isolate a number of culturable endophytes from this plant, some of which were capable of producing bioactive compounds on applying the OSMAC approach.[42,43] Interestingly, we could isolate an endophytic fungus from the healthy leaves of this plant, which was identified and characterized as *Phyllosticta capitalensis* (anamorph of *Guignardia mangiferae*) (unpublished data). Strikingly, however, PCR analysis of the fungal metagenome using universal primers specific to bacterial 16S rRNA revealed the presence of an endosymbiotic bacterium in the isolated fungus. In order to unambiguously confirm the presence and identity of the endosymbiotic bacterium, we employed several parallel approaches.[77] First, the amplified 16S rRNA sequence, which is specific for bacteria (prokaryotes) and not for fungi (eukaryotes), was matched against the nucleotide database of the U.S. National Centre for Biotechnology Information (NCBI) using the Basic Local Alignment Search Tool (BLASTn), and aligned using the EMBOSS-Pairwise Sequence Alignment of the EMBL Nucleotide Database. The presence of an endosymbiotic bacterium, identified as *Herbaspirillum* sp., was thus confirmed by the amplification of the highly conserved 16S rRNA region.[80] Second, we obtained only identical 16S rDNA sequences (1500 base pairs fragment) using the fungal metagenome as template, indicating that only a single type of endosymbiont was present. Third, we replaced the template DNA with sterile double-distilled water as a negative control, which did not amplify any sequence. This confirmed that the identified bacterium was an endosymbiont of the fungus and not a contaminant. These results clearly demonstrated that the isolated endophytic fungus *P. capitalensis* contains an endosymbiotic bacterium, *Herbaspirillum* sp. (unpublished data).

Many species in the genus *Phyllosticta* (Botryosphaeriaceae) are known to be endophytic, or even plant pathogens. In previous reports, *P. capitalensis* was generally considered to be an endophytic fungus with weak pathogenicity,[81,82] even though it can cause leaf blight of *Elaeocarpus glabripetalus*, and spots on foliage and fruits of mango and guava.[83] Moreover, meroterpenes were isolated as bioactive secondary metabolites by fermenting another strain of this fungus.[84–86] Although we confirmed the presence of an endosymbiont within the fungus, we could not cultivate this endosymbiotic bacterium as a free-living (axenic) microbe using either the OSMAC approach[43] or other established methods.[77] Moreover, it was observed that the amplification of 16S rRNA from the fungal metagenome was related to the subculture generation of fungus. The positive band was amplified only when PCR analysis was performed using template DNA of first-generation fungal isolate (mother plate). Repeated subculturing resulted in loss of the desired fragment, which pointed toward the loss of the endosymbiotic bacterium when culturing the fungus under *in vitro* conditions different to that of its ecological niche. The genus *Herbaspirillum* (Oxalobacteraceae) typically consists of Gram-negative diazotrophic bacterial species, many of which have been shown to be associated with plants.[87] Systemic colonization of both above-ground and below-ground tissues and apoplastic compartments by plant-associated *Herbaspirillum* species has been shown to be a unique feature of the bacteria exhibiting endophytic nature.[88] Bacteria belonging to the genus *Herbaspirillum* are well known, not only as endophytes harbored in plants ranging from maize and sorghum to sugarcane, with a capacity of fixing nitrogen and promoting the growth of host plants, but, in many cases, also as uncultivable bacteria.[89–92]

We employed the OSMAC approach on this endophytic fungus containing the endosymbiont. Interestingly, when growing the endophyte in potato dextrose broth (PDB) under static conditions, two new lactam-fused 4-pyrones were produced (unpublished data). Thus far, from the structural point of view, the lactam-fused 4-pyrones produced by the endophytic fungus containing the endosymbiotic bacterium are plausibly biosynthesized via a hybrid PKS/NRPS pathway. Using this background information, we are currently performing *de novo* sequencing and molecular analyses of the endophytic fungus *P. capitalensis* along with its endosymbiotic bacterium, *Herbaspirillum* sp., in order to elucidate the complete biosynthetic pathway of the new lactam-fused 4-pyrones.

7.3.3 Endophytic Bacterial Interactions with Associated and Invading Bacteria

The *in planta* apoplastic environment forms a unique niche for endophytic microorganisms, wherein the internal environment of the host plants plays a critical role in both establishing and maintaining their mutualistic association with resident endophytes.

It is now evident that production of complex bioactive compounds by plants and associated microorganisms are direct or indirect results of their complex and dynamic ecological interactions in Nature. The biosynthetic pathways of these compounds might have gradually evolved over time to benefit the host organisms and contribute to maintaining or improving their fitness in the environment. Thus far, in almost every ecological niche, the coevolution of plants and associated endophytes enables them to recognize invading specialist and/or generalist pathogens. As their primary line of defense against pathogens, endophytes (even plants themselves in many cases) might develop chemical defense strategies by producing bioactive (antimicrobial) compounds *in situ*. Such tactics, commonly known as antimicrobial strategies (Figure 7.8a), are quite effective in warding off a plethora of invading microbes, either by killing them or inhibiting their growth. However, it is compelling that in a selected ecological niche, the result of continued exposure to a given bioactive compound produced by endophytes over a

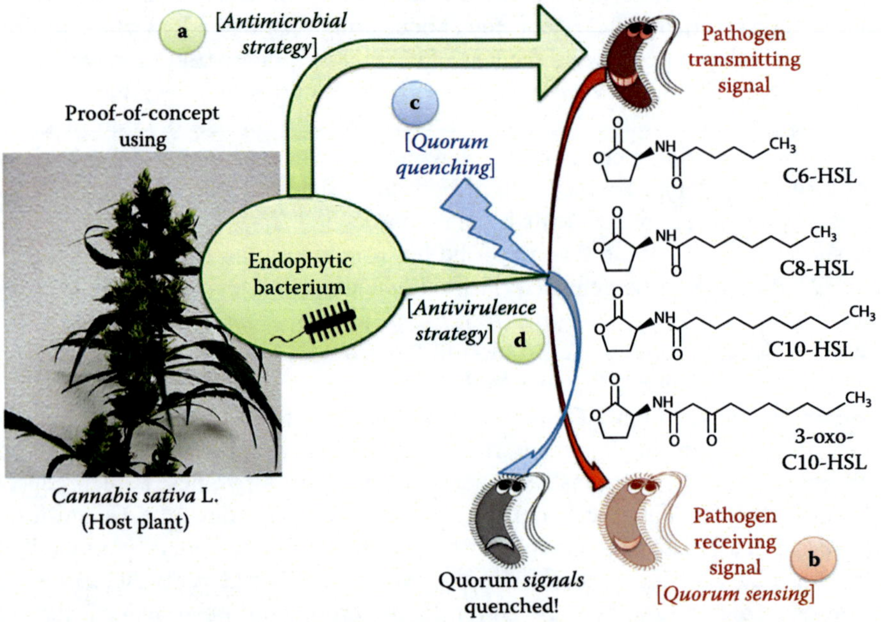

FIGURE 7.8 Schematic representation of the antivirulence strategy of endophytes harbored in *Cannabis sativa* L. plants. **(a)** Antimicrobial strategy as a primary line of chemical defense for endophytes against invading pathogens by producing bioactive (antimicrobial) compounds *in situ* in the host plants. **(b)** Quorum sensing, the cell-to-cell communication system mediated by autoinducers, which is responsible for the regulation of virulence factors, infections, invasion, colonization, biofilm formation, and antimicrobial resistance within bacterial populations invading plants. **(c)** Quorum quenching, the destruction of quorum-sensing cascades which prevents colonization of plants by invading pathogens. **(d)** Antivirulence strategy, a second line of defense coevolved by endophytes whereby they inhibit communication between invading pathogens or prevent pathogens from recognizing each other, thereby preventing them from colonizing plants.

period of time delivers a form of evolutionary pressure on the invading pathogens, leading to the emergence of antimicrobial resistance. For instance, many bacterial pathogens which were previously susceptible to the antibacterial compounds produced by a plant or its associated endophyte might gradually devise means to resist the mode of action of the compounds. Therefore, even though such an antimicrobial strategy aids in inhibiting pathogenic growth, it is also one of the foremost causes of inducing drug resistance in bacteria due to the selection pressure on their growth and survival.[93]

As opposed to the aforementioned antimicrobial strategy, interference with pathogenic bacterial virulence and/or cell-to-cell signaling pathways without killing them or preventing their growth is emerging as an alternative, powerful, and effective mechanism, commonly known as the "antivirulence strategy" (Figure 7.8).[94,95] As detailed in the earlier sections of this chapter, several examples have demonstrated the coevolution of certain functional traits, such as production of bioactive natural products, in order to aid in plant defense responses. However, during their coexistence with host plants over long evolutionary periods of time, endophytes encounter invasion by a plethora of specific and/or generalist pathogens. Consequently, in order to persist in their ecological niches, endophytes might evolve supplementary protection strategies that inhibit the pathogens from developing resistance against endophytic and host plant defense compounds. Thus, endophytes can sustain their persistence within the host plants. One such second line of defense evolved by endophytes is called "quorum quenching" (Figure 7.8), whereby the endophytes disrupt the communication between the invading pathogenic bacterial cells (e.g., quorum-sensing cascades) instead of asserting any selective pressure on their growth.[96–98] Studies on quorum sensing, an important cell-to-cell communication system enabling microbe-microbe interaction, colonization, bacterial pathogenesis, and invasion across populations (Figure 7.8b), have been reported to have massive biotechnological implications, not only in disease management, but also against emerging antibiotic resistance.[99–101] Although oligopeptides are released by Gram-positive bacteria to communicate with each other, *N*-acylated L-homoserine lactones (AHLs) are released as the quorum signals in Gram-negative bacteria. These autoinducers further synchronize communication across pathogenic microbial populations for invasion, colonization, pathogenesis, and thwarting chemical defense like antibiotics of other microorganisms including endophytes.[102]

As a proof of concept, we used the well-known medicinal plant, *C. sativa*, and a plethora of endophytes we isolated and characterized from the plant,[96] to test the endophyte-mediated antivirulence strategy, that is, the attenuation of virulence factors released by pathogens without killing them or inhibiting their growth (Figure 7.8d).[96,97] Our goal was to unravel an important facet of plant-endophyte-pathogen tripartite interaction; we explored the association of *C. sativa* plants with their endophytes under various abiotic and biotic selection pressures that might have led to the development of quorum-quenching ability in the endophytes. As a proxy for invading phytopathogens, we used the well-known Gram-negative biosensor strain, *Chromobacterium violaceum*, known to produce the purple pigment violacein as a result of quorum sensing by means of the *CviI/CviR* synthase-receptor

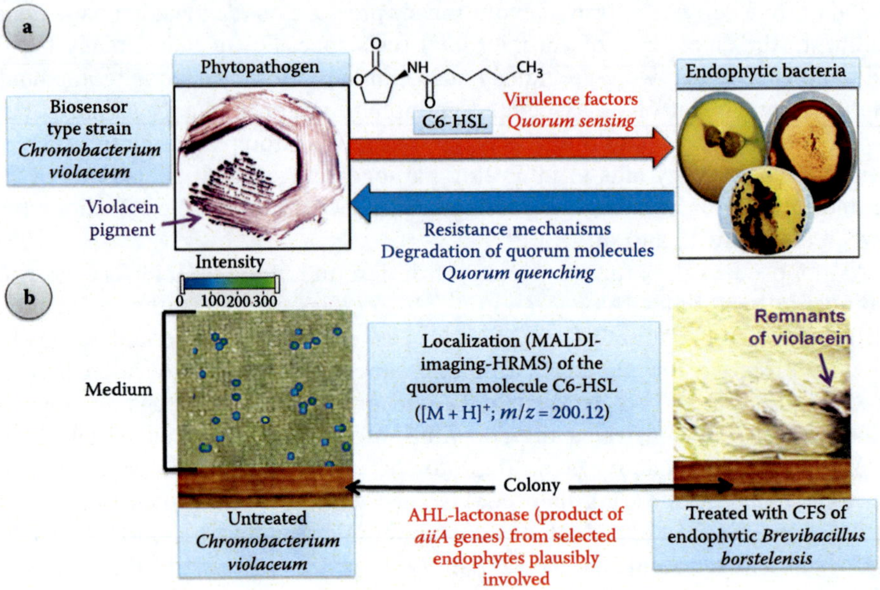

FIGURE 7.9 Quorum quenching by a selected endophytic bacterium isolated from *Cannabis sativa*. **(a)** Schematic representation of the cross talk between endophytes and pathogens using the Gram-negative biosensor strain *Chromobacterium violaceum*, which produces the purple pigment violacein (virulence factor) as a result of *CviI/CviR* synthase-receptor-mediated quorum sensing. This is countered by degradation of the virulence factor by selected endophytic bacteria harboring *C. sativa* plants in our study. **(b)** MALDI-imaging-HRMS shows the production, release, and extracellular accumulation of the quorum signal C6-HSL by untreated *C. violaceum*, and remnants of violacein after C6-HSL has been quenched by the cell-free supernatant of endophytic *Brevibacillus borstelensis*.

signaling (Figure 7.9a).[96] We employed a combination of high-performance liquid chromatography, electrospray ionization, high-resolution mass spectrometry (HPLC-ESI-HRMS^n) and MALDI-imaging-HRMS to quantify and visualize the spatial distribution of cell-to-cell quorum-sensing signals of *C. violaceum*. We further showed that potent endophytic bacteria harbored in *C. sativa* plants can selectively and differentially quench the quorum-sensing molecules of *C. violaceum*. Therefore, using combinations of HPLC-HRMS and MALDI-imaging-HRMS, we ascertained that potent endophytic bacterial isolates selectively mitigate four different quorum signals: the AHLs [*N*-hexanoyl-L-homoserine lactone (C6-HSL), *N*-octanoyl-L-homoserine lactone (C8-HSL), *N*-decanoyl-L-homoserine lactone (C10-HSL), and *N*-(3-oxodecanoyl)-L-homoserine lactone (3-oxo-C10-HSL)] used by *C. violaceum* (Figure 7.9b, representative for C6-HSL for a selected endophytic bacterial strain). Furthermore, we performed additional assays in parallel to further validate the fact that the studied endophytes were capable of significantly reducing the virulence factor (violacein in this case) without inhibiting the growth of *C. violaceum*.[96] The potential of endophytic bacteria as biocontrol agents against bacterial pathogens, as well as antivirulence agents that might be beneficial in quorum-inhibiting therapies, could be fundamentally elaborated as a proof of concept in our study. In particular, diminution of these communication signals will lead to suppression of pathogen virulence without introducing additional resistance-inducing selection pressures,[103] which in turn will prevent the pathogens colonizing the host plants already inhabited by the endophytes. Quorum quenching is, thus far, one of many different antivirulence strategies that are developed by selected endophytic bacteria to thrive and function in their natural ecological niches.

Taken together, this work demonstrates an important survival strategy used by endophytes to circumvent invasion by a plethora of generalist and specific pathogens. This study exemplifies a significant biological role played by the endophytes in different ecological niches, by acting as antivirulence agents, not only aiding host plant defense but also for maintaining colonization and their own survival inside plants.

7.3.4 Endophytic Bacterial Communities

In every ecological system, microorganisms coexist in the form of a *microbiota* in order to obtain a permanent niche for existence and collectively function as a consortium rather than as axenic organisms. This is also true for plant-associated microorganisms, including endophytes (see Figure 7.1).[10,104] Several recent studies on plant-microbe interactions have unambiguously substantiated that *the functional capacity of the plant microbiome is not equal to the sum of its individual components, since microbial species strongly and frequently interact with each other and form a complex network.*[105] For example, Reinhold-Hurek and colleagues studied the root microbiome of rice under field conditions to demonstrate that not only were the bacterial endophytic communities adapted to the root niche conditions, but also that the communities had the potential to deliver succinct ecological functions, such as promotion of host plant growth and reduction of stress, protection against pathogens, and even root-mediated bioremediation.[106] More recently, it was shown using rice plants that even the host plant has an effect on structuring and restructuring the endophytic as well as rhizosphere microbial communities, such that multiple functions can be delivered by different microcommunities within the whole microbiome.[107] In another study, Santhanam et al.[108] established that in tobacco plants, the native root-associated bacterial community (and not any single endophytic bacterium) considerably improved the survivability and defense of the community-containing plants against black root formation and sudden tissue collapse. Yet another example of endophytes providing host plant fitness as a consortium is that of *Parkinsonia aculeata* L. trees, where endophytic archaeal as well as fungal community composition and structure is correlated to the incidence of the dieback disease.[109] Thus far, it has been firmly established that we "*need a consortium of two, three, or even five or more collaborating strains that can withstand the ecological forces*" and function efficiently in their habitat niche.[110]

We investigated the fundamental role of chemical cross talk in plants and associated endophytic microbiota, using the important anticancer and cytotoxic compound, maytansine, in Celastraceae plants such as *Putterlickia verrucosa* and *Putterlickia retrospinosa*. We intended to elucidate the actual producer(s) responsible for maytansine biosynthesis in plants, which was a conundrum since its discovery in the 1970s.[111–113] Maytansine (Figure 7.10) is a benzoansamacrolide that was first isolated by Kupchan et al. in the 1970s using bioactivity-guided fractionation of extracts of Celastraceous plants such as *Putterlickia* spp. and *Maytenus* spp.[114,115] Maytansine is highly toxic and one of the most potent microtubule-targeting compounds known today.[116–118] This cytotoxic agent binds to tubulin that is located at the ends of microtubules, which leads to the suppression of microtubule dynamics and further causes cells to arrest in the G2/M phase of the cell cycle. In order to bypass the high cytotoxicity, a maytansinoid-antibody conjugate is used in therapy to specifically target mamma carcinoma cells for the treatment of breast cancer. Notably, two maytansine derivatives DM1and DM4 are the main building blocks of a new class of drugs to treat breast cancer.[119] Trastuzumab Emtansine (T-DM1), developed by Roche in partnership with ImmunoGen and which received marketing approval in 2013, is a human

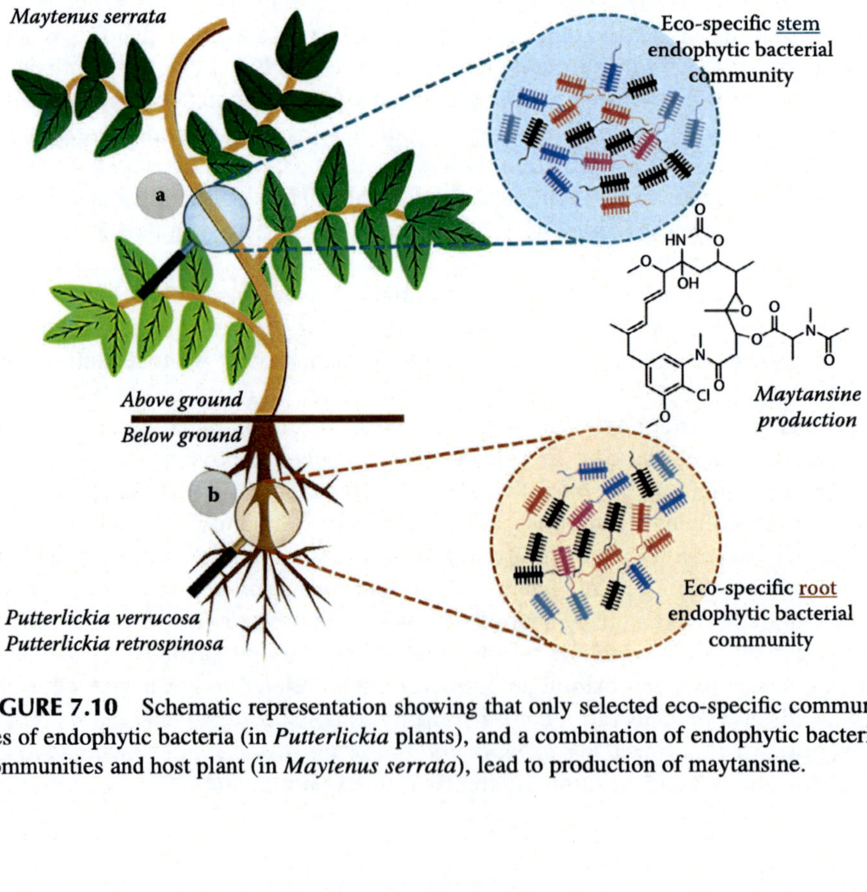

FIGURE 7.10 Schematic representation showing that only selected eco-specific communities of endophytic bacteria (in *Putterlickia* plants), and a combination of endophytic bacterial communities and host plant (in *Maytenus serrata*), lead to production of maytansine.

epidermal growth factor receptor (herceptin2)-targeted antibody drug conjugate composed of trastuzumab, a stable thioether linker, and the potent cytotoxic agent DM1.[120,121] Furthermore, the drug SAR3419, developed by Sanofi in collaboration with ImmunoGen, is another representative example of a maytansine-based anticancer drug currently in the market. Contrary to DM1, this drug is composed of a humanized IgG1 monoclonal anti-CD19 antibody (huB4) conjugated via a cleavable disulfide linker to DM4.[122]

After isolating the endophytic bacterial community from different tissues of *Putterlickia* plants, we used the combination of bioanalytical and genome-mining methods to elucidate the source and site of maytansine biosynthesis in plants. Our study demonstrated that maytansine is actually a biosynthetic product of root-associated endophytic bacterial consortia (Figure 7.10).[123] The knowledge gained from this study provides fundamental insights into the biosynthesis of so-called plant metabolites by endophytic microbiota residing in distinct ecological niches. Following up from these interesting results, endophytic bacterial communities harboring different tissues of *M. serrata* originating from Cameroon were investigated using targeted genome-mining techniques coupled with bioanalytical approaches to elucidate the source of maytansine biosynthesis. Notably, in *M. serrata*, maytansine was found in the above ground tissues as opposed to what was observed in *P. verrucosa* and *P. retrospinosa* plants, where maytansine was found in the roots. In *M. serrata*, it was revealed that the host plant, along with its cryptic endophytic microbiota, produces the biosynthetically unique core structural moiety AHBA, which serves as the unique starter unit for maytansine biosynthesis. However, the biosynthetic step of halogenase-mediated incorporation of chlorine, which is missing in the host plant, is accomplished by the culturable stem endophytic bacterial community (Figures 7.10 and 7.11a).[124] The remarkable subtleties and differences in the maytansine biosynthetic pathway in *Putterlickia* plants compared to *Maytenus* plants vis-à-vis their endophytic microbiota points toward coevolution of multiple parallel pathways in different ecosystems, with plausibly diverse post-PKS modifications of the intermediates in the pathway.[125] Recently, the seeds of *P. verrucosa* plants were investigated for the presence and localization of maytansinoids by MALDI-imaging-HRMS in order to test the hypothesis that maytansinoids indeed play an ecological role as chemical defense compounds during germination of *Putterlickia* plants (unpublished data). Our work demonstrates that maytansine is biosynthesized only when a niche-specific endophytic microbiota interacts with their host plants over an evolutionary time scale (Figure 7.10). More recently, we investigated another unexplored *Putterlickia* plant (*Putterlickia pyracantha*) to study the occurrence and spatial distribution of maytansinoids in addition to maytansine, such as maytanprine, maytanbutine, maytanvaline, and others.[126] It is intriguing how the biosynthetic pathways of the different maytansinoids might have coevolved, and how are they produced *in planta* (Figure 7.11b). Taken together, our aforementioned results have opened up further fundamental questions with regard to the occurrence, distribution, biosynthesis, and ecological relevance of maytansinoids in Celastraceae plants.

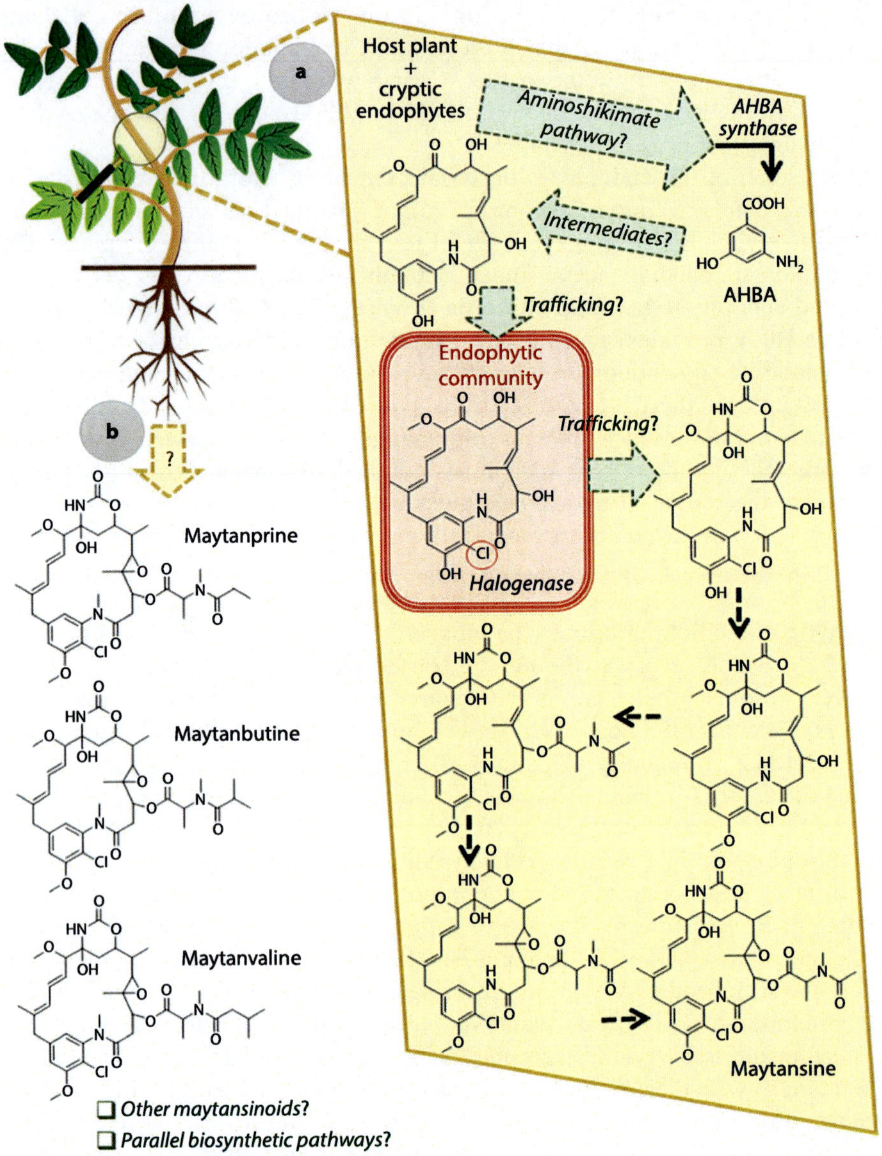

FIGURE 7.11 Plant-endophyte cross-species biosynthetic pathway of maytansine. **(a)** Proposed biosynthetic pathway in *M. serrata*. Plausible steps of the aminoshikimate pathway and of trafficking of maytansine precursors between host plant and its endophytic bacterial community leading to the cross-species biosynthesis of maytansine are marked with "?". Dotted arrows represent proposed hypothetical steps. **(b)** Other important maytansinoids discovered in the plant.

7.4 OUTLOOK

Endophytes (see Figure 7.1) are intrinsic and crucial partners within the ecological marketplace of plant-associated macro- and microbiota. It has now been decisively recognized by a plethora of studies on different facets of endophytes that any plant ecosystem would be rendered virtually nonfunctional without the ecologically meaningful functional traits of endophytes. Notably, expression of such functional traits (e.g., production of secondary metabolites and communication molecules) confers fitness costs, and is, therefore, subject to regulation. It is gradually becoming clear that a comprehensive understanding of plant-endophyte crosstalk requires not only elucidation of host plant signal-mediated triggers for elicitation of endophytic functional traits and vice versa, but also how crosstalk (e.g., chemical and physiological cues) between the interacting partners changes their metabolic and ecological interactions. The multifaceted interaction networks and signaling systems among coexisting endophytes, and between endophytes and associated macro- and microorganisms, are important ecological drivers of any plant ecosystem. Comprehending these functions is vital for us to exploit the potential of endophytes, and divulge their ecological roles in different environmental setups. Admittedly, we have only started *scratching the surface* in our understanding of plant-endophyte interactions and the fascinating impact of endophytes.

REFERENCES

1. Hardoim, P. R.; van Overbeek, L. S.; Berg, G.; Pirttilä, A. M.; Compant, S.; Campisano, A.; Döring, M.; Sessitsch, A. The hidden world within plants: Ecological and evolutionary considerations for defining functioning of microbial endophytes. *Microbiol. Mol. Biol. Rev.* **2015**, 79, 293–320.
2. Newman, D. J.; Cragg, G. M. Endophytic and epiphytic microbes as "sources" of bioactive agents. *Front. Chem.* **2015**, 3, 34.
3. Mithöfer, A.; Boland, W. Do you speak chemistry? *EMBO Rep.* **2016**, 17, 626–629.
4. Jia, M.; Chen, L.; Xin, H.-L.; Zheng, C.-J.; Rahman, K.; Han, T.; Qin, L.-P. A friendly relationship between endophytic fungi and medicinal plants: A systematic review. *Front. Microbiol.* **2016**, 7, 906.
5. Phelan, V. V.; Liu, W. T.; Pogliano, K.; Dorrestein, P. C. Microbial metabolic exchange—The chemotype-to-phenotype link. *Nat. Chem. Biol.* **2011**, 8, 26–35.
6. Kusari, S.; Singh, S.; Jayabaskaran, C. Biotechnological potential of plant-associated endophytic fungi: Hope versus hype. *Trends Biotechnol.* **2014**, 32, 297–303.
7. West, S. A.; Griffin, A. S.; Gardner, A.; Diggle, S. P. Social evolution theory for microorganisms. *Nat. Rev. Microbiol.* **2006**, 4, 597–607.
8. Werner, G. D. A.; Strassmann, J. E.; Ivens, A. B.; Engelmoer, D. J.; Verbruggen, E.; Queller, D. C.; Noë, R.; Johnson, N. C.; Hammerstein, P.; Kiers, E. T. Evolution of microbial markets. *Proc. Natl. Acad. Sci. USA* **2014**, 111, 1237–1244.
9. Kusari, S.; Spiteller, M. Are we ready for industrial production of bioactive plant secondary metabolites utilizing endophytes? *Nat. Prod. Rep.* **2011**, 28, 1203–1207.
10. Kusari, S.; Hertweck, C.; Spiteller, M. Chemical ecology of endophytic fungi: Origins of secondary metabolites. *Chem. Biol.* **2012**, 19, 792–798.
11. Newman, D. J.; Cragg, G. M. Natural products as sources of new drugs from 1981 to 2014. *J. Nat. Prod.* **2016**, 79, 629–661.

12. Wang, W.-X.; Kusari, S.; Spiteller, M. Unraveling the chemical interactions of fungal endophytes for exploitation as microbial factories. In *Fungal Applications in Sustainable Environmental Biotechnology*, Purchase, D., Ed. Springer International, Geneva, Switzerland, **2016**; pp. 353-370.
13. Kusari, S.; Pandey, S. P.; Spiteller, M. Untapped mutualistic paradigms linking host plant and endophytic fungal production of similar bioactive secondary metabolites. *Phytochemistry* **2013**, 91, 81–87.
14. Rico, A.; McCraw, S. L.; Preston, G. M. The metabolic interface between *Pseudomonas syringae* and plant cells. *Curr. Opin. Microbiol.* **2011**, 14, 31–38.
15. Kim, K.-T.; Jeon, J.; Choi, J.; Cheong, K.; Song, H.; Choi, G.; Kang, S.; Lee, Y.-H. Kingdom-wide analysis of fungal small secreted proteins (SSPs) reveals their potential role in host association. *Front. Plant Sci.* **2016**, 7, 186.
16. Hacquard, S.; Kracher, B.; Hiruma, K.; Münch, P. C.; Garrido-Oter, R.; Thon, M. R.; Weimann, A. et al. Survival trade-offs in plant roots during colonization by closely related beneficial and pathogenic fungi. *Nat. Commun.* **2016**, 7, 11362.
17. Tanaka, A.; Christensen, M. J.; Takemoto, D.; Park, P.; Scott, B. Reactive oxygen species play a role in regulating a fungus-perennial ryegrass mutualistic interaction. *Plant Cell* **2006**, 18, 1052–1066.
18. Pandey, S. P.; Shahi, P.; Gase, K.; Baldwin, I. T. Herbivory-induced changes in the small-RNA transcriptome and phytohormone signaling in *Nicotiana attenuata*. *Proc. Natl. Acad. Sci. USA* **2008**, 105, 4559–4564.
19. Pieterse, C. M.; Leon-Reyes, A.; Van der Ent, S.; Van Wees, S. C. Networking by small-molecule hormones in plant immunity. *Nat. Chem. Biol.* **2009**, 5, 308–316.
20. Camehl, I.; Sherameti, I.; Venus, Y.; Bethke, G.; Varma, A.; Lee, J.; Oelmüller, R. Ethylene signalling and ethylene-targeted transcription factors are required to balance beneficial and nonbeneficial traits in the symbiosis between the endophytic fungus *Piriformospora indica* and *Arabidopsis thaliana*. *New Phytol.* **2010**, 185, 1062–1073.
21. Fones, H. N.; Preston, G. M. Trade-offs between metal hyperaccumulation and induced disease resistance in metal hyperaccumulator plants. *Plant Pathol.* **2013**, 62, 63–71.
22. Oldroyd, G. E. Speak, friend, and enter: Signalling systems that promote beneficial symbiotic associations in plants. *Nat. Rev. Microbiol.* **2013**, 11, 252–263.
23. O'Leary, B.; Preston, G. M.; Sweetlove, L. J. Increased β-cyanoalanine nitrilase activity improves cyanide tolerance and assimilation in *Arabidopsis*. *Mol. Plant* **2014**, 7, 231–243.
24. Busby, P. E.; Ridout, M.; Newcombe, G. Fungal endophytes: Modifiers of plant disease. *Plant Mol. Biol.* **2016**, 90, 645–655.
25. Lery, L. M.; Hemerly, A. S.; Nogueira, E. M.; von Krüger, W. M.; Bisch, P. M. Quantitative proteomic analysis of the interaction between the endophytic plant-growth-promoting bacterium *Gluconacetobacter diazotrophicus* and sugarcane. *Mol. Plant Microbe Interact.* **2011**, 24, 562–576.
26. Arencibia, A. D.; Vinagre, F.; Estevez, Y.; Bernal, A.; Perez, J.; Cavalcanti, J.; Santana, I.; Hemerly, A. S. *Gluconacetobacter diazotrophicus* elicits a sugarcane defense response against a pathogenic bacteria *Xanthomonas albilineans*. *Plant Signal. Behav.* **2006**, 1, 265–273.
27. Wu, S.; Shan, L.; He, P. Microbial signature-triggered plant defense responses and early signaling mechanisms. *Plant Sci.* **2014**, 228, 118–126.
28. Pieterse, C. M.; Zamioudis, C.; Berendsen, R. L.; Weller, D. M.; Van Wees, S. C.; Bakker, P. A. Induced systemic resistance by beneficial microbes. *Annu. Rev. Phytopathol.* **2014**, 52, 347–375.
29. Li, B.; Meng, X.; Shan, L.; He, P. Transcriptional regulation of pattern-triggered immunity in plants. *Cell Host Microbe* **2016**, 19, 641–650.
30. Vadassery, J.; Oelmüller, R. Calcium signaling in pathogenic and beneficial plant microbe interactions: What can we learn from the interaction between *Piriformospora indica* and *Arabidopsis thaliana*. *Plant Signal. Behav.* **2009**, 4, 1024–1027.

31. Vadassery, J.; Ranf, S.; Drzewiecki, C.; Mithöfer, A.; Mazars, C.; Scheel, D.; Lee, J.; Oelmüller, R. A cell wall extract from the endophytic fungus *Piriformospora indica* promotes growth of *Arabidopsis* seedlings and induces intracellular calcium elevation in roots. *Plant J.* **2009**, 59, 193–206.
32. Ranf, S.; Eschen-Lippold, L.; Pecher, P.; Lee, J.; Scheel, D. Interplay between calcium signalling and early signalling elements during defence responses to microbe- or damage-associated molecular patterns. *Plant J.* **2011**, 68, 100–113.
33. Khan, A. L.; Hamayun, M.; Kim, Y.-H.; Kang, S.-M.; Lee, J.-H.; Lee, I.-J. Gibberellins producing endophytic *Aspergillus fumigatus* sp. LH02 influenced endogenous phytohormonal levels, isoflavonoids production and plant growth in salinity stress. *Processes Biochem.* **2011**, 46, 440–447.
34. Lahrmann, U.; Zuccaro, A. Opprimo ergo sum-evasion and suppression in the root endophytic fungus *Piriformospora indica. Mol. Plant Microbe Interact.* **2012**, 25, 727–737.
35. Kusari, S.; Zühlke, S.; Spiteller, M. An endophytic fungus from *Camptotheca acuminata* that produces camptothecin and analogues. *J. Nat. Prod.* **2009**, 72, 2–7.
36. Kusari, S.; Zühlke, S.; Spiteller, M. Effect of artificial reconstitution of the interaction between the plant *Camptotheca acuminata* and the fungal endophyte *Fusarium solani* on camptothecin biosynthesis. *J. Nat. Prod.* **2011**, 74, 764–775.
37. Kusari, S.; Kosuth, J.; Cellarova, E.; Spiteller, M. Survival strategies of endophytic *Fusarium solani* against indigenous camptothecin biosynthesis. *Fungal Ecol.* **2011**, 4, 219–223.
38. Rovenich, H.; Zuccaro, A.; Thomma, B. P. Convergent evolution of filamentous microbes towards evasion of glycan-triggered immunity. *New Phytol.* **2016**, 212, 896–901.
39. Selin, C.; de Kievit, T. R.; Belmonte, M. F.; Fernando, W. G. D. Elucidating the role of effectors in plant-fungal interactions: Progress and challenges. *Front. Microbiol.* **2016**, 7, 600.
40. Veneault-Fourrey, C.; Commun, C.; Kohler, A.; Morin, E.; Balestrini, R.; Plett, J.; Danchin, E. et al. Genomic and transcriptomic analysis of *Laccaria bicolor* CAZome reveals insights into polysaccharides remodelling during symbiosis establishment. *Fungal Genet. Biol.* **2014**, 72, 168–181.
41. Oliveira-Garcia, E.; Deising, H. B. Attenuation of PAMP-triggered immunity in maize requires down-regulation of the key β-1,6-glucan synthesis genes KRE5 and KRE6 in biotrophic hyphae of *Colletotrichum graminicola. Plant J.* **2016**, 87, 355–375.
42. Wang, W.-X.; Kusari, S.; Laatsch, H.; Golz, C.; Kusari, P.; Strohmann, C.; Kayser, O.; Spiteller, M. Antibacterial azaphilones from an endophytic fungus *Colletotrichum* sp. BS4. *J. Nat. Prod.* **2016**, 79, 704–710.
43. Bode, H. B.; Bethe, B.; Höfs, R.; Zeeck, A. Big effects from small changes: Possible ways to explore nature's chemical diversity. *ChemBioChem* **2002**, 3, 619–627.
44. Gao, J.; Yang, S.; Qin, J. C. Azaphilones: Chemistry and biology. *Chem. Rev.* **2013**, 113, 4755–4811.
45. Stadler, M.; Ju, Y.-M.; Rogers, J. D. Chemotaxonomy of *Entonaema, Rhopalostroma* and other Xylariaceae. *Mycol. Res.* **2004**, 108(Pt 3), 239–256.
46. Stadler, M.; Fournier, J. Pigment chemistry, taxonomy and phylogeny of the Hypoxyloideae (Xylariaceae). *Rev. Iberoam. Micol.* **2006**, 23, 160–170.
47. Frisvad, J. C.; Andersen, B.; Thrane, U. The use of secondary metabolite profiling in chemotaxonomy of filamentous fungi. *Mycol. Res.* **2008**, 112, 231–240.
48. Matsuzaki, K.; Tahara, H.; Inokoshi, J.; Tanaka, H.; Masuma, R.; Omura, S. New brominated and halogen-less derivatives and structure-activity relationship of azaphilones inhibiting gp120-CD4 binding. *J. Antibiot.* **1998**, 51, 1004–1011.
49. Nam, J. Y.; Kim, H. K.; Kwon, J. Y.; Han, M. Y.; Son, K. H.; Lee, U. C.; Choi, J. D.; Kwon, B. M. 8-*O*-Methylsclerotiorinamine, antagonist of the Grb2-SH2 domain, isolated from *Penicillium multicolor. J. Nat. Prod.* **2000**, 63, 1303–1305.

50. Clark, R. C.; Lee, S. Y.; Searcey, M.; Boger, D. L. The isolation, total synthesis and structure elucidation of chlorofusin, a natural product inhibitor of the p53-mDM2 protein-protein interaction. *Nat. Prod. Rep.* **2009**, 26, 465–477.
51. Cano, J.; Guarro, J.; Gené, J. Molecular and morphological identification of *Colletotrichum* species of clinical interest. *J. Clin. Microbiol.* **2004**, 42, 2450–2454.
52. Photita, W.; Taylor, P. W. J.; Ford, R.; Hyde, K. D.; Lumyong, S. Morphological and molecular characterization of *Colletotrichum* species from herbaceous plants in Thailand. *Fungal Divers.* **2005**, 18, 117–133.
53. Kogel, K. H.; Franken, P.; Hückelhoven, R. Endophyte or parasite-what decides? *Curr. Opin. Plant Biol.* **2006**, 9, 358–363.
54. Fesel, P. H.; Zuccaro, A. Dissecting endophytic lifestyle along the parasitism/mutualism continuum in *Arabidopsis*. *Curr. Opin. Microbiol.* **2016**, 32, 103–112.
55. Youk, H.; Lim, W. A. Secreting and sensing the same molecule allows cells to achieve versatile social behaviors. *Science* **2014**, 343, 1242782.
56. Lee, A. J.; You, L. Cell biology. Cells listen to their inner voice. *Science* **2014**, 343, 624–625.
57. Sawai, S.; Thomason, P. A.; Cox, E. C. An autoregulatory circuit for long-range self-organization in *Dictyostelium* cell populations. *Nature* **2005**, 433, 323–326.
58. De Monte, S.; d'Ovidio, F.; Danø, S.; Graae Sørensen, P. Dynamical quorum sensing: Population density encoded in cellular dynamics. *Proc. Natl. Acad. Sci. USA* **2007**, 104, 18377–18381.
59. Ng, W. L.; Bassler, B. L. Bacterial quorum-sensing network architectures. *Annu. Rev. Genet.* **2009**, 43, 197–222.
60. Gregor, T.; Fujimoto, K.; Masaki, N.; Sawai, S. The onset of collective behavior in social amoebae. *Science* **2010**, 328, 1021–1025.
61. Danino, T.; Mondragón-Palomino, O.; Tsimring, L.; Hasty, J. A synchronized quorum of genetic clocks. *Nature* **2010**, 463, 326–330.
62. Cantrell, D. A.; Smith, K. A. The interleukin-2 T-cell system: A new cell growth model. *Science* **1984**, 224, 1312–1316.
63. Fallon, E. M.; Lauffenburger, D. A. Computational model for effects of ligand/receptor binding properties on interleukin-2 trafficking dynamics and T cell proliferation response. *Biotechnol. Prog.* **2000**, 16, 905–916.
64. Feinerman, O.; Jentsch, G.; Tkach, K. E.; Coward, J. W.; Hathorn, M. M.; Sneddon, M. W.; Emonet, T.; Smith, K. A.; Altan-Bonnet, G. Single-cell quantification of IL-2 response by effector and regulatory T cells reveals critical plasticity in immune response. *Mol. Syst. Biol.* **2010**, 6, 437.
65. Savir, Y.; Waysbort, N.; Antebi, Y. E.; Tlusty, T.; Friedman, N. Balancing speed and accuracy of polyclonal T cell activation: A role for extracellular feedback. *BMC Syst. Biol.* **2012**, 6, 111.
66. Aspinwall, C. A.; Lakey, J. R. T.; Kennedy, R. T. Insulin-stimulated insulin secretion in single pancreatic beta cells. *J. Biol. Chem.* **1999**, 274, 6360–6365.
67. Leibiger, I. B.; Leibiger, B.; Berggren, P. O. Insulin signaling in the pancreatic beta-cell. *Annu. Rev. Nutr.* **2008**, 28, 233–251.
68. Wang, W.-X.; Kusari, S.; Sezgin, S.; Lamshöft, M.; Kusari, P.; Kayser, O.; Spiteller, M. Hexacyclopeptides secreted by an endophytic fungus *Fusarium solani* N06 act as crosstalk molecules in *Narcissus tazetta*. *Appl. Microbiol. Biotechnol.* **2015**, 99, 7651–7662.
69. Schulz, B.; Römmert, A.-K.; Dammann, U.; Aust, H. J.; Strack, D. The endophyte-host interaction: A balanced antagonism. *Mycol. Res.* **1999**, 103, 1275–1283.
70. Adame-Álvarez, R. M.; Mendiola-Soto, J.; Heil, M. Order of arrival shifts endophyte-pathogen interactions in bean from resistance induction to disease facilitation. *FEMS Microbiol. Lett.* **2014**, 355, 100–107.

71. Li, G.; Kusari, S.; Golz, C.; Strohmann, C.; Spiteller, M. Three cyclic pentapeptides and a cyclic lipopeptide produced by endophytic *Fusarium decemcellulare* LG53. *RSC Adv.* **2016**, 6, 54092–54098.
72. Wang, X. M.; Yang, B.; Wang, H. W.; Yang, T.; Ren, C. G.; Zheng, H. L.; Dai, C. C. Consequences of antagonistic interactions between endophytic fungus and bacterium on plant growth and defense responses in *Atractylodes lancea*. *J. Basic Microbiol.* **2015**, 55, 659–670.
73. Kobayashi, D. Y.; Crouch, J. A. Bacterial/fungal interactions: From pathogens to mutualistic endosymbionts. *Annu. Rev. Phytopathol.* **2009**, 47, 63–82.
74. Moebius, N.; Üzüm, Z.; Dijksterhuis, J.; Lackner, G.; Hertweck, C. Active invasion of bacteria into living fungal cells. *eLife* **2014**, 3, e03007.
75. Leveau, J. H.; Preston, G. M. Bacterial mycophagy: Definition and diagnosis of a unique bacterial-fungal interaction. *New Phytol.* **2008**, 177, 859–876.
76. Castillo, D. M.; Pawlowska, T. E. Molecular evolution in bacterial endosymbionts of fungi. *Mol. Biol. Evol.* **2010**, 27, 622–636.
77. Partida-Martinez, L. P.; Hertweck, C. Pathogenic fungus harbours endosymbiotic bacteria for toxin production. *Nature* **2005**, 437, 884–888.
78. Lackner, G.; Hertweck, C. Impact of endofungal bacteria on infection biology, food safety, and drug development. *PLoS Pathog.* **2011**, 7, e1002096.
79. Scherlach, K.; Busch, B.; Lackner, G.; Paszkowski, U.; Hertweck, C. Symbiotic cooperation in the biosynthesis of a phytotoxin. *Angew. Chem. Int. Ed.* **2012**, 51, 9615–9618.
80. Hoffman, M. T.; Gunatilaka, M. K.; Wijeratne, K.; Gunatilaka, L.; Arnold, A. E. Endohyphal bacterium enhances production of indole-3-acetic acid by a foliar fungal endophyte. *PLoS One* **2013**, 8, e73132.
81. Romão, A. S.; Spósito, M. B.; Andreote, F. D.; Azevedo, J. L.; Araújo, W. L. Enzymatic differences between the endophyte *Guignardia mangiferae* (Botryosphaeriaceae) and the citrus pathogen *G. citricarpa*. *Genet. Mol. Res.* **2011**, 10, 243–252.
82. Wikee, S.; Lombard, L.; Crous, P. W.; Nakashima, C.; Motohashi, K.; Chukeatirote, E.; Alias, S. A.; McKenzie, E. H. C.; Hyde, K. D. *Phyllosticta capitalensis*, a widespread endophyte of plants. *Fungal Divers.* **2013**, 60, 91–105.
83. Wickert, E.; de Macedo Lemos, E.; Takeshi Kishi, L.; de Souza, A.; de Goes, A. Genetic diversity and population differentiation of *Guignardia mangiferae* from "Tahiti" acid lime. *Sci. World J.* **2012**, 2012, 125654.
84. Yuan, W. H.; Liu, M.; Jiang, N.; Guo, Z. K.; Ma, J.; Zhang, J.; Song, Y. C.; Tan, R. X. Guignardones A-C: Three meroterpenes from *Guignardia mangiferae*. *Eur. J. Org. Chem.* **2010**, 2010, 6348–6353.
85. Guimarães, D. O.; Lopes, N. P.; Pupo, M. T. Meroterpenes isolated from the endophytic fungus *Guignardia mangiferae*. *Phytochem. Lett.* **2012**, 5, 519–523.
86. Han, W. B.; Dou, H.; Yuan, W. H.; Gong, W.; Hou, Y. Y.; Ng, S. W.; Tan, R. X. Meroterpenes with toll-like receptor 3 regulating activity from the endophytic fungus *Guignardia mangiferae*. *Planta Med.* **2015**, 81, 145–151.
87. Spilker, T.; Uluer, A. Z.; Marty, F. M.; Yeh, W. W.; Levison, J. H.; Vandamme, P.; LiPuma, J. J. Recovery of *Herbaspirillum* species from persons with cystic fibrosis. *J. Clin. Microbiol.* **2008**, 46, 2774–2777.
88. Olivares, F. L.; James, E. K.; Baldani, J. I.; Döbereiner, J. *Herbaspirillum frisingense* sp. nov., a new nitrogen-fixing bacterial species that occurs in C4-fibre plants. *New Phytol.* **1997**, 135, 723–737.
89. Mano, H.; Morisaki, H. Endophytic bacteria in the rice plant. *Microbes Environ.* **2008**, 23, 109–117.
90. Zhu, B.; Ye, S.; Chang, S.; Chen, M.; Sun, L.; An, Q. Genome sequence of the pathogenic *Herbaspirillum seropedicae* strain Os45, isolated from rice roots. *J. Bacteriol.* **2012**, 194, 6995–6996.

91. Straub, D.; Rothballer, M.; Hartmann, A.; Ludewig, U. The genome of the endophytic bacterium *H. frisingense* GSF30(T) identifies diverse strategies in the *Herbaspirillum* genus to interact with plants. *Front. Microbiol.* **2013**, 4, 168.
92. Chen, L.; Jia, R.B.; Li, L. Bacterial community of iron tubercles from a drinking water distribution system and its occurrence in stagnant tap water. *Environ. Sci. Process. Impacts* **2013**, 15, 1332–1340.
93. Clatworthy, A. E.; Pierson, E.; Hung, D. T. Targeting virulence: A new paradigm for antimicrobial therapy. *Nat. Chem. Biol.* **2007**, 3, 541–548.
94. Rasko, D. A.; Sperandio, V. Anti-virulence strategies to combat bacteria-mediated disease. *Nat. Rev. Drug Discov.* **2010**, 9, 117–128.
95. LaSarre, B.; Federle, M. J. Exploiting quorum sensing to confuse bacterial pathogens. *Microbiol. Mol. Biol. Rev.* **2013**, 77, 73–111.
96. Kusari, P.; Kusari, S.; Lamshöft, M.; Sezgin, S.; Spiteller, M.; Kayser, O. Quorum quenching is an antivirulence strategy employed by endophytic bacteria. *Appl. Microbiol. Biotechnol.* **2014**, 98, 7173–7183.
97. Kusari, P.; Spiteller, M.; Kayser, O.; Kusari, S. Recent advances in research on *Cannabis sativa* L. endophytes and their prospect for the pharmaceutical industry. In *Microbial Diversity and Biotechnology in Food Security*, Kharwar, R. N.; Upadhyay, R.; Dubey, N.; Raghuwanshi, R., Eds. Springer, New Delhi, India, **2014**, pp. 3–15.
98. Kusari, P.; Kusari, S.; Spiteller, M.; Kayser, O. Implications of endophyte-plant cross-talk in light of quorum responses for plant biotechnology. *Appl. Microbiol. Biotechnol.* **2015**, 99, 5383–5390.
99. Hartmann, A.; Rothballer, M.; Hense, B. A.; Schröder, P. Bacterial quorum sensing compounds are important modulators of microbe-plant interactions. *Front. Plant Sci.* **2014**, 5, 131.
100. Cornforth, D. M.; Popat, R.; McNally, L.; Gurney, J.; Scott-Phillips, T. C.; Ivens, A.; Diggle, S. P.; Brown, S. P. Combinatorial quorum sensing allows bacteria to resolve their social and physical environment. *Proc. Natl. Acad. Sci. USA* **2014**, 111, 4280–4284.
101. Safari, M.; Amache, R.; Esmaeilishirazifard, E.; Keshavarz, T. Microbial metabolism of quorum-sensing molecules acylhomoserine lactones, γ-heptalactone and other lactones. *Appl. Microbiol. Biotechnol.* **2014**, 98, 3401–3412.
102. Teplitski, M.; Mathesius, U.; Rumbaugh, K. P. Perception and degradation of *N*-acyl homoserine lactone quorum sensing signals by mammalian and plant cells. *Chem. Rev.* **2011,** 111, 100–116.
103. Cegelski, L.; Marshall, G. R.; Eldridge, G. R.; Hultgren, S. J. The biology and future prospects of antivirulence therapies. *Nat. Rev. Microbiol.* **2008**, 6, 17–27.
104. Gaiero, J. R.; McCall, C. A.; Thompson, K. A.; Day, N. J.; Best, A. S.; Dunfield, K. E. Inside the root microbiome: Bacterial root endophytes and plant growth promotion. *Am. J. Bot.* **2013**, 100, 1738–1750.
105. van der Heijden, M. G. A.; Hartmann, M. Networking in the plant microbiome. *PLoS Biol.* **2016**, 14, e1002378.
106. Sessitsch, A.; Hardoim, P.; Döring, J.; Weilharter, A.; Krause, A.; Woyke, T.; Mitter, B. et al. Functional characteristics of an endophyte community colonizing rice roots as revealed by metagenomic analysis. *Mol. Plant Microbe Interact.* **2012**, 25, 28–36.
107. Breidenbach, B.; Pump, J.; Dumont, M. G. Microbial community structure in the rhizosphere of rice plants. *Front. Microbiol.* **2016**, 6, 1537.
108. Santhanam, R.; Luu, V. T.; Weinhold, A.; Goldberg, J.; Oh, Y.; Baldwin, I. T. Native root-associated bacteria rescue a plant from a sudden-wilt disease that emerged during continuous cropping. *Proc. Natl. Acad. Sci. USA* **2015**, 112, E5013–E5020.
109. Steinrucken, T. V.; Bissett, A.; Powell, J. R.; Raghavendra, A. K. H.; van Klinken, R. D. Endophyte community composition is associated with dieback occurrence in an invasive tree. *Plant Soil* **2016**, 405, 311–323.

110. de Vrieze, J. The littlest farmhands. *Science* **2015**, 349, 680–683.
111. Yu, T. W.; Bai, L.; Clade, D.; Hoffmann, D.; Toelzer, S.; Trinh, K. Q.; Xu, J.; Moss, S. J.; Leistner, E.; Floss, H. G. The biosynthetic gene cluster of the maytansinoid antitumor agent ansamitocin from *Actinosynnema pretiosum*. *Proc. Natl. Acad. Sci. USA* **2002**, 99, 7968–7973.
112. Pullen, C. B.; Schmitz, P.; Hoffmann, D.; Meurer, K.; Boettcher, T.; von Bamberg, D.; Pereira, A. M. et al. Occurrence and non-detectability of maytansinoids in individual plants of the genera *Maytenus* and *Putterlickia*. *Phytochemistry* **2003**, 62, 377–387.
113. Wings, S.; Müller, H.; Berg, G.; Lamshöft, M.; Leistner, E. A study of the bacterial community in the root system of the maytansine containing plant *Putterlickia verrucosa*. *Phytochemistry* **2013**, 91, 158–164.
114. Kupchan, S. M.; Komoda, Y.; Court, W. A.; Thomas, G. J.; Smith, R. M.; Karim, A.; Gilmore, C. J.; Haltiwanger, R. C.; Bryan, R. F. Maytansine, a novel antileukemic ansa macrolide from *Maytenus ovatus*. *J. Am. Chem. Soc.* **1972**, 94, 1354–1356.
115. Kupchan, S. M.; Komoda, Y.; Branfmann, A. R.; Sneden, A. T.; Court, W. A.; Thomas, G. J.; Hintz, H. P. et al. The maytansinoids. Isolation, structural elucidation, and chemical interrelation of novel ansa macrolides. *J. Org. Chem.* **1977**, 42, 2349–2357.
116. Wolpert-DeFilippes, M. K.; Adamson, R. H.; Cysyk, R. L.; Johns, D. G. Initial studies on the cytotoxic action of maytansine, a novel ansa macrolide. *Biochem. Pharmacol.* **1975**, 24, 751–754.
117. Issel, B. F.; Crooke, S. T. Maytansine. *Cancer Treat. Rev.* **1978**, 5, 199–207.
118. Nakao, H.; Senokuchi, K.; Umebayashi, C.; Masuda, T.; Oyama, Y.; Yonemori, S. Cytotoxic activity of maytanprine isolated from *Maytenus diversifolia* in human leukemia K562 cells. *Biol. Pharm. Bull.* **2004**, 27, 1236–1240.
119. Lopus, M. Antibody-DM1 conjugates as cancer therapeutics. *Cancer Lett.* **2011**, 307, 113–118.
120. Santi, D.; Myles, D. C.; Metcalf, B.; Hutchinson, R.; Ashley, G. Maytansines and maytansine conjugates. US Patent 0109682 A1, **2003**.
121. LoRusso, P. M.; Weiss, D.; Guardino, E.; Girish, S.; Sliwkowski, M. X. Trastuzumab emtansine: A unique antibody-drug conjugate in development for human epidermal growth factor receptor 2-positive cancer. *Clin. Cancer Res.* **2011**, 17, 6437–6447.
122. Blanc, V.; Bousseau, A.; Caron, A.; Carrez, C.; Lutz, R. J.; Lambert, J. M. SAR3419: An anti-CD19-Maytansinoid Immunoconjugate for the treatment of B-cell malignancies. *Clin. Cancer Res.* **2011**, 17, 6448–6458.
123. Kusari, S.; Lamshöft, M.; Kusari, P.; Gottfried, S.; Zühlke, S.; Louven, K.; Hentschel, U.; Kayser, O.; Spiteller, M. Endophytes are hidden producers of maytansine in *Putterlickia* roots. *J. Nat. Prod.* **2014**, 77, 2577–2584.
124. Kusari, P.; Kusari, S.; Eckelmann, D.; Zühlke, S.; Kayser, O.; Spiteller, M. Cross-species biosynthesis of maytansine in *Maytenus serrata*. *RSC Adv.* **2016**, 6, 10011–10016.
125. Spiteller, P.; Bai, L.; Shang, G.; Carroll, B. J.; Yu, T. W.; Floss, H. G. The post-polyketide synthase modification steps in the biosynthesis of the antitumor agent ansamitocin by *Actinosynnema pretiosum*. *J. Am. Chem. Soc.* **2003**, 125, 14236–14237.
126. Eckelmann, D.; Kusari, S.; Spiteller, M. Occurrence and spatial distribution of maytansinoids in *Putterlickia pyracantha*, an unexplored resource of anticancer compounds. *Fitoterapia* **2016**, 113, 175–181.

8 Microbial Coculture and OSMAC Approach as Strategies to Induce Cryptic Fungal Biogenetic Gene Clusters

Georgios Daletos, Weaam Ebrahim,**
Elena Ancheeva, Mona El-Neketi,
Wenhan Lin, and Peter Proksch

CONTENTS

8.1 INTRODUCTION

Natural products continue to provide important lead structures in the search for new drugs.[1] It is estimated that more than 50% of all drugs that are currently on the market are either natural products or natural product–derived analogs.[2] The share of natural products is especially high for cytostatic compounds that are used for cancer chemotherapy and for antibiotics. For antibiotics alone close to two-thirds of all drugs available today are natural products or natural product derived.[3] Microbial natural compounds contribute strongly in this count, and among them fungal secondary metabolites play pivotal roles as evidenced, for example, by the recent introduction of the echinocandins to the drug market as new antifungal agents.[4]

* These two authors contributed equally to this work. Georgios Daletos contributed to the coculture part. Weaam Ebrahim contributed to the OSMAC part.

Nevertheless, biodiscovery of new bioactive metabolites from fungi (as well as from bacteria such as actinomycetes) suffers from a high rate of reisolation of already known compounds, which often makes the search for new leads from nature tiresome and expensive.

An obvious reason for the high rediscovery rate of known compounds from fungi lies in the fact that many microbial biosynthetic gene clusters are apparently not transcribed under standard laboratory conditions but remain silent. As an outcome, just a small amount of the genuine biosynthetic diversity of microorganisms is translated to secondary metabolites, which constitutes the currently available microbial metabolites that form the backbone of drug discovery. Several approaches are present to overcome these drawbacks during microbial fermentation. These include the one strain many compounds (OSMAC) technique where productive strains are cultured in various culture media and under specific culture conditions in order to obtain a maximum diversity of fungal metabolites.[5]

The term OSMAC was first introduced by the group of Zeeck in 2000.[6] Since then numerous studies have shown that the chemical diversity of fungi grown under laboratory conditions can be greatly enhanced by changing the media composition or other factors such as the temperature, light conditions, the shape of the culture vessels, aeration, and others. Even minor changes such as the exchange of tap water versus distilled water for media preparation have been shown to influence the natural products that are produced by fungi as highlighted in a study by Paranagama et al.[7]

Another approach that has proven highly successful in triggering the activation of silent biosynthetic gene clusters is the cocultivation (also called mixed fermentation) of two or more different microbes together in one culture vessel rather than maintaining pure cultures. This strategy mimics the natural microbial ecosystem, defined as microbiome, where the microorganisms interact with each other in synergistic and/or antagonistic relationships, which are the driving ecological forces for triggering secondary metabolite biosynthesis. Thus, culturing different microbial strains together may lead to the production of *cryptic* natural products that are not observed when the respective microorganisms are grown axenically.[8,9]

This chapter provides an overview of successful examples that highlight the power of the OSMAC approach or of microbial cocultivation for an elicitation of bioactive fungal natural products that resulted either in an enhanced accumulation of constitutively present compounds (= compounds that are also detected in fungal cultures grown under conventional culture conditions), and/or in an accumulation of new natural compounds missing in axenic control cultures. The examples that were chosen for this review chapter cover the time period from 2000 to 2016. Structures of novel and/or significantly bioactive lead compounds are given in Figures 8.1 through 8.14 later in the chapter. Next to providing an overview on published work that follows either the OSMAC or the cocultivation approach, it is the aim of this review to stimulate others who are active in the field to fully explore the biosynthetic potential of fungi with regard to new bioactive leads by developing innovative cultivation techniques.

8.2 PRODUCTION OF NATURAL PRODUCTS FOLLOWING THE OSMAC APPROACH

The OSMAC concept is applied to explore changes of microbial metabolite profiles influenced by changing fermentation conditions, such as different media, different temperature regimes, different culture vessels, and other factors. The original concept dates back to a study employing the fungus *Sphaeropsidales* sp. (strain F-24′707).[6] Following exposure of the wild-type fungus to ultraviolet (UV) irradiation, 25 different fungal morphotypes that differed in coloration were isolated. These mutants were cultivated employing two different liquid media on shakers at 28°C for 3 days. Analysis of the wild-type fungus and of the various mutants revealed striking differences. Whereas the wild type produced mainly the spirobisnaphthalene cladospirone bisepoxide palmarumycin C_{13} (**1**),[10] the mutants accumulated the new antibacterial macrolide compound, mutolide (**2**) (Figure 8.1).

The F-actin inhibitor jaspaklinolide, which is a cyclodepsipeptide that was originally obtained from the sponge *Jaspis splendens*,[11] was added to a liquid Czapek-Dox culture medium that was used for fermentation of the marine-derived fungus *Phomopsis asparagi* isolated from the sponge *Rhaphidophlus juniperina*. Addition of jasplakinolide to the culture medium induced the accumulation of three new bioactive chaetoglobosine derivatives, including chaetoglobosin-510 (**3**), -540 (**4**), and -542 (**5**) (Figure 8.1). Cytochalasin derivatives are well-known actin depolymerizers,[12] and are usually accumulated only by members of the fungal genera *Phomopsis* and *Chaetomium* sp.[13] Addition of the F-actin inhibitor jaspaklinolide to the culture medium thus resulted in the accumulation of actin-depolymerizing cytochalasins.[13]

Fermentation of fungi in microtiter plates (1 mL volume per well) provided a low-cost biodiscovery strategy that allows rapid screening of a large number of fungal strains.[14] When this experiment was conducted using over 2000 different fungal strains growing in microtiter plates, several antifungal metabolites were detected that were absent when the respective fungi were grown in 1 L culture flasks. These metabolites included brefeldin A (**6**) from *Eupenicillium brefeldianum* (strain F-146,140), cerulenin (**7**) and helvolic acid (**8**) from *Sarocladium oryzae*, moriniafungin from *Morinia pestalozzioides* (F-090,354), arundifungin from *Arthrinium arundinis* (F-142,740), furanocandin from *Monocillium* sp. (F-210,948), xylarin from *Xylaria* sp. (F-212,836), aspirochlorine from *Leiothecium* sp. (F-215,757), leucinostatin from *Paecilomyces lilacinus* (F-236,792), pneumocandin from *Glarea lozoyensis* (F-239,379), 9-methoxystrobilurin K from *Favolaschia pustulata* (F-242,456), lovastatin (**9**) from *Monascus purpureus* (F-242,596), BE-49385 from *Paecilomyces inflatus* (F-253,279), and echinocandin B and sterigmatocystin (**10** and **11**) from *Emericella rugulosa* (F-173,113) (Figure 8.1).[14]

Modifying the carbon source of the culture medium is often found to have a distinct effect on the patterns of the compounds produced.[15] Following this approach, 1% glucose in type B media (1% malt extract, 1% glucose, and 0.05% peptone in artificial seawater adjusted to pH 7.5) was replaced by 1% soluble starch in type A medium (1% malt extract, 1% soluble starch, and 0.05% peptone in artificial seawater adjusted to pH 7.5) for fermentation of the sponge-derived fungus *Gymnascella dankaliensis*.[15–18] This resulted in major differences in the high-performance

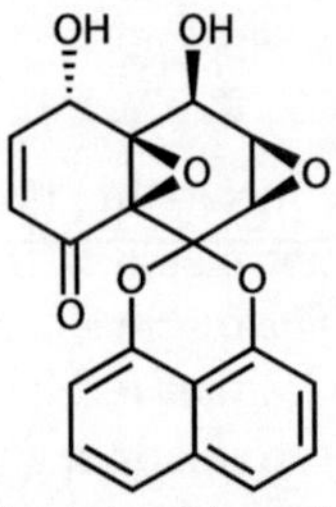

Cladospirone bisepoxide = Palmarumycin C_{13} (**1**)

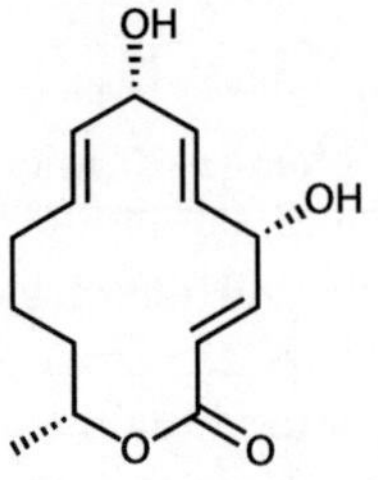

Mutolide (**2**)

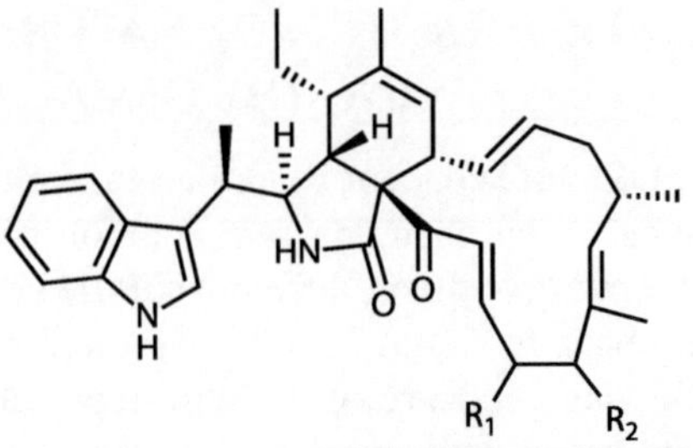

$R_1 = R_2 = H$: Chaetoglobosin-510 (**3**)
$R_1 = {=}O$, $R_2 = $ —OH : Chaetoglobosin-540 (**4**)
$R_1 = R_2 = $ —OH : Chaetoglobosin-542 (**5**)

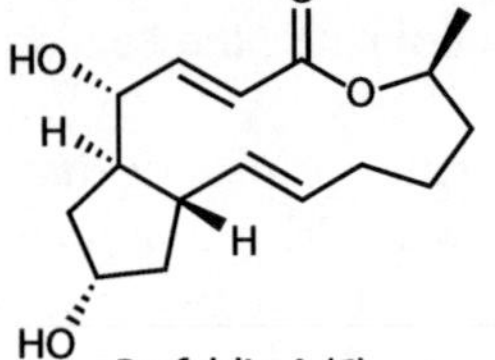

Brefeldin A (**6**)

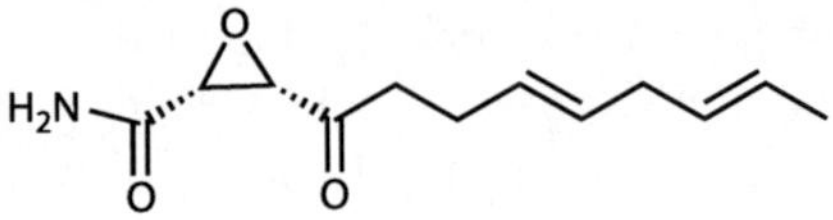

Cerulenin (**7**)

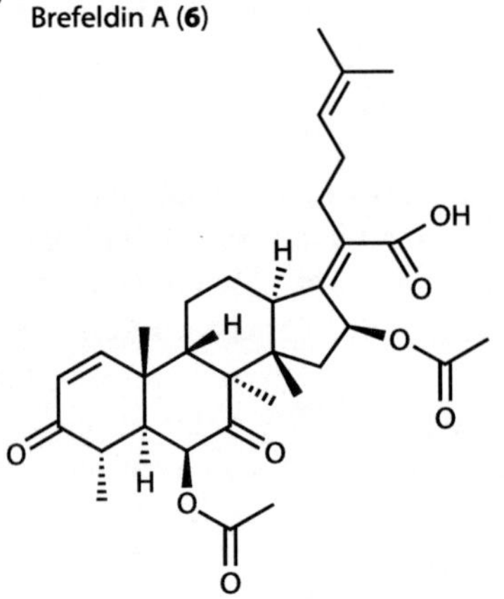

Helvolic acid (**8**)

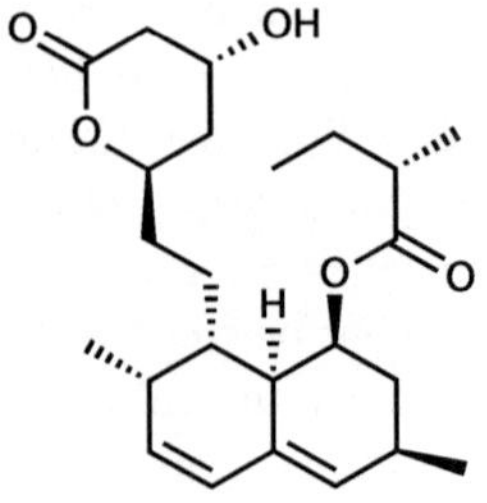

Lovastatin (**9**)

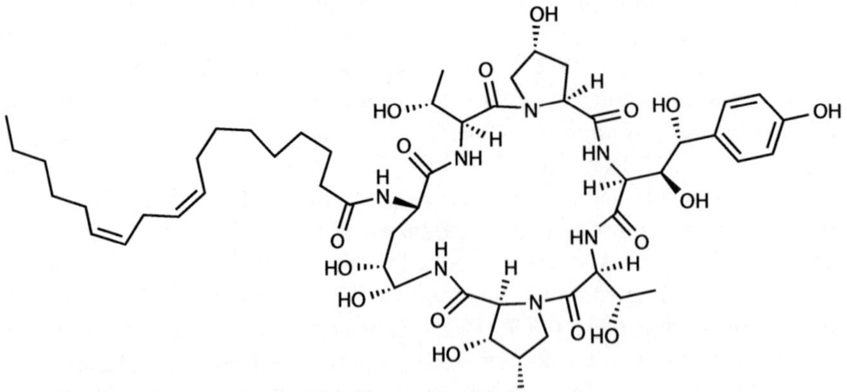

Echinocandin B (**10**)

FIGURE 8.1 Structures **1–10**. *(Continued)*

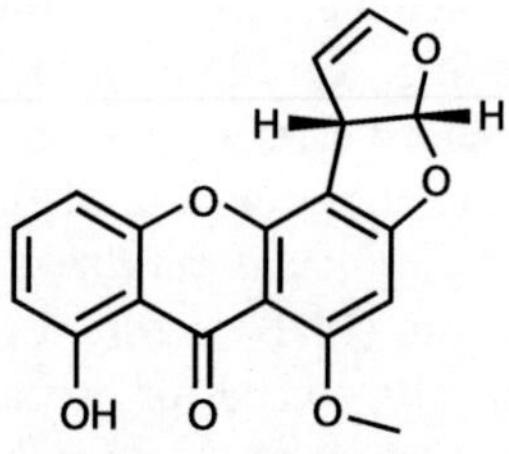

Sterigmatocystin (**11**)

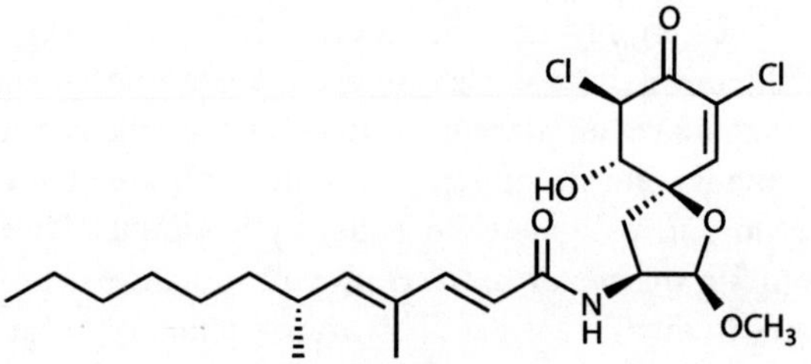

Gymnastatin B (**12**)

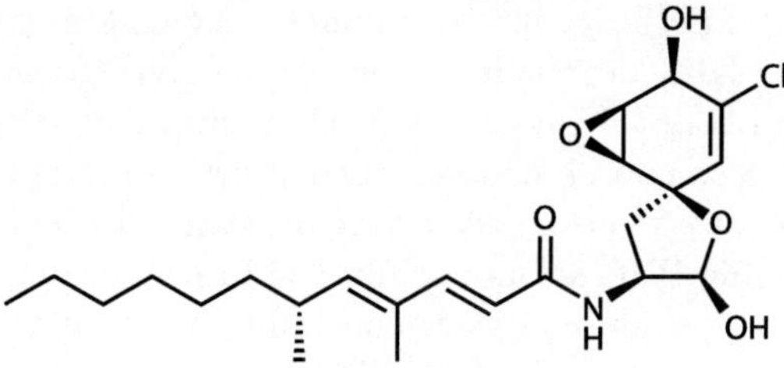

Gymnastatin E (**13**)

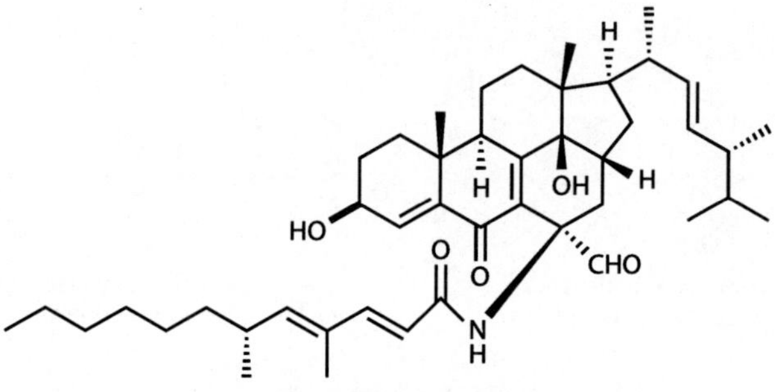

Gymnasterone A (**14**)

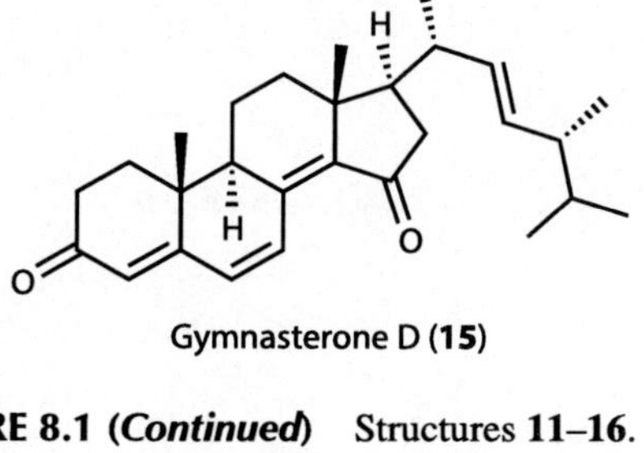

Gymnasterone D (**15**)

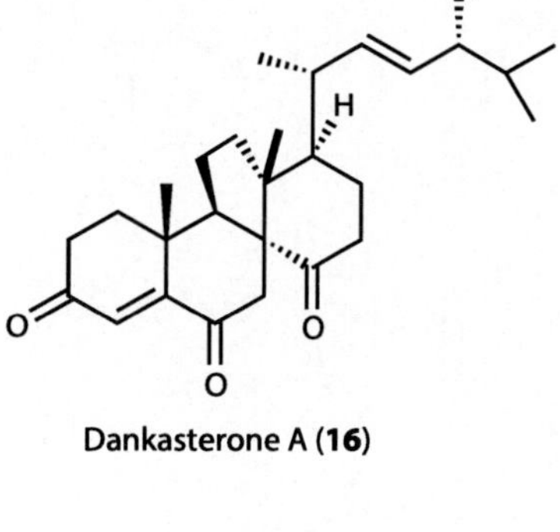

Dankasterone A (**16**)

FIGURE 8.1 (*Continued*) Structures **11–16**.

liquid chromatography (HPLC) patterns of the fungus in both culture media. Original type B media produced the known potent cytotoxic compounds gymnastatins A–H (e.g., **12** and **13**; Figure 8.1)[16–18] in addition to the new cytotoxic steroids gymnasterones A–D (e.g., **14** and **15**; Figure 8.1)[15] whereas type A media induced the production of the novel cytotoxic steroids dankasterones A and B (e.g., **16**; Figure 8.1).[15]

Paranagama et al.[7] demonstrated that even the type of water that is used to prepare culture media and its mineral content can influence the metabolic profile of fungi. The authors investigated the endophytic fungus *Paraphaeosphaeria quadriseptata* by changing the water used to prepare potato dextrose broth (PDB) medium from tap water to distilled water. This minor change in the water type used for preparing the culture media caused striking changes in the metabolite profiles and induced the biosynthesis of several new compounds. From the tap water culture, the major compound monocillin I (**17**) together with minor isocoumarins paraphaeosphaerins A–C, aposphaerin C (**20**), eugenetin, 6-methoxymethyleugenin, and 6-hydroxymethyleugenin were isolated,[19,20] whereas from cultures that were grown in media containing distilled water, new compounds such as cytosporones F–I (**23**–**26**), quadriseptin A, and the antiproliferative agents 5′-hydroxymonocillin III (**27**), in addition to the known monocillin III (**28**), monocillin I (**17**), and aposphaerin B (**19**) were obtained (Figure 8.2). In addition, water containing heavy metals such as $CuSO_4$,

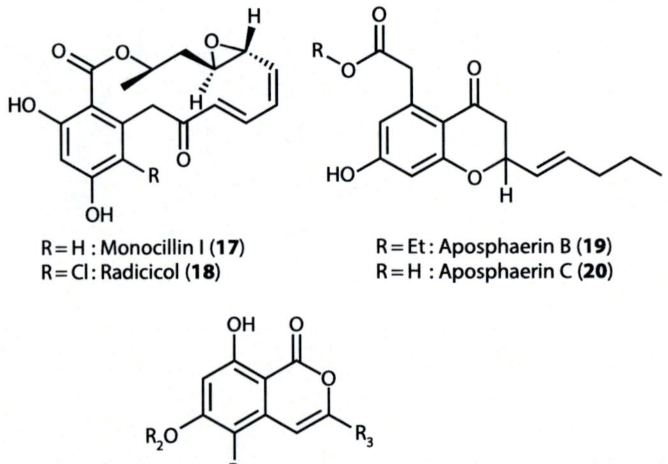

R = H : Monocillin I (**17**)
R = Cl : Radicicol (**18**)

R = Et : Aposphaerin B (**19**)
R = H : Aposphaerin C (**20**)

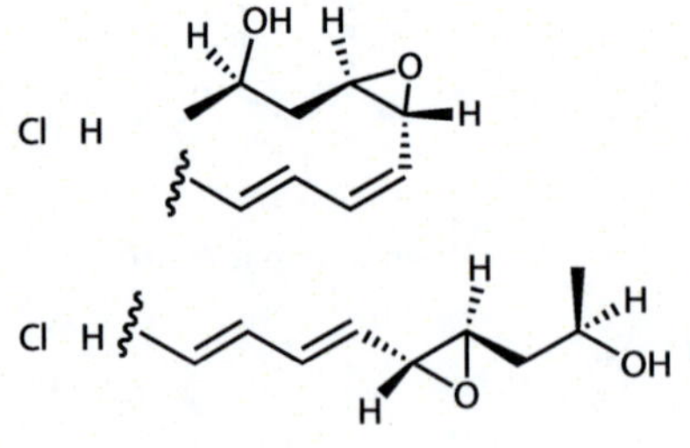

: Chaetochiversin A (**21**)

: Chaetochiversin B (**22**)

FIGURE 8.2 Structures 17–22. *(Continued)*

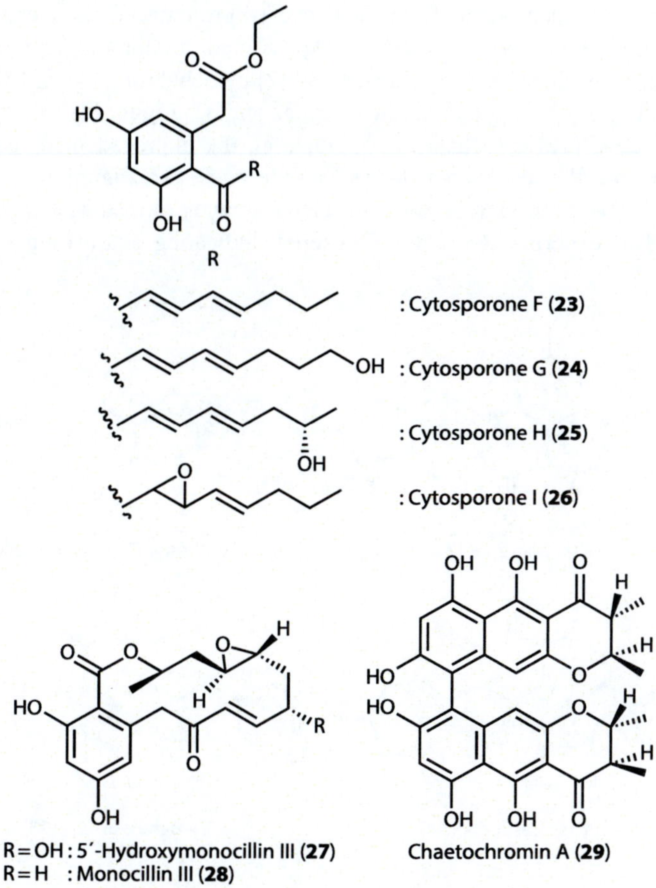

FIGURE 8.2 (*Continued*) Structures 23–29.

$ZnSO_4$, $Cd(NO3)_2$, and $K_2Cr_2O_7$ in the PDB culture medium used for fermentation of *P. quadriseptata* induced mainly production of monocillin I (**17**).[7]

Chaetomium chiversii was investigated by the same research group using liquid cultures on shakers instead of cultures on solid media. When grown on solid media, the fungus produced mainly radicicol (**18**) and chaetochiversins A and B (**21** and **22**),[7] while culturing *C. chiversii* in liquid PDB medium with shaking afforded the known compound chaetochromin A (**29**) as a major metabolite instead (Figure 8.2).

Cytochalasins are bioactive fungal compounds biosynthesized by different fungi including taxa of the genera *Phomopsis*, *Chalara*, *Hyposylon*, *Xylaria*, *Daldinia*, *Pseudeurotium*, and *Phoma*.[21] The diverse structures and pronounced biological activities of these compounds encouraged numerous groups to search for new cytochalasin derivatives employing the OSMAC approach.[22] Changing the composition of culture media used for fermentation of the deep sea–derived fungus *Spicaria elegans* KLA03 (obtained from marine sediments collected in Jiaozhou Bay, China) from type A liquid medium (glucose, peptone, malt extract, yeast extract, and seawater after adjusting its pH to 7.0) to type B liquid culture (2% soluble starch, 1.5% soybean flour, 0.5%

yeast extract, 0.2% peptone and seawater) caused pronounced changes of the respective HPLC profiles.[21–23] When cultured in type A medium, the fungus accumulated the new cytochalasins E and K (**30**), 10-phenyl-[12]-cytochalasins Z_7, Z_8 (**31**), and Z_9, as well as novel open-chain cytochalasins (e.g., **32** and **33**; Figure 8.3), in addition to one known derivative [12]-cytochalasin[21,22]; however, the unprecedented spicochalasin A (**34**), five chemically new aspochalasins M–Q (e.g., **35–37**, Figure 8.3), in addition to known aspochalasin derivatives, were detected following fermentation in type B culture medium.[23] This experiment was the first report indicating that a fungal strain is able

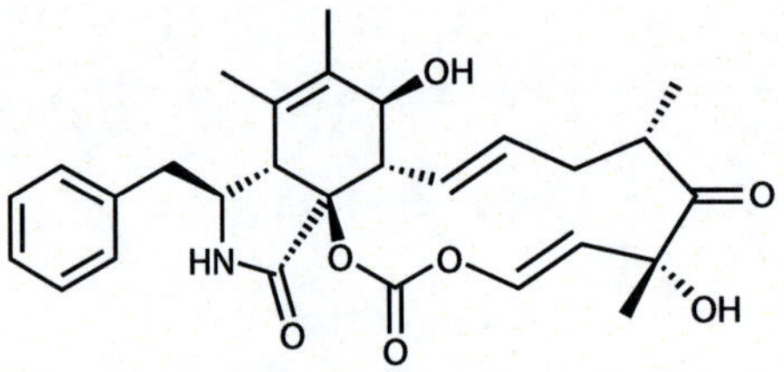

Cytochalasin K (**30**)

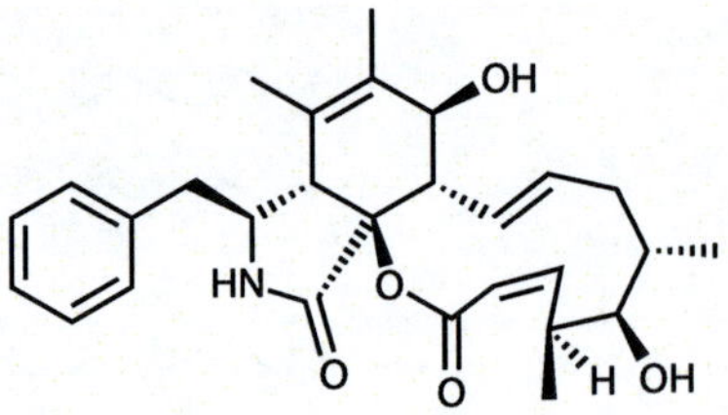

10-Phenyl-[12]-cytochalasin Z_8 (**31**)

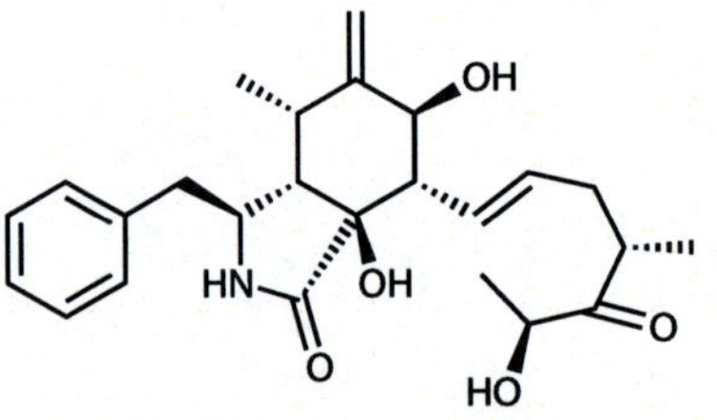

Cytochalasin Z_{11} (**32**)

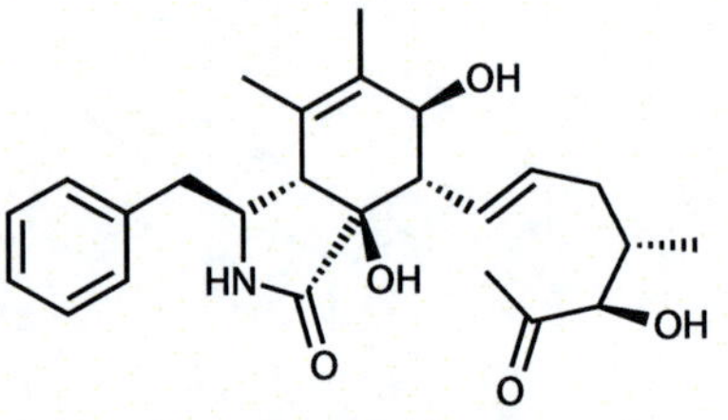

Cytochalasin Z_{14} (**33**)

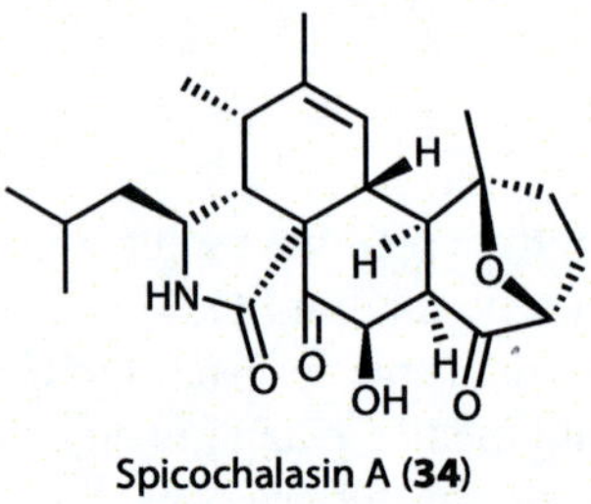

Spicochalasin A (**34**)

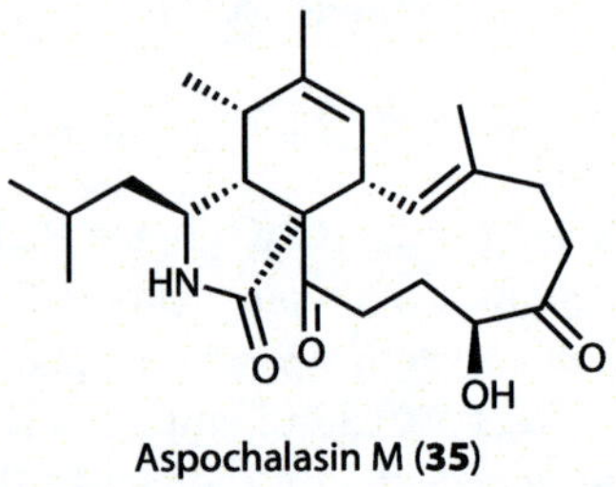

Aspochalasin M (**35**)

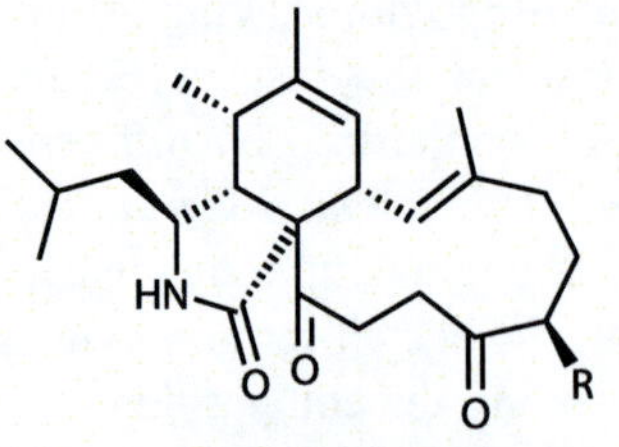

R=OH : Aspochalasin P (**36**)
R=H : Aspochalasin Q (**37**)

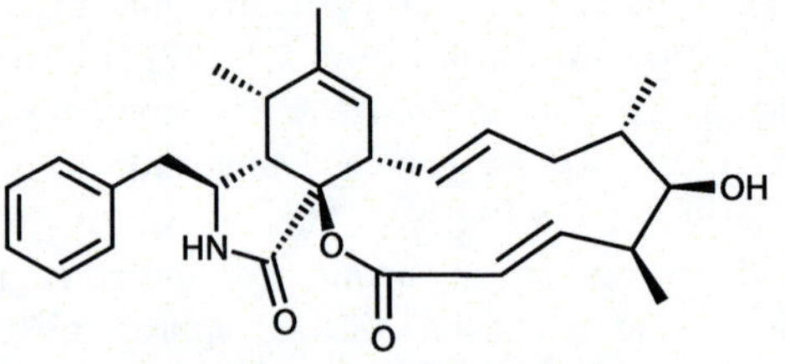

7-Deoxycytoochalasin Z_7 (**38**)

FIGURE 8.3 Structures **30–38**. *(Continued)*

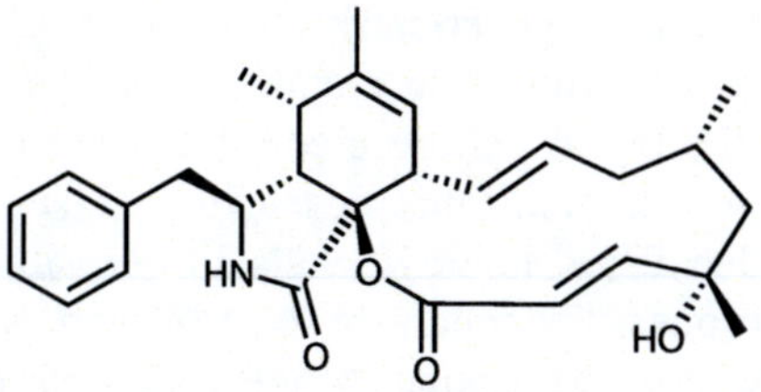

7-Deoxycytoochalasin Z_9 (**39**)

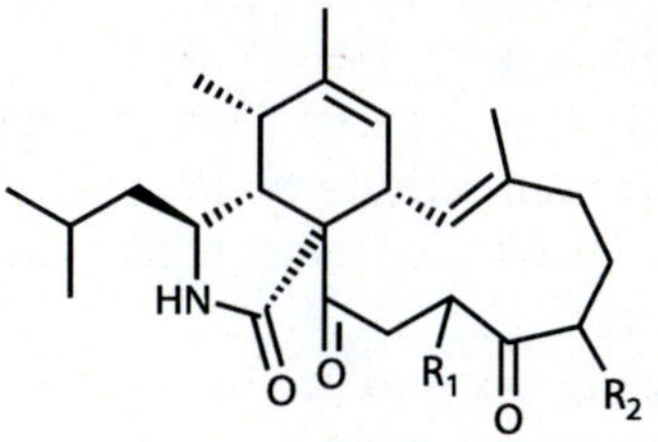

$R_1 = OCH_3$, $R_2 = OH$: Aspochalasin R (**40**)
$R_1 = OH$, $R_2 = H$: Aspochalasin S (**41**)
$R_1 = OH$, $R_2 = OH$: Aspochalasin T (**42**)

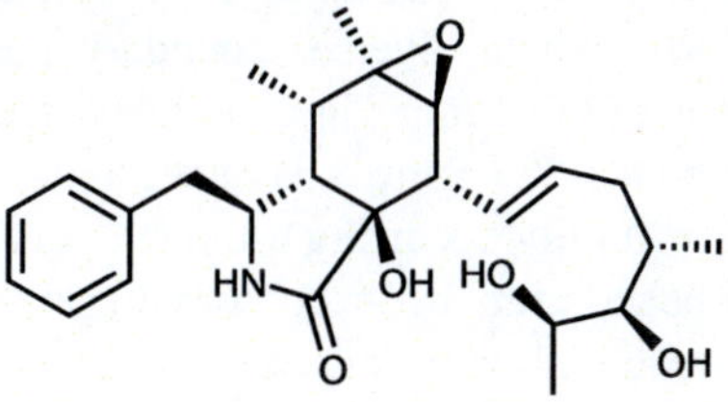

Cytochalasin Z_{21} (**43**)

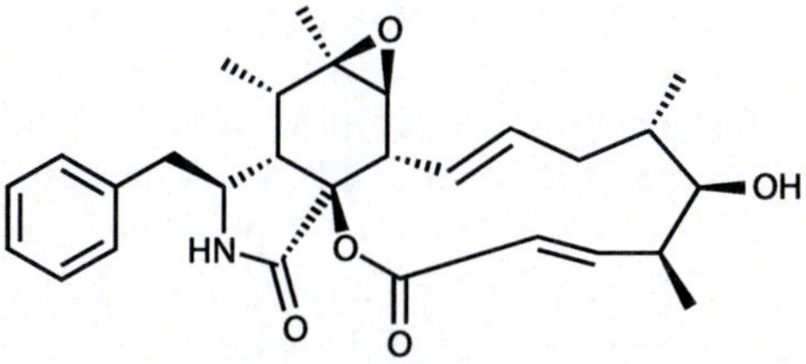

22-Oxa-[12]-cytochalasin 1 (**44**)

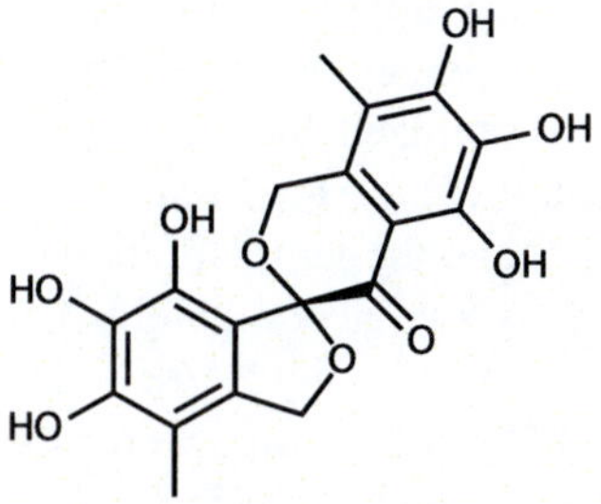

Eleganketal A (**45**)

FIGURE 8.3 (*Continued*) Structures **39–45**.

to biosynthesize two biosynthetically different types of cytochalasin derivatives, which involve either a phenylalanine or a leucine precursor.

Using the same fungus *S. elegans* KLA03, Lin et al.[24] investigated the effect of metyrapone (a cytochrome P-450 inhibitor), which was added on day 6 of the fermentation to the liquid culture medium (glucose, peptone, malt extract, yeast extract, and seawater, pH 7). Metyrapone inhibits P450 mono-oxygenases that catalyze oxidations of cytochalasins. The presence of metapyrone in the medium led to accumulation of two new deoxy-cytochalasins, including the moderately cytotoxic 7-deoxy-cytochalasin Z_7 (**38**) and 7-deoxy-cytochalasin Z_9 (**39**) (Figure 8.3). Interestingly, compounds **38** and **39** are precursors of cytochalasins Z_7 and Z_9, respectively.[24]

When *S. elegans* KLA03 was cultured in liquid medium that was composed of 2% soluble starch, 1.5% soybean flour, 0.5% yeast extract, 0.2% peptone, and seawater at 28°C on shakers for 14 days, new apochalasin derivatives including aspochalasins R–T (**40–42**; Figure 8.3) were accumulated.[25]

Wang et al.[26] were able to obtain additional new cytochalasin derivatives from *S. elegans* KLA03 through further modifications of the culture conditions.

Cytochalasin derivatives are biogenetically obtained from an acetate moiety, a methionine-derived octa- or nonaketide chain, and an amino acid. Addition of L- and D-tryptophan to the liquid culture media (static or on shakers) was found to cause the accumulation of additional new cytochalasins. These amino acid–enriched cultures afforded three new cytochalasin derivatives including (**43**), in addition to three known ones. Cytochalasin Z_{21} (**43**) and 22-oxa-12-cytochalasin 1 (**44**) showed strong cytotoxicity against A-549 cells with IC_{50} values of 8.2 and 3.1 mM, respectively (Figure 8.3).[26]

Luan et al.[27] were attracted by the amazing ability of the fungal strain *S. elegans* KLA03 to produce diverse secondary metabolites using the OSMAC method.[21–23] Cultivation of this fungus in a modified mannitol-based medium led to the isolation of eleganketal A (**45**; Figure 8.3) rather than cytochalasin derivatives. Compound **45** represents a naturally occurring aromatic polyketide possessing a rare highly oxygenated spiro[isobenzofuran-1,3′-isochroman] ring system.[27]

The fractional factorial design approach, which is considered to be a novel experimental design, was applied for induction of new compounds by two *Penicillium* species including *Penicillium oxalicum* and *Penicillium citrinum*.[28] Following this experimental approach, the disadvantage of obtaining only trace amounts of extracts from the original culture media of *P. oxalicum* and *P. citrinum* was eliminated. The authors optimized five main growth parameters including the total medium saline concentration, the total nutrient concentration, the time of incubation, the pH of the medium, and the temperature used for fermentation. Under these optimized culture conditions, two new alkaloids, citrinalins A and B (**46** and **47**), in addition to several known compounds, including the polyketides 3 and 4, cyclopiamines A and B (**48** and **49**), and citrinin, were isolated from *P. citrinum*. From *P. oxalicum* the known compounds meleagrin (**50**) and oxaline (**51**) were obtained following optimization of the culture conditions (Figure 8.4). Yields of compounds were enhanced up to 500% in both *Penicillium* taxa following use of this experimental approach.

Addition of 50 mM $CaBr_2$ to liquid SWS growth medium [soytone (0.1%), soluble starch (1.0%), and seawater (100%)] used for fermentation of the marine-derived *Fusarium tricinctum* resulted in the production of the new halogenated chlamydosporol analogs bromomethylchlamydosporols A and B (**52** and **53**; Figure 8.4), as well as the known chlamydosporol (obtained as an inseparable epimeric mixture of (7*R*:7*S*) at a ratio of 1:1) and the known fusarielin A, instead of peptides that are normally produced by this fungus in halide-free culture medium.[29]

Self-toxic fungal metabolites are natural products that either slow the growth of their producers or even kill them.[30] This is exemplified by cercosporamide (**54**; Figure 8.5), which is produced by the slow-growing unidentified fungus LV-2841. In an effort to increase the production of this broad-spectrum antifungal agent, Singh et al. added different types of resins to the fungal culture medium, thereby absorbing the compound and lowering its toxicity to the producing fungus. This was found to increase the growth of the fungus as well as the production of cercosporamide (**54**). Addition of 3% Diaion® HP20 to Difco™ dehydrated PDB culture medium resulted in a 100-fold enhanced accumulation of cercosporamide (**54**).[30]

Addition of agar to the liquid culture medium used for fermentation of the endolichenic fungus *Corynespora* sp. BA-10763[31] resulted in the induction of

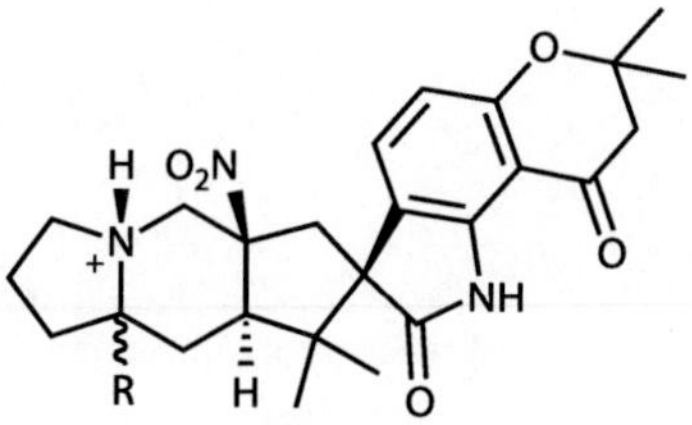

R = ▬ H: Citrinalin A (**46**)
R = ····· H: Citrinalin B (**47**)

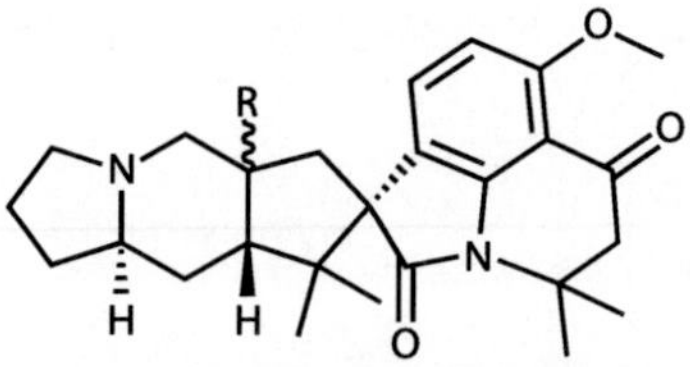

R = ····· NO_2: Cyclopiamine A (**48**)
R = ▬ NO_2: Cyclopiamine B (**49**)

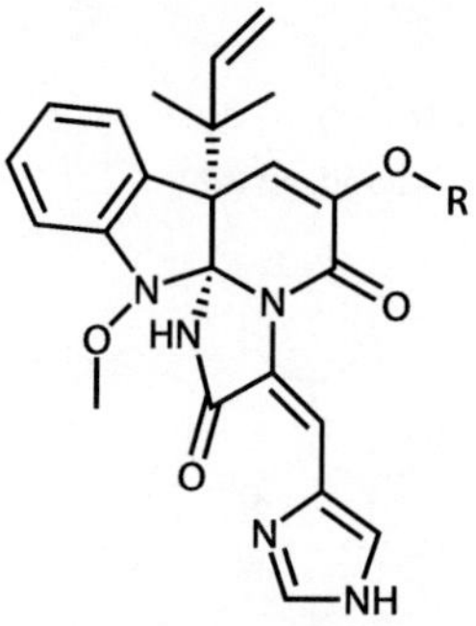

R = H : Meleagrin (**50**)
R = CH_3 : Oxaline (**51**)

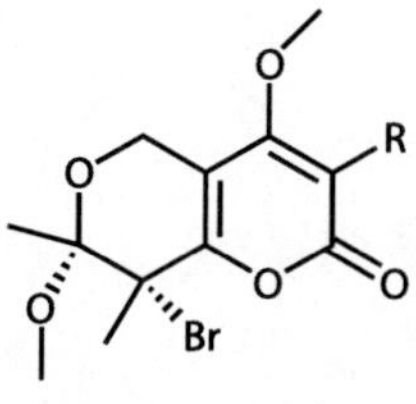

R = H : Bromomethylchlamydosporol A (**52**)
R = Br : Bromomethylchlamydosporol B (**53**)

FIGURE 8.4 Structures **46–53**.

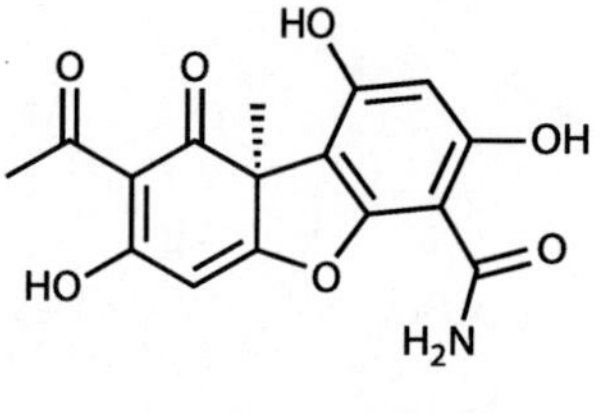

Cercosporamide (**54**)

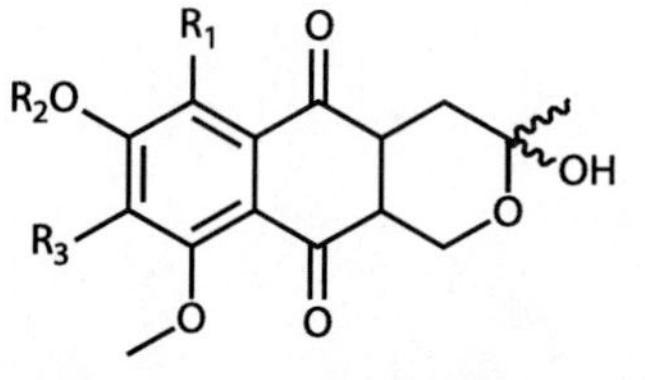

$R_1 = R_3 = H, R_2 = CH_3$: Herbarin (**55**)
$R_1 = R_2 = R_3 = H$: 7-Desmethylherbarin (**56**)
$R_1 = H, R_2 = CH_3, R_3 = OH$: 8-Hydroxyherbarin (**57**)
$R_1 = OH, R_2 = CH_3, R_3 = H$: 8-*O*-Methylfusarubin (**58**)

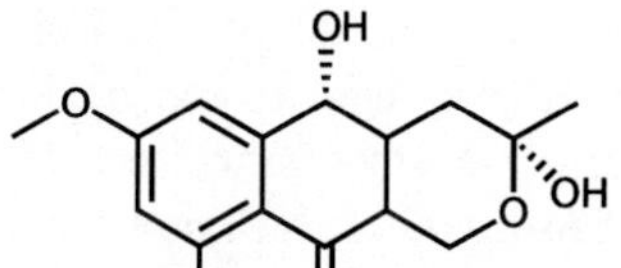

R = CH_3 : 9-*O*-Methylscytalol A (**59**)
R = H : Scytalol A (**60**)

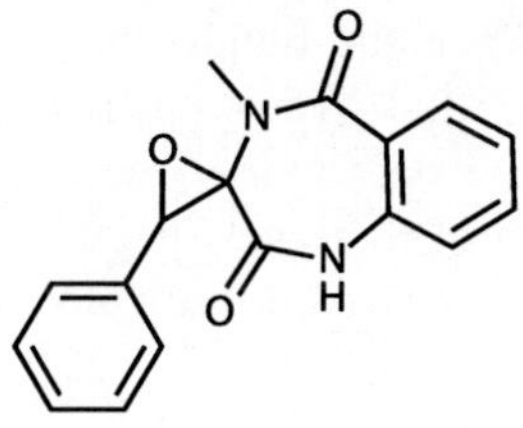

Cyclopenin (**61**)

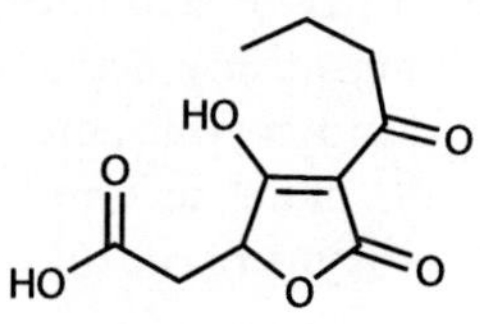

Carlosic acid (**62**)

FIGURE 8.5 Structures **54–62**.

(*Continued*)

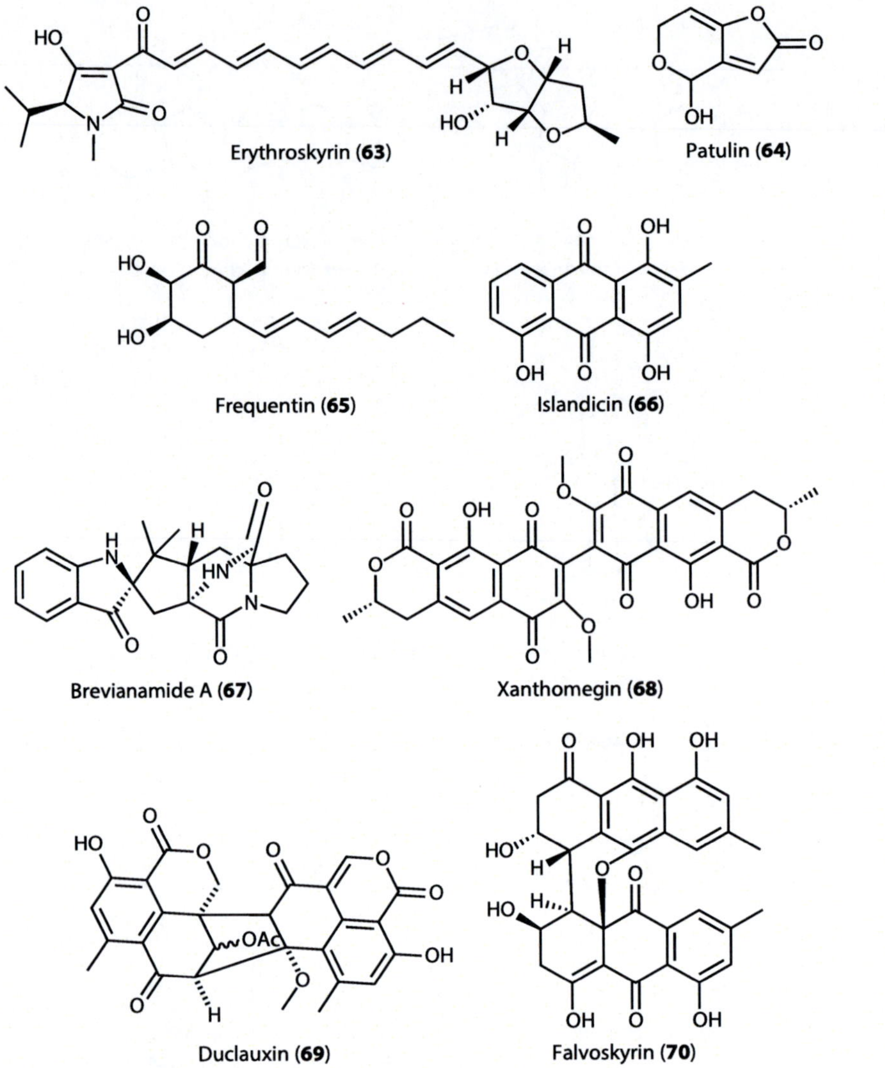

FIGURE 8.5 (*Continued*) Structures **63–70**.

several new metabolites. When the fungus was grown in the original liquid PDB medium, mainly herbarin (**55**), 1-hydroxydehydroherbarin, 1-methoxydehydroherbarin, and corynesporol were detected,[32] whereas addition of agar to the medium resulted in the accumulation of three new heptaketides—7-desmethylherbarin (**56**), 8-hydroxyherbarin (**57**), and 9-*O*-methylscytalol A (**59**)—and of the known compounds herbarin (**55**), 8-*O*-methylfusarubin (**58**), scytalol A (**60**), scorpinone, and 8-*O*-methylbostrycoidin (Figure 8.5).[31]

Trace elements, salts, sugars, and heavy metals are components of culture media that play a pivotal role in the fungal secondary metabolism.[33] This fact was further supported by addition of sucrose and several metal ions [NaCl and $Cd(NO_3)_2$] to yeast extract sucrose (YES) liquid and solid media used for fermentation of *Penicillium janthinellum* and *Penicillium duclauxii*. Addition of $Cd(NO_3)_2$ to the culture medium of *P. janthinellum* resulted in a striking change of the HPLC profile of the fungal extract. Cyclopenin (**61**), carlosic acid (**62**), erythroskyrin (**63**), kojic acid, and patulin (**64**) were produced by *P. janthinellum* in cadmium nitrate-free medium, while cyclopenin (**61**), carlosic acid (**62**), frequentin (**65**), and islandicin (**66**) were induced in media that contained 100 or 500 ppm $Cd(NO_3)_2$. Similar effects were observed for *P. duclauxii* where patulin (**64**), brevianamide A (**67**), xanthomegin (**68**), and duclauxin (**69**) were produced in $Cd(NO_3)_2$-free medium, whereas only brevianamide A (**67**) was detected in $Cd(NO_3)_2$-based medium (Figure 8.5).

When NaCl was added to the culture medium used for fermentation of *P. janthinellum*, carlosic acid (**62**), kojic acid, frequentin (**65**), and islandicin (**66**) were detected, whereas in NaCl-free media, only carlosic acid (**62**), erythroskyrin (**63**), and patulin (**64**) were found.[33] Similar effects were found for *P. duclauxii*, which only produced patulin (**64**) in NaCl-free medium, while citrinin, brevianamide A (**67**), xanthomegin (**68**), mycophenolic acid, and flavoskyrin (**70**) were obtained in NaCl-containing media (Figure 8.5).[33]

In addition to NaCl, sucrose was also found to affect the metabolite patterns of *P. janthinellum* and of *P. duclauxii*. Cyclopenin (**61**), carlosic acid (**62**), erythroskyrin (**63**), kojic acid, and patulin (**64**) were accumulated only in the presence of sucrose in culture media used for fermentation of *P. janthinellum*, but not in sucrose-free cultures. Sucrose added to the culture medium of *P. duclauxii* resulted in the production of citrinin, patulin (**64**), brevianamide A (**67**), xanthomegin (**68**), duclauxin (**69**), flavoskyrin (**70**), mycophenolic acid, and gentisic acid, which were not detected in sucrose-free medium (Figure 8.5).

Spirobisnaphthalenes are naphthoquinone derivatives which are known for their antileishmanial, antibacterial, allelochemical, antitumor, and antifungal activities. To date, over 80 different spirobisnaphthalenes have been identified as natural products.[34] Diepoxin ζ (**1**) (also known as palmarumycin C_{13}, cladospirone bisepoxide, or Sch53514) was the first member of this group of compounds.[34] Addition of yeast extract to liquid modified Sabouraud broth medium was found to enhance the production of palmarumycin C_{13} (**1**) by the endophytic fungus Dzf12 (later identified as *Berkleasmium* sp. Dzf12). From fungal cultures that lacked yeast in their media, only trace amounts of palmarumycin C_{13} (**1**), in addition to palmarumycin CP17, diepoxin kappa, diepoxin eta (**71**; Figure 8.6), and diepoxin gamma, were isolated.[35]

In a follow-up study, Li et al.[36] induced the accumulation of palmarumycin C_{13} (**1**) by the same fungal strain through addition of polysaccharides, which are produced by the fungal host plant *Dioscorea zingiberensis* as stimulating factors. Three polysaccharides, including sodium hydroxide–extracted polysaccharide (SEP), water-extracted polysaccharide (WEP), and acid-extracted polysaccharide (AEP) that had been obtained from the rhizomes of *D. zingiberensis*, were added to the liquid culture of the fungus. Among these polysaccharides, WEP was found to be the most effective inducer to enhance palmarumycin C_{13} (**1**) production by 2.7-fold in comparison with the control.[36]

In 2012, Ding and coworkers applied a similar protocol, but used crude oligosaccharides instead of polysaccharides derived from the host plant *D. zingiberensis*

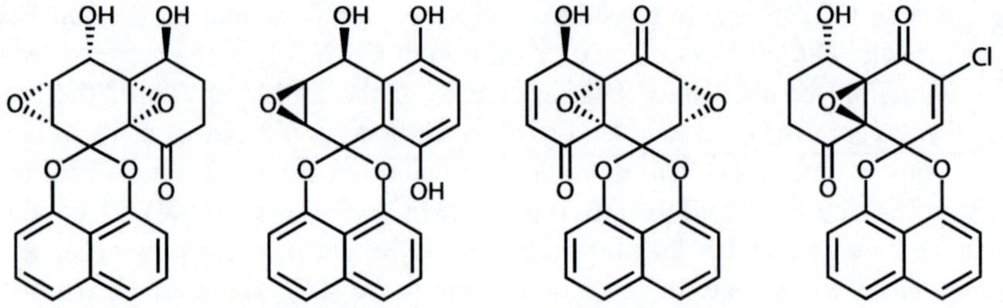

Diepoxin eta (**71**) Palmarumycin C_{12} (**72**) Diepoxin sigma (**73**) Palmarumycin C_8 (**74**)

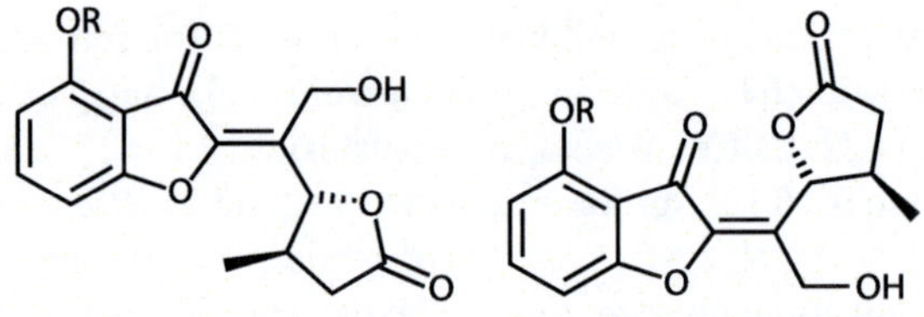

R = Me : Photinide A (**75**)
R = H : Photinide C (**76**)

R = Me : Photinide B (**77**)
R = H : Photinide D (**78**)

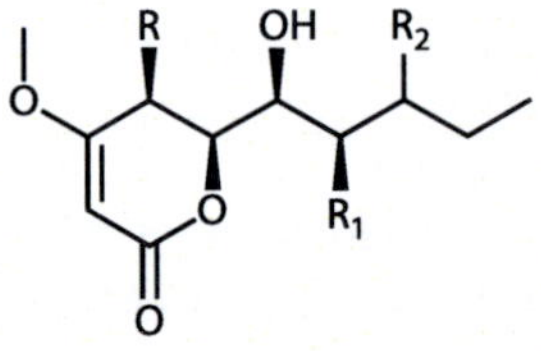

R = OH, $R_1 = R_2 = H$: Photipyrone A (**79**)
R = R_1 = H, R_2 = OH : Photipyrone B (**80**)

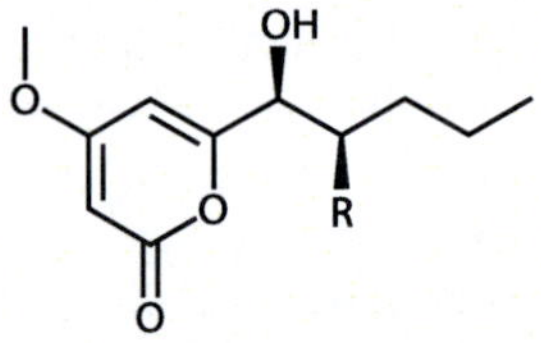

R = H : 1′-Hydroxy-4-methoxy-6-pentyl-2*H*-pyran-2-one (**81**)
R = OH : 1′,2′-Dihydroxy-4-methoxy-6-pentyl-2*H*-pyran-2-one (**82**)

FIGURE 8.6 Structures **71–82**.

for inducing compound production by the fungus *Berkleasmium* sp. Dzf12.[37] Oligosaccharides were prepared by acid hydrolysis of their corresponding polysaccharides (WEP), (SEP), and (AEP). Screening results indicated that addition of WEP oligosaccharide to liquid PDB culture medium induced palmarumycins C_{12} (**72**; Figure 8.6) and C_{13} (**1**). An almost tenfold upregulation was found for palmarumycin C_{12} (**72**) and a threefold upregulation for palmarumycin C_{13} (**1**) compared to controls lacking oligosaccharides. The authors concluded that oligosaccharides (i.e., OW) rather than their corresponding polysaccharides (i.e., WEP) exert the highest inducing effect on palmarumycin production.[37]

In a further attempt to maximize the yield of palmarumycin C_{12} (**72**) and palmarumycin C_{13} (**1**), Mou et al.[38] added three metal ions, including Cu^{2+}, Al^{3+}, and Ca^{2+}, in concentrations of 1.36, 2.5, and 7.85 mmol/L, respectively, to liquid PDB medium used for fermentation of the fungus *Berkleasmium* sp. Dzf12 and compared the yields of palmarumycins in treated cultures versus untreated controls. A three- to fourfold enhancement of the production of palmarumycins C_{13} (**1**) and C_{12} (**72**) was obtained following addition of (Ca^{2+}), (Cu^{2+}), and (Al^{3+}).[38]

Shan et al. investigated further minor spirobisnaphthalenes by optimizing the culture conditions of the fungus *Berkleasmium* sp. Dzf12 in PDB medium (potato 200 g/L and dextrose 20 g/L) on rotary shakers (150 rpm) at a temperature regime of 25°C (7 days cultivation time).[39] These optimized culture conditions led to an accumulation of further minor spirobisnaphthalenes including (**73**) and (**74**) (Figure 8.6), in addition to previously isolated compounds (**1**) and (**72**). In particular, palmarumycin C_8 (**74**) exhibited both antibacterial and antifungal activities, while diepoxin δ (**73**) and palmarumycin C_8 (**74**) showed potent cytotoxicity against five different human cancer cell lines (HCT-8, Bel-7402, BGC-823, A 549, A 2780) with IC_{50} values of 1.3–5.8 μM.[39]

Mou et al. studied the effects of organic-aqueous two-phased cultures on the metabolism of *Berkleasmium* sp. Dzf12 with regard to the production of bioactive deoxypreussomerin spirobisnaphthalenes (**1**) and (**72**).[40] Structurally, (**1**) and (**72**) are nonpolar metabolites which can be easily dissolved in the organic layer of the two-phased cultures. Addition of 5% (v/v) butyl oleate as organic phase to PDB medium led to a yield of 191.6 mg/L palmarumycin C_{12} (**72**), which is almost 35 times higher than that observed for control cultures lacking butyl oleate as organic phase.[40]

Fungi from the genus *Pestalotiopsis* are known to be prolific sources of many novel bioactive metabolites.[41] Changing the culture medium from SA medium to solid rice medium led to major changes in the HPLC profiles of the fungal endophyte *Pestalotiopsis photiniae*. Photinides A–F (e.g., **75–78**; Figure 8.6) were isolated when the fungus was grown in SA culture medium,[42] whereas the two new γ-lactone derivatives photipyrones A (**79**) and B (**80**), along with four known analogs, namely LL-P880a, LL-P880b, 1′-hydroxy-4-methoxy-6-pentyl-2*H*-pyran-2-one (**81**), and 1′,2′-dihydroxy-4-methoxy-6-pentyl-2*H*-pyran-2-one (**82**), were isolated from solid rice cultures of *P. photiniae* (Figure 8.6).[41]

El-Neketi et al. investigated the endophyte *P. citrinum* isolated from the Moroccan plant *Ceratonia siliqua* by culturing it either on solid rice medium or on white beans.[43] Rice cultures of *P. citrinum* yielded mainly the isochromans (3*S*)-6-hydroxy-8-methoxy-3,5-dimethylisochroman and arohynapene D (**83**), in addition to quinolactacide derivatives, quinolactacide (**84**), tetrahydroquinolactacide, and methylpenicinoline (**85**), and their mixed quinolactacide-isochroman product, citriquinochroman (**86**), as well as 6-methylcurvulinic acid (Figure 8.7). When the fungus was grown on white beans, mainly tanzawaic acid derivatives (e.g., **87**, **88**; Figure 8.7), the alkaloid 8-methoxy-3,5-dimethylisoquinoline-6-ol, (3*S*)-4,6-dihydro-8-methoxy-3,5-dimethyl-6-oxo-3*H*-2-benzopyran, citrinamide A (**89**), vanillic acid, trichodermamide C, and indole acetic acid methyl ester were obtained. Citriquinochroman (**86**) exhibited pronounced cytotoxic activity against

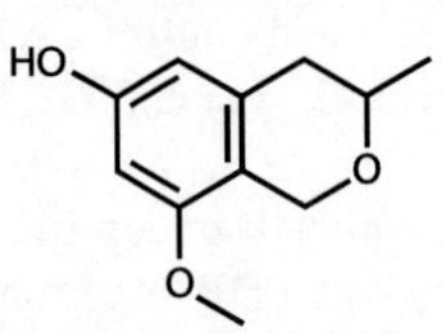
Arohynapene D (**83**)

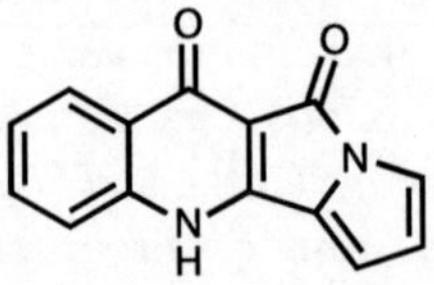
Quinolactacide (**84**)

Methylpenicinoline (**85**)

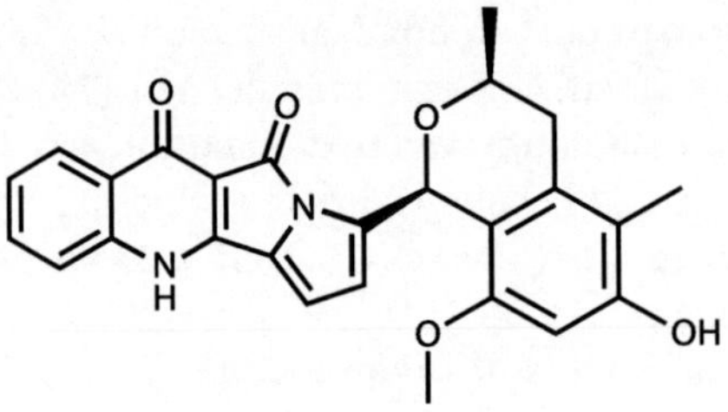
Citriquinochroman (**86**)

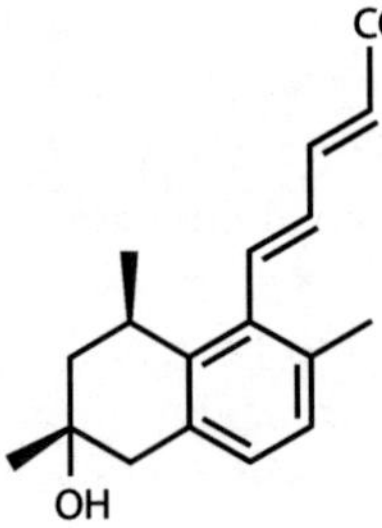
Tanzawaic acid D (**87**)

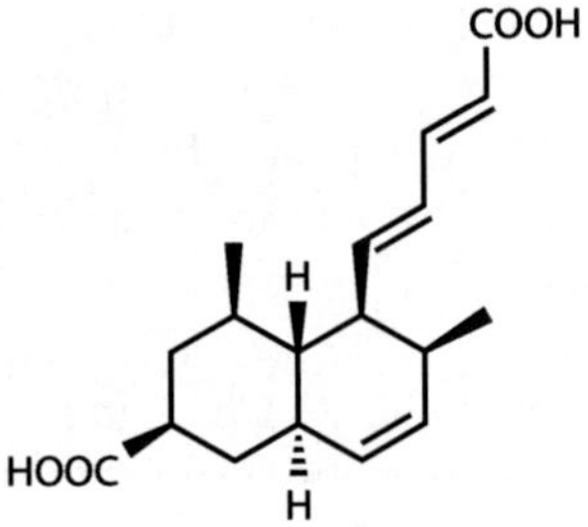
Tanzawaic acid F (**88**)

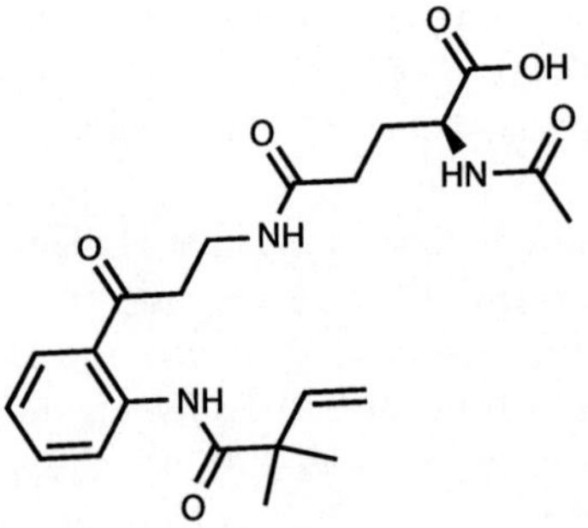
Citrinamide A (**89**)

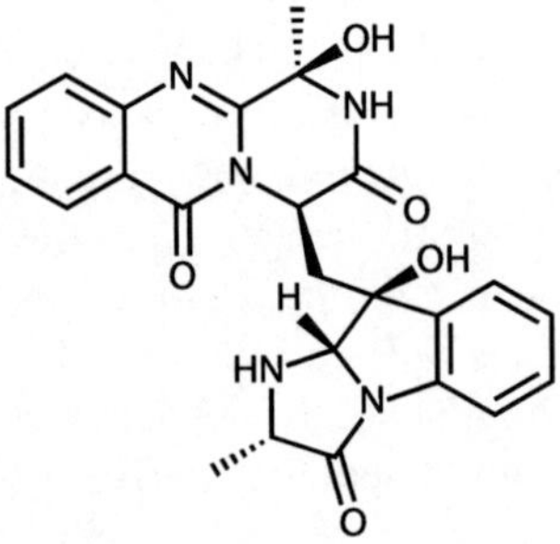
3-Hydroxyfumiquinazoline A (**90**)

FIGURE 8.7 Structures **83–90**. *(Continued)*

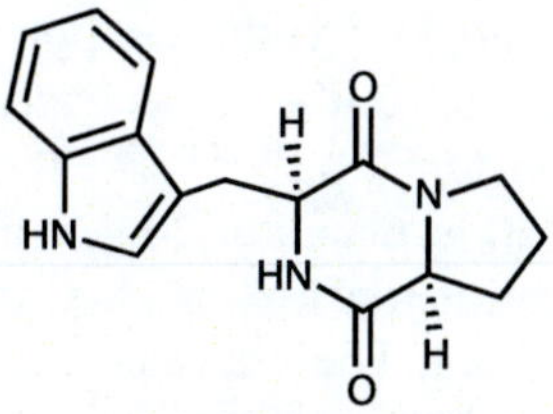

Brevianamide F (**91**)

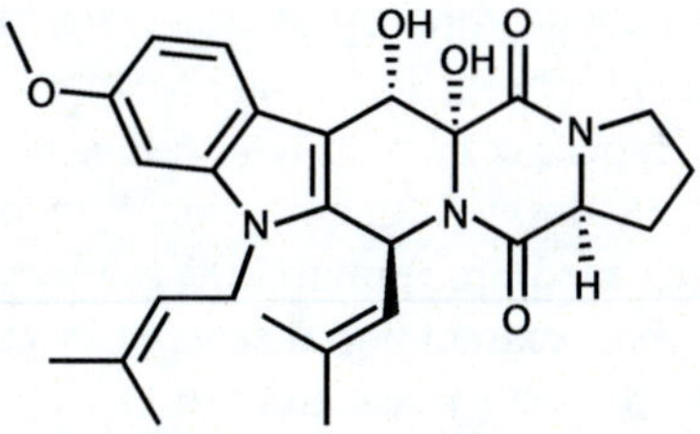

Fumitremorgin B (**92**)

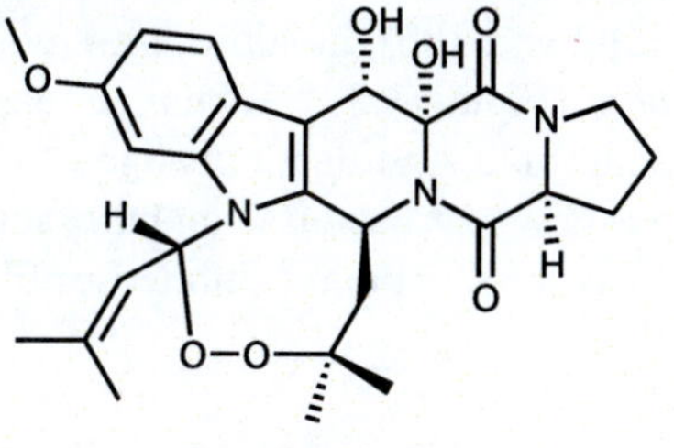

Verruculogen (**93**)

FIGURE 8.7 (*Continued*) Structures **91–93**.

the mouse lymphoma L5178Y cell line with an IC_{50} value of 6.1 μM; however, its monomers were inactive in the same cell line.[43]

The fungal genus *Aspergillus* is well known for its production of allelochemicals, which may have either beneficial (positive allelopathy) or detrimental (negative allelopathy) effects on target organisms.[44] In 2012, Gao and coworkers isolated the allelochemical-producing endophytic *Aspergillus fumigatus* strain LN-4 from the stem bark of *Melia azedarach*. When *A. fumigatus* LN-4 was cultured on SP solid medium (sucrose 20 g/L, peptone 30 g/L, KH_2PO_4 0.5 g/L, $MgSO_4$ 0.5 g/L, and agar 15 g/L and flask cultures were incubated at 28 ± 0.5°C for 21 days),[45] as well as on other culture media like PDB and rice, strikingly different HPLC profiles were obtained. SP solid medium yielded 39 fungal metabolites that possess antifungal, antifeedant, and toxic activities, including a new alkaloid, 3-hydroxyfumiquinazoline A (**90**).[45] PDB and rice media, however, yielded a series of indole diketopiperazines including (**91–93**) (Figure 8.7) known as potential plant growth inhibitors.[46] The tryprostatin-type compound brevianamide F (**91**) caused a stronger inhibition of the growth of lettuce seedlings than the positive control glyphosate. Compound **91** also inhibited the growth of seedlings of other plants such as *Raphanus sativus* or *Amaranthus mangostanus*, indicating that it could be possibly used as a natural eco-friendly herbicide in the future.

Wang et al.[47] investigated changes of the culture medium used for fermentation of the lichen-derived fungus *Ulocladium* sp., which was derived from an *Everniastrum* sp. The fungus was cultivated using either Czapek's liquid medium or under static conditions using PDB culture medium. When grown in Czapek's liquid medium, *Ulocladium* sp. afforded tricycloalternarenes including (**94–96**)[47] while, under static culture conditions on PDB medium, new ophiobolin derivatives including (**97**), in addition to known strong cytotoxic congeners including (**98**), were accumulated (Figure 8.8). Ophiobolin T (**97**) and 6-epi-ophiobolin G (**98**) showed potent cytotoxicity against HepG2 cells with IC_{50} values of 0.2 and 0.3 μM, respectively.[47]

The deep sea–derived fungal strain *Penicillium* sp. F23-2[48,49] was either grown in seawater-based medium (potato 200 g, glucose 20 g, mannitol 20 g, maltose 10 g, peptone 5 g, yeast extract 3 g, dissolved in 1:l seawater, pH 6.0 at 28°C for 45 days) or in agitated peptone-yeast-glucose (PYG) medium. The seawater-based medium yielded meleagrin alkaloids, diketopiperazines, and terpenoids including (**99–101**), whereas the PYG medium induced the accumulation of new nitrogen-containing sorbicillamines including (**102** and **103**), in addition to two known compounds, bisvertinolone and rezishanone C (Figure 8.8).[50] When the fungus *Penicillium* sp. F23-2 was grown on solid rice medium, a significant change of the metabolite profile was observed that included five new antibacterial ambuic acid analogs, penicyclones A–E (e.g., **104**, **105**; Figure 8.8), which exhibited antibacterial activity against *Staphylococcus aureus*.[51]

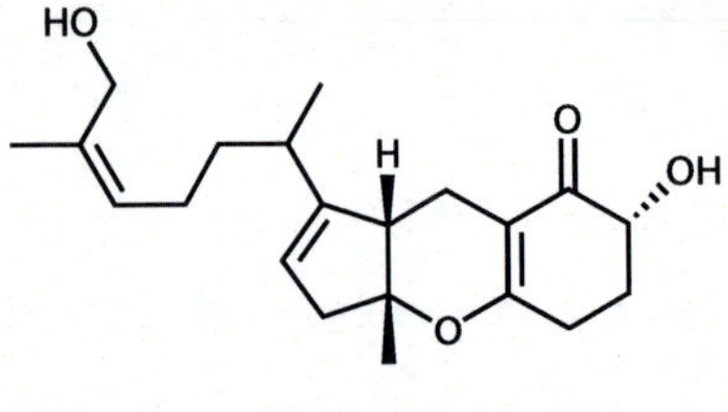

Tricycloalternarene G (**94**)

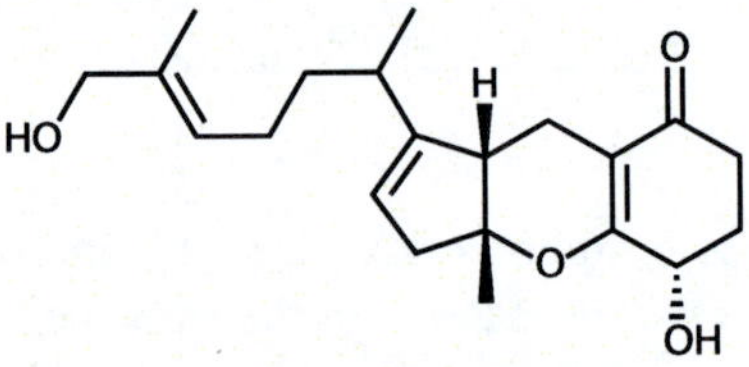

Tricycloalternarene 2a (**95**)

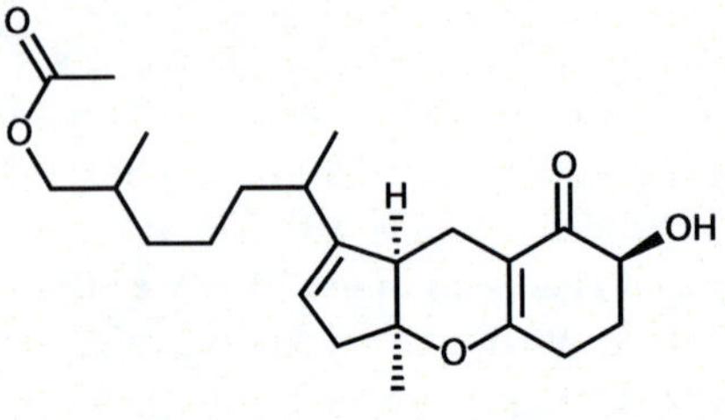

Tricycloalternarene D (**96**)

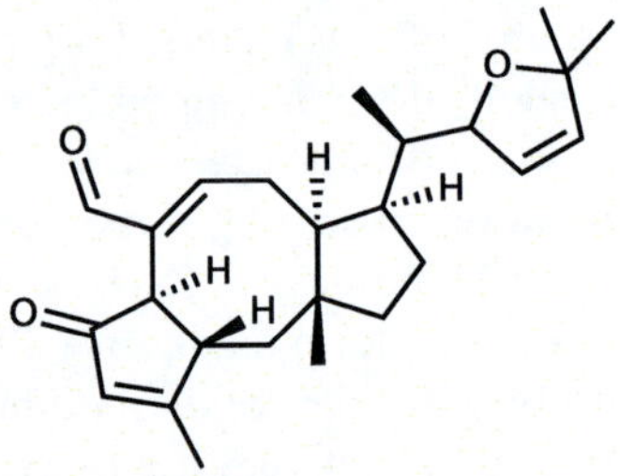

Ophiobolin T (**97**)

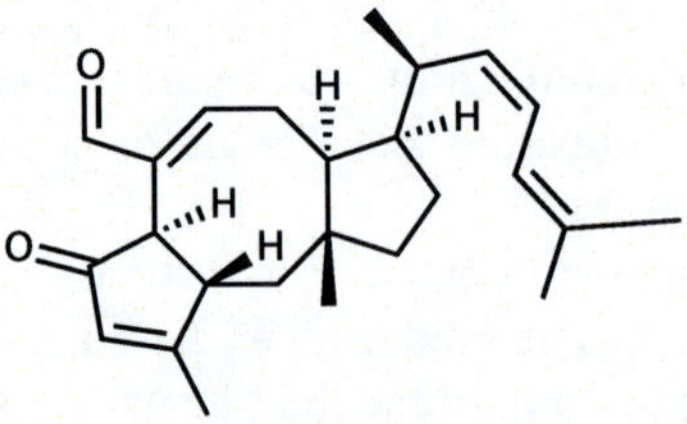

6-Epi-Ophiobolin G (**98**)

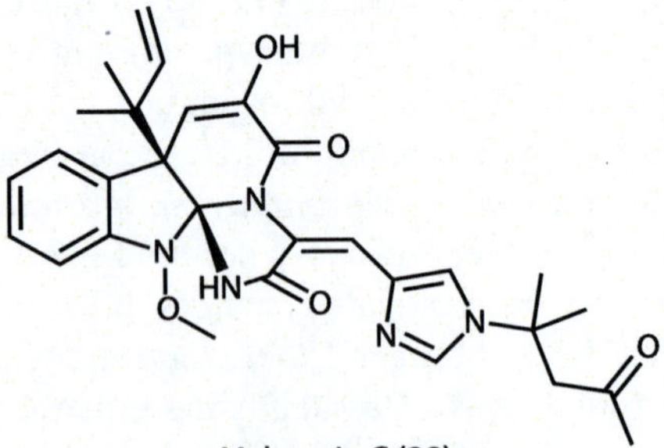

Meleagrin C (**99**)

FIGURE 8.8 Structures 94–99. *(Continued)*

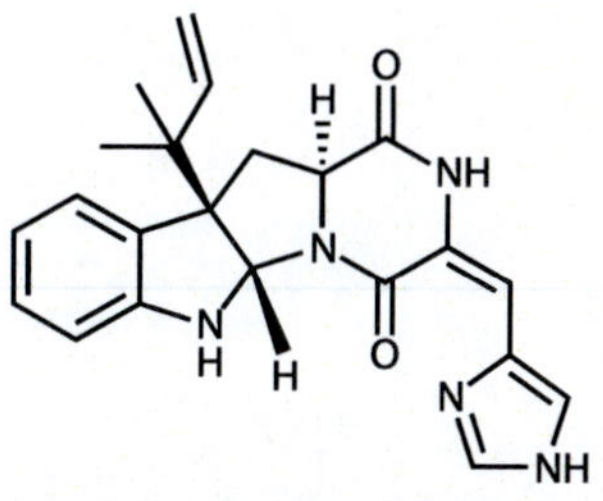

Roquefortine C (**100**)

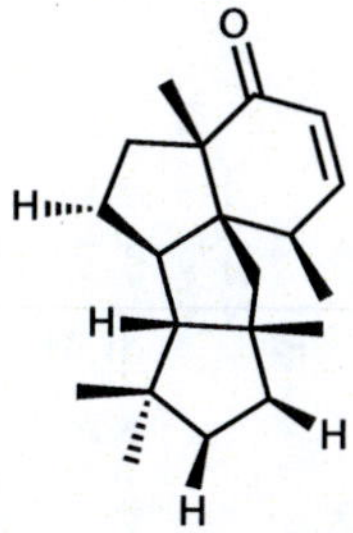

Conidiogenone B (**101**)

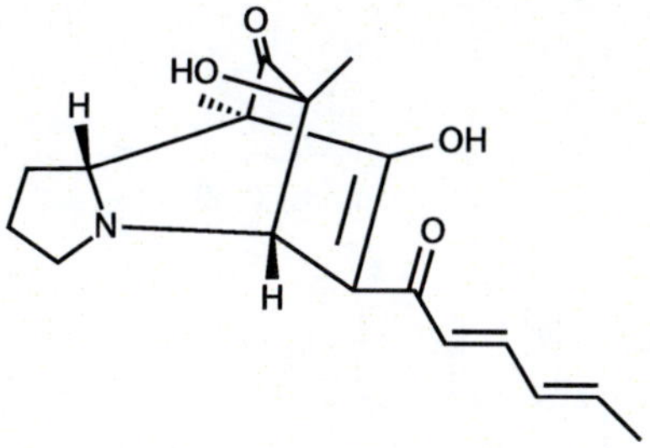

Sorbicillamine A (**102**)

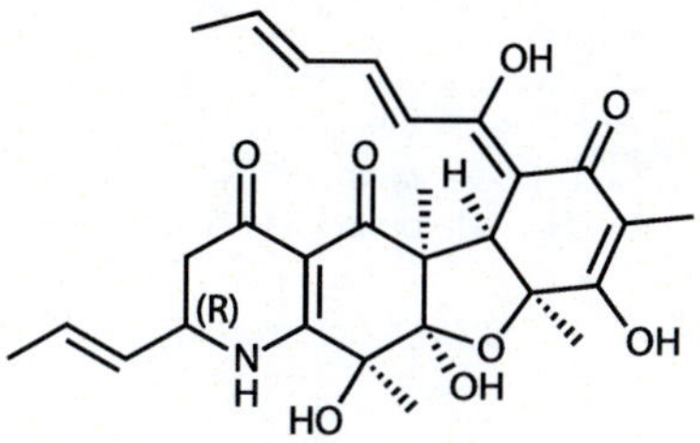

Sorbicillamine C (**103**)

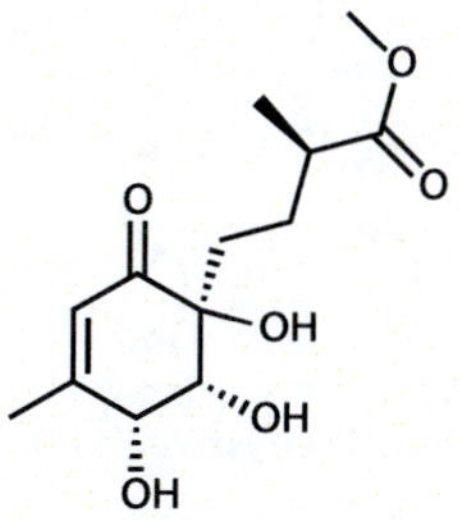

Penicyclone C (**104**)

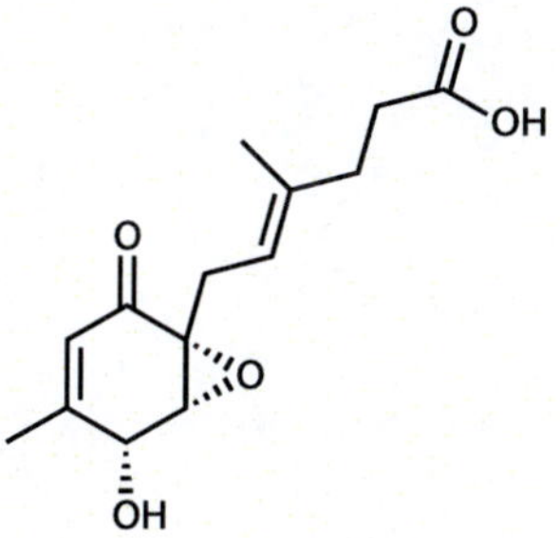

Penicyclone E (**105**)

FIGURE 8.8 (*Continued*) Structures **100–105**.

The endophyte *Talaromyces wortmannii* that was isolated from the Egyptian medicinal plant *Aloe vera* yielded strikingly different metabolite patterns when grown on solid rice medium or on white beans.[52–54] Two cyclic peptides talaromins A and B (**106** and **107**) were isolated from fungal cultures fermented on white beans,[53] whereas the same fungus grown on rice medium yielded a new atropisomer (**108**), three wortmannin derivatives including (**109**), several bisdihydroanthracenone atropisomers including the homodimeric flavomannin A, the previously unreported flavomannin B, two new unsymmetrical dimers, two new mixed dihydroanthracenone/anthraquinone dimers, as well as further known metabolites including (**110**) and (**111**) (Figure 8.9). Compounds (**108**), (**110**), and (**111**) exhibited considerable antibiotic activities against gram-positive human pathogenic bacteria including methicillin-resistant *S. aureus* (MRSA), with minimum inhibitory concentration (MIC) values ranging between 4 and 16 μg/mL.[54]

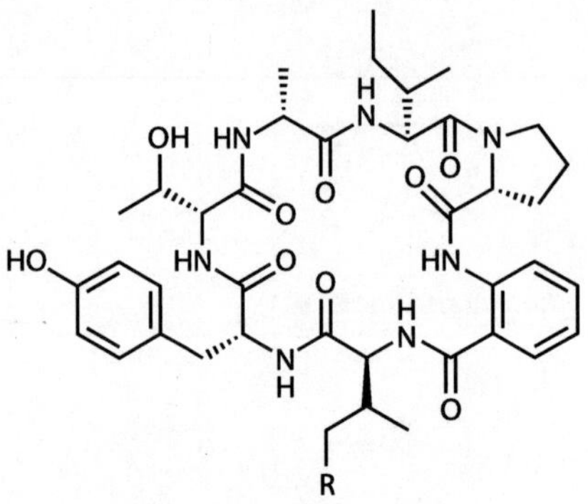

R=H : Talaromin A (**106**)
R=CH_3 : Talaromin B (**107**)

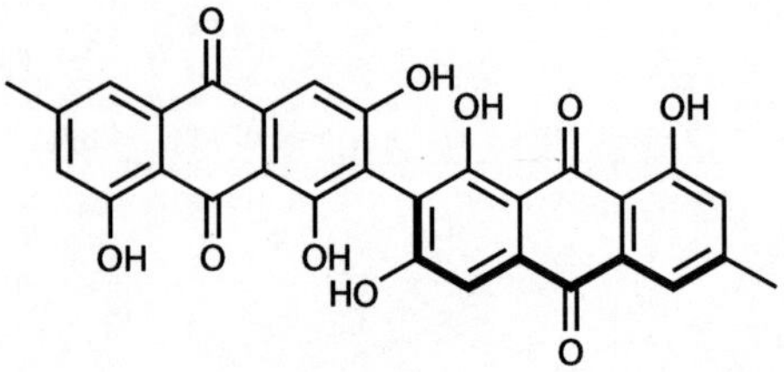

Biemodin (**108**)

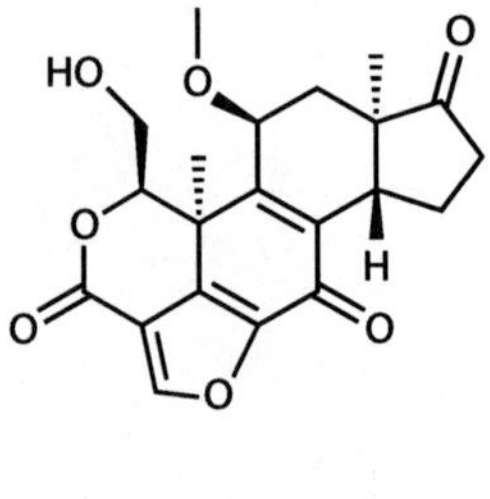

Wortmannin B (**109**)

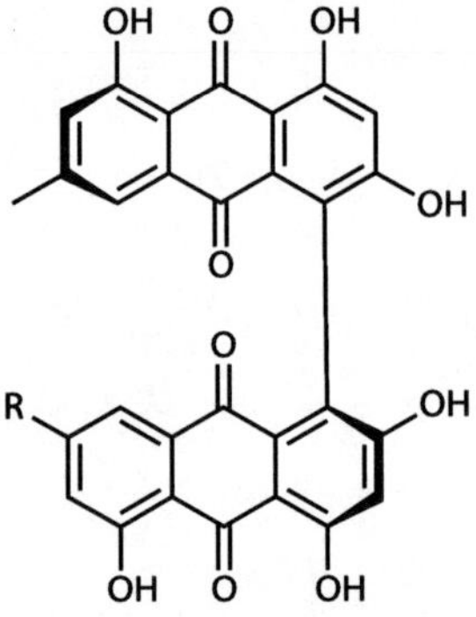

R=CH_3 : Skyrin (**110**)
R=CH_2OH : Oxyskyrin (**111**)

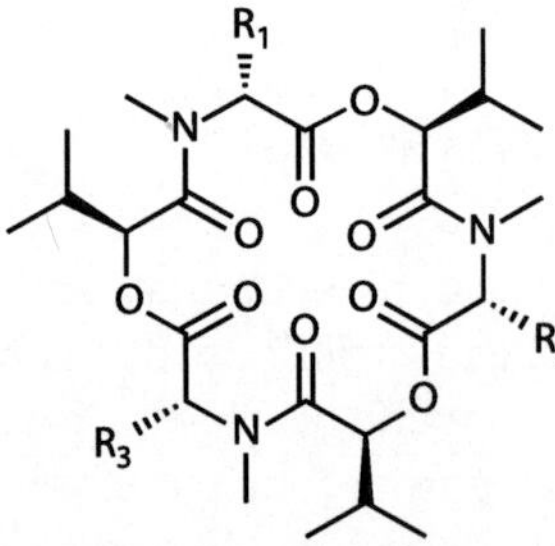

R_1=$CH(CH_3)CH_2CH_3$, R_2=$CH(CH_3)CH_2CH_3$, R_3=$CH(CH_3)CH_2CH_3$: Enniatin A (**112**)
R_1=$CH(CH_3)CH_2CH_3$, R_2=$CH(CH_3)CH_2CH_3$, R_3=$CH(CH_3)_2$: Enniatin A_1 (**113**)
R_1=$CH(CH_3)_2$, R_2=$CH(CH_3)_2$, R_3=$CH(CH_3)_2$: Enniatin B (**114**)
R_1=$CH(CH_3)_2$, R_2=$CH(CH_3)_2$, R_3=$CH(CH_3)CH_2CH_3$: Enniatin B_1 (**115**)

FIGURE 8.9 Structures **106–115**.

Enniatins are cyclic hexadepsipeptides that consist of three D-2-hydroxyisovaleric acid (D-Hiv) residues that are linked alternatively to *N*-methyl-L-amino acid residues (*N*-methyl-L-valine, *N*-methyl-L-isoleucine, and *N*-methyl-L-leucine).[55] Due to their ionophoric properties, enniatins exhibit insecticidal, antimycobacterial, and anticancer activities, in addition to the inhibition of various enzymes, for example, acyl-CoA-cholesterol-acyl transferase and cyclic nucleotide phosphodiesterase.[56] Wang et al. investigated the fungal endophyte *F. tricinctum* by culturing it on seven solid media in addition to one liquid culture medium with or without addition of peptone or of amino acids, with the aim of maximizing enniatin production.[57] A fermentation time of 18 days was found to give the highest enniatin yield on white beans (*Phaseolus vulgaris*, solid medium). This culture medium proved to be the optimum fermentation medium to induce enniatins A, A_1, B, and B_1 (**112–115**; Figure 8.9), which in total reached yields of about 1.3 g in 1 L culture.[57]

A similar sensitivity to culture medium alteration was observed for the deep sea–derived fungus *Penicillium paneum* SD-44.[58] This fungus was originally cultivated on solid rice medium for 30 days at room temperature to yield triazole alkaloids including (**116**) and dihydroimidazole ones including (**117**) (Figure 8.10).[59] Li et al. cultured the fungus in a bioreactor in liquid medium (consisting of mannitol 20 g, maltose 20 g, monosodium glutamate 10 g, glucose 10 g, yeast extract 3 g, corn steep liquor 1 g, KH_2PO_4 0.5 g, and $MgSO_4{\cdot}7H_2O$ 0.3 g, in 1 L filtered seawater at 28°C and 150 rpm/min).[58] This modification in culture conditions resulted in the isolation of five new anthranilic acid derivatives, penipacids A–E (e.g., **118**, **119**; Figure 8.10), and one known analog, 2-[(1-methyl-2-oxopropylidene)amino]-benzoic acid. Penipacids A and E (**118**) and (**119**) inhibited the activity of the human colon cancer RKO cell line with IC_{50} values of 8.4 and 9.7 μM, respectively, while the compound 2-[(1-methyl-2-oxopropylidene)amino]-benzoic acid was cytotoxic to the Hela cell line with an IC_{50} value of 6.6 μM.

Zhou et al.[60] and Gao et al.[61] investigated the marine mud–derived fungus *Aspergillus versicolor* strain ZLN-60 by culturing it under static conditions in 130 Erlenmeyer flasks of 1 L containing a liquid medium composed of mannitol (20 g/L), maltose (20 g/L), glucose (10 g/L), monosodium glutamate (10 g/L), KH_2PO4 (0.5 g/L), $MgSO_4 \cdot 7H_2O$ (0.3 g/L), yeast extract (3 g/L), corn steep liquor (1 g/L), and seawater after adjusting its pH to 6.5.[60,61] Anthranilic acid (ATA)-containing pentapeptides, versicotides A (**120**) and B, and prenylated diphenyl ethers (diorcinols B–E) (e.g., **121**) were isolated from this culture (Figure 8.10). This chemo-diversity encouraged Peng et al. to culture the same fungal strain on solid rice medium, which resulted in the isolation of five new cyclic peptides, including psychrophilin E (**122**) and versicotide C (**123**) (Figure 8.10).[62]

Li and coworkers in 2014 investigated the influence of eutrophic (nutrient-rich) and oligotrophic (nutrient-poor) culture media, and the impact of the addition of Mg^{+2}, on the metabolic profile of marine-derived fungus *Ascotricha* sp. ZJ-M-5. *Ascotricha* sp. ZJ-M-5 produces cyclonerodiol sesquiterpenoid derivatives including cyclonerodiol (**124**; Figure 8.10) when fermented in a medium containing complex nutrients, including yeast extract, peptone, and corn syrup, with a yield of cyclonerodiol (**124**) higher than 23 mg/L. The impact of Mg^{2+} ($MgCl_2$ instead of $MgSO_4$) on the secondary metabolite production of this fungus was investigated using modified Czapek

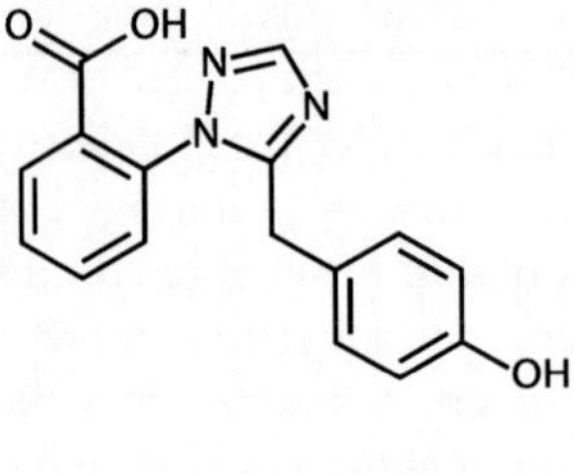

Penipanoid A (**116**)

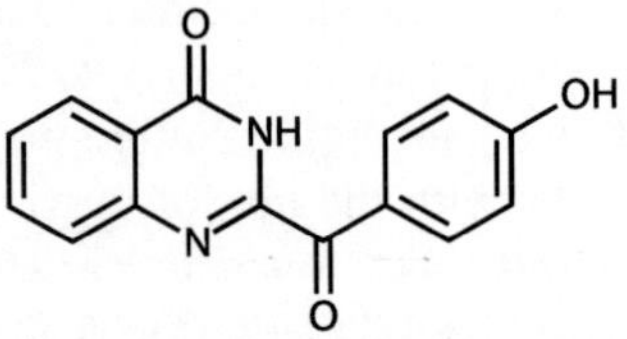

Penipanoid C (**117**)

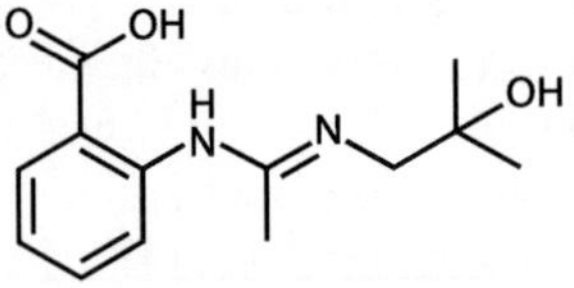

Penipacid A (**118**)

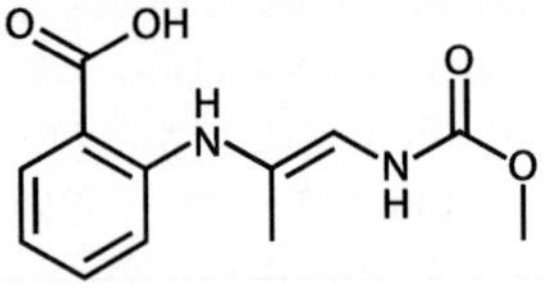

Penipacid E (**119**)

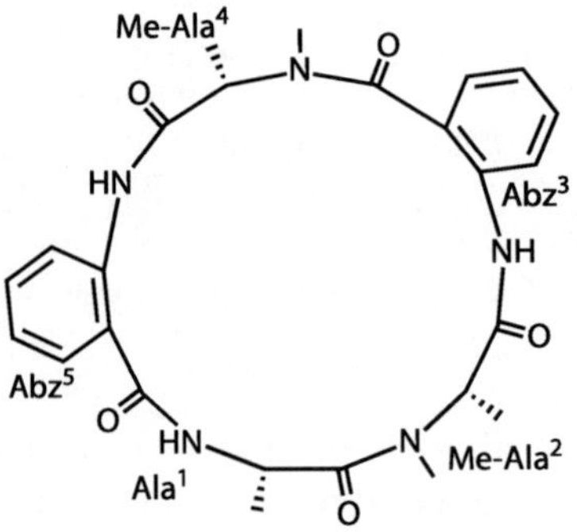

Versicotide A (**120**)

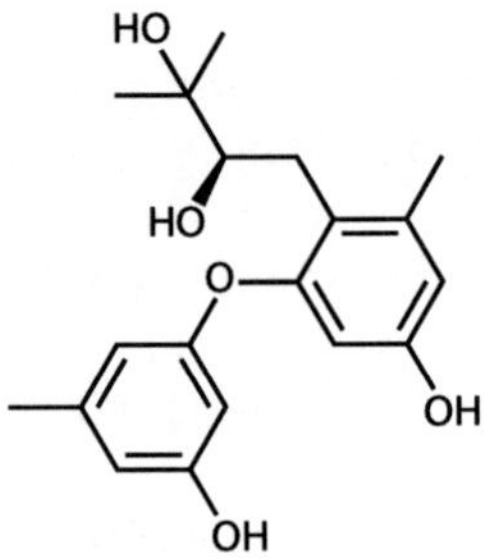

Diorcinol B (**121**)

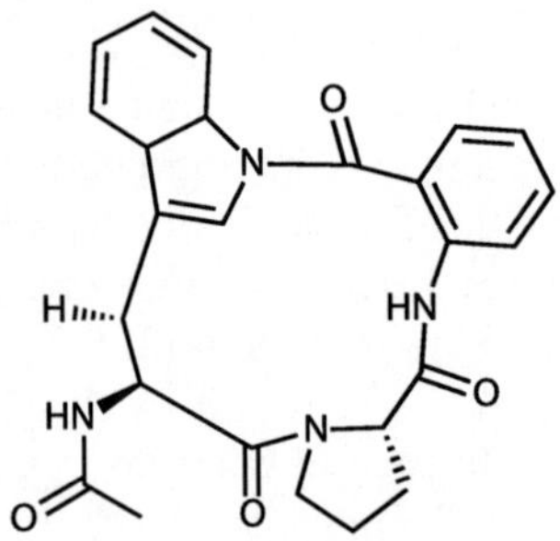

Psychrophilin E (**122**)

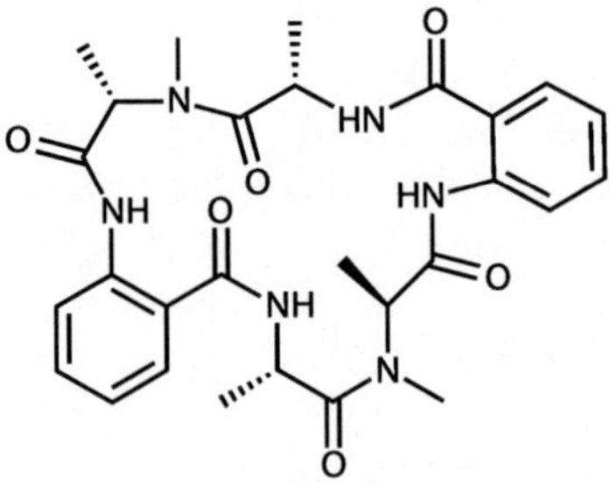

Versicotide C (**123**)

FIGURE 8.10 Structures **116–123**. *(Continued)*

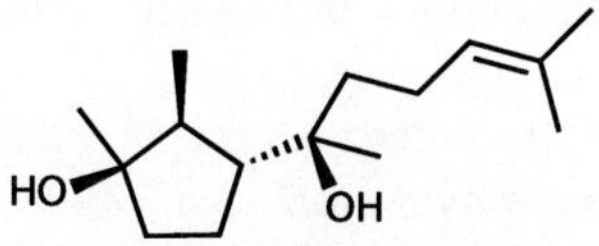

Cyclonerodiol (**124**)

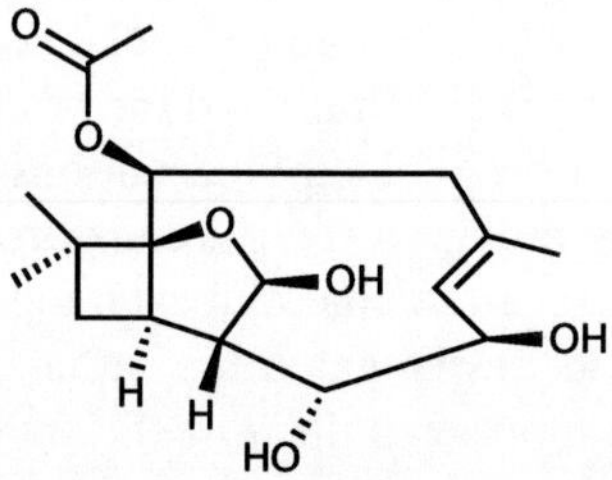

(+)-6-*O*-Demethylpestaltiopsin A (**125**)

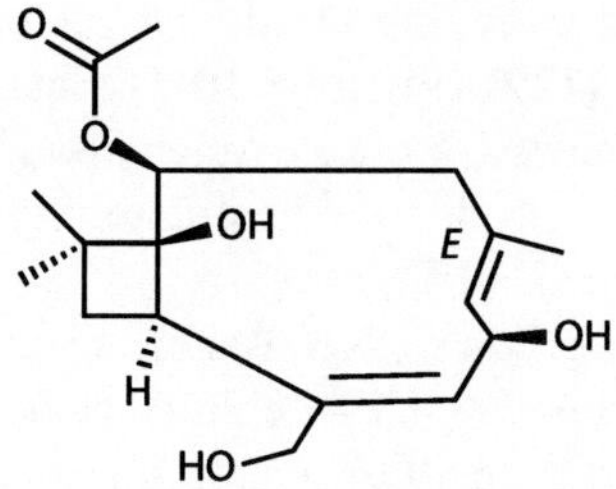

(–)-6-*O*-Demethylpestaltiopsin B (**126**)

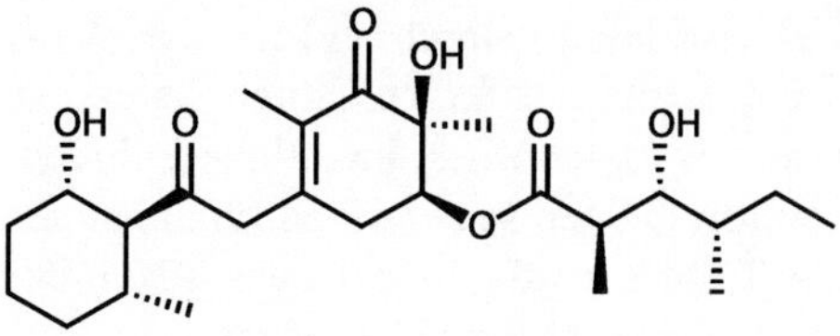

Dothideomycetone A (**127**)

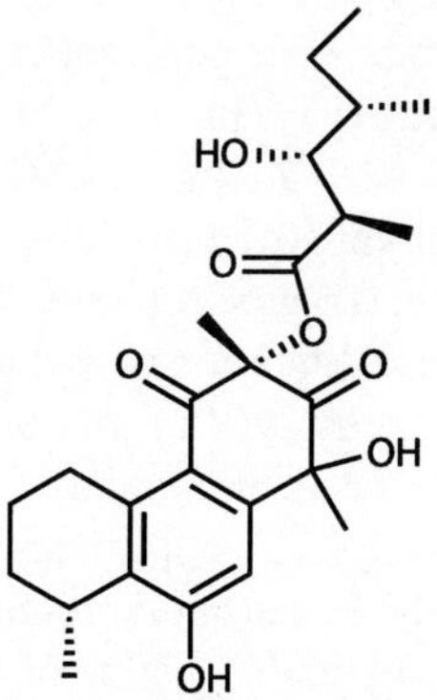

Dothideomycetide A (**128**)

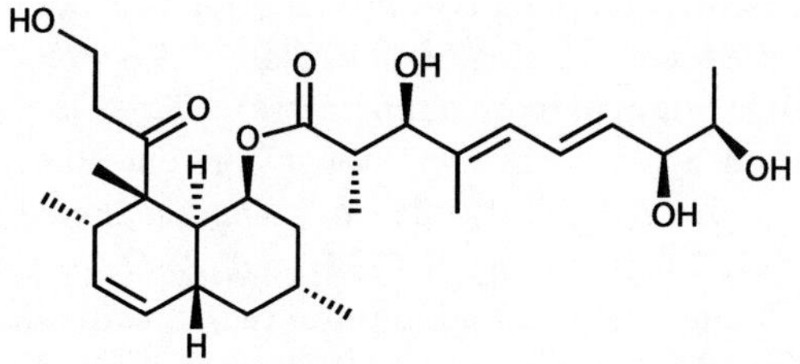

Calbistrin F (**129**)

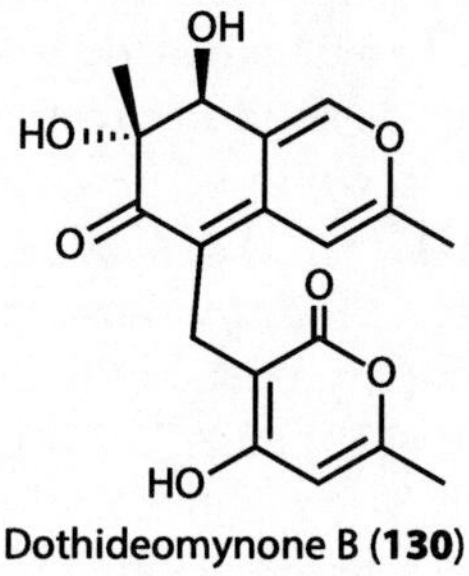

Dothideomynone B (**130**)

FIGURE 8.10 (*Continued*) Structures **124–130**.

Dox broth composed of sucrose and $NaNO_3$ as sole sources of carbon and nitrogen, respectively (oligotrophic medium). The presence of Mg^{2+} caused a weak induction of the biosynthesis of cyclonerodiol (**124**) and (−)-6-*O*-demethylpestalotiopsin B (**126**), while it markedly inhibited the production of (+)-6-*O*-demethylpestalotiopsin A (**125**) and (+)-6-*O*-demethylpestaltiopsin C (Figure 8.10). Higher concentrations of $MgCl_2$ in the modified Czapek Dox broth inhibited the production of compounds **125**, **126**, and (+)-6-*O*-demethylpestaltiopsin C.[63]

Hewage et al. investigated the influence of the potato source in PDB medium with regard to the metabolism of the fungal endophyte *Dothideomycete* sp. CRI7, which was isolated from roots of the Thai medicinal plant *Tiliacora triandra*.[64] Previous fermentation of the same fungal strain in PDB medium that had been prepared from fresh potato tubers yielded compounds including dothideomycetone A (**127**), dothideomycetide A (**128**), and calbistrin F (**129**) (Figure 8.10).[65] Changing the source of potato starch and the use of two commercial malt extracts, one from Spain and the other from the United States, caused the production of structurally very different new metabolites, including dothideomynone B (**130**).

The prolific deep sea–derived fungal strain *Aspergillus* sp. XS-20090B15 proved likewise to be sensitive to changes of the culture media.[66] Upon culturing the fungus for 5 weeks on potato glucose L-methionine-free medium under static conditions, the lumazine-type peptides penilumamide (**131**) and penilumamide D, together with the cyclic pentapeptide asperpeptide A were obtained, while cultivation in a L-methionine-based culture medium yielded the new compounds penilumamides B (**132**) and C (**133**) in addition to the known penilumamide (**131**) (Figure 8.11).

A further study was conducted by Chen et al. in 2014 with an aim to obtain biologically active compounds from the same productive fungal strain by cultivating it on solid rice medium (12 g of natural sea salt, 100 g of rice, 0.6 g of peptone, 100 mL of H_2O for 35 days at room temperature).[67] This OSMAC modification of the medium gave rise to a fungal extract showing potent anti-RSV activity. Chromatographic analysis of this extract afforded two unprecedented prenylated alkaloids, 22-*O*-(*N*-Me-L-valyl) aflaquinolone B (**134**) and 22-*O*-(*N*-Me-L-valyl)-21-epi-aflaquinolone B (**135**) (Figure 8.11), in addition to two known alkaloids. In particular, compound **135** exhibited the strongest anti-RSV activity as indicated by its IC_{50} value of 42 nM.

Two new 4-methyl-progesteroids, nodulisporisteroid A (**136**) and nodulisporisteroid B, were obtained from the endolichenic fungal strain *Nodulisporium* sp. (No. 65-17-2-1), along with two related metabolites, demethoxyviridin and inoterpene B, following cultivation of the fungus on solid rice medium.[68] Cultivation of the fungus on PDB medium resulted in the isolation of new cytotoxic 4-methyl-progesteroid derivatives nodulisporisteroids C–L (e.g., **137**, **138**; Figure 8.11).[69]

The deep sea–derived fungus *Cladosporium sphaerospermum* 2005-01-E3[70] when cultured on solid rice medium produced the known cladosins A–E, whereas two new polyketides, cladosins F (**139**) and G (**140**) (Figure 8.11), both exhibiting a rare 6(3)-enamino-8,10-dihydroxy-tetraketide core structure, were obtained when the fungus was cultured on soybean-based medium instead.

The Palauan marine-derived fungus *Trichoderma* sp. TPU199 (cf. T. *brevicompactum*) was found to produce two known epidithiodiketopiperazines, pretrichodermamide A and DC1149B (**141**), when grown in bacto-malt extract

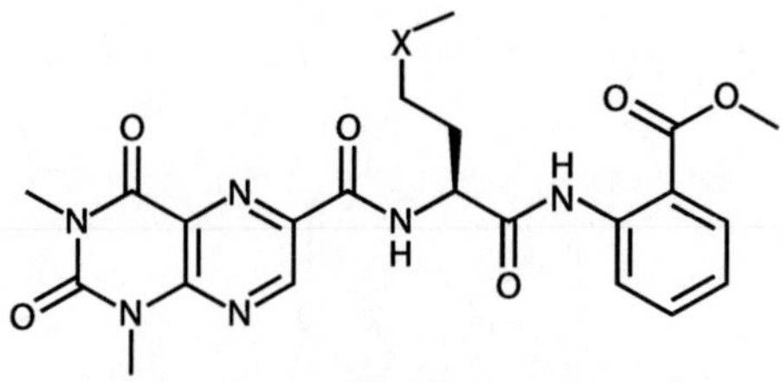

X = S = O : Penilumamide (**131**)
X = S : Penilumamide B (**132**)
X = S(=O)$_2$: Penilumamide C (**133**)

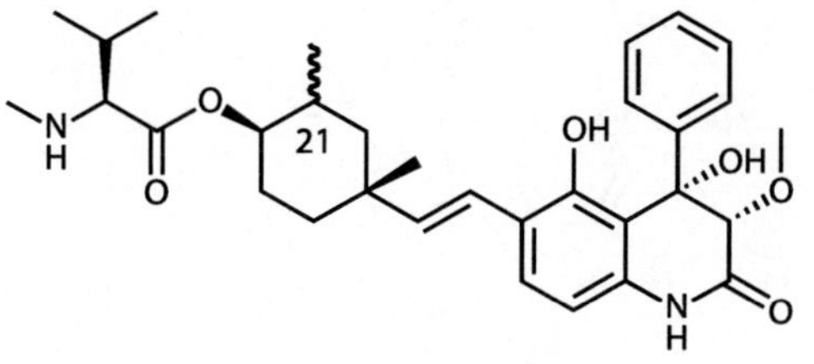

21*S*: 22-*O*-(*N*-Me-L-valyl) aflaquinolone B (**134**)
21*R*: 2-*O*-(*N*- Me-L-valyl)-21-epi-aflaquinolone B (**135**)

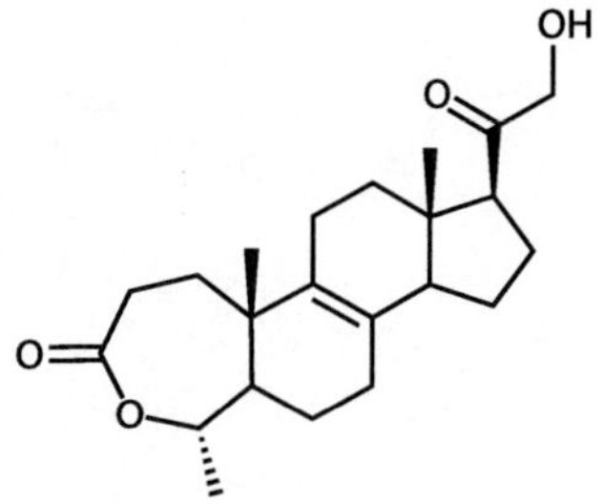

Nodulisporisteriod A (**136**)

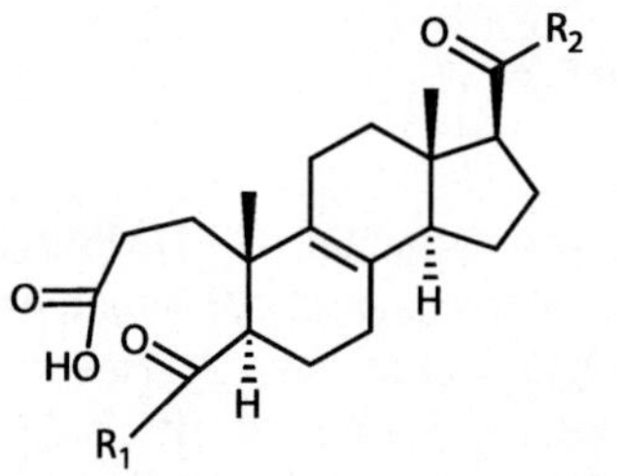

R$_1$ = OH, R$_2$ = Me : Nodulisporisteroid J (**137**)
R$_1$ = Me R$_2$ = OH : Nodulisporisteroid K (**138**)

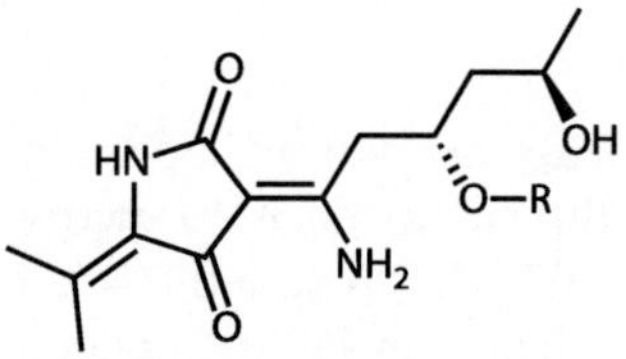

R = H : Cladosin F (**139**)
R = Me : Cladosin G (**140**)

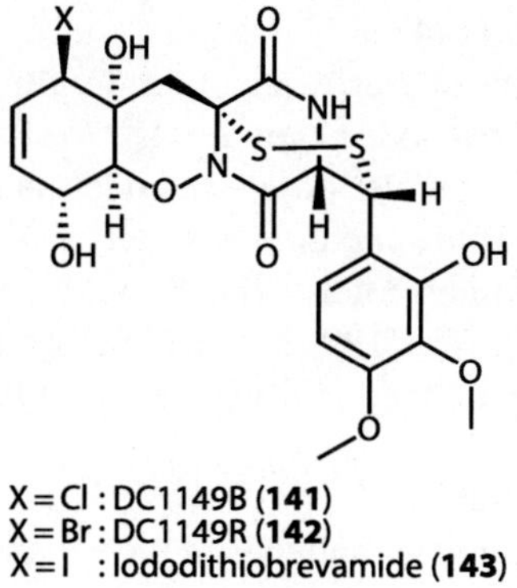

X = Cl : DC1149B (**141**)
X = Br : DC1149R (**142**)
X = I : Iododithiobrevamide (**143**)

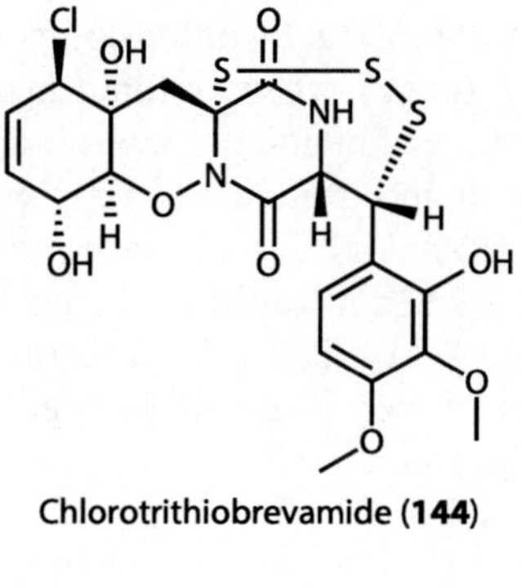

Chlorotrithiobrevamide (**144**)

FIGURE 8.11 Structures **131–144**. (*Continued*)

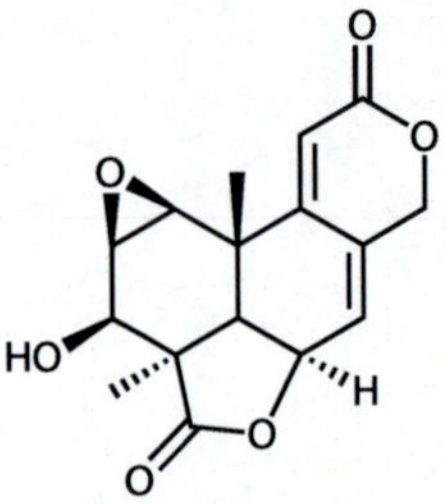

WA (**145**)

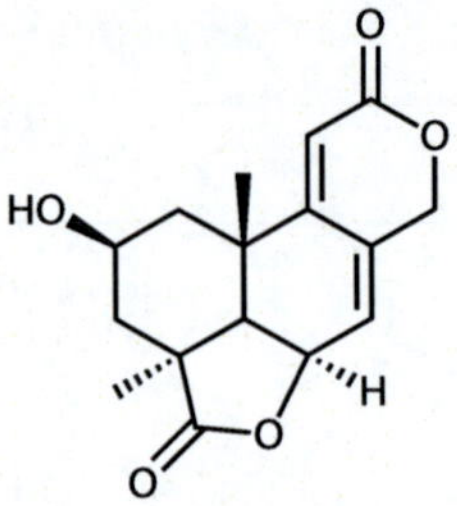

WB (**146**)

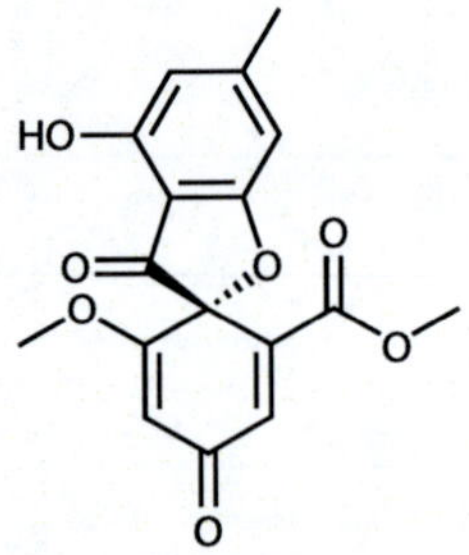

(+)-Bisdechlorogeodin (**147**)

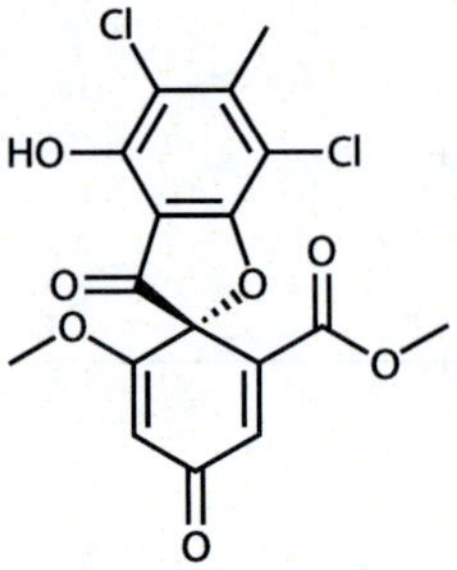

(+)-Geodin (**148**)

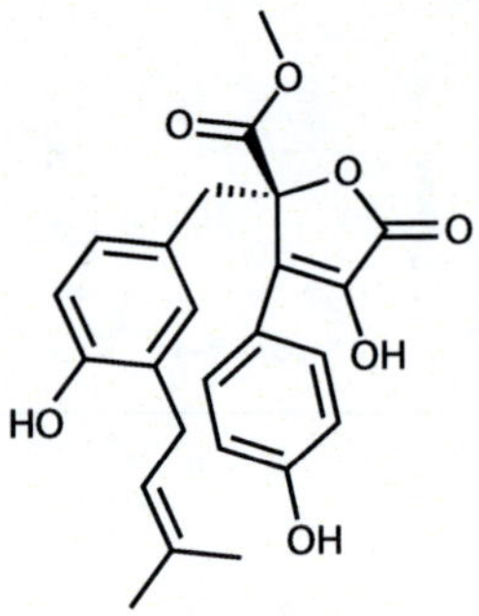

Butyrolactone I (**149**)

FIGURE 8.11 ***(Continued)*** Structures **145–149**.

broth (MEB) medium.[71] Namikoshi and coworkers added 1% DMSO to the culture medium that contained 3.0% sucrose, 3.0% soluble starch, 1.0% malt extract, 0.30% Ebios, 0.50% KH_2PO_4, and 0.050% $MgSO_4 \cdot 7H_2O$ in natural seawater adjusted to pH 6.0, which induced the production of the unprecedented trithio-derivative (**144**; Figure 8.11).[72] Compound **144** was only detected when the fungus was grown in seawater medium, and not in freshwater medium. Apparently the combination of DMSO and NaCl in seawater activated cryptic biosynthetic fungal genes to induce the accumulation of the trisulfide compound (**144**).

The same fungus when cultured in agitated freshwater liquid culture medium was found to accumulate gliovirin and trichodermamide A.[73] Seawater-based medium was found to induce mainly the accumulation of trichodermamide B and DC1149B (**141**). Addition of NaI to the culture medium resulted in the isolation of DC1149R (**142**), in addition to a new iodinated derivative, iododithiobrevamide (**143**) (Figure 8.11). Thus, *Trichoderma* sp. TPU199 is a remarkable fungus that produces rare types of epidithiodiketopiperazines and efficiently utilizes halide ions in the medium.

Xu et al. isolated the deep sea sediment–derived fungus *Aspergillus dimorphicus*, which produces low amounts of potent antitumor compounds of the wentilactone type (**145** and **146**; Figure 8.11) in modified PDB liquid medium under static conditions.[74] Various culture conditions, including different degrees of salinity, pH, temperature, and culture times, were explored by the authors with regard to their influence on the induction of wentilactone derivatives. Optimized fermentation conditions included static cultivation of the fungus in PDB with a salinity between 17.5% and 35.0%, pH 6–8, temperature regimes of 23°C–25°C, and fermentation times ranging from 25 to 30 days. Small-molecule elicitors such as 3% methanol were found to further induce the production of wentilactones. These optimized conditions were found to induce WA (**145**) and WB (**146**) to concentrations of 13.4 and 6.5 mg/L, respectively.

Boruta et al. studied the highly productive fungus *Aspergillus terreus* ATCC 20542, which accumulated several bioactive secondary metabolites, including (**147–149**) albeit in small amounts (Figure 8.11).[75] Several parameters were explored in order to increase the yield of compounds. These included the addition of rapeseed oil or inulin, different nitrogen sources, a reduction of the chlorine content, different salinity, and different aeration. Addition of rapeseed oil to chlorine-deficient medium and high aeration proved to be important to maximize the yield of compounds. Under these conditions, the highest levels of (+)-bisdechlorogeodin (**147**), (+)-geodin (**148**), asterric acid, butyrolactone I (**149**), mevinolinic acid, and terrein were obtained, which were present only in trace amounts in the original culture lacking rapeseed oil.

The endolichenic fungus *Myxotrichum* sp. afforded a unique series of citromycetin and fulvic acid derivatives including (**150–152**; Figure 8.12) when cultured in PDB medium under shaking.[76] When this fungal strain was cultured on solid rice medium instead, the resulting fungal extract showed a very different HPLC. The new myxotritones A–C (e.g., **152**; Figure 8.12), along with the known 7,8-dihydro-7*R*,8*S*-dihydroxy-3,7-dimethyl-2-benzopyran-6-one, were obtained from solid rice cultures of the fungus. The 7,8-dihydro-7*R*,8*S*-dihydroxy-3,7-dimethyl-2-benzopyran-6-one potently inhibited root elongation of *Arabidopsis thaliana*, which confirmed that this fungus exerts a protective function for the harboring lichen against competing higher plants.[77]

The fungal endophyte *Aspergillus flavus* SNFSt, isolated from *Solanum nigrum*,[78] was found to produce trace amounts of the cytotoxic alkaloid solamargine (**153**;

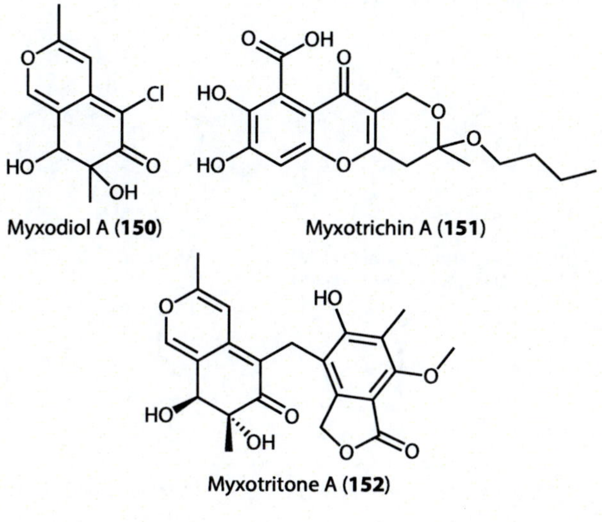

FIGURE 8.12 Structures **150–152**. *(Continued)*

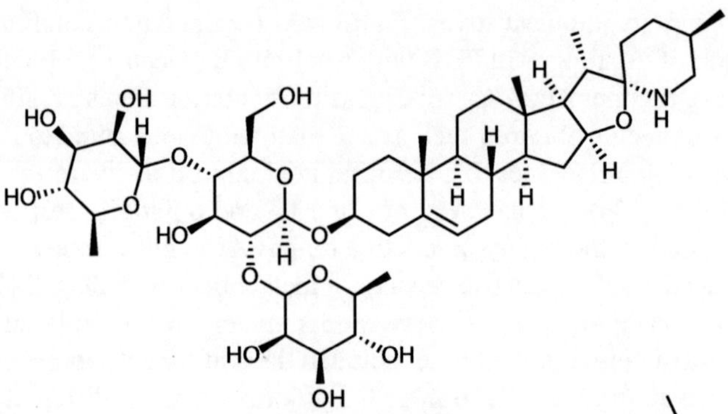

Solamargine (**153**)

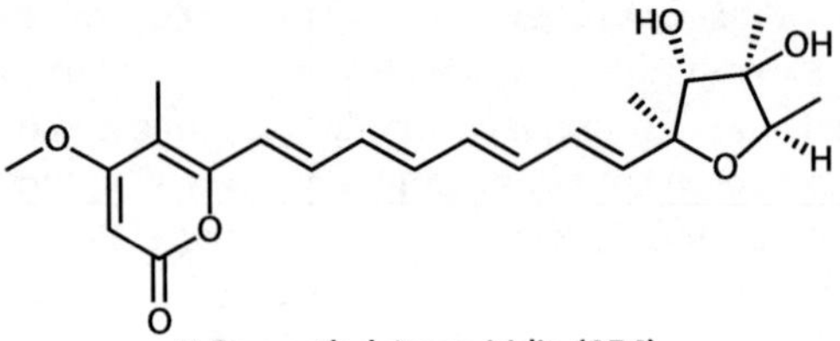

7-Desmethylcitreoviridin (**154**)

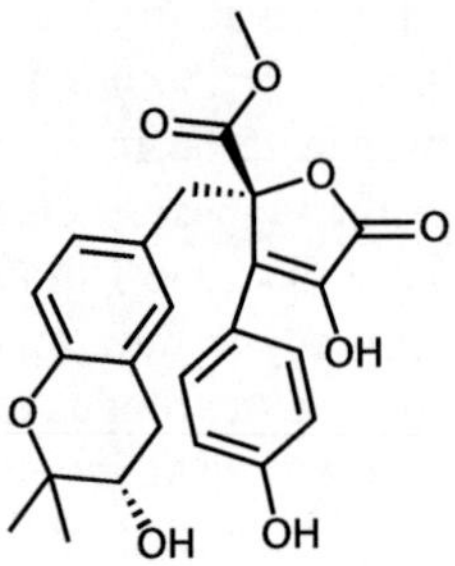

Butyrolactone V (**155**)

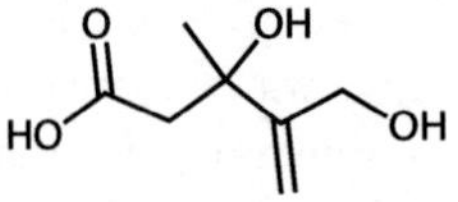

Mevalocidin (**157**)

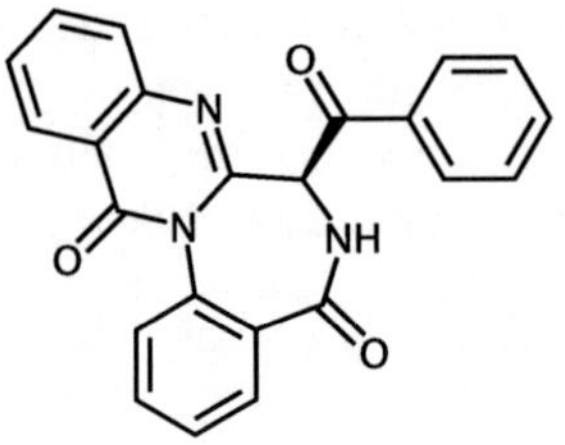

Novobenzomalvin C (**156**)

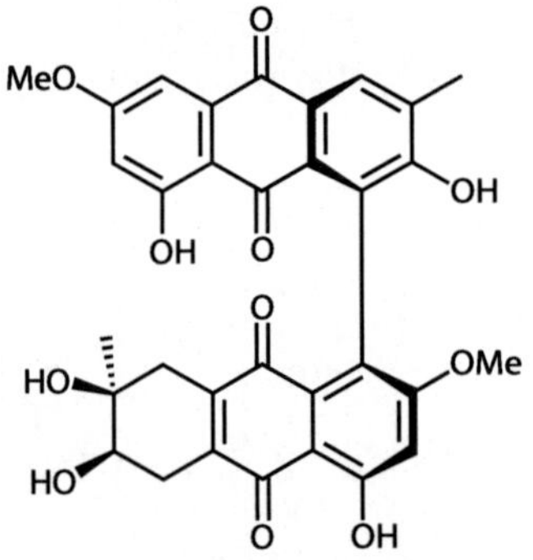

Alterporriol X (**160**)

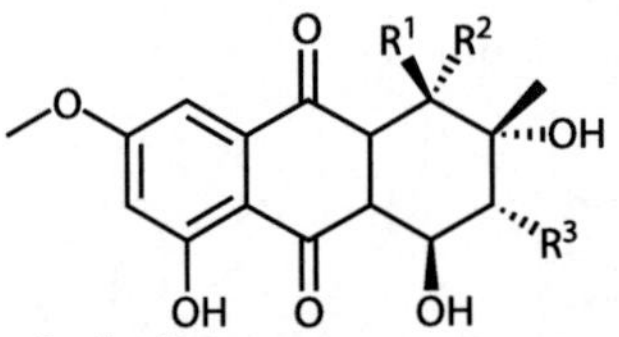

R_1	R_2	R_3	
OH	H	OAc	Altersolanol N (**158**)
OH	H	OH	Altersolanol A (**159**)

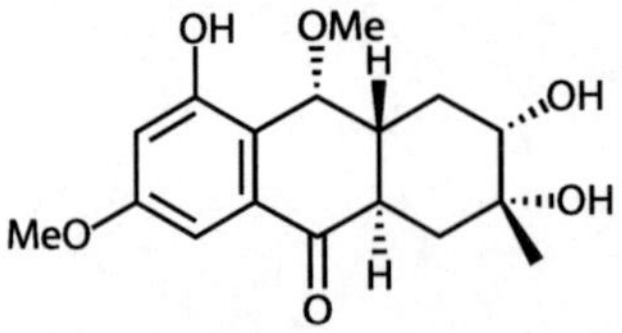

10-Methylaltersolanol Q (**161**)

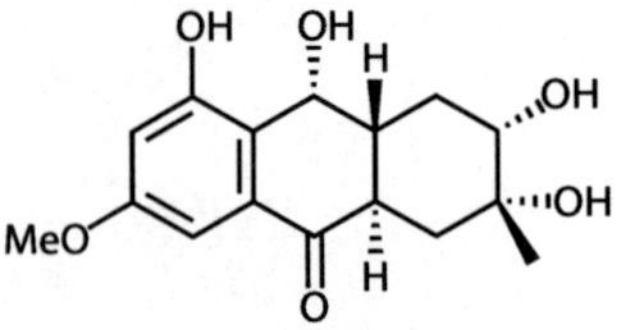

Altersolanol Q (**162**)

FIGURE 8.12 (*Continued*) Structures **153–162**.

Figure 8.12), a common metabolite of its host plant. Repeated subculturing of the fungus was found to increase the yield of this compound, resulting in the production of 250–300 µg/L solamargine till the eighth generation. At this concentration, the fungal endophyte can be considered as an alternative source for production of this compound.[78]

A comparative metabolomic OSMAC study was conducted with the marine-derived fungus *A. terreus*, which was grown under various culture conditions such as on malt agar, malt extract, glycerol, oatmeal, trace element solution, barley-spelt solid, or PDB media.[79] Malt agar and barley-spelt solid media proved to be the best. Multivariate analysis of liquid chromatography/mass spectrometry (LC/MS) data obtained from organic extracts indicated the presence of the new cytotoxic compound 7-desmethylcitreoviridin (**154**), which is an analog of the known compound citreoviridin, as well as of other known fungal metabolites, including butyrolactone V (**155**) and novobenzomalvin C (**156**) (Figure 8.12).

The fungal secondary metabolite mevalocidin (**157**; Figure 8.12) is a well-known herbicidal metabolite that is produced by the fungus *Coniolariella* sp. when cultured on malt extract agar (MEA).[80] Mevalocidin is easily absorbed by plants. Moreover, it is transported via the phloem and xylem mobile, thus easing distribution throughout the meristem.[81] Due to the high demand for new herbicidal metabolites as weed-controlling agents for organic farming, Oberlies and coworkers attempted to maximize the yield of this highly active metabolite and to possibly detect other related derivatives.[80] Several growth conditions, including culture time, temperature, and a variation of the composition of media, were studied for their effects on two mevalocidin-producing patented fungal strains, *Coniolariella* sp.-MSX56446 and -MSX92917, respectively. The highest yield of **157** was detected when the fungus was cultured on solid rice medium. Moreover, addition of agar to the medium was likewise found to enhance the yield of **157** compared to a liquid medium lacking agar. These optimized conditions were able to induce **157** with up to 2.2 mg per each gram of rice medium. The optimum culturing time was 20 days, whereas the optimum temperature was between 30°C and 35°C. *n*-Butanol was found to be the best extraction solvent for obtaining gram quantities of **157**.

The endophytic fungus *Stemphylium globuliferum* is well known for its production of anthraquinone or tetrahydroanthraquinone derivatives including monomers or dimers when cultured on solid rice medium.[82–84] Moussa et al.[85] attempted to obtain further new derivatives from the same fungus by culturing it on white beans. Whereas on solid rice medium, the fungus was shown to produce known anthraquinones and tetrahydroanthraquinones derivatives including altersolanol N (**158**) and altersolanol A (**159**), cultivation on white beans resulted in the production of new derivatives, including alterporriol X (**160**) and 10-methylaltersolanol Q (**161**), in addition to altersolanol Q (**162**) (Figure 8.12).[85]

8.3 INDUCTION OF NATURAL PRODUCTS THROUGH COCULTIVATION

In 2001, Cueto et al. reported the induction of a new chlorinated benzophenone antibiotic, namely pestalone (**163**; Figure 8.13), which was obtained through mixed fermentation of a marine-derived fungus of the genus *Pestalotia* with a unicellular marine α-proteobacterium (later identified as *Thalassospira* sp. CNJ-328).[86]

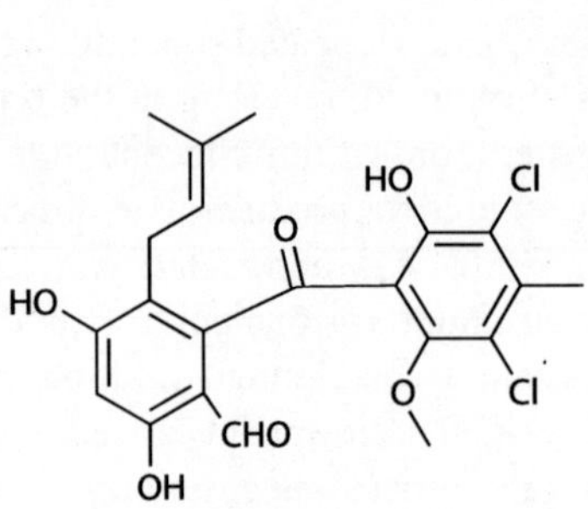

Pestalone (**163**)

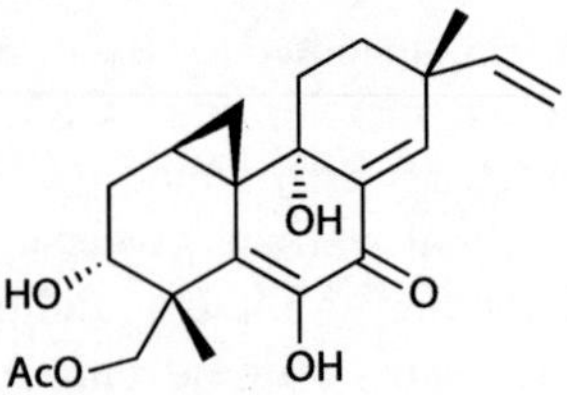

Libertellenone D (**164**)

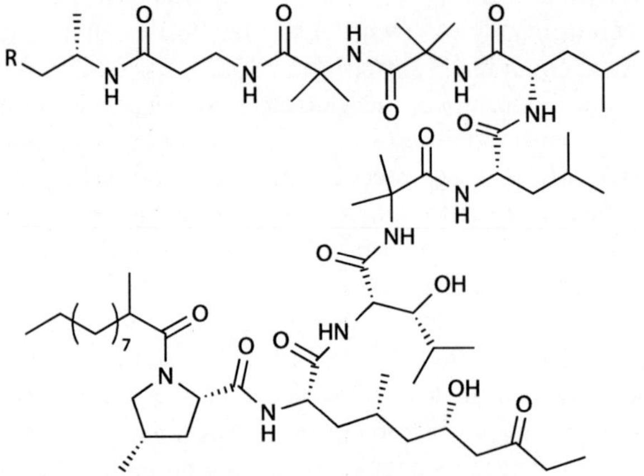

R = $NHCH_3$: Acremostatin A (**165**)
R = $N(CH_3)_2$: Acremostatin B (**166**)
R = $NO(CH_3)_2$: Acremostatin C (**167**)

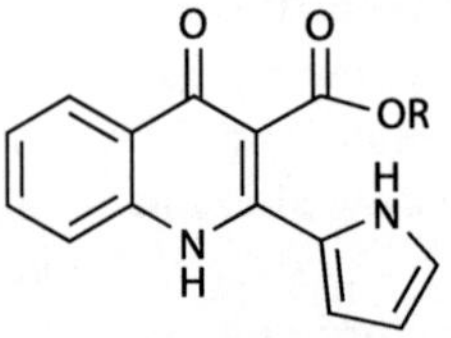

R = H : Marinamide (**168**)
R = CH_3 : Marinamide methyl ester (**169**)

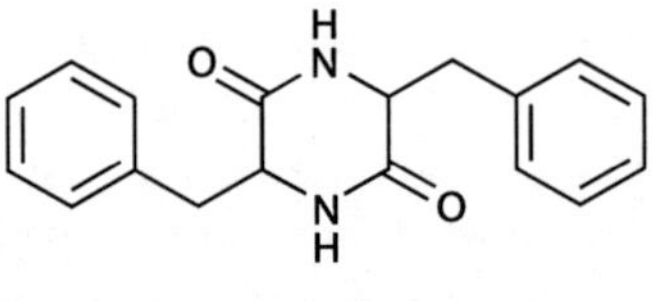

Cyclo-(Phe-Phe) (**170**)

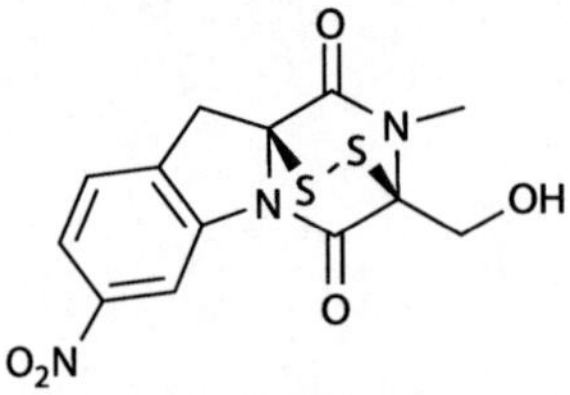

Glionitrin A (**171**)

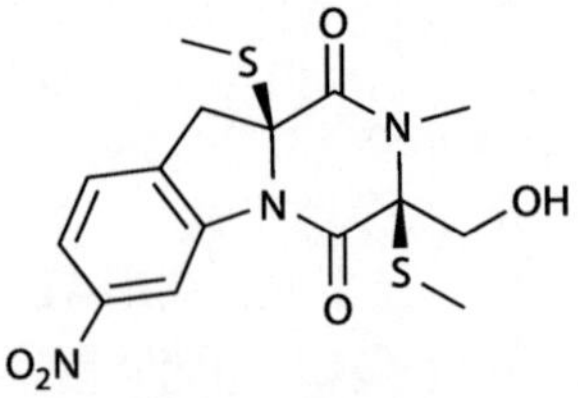

Glionitrin B (**172**)

FIGURE 8.13 Structures **163–172**. *(Continued)*

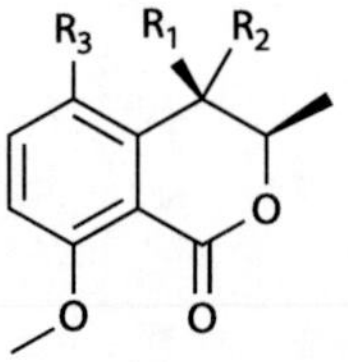

$R_1 = H, R_2 = H, R_3 = H$: *O*-methylmellein (**173**)
$R_1 = OH, R_2 = H, R_3 = H$: 4*β*-Hydroxy-8-*O*-methylmellein (**174**)
$R_1 = H, R_2 = OH, R_3 = H$: 4*α*-Hydroxy-8-*O*-methylmellein (**175**)
$R_1 = H, R_2 = H, R_3 = OH$: 5-Hydroxy-8-*O*-methylmellein (**176**)

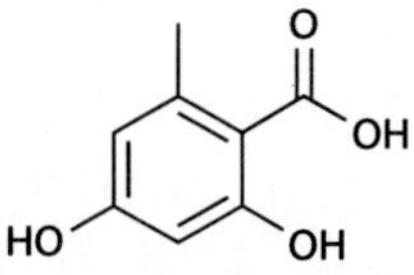

Orsellinic acid (**177**)

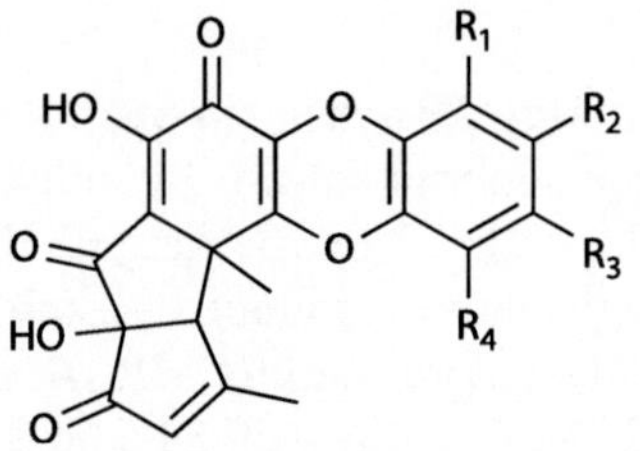

$R_1 = OH, R_2 = H, R_3 = CH_3, R_4 = H$: F-9775A (**178**)
$R_1 = H, R_2 = CH_3, R_3 = H, R_4 = OH$: F-9775B (**179**)

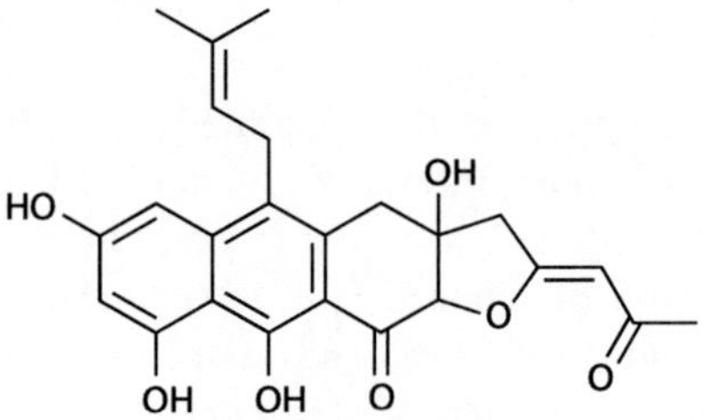

Fumicycline A (**180**)

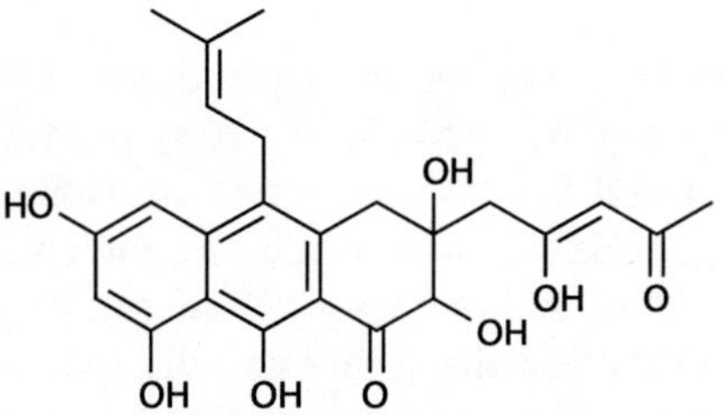

Fumicycline B (**181**)

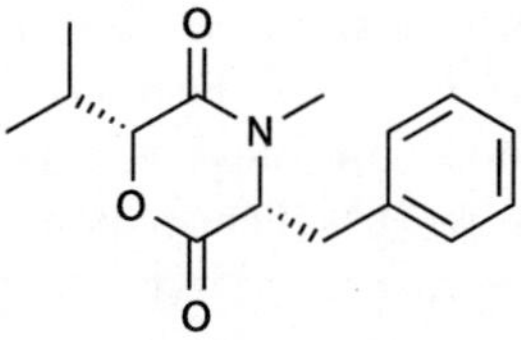

Lateritin (**182**)

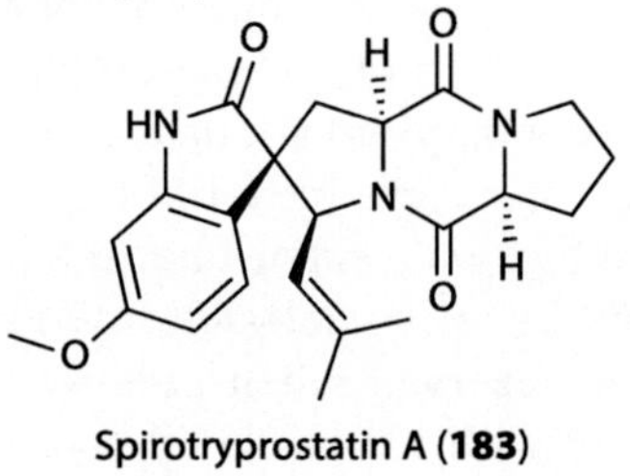

Spirotryprostatin A (**183**)

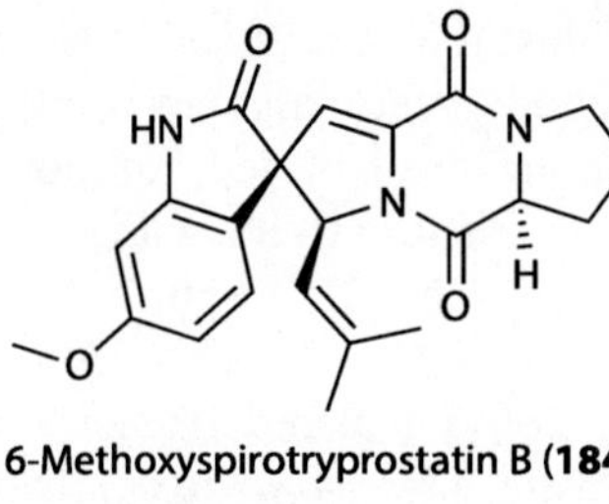

6-Methoxyspirotryprostatin B (**184**)

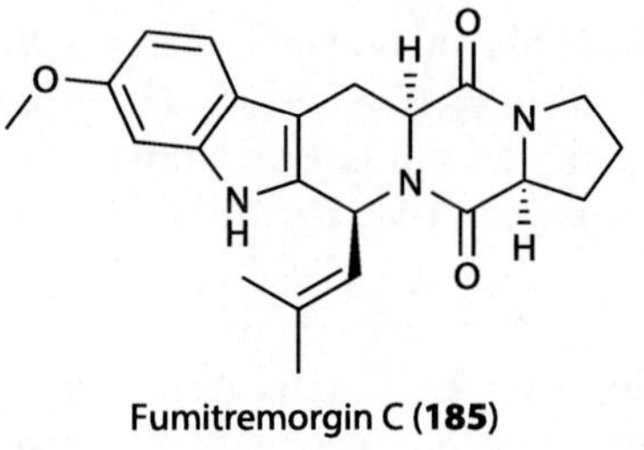

Fumitremorgin C (**185**)

FIGURE 8.13 (*Continued*) Structures **173–185**. (*Continued*)

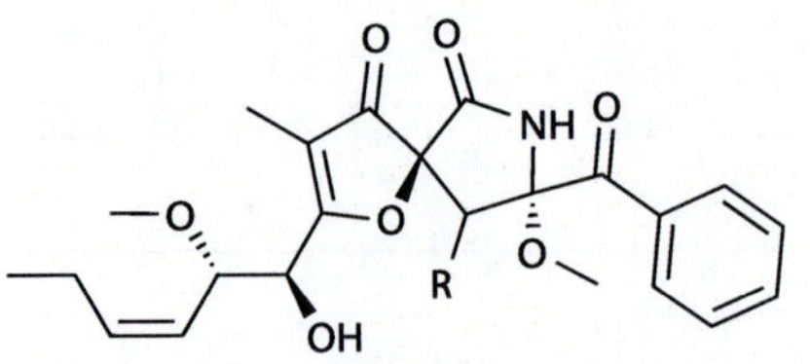

R = α-OH : 11-*O*-Methylpseurotin A (**186**)
R = β-OH : 11-*O*-Methylpseurotin A_2 (**187**)

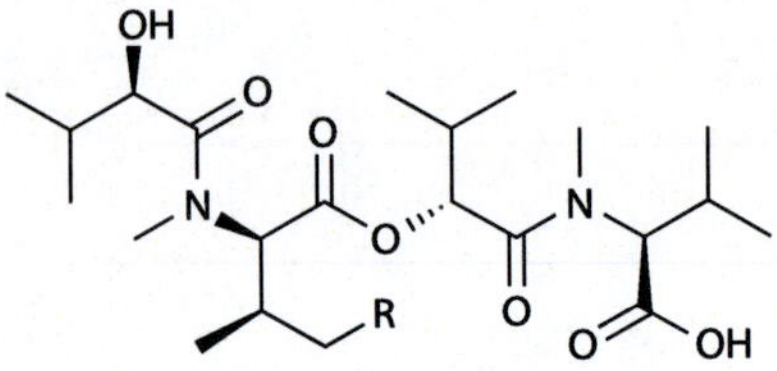

R = H : Subenniatin A (**188**)
R = CH_3 : Subenniatin B (**189**)

FIGURE 8.13 (*Continued*) Structures **186–189**.

The fungus had been isolated from the brown alga *Rosenvingea* sp. collected in the Bahamas. Interestingly, neither the cell-free supernatant nor the ethyl acetate extract of the bacterium triggered pestalone biosynthesis by the fungus. However, addition of ethanol (1% v/v) to the pure fungal culture induced pestalone production. Pestalone (**163**) displayed pronounced activity against MRSA and vancomycin-resistant *Enterococcus faecium* with MIC values of 84 and 178 nM, respectively. In addition, it exhibited moderate cytotoxicity against a panel of 60 human tumor cell lines (mean IC_{50} 6.0 μM), indicating its potential as an antimicrobial lead compound.[86]

In a subsequent study, addition of the same α-proteobacterium strain (*Thalassospira* sp. CNJ-328) that triggered pestalone (**163**) production in a 3-day-old culture of the marine-derived fungus *Libertella* sp. rapidly (within 24 h) induced the biosynthesis of a series of new pimarane diterpenes, libertellenones A–D.[87] These compounds were not observed in axenic fungal or bacterial controls. Taking into consideration that pimarane diterpenes have not been encountered in bacteria, there is a high probability for the production of these compounds by the fungus. In a similar manner to pestalone (**163**) production, addition of the cell-free broth or the organic extract of the bacterium showed no effect on the secondary metabolism of the fungus. The same result was observed after adding autoclaved bacterial cells to the fungal culture, thus indicating that the induced biosynthesis of libertellenones A–D is not regulated by signaling molecules, but rather by cell-cell interactions. Interestingly, diterpene biosynthesis was not observed when either *Libertella* or *Thalassospira* sp. CNJ-328 was cocultured with various combinations of different bacteria or fungi, respectively. Despite this specific fungal-bacterial interaction, libertellenones A–D displayed no significant activity against the cocultivated bacterial strain (*Thalassospira* sp. CNJ-328). These compounds were also evaluated against HCT-116 human colon carcinoma cells. Among the isolated compounds, only libertellenone D (**164**; Figure 8.13) exhibited potent activity with an IC_{50} value of 0.8 μM, indicating that the cyclopropane ring plays an important role for the cytotoxicity of these derivatives.[87]

Coculture of the fungus *Acremonium* sp. Tbp-5 isolated from the European yew (*Taxus baccata* L.), with the mycoparasite *Mycogone rosea* DSM 12973 led to the isolation of three new peptaibiotics, acremostatins A (**165**), B (**166**), and C (**167**) (Figure 8.13).[88] The same strain of *Acremonium* sp. was previously shown to produce a mixture of lipopeptides and leucinostatins, which possess a 4-methylhex-2-enoic acid instead of a 2-methyldecanoic acid (MDA) moiety, as in the case of acremostatins (**165–167**). Interestingly, the MDA moiety is encountered in related lipopeptides, roseoferins, which are constituents of *M. rosea*, thus indicating that the mixed biogenesis of acremostatins in *Acremonium* sp. is regulated by the cocultivation partner *M. rosea*.[88]

Mixed fermentation of two mangrove-derived endophytic fungi (strain Nos. 1924# and 3893#) from the South China Sea yielded two novel 4-quinolone analogs, marinamide (**168**) and its methyl ester (**169**) (Figure 8.13), which were not produced when either microorganism was cultured axenically.[89,90] Compounds **168** and **169** displayed comparable activities against *Escherichia coli*, *Pseudomonas pyocyanea*, and *S. aureus* (diameter of bacteriostatic ring from 0.9 to 2 cm) in the agar plate diffusion assay at a concentration of 3.9 and 3.7 mM, respectively. Moreover, both compounds exhibited potent cytotoxicity toward HepG2 (hepatocellular carcinoma), 95-D (lung cancer), MGC832 (gastric cancer), and HeLa (cervical carcinoma) cells with IC_{50} values in the range of 0.4 nM to 2.5 μM.[90] In a subsequent study, mixed fermentation of the same fungal strains (strain Nos. $1924^{\#}$ and $3893^{\#}$) afforded two further *de novo* induced metabolites, including 6-methylsalicylic acid and *cyclo*-(Phe-Phe) dipeptide (**170**; Figure 8.13). Interestingly, 6-methylsalicylic acid was found to be responsible for the observed insecticidal activity of the ethyl acetate extract against *Sinergasilus* sp. and *Heliothis armigera* (Huehner).[91]

Coculture of a *Sphingomonas* bacterial strain KMK-001 with the fungus *A. fumigatus* strain KMC-901 induced the production of a new diketopiperazine, glionitrin A (**171**; Figure 8.13).[92] Both microbes were collected from a coal mine drainage, which was contaminated with heavy metals and sulfuric acid. The metabolite was detected only after eight days of coculture, followed by a decline in the bacterial density. Thus, it could be assumed that glionitrin A (**171**) was produced by the fungus as a response to cocultivation with the bacterium. The fungal origin of **171** was further corroborated by HPLC analysis of the pure culture broth of *A. fumigatus*, in which the related diketopiperazine disulfides gliotoxin and dehydrogliotoxin were detected. On the other hand, the pure culture of *Sphingomonas* did not produce any notable secondary metabolites. In order to further investigate the factors that elicit the induction of **171**, the fungal extract or pure gliotoxin, which was the major constituent of *A. fumigatus*, was added to the bacterial culture broth. However, both cultures failed to produce **171**, suggesting that its induction during cocultivation was not the result of bacterial enzymatic postmodification. In a second set of experiments, the bacterium and the fungus were separated from the coculture broth and were grown axenically under the same culture conditions. However, neither culture produced **171**. Furthermore, it was investigated whether external factors could elicit the fungal production of **171**. Accordingly, the cellular lysate, the organic extract of the bacterial culture, as well as different chemicals, such as phenol, sulfuric acid, and nystatin, were added to the fungal culture. Notably, none of these experiments afforded **171**, thus suggesting that its production may be induced through long-term fungal-bacterial interactions.[92] At this point, it is worth mentioning that in the coculture medium, in addition to **171**, a further new minor diketopiperazine derivative was detected, which was named glionitrin B (**172**; Figure 8.13).[93] The production of the latter was induced only after long-term mixed fermentation (lasting more than 18 days) or when the bacterium was densely growing on the medium. Glionitrin A (**171**) showed potent cytotoxicity against a series of human cancer cell lines, including HCT-116 (colorectal carcinoma), A549 (lung carcinoma), AGS (gastric adenocarcinoma), DU145 (prostate carcinoma), MCF-7 (breast adenocarcinoma), and HepG2 (hepatocellular carcinoma) cells with IC_{50} values ranging from 0.2 to 2.3 μM. Moreover, **171** exhibited potent activity against three MRSA strains with an MIC value of 2.2 μM.[92] On the other hand, glionitrin B (**172**) exhibited no activity toward the DU145 cell line, indicating that the disulfide bridge is important for the cytotoxicity of these derivatives. Nevertheless, **172** caused 46% inhibition of DU145 cell invasion at a concentration of 60 μM, by decreasing the mRNA levels of the proteolytic enzymes metalloproteinases-2 and -9 (MMP-2 and MMP-9, respectively), highlighting its antimetastatic potential.[93]

Esca is a grapevine disease which is considered to be caused by a physiological misbalance of different fungi colonizing the plant. Most of these species are considered latent pathogens leading to the degradation of wood constituents (i.e. lignin); however, the exact process of this disease is not fully understood.[94,95] Two esca-associated fungi, *Eutypa lata* and *Botryosphaeria obtusa*, were cocultured on agar plates and their metabolic profiles were investigated in order to get a further insight into the processes implicated in esca disease.[96] Interestingly, analysis of the confrontation zone extract (between *E. lata* and *B. obtusa*) using ultra-high-performance liquid chromatography coupled to time-of-flight mass spectrometry (UHPLC-TOFMS) revealed approximately 60% metabolites that were not detected in the pure strains. Microisolation of the mainly induced peaks by MS-monitored semipreparative LC, and subsequent identification by capillary NMR (CapNMR), afforded compounds **173–176** (Figure 8.13). Notably, *O*-methylmellein (**173**) showed inhibitory activity against *B. obtusa* (25 μg spotted), but was inactive toward *E. lata*, thus suggesting that it might be produced by the latter as a result of the confrontation during coculture. On the other hand, the hydroxylated derivatives (**174–176**) showed no inhibition effect on the growth of *B. obtusa* or *E. lata*, leading to the speculation that these metabolites might be metabolic products of *O*-methylmellein (**173**) via activation of detoxifying enzymes released by *B. obtusa*, as previously reported in the case of wood-decaying fungi.[95] Moreover, compounds **173–176** were evaluated for their inhibitory activity on the germination rate of garden cress (*Lepidium sativum* L.) on agar medium. Similarly, only *O*-methylmellein (**173**) exhibited a significant antigerminative effect at 52 μM, which was in agreement with the observed phytotoxicity of the confrontation zone extract.[96] Thus, the strong induction of **173** during coculture between *E. lata* and *B. obtusa* suggests its putative role in esca disease, and provides new insights into the study of plant-fungus interactions.

Individual coculture of the fungus *Aspergillus nidulans* with a collection of 58 soil-dwelling actinomycetes was probed by microarray-based analysis. Notably, only *Streptomyces hygroscopicus* (later renamed *Streptomyces rapamycinicus*) caused a significant effect on fungal secondary biosynthesis with orsellinic acid (**177**), lecanoric acid (a typical lichen metabolite), and the cathepsin K inhibitors F-9775A (**178**) and F-9775B (**179**) (Figure 8.13) being produced *de novo* by the cocultured fungus.[97] Moreover, during coculture of *A. nidulans* with *S. rapamycinicus*, transcriptome analyses were conducted to monitor the impact of the expression of cryptic fungal biogenetic gene clusters. Accordingly, Northern blot analysis indicated that expression of the polyketide synthase (PKS) gene (*orsA*) is required for the production of **177–179**. This was further confirmed by targeted gene inactivation experiments of the respective PKS gene in the genome of *A. nidulans*, in which the biosynthesis of **177–179** was abolished during coculture of the resulting gene knockout mutant with *S. rapamycinicus*. Interestingly, a phylogenetic analysis revealed that orthologs of this PKS gene cluster are widespread in all major fungal groups, providing evidence of specific interactions among microorganisms. To further investigate whether the fungal gene expression is triggered by low–molecular weight bacterial metabolites that are diffused into the coculture environment, the axenic fungal culture was treated with the supernatant of the bacterial culture, with the coculture extract, or with heat-inactivated bacteria. Moreover, a mixed fermentation experiment was carried out, in which *A. nidulans* and *S. rapamycinicus* were separated by a dialysis tube membrane. The impact of the supernatant from the coculture of *S. rapamycinicus* with the gene (*orsA*) knockout mutant of *A. nidulans* was likewise investigated to exclude the involvement of signaling molecules that are only produced in the coculture, or that cannot pass through the dialysis tube membrane. Surprisingly, qRT-PCR analyses showed no fungal response in PKS gene expression, thus indicating that only direct physical interaction between *S. rapamycinicus* and *A. nidulans* triggers the production of polyketide orsellinic acid (**177**) and its derivatives (**178**, **179**, and lecanoric acid). This assumption was further corroborated by scanning electron microscopy of the coculture, which revealed that the bacterial mycelium is nested within the fungal mycelium and is attached to fungal hyphae, thus establishing an intimate interaction with *A. nidulans*. Interestingly, in a subsequent study, it was demonstrated that upon direct physical contact, the streptomycete triggers targeted histone modification in *A. nidulans*, which is catalyzed by the histone acetyltransferase (HAT) Saga/Ada complex.[98] It was further shown that this posttranslational regulation mediated by the bacterium is required for a specific fungal response, including the induction of the orsellinic acid gene cluster. Consistently, addition of the histone deacetylase (HDAC) inhibitor, suberoylanilide hydroxamic acid (SAHA), to the fungal culture resulted in activation of the *orsA* gene without the need for mixed fermentation of *A. nidulans* with *S. rapamycinicus*. These findings shed light on histone modification as a potential molecular basis of microbial cross talk.[97,98]

In a subsequent study, cocultivation of the human pathogenic fungus *A. fumigatus* with the same actinomycete, *S. rapamycinicus*, led to the induction of a previously unreported prenylated polyphenol, fumicycline A (**180**), as well as its hydrated congener fumicycline B (**181**), which were not produced in the axenic fungal culture (Figure 8.13).[99] A full genome microarray assay revealed that in the coculture a large number of fungal genes were upregulated. However, no activation of the PKS gene cluster was observed when the axenic fungal culture was supplemented with either pure bacterial culture medium or the sterile filtered supernatant of the bacterial culture. Likewise, dialysis experiments and scanning electron microscopy unveiled that only a direct physical contact between the bacterial and fungal cells could influence induction of fumicyclines, as previously observed between *S. rapamycinicus* and *A. nidulans*.[97] Consistently, targeted deletion of the PKS gene (*fccA*) in the genome of *A. fumigatus* abolished fumicycline biosynthesis, thus verifying the involvement of the cryptic gene locus in the production of **180** and **181**. This was further corroborated by overexpression of a pathway-specific regulatory gene (*fccR*) in mutant strains, resulting in the activation of the PKS gene cluster and subsequent enhancement of the production of these derivatives. However, when the respective mutant strains were cocultured with the bacterium, the transcript level of most cluster genes was further upregulated, suggesting that *S. rapamycinicus* interferes with gene expression in *A. fumigatus* by modulating other, as yet unknown, regulatory processes. To further support this hypothesis, the HDAC inhibitor SAHA was added to an axenic culture of *A. fumigatus*. As in the case of the related fungus *A. nidulans*,[98] the expression of the PKS gene cluster was induced, thus providing evidence that the bacterium modulates the epigenetic regulation of fungal gene expression during mixed fermentation. Furthermore, it was found that fumicyclines A (**180**) and B (**181**) exhibit moderate activity against *S. rapamycinicus*, suggesting that these metabolites might contribute to fungal defense as a stress response during mixed fermentation.[99]

Lateritin (**182**; Figure 8.13) was obtained from a coculture of five different filamentous fungi, including *Ovadendron sulphureoochraceum*, *Ascochyta pisi*, *Emericellopsis minima*, *Cylindrocarpon destructans*, and *Fusarium oxysporum*, whereas none of these fungi produced **182** when cultured alone.[100] The production of lateritin (**182**) was found to be the highest when the fungi were cultured for 21 days on solid media, followed by mixed fermentation for 14 days. Interestingly, the induction of **182** during coculture was found to be reproducible, denoting the feasibility of mixed fermentation with more than two microorganisms. At the end of the fermentation period, only *F. oxysporum* remained, and thus the suppression of the cocultivated fungi, as well as the production of the cryptic metabolite **182** might be the result of fungal competition for limited nutrients. Lateritin (**182**) inhibited the growth of *Candida albicans* and of gram-positive bacteria, including *Micrococcus luteus* Presque Isle 456, *S. aureus* ATCC 29213, *Enterococcus faecalis* ATCC 29212, and *Streptococcus pneumoniae* ATCC 6303 with MIC values in the range of 7.6–61.2 μM. However, lateritin (**182**) also showed significant cytotoxicity in a panel of different human cancer cell lines, displaying lack of selectivity with regard to its biological activity.[100]

Addition of *Streptomyces bullii* strain C2, isolated from hyper-arid Atacama Desert soil, to a 3-day-old culture of *A. fumigatus* MBC-F1-10 resulted in the formation of a complex set of metabolites, including seven diketopiperazine (DKP) metabolites (brevianamide F (**91**), spirotryprostatin A (**183**), 6-methoxyspirotryprostatin B (**184**), fumitremorgin B (**92**), fumitremorgin C (**185**), 12,13-dihydroxyfumitremorgin C, and verruculogen (**93**)), two pseurotin-type metabolites (11-*O*-methylpseurotin A (**186**) and its new isomer 11-*O*-methylpseurotin A_2 (**187**)), as well as ergosterol (Figure 8.13).[101] Surprisingly, LC-MS analysis of the axenic bacterial and fungal controls revealed only sparse metabolic profiles with hardly any dominating peaks. Nevertheless, all induced compounds during the fungal-bacterial coculture are assumed to be of fungal origin based on their structural similarity to known fungal metabolites. When a bacteria-free medium was added to the axenic culture of *A. fumigatus* MBC-F1-10, no *de novo* production of compounds was observed, thus ruling out the influence of media components on the induction of fungal biosynthesis. Likewise, no effect was observed after addition of the bacterial organic extract or the autoclaved bacterial culture broth. These findings suggest that a long-term simulated competitive environment and/or direct physical contact might be possible mechanisms of the induced biosynthesis of **91–93** and **183–187** through microbial coculture. This is in agreement with previous fungal-bacterial mixed fermentation experiments, likewise involving fungi of the genus *Aspergillus*.[97,99] Despite the fact that these compounds were produced by the fungus during coculture with the bacterium, none of them exhibited significant antibacterial activity when tested against *S. aureus* or *E. coli*. All induced compounds were further evaluated for their potential trypanocidal and leishmanicidal activities. Fumitremorgin B (**92**) and verruculogen (**93**) showed strong growth inhibitory activity toward *Trypanosoma brucei brucei* and *Leishmania donovani* with EC_{50} values of 0.2 and 3–4 μM, respectively, whereas 12,13-dihydroxyfumitremorgin C displayed moderate activity against *T. brucei brucei* (EC_{50} = 7.4 μM). However, compounds **91–93** also exhibited potent cytotoxicity against normal human diploid fetal lung fibroblast (MRC-5) cells, which precluded further investigation of their antiprotozoal activities due to lack of selectivity.[101]

Mixed fermentation of the endophytic fungi *Fusarium begoniae* and *F. tricinctum* (from the plant *Aristolochia paucinervis*) on white bean medium resulted in the production of two new linear depsipeptides, subenniatins A (**188**) and B (**189**) (Figure 8.13), which were not detected in either of the pure cultures.[102] Compounds **188** and **189** originate from *F. tricinctum*, as they are considered to be biogenetic building blocks of the archetypal cyclic peptides enniatins, which were detected in the respective axenic fungal culture. However, in contrast to enniatins, subenniatins A and B (**188** and **189**) were found to be inactive against the mouse lymphoma (L5178Y) cell line (IC_{50} > 10 μM). Interestingly, when *F. tricinctum* was cocultured with *Fusarium equiseti* no significant change in the metabolite production was observed, indicating a specific rather than a general response of *F. tricinctum* during interaction with other *Fusarium* species.[102]

In a subsequent study, mixed fermentation of the endophytic fungus *F. tricinctum* with the bacterium *Bacillus subtilis* 168 trpC2 on solid rice medium led to the induction of three new metabolites, macrocarpon C (**190**), *N*-(carboxymethyl)anthranilic acid, and (−)-citreoisocoumarinol (**191**), in addition to the known (−)-citreoisocoumarin (**192**), which were not detected in axenic fungal or bacterial controls (Figure 8.14).[103] Moreover, HPLC analysis of the coculture extract revealed up to a 78-fold increase of enniatin accumulation compared to axenic cultures of *F. tricinctum*. The induction of fungal metabolites was found to correlate with time, with the strongest effect observed after a 6-day preincubation time of the solid medium with *B. subtilis* prior to inoculation with *F. tricinctum*. Interestingly, when *F. tricinctum* was cocultured with the bacterium *Streptomyces lividans*, compounds **190–192** were not detected. Instead the production of other, yet unknown, compounds was induced that were lacking in axenic fungal and bacterial controls, as well as in cocultures of *F. tricinctum* with *B. subtilis*, hinting at a specificity of *F. tricinctum* toward different prokaryotes. Notably, in a recent study it was shown that fungi of the genus *Fusarium* redirect the aurofusarin pathway to produce citreoisocoumarin as a response to low nitrogen content of the medium.[104]

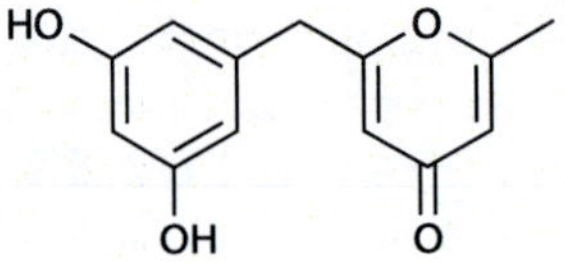

Macrocarpon C (**190**)

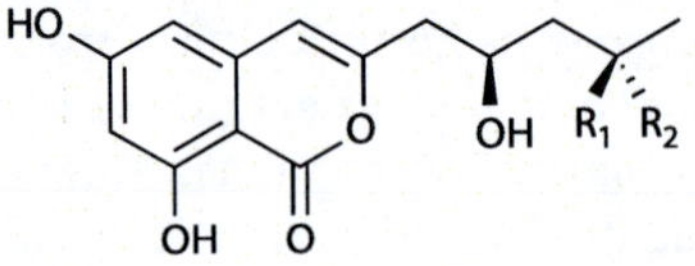

$R_1 = OH; R_2 = H$: (–)-Citreoisocoumarinol (**191**)
$R_1 + R_2 = O$: (–)-Citreoisocoumarin (**192**)

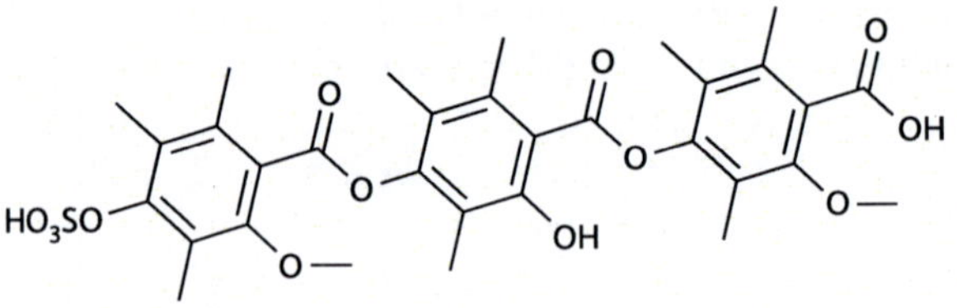

4″-Hydroxysulfoxy-2,2″-dimethylthielavin P (**193**)

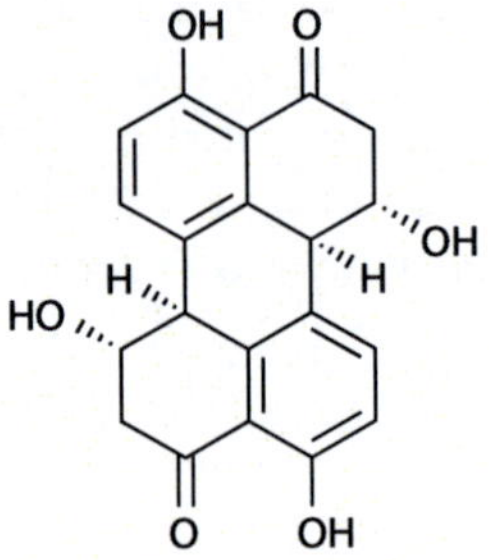

Stemphyperylenol (**194**)

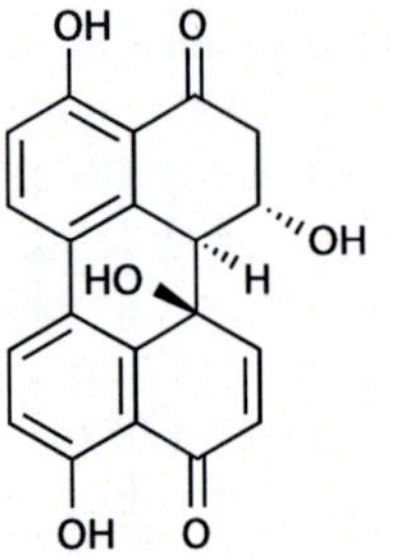

Alterperylenol (**195**)

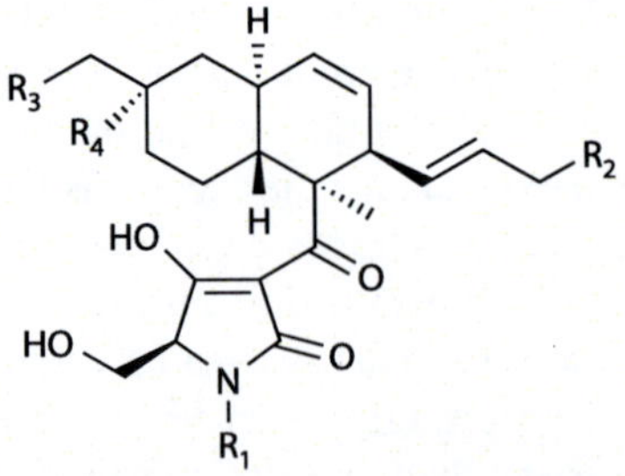

$R_1 = CH_3, R_2 = H, R_3 = H, R_4 = OH$: Pallidorosetin A (**196**)
$R_1 = CH_3, R_2 = OH, R_3 = OH, R_4 = H$: Pallidorosetin B (**197**)
$R_1 = H, R_2 = H, R_3 = OH, R_4 = H$: *N*-Demethylophiosetin (**198**)
$R_1 = CH_3, R_2 = H, R_3 = OH, R_4 = H$: Ophiosetin (**199**)
$R_1 = CH_3, R_2 = H, R_3 = H, R_4 = H$: Equisetin (**200**)

FIGURE 8.14 Structures 190–200. *(Continued)*

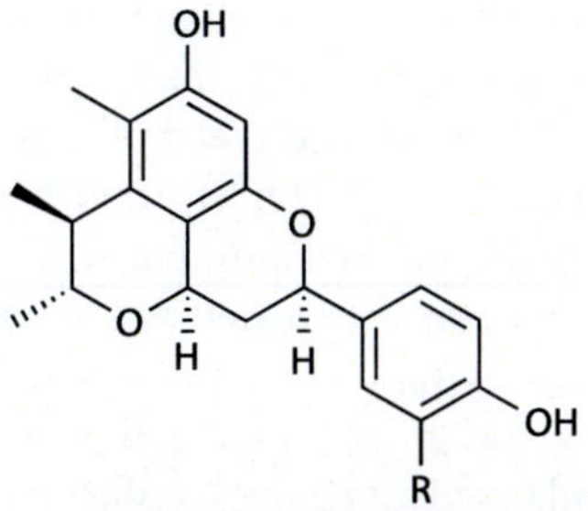

R = H : Citrifelin A (**201**)
R = OCH_3 : Citrifelin B (**202**)

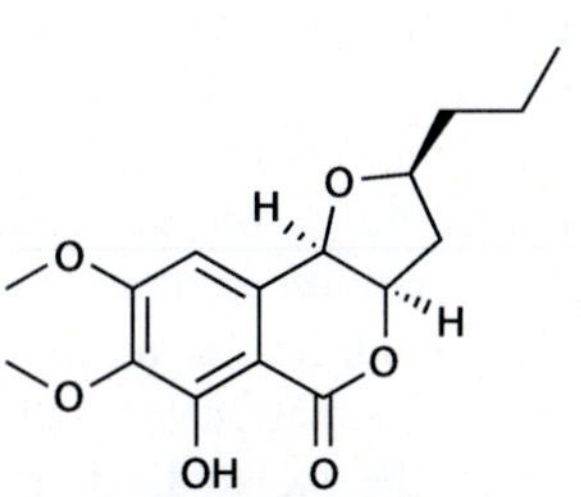

Monocerin (**203**)

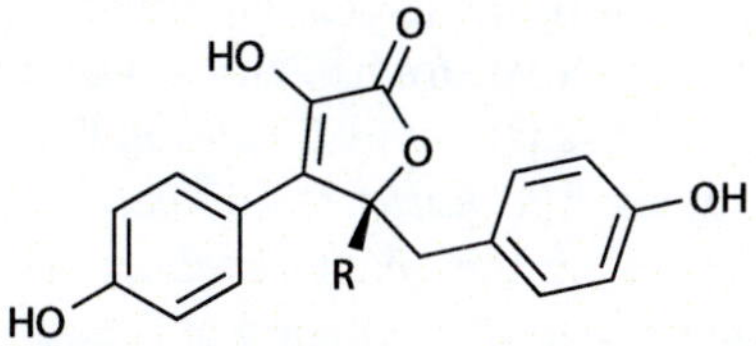

R = OCH_3 : Isobutyrolactone II (**204**)
R = OH : 4-*O*-Demethylisobutyrolactone II (**205**)

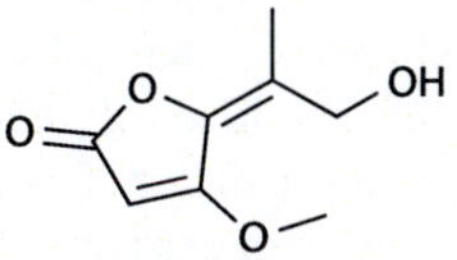

Coculnol (**206**)

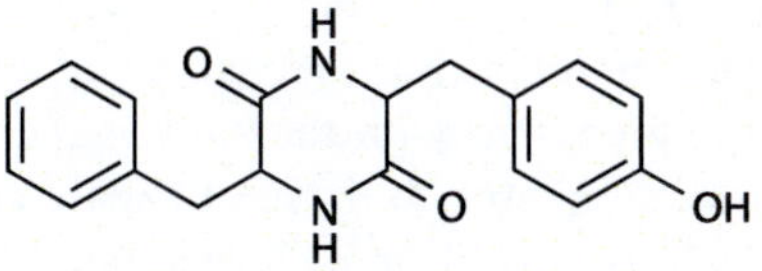

Cyclo-(Phe-Tyr) (**207**)

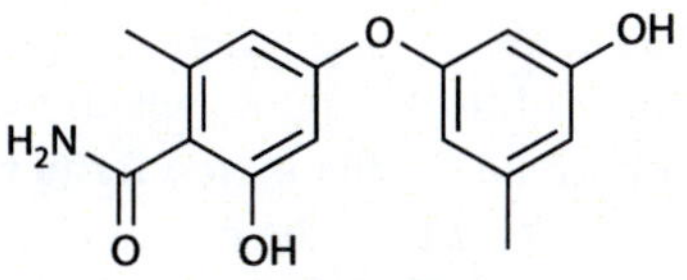

Austramide (**208**)

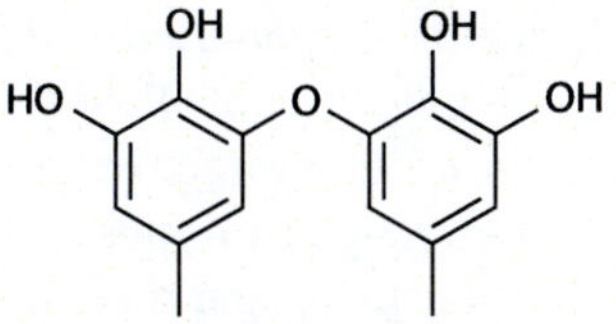

Violaceol I (**209**)

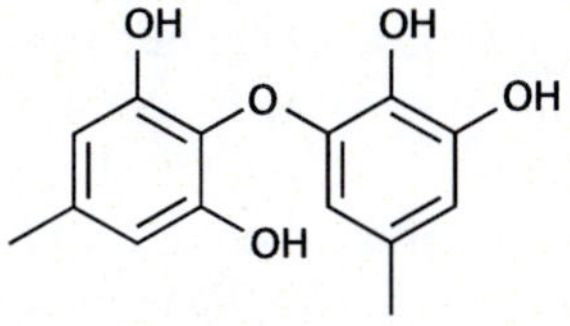

Violaceol II (**210**)

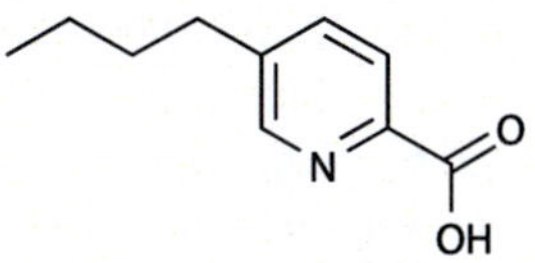

Fusaric acid (**211**)

FIGURE 8.14 (*Continued*) Structures **201–211**.

Thus, the *de novo* production of (−)-citreoisocoumarin (**192**) and the related compound (−)-citreoisocoumarinol (**191**) during coculture might be a result of competition for limited nutrients between *F. tricinctum* and *B. subtilis* 168 trpC2. In addition, enniatins A_1 (**113**) and B_1 (**115**), which showed an enhanced production during coculture, exhibited strong activity against *S. pneumoniae*, *E. faecalis*, and *S. aureus*, as well as against multidrug-resistant staphylococcal and enterococcal clinical isolates, with MIC values in the range of 3–12 μM. For a further investigation of the role of enniatins (**113** and **115**) during coculture, the antibacterial activity of these metabolites was evaluated toward the cocultivated bacterial strain *B. subtilis* 168 trpC2. Interestingly, these peptides inhibited the bacterial growth (MIC_{50} values 12 and 24 μM, respectively), indicating that *F. tricinctum* enhanced their production as a chemical defense against its competitor.[103]

An unusual long-distance microbial interaction was observed during mixed fermentation of the dermatophyte *Trichophyton rubrum* with the filamentous fungus *Bionectria ochroleuca* (Sin80).[105] In order to investigate metabolite induction during coculture, analytical strategies based on UHPLC-TOFMS metabolomics were employed. Comparison of the resulting coculture metabolome fingerprint in the confrontation zone with those of the axenic cultures revealed five *de novo* induced metabolites, which were further purified employing software-oriented semipreparative HPLC-MS microisolation after chromatographic optimization. However, due to low amounts of the isolated compounds, only one metabolite could successfully be identified based on microflow NMR analysis as 4″-hydroxysulfoxy-2,2″-dimethylthielavin P (**193**; Figure 8.14). Interestingly, its nonsulfated form could be detected in the UHPLC-TOFMS fingerprint of the axenic culture of *B. ochroleuca*, indicating that **193** is produced by the respective strain. Thus, it was suggested that compound **193** is further sulfated during cross talk with *T. rubrum*.[105]

Mixed fermentation of the endophytes *Alternaria tenuissima* and *Nigrospora sphaerica*, both isolated from the stems of the plant *Smallanthus sonchifolius* (Asteraceae) in malt extract broth led to increased production of the polyketide stemphyperylenol (**194**; Figure 8.14), which was not detected in axenic fungal cultures.[106] Moreover, the production of its analog, alterperylenol (**195**), was enhanced in the mixed culture compared to the pure culture of *A. tenuissima*. The induced production of **194** and **195** suggested the presence of inhibition mechanisms caused by the confronted fungal strains, which was in agreement with the slower growth rate that was observed for the mixed fermentation compared to those of axenic cultures. The metabolic profile of the mixed fungal culture was likewise investigated in semisolid medium (potato dextrose agar). During coculture, a growth inhibitory effect of *N. sphaerica* could be visualized, thus suggesting that *A. tenuissima* produced the polyketides **194** and **195** as a response to the fast growth rate of *N. sphaerica* in the medium. On the other hand, the slow growth rate of *A. tenuissima* did not allow any observation of colony deformation due to diffusible metabolites that might be produced by *N. sphaerica* into the coculture medium. To further investigate this possibility, *N. sphaerica* was inoculated on a cellophane membrane above the agar medium. The membrane allows the diffusion of nutrients through it, without fungal growth into the agar medium.[107] However, when the cellophane with *N. sphaerica* was removed from the plate, followed by inoculation of *A. tenuissima* on the same medium, no effect on the growth of the latter was observed. This result suggested that no diffusible metabolites were produced by *N. sphaerica* with inhibitory activity against the challenging fungus *A. tenuissima*. Moreover, to further investigate whether volatile antifungal compounds were induced during coculture, two Petri dishes, each containing a pure strain of *A. tenuissima* or *N. sphaerica*, were sealed together with Parafilm M®. Likewise, no effect on the growth rate was observed, indicating that the resulting fungal inhibition during coculture is only caused by diffusible metabolites that are released into the medium by *A. tenuissima*, presumably in order to control the faster growth rate of *N. sphaerica*. It should be noted, however, that the production of stemphyperylenol (**194**) was particularly enhanced during liquid mixed fermentation, compared to the inhibition zone of the semisolid mixed culture, in which there is no fungal contact. This finding suggested that direct interactions between *A. tenuissima* and *N. sphaerica* may likewise play an important role in triggering the production of the polyketides **194** and **195**. Compound **194** inhibited the growth of *N. sphaerica* at 200 μM. However, no phytotoxicity was observed when **194** was exogenously applied to leaves of the host plant *S. sonchifolius*, even at much higher concentrations (2 mM) after 7 days of treatment, suggesting a putative role of these polyketides (**194** and **195**) in the fungal endophyte–host plant mutualistic relationships.[106]

Mixed fermentation of the fungus *Fusarium pallidoroseum* with the bacterium *Saccharopolyspora erythraea* resulted in the *de novo* production of four decalin-type tetramic acid derivatives (**196–199**), including three new metabolites pallidorosetins A (**196**) and B (**197**), and *N*-demethylophiosetin (**198**), as well as the known compound ophiosetin (**199**) (Figure 8.14).[108] These metabolites are assumed to be derived from *F. pallidoroseum* as they are biogenetically related to equisetin (**200**; Figure 8.14), a known metabolite of several fungi of the genus *Fusarium*. Interestingly, **200** was detected in the axenic fungal extract; however, during coculture there was a 30-fold increase in the accumulation of this metabolite. Equisetin (**200**) was active against *S. erythraea* and *S. aureus* at ≤2.5 μg/disk in the Kirby-Bauer disk diffusion assay, leading to the hypothesis that the fungus enhanced production of this metabolite as a defense mechanism against its competitor. This was further corroborated by the observation that the production of **200** was likewise enhanced during coculture of *F. pallidoroseum* with other gram-positive bacteria, including *Bacillus* or *Staphylococcus* strains. Moreover, **200** displayed pronounced cytotoxicity toward the leukemia cell line CCRF-CEM with an IC_{50} value of 144 nM, as well as against a panel of different human cancer cell lines in the low micromolar range, whereas the remaining compounds (**196–199**) were inactive (IC_{50} > 20 μM).[108] It is worth mentioning that a further tetramic acid analog, trichosetin, was produced *de novo* during coculture of the fungus *Trichoderma harzianum* with the callus of the medicinal plant *Catharanthus roseus*, denoting that these metabolites likely play an important role in microbial communication.[109]

Mixed fermentation of the fungus *P. citrinum* MA-197, isolated from the mangrove plant *Lumnitzera racemosa*, with the fungus *Beauveria felina* EN-135, isolated from an unidentified bryozoan, afforded two new citrinin adducts, citrifelins A (**201**) and B (**202**) (Figure 8.14), featuring a unique tetracyclic carbon skeleton that was not present in the axenic fungal controls. The biological origin of these compounds probably goes back to *P. citrinum* MA-197 due to their biogenetic relationship with citrinin isolated from the same fungus. Both compounds (**201** and **202**) showed inhibitory activity toward *E. coli* and *S. aureus* with MIC values ranging between 5.6 and 24.5 μM.[110]

Coculture of two *Penicillium* sp. strains (IO1 and IO2), both isolated from the Mediterranean sponge *Ircinia oros*, resulted in a vastly different metabolite pattern compared to those of the axenic fungal controls grown on solid rice medium, as shown by HPLC analysis of the respective extracts.[111] Accordingly, two known polyketides, norlichexanthone and monocerin (**203**; Figure 8.14), were induced that were lacking in either of the axenic fungal controls. Monocerin (**203**) exhibited strong cytotoxicity against L5178Y (murine lymphoma) cells with an IC_{50} value of 8.4 μM, leading to the speculation that the *de novo* production of these polyketides might be triggered by one of these fungi as a defense mechanism against its competitor.[111]

Coculture of the soil-dwelling fungus *A. terreus*, isolated from the sediment of a hypersaline lake in Wadi EI Natrun in Egypt, either with *B. subtilis* or with *Bacillus cereus* on solid rice medium yielded the new isobutyrolactone II (**204**) and 4-*O*-demethylisobutyrolactone II (**205**) (Figure 8.14), in addition to the known *N*-(carboxymethyl)anthranilic acid, which were not observed in either of the axenic fungal or bacterial controls.[112] Moreover, cocultivation resulted in a 34-fold enhancement in the accumulation of terrein and butyrolactones I–III and VI, which are typical metabolites of *A. terreus*, compared to the axenic fungal control. Interestingly, the production of **204** and **205** was likewise induced when *A. terreus* was cultured in the presence of autoclaved *B. subtilis*, indicating that the induction of fungal cryptic metabolites is not due to heat-labile bacterial signaling molecules. On the other hand, *N*-(carboxymethyl)anthranilic acid was only induced during mixed fermentation of *A. terreus* with live *B. subtilis*. Recently, its production was also reported during coculture of *F. tricinctum* with *B. subtilis*.[103] In addition, this metabolite is derived from anthranilic acid, which has been reported as a constituent of several bacterial species. Thus, the possibility of *B. subtilis* being the biological origin of *N*-(carboxymethyl)anthranilic acid cannot be excluded. Surprisingly, mixed fermentation of *A. terreus* either with *Streptomyces coelicolor* or with *S. lividans* caused no induction of secondary metabolites, indicating a specific fungal response toward different bacteria, which is in agreement with previously reported results.[103,112]

The mixed fermentation broth of the fungi *Fusarium solani* FKI-6853 and *Talaromyces* sp. FKA-65, obtained from soil samples collected from Haha-jima and Kouzu islands in Japan, exhibited anti-influenza virus activity toward the strain A/PR/8/34 (H1N1), whereas the culture broths originating from axenic cultivation showed no activity in the respective assay, suggesting a change in the metabolite pattern during coculture.[113] Subsequent chemical investigation of the coculture extract led to the isolation of a new penicillic acid derivative, coculnol (**206**; Figure 8.14). Nevertheless, compound **206** displayed no inhibitory activity toward the A/PR/8/34 (H1N1) virus (IC_{50} > 100 μM). In addition, compound **206** was further evaluated for its cytotoxicity against Madin-Darby canine kidney (MDCK) cells, but was found to be largely inactive.[113]

The mixed culture of the fungus *Aspergillus niger* N402 with the bacterium *S. coelicolor* A3(2) M145 was investigated for the induction of new metabolites compared to the axenic fungal or bacterial controls.[114] Preliminary experiments revealed that the pH of the medium had a great impact on the growth of both microbes, and therefore the medium was optimally adjusted to pH 5 (by adding phosphate buffer) for a mutual interaction to occur. Moreover, due to the higher growth rate of *A. niger*, the fungal spores were inoculated into liquid normal minimal medium with phosphate (NMMP) following a 3-day preincubation with *S. coelicolor* A3(2) M145. In order to visualize the viable fungal biomass during mixed fermentation, *A. niger* strain AR19#1, which expresses enhanced green fluorescent protein (eGFP), was utilized. Notably, *A. niger* AR19#1 was rapidly (within 24 h) colonized by *S. coelicolor*, which was shown to degrade the fungal mycelium, as evident from the loose fungal hyphae in the culture broth. Subsequent metabolite profiling of large-scale mixed cultures between *A. niger* N402 (strain not expressing eGFP) and *S. coelicolor* A3(2) M145, employing UHPLC-TOFMS and NMR-based metabolomics, revealed five *de novo* induced metabolites: *cyclo*-(Phe-Phe) (**170**), *cyclo*-(Phe-Tyr) (**207**; Figure 8.14), phenylacetic acid, 2-hydroxyphenylacetic acid, and furan-2-carboxylic acid. The highly induced compounds **170** and **207** are assumed to be derived from *A. niger*, as fungi of the genus *Aspergillus* are well known for the production of 2,5-diketopiperazine-type metabolites. When *A. niger* was inoculated into the culture, containing cell-free supernatant of a 5-day-old bacterial culture, the production of **170** and phenylacetic acid was induced, indicating that metabolites in the bacterial culture filtrate may act as biotransformation substrates or may trigger silent biosynthetic pathways of *A. niger*.[114] Interestingly enough, this finding stands in contrast to previously reported coculture studies, in which an intimate fungal-bacterial interaction is required for the induction of cryptic metabolites.[97,101]

A new diphenyl ether, austramide (**208**; Figure 8.14), was obtained from the coculture of the endophytic fungus *Aspergillus austroafricanus*, isolated from leaves of the aquatic plant *Eichhornia crassipes*, either with *B. subtilis* or with *S. lividans*, while the compound was lacking in axenic fungal or bacterial controls.[115] In addition, the accumulation of several diphenyl ether derivatives, including violaceols I (**209**) and II (**210**) (Figure 8.14), as well as diorcinol was strongly enhanced up to 29-fold when *A. austroafricanus* was cocultured with *B. subtilis*, compared to the respective axenic controls. However, these metabolites were only induced up to fourfold when *A. austroafricanus* was cocultured with *S. lividans*, indicating a specific response of the fungus toward different prokaryotes, as previously reported in coculture studies involving *A. terreus* or *F. tricinctum* with different bacteria.[103,112] Interestingly, compounds **209** and **210** showed activity against *S. aureus* ATCC 700699 with MIC values of 25 μM for both compounds, whereas diorcinol displayed activity toward *B. subtilis* 168 trpC2, which was employed during coculture, with an MIC value of 34.8 μM. Thus, the accumulation of diphenyl ethers might be seen as a defense mechanism of *A. austroafricanus* in the presence of the cocultured bacteria.[115]

Several pairs of human pathogenic nondermatophyte fungi, mainly *Fusarium* sp., were cocultured on Petri dishes and screened for morphological changes and metabolomic modifications.[116] These strains were obtained from nails of patients suffering from azole drug-resistant onychomycosis with the aim of simulating interactions that might occur in the nail mycobiome. Interestingly, during coculture of *F. cf. oxysporum* SIN2 with *Sarocladium cf. strictum* SIN29, a distance inhibition phenotype was observed, suggesting a release of low–molecular weight antifungal metabolites into the coculture medium. In addition, for the coculture involving *F. cf. oxysporum* SIN17 and *S. cf. strictum* SIN29, it was observed that in the interaction area both strains were in contact, whereas the mycelial morphology was altered, and thus the confrontation zone of the two fungi was excised for further chemical analysis. Metabolite profiling, employing UHPLC-TOFMS, revealed the production of the mycoalexin fusaric acid (**211**; Figure 8.14) along with two other peaks of lower intensity, which were absent in both pure culture extracts. However, the latter compounds could not be identified due to small amounts. Notably, a targeted metabolomics study of **211** as biomarker in a large set of fungal strains (106 cocultures) revealed that *Fusarium* strains, lacking fusaric acid when cultured axenically, could produce this compound *de novo* during coculture, thus raising the question of a possible production and role of **211** in fungal virulence associated with onychomycosis.[116]

8.4 CONCLUSION

Filamentous fungi have an enormous potential as valuable sources of a multitude of bioactive natural products, many of which play an eminent role in drug discovery.[117,118] Interestingly, recent advances in genome sequencing have shown that the biogenetic potential of fungi is greatly underestimated. A reason for this lies in the fact that fungi express only a subset of biogenetic gene clusters, and thus are capable of producing a far larger number of chemical entities that outnumber the secondary metabolites observed under standard laboratory growth conditions.[119] Therefore, a core area of interest in microbiology is to exploit biogenetic pathways that remain *silent* in the absence of particular triggers. To address this challenge, various genome sequence–independent strategies have been employed, such as selective variation of abiotic culture conditions (OSMAC approach) or cultivation of two or more microbes in a single confined environment (coculture or mixed fermentation) with regard to the elicitation of the cryptic metabolome in a competitive environment.[5,120] In contrast to the OSMAC approach, coculture is still in its infancy, perhaps due to reproducibility issues.

One of the major challenges is simulating the growth conditions under which fungal cryptic biogenetic pathways are activated. In the OSMAC approach, many cultivation parameters can be changed individually or in combination involving, for example, temperature, pH, culture container and its volume, aeration, cultivation time, light intensity, and media composition, including carbon and nitrogen sources, as well as salt concentration. These parameters are found to affect the biosynthesis of secondary metabolites on several levels, such as transcription, translation, and enzyme inhibition or activation.[121] However, the molecular mechanisms required for activation of fungal cryptic metabolites, either via such culture modifications or via natural mutualistic and/or antagonistic microbial relationships during coculture, remain largely underexplored.[122] In a recent study, it was demonstrated that *S. rapamycinicus* triggered secondary metabolite production in *A. nidulans* through targeted histone modification catalyzed by the histone acetyltransferase (HAT) Saga/Ada complex, thus shedding light on histone modification as a potential molecular basis of microbial cross talk.[98] Cocultivation of microbes has also been shown to provoke exchange of whole gene fragments (horizontal gene transfer), or has been used to explore quorum-sensing (QS) molecules which modulate the microbial phenotype (e.g., via virulence attenuation), without inhibiting the growth of a microbial strain.[101,123,124] Thus, a deep understanding of the molecular mechanisms and growth parameters that trigger cryptic secondary metabolites holds promise not only with regard to discovery of new leads, but also as an effective tool for elucidating the complexity of interspecies cross talk occurring in the natural environment.

Accessing of cryptic metabolites is extremely challenging due to the complexity of microbial extracts. However, recent progress in analytical methods, mainly based on state-of-the-art mass spectrometry, including MALDI-MS/MS and nanoDESI-MS imaging approaches, as well as sensitive microflow NMR, requiring only microgram amounts of compounds for structure identification, has brought great improvements in the analysis of the microbial metabolome.[125] Moreover, improvements in chromatographic techniques allow a fast and efficient targeted isolation at the microgram level of metabolites from complex mixtures. A notable example is the structure identification of the mycoalexin fusaric acid, employing UHPLC-TOFMS, which required only three 9 cm Petri dishes of fungal coculture between *F. cf. oxysporum* SIN17 and *S. cf. strictum* SIN29.[116] Moreover, metabolite-profiling technology and data-mining approaches have also been successfully applied to highlight the compounds of interest (i.e., cryptic metabolites) resulting from such interactions, thus overcoming the major challenge of analysis of complex biological systems or the frequent rediscovery of previously identified natural products.[122]

Conclusively, the various recent papers described in the present chapter highlight the potential of OSMAC and coculture approaches as powerful strategies for inducing previously unexpressed biogenetic pathways, leading to the induction of new compounds. Taking into account the practically infinite combinations of biotic and/or abiotic stress factors that could be applied to the fungal cultures, in addition to advances in molecular and analytical methods, it is expected that these approaches will continue to play an important role as viable avenues for harvesting the *cryptic* secondary metabolome of fungi.

ACKNOWLEDGMENTS

Financial support of the DFG (GRK 2158) and of the Manchot Foundation to P.P. is gratefully acknowledged.

REFERENCES

1. Newman, D. J.; Cragg, G. M.; Snader, K. M. Natural products as sources of new drugs over the period 1981–2002. *J. Nat. Prod.* **2003**, *66*, 1022–1037.
2. Baker, D. D.; Chu, M.; Oza, U.; Rajgarhia, V. The value of natural products to future pharmaceutical discovery. *Nat. Prod. Rep.* **2007**, *24*, 1225–1244.
3. Lucas, X.; Senger, C.; Erxleben, A.; Grüning, B. A.; Döring, K.; Mosch, J.; Flemming, S.; Günther, S. StreptomeDB: A resource for natural compounds isolated from *Streptomyces* species. *Nucleic Acids Res.* **2013**, *41*, D1130–D1136.
4. Balashov, S. V.; Park, S.; Perlin, D. S. Assessing resistance to the echinocandin antifungal drug caspofungin in *Candida albicans* by profiling mutations in FKS1. *Antimicrob. Agents Chemother.* **2006**, *50*, 2058–2063.
5. Marmann, A.; Aly, A. H.; Lin, W.; Wang, B.; Proksch, P. Co-cultivation—A powerful emerging tool for enhancing the chemical diversity of microorganisms. *Mar. Drugs* **2014**, *12*, 1043–1065.
6. Bode, H. B.; Walker, M.; Zeeck, A. Structure and biosynthesis of mutolide, a novel macrolide from a UV mutant of the Fungus F-24′707. *Eur. J. Org. Chem.* **2000**, *2000*, 1451–1456.
7. Paranagama, P. A.; Wijeratne, E. M. K.; Gunatilaka, A. A. L. Uncovering biosynthetic potential of plant-associated fungi: Effect of culture conditions on metabolite production by *Paraphaeosphaeria quadriseptata* and *Chaetomium chiversii*. *J. Nat. Prod.* **2007**, *70*(12), 1939–1945.
8. Wiener, P. Experimental studies on the ecological role of antibiotic production in bacteria. *Evol. Ecol.* **1996**, *10*, 405–421.

9. Tarkka, M. T.; Sarniguet, A.; Frey-Klett, P. Inter-kingdom encounters: Recent advances in molecular bacterium–fungus interactions. *Curr. Genet.* **2009**, *55*, 233–243.
10. Petersen, F.; Moerker, T.; Vanzanella, F.; Peter, H. H. Production of cladospirone bisepoxide, a new fungal metabolite. *J. Antibiot.* **1994**, *47*, 1098–1103.
11. Crews, P.; Manes, L. V.; Boehler, M. Jasplakinolide, a cyclodepsipeptide from the marine sponge, *Jaspis* sp.. *Tetrahedron Lett.* **1986**, *27*, 2797–2800.
12. Mills, J. W.; Pedersen, S. F.; Walmod, P. S.; Hoffmann, E. K. Effect of cytochalasins on F-actin and morphology of Ehrlich ascites tumor cells. *Exp. Cell Res.* **2000**, *261*, 209–219.
13. Christian, O. E.; Compton, J.; Christian, K. R.; Mooberry, S. L.; Valeriote, F. A.; Crews, P., Using Jasplakinolide to turn on pathways that enable the isolation of new chaetoglobosins from *Phomospis asparagi. J. Nat. Prod.* **2005**, *68*, 1592–1597.
14. Bills, G.; Platas, G.; Fillola, A.; Jimenez, M.; Collado, J.; Vicente, F.; Martin, J.; Gonzalez, A.; Bur-Zimmermann, J.; Tormo, J. Enhancement of antibiotic and secondary metabolite detection from filamentous fungi by growth on nutritional arrays. *J. Appl. Microbiol.* **2008**, *104*, 1644–1658.
15. Amagata, T.; Tanaka, M.; Yamada, T.; Doi, M.; Minoura, K.; Ohishi, H.; Yamori, T.; Numata, A. Variation in cytostatic constituents of a sponge-derived *Gymnascella dankaliensis* by manipulating the carbon source. *J. Nat. Prod.* **2007**, *70*, 1731–1740.
16. Numata, A.; Amagata, T.; Minoura, K.; Ito, T. Gymnastatins, novel cytotoxic metabolites produced by a fungal strain from a sponge. *Tetrahedron Lett.* **1997**, *38*, 5675–5678.
17. Amagata, T.; Doi, M.; Ohta, T.; Minoura, K.; Numata, A. Absolute stereostructures of novel cytotoxic metabolites, gymnastatins A-E, from a *Gymnascella* species separated from a *Halichondria* sponge. *J. Chem. Soc., Perkin Trans.* **1998**, *1*(21), 3585–3600.
18. Amagata, T.; Minoura, K.; Numata, A. Gymnastatins F-H, cytostatic metabolites from the sponge-derived fungus *Gymnascella dankaliensis*. *J. Nat. Prod.* **2006**, *69*, 1384–1388.
19. Wijeratne, E. K.; Carbonezi, C. A.; Takahashi, J. A.; Seliga, C. J.; Turbyville, T. J.; Pierson, E. E.; Pierson III, L. S. et al. Optimization of production and structure-activity relationship studies of monocillin I, the cytotoxic constituent of *Paraphaeosphaeria quadriseptata*. *J. Antibiot.* **2004**, *57*, 541–546.
20. Wijeratne, E. K.; Paranagama, P. A.; Gunatilaka, A. L. Five new isocoumarins from Sonoran desert plant-associated fungal strains *Paraphaeosphaeria quadriseptata* and *Chaetomium chiversii*. *Tetrahedron* **2006**, *62*, 8439–8446.
21. Liu, R.; Gu, Q.; Zhu, W.; Cui, C.; Fan, G.; Fang, Y.; Zhu, T.; Liu, H. 10-Phenyl-[12]-cytochalasins Z_7, Z_8, and Z_9 from the marine-derived fungus *Spicaria elegans*. *J. Nat. Prod.* **2006**, *69*, 871–875.
22. Liu, R.; Lin, Z.; Zhu, T.; Fang, Y.; Gu, Q.; Zhu, W. Novel open-chain cytochalsins from the marine-derived fungus *Spicaria elegans*. *J. Nat. Prod.* **2008**, *71*, 1127–1132.
23. Lin, Z.; Zhu, T.; Wei, H.; Zhang, G.; Wang, H.; Gu, Q. Spicochalasin A and new aspochalasins from the marine-derived fungus *Spicaria elegans*. *Eur. J. Org. Chem.* **2009**, *2009*, 3045–3051.
24. Lin, Z.-J.; Zhu, T.-J.; Zhang, G.-J.; Wei, H.-J.; Gu, Q.-Q. Deoxy-cytochalasins from a marine-derived fungus *Spicaria elegans*. *Can. J. Chem.* **2009**, *87*, 486–489.
25. Lin, Z. J.; Zhu, T. J.; Chen, L.; Gu, Q. Q. Three new aspochalasin derivatives from the marine-derived fungus *Spicaria elegans*. *Chin. Chem. Lett.* **2010**, *21*, 824–826.
26. Wang, F. Z.; Wei, H. J.; Zhu, T. J.; Li, D. H.; Lin, Z. J.; Gu, Q. Q. Three new cytochalasins from the marine-derived fungus *Spicaria elegans* KLA03 by supplementing the cultures with L-and D-tryptophan. *Chem. Biodivers.* **2011**, *8*, 887–894.
27. Luan, Y.; Wei, H.; Zhang, Z.; Che, Q.; Liu, Y.; Zhu, T.; Mándi, A.; Kurtán, T.; Gu, Q.; Li, D. Eleganketal A, a highly oxygenated dibenzospiroketal from the marine-derived fungus *Spicaria elegans* KLA03. *J. Nat. Prod.* **2014**, *77*, 1718–1723.

28. Pimenta, E. F.; Vita-Marques, A. M.; Tininis, A.; Seleghim, M. H.; Sette, L. D.; Veloso, K.; Ferreira, A. G.; Williams, D. E.; Patrick, B. O.; Dalisay, D. S. Use of experimental design for the optimization of the production of new secondary metabolites by two *Penicillium* species. *J. Nat. Prod.* **2010**, *73*, 1821–1832.
29. Nenkep, V.; Yun, K.; Zhang, D.; Choi, H. D.; Kang, J. S.; Son, B. W. Induced production of bromomethylchlamydosporols A and B from the marine-derived fungus *Fusarium tricinctum*. *J. Nat. Prod.* **2010**, *73*, 2061–2063.
30. Singh, M. P.; Leighton, M. M.; Barbieri, L. R.; Roll, D. M.; Urbance, S. E.; Hoshan, L.; McDonald, L. A. Fermentative production of self-toxic fungal secondary metabolites. *J. Ind. Microbiol. Biotechnol.* **2010**, *37*, 335–340.
31. Wijeratne, E. K.; Bashyal, B. P.; Gunatilaka, M. K.; Arnold, A. E.; Gunatilaka, A. L. Maximizing chemical diversity of fungal metabolites: Biogenetically related heptaketides of the endolichenic fungus *Corynespora* sp. *J. Nat. Prod.* **2010**, *73*, 1156–1159.
32. Paranagama, P. A.; Wijeratne, E. K.; Burns, A. M.; Marron, M. T.; Gunatilaka, M. K.; Arnold, A. E.; Gunatilaka, A. L. Heptaketides from *Corynespora* sp. inhabiting the cavern beard lichen, *Usnea cavernosa*: First report of metabolites of an endolichenic fungus. *J. Nat. Prod.* **2007**, *70*(1), 1700–1705.
33. Zain, M.; El-Sheikh, H.; Soliman, H.; Khalil, A. Effect of certain chemical compounds on secondary metabolites of *Penicillium janthinellum* and *P. duclauxii*. *J. Saudi Chem. Soc.* **2011**, *15*, 239–246.
34. Zhao, J.; Zheng, B.; Li, Y.; Shan, T.; Mou, Y.; Lu, S.; Li, P.; Zhou, L. Enhancement of diepoxin ζ production by yeast extract and its fractions in liquid culture of *Berkleasmium*-like endophytic fungus Dzf12 from *Dioscorea zingiberensis*. *Molecules* **2011**, *16*, 847–856.
35. Cai, X.; Shan, T.; Li, P.; Huang, Y.; Xu, L.; Zhou, L.; Wang, M.; Jiang, W. Spirobisnaphthalenes from the endophytic fungus Dzf12 of *Dioscorea zingiberensis* and their antimicrobial activities. *Nat. Prod. Commun.* **2009**, *4*, 1469–1472.
36. Li, Y.; Li, P.; Mou, Y.; Zhao, J.; Shan, T.; Ding, C.; Zhou, L. Enhancement of diepoxin ζ production in liquid culture of endophytic fungus *Berkleasmium* sp. Dzf12 by polysaccharides from its host plant *Dioscorea zingiberensis*. *World J. Microbiol. Biotechnol.* **2012**, *28*, 1407–1413.
37. Li, Y.; Shan, T.; Mou, Y.; Li, P.; Zhao, J.; Zhao, W.; Peng, Y.; Zhou, L.; Ding, C. Enhancement of palmarumycin C_{12} and C_{13} production in liquid culture of the endophytic fungus *Berkleasmium* sp. Dzf12 by oligosaccharides from its host plant *Dioscorea zingiberensis*. *Molecules* **2012**, *17*, 3761–3773.
38. Mou, Y.; Luo, H.; Mao, Z.; Shan, T.; Sun, W.; Zhou, K.; Zhou, L. Enhancement of palmarumycins C_{12} and C_{13} production in liquid culture of endophytic fungus *Berkleasmium* sp. Dzf12 after treatments with metal ions. *Int. J. Mol. Sci.* **2013**, *14*, 979–998.
39. Shan, T.; Tian, J.; Wang, X.; Mou, Y.; Mao, Z.; Lai, D.; Dai, J.; Peng, Y.; Zhou, L.; Wang, M. Bioactive spirobisnaphthalenes from the endophytic fungus *Berkleasmium* sp. *J. Nat. Prod.* **2014**, *77*, 2151–2160.
40. Mou, Y.; Xu, D.; Mao, Z.; Dong, X.; Lin, F.; Wang, A.; Lai, D.; Zhou, L.; Xie, B. Enhancement of palmarumycin c12 and c13 production by the endophytic fungus *Berkleasmium* sp. Dzf12 in an aqueous-organic solvent system. *Molecules* **2015**, *20*, 20320–20333.
41. Ding, G.; Qi, Y.; Liu, S.; Guo, L.; Chen, X. Photipyrones A and B, new pyrone derivatives from the plant endophytic fungus *Pestalotiopsis photiniae*. *J. Antibiot.* **2012**, *65*, 271–273.
42. Ding, G.; Zheng, Z.; Liu, S.; Zhang, H.; Guo, L.; Che, Y. Photinides A–F, cytotoxic benzofuranone-derived γ-lactones from the plant endophytic fungus *Pestalotiopsis photiniae*. *J. Nat. Prod.* **2009**, *72*, 942–945.
43. El-Neketi, M.; Ebrahim, W.; Lin, W.; Gedara, S.; Badria, F.; Saad, H.-E. A.; Lai, D.; Proksch, P. Alkaloids and polyketides from *Penicillium citrinum*, an endophyte isolated from the Moroccan plant *Ceratonia siliqua*. *J. Nat. Prod.* **2013**, *76*, 1099–1104.

44. Adler, M. J.; Chase, C. A. Comparison of the allelopathic potential of leguminous summer cover crops: Cowpea, sunn hemp, and velvetbean. *HortScience* **2007**, *42*, 289–293.
45. Li, X.-J.; Zhang, Q.; Zhang, A.-L.; Gao, J.-M. Metabolites from *Aspergillus fumigatus*, an endophytic fungus associated with *Melia azedarach*, and their antifungal, antifeedant, and toxic activities. *J. Agric. Food Chem.* **2012**, *60*, 3424–3431.
46. Zhang, Q.; Wang, S.-Q.; Tang, H.-Y.; Li, X.-J.; Zhang, L.; Xiao, J.; Gao, Y.-Q.; Zhang, A.-L.; Gao, J.-M. Potential allelopathic indole diketopiperazines produced by the plant endophytic *Aspergillus fumigatus* using the one strain–many compounds method. *J. Agric. Food Chem.* **2013**, *61*, 11447–11452.
47. (a) Wang, Q. X.; Bao, L.; Yang, X. L.; Guo, H.; Ren, B.; Guo, L. D.; Song, F. H.; Wang, W. Z.; Liu, H. W.; Zhang, L. X., Tricycloalternarenes F–H: Three new mixed terpenoids produced by an endolichenic fungus *Ulocladium* sp. using OSMAC method. *Fitoterapia* **2013**, *85*, 8–13; (b) Wang, Q.-X.; Bao, L.; Yang, X.-L.; Liu, D.-L.; Guo, H.; Dai, H.-Q.; Song, F.-H.; Zhang, L.-X.; Guo, L.-D.; Li, S.-J. Ophiobolins P–T, five new cytotoxic and antibacterial sesterterpenes from the endolichenic fungus *Ulocladium* sp. *Fitoterapia* **2013**, *90*, 220–227.
48. Du, L.; Feng, T.; Zhao, B.; Li, D.; Cai, S.; Zhu, T.; Wang, F.; Xiao, X.; Gu, Q. Alkaloids from a deep ocean sediment-derived fungus *Penicillium* sp. and their antitumor activities. *J. Antibiot.* **2010**, *63*, 165–170.
49. Du, L.; Li, D.; Zhu, T.; Cai, S.; Wang, F.; Xiao, X.; Gu, Q. New alkaloids and diterpenes from a deep ocean sediment derived fungus *Penicillium* sp. *Tetrahedron* **2009**, *65*, 1033–1039.
50. Guo, W.; Peng, J.; Zhu, T.; Gu, Q.; Keyzers, R. A.; Li, D. Sorbicillamines A–E, nitrogen-containing sorbicillinoids from the deep-sea-derived fungus *Penicillium* sp. F23-2. *J. Nat. Prod.* **2013**, *76*, 2106–2112.
51. Guo, W.; Zhang, Z.; Zhu, T.; Gu, Q.; Li, D. Penicyclones A–E, antibacterial polyketides from the deep-sea-derived fungus *Penicillium* sp. F23-2. *J. Nat. Prod.* **2015**, *78*, 2699–2703.
52. Bara, R.; Aly, A. H.; Pretsch, A.; Wray, V.; Wang, B.; Proksch, P.; Debbab, A. Antibiotically active metabolites from *Talaromyces wortmannii*, an endophyte of *Aloe vera*. *J. Antibiot.* **2013**, *66*, 491–493.
53. Bara, R.; Aly, A. H.; Wray, V.; Lin, W.; Proksch, P.; Debbab, A. Talaromins A and B, new cyclic peptides from the endophytic fungus *Talaromyces wortmannii*. *Tetrahedron Lett.* **2013**, *54*, 1686–1689.
54. Bara, R.; Zerfass, I.; Aly, A. H.; Goldbach-Gecke, H.; Raghavan, V.; Sass, P.; Mándi, A.; Wray, V.; Polavarapu, P. L.; Pretsch, A. Atropisomeric dihydroanthracenones as inhibitors of multiresistant *Staphylococcus aureus*. *J. Med. Chem.* **2013**, *56*, 3257–3272.
55. Strongman, D. B.; Strunz, G. M.; Giguere, P.; Yu, C.-M.; Calhoun, L. Enniatins from *Fusarium avenaceum* isolated from balsam fir foliage and their toxicity to spruce budworm larvae, *Choristoneura fumiferana* (Clem.) (Lepidoptera: Tortricidae). *J. Chem. Ecol.* **1988**, *14*, 753–764.
56. Tomoda, H.; Huang, X.-H.; Caoa, J.; Nishida, H.; Nagao, R.; Okuda, S.; Tanaka, H.; Omura, S.; Arai, H.; Inoue, K. Inhibition of acyl-CoA: Cholesterol acyltransferase activity by cyclodepsiptide antibiotics. *J. Antibiot.* **1992**, *45*, 1626–1632.
57. Wang, J.-P.; Debbab, A.; Pérez Hemphill, C. F.; Proksch, P. Optimization of enniatin production by solid-phase fermentation of *Fusarium tricinctum*. *Z. Naturforsch. C* **2013**, *68*, 223–230.
58. Li, C.-S.; Li, X.-M.; Gao, S.-S.; Lu, Y.-H.; Wang, B.-G. Cytotoxic anthranilic acid derivatives from deep sea sediment-derived fungus *Penicillium paneum* SD-44. *Mar. Drugs* **2013**, *11*, 3068–3076.
59. Li, C.-S.; An, C.-Y.; Li, X.-M.; Gao, S.-S.; Cui, C.-M.; Sun, H.-F.; Wang, B.-G. Triazole and dihydroimidazole alkaloids from the marine sediment-derived fungus *Penicillium paneum* SD-44. *J. Nat. Prod.* **2011**, *74*, 1331–1334.

60. Zhou, L. N.; Gao, H. Q.; Cai, S. X.; Zhu, T. J.; Gu, Q. Q.; Li, D. H. Two new cyclic pentapeptides from the marine-derived fungus *Aspergillus versicolor*. *Helv. Chim. Acta* **2011**, *94*, 1065–1070.
61. Gao, H.; Zhou, L.; Cai, S.; Zhang, G.; Zhu, T.; Gu, Q.; Li, D. Diorcinols B-E, new prenylated diphenyl ethers from the marine-derived fungus *Aspergillus versicolor* ZLN-60. *J. Antibiot.* **2013**, *66*, 539–542.
62. Peng, J.; Gao, H.; Zhang, X.; Wang, S.; Wu, C.; Gu, Q.; Guo, P.; Zhu, T.; Li, D. Psychrophilins E–H and versicotide c, cyclic peptides from the marine-derived fungus *Aspergillus versicolor* ZLN-60. *J. Nat. Prod.* **2014**, *77*, 2218–2223.
63. Wang, W.-J.; Li, D.-Y.; Li, Y.-C.; Hua, H.-M.; Ma, E.-L.; Li, Z.-L. Caryophyllene sesquiterpenes from the marine-derived fungus *Ascotricha* sp. ZJ-M-5 by the one strain–many compounds strategy. *J. Nat. Prod.* **2014**, *77*, 1367–1371.
64. Hewage, R. T.; Aree, T.; Mahidol, C.; Ruchirawat, S.; Kittakoop, P. One strain-many compounds (OSMAC) method for production of polyketides, azaphilones, and an isochromanone using the endophytic fungus *Dothideomycete* sp. *Phytochemistry* **2014**, *108*, 87–94.
65. Senadeera, S. P.; Wiyakrutta, S.; Mahidol, C.; Ruchirawat, S.; Kittakoop, P. A novel tricyclic polyketide and its biosynthetic precursor azaphilone derivatives from the endophytic fungus *Dothideomycete* sp. *Org. Biomol. Chem.* **2012**, *10*, 7220–7226.
66. Chen, M.; Shao, C.-L.; Fu, X.-M.; Kong, C.-J.; She, Z.-G.; Wang, C.-Y. Lumazine peptides penilumamides B–D and the cyclic pentapeptide asperpeptide A from a gorgonian-derived *Aspergillus* sp. fungus. *J. Nat. Prod.* **2014**, *77*, 1601–1606.
67. Chen, M.; Shao, C.-L.; Meng, H.; She, Z.-G.; Wang, C.-Y. Anti-respiratory syncytial virus prenylated dihydroquinolone derivatives from the gorgonian-derived fungus *Aspergillus* sp. XS-20090B15. *J. Nat. Prod.* **2014**, *77*, 2720–2724.
68. Zheng, Q.-C.; Chen, G.-D.; Kong, M.-Z.; Li, G.-Q.; Cui, J.-Y.; Li, X.-X.; Wu, Z.-Y.; Guo, L.-D.; Cen, Y.-Z.; Zheng, Y.-Z. Nodulisporisteriods A and B, the first 3, 4-seco-4-methyl-progesteroids from *Nodulisporium* sp. *Steroids* **2013**, *78*, 896–901.
69. Zhao, Q.; Wang, G.-Q.; Chen, G.-D.; Hu, D.; Li, X.-X.; Guo, L.-D.; Li, Y.; Yao, X.-S.; Gao, H. Nodulisporisteroids C–L, new 4-methyl-progesteroid derivatives from *Nodulisporium* sp. *Steroids* **2015**, *102*, 101–109.
70. Yu, G.-H.; Wu, G.-W.; Zhu, T.-J.; Gu, Q.-Q.; Li, D.-H. Cladosins F and G, two new hybrid polyketides from the deep-sea-derived *Cladosporium sphaerospermum* 2005-01-E3. *J. Asian Nat. Prod. Res.* **2015**, *17*, 120–124.
71. Seephonkai, P.; Kongsaeree, P.; Prabpai, S.; Isaka, M.; Thebtaranonth, Y. Transformation of an irregularly bridged epidithiodiketopiperazine to trichodermamide A. *Org. Lett.* **2006**, *8*, 3073–3075.
72. Yamazaki, H.; Takahashi, O.; Murakami, K.; Namikoshi, M. Induced production of a new unprecedented epitrithiodiketopiperazine, chlorotrithiobrevamide, by a culture of the marine-derived *Trichoderma cf. brevicompactum* with dimethyl sulfoxide. *Tetrahedron Lett.* **2015**, *56*, 6262–6265.
73. Yamazaki, H.; Rotinsulu, H.; Narita, R.; Takahashi, R.; Namikoshi, M. Induced production of halogenated epidithiodiketopiperazines by a marine-derived *Trichoderma cf. brevicompactum* with sodium halides. *J. Nat. Prod.* **2015**, *78*, 2319–2321.
74. Xu, R.; Xu, G.-M.; Li, X.-M.; Li, C.-S.; Wang, B.-G. Characterization of a newly isolated marine fungus *Aspergillus dimorphicus* for optimized production of the anti-tumor agent wentilactones. *Mar. Drugs* **2015**, *13*, 7040–7054.
75. Boruta, T.; Bizukojc, M. Induction of secondary metabolism of *Aspergillus terreus* ATCC 20542 in the batch bioreactor cultures. *Appl. Microbiol. Biotechnol.* **2016**, *100*(7), 3009–3022.
76. Yuan, C.; Wang, H.-Y.; Wu, C.-S.; Jiao, Y.; Li, M.; Wang, Y.-Y.; Wang, S.-Q.; Zhao, Z.-T.; Lou, H.-X. Austdiol, fulvic acid and citromycetin derivatives from an endolichenic fungus *Myxotrichum* sp. *Phytochem. Lett.* **2013**, *6*, 662–666.

77. Yuan, C.; Guo, Y.-H.; Wang, H.-Y.; Ma, X.-J.; Jiang, T.; Zhao, J.-L.; Zou, Z.-M.; Ding, G. Allelopathic polyketides from an endolichenic fungus *myxotrichum* sp. by using OSMAC strategy. *Sci. Rep.* **2016**, *6*, 19350.
78. El-Hawary, S. S.; Mohammed, R.; AbouZid, S. F.; Bakeer, W.; Ebel, R.; Sayed, A. M.; Rateb, M. E. Solamargine production by a fungal endophyte of *Solanum nigrum*. *J. Appl. Microbiol.* **2016**, *120*, 900–911.
79. Adpressa, D. A.; Loesgen, S. Bioprospecting chemical diversity and bioactivity in a marine derived *Aspergillus terreus*. *Chem. Biodivers.* **2016**, *13*, 253–259.
80. Sica, V. P.; Figueroa, M.; Raja, H. A.; El-Elimat, T.; Darveaux, B. A.; Pearce, C. J.; Oberlies, N. H. Optimizing production and evaluating biosynthesis in situ of a herbicidal compound, mevalocidin, from *Coniolariella* sp. *J. Ind. Microbiol. Biotechnol.* **2016**, *43*(8), 1149–1157.
81. Gerwick, B. C.; Brewster, W. K.; Fields, S. C.; Graupner, P. R.; Hahn, D. R.; Pearce, C. J.; Schmitzer, P. R.; Webster, J. D. Mevalocidin: A novel, phloem mobile phytotoxin from *Fusarium* DA056446 and *Rosellinia* DA092917. *J. Chem. Ecol.* **2013**, *39*, 253–261.
82. Debbab, A.; Aly, A. H.; Edrada-Ebel, R.; Wray, V.; Müller, W. E. G.; Totzke, F.; Zirrgiebel, U. et al. Bioactive metabolites from the endophytic fungus *Stemphylium globuliferum* isolated from *Mentha pulegium*. *J. Nat. Prod.* **2009**, *72*, 626–631.
83. Liu, Y.; Marmann, A.; Abdel-Aziz, M. S.; Wang, C. Y.; Müller, W. E. G.; Lin, W. H.; Mándi, A.; Kurtán, T.; Daletos, G.; Proksch, P. Tetrahydroanthraquinone derivatives from the endophytic fungus *Stemphylium globuliferum*. *Eur. J. Org. Chem.* **2015**, *2015*, 2646–2653.
84. Debbab, A.; Aly, A. H.; Edrada-Ebel, R.; Wray, V.; Pretsch, A.; Pescitelli, G.; Kurtan, T.; Proksch, P. New anthracene derivatives–structure elucidation and antimicrobial activity. *Eur. J. Org. Chem.* **2012**, *2012*, 1351–1359.
85. Moussa, M.; Ebrahim, W.; El-Neketi, M.; Mándi, A.; Kurtán, T.; Hartmann, R.; Lin, W. H., Liu, Z.; Proksch, P. Tetrahydroanthraquinone derivatives from the mangrove-derived endophytic fungus *Stemphylium globuliferum*. *Tetrahedron Lett.* **2016**, *57*, 4074–4078.
86. (a) Cueto, M.; Jensen, P. R.; Kauffman, C.; Fenical, W.; Lobkovsky, E.; Clardy, J. Pestalone, a new antibiotic produced by a marine fungus in response to bacterial challenge. *J. Nat. Prod.* **2001**, *64*, 1444–1446; (b) In a recent study, pestalone was tested against various MRSA strains; however, it did not show the expected degree of activity (7–22 μM): Augner, D.; Krut, O.; Slavov, N.; Gerbino, D. C.; Sahl, H.-G.; Benting, J.; Nising, C. F.; Hillebrand, S.; Krönke, M.; Schmalz, H.-G. On the antibiotic and antifungal activity of pestalone, pestalachloride A, and structurally related compounds. *J. Nat. Prod.* **2013**, *76*, 1519–1522.
87. Oh, D.-C.; Jensen, P. R.; Kauffman, C. A.; Fenical, W. Libertellenones A–D: Induction of cytotoxic diterpenoid biosynthesis by marine microbial competition. *Bioorg. Med. Chem.* **2005**, *13*, 5267–5273.
88. Degenkolb, T.; Heinze, S.; Schlegel, B.; Strobel, G.; Gräfe, U. Formation of new lipoaminopeptides, acremostatins A, B, and C, by co-cultivation of *Acremonium* sp. Tbp-5 and *Mycogone rosea* DSM 12973. *Biosci. Biotechnol. Biochem.* **2002**, *66*, 883–886.
89. Zhu, F.; Lin, Y. Marinamide, a novel alkaloid and its methyl ester produced by the application of mixed fermentation technique to two mangrove endophytic fungi from the South China Sea. *Chin. Sci. Bull.* **2006**, *51*, 1426–1430.
90. Zhu, F.; Chen, G.; Wu, J.; Pan, J. Structure revision and cytotoxic activity of marinamide and its methyl ester, novel alkaloids produced by co-cultures of two marine-derived mangrove endophytic fungi. *Nat. Prod. Res.* **2013**, *27*, 1960–1964.
91. Zhu, F.; Lin, Y.; Ding J.; Wang, X.; Huan, L. Secondary metabolites of two marine-derived mangrove endophytic fungi (strains nos. 1924# and 3893#) by mixed fermentation. *Chem. Ind. Forest Prod.* **2007**, *27*, 8–10.

92. Park, H. B.; Kwon, H. C.; Lee, C.-H.; Yang, H. O. Glionitrin A, an antibiotic—Antitumor metabolite derived from competitive interaction between abandoned mine microbes. *J. Nat. Prod.* **2009**, *72*, 248–252.
93. Park, H. B.; Kim, Y.-J.; Park, J.-S.; Yang, H. O.; Lee, K. R.; Kwon, H. C. Glionitrin B, a cancer invasion inhibitory diketopiperazine produced by microbial coculture. *J. Nat. Prod.* **2011**, *74*, 2309–2312.
94. Graniti, A.; Surico, G.; Mugnai, L. Esca of grapevine: A disease complex or a complex of diseases? *Phytopathol. Mediterr.* **2000**, *39*, 16–20.
95. Bruno, G.; Sparapano, L. Effects of three esca-associated fungi on *Vitis vinifera* L.: II. Characterization of biomolecules in xylem sap and leaves of healthy and diseased vines. *Physiol. Mol. Plant Pathol.* **2006**, *69*, 195–208.
96. Glauser, G.; Gindro, K.; Fringeli, J.; De Joffrey, J.-P.; Rudaz, S.; Wolfender, J.-L. Differential analysis of mycoalexins in confrontation zones of grapevine fungal pathogens by ultrahigh pressure liquid chromatography/time-of-flight mass spectrometry and capillary nuclear magnetic resonance. *J. Agric. Food Chem.* **2009**, *57*, 1127–1134.
97. Schroeckh, V.; Scherlach, K.; Nützmann, H.-W.; Shelest, E.; Schmidt-Heck, W.; Schuemann, J.; Martin, K.; Hertweck, C.; Brakhage, A. A. Intimate bacterial-fungal interaction triggers biosynthesis of archetypal polyketides in *Aspergillus nidulans*. *Proc. Natl. Acad. Sci. USA* **2009**, *106*, 14558–14563.
98. Nützmann, H.-W.; Reyes-Dominguez, Y.; Scherlach, K.; Schroeckh, V.; Horn, F.; Gacek, A.; Schümann, J.; Hertweck, C.; Strauss, J.; Brakhage, A. A. Bacteria-induced natural product formation in the fungus *Aspergillus nidulans* requires Saga/Ada-mediated histone acetylation. *Proc. Natl. Acad. Sci. USA* **2011**, *108*, 14282–14287.
99. König, C. C.; Scherlach, K.; Schroeckh, V.; Horn, F.; Nietzsche, S.; Brakhage, A. A.; Hertweck, C. Bacterium induces cryptic meroterpenoid pathway in the pathogenic fungus *Aspergillus fumigatus*. *ChemBioChem* **2013**, *14*, 938–942.
100. (a) Pettit, R. K.; Pettit, G. R.; Xu, J.-P.; Weber, C. A.; Richert, L. A. Isolation of human cancer cell growth inhibitory, antimicrobial lateritin from a mixed fungal culture. *Planta Med.* **2010**, *76* , 500–501; (b) In a recent study, it was suggested that the reported structure of lateritin should be revised to be identical with the trimeric depsipeptide beauvericin: Ola, A. R. B.; Aly, A. H.; Lin, W.; Wray, V.; Debbab, A. Structural revision and absolute configuration of lateritin. *Tetrahedron Lett.* **2014**, *55*, 6184–6187.
101. Rateb, M. E.; Hallyburton, I.; Houssen, W. E.; Bull, A. T.; Goodfellow, M.; Santhanam, R.; Jaspars, M.; Ebel, R. Induction of diverse secondary metabolites in *Aspergillus fumigatus* by microbial co-culture. *RSC Adv.* **2013**, *3*, 14444–14450.
102. Wang, J.-P.; Lin, W.; Wray, V.; Lai, D.; Proksch, P. Induced production of depsipeptides by co-culturing *Fusarium tricinctum* and *Fusarium begoniae*. *Tetrahedron Lett.* **2013**, *54*, 2492–2496.
103. Ola, A. R. B.; Thomy, D.; Lai, D.; Brötz-Oesterhelt, H.; Proksch, P. Inducing secondary metabolite production by the endophytic fungus *Fusarium tricinctum* through coculture with *Bacillus subtilis*. *J. Nat. Prod.* **2013**, *76*, 2094–2099.
104. Sørensen, J. L.; Nielsen, K. F.; Sondergaard, T. E. Redirection of pigment biosynthesis to isocoumarins in *Fusarium*. *Fungal Genet. Biol.* **2012**, *49*, 613–618.
105. Bertrand, S.; Schumpp, O.; Bohni, N.; Monod, M.; Gindro, K.; Wolfender, J.-L. De novo production of metabolites by fungal co-culture of *Trichophyton rubrum* and *Bionectria ochroleuca*. *J. Nat. Prod.* **2013**, *76*, 1157–1165.
106. (a) Chagas, F. O.; Dias, L. G.; Pupo, M. T. A mixed culture of endophytic fungi increases production of antifungal polyketides. *J. Chem. Ecol.* **2013**, *39*, 1335–1342; (b) The absolute configuration of stemphyperylenol was subsequently revised as shown in Figure 8.14: Podlech, J.; Fleck, S. C.; Metzler, M.; Bürck, J.; Ulrich, A. S. Determination of the absolute configuration of perylene quinone-derived mycotoxins by measurement and calculation of electronic circular dichroism spectra and specific rotations. *Chem. Eur. J.* **2014**, *20*, 11463–11470.

107. Gagnon, C. Fungus culture on cellophane membrane for cytochemical tests. *Stain Technol.* **1966**, *41*, 247–251.
108. Whitt, J.; Shipley, S. M.; Newman, D. J.; Zuck, K. M. Tetramic acid analogues produced by coculture of *Saccharopolyspora erythraea* with *Fusarium pallidoroseum*. *J. Nat. Prod.* **2014**, *77*, 173–177.
109. Marfori E. C.; Kajiyama, S.; Fukusaki, E.; Kobayashi, A. Trichosetin, a novel tetramic acid antibiotic produced in dual culture of *Trichoderma harzianum* and *Catharanthus roseus* callus. *Z. Naturforsch. C* **2002**, *57*, 465–470.
110. Meng, L.-H.; Liu, Y.; Li, X.-M.; Xu, G.-M.; Ji, N.-Y.; Wang, B.-G. Citrifelins A and B, citrinin adducts with a tetracyclic framework from cocultures of marine-derived isolates of *Penicillium citrinum* and *Beauveria felina*. *J. Nat. Prod.* **2015**, *78*, 2301–2305.
111. Chen, H.; Aktas, N.; Konuklugil, B.; Mándi, A.; Daletos, G.; Lin, W.; Dai, H.; Kurtán, T.; Proksch, P. A new fusarielin analogue from *Penicillium* sp. isolated from the Mediterranean sponge *Ircinia oros*. *Tetrahedron Lett.* **2015**, *56*, 5317–5320.
112. Chen, H.; Daletos, G.; Abdel-Aziz, M. S.; Thomy, D.; Dai, H.; Brötz-Oesterhelt, H.; Lin, W.; Proksch, P. Inducing secondary metabolite production by the soil-dwelling fungus *Aspergillus terreus* through bacterial co-culture. *Phytochem. Lett.* **2015**, *12*, 35–41.
113. Nonaka, K.; Chiba, T.; Suga, T.; Asami, Y.; Iwatsuki, M.; Masuma, R.; Ōmura, S.; Shiomi, K. Coculnol, a new penicillic acid produced by a coculture of *Fusarium solani* FKI-6853 and *Talaromyces* sp. FKA-65. *J. Antibiot.* **2015**, *68*, 530–532.
114. Wu, C.; Zacchetti, B.; Ram, A. F. J.; van Wezel, G. P.; Claessen, D.; Hae Choi, Y. Expanding the chemical space for natural products by *Aspergillus-Streptomyces* co-cultivation and biotransformation. *Sci. Rep.* **2015**, *5*, 10868.
115. Ebrahim, W.; El-Neketi, M.; Lewald, L.-I.; Orfali, R. S.; Lin, W.; Rehberg, N.; Kalscheuer, R.; Daletos, G.; Proksch, P. Metabolites from the fungal endophyte *Aspergillus austroafricanus* in axenic culture and in fungal-bacterial mixed cultures. *J. Nat. Prod.* **2016**, *79*, 914–922.
116. Bohni, N.; Hofstetter, V.; Gindro, K.; Buyck, B.; Schumpp, O.; Bertrand, S.; Monod, M.; Wolfender, J.-L. Production of fusaric acid by *Fusarium* spp. in pure culture and in solid medium co-cultures. *Molecules* **2016**, *21*, 370.
117. Cragg, G. M.; Newman, D. J. Natural products: A continuing source of novel drug leads. *Biochim. Biophys. Acta* **2013**, *1830*, 3670–3695.
118. Debbab, A.; Aly, A. H.; Lin, W. H.; Proksch, P. Bioactive compounds from marine bacteria and fungi. *Microb. Biotechnol.* **2010**, *3*, 544–563.
119. Bergmann, S.; Schümann, J.; Scherlach, K.; Lange, C.; Brakhage, A. A.; Hertweck, C. Genomics-driven discovery of PKS-NRPS hybrid metabolites from *Aspergillus nidulans*. *Nat. Chem. Biol.* **2007**, *3*, 213–217.
120. Bode, H. B.; Bethe, B.; Höfs, R.; Zeeck, A. Big effects from small changes: Possible ways to explore nature's chemical diversity. *ChemBioChem* **2002**, *3*, 619–627.
121. Pettit, R. K. Small-molecule elicitation of microbial secondary metabolites. *Microb. Biotechnol.* **2011**, *4*, 471–478.
122. Scherlach, K.; Hertweck, C. Triggering cryptic natural product biosynthesis in microorganisms. *Org. Biomol. Chem.* **2009**, *7*, 1753–1760.
123. Kurosawa, K.; MacEachran, D. P.; Sinskey, A. J. Antibiotic biosynthesis following horizontal gene transfer: New milestone for novel natural product discovery? *Expert Opin. Drug Discov.* **2010**, *5*, 819–825.
124. Kalia, V. C. Quorum sensing inhibitors: An overview. *Biotechnol. Adv.* **2013**, *31*, 224–245.
125. Bertrand, S.; Bohni, N.; Schnee, S.; Schumpp, O.; Gindro, K.; Wolfender, J.-L. Metabolite induction via microorganism co-culture: A potential way to enhance chemical diversity for drug discovery. *Biotechnol. Adv.* **2014**, *32*, 1180–1204.

9 Natural Products of the Rhizosphere and Its Microorganisms

Bioactivities and Implications of Occurrence

Maria C. F. de Oliveira, Jair Mafezoli, and A. A. Leslie Gunatilaka

CONTENTS

9.1 GENERAL INTRODUCTION

The rhizosphere, a term coined and defined by Hiltner in 1904, is the biologically active zone of the soil around roots of terrestrial plants.[1] Recent studies have shown that the rhizosphere is a complex interface between plant roots and soil, harboring a myriad of microorganisms and invertebrates that interact with each other and the plant, affecting growth and stress tolerance of plants and biogeochemical cycling. Plants are known to secrete up to 40% of their photosynthates into the rhizosphere, attracting microbes and making the microbial population densities in the rhizosphere much higher than those of the surrounding bulk soil.[2] These rhizosphere microorganisms then establish a functional diversity, including decomposition of organic matter, nitrogen fixation, solubilization of phosphate, transformation of sulfur and iron, production of siderophores (iron-binding compounds), and release of plant hormones and other natural products useful for biotic control.[3] Thus, the rhizosphere is intriguingly complex and dynamic, and understanding its evolution and ecology is considered key to enhancing plant productivity and ecosystem functioning.[4] The rhizosphere ecosystem is also affected by plant root rhizodeposition products, which are composed of exudates, lysates, mucilage, secretions, and dead cell material, as well as volatile organic compounds (VOCs) and gases, including respiratory carbon dioxide.[5] Some of these rhizodeposition products contain plant primary metabolites such as organic acids, carbohydrates, and amino acids and an array of secondary metabolites (small-molecule natural products).[6] Many small-molecule rhizodeposition products and root exudates have a strong impact in the rhizosphere as they mediate plant–microbial, microbial–microbial, plant–plant, and plant–animal interactions with significant implications in agronomy and ecology (Figure 9.1).

The rhizosphere microbiome (rhizomicrobiome) is highly diverse and includes both bacteria and fungi. This enormous and complex plant-associated microbial community, also referred to as the second genome of the plant, has recently been compared with the microbial communities in the human gut, as similar functions can be ascribed to both these microbiomes.[7] Some common functions of these microbiomes

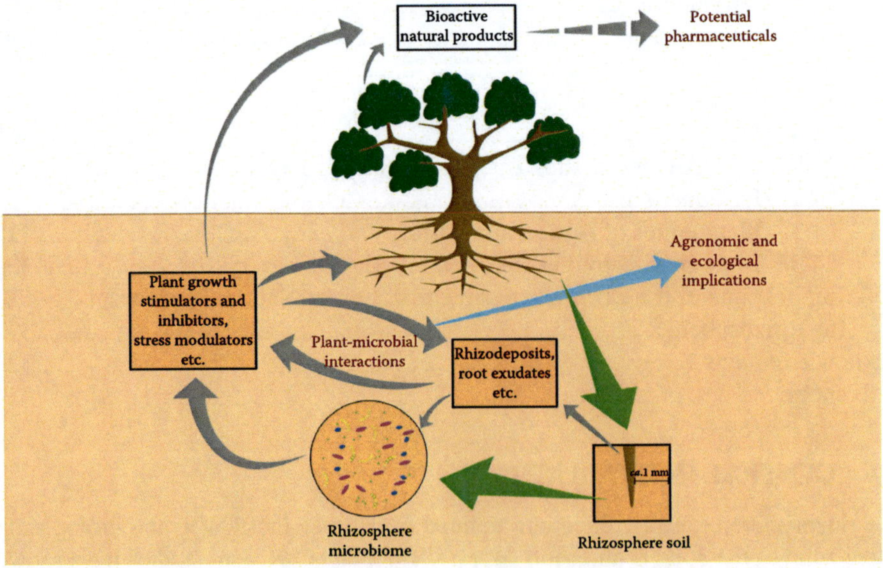

FIGURE 9.1 Natural products–mediated interactions in the rhizosphere, their agronomic and ecological implications, and importance as potential pharmaceuticals.

include nutrient uptake, prevention of colonization by pathogens, modulation of host immunity, and control of microbial density and diversity. These functions are known to be associated with a variety of small-molecule natural products, which may exert these functions by interacting with biological molecules. These small-molecule natural products, in addition to their agronomic and ecological implications, may therefore exhibit a variety of beneficial biological activities with potential applications in human and animal health (Figure 9.1). This chapter summarizes the distribution, structural diversity, biological activity, and implications of the occurrence of natural products of the rhizosphere, which are derived from its associated plants and microorganisms. Also discussed is the importance of these and small-molecule rhizodeposition products in a variety of interactions between the rhizosphere organisms. Soil-borne fungi, especially those that occur in the rhizosphere, are known to develop symbiotic structures (mycorrhiza) with many terrestrial plants, creating another sphere called the mycorrhizosphere. Fungi of the mycorrhizosphere, comprising ectomycorrhizal (ECM) and endomycorrhizal (arbuscular mycorrhizal, AM) fungi, are of immense importance to the functioning and the ecological success of the associated plant. As AM fungi reside within plants similar to endophytic fungi, natural products of these will not be considered in this chapter. However, where relevant, the effect of natural products of the rhizosphere on AM fungi will be considered here.

9.2 RHIZOSPHERE MICROBIAL DIVERSITY

9.2.1 Introduction

The rhizosphere plays host to rich and diverse communities of bacteria and fungi. The term *microbial diversity* can be described as the propensity of microorganisms from different taxa, as well as their distribution within the taxa, including the diversity of fungi and bacteria, in microbial communities of a given ecosystem.[8] It was only about 300 years ago, with the advent of the microscope, that microbial life was known. In addition, the study of culturable microorganisms started only after the development of microbial culture techniques.[9] It has been suggested that the number of microbial species in the rhizosphere may vary from 1,000 to 1,000,000, and that about 98% of these microorganisms may not be culturable.[10] However, thanks to modern culture-independent techniques, the microbial diversity in rhizosphere soils has been shown to be much higher than in bulk soils. Using these techniques, important information about the microbial diversity, such as their richness, abundance, and functions, may also be acquired.[11–13] Among these modern approaches, molecular techniques involving the characterization of the microbial community's isolated nucleic acids are the most popular, as these can be utilized for the identification of both culturable and unculturable microorganisms. A recently developed technique for this purpose includes metagenomic (total genomic DNA) analysis, which is a powerful tool for the determination of the total genetic diversity and strain composition of a given microbial community.[6,14]

The main factors responsible for the microbial diversity of the rhizosphere include the type and zone of the host plant root, released metabolites from the plant, plant cycle (three basic stages of plant life cycle), and movement of microbes in the soil.[4,15] Because of their proximity to nutrient-rich plant root exudates, the rhizosphere microbial communities are distinct from those of the bulk soil, and the number of microorganisms occurring in the rhizosphere is usually 10–20 times greater than that in the nonrhizosphere soil.[16] Investigation of the microbial diversity of the rhizosphere and nonrhizosphere soils associated with four medicinal plants has shown that for all these plants the microbial population (fungi, bacteria, and actinomycetes) was about 1.16–1.48 fold higher in rhizosphere soils compared with bulk soils.[17] These differences in microbial populations in the rhizosphere, compared with the surrounding bulk soil, have been explained by considering the effect of plants in creating a dynamic environment where microbes can develop and interact, known as the rhizosphere effect.[7] Thus, the rhizosphere is believed to represent a natural *microbial seed bank.*

The composition of microbial communities in the rhizosphere is unique and strongly influenced by biotic and abiotic factors. Among the biotic factors, the associated plant plays an important role, and the link between these microorganisms and the associated plant is considered to have resulted from their coevolution.[4,12,14,15] The adaptation of the plant to its habitat is known to be assisted by the rhizosphere microbial diversity.[15] It has been found that the microbial diversity in the rhizosphere can be either similar or different for a given plant species, or different for the same species with different genotypes.[7] Although not conclusive, some studies have suggested that the plant plays a role in determining the microbial diversity of its rhizosphere, based on the taxonomy of the microorganisms and their functions in the rhizosphere. This latter hypothesis was supported by the finding that the diversity of bacteria capable of oxidizing ammonia to nitrogen was higher in the rhizosphere when compared with the bulk soil.[11] The diversity of bacteria and fungi in the rhizosphere soil of *Medicago truncatula* (barrel clover) at five different stages of growth (vegetative and reproductive) has been investigated.[18] It was observed that the microbial communities changed with the developmental stage of the plant. The microbial diversity in the rhizosphere of a plant is not uniform due to the variation in the type and zone of its root and the movement of its actively growing root. The influence of some natural products released by plants in their root exudates on rhizosphere microbial communities will be discussed in detail below (see Section 9.3.2).

9.2.2 Bacterial Diversity

Comparative studies of bacterial diversity in the rhizosphere and bulk soils involving cultivation-based analysis have revealed a higher population of these microorganisms in the rhizosphere. A recent study of bacterial population in the rhizosphere soil from *Origanum vulgare* (oregano) has revealed a higher density in the rhizosphere soil [$4.64 \pm 0.44 \times 10^5$ cfu (colony-forming unit) g^{-1}] than in the bulk soil ($1.71 \pm 0.51 \times 10^5$ cfu g^{-1}).[19] Most of the communities were Gram-negative bacteria, and the predominance of this group in the rhizosphere was corroborated by metagenomics analysis which included both culturable and unculturable microorganisms. These culture-independent approaches have also revealed the presence of many unexplored rhizobacterial communities from different groups such as Proteobacteria, Actinobacteria, and Acidobacteria.[12] Furthermore, recent 16S rDNA– and 16S rRNA–based analyses of the rhizobacterial communities of several agriculturally important crop plants have revealed the predominance of Actinobacteria and Proteobacteria in their rhizospheres.[3] *Pseudomonas* and *Burkholderia* have been reported as the predominant genera of some Proteobacteria (Gram-negative bacteria) in the rhizosphere.[12] The ability of Proteobacteria to use a great variety of root-derived carbon substrates, in addition to their capacity to grow fast, may justify their abundance in the rhizosphere soil. Other examples of the predominance of Gram-negative bacteria in the rhizosphere have also been reported. These include the isolation and identification of: (i) *Pseudomonas* strains from the rhizosphere of four medicinal plants [*Ocimum sanctum* (tulsi, holy basil), *Catharanthus roseus* (Madagascar periwinkle), *Aloe vera* (aloe), and *Coleus forskholii* (coleus)], in addition to *Azospirillum* and *Azotobacter* populations[17]; (ii) 20 strains of *Pseudomonas fluorescens* from the rhizosphere of rice[20]; (iii) more than two hundred bacterial strains from the rhizosphere of *Populus euphratica* (desert poplar), some of which have been identified as Gram-negative bacteria of the genera *Pseudomonas*, *Stenotrophomonas*, and *Serratia*[21]; and (iv) over one hundred culturable and morphologically different bacterial isolates from the rhizosphere of *Origanum vulgare* (oregano), of which Proteobacteria were found to be predominant.[19]

9.2.3 Fungal Diversity

The rhizosphere is known to host a great diversity of fungal species, and this is strongly influenced by the plant species, its developmental stage, and the soil type. Based on their function in the rhizosphere, these fungi may be classified into three groups, namely plant-beneficial, plant-pathogenic, and fungal decomposers.[3] Among the plant-beneficial fungi, those associated with ectomycorrhiza constitute an important group of over 4000 species belonging to the Basidiomycotina and the Ascomycotina. These ECM fungi were found associated with woody plants of the families Betulaceae, Fagaceae, Myrtaceae, and Pinaceae, and include the fungal genera *Laccaria*, *Lactarius*, *Paxillus*, *Pisolithus*, *Scleroderma*, *Suillus*, and *Wilcoxia*.[22] Studies on the isolation and identification of the fungal strains associated with some plants of the family Brassicaceae have revealed *Fusarium* and *Rhizopus* as the predominant genera besides *Trichoderma* and *Mucor*.[23] In addition to these, several fungal strains encountered in the rhizospheres of several plants and mangrove soils have been cultured for the purpose of isolation and characterization of their biologically active secondary metabolites. Fungal strains isolated from the rhizospheres of plants include *Aspergillus cervinus* from *Anicasanthus thurberi* (Thurber's desert honeysuckle; Acanthaceae),[24] *As. flavipes* from *Ericameria laricifolia* (turpentine bush; Asteraceae),[25] *As. taichungensis* from *Acrostichum aureum* (golden leather fern; Pteridaceae),[26] *As. terreus* from *Ambrosia ambrosoides* (canyon ragweed; Asteraceae) and *Brickellia* sp. (Asteraceae),[24] *As. terreus* from *Opuntia versicolor* (staghorn cholla; Cactaceae),[27] *As. tubingensis* and *Penicillium* sp. AH-00-89-F6 from *Fallugia paradoxa* (apache plume; Rosaceae),[28,29] *As. wentii* from *Larrea tridentata* (creosote bush; Zygophyllaceae),[24] *Aspergillus* sp. YIM PH30001 from *Panax notoginseng* (notoginseng; Panaxaceae),[30] *Chaetomium globosum* and *Paraphaeosphaeria quadriseptata* from *Opuntia leptocaulis* (desert Christmas cactus; Cactaceae),[31,32] *Penicillium sumatrense* from *Lumnitzera racemosa* (black mangrove; Combretaceae),[33] and *Talaromyces verruculosus* from *Stellera chamaejasme* (Thymelaeaceae).[34] In addition, *As. effuses* H1-1 isolated from mangrove rhizosphere soil has been cultured and investigated for its constituent metabolites.[35,36] The rhizosphere of the Brazilian medicinal plant *Senna spectabilis* (crown of gold tree; Fabaceae) has been recently reported to be a rich source of fungal strains. Of over 150 fungal strains isolated, the majority were found to be *Fusarium* species, identified as *F. oxysporum* and *F. solani*, and the extracts derived from these have been subjected to nuclear magnetic resonance (NMR)–based metabolomics (see Section 9.3.4).[37]

9.3 NATURAL PRODUCTS OF THE RHIZOSPHERE

9.3.1 Introduction

A range of associations between microbes and plant roots, from pathogenic to symbiotic to commensal, can be established in the rhizosphere. Organisms living in association with each other, such as plants and microorganisms of the rhizosphere, are well-known producers of natural products. These natural products have been shown to mediate a variety of plant–plant, plant–microbial, and microbial–microbial interactions, including attraction of beneficial soil-borne microorganisms to the rhizosphere, stimulation or inhibition of plant growth and stress modulation, and influence on growth and even survival of the plant (Figure 9.1). The rhizosphere is also an environment with complex ecological interactions among the members of its microbial community.[15,38] Thus, rhizosphere microorganisms represent a rich resource of biologically active small-molecule natural products, with chemical structures that have been optimized by evolution, that are produced for communication and in response to changes in their habitats, including environmental stress.[39] Many of these associations may be characterized as symbiotic associations involving chemical interactions, and thus can be termed *chemosymbiosis*.[40] The demonstration that the rhizosphere microbial diversity is strongly influenced by the diversity of plant species and environmental factors suggests that a previously unexploited opportunity exists for harvesting natural products from this group of plant-associated microorganisms. However, the secondary metabolites of rhizosphere microorganisms have not received much attention, compared with those of soil-borne and endophytic microorganisms, although it has been found that Actinomycetes producers of antifungal compounds could be isolated with higher frequency from the rhizosphere bacteria of the big sagebrush (*Artemisia tridentata*) than from bacterial communities of the bulk soil.[41] Depicted in Figure 9.2 are some natural products involved in significant chemicobiological interactions between plant roots and their rhizomicrobiomes, and functional roles of some of these natural products in the rhizosphere.

9.3.2 Plant Metabolites

Plants produce an array of primary and secondary metabolites for their interactions with each other, their associated organisms, and the environment. These include molecules used in signaling and other strategies above ground to interact with

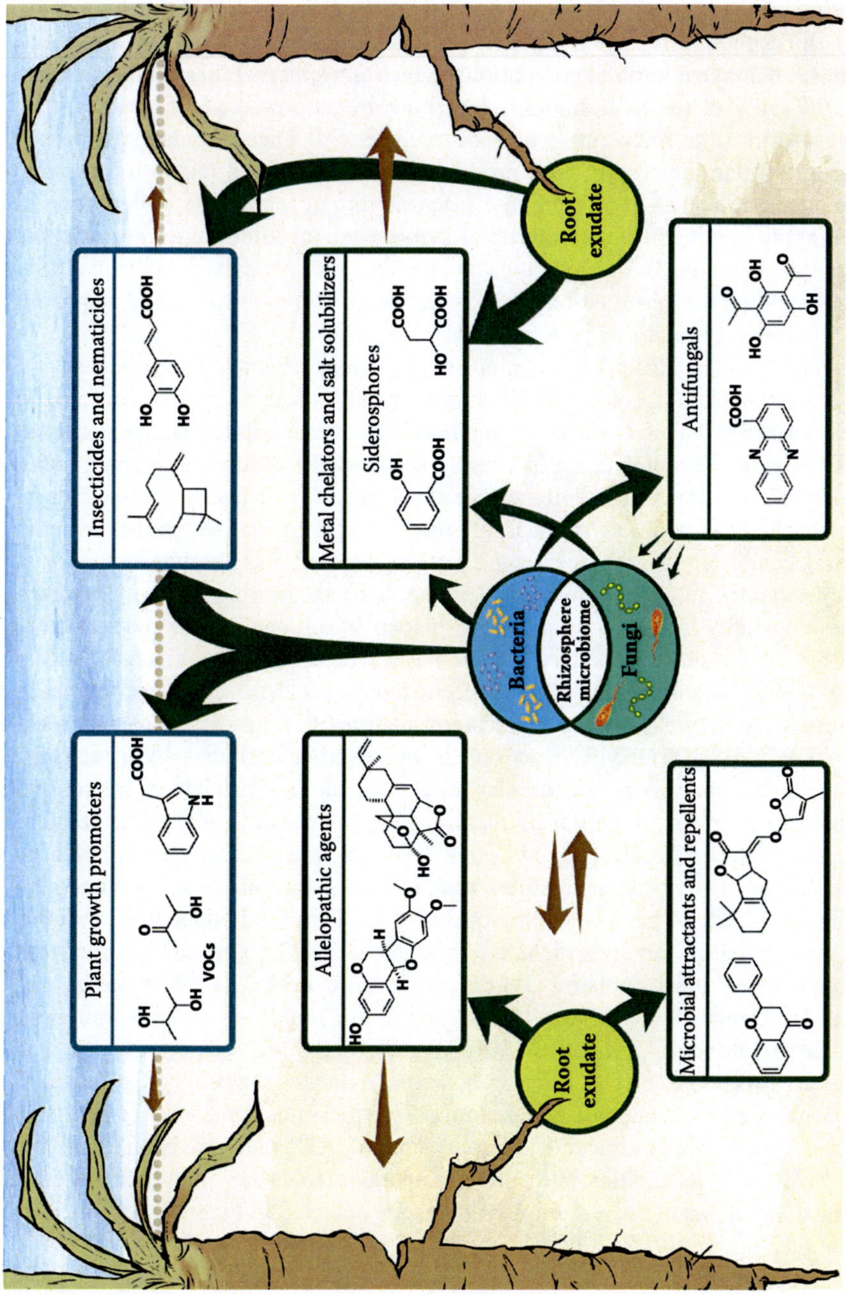

FIGURE 9.2 Some significant chemicobiological interactions mediated by natural products of the rhizosphere involving plant roots and their associated microbiomes.

neighboring plants and influence insects and birds, and below ground to interact with plants and other organisms (microorganisms, soil nematodes, protozoa, microarthropods, etc.) in the rhizosphere. During the germination of seeds and growth of the resulting plants, the release of metabolites in the form of root exudates by the so-called rhizosphere effect of the plants provides the driving force for the initial development of active microbial populations in the rhizosphere. It has been estimated that ca. 10%–50% of the total amount of carbohydrates produced by photosynthesis is released into the rhizosphere as root exudates.[42,43] These exudates contain a wide variety of plant-derived natural products that are useful in regulating the rhizosphere interactions (see Figure 9.2) by: (a) attracting or repelling its inhabitants,[44] (b) encouraging beneficial plant–microbial symbioses, (c) inhibiting the growth of competing plant species,[45] (d) changing the chemical and physical properties of the soil, (e) stimulating the germination of parasitic seeds,[46] (f) mobilizing and acquiring nutrients,[47] and (g) detoxifying heavy metals.[43]

The qualitative and quantitative compositions of root exudates have been shown to depend on the plant species and its developmental stage, as well as on abiotic and biotic factors.[48,49] This is further complicated by the possibility of some chemical constituents in the root exudate binding to soil particles and/or undergoing metabolic modifications (microbial biotransformation) caused by some microbes present in the rhizosphere. In addition, the lack of suitable experimental techniques to emulate natural conditions has hampered the investigation of the involvement of root exudate metabolites in chemicobiological interactions in the rhizosphere. However, recent developments in metabolomics have helped in the analysis of root exudates for their constituent plant secondary metabolites.[50] The application of metabolomics involving liquid chromatography/mass spectroscopy (LC/MS), ultra performance liquid chromatography/electrospray ionization-quadrupole time of flight mass spectroscopy (UPLC/ESI-QTOFMS), and online hydrogen/deuterium (H/D) exchange techniques for the analysis of a semipolar fraction of the root exudate of hydroponically grown Arabidopsis (*Arabidopsis thaliana*) has suggested that it contained 103 different metabolites.[51] In this study, different verification levels were used to identify metabolites. Forty-two of the metabolites were identified by comparison with authentic samples, of which 24 have been previously reported as metabolites of *A. thaliana*. These 24 confirmed metabolites included 13 amino acids, 3 degradation products of methionine-derived glucosinolates, 2,3-dihydrobenzoic acid 3-*O*-*β*-D-xyloside (**1**), esculetin (**2**), esculin (**3**), scopoletin (**4**), lariciresinol (**5**), indol-3-ylmethylamine (**6**), indole-3-carboxylic acid (**7**), and 9,12,13-trihydroxy-10(*E*),15(*Z*)-octadecadienoic acid (**8**). Table 9.1 lists some selected plants whose root exudates/rhizosphere soil extracts have been analyzed by untargeted metabolomics, phytochemical analysis, or bioassay-driven fractionation; small-molecule natural products (**1**–**17**) that have been identified; their known biological activities; and possible functional roles they play in rhizosphere interactions. Structures of these natural products are depicted in Figure 9.3.

Flavonoids, one of the most common groups of phenolic natural products found ubiquitously in plants, and their glycosides have also been found to constitute a large part of plant root exudates.[52] Flavonoids (**18**–**26**; Figure 9.4) are important in rhizosphere interactions and functions, as they have been shown to stimulate or inhibit rhizobial nodulation (*nod*) gene expression, cause chemoattraction of rhizobia

TABLE 9.1
Natural Products of Some Plant Root Exudates Involved in Rhizosphere Interactions, Their Known Biological Activities, and Functional Roles in the Rhizosphere

Plant Species (Common Name)	Natural Products Identified in the Rhizosphere[a]	Known Biological Activity[b]	Possible Functional Role in the Rhizosphere	Reference(s)
Arabidopsis thaliana (Arabidopsis)	2,3-Dihydroxybenzoic acid-3-*O*-*β*-D-xyloside (**1**)		Resistance to biotropic pathogens	[51,102]
	Esculetin (**2**)	Antibacterial		[47,103,104]
	Esculin (**3**)		Acquisition of iron	[105]
	Scopoletin (**4**)	Antibacterial	Acquisition of iron	
	Lariciresinol (**5**)	Antifungal	Acquisition of iron	
	Indol-3-ylmethylamine (**6**)	Antifungal	Inducible preinvasion resistance to fungal pathogens	[106,107]
	Indole-3-carboxylic acid (**7**)			
	9,12,13-Trihydroxy-10(*E*),15(*Z*)-octadecadienoic acid (**8**)	Antifungal		[108,109]
Beta vulgaris (Sugar beet)	Citramalic acid (**9**)		Solubilizing insoluble phosphates	[110,111]
	Salicylic acid (**10**)	Antifungal	Signaling plant–microbe interaction: solubilizing insoluble phosphates	
Eperua falcata (Aublet)	Isoliquiritigenin (**11**)		Acquisition of nitrogen	[112]
	Liquiritigenin (**12**)		Acquisition of nitrogen	
Trifolium pretense (Red clover)	(6a*R*,11a*R*)-Maakiain (**13**)	Phytotoxic	Allelopathic agent	[113]
	(6a*R*,11a*R*)-Trifolirhizin (**14**)	Phytotoxic	Allelopathic agent	
Zea mays (Maize)	2-*β*-D-Glucopyranosyloxy-4-hydroxy-7-methoxy-1,4(2*H*)-benzoxin-3(4*H*)-one (DIMBOA-Glu) (**15**)	Insecticidal	Induction of herbivore resistance	[114,115]
	2-*β*-D-Glucopyranosyloxy-7-methoxy-1,4(2*H*)-benzoxin-3(4*H*)-one (HMBOA-Glu) (**16**)	Insecticidal	Induction of herbivore resistance	[114,115]

(Continued)

TABLE 9.1 (*Continued*)
Natural Products of Some Plant Root Exudates Involved in Rhizosphere Interactions, Their Known Biological Activities, and Functional Roles in the Rhizosphere

Plant Species (Common Name)	Natural Products Identified in the Rhizosphere[a]	Known Biological Activity[b]	Possible Functional Role in the Rhizosphere	Reference(s)
	2-*β*-D-Glucopyranosyloxy-4,7-dimethoxy-3,4-dihydro-1,4(2*H*)-benzoxin-3(4*H*)-one (HDMBOA-Glu) (**17**)	Insecticidal	Induction of herbivore resistance	[114,115]

[a] Technique(s) used for the identification of natural products included LC/MS, UHPLC/ESI-QTOFMS, and online H/D exchange for *A. thaliana*[51]; HPLC/ESI-MS for *B. vulgaris*[110]; UHPLC/DAD (Diode Array Detection) /ESI-QTOFMS for *E. falcata*[111]; bioassay-guided fractionation involving HPLC and LC-MS for *T. pretense*[113]; and UHPLC/TOF-MS and CaPNMR (capillary nuclear magnetic resonance) for *Z. mays*.[114]

[b] Biological activity listed may not necessarily be associated with those encountered only in the rhizosphere of these plants.

toward roots, inhibit root pathogens, stimulate mycorrhizal spore germination and hyphal branching, affect quorum sensing, mediate plant allelopathy, and chelate soil nutrients. These biological functions of flavonoids in the rhizosphere have been found to depend on their structures. Most flavonoids contain a six-membered heterocyclic ring, formed by a Michael-type nucleophilic attack of a phenol group on to the unsaturated ketone of their biosynthetic precursors, chalcones, affording flavanones [e.g., naringenin (**18**)]. Flavanones can then give rise to many structural variants of their basic skeleton, including flavones [e.g., apigenin (**19**) and luteolin (**20**)], flavonols [e.g., kaempferol (**21**) and quercetin (**22**)], and catechins [e.g., (+)-catechin (**23**)]. Isoflavonoids form a distinct subclass of flavonoids having structural variants in which the shikimate-derived aromatic ring has migrated to the adjacent carbon atom of the heterocyclic ring; these include isoflavones [e.g., genistein (**24**)], coumestans [e.g., coumestrol (**25**)], and pterocarpans [e.g., medicarpin (**26**)]. One of the first flavonoids to be discovered to act as a *nod* gene inducer in rhizobia is the flavone luteolin (**20**).[53] The flavanone naringenin (**18**), in addition to acting as a *nod* gene inducer in rhizobia of pea (*Pisum sativum*), has been found to inhibit quorum sensing in *Escherichia coli*, *P. aeruginosa*, and *Vibrio fischeri*.[52] Interestingly, the flavones apigenin (**19**) and luteolin (**20**) have been shown to evoke a strong chemoattractant response from the rhizobia at much lower concentrations than those reported for their activity as *nod* gene inducers. Flavonols, such as kaempferol (**21**) and quercetin (**22**), and the isoflavone genistein (**24**) are some flavonoids involved in altering the soil chemistry of the rhizosphere by acting as antioxidants and metal chelators. Chelation and reduction of metals [e.g., Fe(III) to Fe(II)] can affect the

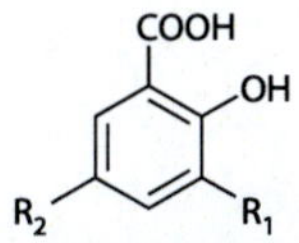

R_1 = O-Xyl, R_2 = OH, 2,3-Dihydroxybenzoic acid 3-*O*-*β*-D-xyloside (**1**)
R_1 = R_2 = H, Salicylic acid (**10**)

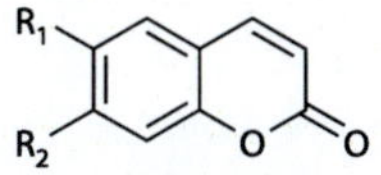

R_1 = R_2 = OH, Esculetin (**2**)
R_1 = O-Glc, R_2 = OH, Esculin (**3**)
R_1 = OMe, R_2 = OH, Scopoletin (**4**)

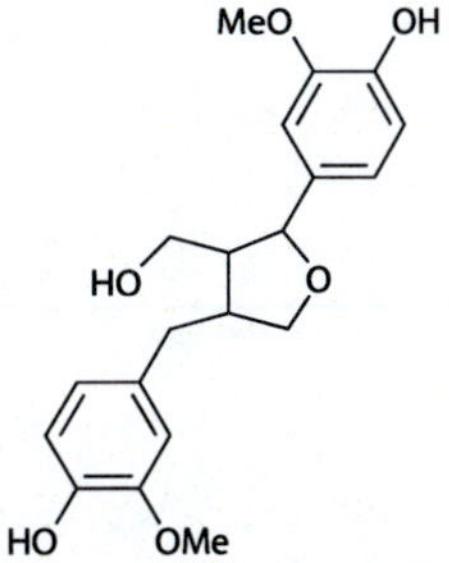

Lariciresinol (**5**)

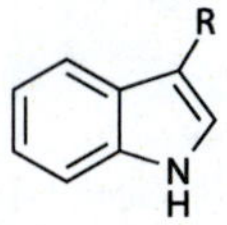

R = CH_2NH_2, Indol-3-ylmethylamine (**6**)
R = CO_2H, Indole-3-carboxylic acid (**7**)

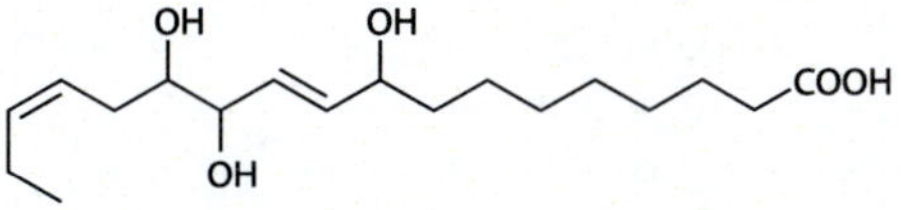

9,12,13-Trihydroxy-10(*E*),15(*Z*)-octadecadienoic acid (**8**)

Citramalic acid (**9**)

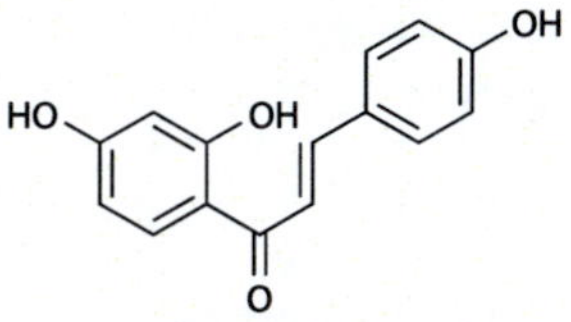

Isoliquiritigenin (**11**)

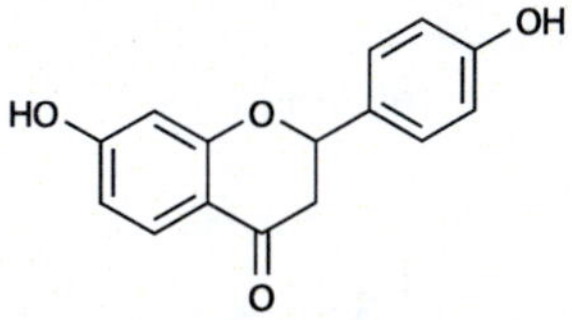

Liquiritigenin (**12**)

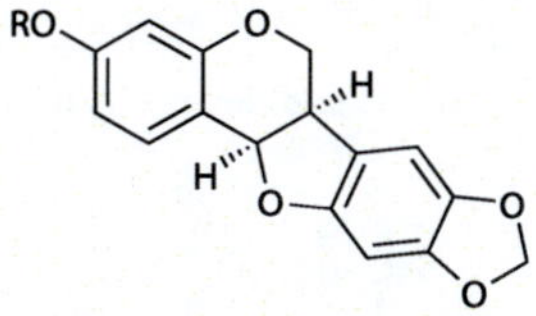

R = H, (6a*R*,11a*R*)-Maakiain (**13**)
R = Glc, (6a*R*,11a*R*)-Trifolirhizin (**14**)

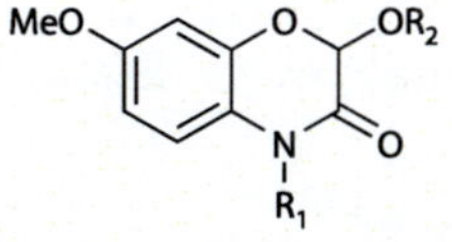

R_1 = OH, R_2 = Glu, DIMBOA-Glu (**15**)
R_1 = H, R_2 = Glu, HMBOA-Glu (**16**)
R_1 = OMe, R_2 = Glu, HDMBOA-Glu (**17**)

FIGURE 9.3 Natural products of plant root exudates and rhizosphere soil extracts identified by untargeted metabolomics, phytochemical analysis, or bioassay-driven fractionation (for possible functional roles of these natural products in the rhizosphere, see Table 9.1).

concentration of nutrients in the rhizosphere, especially the availability of iron and phosphates essential for plant growth. Flavonols [e.g., quercetin (**22**)] are also known to contribute to resistance against microbial pathogens such as *E. coli* and *Neurospora crassa*. The flavonol-derived compound catechin (**23**) has been reported to exhibit dual activity, inhibiting quorum sensing in plant host–related soil bacteria and acting as an allelopathic agent suppressing plant growth. Catechin present in the root exudate of *Centaurea stoebe* spp. *micranthos* (spotted knapweed) has been reported to act as a potent allelochemical.[54] A mutant of *Medicago truncata* (barrel clover) with capacity to hyperaccumulate coumestrol (**25**), an isoflavonoid-derived coumestan, capable of stimulating hyphal growth of fungi, has been found to be hyperinfected by its mycorrhizal symbiont.[55] Interestingly, in contrast to other

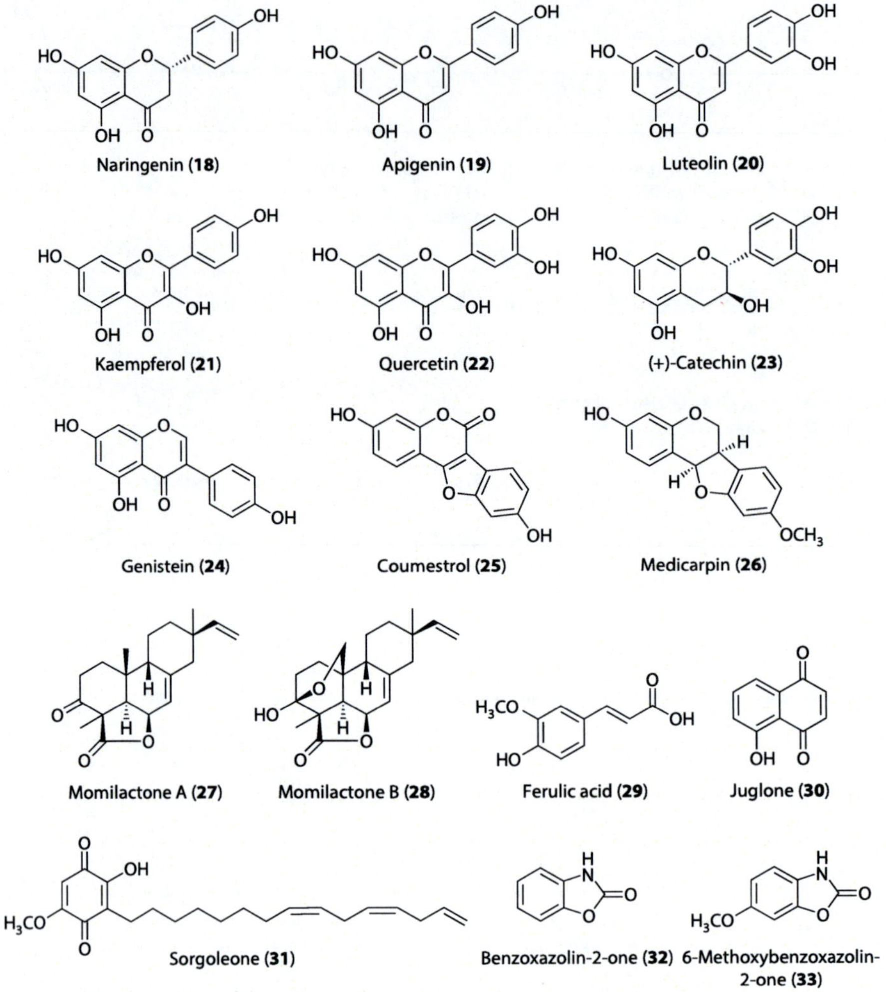

FIGURE 9.4 Plant-derived natural products in root exudates and the rhizosphere.

flavonoids, coumestrol (**25**) and the structurally related pterocarpan medicarpin (**26**) have shown *nod* gene–repressing activity for certain rhizobia. It has been suggested that the *nod* gene activators and repressors act in concert to maintain an optimal level of Nod factor production and prevent the elicitation of defense responses by the host plant.[56]

Plants are known to suppress the growth of their neighbors by producing and exuding specific natural products, termed allelochemicals, into their rhizospheres. Thus, root exudates represent the largest source of allelochemical input into the rhizosphere. Soil microorganisms are also known to produce allelochemicals directly or indirectly through microbial biotransformations. Allelochemicals belong to a variety of structural classes (see **23** and **27–33**; Figure 9.4) including fatty acids [FAs; e.g., palmitic and stearic acids], terpenes [e.g., momilactones A (**27**) and B (**28**)], phenolic acids [e.g., ferulic acid (**29**) and salicylic acid (**10**) (Figure 9.3)], quinones [e.g., juglone (**30**) and sorgoleone (**31**)], flavonoids [e.g., catechin (**23**)], and benzoxazolinones [e.g., benzoxazolin-2-one (**32**) and 6-methoxybenzoxazolin-2-one (**33**)]. Demonstration of the role played by allelochemicals in allelopathy (the production of allelochemicals acting as spacers to keep away the invading neighbors) has been found to be a difficult task, due to the chemical complexity of root exudates, interaction of compounds present within the soil matrix and associated microbes, as well as intraspecific variation in the production of allelochemicals.[57] However, recent approaches involving reverse genetics and application of metabolic profiling, coupled with bioassays, have helped to overcome some of these difficulties. In the former approach, the chemical complexity of root exudates was bypassed via selective genetic manipulation to abrogate the production of putative allelopathic compounds.[58] Application of reverse genetics using knockouts of the diterpene synthases [copyl diphosphate synthase 4 (OsCPS4) and kaurene synthase-like 4 (OsKSL4)] involved in the biosynthesis of momilactones A (**27**) and B (**28**), two allelochemicals produced by the rice plant (*Oryza sativa*) to suppress the growth of the widespread rice paddy weed *Echinochloa crus-galli* (barnyard grass), provided strong genetic evidence for natural products–mediated allelopathy, in addition to furnishing a molecular target for breeding and metabolic engineering of rice. In a more recent study, the composition of phytotoxic root exudates of *Heraceleum mantegazzianum* (giant hogweed) on germination of Arabidopsis and *Plantago lanceolata* (narrow leaf plantain) was investigated by metabolic profiling. This study employed ultra high performance liquid chromatography-time of flight-mass spectroscopy (UHPLC–TOF-MS) analysis, and the relationships between the metabolic profiles and the phytotoxic effects were determined using orthogonal partial least-square (OPLS) analyses.[59] It was found that the inhibition of germination of the native species, *Plantago lanceolata*, cooccurring with *H. mantegazzianum*, could be predicted by the metabolic profiles of the root exudates. This study also led to the tentative identification of 15 compounds associated with this inhibition, and 14 of these were found to belong to three groups according to their structures. These consisted of: (i) two dipeptides and one amino acid (tryptophan), (ii) six C_{18} oxylipins, and (iii) five malonyl monoglycosides, demonstrating that metabolic profiling could provide an efficient approach for studying natural products–mediated plant–plant interactions whenever unknown metabolites are involved.[59]

9.3.3 Bacterial Metabolites

Chemotaxis, the movement of an organism in response to a chemical stimulus, has been studied in detail and characterized at the molecular level in several beneficial soil bacteria. Plant-released root exudates and rhizodeposits form a nutrient gradient in the rhizosphere, and these have been found to chemotactically attract diverse motile bacteria that are beneficial to plant productivity.[7] Most of the metabolites synthesized by rhizobacterial communities are quorum-sensing compounds responsible for cell-to-cell signaling among the rhizosphere microorganisms. Of these, *N*-acyl homoserine lactone (AHL) (**34**) is the most investigated quorum-sensing compound, which is produced by many species of genera including *Pseudomonas*, *Burkholderia*, *Serratia*, *Erwinia*, and *Ralstonia* belonging to Proteobacteria.[15] Rhizobacterial metabolites belong to different chemical classes and are known to perform a variety of functional roles in the rhizosphere. Presented in Figure 9.5 are some examples of rhizobacterial

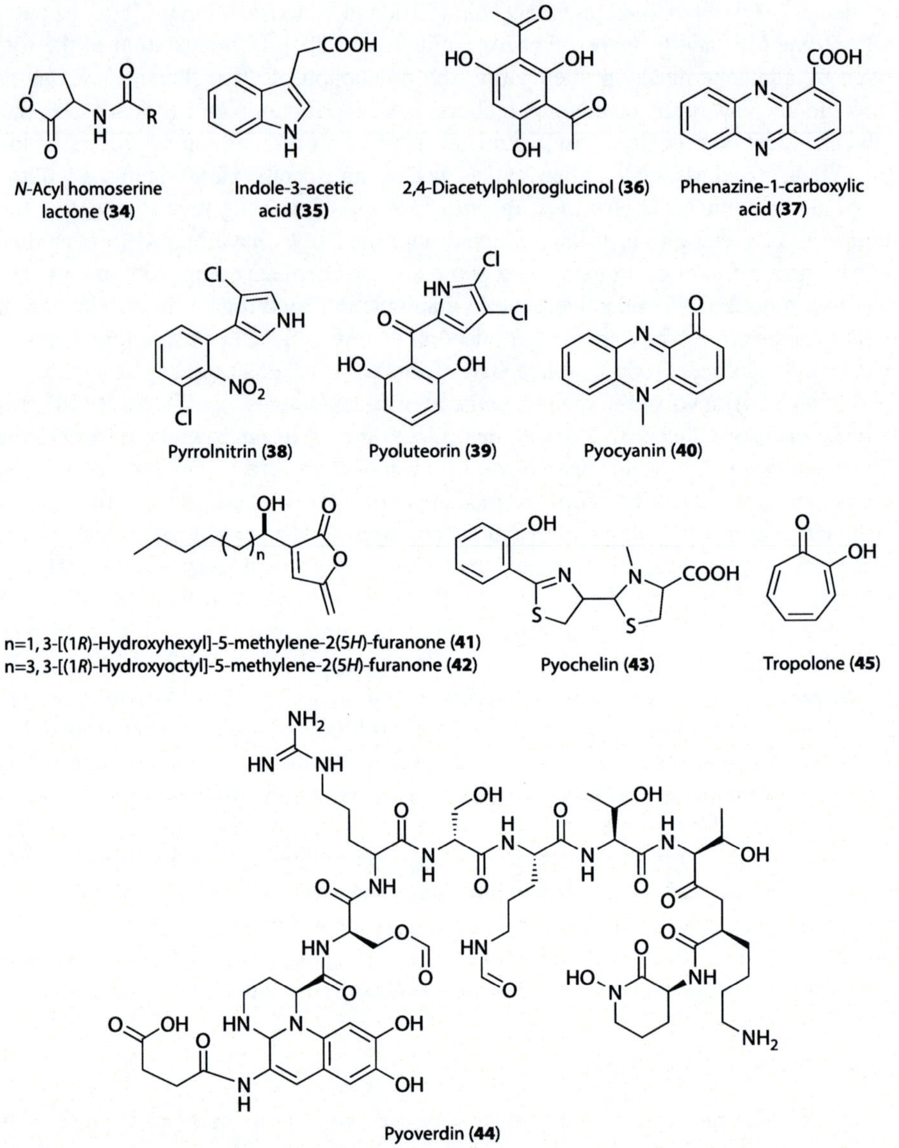

FIGURE 9.5 Natural products of rhizobacterial origin.

metabolites (**34**–**44**) isolated and characterized from cultured organisms. Among these metabolites, indole-3-acetic acid (IAA, **35**) is one of the most important signaling molecules produced by rhizobacteria such as *Pseudomonas* spp. and is involved in the regulation of plant development.[60] The same group of bacteria are also known to produce antifungal metabolites, which include 2,4-diacetylphloroglucinol (2,4-DAPG) (**36**),[61–63] phenazine-1-carboxylic acid (**37**),[63,64] pyrrolnitrin (**38**),[63] pyoluteorin (**39**),[63] pyocyanin (**40**),[63] and the 5-methylenefuranones **41** and **42**.[65] Some rhizosphere pseudomonads are known to produce siderophores, pyochelin (**43**), and pyoverdin (**44**).[3] Recent investigation of the rice rhizobacterium *Burkholderia heleia* PAK1-2, a potent biocontrol agent responsible for preventing bacterial rice seedling blight disease caused by *Burkholderia plantarii*, led to the identification of IAA (**35**) as the active constituent. It was also found that IAA suppressed the production of tropolone (**45**), the phytotoxic metabolite of *B. heleia* PAK1-2.[66] Some rhizobacterial natural products with functional roles in the rhizosphere, together with their reported biological activities, are presented in Table 9.2. It is noteworthy that most of these natural products are of acetate (polyketide) origin. In addition to the above metabolites, both Gram-positive and Gram-negative rhizobacteria have been found to produce VOCs belonging to different organic classes, such as alcohols, aldehydes, ketones, esters, ethers, carboxylic acids, hydrocarbons, and sulfur compounds. Studies on the VOC profiles of plant growth–promoting rhizobacteria (PGPR) have suggested that these compounds may be used as taxonomic markers. The importance of these in rhizosphere interactions is discussed below (see Section 9.4.2).

The FA profiles of rhizobacteria have also been used as chemotaxonomic markers. This is illustrated by the culture-dependent analysis of the FA profiles of 1188 isolates of rhizobacteria associated with barley (*Hordeum vulgare*) roots, leading to

TABLE 9.2
Natural Products from Rhizosphere Bacteria

Bacterial Strain	Natural Product(s)	Reported Biological Activity and/or Functional Role in the Rhizosphere	Reference(s)
Pseudomonas spp.	2,4-Diacetylphloroglucinol (**36**)	Antibiotic	[61–63]
P. aurantiaca	Indole-3-acetic acid (**35**)	Regulator of plant development	[60]
P. fluorensces	Phenazine-1-carboxylic acid (**37**)	Antibiotic	[64]
	Pyrrolnitrin (**38**)	Antibiotic	[63,64]
	Pyoluteorin (**39**)	Antibiotic	[63]
P. jessenii	3-[(1*R*)-Hydroxyhexyl]-5-methylene-2(5*H*)-furanone (= 4,5-Didehydroacaterin) (**41**)	Antibiotic	[65]
	3-[(1*R*)-Hydroxyoctyl]-5-methylene-2(5*H*)-furanone (**42**)	Antibiotic	[65]
P. cepacia	Pyrrolnitrin (**38**)	Antibiotic	[63]
P. aeruginosa	Pyocyanin (**40**)	Antibiotic	[63]
Burkholderia heleia	Indole-3-acetic acid (**35**)	Inhibitor of tropolone (phytotoxin) biosynthesis	[66]
Serratia plymuthica	Pyrrolnitrin (**38**)	Antibiotic	[116]
Acinetobacter calcoaceticus	Acinetobactin-like siderosphore	Siderophore	[79]
Fluorescent pseudomonads	Pyochelin (**43**)	Siderophore	[3]
	Pyoverdin (**44**)	Siderophore	[3]
Burkholderia plantarii	Tropolone (**45**)	Phytotoxin	[66]

their classification into distinct groups. The group containing the highest number of isolates (720 isolates) had a predominance of *Pseudomonas* species and unbranched FAs with an even number of carbon atoms. Branched-chain FAs with an odd number of carbon atoms were predominant in the group of *Cytophaga* species as major rhizobacteria.[67] The predominance of FAs with an even number of carbons (16:0, 16:1, and 18:1) in the rhizosphere of *Pseudomonas* species was corroborated through the analysis of the FA profiles of some rhizobacterial strains collected in Ukraine.[68]

9.3.4 Fungal Metabolites

The metabolome of a fungus consists of two components, namely endometabolomes and exometabolomes, comprising intercellular (mostly primary) and extracellular (mostly secondary) metabolites, respectively. Both primary and secondary metabolites of rhizospheric fungi are known to participate in their interactions with other organisms in the rhizosphere. Similar to plant growth–promoting (PGP) bacteria, a number of plant growth–promoting fungi (PGPFs) occur in rhizosphere soils. The well-known PGPFs belong to the fungal genera *Fusarium*, *Penicillium*, *Phoma*, and *Trichoderma*.[69] In addition to promoting plant growth, these fungi have been found to be beneficial to the host plant, as they elicit induced systemic resistance (ISR) against other fungi, bacteria, viruses, and nematodes. Thus, metabolites produced by rhizosphere fungi may act as signaling molecules, helping to attract or repel other organisms in their interactions with bacteria, other fungi, and other organisms. These signaling molecules also induce metabolic responses in the host plant, inhibiting or killing competitors (such as bacteria and other fungi), or mediating susceptibility in host plants (small-molecule effectors, typically toxins). In addition, nutrition-related metabolites, such as organic acids and metal chelators (siderophores), may be produced and secreted into the rhizosphere to solubilize, bind, and assimilate inorganic nutrients, thereby assisting the host plant. In a recent study to characterize rhizosphere fungi that mediate resistance in tomato plants (*Solanum lycopersicum*) against bacterial wilt disease caused by *Ralstonia solanacearum*, 79 plant growth–promoting fungal strains were isolated from their rhizosphere soil. Among these, nine strains were found to be capable of colonizing the roots of tomato plants, solubilizing phosphate, producing IAA (**35**), and promoting the growth of the plant.[70]

A number of rhizosphere-associated fungi have also been investigated for their constituent natural products. Thus, in a pioneering program to uncover bioactive metabolites in the rhizosphere microbiome of the Sonoran Desert, over 1500 rhizosphere fungal strains have been collected and cultured, and the derived extracts have been screened for their potential anticancer activity.[27] This and other studies have led to the isolation and characterization of rhizosphere fungal–derived natural products, some with unique structures (see below) and promising biological activities (see Section 9.4). Altogether, 18 rhizosphere fungal strains have been investigated to date, including *As. cervinus, As. effuses* H1-1, *As. flavipes*, *As. taichungensis*, *As. terreus* (four strains), *As. tubingensis*, *As. wentii*, *Aspergillus* sp. YIM PH30001, *Chaetomium globosum*, *Paraphaesphaeria quadriseptata*, *Penicillium sumatrense*, *Penicillium* sp. AH-00-89-F6, *Talaromyces verruculosus*, and two unidentified fungal strains inhabiting the wheat rhizosphere and mangrove rhizosphere soil. Presented in Table 9.3 are these fungal strains, together with their plant hosts, isolated natural

TABLE 9.3
Rhizosphere Fungal Strains Investigated, Their Constituent Natural Products, and Reported Biological Activities[a]

Fungal Strain	Plant Host(s) (Family)/ Rhizosphere Soil	Natural Product(s)[b]	Biological Activity[c]	Reference(s)
Aspergillus cervinus (Trichocomaceae)	*Anicasanthus thurberi* (Torr.) Gray (Acanthaceae)	Penicillic acid (**46**)	Selectively cytotoxic	[24]
		6-Methoxy-5(6)-dihydropenicillic acid (**47**)*		
		4*R**,5*S**-Dihydroxy-3-methoxy-5-methylcyclohex-2-enone (**48**)*		
As. effuses H1-1 (Trichocomaceae)	Mangrove rhizosphere soil	Effusin A (**49**)*		[35,36]
		Dihydrocryptoechinulin D (**50**)*	Cytotoxic; Topo I	
		Cryptoechinuline D (**51**)		
		Neoechinulin B (**52**)		
		Dihydroneochinulin B (**53**)*		
		Didehydroechinulin B (**54**)		
		Isodihydroauroglaucin (**55**)	Cytotoxic	
		Aspergin (**56**)	Cytotoxic	
		Auroglaucin (**57**)		
As. flavipes (Trichocomaceae)	*Ericameria laricifolia* Nutt. (Asteraceae)	Aspochalasin C (**58**)	Cytotoxic	[25]
		Aspochalasin D (**59**)	Cytotoxic	
		TMC-169 (**60**)	Cytotoxic	
		Aspochalasin E (**61**)	Cytotoxic	
		Aspochalasin K (**62**)*	Cytotoxic	
		Aspochalasin I (**63**)*	Cytotoxic	
		Aspochalasin J (**64**)*	Cytotoxic	

(Continued)

TABLE 9.3 (*Continued*)
Rhizosphere Fungal Strains Investigated, Their Constituent Natural Products, and Reported Biological Activities[a]

Fungal Strain	Plant Host(s) (Family)/ Rhizosphere Soil	Natural Product(s)[b]	Biological Activity[c]	Reference(s)
As. taichungensis ZHN-7-07 (Trichocomaceae)	*Acrostichum aureum* (Pteridaceae)	Prenylterphenyllin A (**65**)*	Cytotoxic	[26]
		Prenylterphenyllin B (**66**)*		
		Prenylterphenyllin C (**67**)*		
		4″-Dehydro-3-hydroxyterphenyllin (**68**)*	Cytotoxic	
		Prenylterphenyllin (**69**)		
		Terprenin (**70**)	Cytotoxic	
		Deoxyterphenyllin (**71**)	Cytotoxic	
		3-Hydroxyterphenyllin (**72**)		
		Terphenyllin (**73**)		
		3,3′-Dihydroxyterphenyllin (**74**)		
		Prenylcandidusin A (**75**)*		
		Prenylcandidusin B (**76**)*		
		Prenylcandidusin C (**77**)*		
		Candidusin A (**78**)		
		Candidusin C (**79**)		
As. terreus (Trichocomaceae)	*Brickellia* sp. (Asteraceae)	Dehydrocurvularin (**80**)	Cytotoxic; antimitotic; weak Hsp90 inhibitor; p97 and proteasome inhibitor	[24,89]
		11-Methoxycurvularin (**83**)	Cytotoxic; antimitotic; weak Hsp90 inhibitor	
		11-Hydroxycurvularin (**84**)	Selectively cytotoxic; antimitotic; weak Hsp90 inhibitor	

(*Continued*)

TABLE 9.3 (*Continued*)
Rhizosphere Fungal Strains Investigated, Their Constituent Natural Products, and Reported Biological Activities[a]

Fungal Strain	Plant Host(s) (Family)/ Rhizosphere Soil	Natural Product(s)[b]	Biological Activity[c]	Reference(s)
As. terreus (Trichocomaceae)	*Opuntia versicolor* Engelm. (Cactaceae)	Betulinan A (**88**)	Cytotoxic	[27,90]
		Asterriquinone D (**89**)	Cytotoxic	
		Asterriquinone C-1 (**90**)	Cytotoxic	
		(–)-Quadrone (**91**)	Cytotoxic	
		(+)-Terrecyclic acid A (**92**)	Cytotoxic; cell cycle inhibitor; induction of heat shock response; NF-kB inhibitor	
		(+)-5,6-Dihydro-6-hydroxyterrecyclic acid A (**93**)*	Cytotoxic	
		(+)-5,6-Dihydro-6-methoxyterrecyclic acid A (**94**)*	Cytotoxic	
		Asterredione (**95**)*	Cytotoxic	
As. terreus (Trichocomaceae)	*Ambrosia ambrosoides* (Cav.) Payne (Asteraceae)	Terrefuranone (**96**)*		[24]
		N[a]-Acetylaszonalemin (LL-S490β) (**97**)		
		Terrequinone A (**98**)*	Cytotoxic	
As.terreus Gwq-48[d] (Trichocomaceae)	Mangrove rhizosphere soil	Isoaspulvinone E (**99**)*	Antiviral	[92]
		Aspulvinone E (**100**)	Antiviral	
		Pulvic acid (**101**)		
As. tubingensis (Trichocomaceae)	*Fallugia paradoxa* D. Don (Rosaceae)	Asperpyrone D (**102**)*		[28]
		Asperpyrone A (**103**)		
		Fonsecinone A (**104**)		
		Dianhydro-aurasperone C (**105**)		
		Aurasperone A (**106**)		
		Aurasperone E (**107**)		
		Fonsecin (**108**)		
		Fonsecin B (**109**)		
		Rubrofusarin B (**110**)		
		TMC-256A1 (**111**)		
		Funalenone (**112**)		
		Malformin A_1 (**113**)	Cytotoxic	

(*Continued*)

TABLE 9.3 (*Continued*)
Rhizosphere Fungal Strains Investigated, Their Constituent Natural Products, and Reported Biological Activities[a]

Fungal Strain	Plant Host(s) (Family)/ Rhizosphere Soil	Natural Product(s)[b]	Biological Activity[c]	Reference(s)
As. wentii (Trichocomaceae)	*Larrea tridentata* (DC.) Coville (Zygophyllaceae)	Penicillic acid (**46**)	Selectively cytotoxic	[24]
Aspergillus sp. YIM PH30001 (Trichocomaceae)	*Panax notoginseng* (Panaxaceae)	Arugosin C (**114**)	Antibacterial	[30]
		Averantin (**115**)	Antibacterial; antifungal	
		Methylaverantin (**116**)		
		Averufin (**117**)	Antibacterial	
		8-*O*-Methyl averufin (**118**)		
		Averythrin (**119**)	Antibacterial; antifungal	
		Sterigmatocystin (**120**)		
		Demethylsterigmatocystin (**121**)		
		Dihydrosterigmatocystin (**122**)		
		Diorcinol (**123**)		
		Versicolorin B (**124**)	Antifungal	
		Ziganein-1-methyl ether (**125**)	Antibacterial	
		8-*O*-Methylchrysophanol (**126**)		
Chaetomium globosum (Chaetomiaceae)	*Opuntia leptocaulis* DC. (Cactaceae)	Chrysazin (**127**)		[73]
		1,3,6,8-Tetrahydroxyanthraquinone (**128**)		
		Globosuxanthone A (**129**)*	Cytotoxic	
		Globosuxanthone B (**130**)*		
		Globosuxanthone C (**131**)*		
		Globosuxanthone D (**132**)*		
		2-Hydroxyvertixanthone (**133**)		

(*Continued*)

TABLE 9.3 (*Continued*)
Rhizosphere Fungal Strains Investigated, Their Constituent Natural Products, and Reported Biological Activities[a]

Fungal Strain	Plant Host(s) (Family)/ Rhizosphere Soil	Natural Product(s)[b]	Biological Activity[c]	Reference(s)
Paraphaeosphaeria quadriseptata (Montagnulaceae)	*Opuntia leptocaulis* DC. (Cactaceae)	Monocillin I (**134**)	Cytotoxic; Hsp90 inhibitor	[32]
		Monocillin III (**135**)		
		5′-Hydroxymonocillin III (**136**)*		
		Paraphaeosphaerin A (**137**)*		
		Paraphaeosphaerin B (**138**)*		[74]
		Paraphaeosphaerin C (**139**)*		
		Aposphaerin C (**140**)*		
		Aposphaerin B (**141**)		
		Quadriseptin A (**142**)*		
		Cytosporone F (**143**)*		
		Cytosporone G (**144**)*		
		Cytosporone H (**145**)*		
		Cytosporone I (**146**)*		
Penicillium sumatrense MA-92 (Trichocomaceae)	*Lumnitzera racemosa* (Combretaceae)	Dehydrocurvularin (**80**)	Cytotoxic; p97 and proteasome inhibitor	[33,89]
		Curvularin (**81**)		
		Curvularin-7-*O*-β-D-glucopyranoside (**82**)		
		Sumalarin A (**85**)*	Cytotoxic	
		Sumalarin B (**86**)*	Cytotoxic	
		Sumalarin C (**87**)*	Cytotoxic	

(*Continued*)

TABLE 9.3 (*Continued*)
Rhizosphere Fungal Strains Investigated, Their Constituent Natural Products, and Reported Biological Activities[a]

Fungal Strain	Plant Host(s) (Family)/ Rhizosphere Soil	Natural Product(s)[b]	Biological Activity[c]	Reference(s)
Penicillium sp. AH-00-89-F6 (Trichocomaceae)	*Fallugia paradoxa* D. Don (Rosaceae)	Dehydrocurvularin (**80**)	Cytotoxic; antimitotic; weak Hsp90 inhibitor; p97 and proteasome inhibitor	[24,29,89]
		11-Methoxycurvularin (**83**)	Cytotoxic; antimitotic; weak Hsp90 inhibitor	
		11-Hydroxycurvularin (**84**)	Selectively cytotoxic; antimitotic; weak Hsp90 inhibitor	
		1,3-Dihydroxy-6-methyl-7-methoxyanthraquinone (**147**)		
		1,3-Dihydroxy-6-hydroxymethyl-7-methoxyanthraquinone (**148**)*		
Talaromyces verruculosus (Trichocomaceae)	*Stellera chamaejasme* L. (Thymelaeaceae)	(–)-8-Hydroxy-3-(4-hydroxypentyl)-3,4-dihydroisocoumarin (**149**)*		[34]
		(*E*)-3-(2,5-Dioxo-3-(propan-2-ylidene)pyrrolidin-1-yl)acrylic acid (**150**)	Weakly antifungal	
Unidentified fungal strain	*Triticum* sp. (Wheat)	(*E*)-5-(3-Carboxy-2-butenyloxy)-7-hydroxy-4,6-dimethylphthalide (**151**)*	Antifungal	[75]
		(+)-5-(3-Carboxy-butoxy)-7-hydroxy-4,6-dimethylphthalide (**152**)*		
Unidentified fungal strain E33[d]	Mangrove rhizosphere soil	3,5′-Dihydroxy-4′,5-dimethoxy-2′-methyl-(1,1′-biphenyl)-2-carboxylic acid methyl ester (**153**)*		[76]

[a] See Figures 9.6 through 9.10 for structures of these natural products.
[b] New natural products are indicated with an asterisk (*).
[c] Only those found to be active in assays used are indicated (for additional details of bioactivity, see Section 9.4).
[d] Although this fungal strain was isolated from a mangrove rhizosphere soil sample collected from the coast, authors have claimed it to be a marine-derived fungus.

products, their biological activities, and relevant literature references. Structures of natural products (**46–153**) isolated and characterized from these fungal strains to date are depicted in Figures 9.6 through 9.10. Of the 107 metabolites encountered, 44 were new natural products, and the discussion below will focus mainly on those with novel structures.

Small-molecule natural products (**46–153**) encountered in cultured rhizosphere fungal strains include those derived from the acetate (polyketide), mevalonate, shikimate, and mixed biosynthetic pathways. In addition, a few nitrogen-containing metabolites, including alkaloids, have also been isolated and characterized from these microorganisms. The majority of rhizosphere fungal–derived natural products were of polyketide origin. This is not surprising, as polyketide synthase (PKS) genes have been found to be abundant in the genomes of several fungal species.[71] Recent analysis of 17 sugarcane-derived rhizosphere and endophytic fungi has revealed that their genomes encode 36 putative PKS sequences and 26 shared sequence homologies with *β*-ketoacyl synthase domains, while 10 sequences show homology to known C-methyltransferase domains.[72]

Investigation of *As. effuses* H1-1, a rhizosphere fungal strain collected from the mangrove soil of the coast of Fujian province in China, has afforded three new diketopiperazine alkaloids, effusin A (**49**), dihydrocryptoechinulin D (**50**), and dihydroneoechinulin B (**53**), together with cryptoechinuline D (**51**), neoechinulin B (**52**), didehydroechinulin B (**54**), isodihydroauroglaucin (**55**), aspergin (**56**), and auroglaucin (**57**) (see Figure 9.6).[35,36] Interestingly, **49** and **50** were found to occur as racemic mixtures, and their corresponding enantiomers **49a**, **49b** and **50a**, **50b** have been separated by chiral HPLC. Possible biosynthetic pathways to effusin A (**49**) from **52** and **56**, and to dihydrocryptoechinulin D (**50**) from **55** and **56**, have been proposed. A series of new cytochalasins, aspochalasins I (**63**), J (**64**), and K (**62**), together with four known members of this series (**58–61**) (Figure 9.6) have been found in *As. flavipes* inhabiting the rhizosphere of turpentine bush (*Ericameria laricifolia*).[25] Investigation of the fungus *As. taichungensis*, collected from the root soil of the mangrove fern *Acrostichum aureum*, has afforded seven new metabolites, including six prenylated polyhydroxy-*p*-terphenyl analogs, prenylterphenyllins A–C (**65–67**), prenylcandidusins A–C (**75–77**), and 4″-dehydro-3-hydroxyterphenyllin (**68**), together with eight of their known analogs (**69–74**, **78**, and **79**) (see Figure 9.7).[26] Two new derivatives of the sesquiterpenoid (+)-terrecyclic acid A (**92**), namely (+)-5,6-dihydro-6-hydroxyterrecyclic acid A (**93**) and (+)-5,6-dihydro-6-methoxyterrecyclic acid A (**94**), together with **92** and the derived cyclic lactone (–)-quadrone (**91**), were found to occur as major constituents of *As. terreus* isolated from the rhizosphere of the cactus staghorn cholla (*Opuntia versicolor*),[27] whereas *As. terreus* isolated from the rhizosphere of canyon ragweed (*Ambrosia ambrosoides*) afforded the new alkaloids terrefuranone (**96**) and terrequinone A (**98**), together with N^a-acetyl aszonalemin (LL-S490β) (**97**) (see Figure 9.8).[24] A new dimeric naphtho-γ-pyrone, asperpyrone D (**102**), nine known naphtho-γ-pyrones (**103–111**), funalenone (**112**), and the cyclic penta-peptide malformin A_1 (**113**) were found in *As. tubingensis* occurring in the rhizosphere of apache plume (*Fallugia paradoxa*) (see Figure 9.8).[28]

Isolation of a series of xanthone analogs, globosuxanthones A–D (**129–132**) and the known metabolites **127**, **128**, and **133** (Figure 9.9), has been reported from

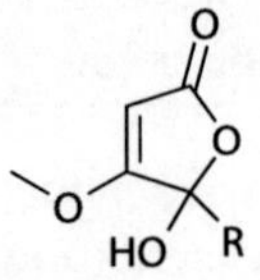

R= , Penicillic acid (**46**)

R= OMe, 6-Methoxy-5(6)-dihydropenicillic acid (**47**)

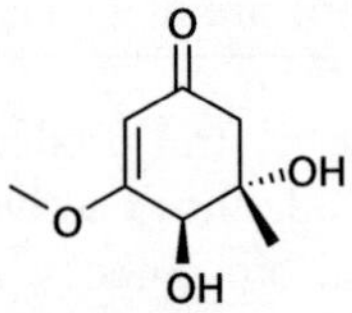

4*R**,5*S**-Dihydroxy-3-methoxy-5-methylcyclohex-2-enone (**48**)

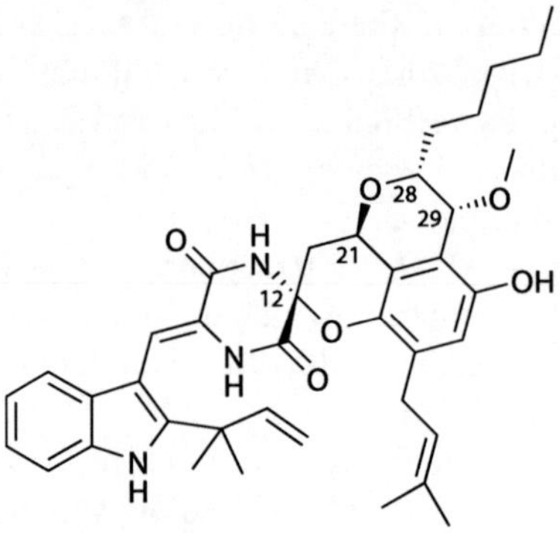

Effusin A (**49**)

(**49a**) 12*R*,21*R*,28*R*,29*R*

(**49b**) 12*S*,21*S*,28*S*,29*S*

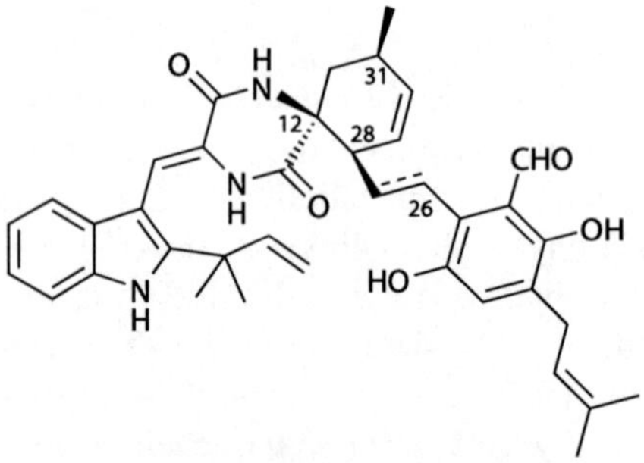

Dihydrocryptoechinulin D (**50**)

(**50a**) 12*R*,28*S*,31*S*

(**50b**) 12*S*,28*R*,31*R*

Δ^{26}, Cryptoechinuline D (**51**)

(**51a**) 12*R*,28*S*,31*S*

(**51b**) 12*S*,28*R*,31*R*

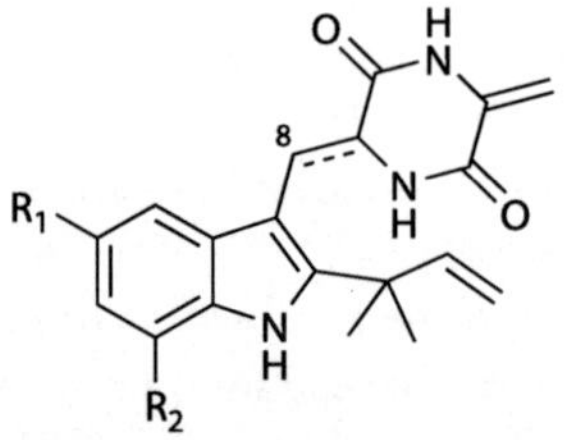

$R_1 = R_2 = H$, Δ^8, Neoechinulin B (**52**)

$R_1 = R_2 = H$, Dihydroneochinulin B (**53**)

$R_1 = R_2 =$, Didehydroechinulin B (**54**)

FIGURE 9.6 Natural products encountered in rhizospheric *Aspergillus cervinus*, *A. effuses* H1-1, *A. flavipes*, and *A. wentii*. (*Continued*)

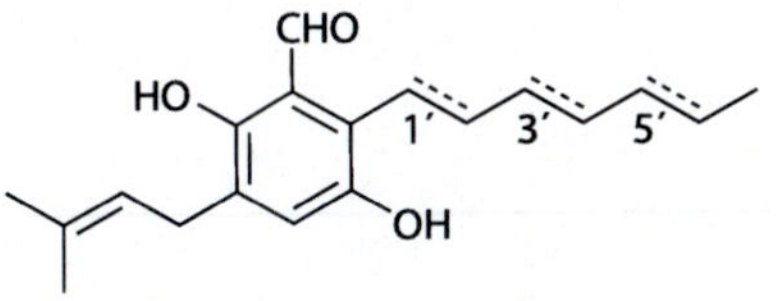

$\Delta^{3',5'}$, Isodihydroauroglaucin (**55**)
$\Delta^{1'}$, Aspergin (**56**)
$\Delta^{1',3',5'}$, Auroglaucin (**57**)

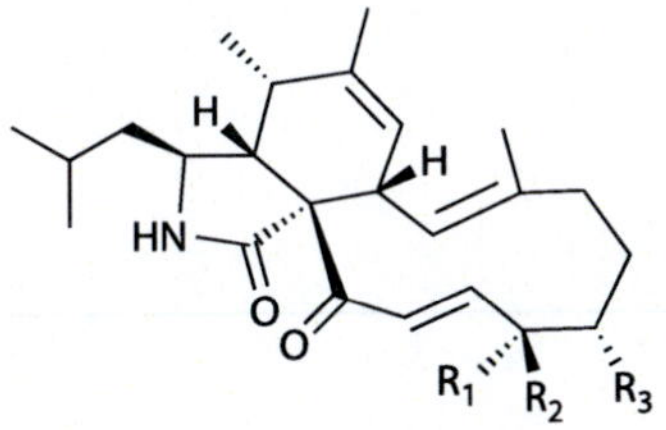

R_1 = H, R_2 = R_3 = OH, Aspochalasin C (**58**)
R_1 = R_3 = OH, R_2 = H, Aspochalasin D (**59**)
R_1 or R_2 = OH, R_3 = H, TMC-169 (**60**)

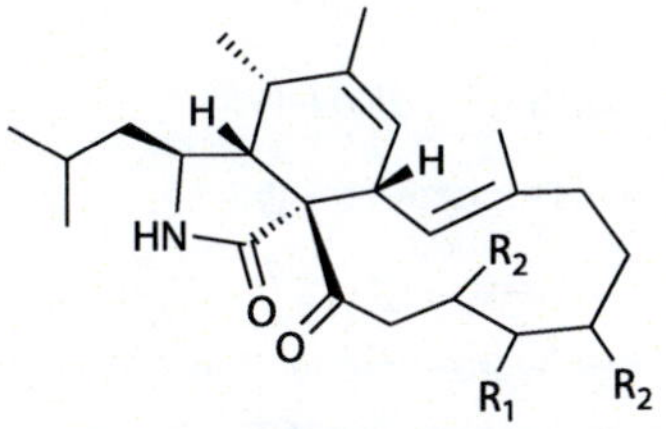

R_1 = R_2 = OH, Aspochalasin E (**61**)
R_1 = OMe R_2 = OH, Aspochalasin K (**62**)

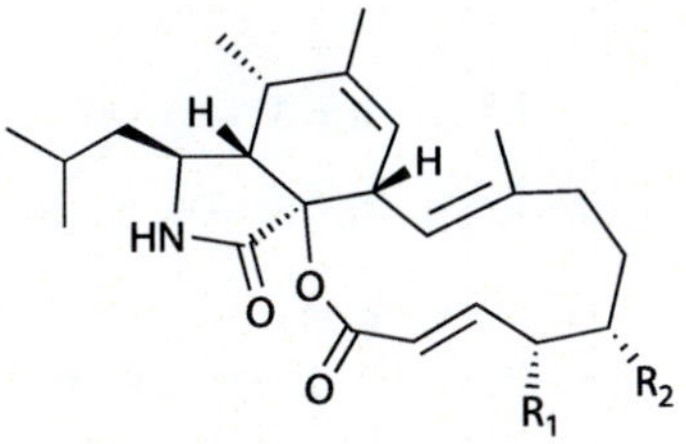

R_1 = R_2 = OH, Aspochalasin I (**63**)
R_1 = OH, R_2 = H, Aspochalasin J (**64**)

FIGURE 9.6 (*Continued*) Natural products encountered in rhizospheric *Aspergillus cervinus*, *A. effuses* H1-1, *A. flavipes*, and *A. wentii.*

the fungal strain *Chaetomium globosum*, collected from the rhizosphere of the Christmas cactus *Opuntia leptocaulis*.[73] Investigation of a solid potato dextrose agar (PDA) culture of another fungal strain, *Paraphaeosphaeria quadriseptata*, collected from the rhizosphere of the same cactus led to the isolation of three novel isocoumarins, paraphaeosphaerins A–C (**137–139**) biogenetically related to the resorcylic acid lactone, monocillin I (**134**), and a new chroman-4-one aposphaerin C (**140**) (see Figure 9.10).[31] However, when this fungus was cultured in liquid potato dextrose broth (PDB) medium made up in distilled water, new metabolites, 5′-hydroxymonocillin III (**136**), quadriseptin A (**142**), and cytosporones F–G (**143–144**), were produced. Interestingly, culture in PDB made up in tap water, or incorporation of heavy metal ions (e.g., Cu^{2+}, Zn^{2+} etc.) into the PDB medium made up in distilled water, induced the production of monocillin I (**134**) as the major constituent.[74] Three new thiol adducts of dehydrocurvularin (**80**), sumalarins A–C (**85–87**) (Figure 9.7), have been found in *Penicillium sumatrense* MA-92 isolated from the rhizosphere of the mangrove plant *Lumnitzera racemosa*,[33] whereas the occurrence of **80** and its Michael adducts with water and methanol, 11-hydroxycurvularin (**84**) and 11-methoxycurvularin (**83**), respectively, together with a new anthraquinone (**147**) (Figure 9.10), has been reported from *Penicillium* sp. AH-00-89-F6 isolated from the rhizosphere of the

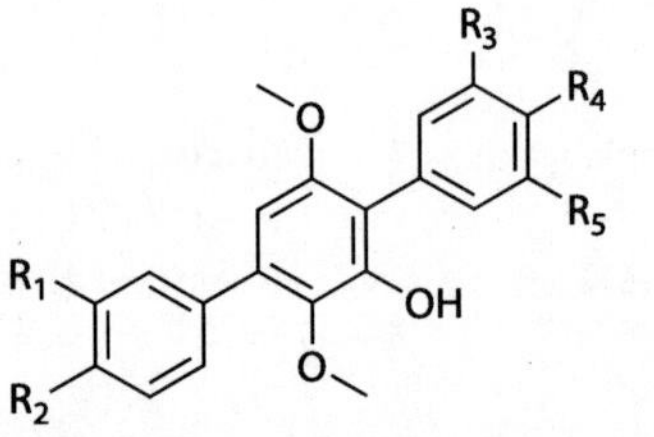

R_1 = , $R_2=R_3=R_4$=OH, R_5=H, Prenylterphenyllin A (**65**)

R_1 = , $R_2=R_4$=OH, $R_3=R_5$=H, Prenylterphenyllin B (**66**)

R_1=H, $R_2=R_3=R_4$=OH, R_5 = , Prenylterphenyllin C (**67**)

$R_1=R_2=R_5$=H, $R_3=R_4$=OH, 4″-Dehydro-3-hydroxyterphenyllin (**68**)

$R_1=R_5$=H, $R_2=R_4$=OH, R_3 = , Prenylterphenyllin (**69**)

$R_1=R_5$=H, $R_2=R_3$=OH, R_4 = O , Terprenin (**70**)

$R_1=R_2=R_3=R_5$=H, R_4=OH, Deoxyterphenyllin (**71**)

$R_1=R_5$=H, $R_2=R_3=R_4$=OH, 3-Hydroxyterphenyllin (**72**)

$R_1=R_3=R_5$=H, $R_2=R_4$=OH, Terphenyllin (**73**)

$R_1=R_2=R_3=R_4$=OH, R_5=H, 3,3′-Dihydroxyterphenyllin (**74**)

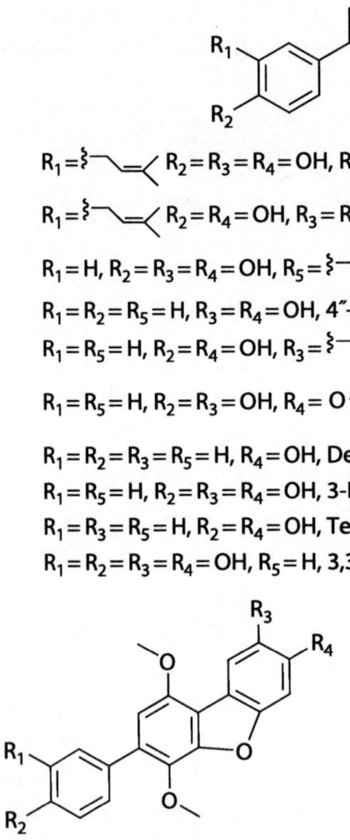

R_1 = , $R_2=R_3=R_4$=OH, Prenylcandidusin A (**75**)

R_1 = , R_2=OH, $R_3=R_4$=OMe, Prenylcandidusin B (**76**)

R_1 = , $R_2=R_3$=OH, R_4=OMe, Prenylcandidusin C (**77**)

R_1=H, $R_2=R_3=R_4$=OH, Candidusin A (**78**)

R_1=H, R_2=OMe, $R_3=R_4$=OH, Candidusin C (**79**)

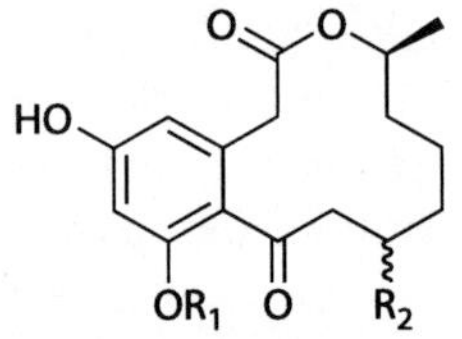

$R_1=R_2$=H, Δ^{10}, Dehydrocurvularin (**80**)

$R_1=R_2$=H, Curvularin (**81**)

R_1=Glu, R_2=H, Curvularin-7-*O*-*β*-D-glucopyranoside (**82**)

R_1=H, R_2=OMe, 11-Methoxycurvularin (**83**)

R_1=H, R_2=OH, 11-Hydroxycurvularin (**84**)

FIGURE 9.7 Natural products encountered in rhizospheric *Aspergillus taichungensis*, *A. terreus*, *Penicillium sumatrense* MA-92, and *Penicillium* sp. AH-00-89-F6. *(Continued)*

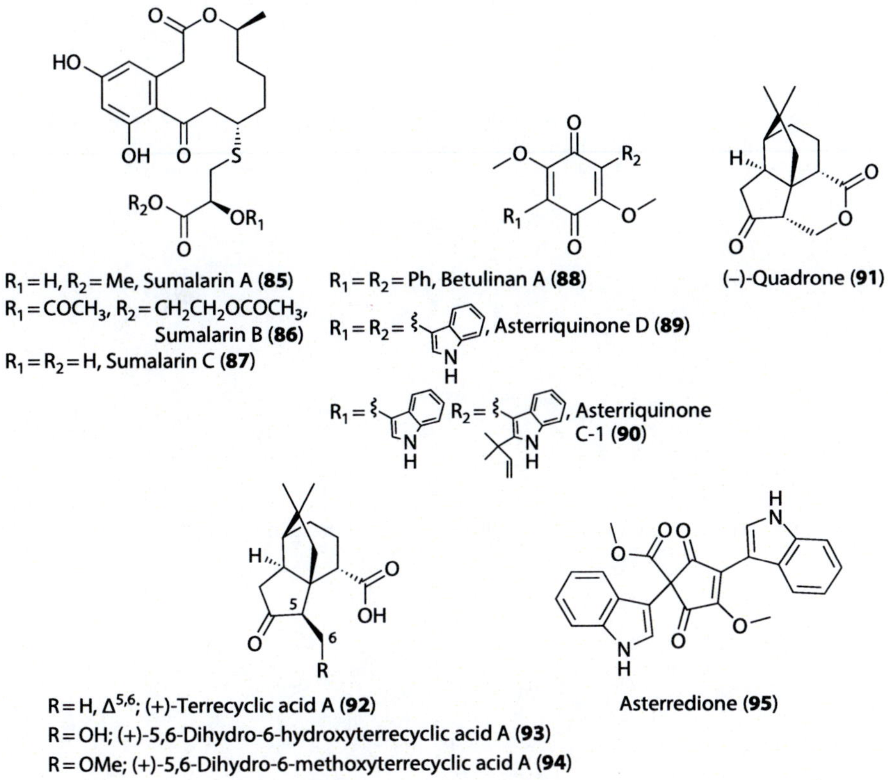

FIGURE 9.7 (*Continued*) Natural products encountered in rhizospheric *Aspergillus taichungensis*, *A. terreus*, *Penicillium sumatrense* MA-92, and *Penicillium* sp. AH-00-89-F6.

Sonoran desert plant *Fallugia paradoxa*.[29] Additional new compounds isolated and characterized from rhizosphere fungal strains include (–)-8-hydroxy-3-(4-hydroxypentyl)-3,4-dihydroisocoumarin (**149**) from *Talaromyces verruculosus* of the rhizosphere of *Stellera chamaejasme*,[34] 7-hydroxy-4,6-dimethylphthalide analogs **151** and **152** from an unidentified fungal strain of the rhizosphere of wheat (*Triticum* sp.),[75] and the (1,1′-biphenyl)-2-carboxylic acid methyl ester analog (**153**) (Figure 9.10) from the unidentified fungal strain E33 collected from mangrove rhizosphere soil of the South China Sea.[76]

Application of proton-NMR–based metabolomics to identify mycotoxin natural products in extracts of *Fusarium oxysporum* and *F. solani*, predominant fungi in the rhizosphere of *Senna spectabilis*, has been recently reported. Analysis of proton-NMR spectra by the application of principal component analysis (PCA) as a dereplication tool led to the identification of beauvericin (**154**) and fusaric acid (**155**) as major metabolites in extracts of several *F. oxysporum* strains, and of the depsipeptide HA23 (**156**) (Figure 9.11) in extracts of a number of strains of *F. solani*.[37]

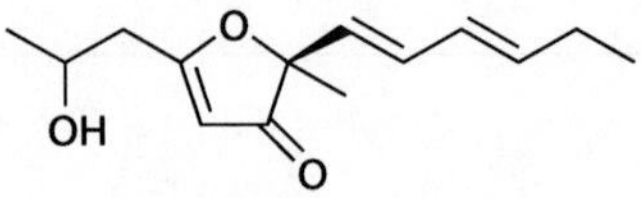

Terrefuranone (**96**)

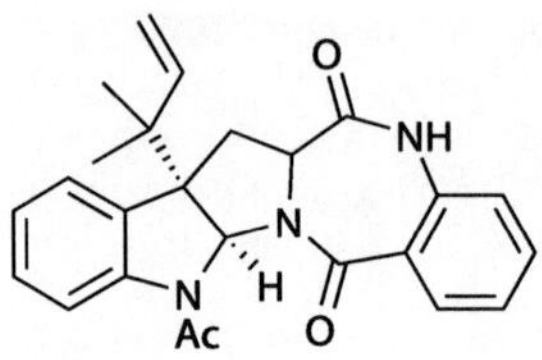

N^a-Acetyl aszonalemin [LL-S490β] (**97**)

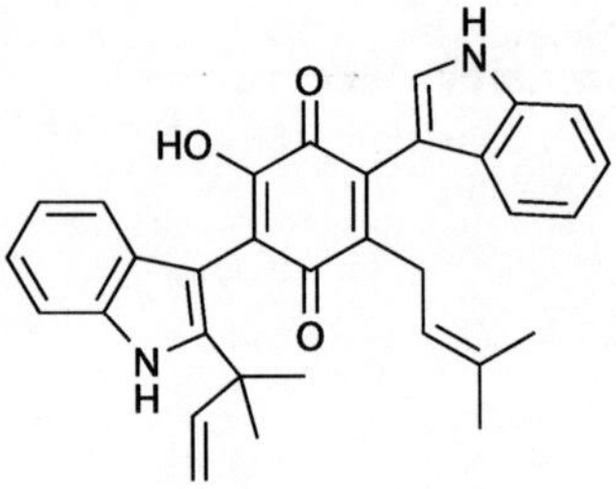

Terrequinone A (**98**)

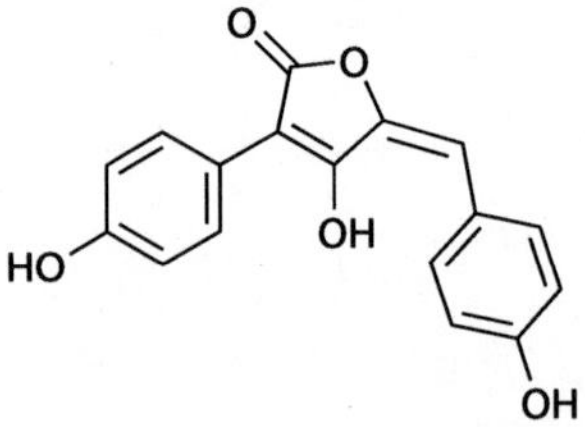

Isoaspulvinone E (**99**)

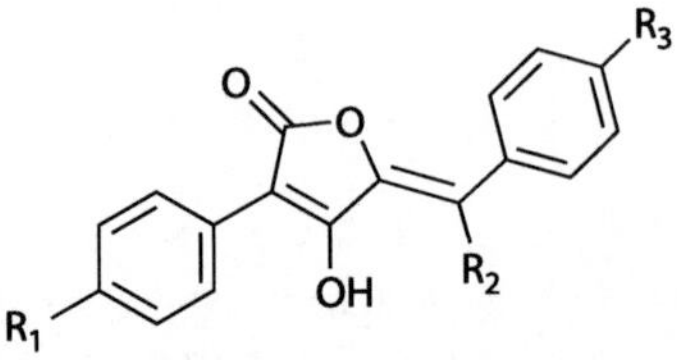

$R_1 = R_3 = OH$, $R_2 = H$, Aspulvinone E (**100**)
$R_1 = R_3 = H$, $R_2 = CO_2H$, Pulvic Acid (**101**)

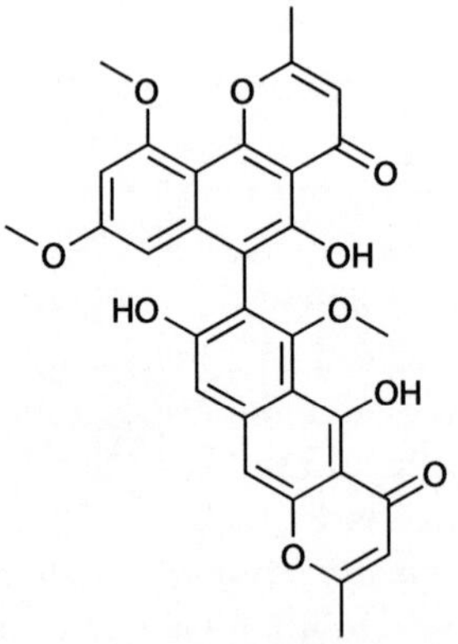

Asperpyrone D (**102**)

FIGURE 9.8 Natural products encountered in rhizospheric *Aspergillus terreus*, *A. terreus* Gwq-48, and *A. tubingensis*. *(Continued)*

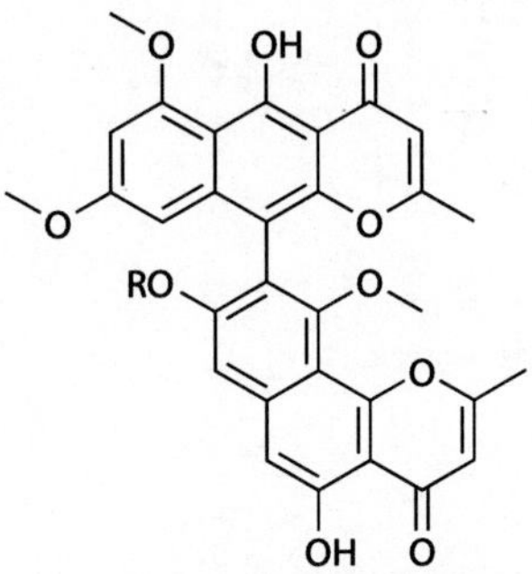

R = H, Asperpyrone A (**103**)
R = Me, Fonsecinone A (**104**)

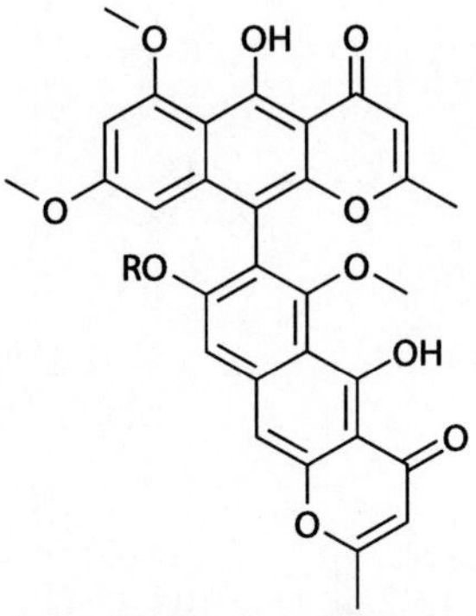

R = H, Dianhydro-aurasperone C (**105**)
R = Me, Aurasperone A (**106**)

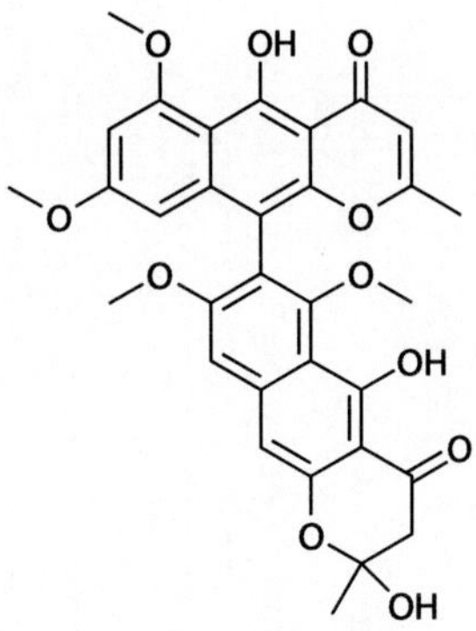

Aurasperone E (**107**)

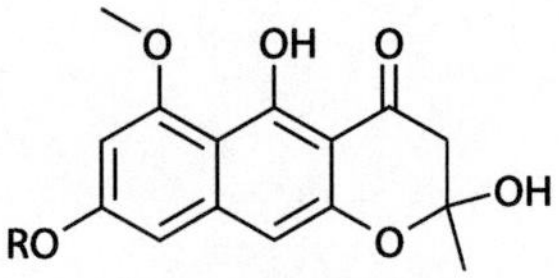

R = H, Fonsecin (**108**)
R = Me, Fonsecin B (**109**)

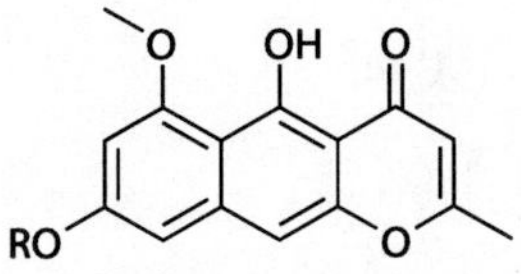

R = H, Rubrofusarin B (**110**)
R = Me, TMC-256A1 (**111**)

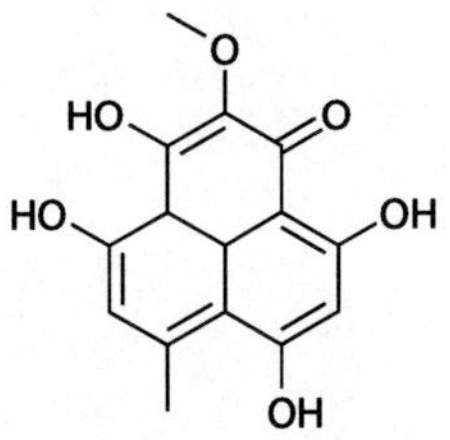

Funalenone (**112**)

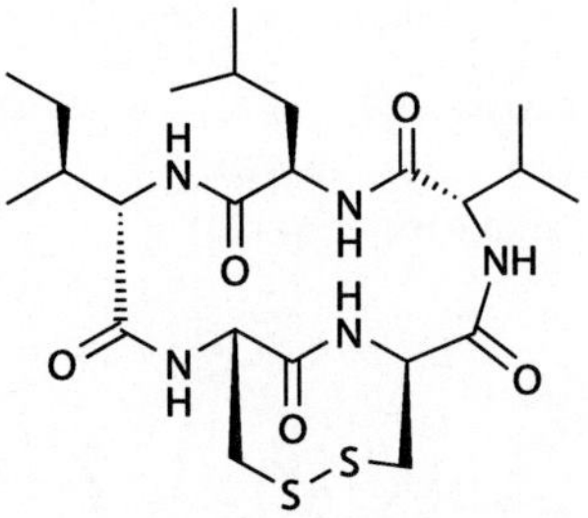

Malformin A_1 (**113**)

FIGURE 9.8 (*Continued*) Natural products encountered in rhizospheric *Aspergillus terreus*, *A. terreus* Gwq-48, and *A. tubingensis*.

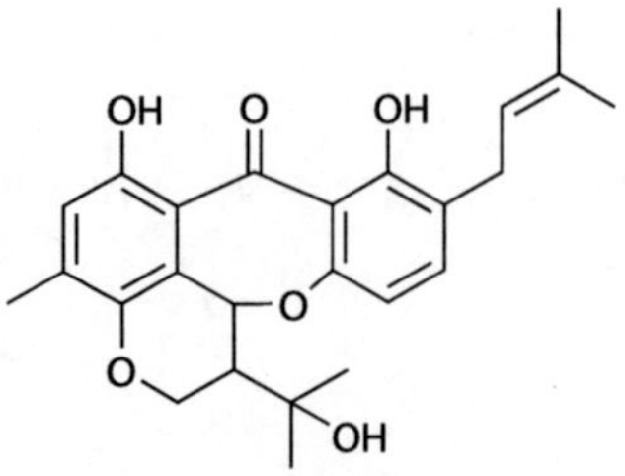

Arugosin C (**114**)

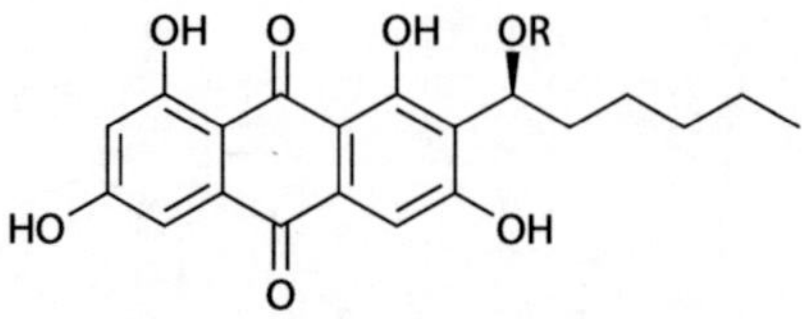

R = H, Averantin (**115**)
R = Me, Methylaverantin (**116**)

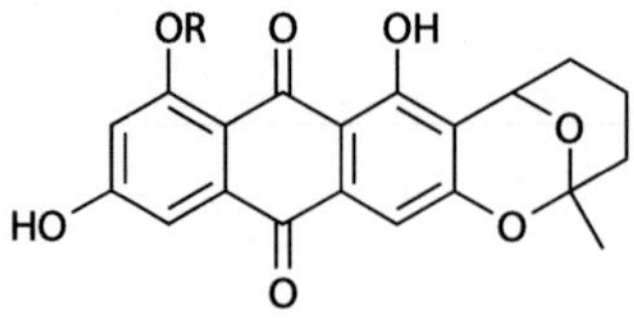

R = H, Averufin (**117**)
R = Me, 8-*O*-Methylaverufin (**118**)

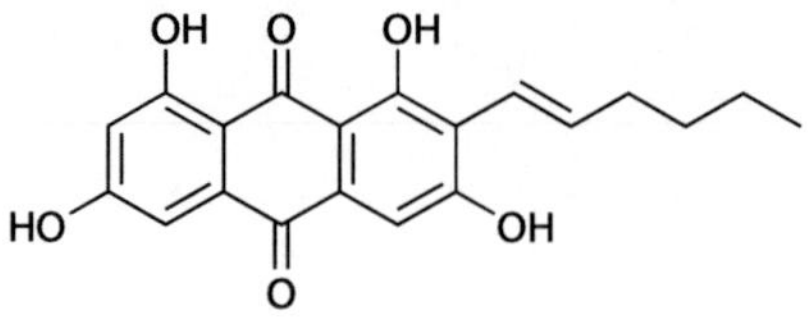

Averythrin (**119**)

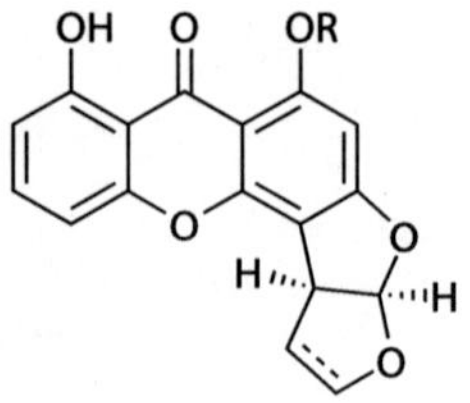

R = Me, $\Delta^{2'}$, Sterigmatocystin (**120**)
R = H, $\Delta^{2'}$, Demethylsterigmatocystin (**121**)
R = Me, 2′,3′-Dihydrosterigmatocystin (**122**)

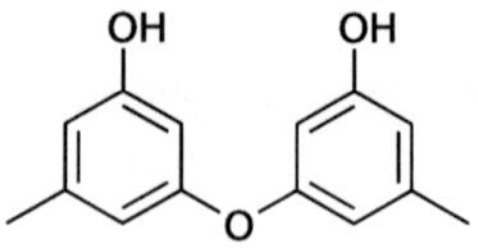

Diorcinol (**123**)

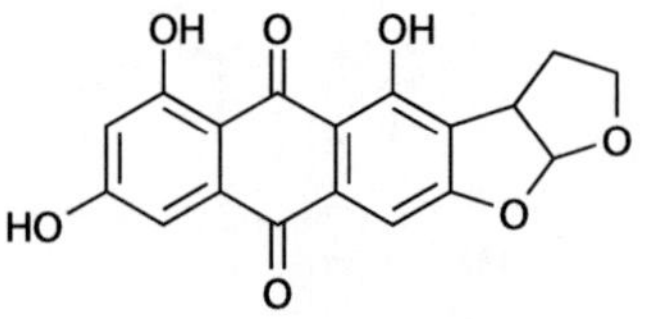

Versicolorin B (**124**)

FIGURE 9.9 Natural products encountered in rhizospheric *Aspergillus* sp. YIM PH30001 and *Chetomium globosum*. *(Continued)*

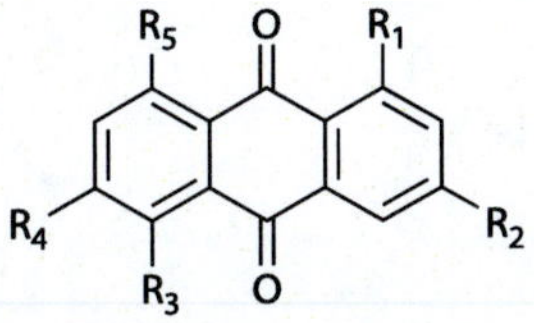

$R_4 = R_5 = H$, $R_1 = OMe$, $R_2 = Me$, $R_3 = OH$, Ziganein-1-methyl ether (**125**)
$R_3 = R_4 = H$, $R_1 = OH$, $R_2 = Me$, $R_5 = OMe$, 8-*O*-Methylchrysophanol (**126**)
$R_2 = R_3 = R_4 = H$, $R_1 = R_5 = OH$, Chrysazin (**127**)
$R_3 = H$, $R_1 = R_2 = R_4 = R_5 = OH$, 1,3,6,8-Tetrahydroxyanthraquinone (**128**)

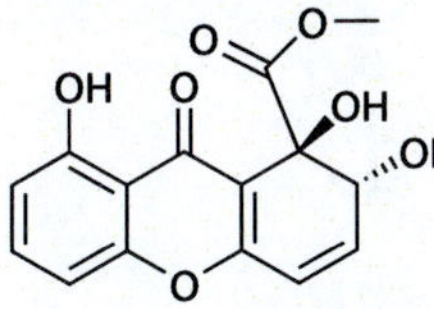

Globosuxanthone A (**129**)

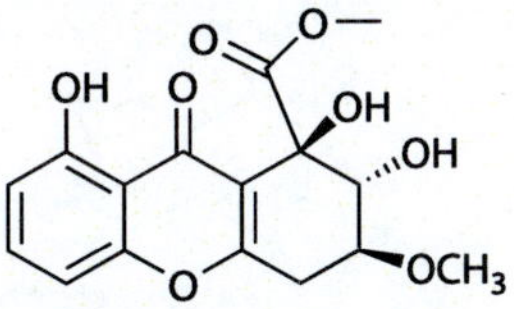

Globosuxanthone B (**130**)

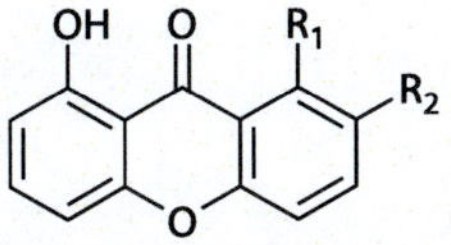

$R_1 = OH$, $R_2 = OMe$, Globosuxanthone C (**131**)
$R_1 = CO_2H$, $R_2 = H$, Globosuxanthone D (**132**)
$R_1 = CO_2Me$, $R_2 = OH$, 2-Hydroxyvertixanthone (**133**)

FIGURE 9.9 (*Continued*) Natural products encountered in rhizospheric *Aspergillus* sp. YIM PH30001 and *Chetomium globosum.*

9.4 BIOLOGICAL ACTIVITIES OF RHIZOSPHERE MICROBIAL METABOLITES

9.4.1 Introduction

As discussed above, the rhizosphere provides a resource-rich environment for its inhabitants. In order to survive in this extremely competitor-rich and predator-rich environment, rhizosphere microorganisms are likely to produce biologically active metabolites, in addition to those produced for their interaction with each other and the plant host. Apart from the anticipated biological activities due to their associations in the rhizosphere, such as plant growth–regulating, phytotoxic, and antifungal activities, natural products of some rhizosphere microorganisms have also been investigated for their applications in agriculture and drug discovery.

9.4.2 Metabolites of Agricultural Utility

PGPR are known to help plant growth by inhibiting pathogenic microorganisms such as bacteria, fungi, and also nematodes. Studies on PGPR metabolites (Table 9.2) have suggested that most of these have antibiotic activity and corroborate their importance in agriculture. One of the most studied disease-suppressing antibiotics is 2,4-DAPG

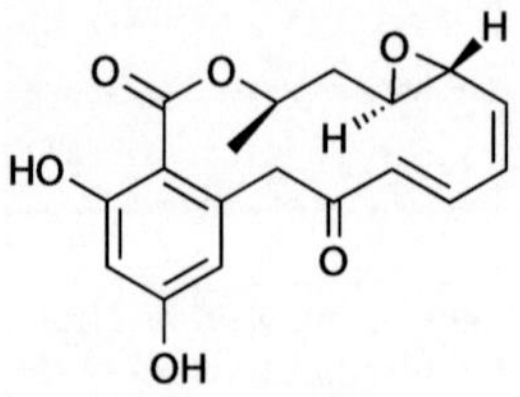

Monocillin I (**134**)

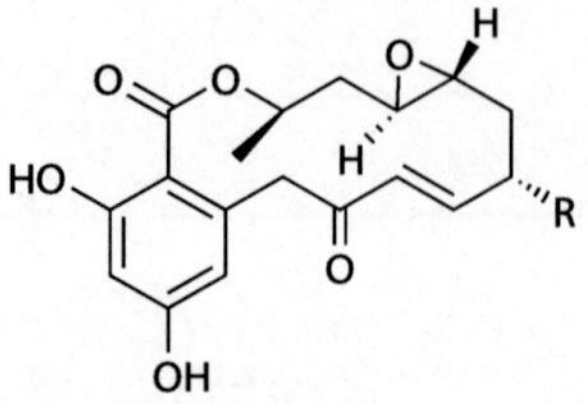

R = H, Monocillin III (**135**)
R = OH, 5′-Hydroxymonocillin III (**136**)

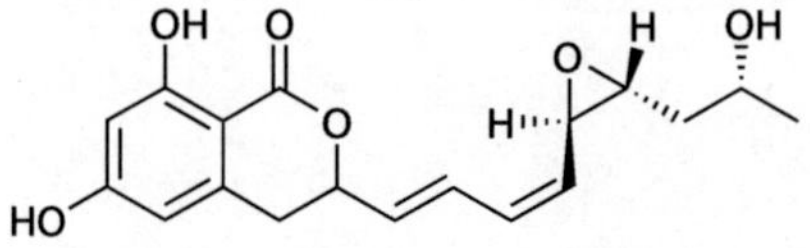

Paraphaeosphaerin A (**137**)

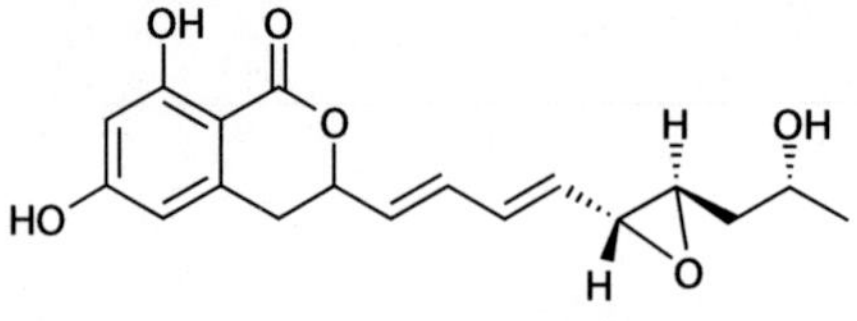

Paraphaeosphaerin B (**138**)

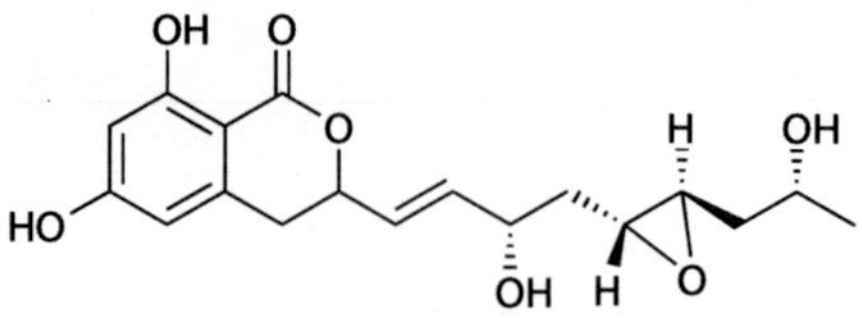

Paraphaeosphaerin C (**139**)

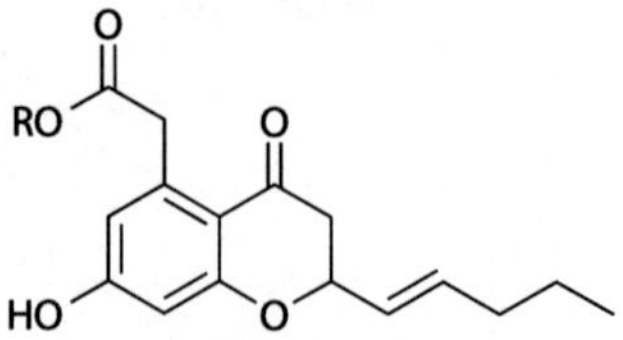

R = H, Aposphaerin C (**140**)
R = Et, Aposphaerin B (**141**)

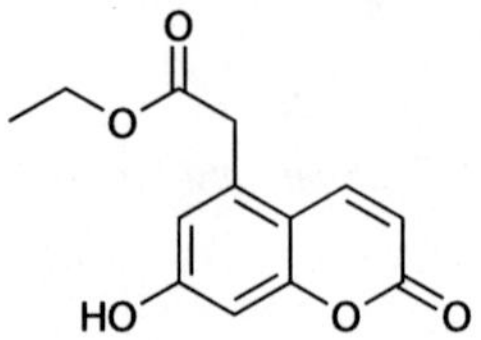

Quadriseptin A (**142**)

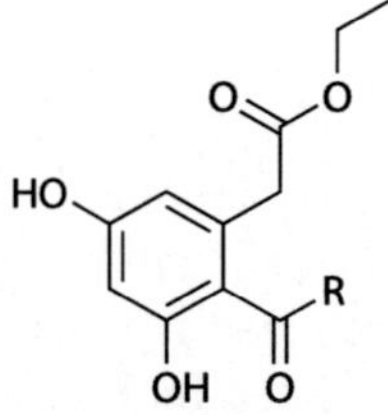

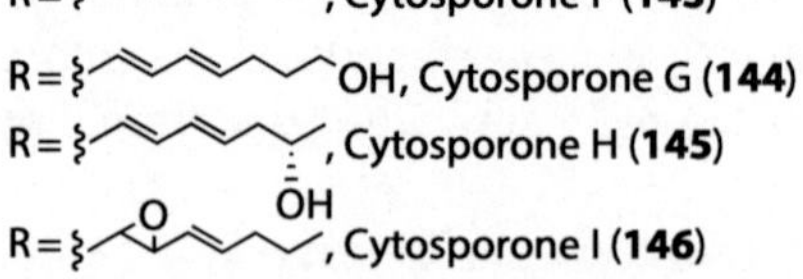

R = , Cytosporone F (**143**)
R = OH, Cytosporone G (**144**)
R = , Cytosporone H (**145**)
R = , Cytosporone I (**146**)

FIGURE 9.10 Natural products encountered in rhizospheric *Parasphaeosphaeria quadriseptata*, *Penicillium* sp. AH-00-89-F6, *Talaromyces verruculosus*, and two unidentified fungal strains. *(Continued)*

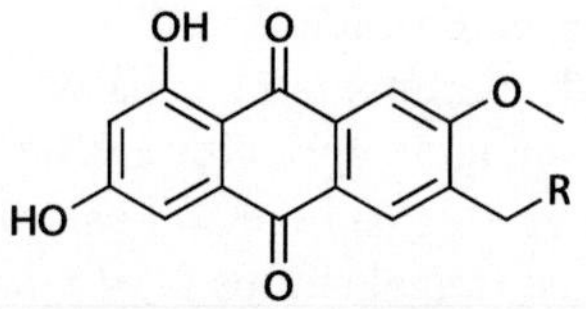

R = H, 1,3-Dihydroxy-6-methyl-7-methoxyanthraquinone (**147**)
R = OH, 1,3-Dihydroxy-6-hydroxymethyl-7-methoxyanthraquinone (**148**)

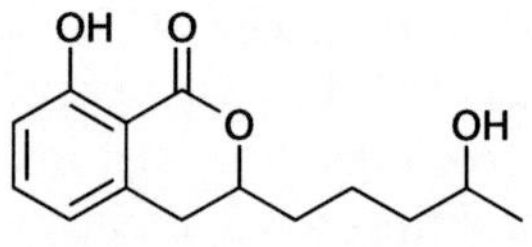

(−)-8-Hydroxy-3-(4-hydroxypentyl)-3,4-dihydroisocoumarin (**149**)

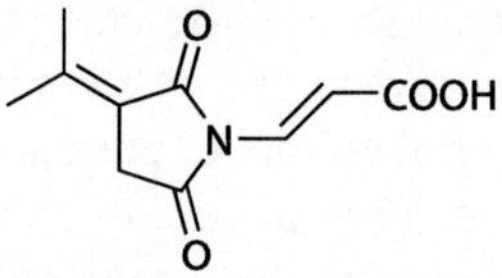

(*E*)-3-(2,5-Dioxo-3-(propan-2-ylidene) pyrrolidin-1-yl)acrylic acid (**150**)

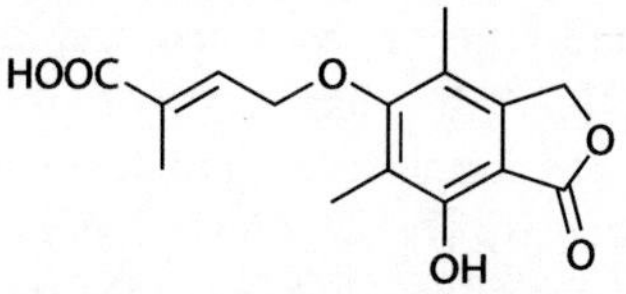

(*E*)-5-(3-Carboxy-2-butenyloxy)-7-hydroxy-4,6-dimethylphthalide (**151**)

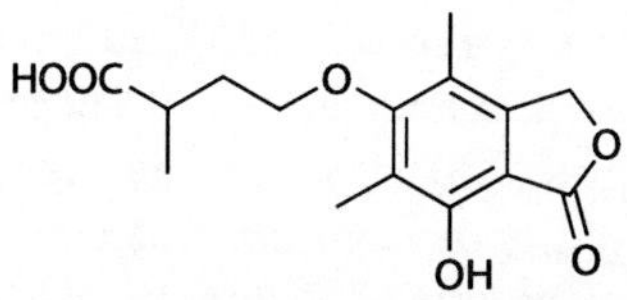

(+)-5-(3-Carboxy-butoxy)-7-hydroxy-4,6-dimethylphthalide (**152**)

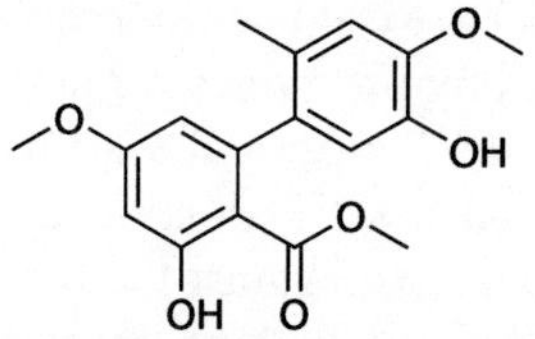

3,5′-Dihydroxy-4′,5-dimethoxy-2′-methyl-(1,1′-biphenyl)-2-carboxylic acid methyl ester (**153**)

FIGURE 9.10 (*Continued*) Natural products encountered in rhizospheric *Parasphaeosphaeria quadriseptata*, *Penicillium* sp. AH-00-89-F6, *Talaromyces verruculosus*, and two unidentified fungal strains.

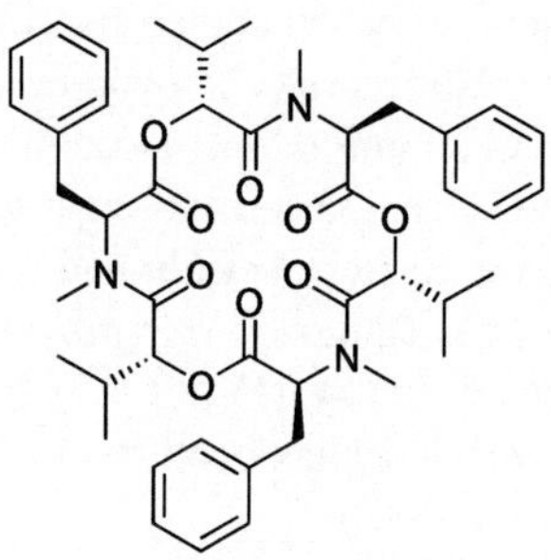

Beauvericin (**154**)

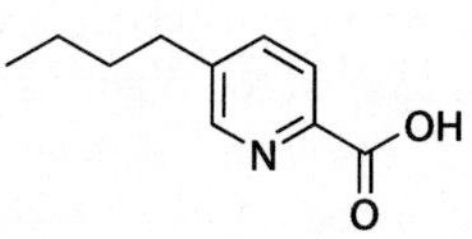

Fusaric acid (**155**)

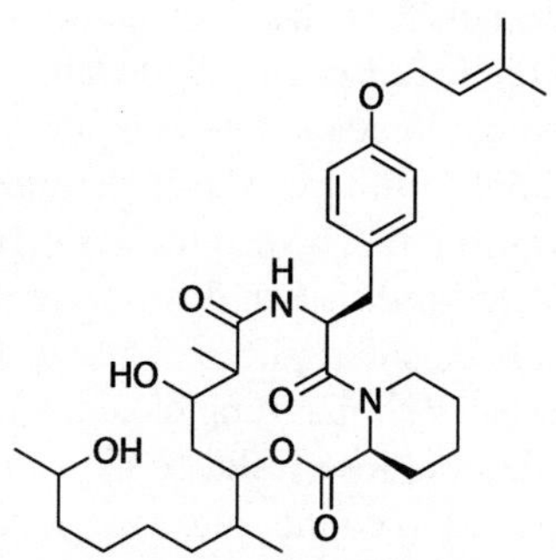

Depsipeptide HA23 (**156**)

FIGURE 9.11 Natural products identified in extracts of rhizospheric *Fusarium oxysporum* and *F. solani* by NMR-based metabolomics.

(**36**) (Figure 9.5) produced by *Pseudomonas* species, including *P. aureofaciens* and *P. fluorescens*.[61,62] It has been suggested that bacterial strains capable of producing 2,4-DAPG show great potential for use in sustainable agriculture. The antagonistic effect of 20 *P. fluorescens* strains, isolated from rhizosphere soil samples collected from rice seedlings, was investigated against the fungal strains of *Magnaporthe oryzae* (rice blast pathogen) and *Rhizoctonia solani* (rice sheath blight pathogen), and resulted in the isolation and characterization of 2,4-DAPG (**36**) as the metabolite responsible for the fungicidal activity of this bacterial strain.[61] IAA (**35**) (Figure 9.5), a well-known plant growth–promoting hormone, has been found to be produced by the rhizobacterium *P. aurantiaca*, and its biosynthesis was correlated with 2,4-DAPG (**36**) production.[60] It is noteworthy that IAA was not detected when the production of 2,4-DAPG was suppressed by exposing this bacterial culture to permanent light. IAA (**35**) produced by *Burkholderia heleia* (Gram-negative bacterium) associated with the rhizosphere of rice has been demonstrated to inhibit the virulence caused by the phytopathogenic bacterium *B. plantarii* by suppressing tropolone (**45**; Figure 9.5) biosynthesis in the latter microorganism.[66] Phenazine-type rhizobacterial metabolites were also found to exhibit antimicrobial activity and play an important role in biological control of phytopathogenic fungi, such as *F. oxysporum* and *Gaeumannomyces graminis* var. *tritici*, in agricultural crops such as wheat and barley.[63,77] The antifungal activity of phenazine-1-carboxylic acid (**37**; Figure 9.5), produced by the rhizobacterium *P. fluorescens*, against the fungal pathogen *F. oxysporum*, was confirmed by *in vitro* studies involving both normal and mutant strains of *P. fluorescens*, the latter lacking the gene responsible for producing **37**. Only the wild-type strain produced this metabolite and showed antifungal activity.[64] Two 5-methylenefuranones (**41** and **42**; Figure 9.5) produced by the rhizobacterium *P. jessenii*, associated with spinach (*Spinacia oleracea*), have been shown to have antifungal activity against the phytopathogens *Aphanomyces cochlioides* and *Pythium aphanidermatum*.[65]

Siderophores are small-molecule natural products that are produced by microorganisms and plants for the purpose of chelating iron [Fe(III)] in their habitats and making it available for themselves.[78] Thus, siderophores play an important role in rhizosphere soils. The production of siderophores (iron-chelating compounds) by PGPR is one of the indirect mechanisms used by some bacteria to promote plant growth. This class of metabolites has also been shown to display strong antimicrobial properties, which are associated with their metal-chelating activity. An acinetobactin-like and catecholate type of siderophore containing the 2,3-dihydroxybenzoic acid moiety, with hydroxyhistamine and threonine as amino acid subunits in its structure, was found to be produced by *Acinetobacter calcoaceticus* H1RFA32 (Gram-negative bacterium), isolated from the rhizosphere of wheat. Both the bacterial strain and the isolated siderophore, when tested against the phytopathogenic fungus *F. oxysporum*, revealed similar fungicidal activity.[79] Pyochelin (**43**) and pyoverdin (**44**) (Figure 9.5) are siderophores produced by pseudomonads. It is noteworthy that **43** has been reported to bind to copper and zinc and display strong antimicrobial activity.[3]

Rhizosphere microorganisms produce a wide range of VOCs that constitute lipophilic, small-molecule natural products, with low boiling point and high vapor pressure.[80] These characteristics facilitate the evaporation and diffusion of VOCs through both water- and gas-filled pores in rhizosphere soils. Release of diffusible

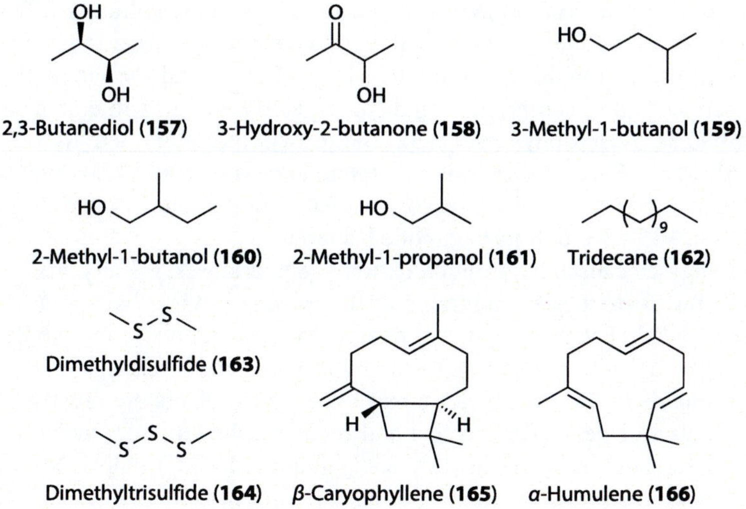

FIGURE 9.12 Volatile organic compounds of rhizosphere microbial origin.

VOCs is one of the mechanisms by which certain microbes influence the growth and fitness of their associated plants. Structures of some VOCs of rhizosphere microbial origin (**157–166**) are depicted in Figure 9.12. Some examples of VOCs produced by PGPR include 2,3-butanediol (**157**), 3-hydroxy-2-butanone (acetoin) (**158**), 3-methyl-1-butanol (**159**), 2-methyl-1-butanol (**160**), 2-methyl-1-propanol (**161**), tridecane (**162**), dimethyldisulfide (DMDS) (**163**), and dimethyltrisulfide (DMTS) (**164**).[81] The ability of VOCs produced by PGPR to trigger ISR in Arabidopsis *thaliana* has been reported.[82] Since the appearance of this report, many PGPR strains were found to elicit ISR and promote plant growth via emission of VOCs. It is interesting that VOCs produced by the Gram-positive strains *Bacillus subtilis* GB03 and *B. amyloliquefaciens* IN937a have been shown to play a role in the plant immune response. The two major VOCs produced by both these strains were identified as **157** and **158**, which were responsible for plant growth promotion. Subsequent studies with commercially available **157** and **158** confirmed their ability to promote plant growth in a dose-dependent manner. In addition, similar experiments with mutant strains of *Bacillus subtilis* where the production of 2,3-butanediol (**157**) was blocked had no effect on plant growth. Studies on VOCs from another PGPR strain, *Paenibacillus polymyxa* E681, revealed the presence of long-chain volatile compounds [e.g., tridecane (**162**)], which were also able to trigger ISR in Arabidopsis.[81] Like rhizobacteria, rhizosphere fungi are also known to produce VOCs exhibiting plant growth–enhancing, antibacterial, and antifungal activity.[83] *F. oxysporum* is a soil-borne fungal species complex often found in the rhizosphere of a variety of plants, some of which are responsible for vascular wilt disease of over 100 plant species. The nonpathogenic strain of *F. oxysporum* MSA35, harboring a consortium of ectosymbiotic bacterial species, has been found to protect plants from pathogenic *F. oxysporum* isolates. The MSA35 strain, in addition to inhibiting the growth of pathogenic *F. oxysporum*, was shown to enhance the growth of lettuce (*Lactuca sativa*) via production of VOCs.[84]

Interestingly, after removing its symbiotic bacteria by serially culturing MSA35 in a medium containing antibiotics, the resulting cured strain was found to be incapable of promoting plant growth. VOCs produced by MSA35 and the cured strain differed in their sesquiterpene profiles, and subsequently, the growth-enhancing VOC was identified as β-caryophyllene (**165**). The growth inhibitory activity of MSA35 against pathogenic *F. oxysporum* was also found to be due to a VOC, which was tentatively identified as α-humulene (**166**), another sesquiterpene emitted by MSA35, but not by MSA35 devoid of its symbiotic bacteria.[84]

Abiotic stresses caused by drought, extreme temperatures, salinity, and so on are well-known threats to the agricultural industry, leading to a loss of over 50% of crop plants worldwide.[85] For the purpose of maintaining growth and productivity, plants must adapt to stress conditions and/or employ enhanced stress tolerance mechanisms. Current strategies to enhance stress tolerance in plants are mostly based on the manipulation of genes that protect and maintain the structure and function of cellular components, resulting in genetically modified (GM) plants. Recent studies have demonstrated that the molecular chaperone heat-shock protein 90 (Hsp90) is essential for the evolutionarily conserved response of an organism to temperature stress. Since plants are stationary, during evolution, they must have acquired a mechanism for adaptation and survival in harsh environments. For those plants growing in deserts, the problem of surviving high temperatures could be more acute. Our observation that the rhizosphere fungus *Paraphaeosphaeria quadriseptata*, which inhabits the Sonoran Desert Christmas cactus (*Opuntia leptocaulis*), contained the mammalian Hsp90 inhibitor monocillin I (**134**; Figure 9.10) in surprisingly high amounts (ca. 30% by weight of total extract)[32] led to the hypothesis that the survival of plants with no morphological adaptations and growing in harsh environments may require physiological adaptations mediated by small-molecule natural products capable of activating heat shock proteins. This hypothesis was tested by addressing the question, does elaboration of monocillin I (**134**) by the rhizosphere fungus *P. quadriseptata* affects plant Hsp90 and hence environmental responsiveness of plants? It led to the demonstration that **134** was capable of binding Arabidopsis Hsp90 and also inhibiting the function of Hsp90 in lysates of wheat (*Triticum aestivum*) germ. In addition, monocillin I treatment of Arabidopsis seedlings induced Hsp101 and Hsp70, known to be conserved components of the stress response. Application of 0.1 μM monocillin I, or growth in the presence of this natural product, allowed Arabidopsis wild-type seedlings, but not Arabidopsis Hsp101 knockout mutant (*AtHSP101*) seedlings, to survive at 45°C for 75 min, otherwise a lethal temperature stress. Intriguingly, cocultivation of *P. quadriseptata* with Arabidopsis was found to enhance plant heat stress tolerance, suggesting the possible functional role of rhizosphere microorganisms in conferring thermotolerance to host plants.[86] The significance of this finding was highlighted in *Science* under editor's choice and in the research news section of *Current Science*.[87,88] The latter news article stated: "This finding opens up many exciting new avenues of investigation. Preconditioning of crop plants by spraying, based on weather forecast, of a fungal-derived natural product may be a new means of protecting crops by induction of Hsp101 for protection against abiotic stresses. This study will undoubtedly stimulate interest in extending the isolation of new fungi from the rhizosphere of plants (where fungi are numerically more because of root exudates) growing in a variety of environments, their culturing, and screening natural products that induce Hsp101 in test plants for protection from heat stress."

9.4.3 Biological Activities Relevant to Drug Discovery

Microorganisms represent a rich source of biologically active natural products with wide-ranging applications such as agrochemicals, antibiotics, immunosuppressants, anticancer drugs, and antiparasitic agents. Compared with microorganisms of the bulk soil, those of the rhizosphere microbiome are expected to have a higher capacity to produce biologically active metabolites, as these are implicated in their rhizosphere interactions (see Figure 9.1). Thus, extracts derived from cultured rhizosphere microorganisms, especially fungi, have exhibited biological activities relevant to drug discovery, and their bioactivity-guided fractionation has resulted in a variety of natural products with interesting structures and potential anticancer, antimicrobial, and antiviral activities. As apparent from Table 9.3, the majority of these studies have been conducted in programs involving the discovery of potential anticancer drugs. Out of the total of 18 rhizosphere fungal strains investigated, 13 have been subjected to cytotoxicity-guided fractionation, resulting in the isolation and characterization of over 30 natural products with potential anticancer activity. The discussion below will focus on those natural products with biological activities relevant to drug discovery, and for cytotoxic natural products, only those for which target identification studies have been conducted and/or possible mechanisms of action have been evaluated.

The new spiro-polyketide-diketopiperazine alkaloid, (±)-dihydrocryptoechinulin D (**50**; Figure 9.6), isolated from *As. effuses* H1-1, was found to have selective cytotoxic activity against the P388 leukemia cell line, with an IC_{50} of 1.83 μM, and its possible molecular target has been determined to be topoisomerase I, using an assay involving the relaxation of the supercoiled plasmid DNA.[35] Interestingly, when the two enantiomers (+)-dihydrocryptoechinulin D and (–)-dihydrocryptoechinulin D (**50a** and **50b**, respectively) were separated and tested, **50a** exhibited enhanced activity, whereas **50b** was found to be inactive in this assay. The macrocyclic resorcylic acid lactone dehydrocurvularin (**80**; Figure 9.7), isolated from *As. terreus*,[24] *Penicillium* sp. AH-00-89-F6,[29] and *Penicillium sumatrense*,[33] has recently been shown to inhibit the AAA+ (ATPase associated with diverse cellular activities) chaperone p97 and the proteasome, suggesting that the previously observed cytotoxic activity may be associated with these inhibitory activities.[89] It is noteworthy that the thiol adducts of **80**, sumalarins A–C (**85–87**; Figure 9.7), isolated from *P. sumatrense* and lacking the Michael acceptor enone moiety, were found to be as cytotoxic as **80**.[33] The possible mechanism of cytotoxic activity of the sesquiterpenoid (+)-terrecyclic acid A (**92**; Figure 9.7), isolated from *As. terreus*,[27] has been determined to be due to the modulation of multiple stress pathways, namely the oxidative, heat shock, and inflammatory responses in tumor cells that promote their survival.[90] A new, strongly cytotoxic dihydroxanthenone, globosuxanthone A (**129**; Figure 9.9), isolated from *Chaetomium globosum*, was found to impair cell cycle progression of human non-small-cell carcinoma (NCI-H460) and metastatic prostate cancer (PC-3M) cell lines, leading to the accumulation of cells in the G_2/M and S phases, and also to induce classic signs of apoptosis in NCI-H460.[73] Monocillin I (**134**; Figure 9.10), the known macrocyclic resorcylic acid lactone, was isolated by heat shock induction assay–guided fractionation of an extract of *Paraphaeosphaeria quadriseptata*. Monocillin I was found to induce a strong heat shock response and exhibit strong cytotoxic activity against the human breast adenocarcinoma cell line MCF-7.[91]

Natural products encountered in some rhizosphere fungal strains were also reported to have antibacterial, antifungal, and antiviral activities. It is interesting that the anthraquinones averantin (**115**) and averythrin (**119**) (Figure 9.9), isolated from *Aspergillus* sp. YIM PH30001, showed antibacterial activity against *Bacillus subtilis* and antifungal activity against *F. solani*, whereas the anthraquinones averufin (**117**), versicolorin B (**124**), and ziganein-1-methyl ether (**125**), and the phenolic metabolite arugosin C (**114**) (Figure 9.9) exhibited antibacterial and antifungal activities, respectively.[30] Other rhizosphere fungal metabolites with antifungal activity included (*E*)-3-(2,5-dioxo-3-(propan-2-ylidene)pyrrolidin-1-yl)acrylic acid (**150**; Figure 9.10) from *Talaromyces verruculosus*[34] and (*E*)-5-(3-carboxy-2-butoxy)-7-hydroxy-4,6-dimethylphthalide (**152**) from an unidentified fungal strain isolated from wheat.[75] Isoaspulvinone E (**99**; Figure 9.8), a new butenolide, and two of its known analogs, aspulvinone E (**100**) and pulvic acid (**101**), isolated from a mangrove rhizosphere fungus *As. terreus* Gwq-48 were found to exhibit antiviral activity against the influenza virus H1N1.[92] Of these, only isoaspulvinone E (**99**) was capable of inhibiting H1N1 viral neuraminidase, and docking of **99** and its *E* isomer **100** into the active site of neuraminidase suggested that the *E* double bond at $\Delta^{5(10)}$ was essential for activity. The cyclic hexadepsipeptide beauvericin (**154**; Figure 9.11), identified by NMR-based metabolomics as a major metabolite of several *F. oxysporum* strains, and isolated from the rhizosphere of *Senna spectabilis*,[37] has been reported to exhibit cytotoxic, antimicrobial, antiviral, and insecticidal activities,[93] and to inhibit the migration of metastatic breast cancer (MDA-MB-231) and prostate cancer (PC-3M) cells.[28] In an attempt to discover natural products–based drugs to treat invasive fungal infections, a leading cause of human mortality, it was recently found that beauvericin (**154**) effectively potentiated the activity of azoles, the most widely deployed class of antifungals, blocked the emergence of drug resistance, and rendered resistant pathogens responsive to treatment.[94] In this study, a novel mechanism for beauvericin (**154**) involving the modulation of signaling through global cellular regulators TOR (target of rapamycin), CK2 (casein kinase 2), and Hsp90 was established, which is of significance, as inhibitors of these are known to have potent activity against protozoan parasites, including those causing malaria, trypanosomiasis, and toxoplasmosis.

9.5 IMPLICATIONS OF THE OCCURRENCE AND UTILIZATION OF NATURAL PRODUCTS OF THE RHIZOSPHERE

Compared with other ecosystems on earth, the rhizosphere ecosystem is one of the most complex, as every plant root on the planet is expected to have a rhizosphere that is chemically and biologically unique. In addition, the rhizosphere is intriguingly complex and dynamic, and understanding its ecology is considered important for enhancing ecosystem functioning and plant productivity in agriculture. Although considerable progress has been made in understanding the microbial ecology of the rhizosphere, rhizosphere ecologists face numerous challenges, including the development of new crops and cropping systems to maximize food, feed, fiber, and bioenergy production at low environmental costs.[4] The ability of organisms to form long-term, intimate, and diverse relationships with each other, as is the case with plants and their rhizosphere microbiomes, has been recognized as a common ecological phenomenon. Direct and indirect interactions between hosts and their associated microbiota are known to involve constitutive and inducible changes in the secondary metabolism of these organisms.[95] Thus, for fruitful interactions, plants and their rhizosphere-associated microorganisms must have developed a rich chemical ecology, and natural products of the rhizosphere play a significant role in these ecological interactions. However, passage of these natural products within the rhizosphere ecosystem can occur only by diffusion or by transport via vesicles or transporters, and during the passage, these compounds may undergo oxidation, degradation, microbial biotransformation, or immobilization as a result of binding to soil particles. As highlighted in foregoing sections, rhizosphere natural products with ecological implications and agricultural utility include those involved in communication, symbiotic interactions, chemical defense (using toxic compounds or knocking out the host's defense mechanisms), and allelopathy of rhizosphere-associated organisms; in addition, siderophores with metal-chelating ability produced by rhizosphere organisms have been suggested to be an environmentally friendly alternative to some harmful pesticides, as they are known to improve plant growth by protecting them from pathogens.[96]

In agroecosystems, crop productivity is closely related to plant health. In these ecosystems, crop plants encounter a wide range of pathogenic microorganisms, some of which colonize the rhizosphere and infect plant roots.[97] Some rhizobacterial strains capable of producing antimicrobial natural products are considered potential agents in biocontrol of a variety of soil-borne fungal pathogens, and can be used to complement or even replace the application of chemical fungicides in agricultural practices. However, utilization of these bacterial strains is still a challenge because the strategy involved in each crop protection may vary with differences in cultivars, soil chemistries, and environmental conditions.[98] Take-all disease (TAD) is a worldwide root disease of wheat (*Triticum* spp.) caused by the fungus *Gaeumannomyces graminis* var. *tritici*. The spontaneous decline in the incidence and severity of TAD after extended monoculture of wheat can be considered a very good example of a natural biocontrol phenomenon. This TAD decline has been associated with the buildup of bacterial populations of fluorescent *Pseudomonas* spp. during monoculture of wheat, which consequently increased the concentration of the bacterial natural product 2,4-DAPG (**36**; Figure 9.5), which was shown to be active against some phytopathogenic fungi.[63,99] The black root rot of tobacco caused by the fungus *Thielaviopsis basicola* and the damping-off disease of sugar beet due to *Pythium ultimum* infection were also biocontrolled by *P. fluorescens* strains, and 2,4-DAPG (**36**) was found to be responsible for the ability of *P. fluorescens* strains to control these agriculturally important plant diseases.[63,97]

Rice (*Oryza sativa*) seedling blight is another example of economically costly agricultural diseases which is caused by the bacterium *Burkholderia plantarii*, as a result of its ability to produce the phytotoxic natural product tropolone (**45**; Figure 9.5).[100] It has been suggested that the potent iron-chelating property of tropolone not only accounted for its broad-spectrum antibacterial and antifungal activities, but also contributed to its virulence and the symptoms associated with the onset of bacterial blight disease in rice seedlings.[101] It has been recently shown that the related bacterium *B. heleia* PAK1-2, also occurring in the rice rhizosphere, was able to act as a potent biocontrol agent, preventing rice seedling blight disease.[66] GC-MS and proton NMR-based metabolic profiling of *B. plantarii* suggested that phenylacetic acid (PAA) was a dominant metabolite during rice plant's early growth stage and acted as a direct biosynthetic precursor of tropolone. This work also demonstrated that a culture of *B. heleia* PAK1-2 contained IAA (**35**; Figure 9.5), which was capable of suppressing *B. plantarii* virulence by inhibiting its production of tropolone by disrupting the biosynthetic conversion of PAA to tropolone. More significantly, this study provided insight into microbial interspecies interactions in the rhizosphere ecosystem, understanding of which may lead to the development of novel biocontrol agents for utilization in agriculture.

ACKNOWLEDGMENTS

The authors are thankful to all their collaborators and coworkers for their valuable contributions to the research programs on plant-associated microorganisms. We are also thankful to Dr. E.M.K. Wijeratne and Ms. A.M. Bagley of the University of Arizona for their help in drawing some structures and typing the list of references, the staff of Co-ordination of Social Communication and Institutional Marketing of Federal University of Ceará and the graphic designer Daniel L. Cabral for their assistance in preparing some figures, and Ms. M.K. Gunatilaka for critically reading this chapter. The authors express their appreciation to the Brazilian funding agency Conselho Nacional de Desenvolvimento Científico e Tecnológico (CNPq) (Process 303365/2014-5 to M.C.F. de O and Process 405001/2013-4 to M.C.F de O and A.A.L.G.) and the U.S. National Institutes of Health (Grants R01 CA90265 and P41 GM094060 to A.A.L.G.) for financial support for their research programs.

REFERENCES

1. Hartmann, A.; Rothballer, M.; Schmid, M. Lorenz Hiltner, a pioneer in rhizosphere microbial ecology and soil bacteriology research. *Plant Soil* **2008**, *312*, 7–14.
2. Bais, H. P.; Weir, T. L.; Perry, L. G.; Gilroy, S.; Vivanco, J. M. The role of root exudates in rhizosphere interactions with plants and other organisms. *Annu. Rev. Plant Biol.* **2006**, *57*, 233–266.
3. Schrey, S. D.; Hartmann, A.; Hampp, R. Rhizosphere interactions. In *Ecological Biochemistry: Environmental and Interspecies Interactions* 1st edn.; Krauss, G. J.; Nies, D. H., Eds. Wiley-VCH Verlag GmbH & Co. KGaA, Weinheim, Germany, **2015**, pp. 293–310.
4. Philippot, L.; Raaijmakers, J. M.; Lemanceau, P.; van der Putten, W. H. Going back to the roots: The microbial ecology of the rhizosphere. *Nat. Rev. Microbiol.* **2013**, *11*, 789–799.

5. Lynch, J. M.; Whipps, J. M. Substrate flow in the rhizosphere. *Plant Soil* **1990**, *129*, 1–10.
6. Zhang, Y.; Ruyter-Spira, C.; Bouwmeester, H. J. Engineering the plant rhizosphere. *Curr. Opin. Biotechnol.* **2015**, *32*, 136–142.
7. Berendsen, R. L.; Pieterse, C. M. J.; Bakker, P. A. H. M. The rhizosphere microbiome and plant health. *Trends Plant Sci.* **2012**, *17*, 478–486.
8. Lynch, J. M.; Benedetti, A.; Insam, H.; Nuti, M. P.; Smalla, C.; Torsvik, V.; Nannipieri, P. Microbial diversity in soil: Ecological theories, the contribution of molecular techniques and the impact of transgenic plants and transgenic microorganisms. *Biol. Fertil. Soils* **2004**, *40*, 363–385.
9. Pace, N. R. A molecular view of microbial diversity and the biosphere. *Science* **1997**, *276*, 734–740.
10. Nihorimbere, V.; Ongena, M.; Smargiassi, M.; Thonart, P. Beneficial effect of the rhizosphere microbial community for plant growth and health. *Biotechnol. Agron. Soc. Environ.* **2011**, *15*, 327–337.
11. Singh, B. K.; Millard, P.; Whiteley, A. S.; Murrell, J. C. Unravelling rhizosphere-microbial interactions: Opportunities and limitations. *Trends Microbiol.* **2004**, *12*, 386–393.
12. Buée, M.; De Boer, W.; Martin, F.; van Overbeek, L.; Jurkevitch, E. The rhizosphere zoo: An overview of plant-associated communities of microorganisms, including phages, bacteria, archaea, and fungi, and some of their structuring factors. *Plant Soil* **2009**, *321*, 189–212.
13. Gardner, T.; Acosta-Martinez, V.; Senwo, Z.; Dowd, S. E. Soil rhizosphere microbial communities and enzyme activities under organic farming in Alabama. *Diversity* **2011**, *3*, 308–328.
14. Nannipieri, P.; Ascher, J.; Ceccherini, M. T.; Landi, L.; Pietramellara, G.; Renella, G. Microbial diversity and soil functions. *Eur. J. Soil Sci.* **2003**, *54*, 655–670.
15. Venturi, V.; Keel, C. Signaling in the rhizosphere. *Trends Plant Sci.* **2016**, *21*, 187–198.
16. Whipps, J. M.; Lynch, J. M. The influence of the rhizosphere on crop productivity. *Adv. Microb. Ecol.* **1986**, *9*, 187–244.
17. Karthikeyan, B.; Jaleel, C. A.; Lakshmanan, G. M. A.; Deiveekasundaram, M. Studies on rhizosphere microbial diversity of some commercially important medicinal plants. *Colloids Surf. B: Biointerfaces* **2008**, *62*, 143–145.
18. Mougel, C.; Offre, P.; Ranjard, L.; Corberand, T.; Gamalero, E.; Robin, C.; Lemanceau, P. Dynamic of the genetic structure of bacterial and fungal communities at different developmental stages of *Medicago truncatula* Gaertn. cv. Jemalong line J5. *New Phytol.* **2006**, *170*, 165–175.
19. Bafana, A. Diversity and metabolic potential of culturable root-associated bacteria from *Origanum vulgare* in sub-Himalayan region. *World J. Microbiol. Biotechnol.* **2013**, *29*, 63–74.
20. Battu, P. R.; Reddy, M. S. Isolation of secondary metabolites from *Pseudomonas fluorescens* and its characterization. *Asian J. Res. Chem.* **2009**, *2*, 26–29.
21. Wang, S.; Ouyang, L.; Ju, X.; Zhang, L.; Zhang, Q.; Li, Y. Survey of plant drought-resistance promoting bacteria from *Populus euphratica* tree living in arid area. *Indian J. Microbiol.* **2014**, *54*, 419–426.
22. Reddy, M. S.; Satyanarayana, T. Interactions between ectomycorrhizal fungi and rhizospheric microbes. In *Soil Biology*; Mukerji, K. G.; Manoharachary, C.; Singh, J., Eds. Springer-Verlag, Heidelberg, Germany, **2006**, pp. 245–263.
23. Ishimoto, H.; Fukushi, Y.; Yoshida, T.; Tahara, S. *Rhizopus* and *Fusarium* are selected as dominant fungal genera in rhizospheres of Brassicaceae. *J. Chem. Ecol.* **2000**, *26*, 2387–2399.

24. He, J.; Wijeratne, E. M. K.; Bashyal, B. P.; Zhan, J.; Seliga, C. J.; Liu, M. X.; Pierson, E. E.; Pierson, L. S., III; Van Etten, H. D.; Gunatilaka, A. A. L. Cytotoxic and other metabolites of *Aspergillus* inhabiting the rhizosphere of Sonoran desert plants. *J. Nat. Prod.* **2004**, *67*, 1985–1991.
25. Zhou, G.-X.; Wijeratne, E. M. K., Bigelow, D. ; Pierson, L. S., III; VanEtten, H. D.; Gunatilaka, A. A. L. Aspochalasins I, J, and K: Three new cytotoxic cytochalasans of *Aspergillus flavipes* from the rhizosphere of *Ericameria laricifolia* of the Sonoran desert. *J. Nat. Prod.* **2004**, *67*, 328–332.
26. Cai, S.; Sun, S.; Zhou, H.; Kong, X.; Zhu, T.; Li, D.; Gu, Q. Prenylated polyhydroxy-*p*-terphenyls from *Aspergillus taichungensis* ZHN-7-07. *J. Nat. Prod.* **2011**, *74*, 1106–1110.
27. Wijeratne, E. M. K.; Turbyville, T. J.; Zhang, Z.; Bigelow, D.; Pierson, L. S., III; VanEtten, H. D.; Whitesell, L.; Canfield, L. M.; Gunatilaka, A. A. L. Cytotoxic constituents of *Aspergillus terreus* from the rhizosphere of *Opuntia versicolor* of the Sonoran desert. *J. Nat. Prod.* **2003**, *66*, 1567–1573.
28. Zhan, J.; Gunaherath, G. M. K. B.; Wijeratne, E. M. K.; Gunatilaka, A. A. L. Asperpyrone D and other metabolites of the plant-associated fungal strain *Aspergillus tubingensis*. *Phytochemistry* **2007**, *68*, 368–372.
29. Zhan, J.; Wijeratne, E. M. K.; Seliga, C. J.; Zhang, J.; Pierson, E. E.; Pierson, L. S., III; Vanetten, H. D.; Gunatilaka, A. A. L. A new anthraquinone and cytotoxic curvularins of a *Penicillium* sp. from the rhizosphere of Fallugia paradoxa of the Sonoran desert. *J. Antibiot.* **2004**, *57*, 341–344.
30. Liu, K.; Zheng, Y.; Miao, C.; Xiong, Z.; Xu, L.; Guan, H.; Yang, Y.; Zhao, L. The antifungal metabolites obtained from the rhizospheric *Aspergillus* sp. YIM PH30001 against pathogenic fungi of *Panax notoginseng*. *Nat. Prod. Res.* **2014**, *28*, 2334–2337.
31. Wijeratne, E. M. K.; Paranagama, P. A.; Gunatilaka, A. A. L. Five new isocoumarins from Sonoran desert plant-associated fungal strains *Paraphaeosphaeria quadriseptata* and *Chaetomium chiversii*. *Tetrahedron* **2006**, *62*, 8439–8446.
32. Wijeratne, E. M. K.; Carbonezi, C. A.; Takahashi, J. A.; Seliga, C. J.; Turbyville, T. J.; Pierson, E. E.; Pierson, L. S., III et al. Isolation, optimization of production and structure-activity relationship studies of monocillin I, the cytotoxic constituent of *Paraphaeosphaeria quadriseptata*. *J. Antibiot.* **2004**, *57*, 541–546.
33. Meng, L.-H.; Li, X.-M.; Lv, C.-T.; Li, C.-S.; Xu, G.-M.; Huang, C.-G.; Wang, B.-G. Sulfur-containing cytotoxic curvularin macrolides from *Penicillium sumatrense* MA-92, a fungus obtained from the rhizosphere of the mangrove *Lumnitzera racemosa*. *J. Nat. Prod.* **2013**, *76*, 2145–2149.
34. Miao, F.; Yang, R.; Chen, D.-D.; Wang, Y.; Qin, B.-F.; Yang, X.-J.; Zhou, L. Isolation, identification and antimicrobial activities of two secondary metabolites of *Talaromyces verruculosus*. *Molecules* **2012**, *17*, 14091–14098.
35. Gao, H.; Liu, W.; Zhu, T.; Mo, X.; Mándi, A.; Kurtán, T.; Li, J.; Ai, J.; Gu, Q.; Li, D. Diketopiperazine alkaloids from a mangrove rhizosphere soil derived fungus *Aspergillus effuses* H1-1. *Org. Biomol. Chem.* **2012**, *10*, 9501–9506.
36. Gao, H.; Zhu, T.; Li, D.; Gu, Q.; Liu, W. Prenylated indole diketopiperazine alkaloids from a mangrove rhizosphere soil dervied fungus *Aspergillus effuses* H1-1. *Arch. Pharm. Res.* **2013**, *36*, 952–956.
37. Selegato, D. M.; Freire, R. T.; Tannus, A.; Castro-Gamboa, I. New dereplication method applied to NMR-based metabolomics on different *Fusarium* species isolated from rhizosphere of *Senna spectabilis*. *J. Braz. Chem. Soc.* **2016**, *27*, 1421–1431.
38. Kent, A. D.; Triplett, E. W. Microbial communities and their interactions in soil and rhizosphere ecosystems. *Annu. Rev. Microbiol.* **2002**, *56*, 211–236.

39. Gunatilaka, A. A. L. Natural products from plant-associated microorganisms: Distribution, structural diversity, bioactivity, and implications of their occurrence. *J. Nat. Prod.* **2006**, *69*, 509–526.
40. Flachshaar, D.; Piel, J. Chemical biology of symbiosis. In *Wiley Encyclopedia of Chemical Biology*. Begley, T. P., Ed. John Wiley & Sons, Inc., Hoboken, NJ, Published Online, **2008**, pp. 423–435.
41. Basil, A. J.; Strap, J. L.; Knotek-Smith, H. M.; Crawford, D. L. Studies on the microbial populations of the rhizosphere of big sagebrush (*Artemisia tridentata*). *J. Ind. Microbiol. Biotechnol.* **2004**, *31*, 278–288.
42. Jones, D. L.; Nguyen, C.; Finlay, R. D. Carbon flow in the rhizosphere: Carbon trading at the soil-root interface. *Plant Soil* **2009**, *321*, 5–33.
43. van Dam, N. M.; Bouwmeester, H. J. Metabolomics in the rhizosphere: Tapping into belowground chemical communication. *Trends Plant Sci.* **2016**, *21*, 256–265.
44. Rudrappa, T.; Czymmek, K. J.; Paré, P. W.; Bais, H. P. Root-secreted malic acid recruits beneficial soil bacteria. *Plant Physiol.* **2008**, *148*, 1547–1556.
45. Biedrzycki, M. L.; Jilany, T. A.; Dudley, S. A.; Bais, H. P. Root exudates mediate kin recognition in plants. *Commun. Integr. Biol.* **2010**, *3*, 28–35.
46. Auger, B.; Pouvreau, J.-B.; Pouponneau, K.; Yoneyama, K.; Montiel, G.; Le Bizec, B.; Yoneyama, K.; Delavault, P.; Delourme, R.; Simier, P. Germination stimulants of *Phelipanche ramosa* in the rhizosphere of *Brassica napus* are derived from the glucosinolate pathway. *Mol. Plant Microbe Interact.* **2012**, *25*, 993–1004.
47. Schmid, N. B.; Giehl, R. F.; Döll, S.; Mock, H. P.; Strehmel, N.; Scheel, D.; Kong, X.; Hider, R. C.; von Wirén, N. Feruloyl-CoA 6′-hydroxylase1-dependent coumarins mediate iron acquisition from alkaline substrates in *Arabidopsis*. *Plant Physiol.* **2014**, *164*, 160–172.
48. Chaparro, J. M.; Badri, D. V.; Bakker, M. G.; Sugiyama, A.; Manter, D. K.; Vivanco, J. M. Root exudation of phytochemicals in Arabidopsis follows specific patterns that are developmentally programmed and correlate with soil microbial functions. *PLoS One* **2013**, *8*, e55731.
49. Bouwmeester, H. J.; Roux, C.; Lopez-Raez, J. A.; Bécard, G. Rhizosphere communication of plants, parasitic plants and AM fungi. *Trends Plant Sci.* **2007**, *12*, 224–230.
50. Reuben, S.; Bhinu, V. S.; Swarup, S. Rhizosphere metabolomics: Methods and applications. In *Secondary Metabolites in Soil Ecology: Soil Biology*; Karlovsky, P., Ed. Springer-Verlag, Heidelberg, Germany, **2008**, pp. 37–68.
51. Strehmel, N.; Böttcher, C.; Schmidt, S.; Scheel, D. Profiling of secondary metabolites in root exudates of *Arabidopsis thaliana*. *Phytochemistry* **2014**, *108*, 35–46.
52. Hassan, S.; Mathesius, U. The role of flavonoids in root-rhizosphere signalling: Opportunities and challenges for improving plant-microbe interactions. *J. Exp. Bot.* **2012**, *63*, 3429–3444.
53. Peters, N. K.; Frost, J. W.; Long, S. R. A plant flavone, luteolin, induces expression of *Rhizobium meliloti* nodulation genes. *Science* **1986**, *233*, 977–980.
54. Weir, T. L.; Bais, H. P.; Vivanco, J. M. Intraspecific and interspecific interactions mediated by a phytotoxin, (–)-catechin, secreted by the roots of *Centaurea maculosa* (spotted knapweed). *J. Chem. Ecol.* **2003**, *29*, 2397–2412.
55. Morandi, D.; le Signor, C.; Gianinazzi-Pearson, V.; Duc, G. A *Medicago truncatula* mutant hyper-responsive to mycorrhiza and defective for nodulation. *Mycorrhiza* **2009**, *19*, 435–441.
56. Zuanazzi, J. A. S.; Clergeot, P. H.; Quirion, J.-C.; Husson H.-P.; Kondorosi, A.; Ratet, P. Production of *Sinorhizobium meliloti* nod gene activator and repressor flavonoids from *Medicago sativa* roots. *Mol. Plant Microbe Interact.* **1998**, *11*, 784–794.
57. Belz, R. G. Allelopathy in crop/weed interactions—An update. *Pest Manag. Sci.* **2007**, *63*, 308–326.

58. Xu, M.; Galhano, R.; Wiemann, P.; Bueno, E.; Tiernan, M.; Wu, W. et al. Genetic evidence for natural product-mediated plant-plant alleopathy in rice (*Oryza sativa*). *New Phytol.* **2012**, *193*, 570–575.
59. Jandová, K.; Dostál, P.; Cajthaml, T.; Kameník, Z. Intraspecific variability in allelopathy of *Heracleum mantegazzianum* is linked to the metabolic profile of root exudates. *Ann. Bot.* **2015**, *115*, 821–831.
60. Rovera, M.; Carlier, E.; Pasluosta, C.; Avanzini, G.; Andrés, J.; Rosas, S. *Pseudomonas aurantiaca* SR1: Plant growth promoting traits, secondary metabolites and crop inoculation response. In *Plant-Bacteria Interactions: Strategies and Techniques to Promote Plant Growth*; Ahmad, I.; Pichtel, J.; Hayat, S., Eds. Wiley-VCH Verlag GmbH & Co. KGaA, Weinheim, Germany, **2008**, pp. 155–163.
61. Reddy, B. P.; Reddy, M. S.; Kumar, K. V. K. Characterization of antifungal metabolites of *Pseudomonas fluorescens* and their effect on mycelial growth of *Magnaporthe grisea* and *Rhizoctonia solani*. *Int. J. Pharm Tech Res.* **2009**, *1*, 1490–1493.
62. Sang-Dal, K.; Fuente, L. D. L.; Weller, D. M.; Thomashow, L. S. Colonizing ability of *Pseudomonas fluorescens* 2112, among collections of 2,4-diacetylphloroglucinol-producing *Pseudomonas fluorescens* spp. in pea rhizosphere. *J. Microbiol. Biotechnol.* **2012**, *22*, 763–770.
63. Dwivedi, D.; Johri, B. N. Antifungals from fluorescent pseumonads: Biosynthesis and regulation. *Curr. Sci.* **2003**, *85*, 1693–1703.
64. Upadhyay, A.; Srivastava, S. Phenazine-1-carboxylic acid is a more important contributor to biocontrol *Fusarium oxysporum* than pyrrolnitrin in *Pseudomonas fluorescens* strain Psd. *Microbiol. Res.* **2011**, *166*, 323–335.
65. Deora, A.; Hatano, E., Tahara, S.; Hashidoko, Y. Inhibitory effects of furanone metabolites of a rhizobacterium, *Pseudomonas jessenii*, on phytopathogenic *Aphanomyces cochlioides* and *Pythium aphanidermatum*. *Plant Pathol.* **2010**, *59*, 84–99.
66. Wang, M.; Tachibana, S.; Murai, Y.; Li, L.; Lau, S. Y. L.; Cao, M.; Zhu, G.; Hashimoto, M.; Hashidoko, Y. Indole-3-acetic acid produced by *Burkholderia heleia* acts as a phenylacetic acid antagonist to disrupt tropolone biosynthesis in *Burkholderia plantarii*. *Sci. Rep.* *6*, 22596. DOI: 10.1038/srep22596. Published Online: March 3, **2016**.
67. Olsson, S.; Alström, S.; Persson, P. Barley rhizobacterial population characterized by fatty acid profiling. *Appl. Soil Ecol.* **1999**, *12*,197–204.
68. Ivanova, E. P.; Christen, R.; Bizet, C.; Clermont, D.; Motreff, L.; Bouchier, C.; Zhukova, N. V.; Crawford, R. J.; Kiprianova, E. A. *Pseudomonas brassicacearum* subsp. *neoaurantiaca* subsp. nov., orange-pigmented bacteria isolated from soil and the rhizosphere of agricultural plants. *Int. J. Syst. Evol. Microbiol.* **2009**, *59*, 2476–2481.
69. Shoresh, M.; Harman, G. E.; Mastouri, F. Induced systemic resistance and plant responses to fungal biocontrol agents. *Annu. Rev. Phytopathol.* **2010**, *48*, 21–43.
70. Jogaiah, S.; Abdelrahman, M.; Tran, L.-S. P.; Shin-ichi, I. Characterization of rhizosphere fungi that mediate resistance in tomato against bacterial wilt disease. *J. Exp. Bot.* **2013**, *64*, 3829–3842.
71. Brakhage, A. A. Regulation of fungal secondary metabolism. *Nat. Rev. Microbiol.* **2013**, *11*, 21–32.
72. Rojas, J. D.; Sette, L. D.; de Araujo, W. L.; Lopes, M. S. G.; da Silva, L. F.; Furlan, R. L. A.; Padilla, G. The diversity of polyketide synthase genes from sugarcane-derived fungi. *Microb. Ecol.* **2012**, *63*, 565–577.
73. Wijeratne, E. M. K.; Turbyville, T. J.; Fritz, A.; Whitesell, L.; Gunatilaka, A. A. L. A new dihydroxanthenone from a plant-associated strain of the fungus *Chaetomium globosum* demonstrates anticancer activity. *Bioorg. Med. Chem.* **2006**, *14*, 7917–7923.
74. Paranagama, P. A.; Wijeratne, E. M. K.; Gunatilaka, A. A. L. Uncovering biosynthetic potential of plant-associated fungi: Effect of culture conditions on metabolite production by *Paraphaeosphaeria quadriseptata* and *Chaetomium chiversii*. *J. Nat. Prod.* **2007**, *70*, 1939–1945.

75. Takahashi, K.; Koshino, H.; Narita, Y.; Yoshihara, T. Novel antifungal compounds produced by sterile dark, an unidentified wheat rhizosphere fungus. *Biosci. Biotechnol. Biochem.* **2005**, *69*, 1018–1020.
76. Li, C.; Ding, W.; She, Z.; Lin, Y. A new biphenyl derivative from an unidentified marine fungus E33. *Chem. Nat. Compd.* **2008**, *44*, 163–165.
77. Chin-A-Woeng, T. F. C.; Bloemberg, G. V.; Lugtenberg, B. J. J. Phenazines and their role in biocontrol by *Pseudomonas* bacteria. *New Phytol.* **2003**, *157*, 503–523.
78. Ahmed, E.; Holmstrom, S. J. M. Siderophores in environmental research: Roles and applications. *J. Microbial. Biotechnol.* **2014**, *7*, 196–208.
79. Maindad, D. V.; Kasture, V. M.; Chaudhari, H.; Dhavale, D. D.; Chopade, B. A.; Sachdev, D. P. Characterization and fungal inhibition activity of siderophore from wheat rhizosphere associated *Acinetobacter calcoaceticus* strain HIRFA32. *Indian J. Microbiol.* **2014**, *54*, 315–322.
80. Schmidt, R.; Cordovez, V.; de Boer, W.; Raaijmakers, J.; Garbeva, P. Volatile affairs in microbial interactions. *ISME J.* **2015**, *9*, 2329–2335.
81. Farag, M. A.; Zhang, H.; Ryu, C.-M. Dynamic chemical communication between plants and bacteria through airborne signals: Induced resistance by bacterial volatiles. *J. Chem. Ecol.* **2013**, *39*, 1007–1018.
82. Ryu, C.-M.; Farag, M. A.; Hu, C.-H.; Reddy, M. S.; Kloepper, J. W.; Paré, P. W. Bacterial volatiles induce systemic resistance in Arabidopsis. *Plant Physiol.* **2004**, *134*, 1017–1026.
83. Bitas, V.; Kim, H.-S.; Bennett, J. W.; Kang, S. Sniffing on microbes: Diverse roles of microbial volatile organic compounds in plant health. *Mol. Plant Microbe Interact.* **2013**, *26*, 835–843.
84. Minerdi, D.; Bossi, S.; Maffei, M. E.; Gullino, M. L.; Garibaldi, A. *Fusarium oxysporum* and its bacterial consortium promote lettuce growth and expansin A5 gene expression through microbial volatile organic compound (MVOC) emission. *FEMS Microbiol. Ecol.* **2011**, *76*, 342–351.
85. Bray, E. A.; Bailey-Serres, J.; Weretilnyk, E. Responses to abiotic stresses. In *Biochemistry and Molecular Biology of Plants*; Buchanan, B.; Gruissem, W.; Jones, R. L., Eds. American Society of Plant Physiologists, Rockville, MD, **2000**, pp. 1158–1203.
86. McLellan, C. A.; Turbyville, T. J.; Wijeratne, E. M. K.; Kerschen, A.; Vierling, E.; Queitsch, C.; Whitesell, L.; Gunatilaka, A. A. L. A rhizosphere fungus enhances arabidopsis thermotolerance through production of an HSP90 inhibitor. *Plant Physiol.* **2007**, *145*, 174–182.
87. Gough, N. R. Helping plants survive heat stress. *Sci. STKE. 403*, pp. tw333. DOI: 10.1126/stke.4032007tw333. Published Online: September 11, **2007**.
88. Maheshwari, R. Small molecule natural product from a fungus induces thermotolerance in plants. *Curr. Sci.* **2008**, *94*, 16.
89. Tillotson, J.; Bashyal, B. P.; Kang, M.; Shi, T., De La Cruz, F.; Gunatilaka, A. A. L.; Chapman, E. Selective inhibition of p97 by chlorinated analogues of dehydrocurvularin. *Org. Biomol. Chem.* **2016**, *14*, 5918–5921.
90. Turbyville, T. J.; Wijeratne, E. M. K.; Whitesell, L.; Gunatilaka, A. A. L. The anticancer activity of the fungal metabolite terrecyclic acid A is associated with modulation of multiple cellular stress response pathways. *Mol. Cancer Ther.* **2005**, *4*, 1569–1576.
91. Turbyville, T. J.; Wijeratne, E. M. K.; Liu, M. X.; Burns, A. M.; Seliga, C. J.; Luevano, L. A.; David, C. L.; Faeth, S. H.; Whitesell, L.; Gunatilaka, A. A. L. Search for Hsp90 inhibitors with potential anticancer activity: Isolation and SAR studies of radicicol and monocillin I from two plant-associated fungi of the Sonoran desert. *J. Nat. Prod.* **2006**, *69*, 178–184.
92. Gao, H.; Guo, W.; Wang, Q.; Zhang, L.; Zhu, M.; Zhu, T.; Gu, Q.; Wang, W.; Li, D. Aspulvinones from a mangrove rhizosphere soil-derived fungus *Aspergillus terreus* Gwq-48 with anti-influenza A viral (H1N1) activity. *Bioorg. Med. Chem. Lett.* **2013**, *23*, 1776–1778.

93. Wang, Q.; Xu, L. Beauvericin, a bioactive compound produced by fungi: A short review. *Molecules* **2012**, *17*, 2367–2377.
94. Shekhar-Guturja, T.; Gunaherath, G. M. K. B.; Wijeratne, E. M. K.; Lambert, J.-P.; Averette, A. F.; Lee, S. C.; Kim, T. et al. Dual action antifungal small molecule modulates multidrug efflux and TOR signaling. *Nat. Chem. Biol.* **2016**, *12*, 867–875.
95. Oldroyd, G. E. D. Speak, friend, and enter: Signaling systems that promote beneficial symbiotic associations in plants. *Nat. Rev. Microbiol.* **2013**, *11*, 252–263.
96. Schenk, P. M.; Carvalhais, L. C.; Kazan, K. Unraveling plant-microbe interactions: Can multi-species transcriptomics help? *Trends Biotechnol.* **2012**, *30*, 177–184.
97. Almario, J.; Muller, D.; Défago, G.; Moënne-Loccoz, Y. Rhizosphere ecology and phytoprotection in soils naturally suppressive to Thielaviopsis black root rot of tobacco. *Environ. Microbiol.* **2014**, *16*, 1949–1960.
98. Frey-Klett, P.; Burlinson, P.; Deveau, A.; Barret, M.; Tarkka, M.; Sarniguet, A. Bacterial-fungal interactions: Hyphens between agricultural, clinical, environmental, and food microbiologists. *Microbiol. Mol. Biol. Rev.* **2011**, *75*, 583–609.
99. Kwak, Y. S.; Weller, D. M. Take-all of wheat and natural disease suppression: A review. *Plant Pathol. J.* **2013**, *29*, 125–135.
100. Azegami, K.; Nishiyama, K.; Watanabe, Y.; Suzuki, T.; Yoshida, M.; Nose, K.; Toda, S. Tropolone as a root growth-inhibitor produced by a plant pathogenic *Pseudomonas* sp. causing seedling blight of rice. *Jpn. J. Phytopathol.* **1985**, *51*, 315–317.
101. Azegami, K.; Nishiyama, K.; Kato, H. Effect of iron limitation on "*Pseudomonas plantarii*" growth and tropolone and protein production. *Appl. Environ. Microbiol.* **1988**, *54*, 844–847.
102. Bartsch, M.; Bednarek, P.; Vivancos, P. D.; Schneider, B.; von Roepenack-Lahaye, E.; Foyer, C. H.; Kombrink, E.; Scheel, D.; Parker, J. E. Accumulation of isochorismate-derived 2,3-dihydroxybenzoic 3-O-β-D-xyloside in Arabidopsis resistance to pathogens and ageing of leaves. *J. Biol. Chem.* **2010**, *285*, 25654–25665.
103. Cespedes, C. L.; Avila, J. G.; Garcia, A. M.; Becerra, J.; Flores, C.; Aqueveque, P.; Bittner, M.; Hoeneisen, M.; Martinez, M.; Silva, M. Antifungal and antibacterial activities of *Araucaria araucana* (Mol.) K. Koch heartwood lignans. *Z. Naturforsch. C.* **2006**, *61*, 35–43.
104. Park, H.; Son, D. H.; Kim, Y. C.; Kwon, D. R. Antibacterial compositions based on organic acids and coumarins. In Repub. Korean Kongkae Taeho Kongbo, Patent Application No. KR 2007096688 A20071002. Assignee, Rowett Research Institute, Aberdeen, Scotland, **2007**.
105. Hwang, B.; Cho, J.; Hwang, I.-s.; Jin, H.-G.; Woo, E.-R.; Lee, D. G. Antifungal activity of lariciresinol derived from *Sambucus williamsii* and their membrane-active mechanisms in *Candida albicans*. *Biochem. Biophys. Res. Commun.* **2011**, *410*, 489–493.
106. Kayser, O.; Kolodziej, H. Antibacterial activity of extracts and constituents of *Pelargonium sidoides* and *Pelargonium reniforme*. *Planta Med.* **1997**, *63*, 508–510.
107. Consonni, C.; Bednarek, P.; Humphry, M.; Francocci, F.; Ferrari, S.; Harzen, A.; Ver Loren van Themaat, E.; Panstruga, R. Tryptophan-derived metabolites are required for antifungal defense in the Arabidopsis *mlo2* mutant. *Plant Physiol.* **2010**, *152*, 1544–1561.
108. Bednarek, P.; Piślewska-Bednarek, M.; Svatoš, A.; Schneider, B.; Doubský, J.; Mansurova, M.; Humphry, M. et al. A glucosinolate metabolism pathway in living plant cells mediates broad-spectrum antifungal defense. *Science* **2009**, *323*, 101–106.
109. Kato, T.; Yamaguchi, Y.; Abe, N.; Uyehara, T.; Namai, T.; Kodama, M.; Shiobara, Y. Structure and synthesis of unsaturated trihydroxy C_{18} fatty acids in rice plants suffering from rice blast disease. *Tetrahedron Lett.* **1985**, *26*, 2357–2360.

110. Khorassani, R.; Hettwer, U.; Ratzinger, A.; Steingrobe, B.; Karlovsky, P.; Claassen, N. Citramalic acid and salicylic acid in sugar beet root exudates solubilize soil phosphorus. *BMC Plant Biol.* [Online] **2011**, *11*, 121.
111. Vernooij, B.; Uknes, S.; Ward, E.; Ryals, J. Salicylic acid as a signal molecule in plant-pathogen interactions. *Curr. Opin. Cell Biol.* **1994**, *6*, 275–279.
112. Michalet, S.; Rohr, J.; Warshan, D.; Bardon, C.; Roggy, J.-C.; Domenach, A.-M.; Czarnes, S. et al. Phytochemical analysis of mature tree root exudates *in situ* and their role in shaping soil microbial communities in relation to tree N-acquisition strategy. *Plant Physiol. Biochem.* **2013**, *72*, 169–177.
113. Liu, Q.; Xu, R.; Yan, Z.; Jin, H.; Cui, H.; Lu, L.; Zhang, D.; Qin, B. Phytotoxic allelochemicals from roots and root exudates of *Trifolium pratense*. *J. Agric. Food Chem.* **2013**, *61*, 6321–6327.
114. Marti, G.; Erb, M.; Boccard, J.; Glauser, G.; Doyen, G. R.; Villard, N.; Robert, C. A. M.; Turlings, T. C. J.; Rudaz, S.; Wolfender, J.-L. Metabolomics reveals herbivore-induced metabolites of resistance and susceptibility in maize leaves and roots. *Plant Cell Environ.* **2013**, *36*, 621–639.
115. Ahmad, S.; Veyrat, N.; Gordon-Weeks, R.; Zhang, Y.; Martin, J.; Smart, L.; Glauser, G. et al. Benzoxazinoid metabolites regulate innate immunity against aphids and fungi in maize. *Plant Physiol.* **2011**, *157*, 317–327.
116. Liu, X.; Bimerew, M.; Ma, Y.; Muller, H.; Ovadis, M.; Eberl, L.; Berg, G.; Chernin, L. Quorum-sensing signaling is required for production of the antibiotic pyrrolnitrin in a rhizospheric biocontrol strain of *Serratia plymuthica*. *FEMS Microbiol. Lett.* **2007**, *270*, 299–305.

10 Novel Metabolites from Extremophilic Microbes Isolated from Toxic Waste Sites

Andrea Stierle and Don Stierle

CONTENTS

10.1 INTRODUCTION

For the last 30 years, the Stierle Research Laboratory has explored the secondary metabolism of medicinal plants, endophytic fungi, and marine sponge endosymbionts in its search for compounds with biological activity and pharmaceutical potential. There was a compelling biorationale for studying each of these source organisms for the production of drug-like compounds. Medicinal plants have been used for thousands of years for the treatment of a wide array of diseases, and it is logical to assume that this activity would be associated with particular natural products. Endophytic microbes are generally nonpathogenic in nature, but often produce secondary metabolites that enable them to survive in the competitive world of plant interstitial space without harming their host. The same argument can be applied to the endosymbiotic microbes of marine invertebrates. Indeed, microorganisms in most ecosystems establish and define their ecological niches by their ability to control competitors and pathogens with only their cell walls or membranes and chemical arsenals for defense. These chemical arsenals have provided many of the important chemotherapeutics used to date. These include the fungally derived beta-lactam antibiotics, including penicillins, cephalosporins, and carbapenems, the macrolide polyketide antibiotics, including erythromycin, as well as tetracyclines, and aminoglycosides, most of which are produced by actinomycetes.[1] Microbes are also adept at producing compounds with potent anticancer activity, including the tubulin stabilizer epothilone, which was isolated from myxobacterium *Sorangium cellulosum*,[2] and calicheamicin, which was isolated from a novel subspecies of the streptomycete *Micromonospora echinospora calichensis*.[3] In 1993, we reported the production of the important anticancer agent paclitaxel by an endophytic fungus of the Northwest Pacific yew tree,[4] a discovery that has since been replicated by research labs around the world.[5–7]

10.2 TOXIC WASTE: THE NEXT FRONTIER FOR THE DISCOVERY OF DRUG-LIKE MOLECULES

In 1995, however, we launched a tentative investigation of a most unlikely ecosystem for the discovery of drug-like molecules—an abandoned acid mine waste lake in Butte, Montana. There were several reasons for selecting this particular research arena. Our search for taxol-producing microbes was drawing to a close, and although we continued to search for novel compounds from yew tree[8] and redwood endophytes for several more years, we hoped to find a new population of microbes from an as yet unexplored ecosystem. Unfortunately, our funding was exhausted and, with limited resources, we could not launch an expedition into uncharted territories in remote, exotic ecosystems. Our budget could not accommodate a trip to Antarctica, to a volcanic lake in Peru, or to the temperate rainforests of Western Washington.

However, when colleague Bill Chatham discovered green algae growing on a piece of wood floating just below the surface of Berkeley Pit Lake (BPL) less than one mile from our laboratory, we were intrigued by the possibility that other microorganisms might be found in this acid lake.

Although scientists from the Montana Bureau of Mines and Geology had studied the geology and hydrogeology of the region, the possibility of life in those inhospitable waters had not been considered or studied before 1995. BPL is ground zero for the largest Environmental Protection Agency (EPA) Superfund site in North America. Despite its low pH, high oxidation potential (E_h), and high metal concentration, it harbors a population of extremophilic microbes that have yielded a collection of bioactive natural products with promising drug-like properties. We focused on microbes that could be isolated from the waters and sediment of the Pit and grown in the laboratory in submerged culture. The microbes have been collected from the surface waters down to 275 m, established in pure culture, and grown using an array of physicochemical conditions. Bioassay-guided fractionation has directed the purification of small-molecule inhibitors of enzymes associated with inflammation (caspase-1), epithelial–mesenchymal transition (EMT; matrix metalloproteinase-3, MMP-3), and the collateral damage following ischemic stroke (caspase-3). In these studies, attention has been paid to the isolation and characterization of enzyme inhibitors as well as inactive analogs, using *nuclear magnetic resonance (NMR)–guided* chemotype fractionation to facilitate assessment of structure–activity relationships (SARs) and the development of analogs with increased potency. In this chapter, we will provide an overview of these compounds as well as their activities in both enzyme inhibitory assays and orthogonal cell-based assays.

Fundamental to this work however are the microbes themselves and the environment from which they were isolated, so we will provide a profile of Berkeley Pit Lake and a brief history of its evolution from an abandoned mine into a dynamic ecosystem. These conditions have selected for a population of microbes that have evolved a number of survival strategies that include the production of secondary metabolites that regulate pathways associated with low pH and high metal concentrations. It is our goal to demonstrate that these small molecules might also be used to regulate pathways in mammalian systems that are associated with an acidic microenvironment, including inflammatory diseases and certain cancers.

10.3 EVOLUTION OF AN EXTREME ECOSYSTEM

Butte, Montana, has been a mining mecca for over 130 years. In 1955, the Berkeley Pit was gouged out of the Boulder Batholith and gradually developed into a mile wide, 400 m deep pit that was dewatered through constant pumping. When the pumps were *decommissioned* in 1982, groundwater began percolating through the mineral rich overburden and 10,000 miles of mine tunnels honeycombing the Butte hill. Within 2 years, the water level had risen to the base of the Pit, and proceeded to fill the vast basin, and Berkeley Pit Lake was born.[9]

Thirty-four years later, there are over 152 billion L of water in the Pit, with an inflow rate of 15 million L/day. Ironically, pit lakes formed in non-acid-generating limestone formations may gradually evolve into community recreation areas ideal for boating and swimming. Unfortunately, the 342 snow geese that landed in BPL in 1995 and died shortly thereafter as a result belied any notion that it had potential as a recreational site. Much of the groundwater entering the Pit is already contaminated as it meanders through mine tailings and tunnels. Yet no single tributary or combination of tributaries of the Pit Lake can match it in either metal ion or hydrogen ion concentration.[9,10]

Iron pyrite (FeS_2) dominates the local geology and ultimately determines the nature of Berkeley Pit Lake. Even if all of the influent waters were pristine, the pyrite walls of the Pit continually react with air and water to generate sulfuric acid, which further dissolves the minerals in the surrounding rocky overburden. As oxygen concentration decreases with depth, pyrite oxidation and the resulting acid generation might be expected to decrease. However, the oxidation of pyrite sulfide to sulfate by dissolved ferric iron can take place at a rapid rate in acidic waters, even in the complete absence of oxygen.[10,11] The rate of ferrous iron oxidation by oxygen is known to increase many orders of magnitude in the presence of certain acidophilic bacteria, chiefly *Acidithiobacillus ferrooxidans*. Because of this dynamic interplay, the water is acidic (pH 2.5–2.7) and contaminated with high concentrations of metal sulfates of iron, copper, aluminum, cadmium, and zinc.[9–11]

Although conditions within the Pit Lake System do not support *normal* aquatic biota, the extremophilic microbes have adapted well to the environment. Extremophiles can be more than just an interesting scientific oddity. They can provide a new untapped reservoir of bioactive secondary metabolites waiting to be discovered. Clearly, this work is not a microbial ecology study, nor is it a comprehensive assessment of the total diversity of microbial life in the Pit Lake. These compounds may or may not be produced *in situ* under natural conditions. Instead, this research focuses on bioactive compounds produced in the laboratory under controlled conditions that can be isolated, characterized, and studied for their drug-like potential.

10.4 EXTREME ENVIRONMENTS AS A SOURCE OF EXTREMOPHILIC MICROORGANISMS

10.4.1 Extreme Environment

The earth is actually rich in hostile environments and unlikely ecosystems. These include many different natural systems, including deep-sea vents, salt brines, thermal pools, volcanic lakes, and frigid ice fields. Other sites are anthropogenic, often the result of extractive hard rock, oil, or coal mining; munitions testing; or industrial waste dumps. Nobel laureate Paul Crutzen suggested that we are currently living in the *Anthropocene*, an era in which humans and human activities have become a major geological force on the planet.[12] Whether man-made or natural, extreme environments can harbor life forms called *extremophiles*. Bacteriologist Thomas Brock demonstrated the importance of extremophiles in the 1960s when he isolated bacteria from a 70°C thermal pool in Yellowstone National Park.[13] In the 40 years following Brock's discovery, scientists have explored ecological niches as varied as deep-sea vents and Antarctic ice sheets, and have also found unusual life forms in unexpected places.[13]

For the natural products chemist, unusual microbes hold the potential of novel chemistry, often in the form of small molecules, with important medicinal, industrial, or agrochemical applications. The rapidly developing tools of molecular biology have focused attention on genomics and proteomics. Despite this trend however, the study of secondary metabolites, or *secondary metabolomics*, is still an important means to the discovery of new chemotherapeutic agents. When a population of extremophiles are discovered and established in culture, it is the unique challenge of natural products chemistry to isolate and characterize the bioactive components in these organisms.

Berkeley Pit Lake is our exclusive extreme environment. We have confined our studies to the microbial life in the Pit for the simple reason that microbes are the predominant, and perhaps sole, inhabitants of this toxic lake. These microbes include bacteria, fungi, and both heterotrophic and autotrophic protists (green algae).[14] Unlike most high mountain lakes in Montana, BPL does not harbor trout, grayling, or other blue ribbon fish species. Aside from a solitary water bug photographed resting on the surface of the Pit Lake and a flock of snow geese that landed on the water and subsequently died in 1995, no evidence of macrobial life exists.[15]

10.4.2 Different Categories of Extremophilic Microorganisms

The term *extremophile* was coined by MacElroy in 1974 to describe microorganisms that thrive under conditions that would be considered extreme from a human perspective.[16] Of course, extreme is a relative term. Obligate anaerobic microbes have long been known and are generally not classified as extremophiles, yet life without oxygen would certainly be a challenge for most of us. In essence, the term extremophile is used to describe microbes that thrive in environments where most microbes cannot grow or thrive because of extremes in temperature, salinity, pH, or pressure. Extremophiles can be classified according to the environments in which they thrive and certain organisms may fit into multiple categories.[17,18]

- *Acidophiles* thrive in an acidic environment, usually at an optimum pH of 2–3.
- *Alkaliphiles* (*alkalinophiles*) thrive in an alkaline environment, at a minimum pH of 10.
- *Halophiles* require a salty environment, with a minimum salt concentration of 0.2 M.
- *Piezophiles* (*barophiles*) thrive at high pressures often exceeding 100 atm.
- *Psychrophiles* thrive in a cold environment from –20°C (in sea ice) to 20°C.
- *Thermophiles* thrive in a hot environment, with an optimum growth temperature of 45°C or higher.
- *Hyperthermophiles* flourish at even higher temperatures, between 80°C and 100°C.
- *Endoliths* live in rock fissures or within porous rocks and may actively bore into the rock, creating tunnels that conform to the shape of the organism.
- *Metallotolerant organisms* can survive in solutions that contain high levels of dissolved heavy metals such as copper, cadmium, arsenic, and zinc.
- *Radioresistant organisms* can survive high levels of ionizing radiation, including gamma rays and x-ray.

Brock's *Thermus aquaticus* is perhaps the most famous thermophile discovered to date. His initial discovery of viable bacteria thriving in hot springs surprised many scientists because most organisms known at that time would either die or fail to reproduce in 70°C water.[13] *T. aquaticus* owes its continued fame to more than just its unusual habitat, however. It owes its fame and its ability to survive in hot springs in part to a special variant of the enzyme DNA polymerase.[14] DNA polymerases catalyze the synthesis of deoxyribonucleic acid in a template-dependent process that results in a faithful copy of the original DNA molecule. Consequently, these enzymes are necessary to propagate, maintain, and manipulate the genetic code of living organisms. Although DNA polymerases are usually not heat labile, the *T. aquaticus* DNA polymerase variant (*Taq* polymerase) remains operational at high temperatures,[19] and is now a key ingredient in the polymerase chain reaction (PCR) technique.[20]

10.5 EXTREMOPHILES FROM NATURALLY OCCURRING EXTREME ENVIRONMENTS AS A SOURCE OF NEW NATURAL PRODUCTS

Although the main focus of this chapter is the investigation of the secondary metabolites of extreme microbes from *anthropogenic toxic waste environments*, it is appropriate to briefly discuss recent studies of extremophiles from other unusual environments. Many scientists worldwide are looking for life forms in places considered too inhospitable for conventional life, and much of the search is focused on microorganisms. This is not surprising. In the past 30 years, researchers have compiled an increasingly robust map of evolutionary diversification showing that the main diversity of life is microbial, distributed among three primary domains: Archaea, Bacteria, and Eucarya.[21] During this same period, the number of microbes discovered living in extreme environments has increased dramatically. Although many of these investigations focus on the unique physiological characteristics of the extremophiles themselves, there have been numerous reports of the production of novel secondary metabolites by these microorganisms. Giddings and Newman recently published an excellent overview of the bioactive compounds produced by terrestrial extremophiles.[18] A few examples of secondary metabolite production by microbial extremophiles are described below. These include compounds isolated from extremophiles in deep-sea hydrothermal vents, another extreme aqueous ecosystem, and extremophiles from other mine waste sites. These microbes are relevant to our own work as they are metallotolerant, and in some cases, acidophilic, much like the microbes of Berkeley Pit lake. This is only a small sample of the unusual microbes that have been found in unexpected ecosystems.

10.5.1 Deep-Sea Vent Thermophiles

Thermophiles are not confined to the Greater Yellowstone ecosystem. In 1977, scientists exploring the Galápagos Rift along the mid-ocean ridge in the eastern Pacific found a most unusual ecosystem associated with deep-sea hydrothermal vents. It was populated by organisms that could withstand the extreme temperatures, pressures, toxic minerals, and lack of sunlight that characterized this environment. Scientists later realized that bacteria were converting the toxic vent minerals into usable forms of energy through a process called chemosynthesis, providing food for other vent organisms. The microbes that live in this niche include hyperthermophiles and thermophiles from both the bacterial and archaeal domains.

One of the most extreme thermophilic microorganisms is the iron-reducing archaeon *strain 121*.[22] Kashefi and Lovley isolated an iron-reducing archaeon from a water sample taken from an active black smoker hydrothermal vent along the Endeavor segment of the Juan de Fuca Ridge, in the Northeast Pacific Ocean. Strain 121 could grow in temperatures of 85°C–121°C. *Pyrolobus fumarii*, the former thermophilic record holder, could not meet the challenge. After an hour at 121°C, only 1% of its cells were intact and none appeared viable.[22]

Although *P. fumarii* is no longer the most extreme hyperthermophile, it can multiply in temperatures up to 113°C, which is still a remarkable upper limit for a living organism.[23] It grows on the walls of deep-ocean black smokers. The water is rich in minerals, and these (in combination with carbon dioxide) provide the food source for *P. fumarii*. It was discovered by German scientist Karl Stetter and his colleagues, who also discovered the world's smallest microorganism, *Nanoarchaeum equitans*.[24] The name translates *as ancient dwarf who rides the fire ball* from its tendency to adhere to the surface of the archaeal microbe *Ignicoccus* (*fireball*). The discovery of this nanosized hyperthermophilic archaeon led to the creation of a new phylum, Nanoarchaeota. *N. equitans* was found in a 120 m deep submarine hydrothermal vent north of Iceland, and thrives in temperatures close to 100°C. With less than 500 kb in its genome, *N. equitans* represents the smallest archaeal genome sequenced to date and also the smallest genetic code of any living organism.[24]

10.5.2 Secondary Metabolites from Deep-Sea Vent Thermophiles

Much attention has been paid to deep-sea vent thermophiles, but much of the literature has focused on the organisms themselves—their primary metabolism, ability to flourish under such extreme conditions and biodiversity. Five years ago, there was speculation that these organisms would be a rich source of new natural products, although to date only a few classes of molecules have been reported.[25,26] These include the hydroxyethylamine chroman derivatives ammonificins A (**1**) and B (**2**), which were isolated from the marine hydrothermal vent bacterium *Thermovibrio ammonificans*. The authors claimed that this was the first report of secondary metabolites from the marine hydrothermal vent bacterium *T. ammonificans*.[18,27] Further studies of this microbe yielded ammonificins C (**3**) and D (**4**),[18,28] both of which induced apoptosis in micromolar concentrations (Figure 10.1).

Loihichelins A–F (**5**–**10**), six new amphiphilic peptide siderophores, were isolated from the heterotrophic γ-proteobacteria *Halomonas* LOB-5, collected from a vent at 1714 m.[29] It was grown under microaerophilic conditions, either lithotrophically on Fe^{+2} and CO_2, or heterotrophically in an artificial seawater medium. The loihichelins are distinguished from other amphiphilic siderophores by a longer peptide head group and shorter fatty acid appendages (Figure 10.2).

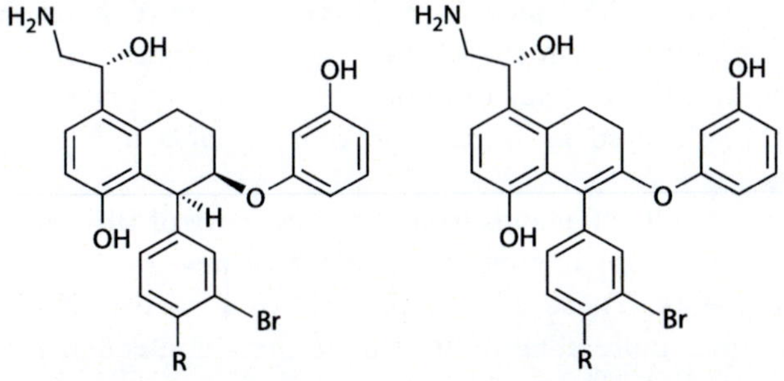

(**1**) Ammonificin A R = OH
(**2**) Ammonificin B R = Br

(**3**) Ammonificin C R = OH
(**4**) Ammonificin D R = Br

FIGURE 10.1 Ammonificins A–D (**1–4**) isolated from deep-sea hydrothermal vent bacterium *Thermovibrio ammonificans*.

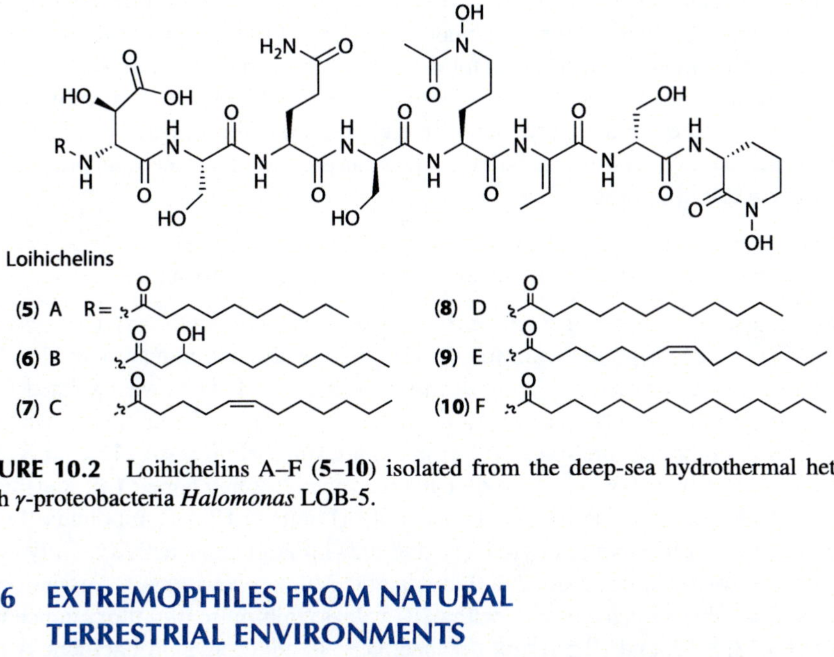

FIGURE 10.2 Loihichelins A–F (**5–10**) isolated from the deep-sea hydrothermal heterotroph γ-proteobacteria *Halomonas* LOB-5.

10.6 EXTREMOPHILES FROM NATURAL TERRESTRIAL ENVIRONMENTS

In recent years, there have also been numerous reports of microorganisms residing in extreme terrestrial conditions. David Boone, a Portland State University microbial ecologist, discovered *Bacillus infernos*, an iron- and sulfate-reducing bacterium and the first anaerobic member of the bacterial genus *Bacillus*. *The Bacillus from hell*, as the name implies, was isolated 2700 m below the land surface. It is thermophilic (60°C), halotolerant (salt concentration 0.6 M), and slightly alkaliphilic.[30]

Deinococcus radiodurans is the most radiation-resistant organism known. It was discovered by Arthur W. Anderson at Oregon Agricultural Experiment Station in Corvallis in 1956 in a can of radiation-sterilized meat. *D. radiodurans* is resistant to genotoxic chemicals, oxidative damage, dehydration, and high levels of ionizing and ultraviolet radiation. It can withstand exposure to radiation levels up to 1.5 million rads (500 rads is lethal to humans). A recombinant strain has been engineered to degrade organopollutants in radioactive, mixed-waste environments.[31]

George Roadcap and colleagues from the University of Illinois at Champaign-Urbana found a microbial community that sets a new record for tolerance to alkaline environments. These alkalinophiles live at a pH of 12.8, tolerating conditions 100 times more alkaline than the closest contender, in contaminated groundwater created by dumping of slag waste from iron ore processing just south of Chicago.[32]

10.7 EXTREMOPHILES FROM ANTHROPOGENIC TERRESTRIAL ENVIRONMENTS

Searching for unusual life forms in an anthropogenic waste site does not engender the same sense of adventure or excitement as an expedition to a naturally occurring extreme environment. Collecting samples in a toxic mine dump or old munitions site is not the stuff of specials on NOVA or National Geographic. But despite their lack of dramatic appeal, coal mine waste heaps, metal-contaminated soils, or an abandoned mine waste lake can harbor a population of organisms with tremendous promise for secondary metabolite production.

10.7.1 Secondary Metabolites from a Coal Mine Acidophile

Although many extremophilic microbes have been isolated from terrestrial ecosystems, very few have been reported from toxic waste that result from human activities. The acidophilic eubacterial strain *Lysinibacillus fusiformis* $KMCOO_3$ was isolated from acidic (pH 3) coal mine drainage contaminated with sulfuric acid and iron-rich, heavy-metal ions.[33] It produces the spiro-cyclopentenones, spirobacillenes A (**11**) and B (**12**), as well as (*Z*)-3-hydroxy-4-(3-indoyl)-1-hydroxyphenyl-2-butenone (**13**), and (*Z*)-3-hydroxy-4-(3-indoyl)-1-phenyl-2-butenone (**14**) (Figure 10.3). Compounds **11–13** were tested for antimicrobial, cytotoxic, and anti-inflammatory activity. Only compound **13** exhibited antimicrobial activity against *Micrococcus luteus*, *Enterococcus hirae*, and *Staphylococcus aureus*, with MIC values (minimum inhibitory concentrations) of 3.13, 3.13, and 12.5 μg/mL, respectively. In addition, spirobacillene A (**11**) weakly inhibited reactive oxygen species (ROS) and NO production, with IC_{50} values of 39 and 43 μM, respectively.[33]

10.7.2 Secondary Metabolites from Tin Mine Alkaliphiles in Yunnan Province, China

Much like our own investigation of Berkeley Pit microbes, in 2007, He and coworkers launched a survey of microorganisms isolated from the alkaline soils of the Datun tin mine tailings in Yunnan province, China, in a search for new metabolites.[34] They screened crude microbial extracts against six tumor cell lines: large-cell lung cancer

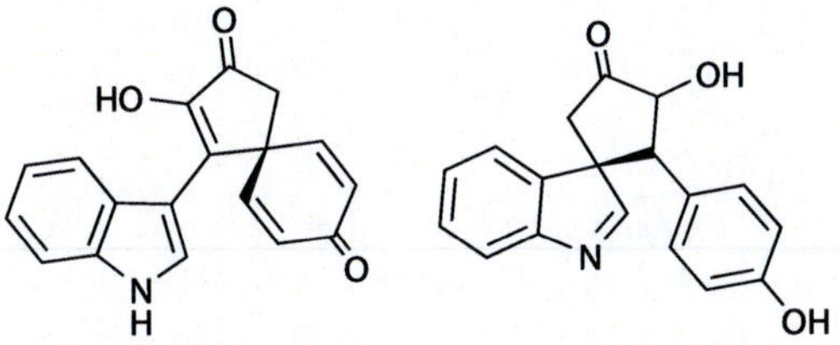

(11) Spirobacillene A **(12)** Spirobacillene B

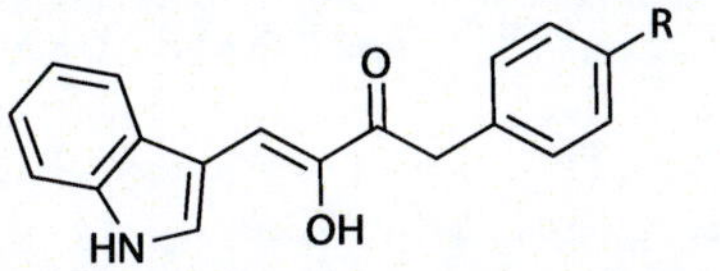

(13) (*Z*)-3-hydroxy-4-(3-indoyl)-1-hydroxyphenyl-2-butenone R = OH
(14) (*Z*)-3-hydroxy-4-(3-indoyl)-1-phenyl-2-butenone R = H

FIGURE 10.3 Spirobacillenes A (**11**) and B (**12**), (*Z*)-3-hydroxy-4-(3-indoyl)-1-hydroxyphenyl-2-butenone (**13**), and (*Z*)-3-hydroxy-4-(3-indoyl)-1-phenyl-2-butenone (**14**) isolated from the acidophilic eubacterial strain *Lysinibacillus fusiformis* KMCOO$_3$.

LXFL 529L, mammary cancer MAXF 401NL, melanoma MEXF 462NL, uterine cancer UXF 1138L, gastric cancer GXF251L, and renal cancer RXF 486L cell lines. Four new pyranonaphthoquinone griseusins, 4′-dehydrodeacetylgriseusin A (**15**), 2*α*,8*α*-epoxyepideacetylgriseusin B (**16**), epideacetylgriseusin A (**17**), and epideacetylgriseusin B (**18**), were produced by the alkaliphilic strains *Nocardiopsis* sp. YIM80133 and DSM1664 (Figure 10.4).

Both **15** and **17** inhibited the growth of all six tumor cell lines, with average IC$_{50}$ values of 0.392 and 5.32 μM, respectively. Compound **15** also exhibited potent cytotoxic activity against monolayer cultures of 37 different human tumor cell lines, with an average IC$_{50}$ value of 430 nM. It was selective against various cell lines, including breast cancer (MDA-MB 231, MDA-MB 68, MCF-7, and MAX7 401; IC$_{50}$, 150–345 nM), renal cancer (RXF 393NL, RXF 468L, and RXF 944L;

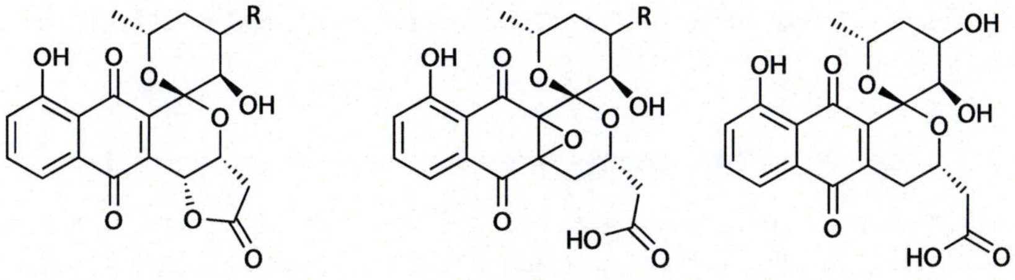

(15) R = O 4′-dehydrodeacetylgriseusin A
(17) R = OH epideacetylgriseusin A
(16) 2*α*,8*α*-epoxyepideacetyl-griseusin B
(18) epideacetylgriseusin B

FIGURE 10.4 Four new pyranonaphthoquinone griseusins, 4′-dehydrodeacetylgriseusin A (**15**), 2*α*,8*α*-epoxyepideacetylgriseusin B (**16**), epideacetylgriseusin A (**17**), and epideacetylgriseusin B (**18**), produced by the alkaliphilic strains *Nocardiopsis* sp. YIM80133 and DSM1664.

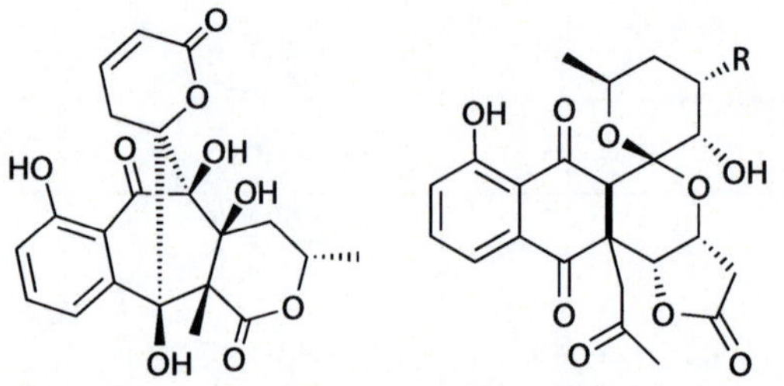

FIGURE 10.5 Naphthospironone A (**19**), as well as griseusins F (**20**) and G (**21**), isolated from the alkaliphilic *Nocardiopsis* sp., strain YIM DT266.

IC_{50}, 95–250 nM), and melanoma (MEXF 276L, MEXF 394NL, MEXF 462NL, MEXF 514L, and MEXF 520L; IC_{50}, 87–280 nM). Compound **15** selectively inhibited the colony formation of several different types of tumors, including colon cancer, breast cancer, melanoma, pancreatic cancer, renal cancer, and leukemia solid tumors, in semisolid medium.[34]

Ding and coworkers isolated the unique cytotoxic agent naphthospironone A (**19**) from a culture of a third *Nocardiopsis* sp. strain, YIM DT266, which was also isolated from the Datun tin mine site (Figure 10.5).[35] This molecule exhibited weak cytotoxic activity against murine fibrosarcoma L929, HeLa, and human lung cancer AGZY cells, with IC_{50} values of 80–221 μM. Naphthospironone A (**19**) also exhibited moderate-to-weak antimicrobial activity against *Staphylococcus aureus*, *Escherichia coli*, *Bacillis subtilis*, and *Aspergillus niger*, with MIC values of 11–25 μg/mL.

Griseusins F (**20**) and G (**21**) (Figure 10.5) were also isolated from the same eubacterium by Ding and coworkers.[36] Both compounds exhibited potent cytotoxicity against human melanoma B16, breast carcinoma MDA-MB-435S, pancreatic cancer CFPAC-1, renal carcinoma ACHN, and colorectal carcinoma HCT 116 cell lines, with IC_{50} values ranging from 0.37 to 0.82 μM, comparable with IC_{50} values of the cisplatin-positive control. Both compounds also exhibited potent antibiotic activity against *S. aureus* ATCC 29213, *Micrococcus luteus*, and *B. subtilis*, with MIC values of 0.80–1.65 μg/mL.[36]

10.7.3 Secondary Metabolites from an Underground Coal Fire Site

Metallotolerant microorganisms are capable of surviving in environments with high levels of heavy metals, such as redox-active iron and copper, and redox-inactive cadmium, arsenic, and zinc. These microbes usually exist near hard rock or coal mines, and are an underexplored resource for new bioactive metabolites. Recently, Wang and coworkers reported the isolation and characterization of frenolicins C–G (**22–26**), new pyranonaphthoquinones from *Streptomyces* sp. RM-4-15, which was isolated from the man-made Appalachian Ruth Mullins underground coal fire site.[37] Compounds **22–25** differ structurally, primarily in the degree of oxygenation of the pyranonaphthoquinone

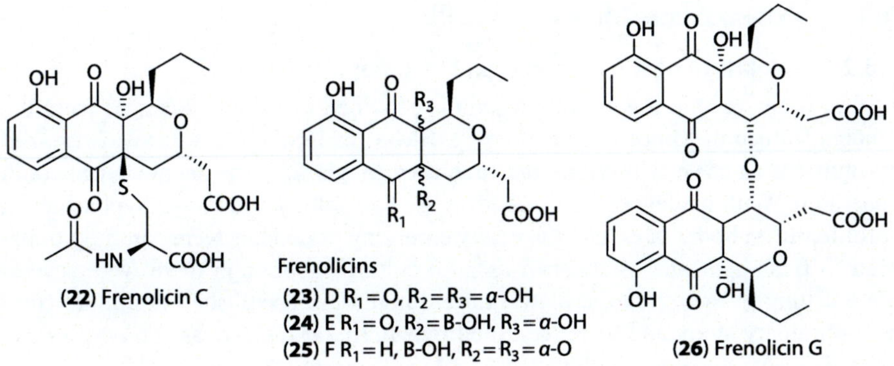

FIGURE 10.6 Frenolicin C–G (**22–26**) produced by metallotolerant *Streptomyces* sp. RM-4-15 isolated from an underground coal fire site.

core and the presence of an *N*-acetyl cysteine residue. However, when scandium was added to the growth medium, *Streptomyces* sp. RM-4-15 produced a new homodimeric analog, frenolicin G (**26**) (Figure 10.6).

10.8 BIOPROSPECTING IN AN EPA SUPERFUND SITE: THE SEARCH FOR EXTREMOPHILES

Thirty years ago, most scientists who studied BPL focused primarily on its geochemistry and on possible remediation strategies. A sample of lake water analyzed by inductively coupled plasma (ICP) in 1995 showed high levels of Fe^{+2}/Fe^{+3} (1200 ppm), Al^{+3} (650 ppm), Cu^{+2} (190 ppm), and many other metal cationic species.[10] It was also very rich in sulfates (8500 ppm), the predominant anionic species that results from oxidation of pyritic sulfides.[10] Although much was known about the geochemistry of the Pit Lake and its surroundings, the biology was largely ignored. It was considered too toxic to support life, so it was not considered a viable ecosystem.

10.8.1 SEARCH FOR ACIDOPHILIC SULFATE-REDUCING BACTERIA

After our initial discovery of microbes in the Pit Lake in 1995, Mitman and coworkers attempted to find acid-tolerant sulfate-reducing bacteria (SRBs) that could mitigate the oxidative processes that contributed to acid generation. Unfortunately, these attempts were not successful.[14] Most of the known SRBs grow optimally at neutral pH.[38] The first acidophilic SRBs isolated and partially characterized were *Desulfosporosinus*-like bacteria isolated from the White River (Monserrat, West Indies) and the abandoned Mynydd Parys copper mine in Wales. Both isolates grew at pH 3.8 using glycerol as electron donor.[39] *Acidophilic* SRBs have been isolated from the Rio Tinto in Spain, a river that has been mined for over 5000 years, but most of these could only grow down to pH 4.0. None of these organisms could survive at pH 2.5, and to date, no SRBs have ever been isolated from water as acidic as BPL.

10.8.2 Extremophilic Microbes of BPL

10.8.2.1 Isolation of the First Fungal Extremophiles

Our first water sample was a surface sample provided to us by colleagues from the Montana Bureau of Mines and Geology (MBMG) in 1995. The water was streaked onto nutrient agar[40] and nutrient agar acidified to pH 2.5, and incubated at room temperature. Within a few days, a smooth yeast-like colony and fungal hyphae began to proliferate on both plates, and the newly emerging organisms were carefully transferred to fresh agar and established as pure cultures. Although there were several different fungi growing on the plates, three fungi were established in culture from this preliminary study and were identified by Microbial ID, Inc. as *Pithomyces* sp., *Penicillium chrysogenum*, and the yeast *Pichia* (*Hansenula*) *anomala*.[40]

10.8.2.2 Initial Pilot Studies with BPL Fungi

In this preliminary study, the organisms were grown in multiple media and harvested as previously described.[40] Each culture was omnimixed and extracted thoroughly with chloroform. The chloroform layer was reduced *in vacuo* to an organic-soluble oil. The water layer was lyophilized, then extracted with chloroform/methanol (1:1, v/v), which was reduced *in vacuo* to generate the freeze-dried extract (FDX). The remaining aqueous material was the freeze-dried residue (FDR). The organic extract, FDX, and FDR were tested for antifungal and antibacterial activity using standard disk assays, and for cytotoxicity using the brine shrimp lethality assay.

For all three organisms, biological activity was concentrated in the organic extract and FDX. Dramatic differences in activity were observed in extracts from the same fungus grown in different media. For instance, the chloroform extract of *Penicillium chrysogenum* exhibited 100% brine shrimp lethality when grown in potato dextrose broth (PDB), but was inactive when grown in any other media. Both *Pithomyces* sp. and the yeast *Pichia anomala* exhibited maximal brine shrimp lethality when grown in acidified $PDBH^+$ broth, and little or no activity in any other nutrient medium.

Although our research was largely focused on the fungal endophytes of *Taxus brevifolia* and *Sequoia sempervirens*, these data eventually helped direct specific fermentation studies of the most promising Pit microbes under conditions that promoted greatest biological activity. In 1996 and 1998, the MBMG provided additional water samples from the surface down to 275 m. It was from these samples that most of the microbial collection was established.

Although it would be fascinating to launch a complete metagenomic survey of the waters and sediments of BPL, our current studies require the isolation and maintenance of a microbial library. (Our microbial isolation methodology has been previously described.[40]) Over the last 20 years, we have isolated and studied over 60 fungi and bacteria from the surface waters down to the lake bottom sediments at 275 m. We are interested primarily in microbes that can be grown in submerged culture using standard methodology (with some nonstandard media) and that flourish at pH 2.5. Of course, many microbes are fastidious and resist standard cultivation techniques, but the BPL microbes that have been established in pure culture and that grow under standard conditions have proven themselves capable of producing interesting new chemistry.

10.8.3 Identification of Berkeley Pit Fungi

Many of the fungi in this collection have been identified by Microbial ID (Delaware) based on sequence analysis of the D_2 variable region of the large-subunit (LSU) rRNA gene. Several species were found at multiple depths, while certain species were isolated from a discreet depth, in sediment samples, or in association with an alga. For example, both *Penicillium rubrum* and *Phialemonium curvatum* were isolated from various depths from surface waters down to 275 m, while *Oidiodendron tenuissimum*, a *Pleurostomophora* sp., and a *Pithomyces* sp. were isolated exclusively from surface water. Bacteria exhibited a similar distribution pattern. *Stenotrophomonas maltophilia* was isolated from sediment samples and from water samples taken at the surface, 10 and 225 m, while *Streptomyces griseoplanus* was isolated exclusively from sediment samples (Figure 10.7).

It was interesting to note that taxonomically and morphologically identical organisms isolated from different depths often produced different secondary metabolites. *Penicillium rubrum* was isolated from surface water, and from depths of 3, 50, and 275 m. All four of these isolates were grown in pilot studies under a variety of physicochemical conditions.[40] At defined harvest times, the cultures were thoroughly extracted with organic solvents and the extracts were tested in several different assays, including antimicrobial, brine shrimp lethality, and enzyme inhibition assays (caspase-1 and MMP-3). Even though these four organisms are morphologically and genetically (at least based on the LSU region) identical, the biological activity of their extracts often differed sharply, with the deep-water strain exhibiting the most

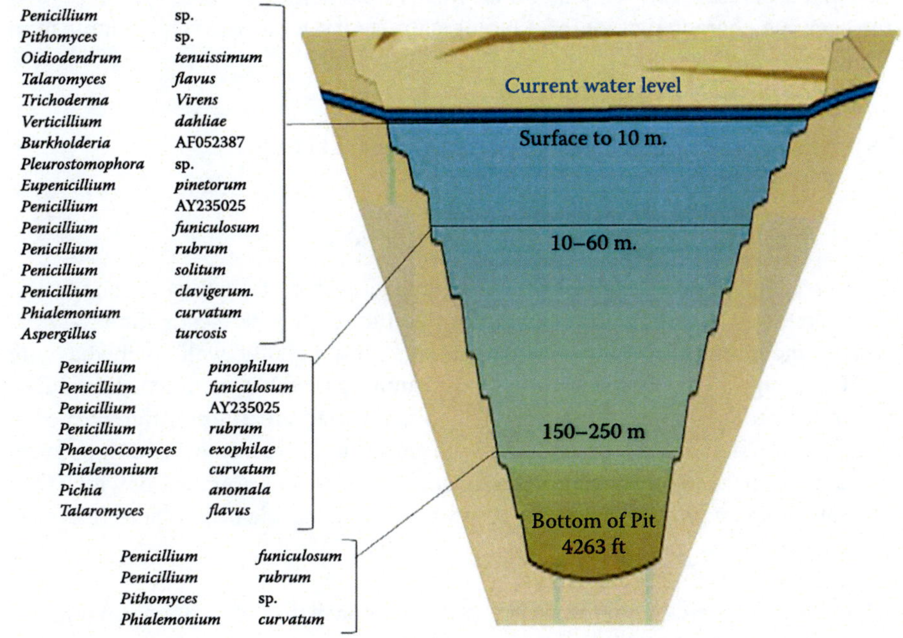

FIGURE 10.7 Overview of microbes isolated from different depths of the Berkeley Pit.

potent enzyme inhibitory activity. We have therefore maintained these four isolates of *P. rubrum* as separate cultures because of the differences in their chemistry. Only the deep-water isolate of *P. rubrum* has been extensively studied to date.[41–44]

10.8.4 Challenges of Fungal Taxonomy

Identification of fungi can be challenging, and taxonomists have used morphology as well as sequence analysis of multiple regions of the fungal rRNA genes to study fungal taxonomy and diversity.[45] These include the small-subunit (SSU) and large-subunit (LSU) rRNA genes and the internal transcribed spacer (ITS) region that separates the two rRNA genes.[46] There is debate among taxonomists regarding which of these regions provides the most robust information for species assignment. The LSU gene has been used extensively for fungal phylogeny and taxonomic placement, including the Assembling the Fungal Tree of Life (AFTOL) Project and environmental surveys.[46]

Although the ITS region provides a useful *bar code* for environmental diversity studies, the extent of sequence variability in this region does not allow for robust sequence alignment. Houbraken and colleagues from the CBS-KNAW Fungal Biodiversity Centre (The Netherlands) propose the use of a secondary identification marker for *Penicillium* and consider β-tubulin (BenA) as the best option. Other proposed secondary markers include calmodulin (CaM) or the RNA polymerase II second largest subunit (RPB2) genes.[47] Many scientific journals now require GenBank accession numbers for microorganisms, as well as submission of a voucher specimen to a culture collection to facilitate access to organisms for other scientists. It can be argued that accurate identification of a secondary metabolite producer is as important as the accurate characterization of a biologically active molecule. The combination contributes to good science.

10.9 Search for Secondary Metabolites from Berkeley Pit Fungi

10.9.1 Prioritizing Microbes for Investigation

When this study began in earnest in 2000, we had over 60 fungal isolates and limited resources, so it was important to prioritize the fungi for investigation. Several microbes, including all of the *Penicillium* sp., grew equally well in media with low pH and high metal concentration, and in more generalist media such as PDB, although the production of secondary metabolites varied greatly in different media. Organisms that demonstrated robust growth under more extreme conditions were designated as *thrivers*—microbes uniquely adapted to the environment in which they were found. Others grew very slowly at low pH and it was clear that these microbes were *extremotolerant* rather than truly extremophilic. Most of our subsequent work focused on the thrivers, the true extremophiles.

The Stierle Research Laboratory is a small lab consisting of two principal investigators (Andrea and Donald Stierle) and undergraduate students, who have been an invaluable component of our research. With their assistance, we were able to grow over half of the microbe collection using 12 different physicochemical conditions and to assess the biological activity of the crude extracts of each fermentation.[40] Extracts that exhibited the most promising biological activity were chromatographed, and each of the resulting column fractions were examined by NMR and assessed for biological activity. Microbes were prioritized for study based on their acid tolerance, activity profiles, and NMR *chemotype*. The most promising microbes were then grown in larger volumes (8–12 L) using conditions associated with the most promising biological activity.

10.9.2 Selection of Appropriate Bioassays: The Path to MMP-3, Caspase-1, and Caspase-3

No matter what source organism is being investigated, it is the unique challenge of the natural product chemist to find suitable assay methods to guide the isolation of secondary metabolites with desirable biological activity. We currently use three different enzyme inhibition assays to guide compound isolation: MMP-3, caspase-1, and caspase-3. Each of these assays is validated by an orthogonal cell-based assay, and each will be discussed in detail later in the chapter. When we first initiated this study however, we used the standard disk antimicrobial assay to guide the search for new antibiotics and antifungal agents, and most of the extracts were either weakly active or inactive.

Rather than abandon the collection, we reconsidered our choice of assay methodologies. Microbes produce secondary metabolites as part of their survival strategies. In essence, part of the driving force for secondary metabolite production is environmental pressure. In the competitive, resource-rich world of plant endophytes, individual microbes carve out territories by the production of antimicrobial compounds.[47–50] In the inhospitable and sparsely populated waters of BPL, there is no need to outcompete other microbes for resources, so there is no selection pressure toward the production of antimicrobial agents.

We therefore turned to the tools of signal transduction to guide our efforts in a different direction, one dictated by low pH and an iron-rich, oxidizing environment. We were most fortunate that the National Institutes of Health introduced the Biomedical Research Infrastructure Network grants program at this time. Funding made possible through this program provided the necessary means to incorporate enzyme inhibition assays into our protocol. It only remained to select the appropriate bioassay.

10.10 Correlations between BPL and Human Pathologies

10.10.1 Correlating Low pH and High Iron Concentration with Inflammation and Carcinogenesis

The bioassay-guided isolation of secondary metabolites is obviously biased by the choice of assay systems. As we evaluated the appropriateness of different assay systems to guide compound isolation, we considered the uniqueness of BPL microbes and their environment, and how that could apply to human pathologies. Two factors seemed particularly relevant: pH and high iron concentration. While these factors clearly dominate the environment of BPL microbes, they have also been implicated in human pathologies.[51–55]

Low pH is associated with both inflammation and carcinogenesis.[51] The tumor microenvironment is usually acidic due to the production of acidic metabolites caused by anaerobic glycolysis under the hypoxic conditions associated with tumors.[51] According to a theory known as the *Warburg effect*, tumors produce lactate because they use the anaerobic glycolytic pathway rather than oxidative phosphorylation for energy production, even in the presence of sufficient oxygen.[55] Low pH in inflammatory tissue is also caused by the production of acids, including quinolinic and retinoic acids, by macrophages.[56,57]

Iron is the most abundant transition metal in BPL—and in the human body as well. In humans, redox cycling of iron is closely associated with the generation of ROS or reactive nitrogen species (RNS) and the induction of oxidative stress, which occurs when the generation of excess free radicals and active intermediates exceeds the system's ability to eliminate them.[52–54] Oxidative stress from increased levels of ROS can lead to damage of proteins, lipids, and DNA, and can promote numerous pathologies, including hepatocarcinoma.[52–54]

The most damaging effects on DNA are due to hydroxyl ions, which are generated via the Fenton reaction,[52] in which intracellular transition metals such as iron donate or accept free electrons and use H_2O_2 to catalyze free radical formation. Hydroxyl radicals attack DNA rapidly due to their high diffusibility, which results in the formation of DNA lesions, including oxidized DNA bases and single-strand and double-strand breaks. Elevated rates of ROS have been detected in almost all cancers, and promote tumor progression and development through a variety of mechanisms.[54]

ROS also play an important role in the progression of inflammatory disorders.[58] One of the mechanisms by which ROS promote inflammation is through the induction of the NLR protein-3 (NLRP3) inflammasome, an important multiprotein complex that binds and subsequently activates the cysteine protease caspase-1.[58] Upon activation, caspase-1 stimulates the production of several proinflammatory cytokines, including interleukin (IL)-1β and IL-18.[58,59] Unfortunately, unresolved, chronic inflammation may lead to tissue damage, cell proliferation, and the generation of ROS and RNS, the hallmarks of a cancer-prone microenvironment.[58]

10.10.2 Enzymes Associated with Inflammation and Carcinogenesis: Caspase-1

As we considered appropriate enzyme assays for our study of BPL extremophiles, it was clear that a case could be made for enzyme pathways associated with inflammation and carcinogenesis. In 2001, researchers had already determined that caspase-1 played a critical role in the regulation of multiple proinflammatory cytokines, and specific caspase-1 inhibitors were already being considered as a new class of anti-inflammatory drugs with multipotent action.[59] Caspase-1 was known to activate IL-1β, a cytokine implicated in a variety of inflammatory diseases, and was also shown to activate interferon (IFN)-gamma-inducing factor (IGIF, IL-18) with equivalent efficiencies *in vitro*.[59] Selective caspase-1 inhibitors were shown to block both lipopolysaccharide (LPS)-induced IL-1β and IFN-gamma production from human mononuclear cells.[59] BioMol (now Enzo Life Sciences) offered a Caspase-1 Assay Kit for Drug Discovery, which became our first biochemical assay to guide the isolation of anti-inflammatory compounds.

10.10.3 Enzymes Associated with Inflammation and Carcinogenesis: MMP-3

Researchers have also demonstrated that matrix metalloproteinases (MMPs) were upregulated in a number of different pathological conditions, including several cancers.[60–65] Extracellular matrices were known to constitute the principal barrier to tumor growth and metastatic spread, and it had been shown that malignant tumors utilized MMPs to overcome these barriers.[60–65] Consequently, inhibitors of MMPs, especially MMP-3 and MMP-9, represented an attractive target for a new class of anticancer agents. MMP-3 (stromelysin) had also been implicated in cartilage damage associated with inflammatory rheumatoid arthritis.[66,67] Although it was not considered a proinflammatory enzyme, MMP-3 was often upregulated in inflammatory diseases. BioMol also offered an MMP-3 Assay Kit for Drug Discovery, so we adopted a binary biochemical approach to our search for compounds that could inhibit enzymes associated with inflammation and carcinogenesis.

10.10.4 Fungal Metacaspases, Apoptosis, and Mammalian Caspases

There is an interesting biochemical correlation between fungal and mammalian cells when exposed to low pH and oxidative stress. In mammalian cells, low pH and ROS are intimately associated with inflammation. Within macrophages, ROS stimulate the activation of the inflammasome, which leads to the activation of caspase-1 and the subsequent activation of proinflammatory cytokines.[58] In an analogous fashion, fungi produce metacaspases in acidic environments when stimulated by ROS-mediated oxidative bursts (Figure 10.8).[68–70] Fungal metacaspases are part of the caspase/paracaspase/metacaspase superfamily that emerged early in eukaryotic evolution as an offshoot of a larger, more ancient class of cysteine proteases.[71] Unlike caspase-1, when metacaspases are upregulated, they promote the progression to apoptosis rather than inflammation. This suggests that metacaspases function more like the mammalian executioner caspases (caspase-3, -6, -7, -8, and -9).[72] Although they are not structurally closely related to caspases, they cleave some of the same substrates, including *Tudor Staphylococcus* nuclease, and are inhibited by the pan-caspase inhibitors zVAD-fmk and zVEID-fmk.[69]

We speculated that a fungus consistently exposed to an oxidative, acidic environment would be a logical source of small molecules that blocked *metacaspase*-induced apoptosis. These same inhibitors might also inhibit *caspase-3*-induced apoptosis (Figure 10.8). While MMP-3 and caspase-1 are *upregulated* in processes associated with carcinogenesis, caspase-3 and apoptosis are generally *downregulated* in carcinogenesis. Caspase-3 catalyzes the terminal step in the biochemical cascade that leads to apoptotic cell death and is important in killing cancer cells.

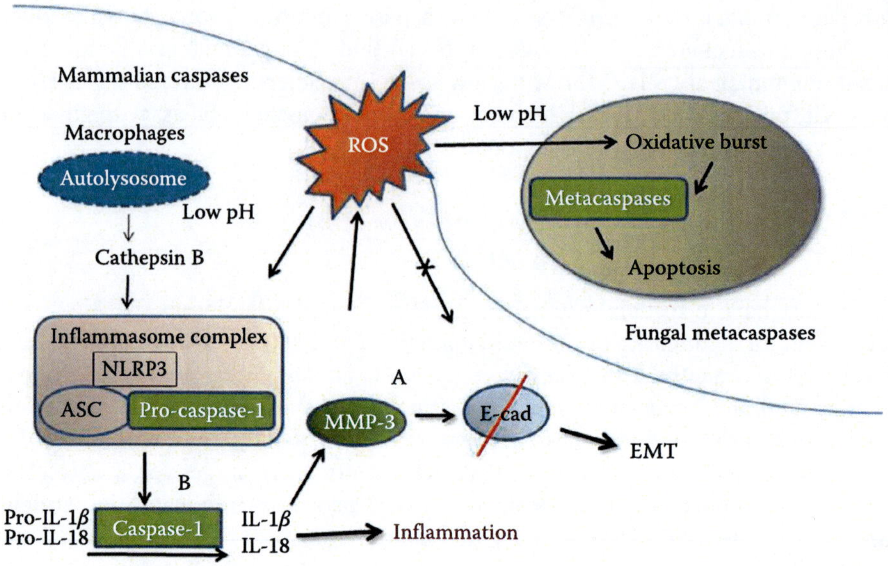

FIGURE 10.8 Low pH and ROS induce caspase-1-mediated responses in mammalian macrophages through the NLRP3 inflammasome, and metacaspase-mediated responses in fungi. Both caspases and metacaspases can be inhibited by the same small molecules.[70]

However, its activation following a cerebral stroke contributes significantly to the delayed neuronal death that occurs following a stroke. If apoptotic processes could be mitigated during this critical period, then neurons in the ischemic penumbra could be salvaged.[73] In 2010, we incorporated BioMol Caspase-3 Inhibition Assays for Drug Discovery into our research protocol to find compounds that could block apoptosis.

10.11 MECHANISTIC PATHWAYS ASSOCIATED WITH INFLAMMATION AND CARCINOGENESIS

We have provided a brief overview of three important physiological phenomena—carcinogenesis, inflammation, and apoptosis—and how they directed our choices of biochemical assays for this research. Each of these processes will now be explored in more detail. In 2002, we assumed that this research would result in two discreet compound libraries: caspase-1 inhibitors that could mitigate inflammation, and MMP-3 inhibitors that inhibited carcinogenesis. Inflammation and carcinogenesis are both complex phenomena and are controlled by numerous and sometimes synergistic mechanisms.[74]

10.11.1 Unresolved Inflammation

When inflammation is initiated by infection, wounding, or irritation, a wide array of immune cells are recruited to the site. This leads to the release of various proinflammatory cytokines and other agents that orchestrate the initiation of an inflammatory cascade that is precisely timed and often resolves itself.[58]

Unfortunately, if unresolved, inflammation may become chronic, which can lead to tissue damage, cell proliferation, and the generation of ROS and RNS, the hallmarks of a cancer-prone microenvironment. Inflammatory cells (macrophages, neutrophils, eosinophils, and basophils) and molecules (cytokines and chemokines) persist after tumor initiation and may infiltrate into the tumor site. They may actually protect tumor cells from the host immune response or directly facilitate angiogenesis, tumor growth, invasion, and metastasis by themselves or by inducing other effector molecules, such as MMPs.[58]

Certain anti-inflammatory agents have also been found to have potent anticancer or chemopreventative activity.[75–77] These include several nonsteroidal anti-inflammatory drugs (NSAIDS), which mitigate inflammation through both cyclooxygenase-2 (COX-2)-dependent and Cox-2-independent pathways. NSAIDs, tolfenamic acid, and sulindac sulfide have been well investigated for their antitumorigenic activity in many different types of cancer.[76] Aspirin has also been shown to have anticancer activity, but its most significant effect might be as a chemopreventive agent against colorectal cancer (CRC). Five cardiovascular prevention randomized controlled trials (RCTs) evaluating the use of aspirin were also analyzed for its effects on cancer outcomes. Daily aspirin use at any dose reduced the risk of CRC by 24% and of CRC-associated mortality by 35%.[77]

Inflammation is an important natural defense reaction induced by tissue damage or the presence of foreign proteins or pathogens. The primary objective of inflammation is to localize and eradicate the irritant and repair the surrounding tissue. Inflammation is a necessary and beneficial process, but it can be difficult to regulate.[78] When the body's own immune responses are directed against its own tissues, autoimmune disorders, characterized by prolonged inflammation and subsequent tissue destruction, can result.[78] Autoimmune disorders can cause immune-responsive cells to attack the linings of the joints, resulting in rheumatoid arthritis, or trigger immune cells to attack the insulin-producing islet cells of the pancreas, leading to insulin-dependent diabetes.[78]

10.11.2 Caspase-1, the Inflammasome, and Inflammation

Caspase-1 plays an important role in inflammation, as well as in chronic inflammation, through the production of specific cytokines. Caspase-1 is activated upon binding to the NLRP3 inflammasome, a multiprotein complex that plays a key role in innate immunity. Once caspase-1 is activated, it cleaves (activates) the precursors of IL-1β, IL-18, and IL-33.[58,79] The production of pro-IL-1β and IL-18 is regulated by *nuclear factor-kappa B* (NF-κB).[79] NF-κB can be activated by exposure of the cell to *tumor necrosis factor* (TNF) and *pathogen-* and *danger-associated molecular patterns* (PAMPS and DAMPS), including bacterial LPS.[79] These same factors, as well as ROS, stimulate the assemblage of the NLRP3 inflammasome (Figure 10.9).[70]

The upregulation of caspase-1 and/or IL-1β has been found in certain breast cancers,[80] acute myelogenous leukemia,[81] melanoma,[82] certain glioblastomas,[83]

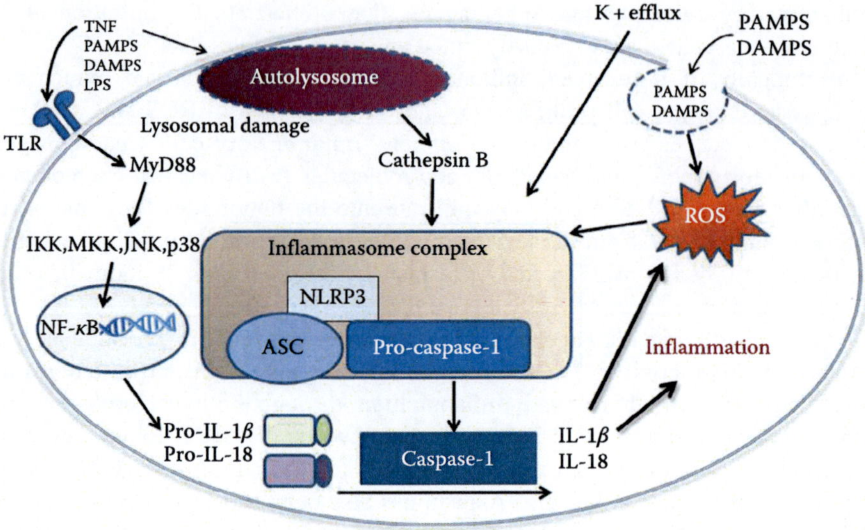

FIGURE 10.9 Inflammasome assembly is stimulated by several factors, including ROS. Caspase-1 is activated upon binding to the NLRP3 inflammasome and in turn activates IL-1β and IL-18. The production of pro-IL-1β and IL-18 is regulated by nuclear factor-kappa B (NF-κB), which is activated by exposure of the cell to tumor necrosis factor (TNF) and pathogen- and danger-associated molecular patterns (PAMPS and DAMPS), including bacterial LPS.[70]

and pancreatic cancers,[84,85] all of which are exacerbated by chronic inflammation associated with activation of the inflammasome. Proinflammatory cytokines can lead to chronic inflammation and the production of ROS, which can induce oxidative damage to DNA and consequently lead to the initiation and progression of carcinogenesis.[79,86] ROS also stimulate the assemblage of inflammasomes within macrophages, which results in higher inflammatory cytokine production and increasing levels of ROS.

10.11.3 Correlating Inflammation and Metastasis

Although inflammation is associated with many aspects of tumorigenesis, its role in metastasis is of particular interest to us and to many other researchers.[87,88] A high concentration of inflammatory cells, particularly tumor-associated macrophages (TAMs), is commonly found at the invasive edge of advanced carcinomas. Macrophages are key cells in chronic inflammation, and TAMs produce a wide variety of growth factors and cytokines (such as TNF-α, IL-6, IL-1, and IFNs) that stimulate the growth, motility, and invasiveness of tumor cells.[88] Mitigating inflammation could therefore play an important role in the prevention of metastasis.

Although treating any cancer can be a challenge, the treatment of metastatic cancer is one of the most daunting tasks facing cancer researchers and oncologists. According to the National Cancer Institute (NCI), the majority of people who die of cancer, die of metastatic cancer.[89] In spite of progress in the understanding of the onset and progression of cancers and in the treatment of localized malignancies, metastatic disease is often incurable. Lung and bronchus cancers still account for more deaths than any other cancer in both men and women.[89] According to the American Cancer Society, the 5-year survival rate for localized disease (invasive cancer confined entirely to the lung and bronchus) is only 54%, but for metastatic disease, the survival rate drops to 4%.[89] The 5-year survival rate for localized breast cancer is 99%, but once the disease metastasizes, the survival rate drops to 24%.[89]

Metastasis takes a similar toll in many other cancers. Current treatment strategies for metastatic cancer largely rely on the use of systemic cytotoxic agents, which often have severe side effects and, in many cases, have limited long-term success.[89] *Preventing* metastatic dissemination would greatly enhance cancer survival rates. Small-molecule inhibitors of the enzymes associated with metastasis would be useful tools to study the phenomenon itself. Moreover, they have the potential to serve as chemotherapies that could effectively prevent the spread of cancers to remote sites.

10.12 CORRELATING METASTASIS WITH EMT AND MMP-3

Researchers have shown the importance of inflammation in metastatic cancer and that certain cytokines enhance the motility of cancer cells. There are other mechanisms that contribute to metastasis however, and one of the most important of these mechanisms is EMT. Metastasis requires both the mobilization of cancer cells and the invasion of healthy tissue, both of which are facilitated by EMT. It is inherent in cells undergoing proliferation and differentiation and is characterized by loss of cell adhesion, repression of E-cadherin (E-cad) expression, and increased cell motility.[86] Although these processes are components of embryogenesis, neural tube development, wound healing, and angiogenesis, they are also the hallmark of tumor cell invasion and metastatic spread of many types of cancer.[86] If EMT could be circumvented, then the risk of metastatic spread would be mitigated. Unfortunately, metastatic disease continues to be intractable and often incurable.

EMT is a complex phenomenon. It involves many different enzymes, but MMP-3 plays a pivotal role in multiple aspects of EMT (which will be described in more detail later) and is upregulated in many cancers, especially metastatic cancers.[86] MMP-3 is part of the family of zinc endopeptidases that are required for the degradation of extracellular matrix components during angiogenesis and wound healing.[61] When these processes become dysregulated however, tumor cell invasion can result. In mammals, endogenous *tissue inhibitors of metalloproteinases* (TIMPs) precisely regulate the levels and metabolic activities of MMPs.[61,62] Disruption of this balance results not only in tumor growth and metastasis, but also in diseases such as arthritis and atherosclerosis, which often have an inflammatory connection.[62] Several studies suggest that the inhibition of MMP activity may prevent tumor cell dissemination.[63]

10.12.1 Prior Clinical Trials with MMP Inhibitors

There were several clinical trials using specific competitive inhibitors of different MMPs throughout the 1990s, all with disappointing results. Two possible reasons for these failures were timing of drug administration and focus on competitive inhibitors:

1. The majority of Phase II and III clinical trials involved late-stage (III and IV) cancers that had already metastasized.[64,65,90] As MMPs often promote carcinogenesis by increasing EMT and subsequent metastasis, chemotherapies that target these enzymes should be prescribed in the early stages of cancer to prevent the progression to metastatic disease.
2. MMPs were assumed to promote EMT only through direct catalytic processes; therefore, all early MMP inhibitors were designed to target the deep-cleft catalytic sites of various MMPs, either selectively or as broad-spectrum therapies.[64,65,90] Recent studies have shown that MMP-3 induces EMT through multiple pathways, and through both its catalytic and noncatalytic hemopexin-like (HPEX) domains.[91,92] Targeting allosteric sites as opposed to the orthosteric site could have different outcomes with reduced toxicity and greater selectivity.[93]

New insights into the complex interactions of MMPs with other molecular pathways, as well as the role of noncatalytic domains in EMT, suggest that MMP inhibitors deserve further consideration.

10.12.2 Catalytic Mechanisms Involved in the Progression of EMT by MMP-3

The catalytic domain of MMP-3 is involved in several aspects of EMT. These include direct cleavage of the cell–cell adhesion protein E-cad, repression of E-cad synthesis, activation of MMP-9, and enhanced expression of Rac1b. Activated MMP-9 also directly promotes EMT through the disruption of the extracellular basal lamina.[92] MMP-3 and MMP-9 work synergistically to dissolve endothelial matrices, an important feature of cells undergoing proliferation. The tumor microenvironment itself is actually a potent carcinogen that facilitates cancer progression and activates dormant cancer cells. It also stimulates tumor formation through induction of even higher levels of enzymes such as MMP-3 and MMP-9.[86]

MMP-3 also induces the expression of a highly activated splice isoform of Rac-1 (called Rac-1b) that was first discovered in colorectal[94] and breast tumors,[86,95] and more recently in pancreatic adenocarcinomas.[96] Rac-1, part of the Rho family of GTPases, is a regulator of many cellular processes, including the cell cycle, epithelial differentiation, and cell–cell adhesion.

Rac-1b stimulates the production and release of mitochondrial superoxide into the cytoplasm. Excess superoxide can have multiple effects: it can induce oxidative DNA damage and genomic instability, and potentiate tumor progression.[86] Superoxide is readily converted to other forms of ROS that stimulate further tumorigenic processes.

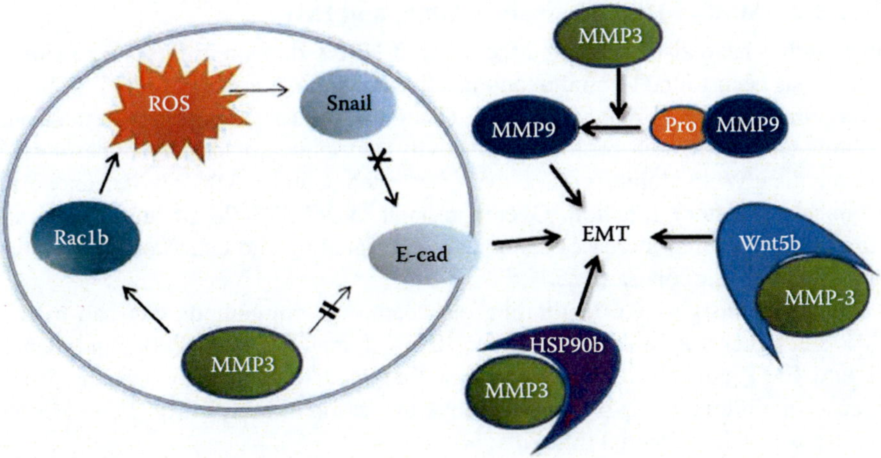

FIGURE 10.10 MMP-3 promotes EMT through multiple mechanisms. It acts catalytically to degrade E-cad, to induce expression of Rac1B, and to activate MMP-9, which subsequently cleaves the basal lamina. Its hemopexin-like domain binds to either extracellular HSP90b or Wnt5b, both of which lead to EMT.[70]

Increased levels of ROS stimulate the expression of the transcription factor *Snail*, which directly represses the synthesis of E-cad. In essence, the tumor itself induces changes that alter cellular structure in culture and tissue structure *in vivo*, leading to EMT and malignant transformation (Figure 10.10).[86]

10.12.3 EMT and the Noncatalytic HPEX Domain of MMP-3

10.12.3.1 MMP-3-HPEX Domain, HSP90b, and EMT

While much attention has been paid to the catalytic domain of MMP-3 and its multiple effects on EMT, Bissell and colleagues demonstrated that the noncatalytic domain of MMP-3 also plays an important role in the promotion of EMT.[91] In mouse models, MMP-3 has been shown to regulate mammary gland branching morphogenesis, and that upregulated expression of MMP-3 in mouse mammary epithelia leads to excessive lateral branching and eventual tumor formation.[91] Bissell's lab developed MMP-3 constructs that lacked catalytic activity and demonstrated that overexpression of these constructs directed mammary epithelial invasion in collagen-I gels. Furthermore, constructs of catalytically active MMP-3 that lacked the HPEX domain did not induce epithelial invasion. Equally surprising, they also found that extracellular *heat shock protein 90b* (HSP90b) was required for the observed effects (Figure 10.10).[91]

HSP90b is often overexpressed in cancer and is associated with more aggressive, larger tumors with more lymph node involvement and decreased patient survival.[91] Preventing the formation of this complex with small-molecule inhibitors that bind either to the MMP-3-HPEX domain or to HSP90b could mitigate EMT and thereby metastasis.

10.12.3.2 MMP-3-HPEX Domain, Wnt5b, and EMT

Other studies have also implicated the MMP-3-HPEX domain in the progression of EMT. It has been found to bind to and inactivate Wnt5b, a noncanonical Wnt ligand that inhibits canonical Wnt signaling.[97] Canonical Wnt is important in embryonic development, but its cell proliferative activity can induce mammary hyperplasia.[97] Noncanonical Wnt5b inhibits this activity, but when bound to MMP-3, it cannot exert its typical regulatory function. Overexpression of MMP-3 (as in tumor cells and mammary stem cells) increases canonical Wnt signaling and concomitant elevation in mammary carcinoma stem cells.[97]

These new insights are facilitating our search for compounds that can bind to MMP-3 at either the catalytic site or the HPEX domain, or to HSP90b, and thereby mitigate EMT and subsequent processes that lead to metastatic cancer. Small-molecule inhibitors of MMP-3 would block the action of both MMP-3 and MMP-9 and help maintain epithelial integrity.

10.12.3.3 MMP-9-HPEX Domain and EMT

The HPEX domains of other MMPs have also been shown to have important functions beyond the catalytic activity of the enzyme. MMP-9-HPEX has also been shown to mediate EMT cell migration.[98] In the only reported study of small-molecule inhibitors of MMP-9-HPEX, specific HPEX inhibitors prevented the homodimerization of MMP-9, which effectively blocked the [CD44–*epithelial growth factor receptor* (EGFR)–*mitogen-activated protein kinase* (MAPK)] signaling pathway and prevented MMP-9–mediated cancer cell proliferation, migration, and invasion.[98] These new insights into the complex interactions of MMPs with other molecular pathways and the role of noncatalytic domains in EMT suggest that MMP inhibitors deserve a second chance.

10.13 CONNECTION BETWEEN INFLAMMATION AND THE EMT MOLECULAR PATHWAY

It is well-documented that caspase-1 activates proinflammatory cytokines, and when upregulated, it can lead to chronic inflammation and metastatic spread of certain cancers. It is equally clear that MMP-3 is instrumental in EMT. These two pathways are also interconnected. Cytokines that are activated upon cleavage by caspase-1 directly induce the production of MMP-3. IL-1β and TNF-α stimulate the production of MMP-3 through the activation of cellular signaling pathways involving MAPKs, NF-κB, and *activating protein-1* (AP-1).[99,100] The resulting upregulation of MMP-3 leads to EMT as a direct result of the inflammatory pathway. The natural product cordycepin, a potent inhibitor of IL-1β production, effectively inhibits MMP-3 expression.[101,102]

In a reciprocal fashion, MMP-3 also stimulates the production of pro-IL-1β and pro-IL-18, which are ultimately activated by caspase-1. Macrophage exposure to MMP-1 or MMP-3 triggers a rapid release of TNF-α, which induces the production of pro-IL-1β and pro-IL-18 through the transcription regulator NF-$\kappa\beta$. MMP-3 inhibitors blocked this activity, effectively reducing the concentration of pro-IL-1β and IL-18, and subsequently preventing synthesis of IL-1β and other cytokines.[103]

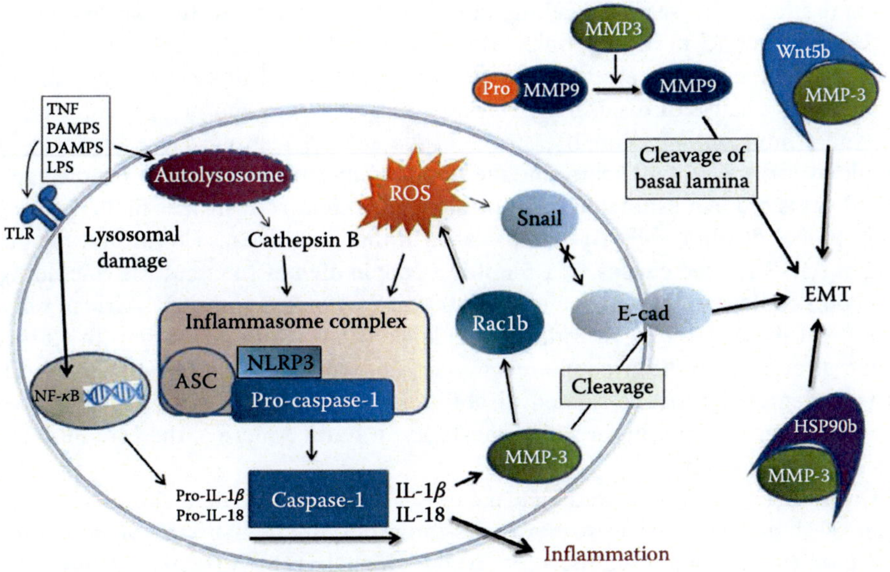

FIGURE 10.11 Inflammation and EMT are two key cellular processes implicated in various pathologies. Several studies have shown that these processes are synergistic. Upregulated MMP-3 leads to the production of ROS, which induces the assemblage of the NLRP3 inflammasome and the subsequent activation (cleavage) of procaspase-1. Caspase-1 activates cytokines 1L-1β and IL-18, which leads to the upregulation of MMP-3. This interplay between MMP-3 and caspase-1 is mediated by ROS and suggests the importance of the *MMP-3–ROS–caspase-1 cycle*.[70]

MMP-3–induced production of ROS also induces the assemblage of inflammasomes in macrophages,[104–106] in essence helping to provide both the raw materials (pro-IL-1β and IL-18) and the machinery (the inflammasome) of inflammation. These data could account for our own observation that MMP-3 inhibitors reduce the production of IL-1β in the *induced inflammasome assay* (IIA; to be described later). MMP-3 not only orchestrates events that promote EMT, but also induces further inflammation in a cell-cyclic series of events (Figure 10.11).[70]

The synergistic relationship between MMP-3 and caspase-1 is mediated by the generation and activities of ROS, a relationship we have termed the *MMP-3–ROS–caspase cycle*.

10.14 CEREBRAL ISCHEMIA, CASPASE-3, AND APOPTOSIS

The biochemical analogies between the proapoptotic fungal metacaspases and caspase-3 were intriguing, but it was not clear how to fit this enzyme into our cancer/inflammation focus.[68–71] Caspase-3 is generally downregulated in cancers, and many cancers produce *X-linked inhibitor of apoptotic protein* (XIAP) to prevent apoptosis of cancer cells.[106] There are several pathologies, however, in which upregulation of caspase-3 can induce severe damage, which might be prevented with an appropriate inhibitor. One of the most important of these pathologies is cerebral stroke, the second leading cause of death and disability worldwide.[107] Each year, over 15 million people suffer a stroke. Of these, over six million die and another five million are left permanently disabled.[108] Stroke is second only to dementia as a cause of disability (including loss of vision and/or, speech, paralysis, and confusion).[108] Globally, stroke is the second leading cause of death in people over the age of 60 years, and the fifth leading cause of death in people aged 15–59 years old. An estimated 6.7 million people died from strokes in 2012 (15% of all global deaths).[107,108] By comparison, in that same year, 1.6 million people died of AIDS-related causes,[109] 1.5 million people died of tuberculosis (including 360,000 people with HIV),[110] and 584,000 died of malaria, mostly African children.[111] In the developing world, the incidence of stroke is increasing. In China, stroke affects 1.3 million people each year, and 75% of the survivors live with varying degrees of stroke-related disabilities. The predictions for the next two decades suggest a tripling in stroke mortality in Latin America, the Middle East, and sub-Saharan Africa.[108]

Despite these alarming statistics, the only FDA-approved medication for treatment of an acute ischemic stroke is the thrombolytic agent *tissue plasminogen activator* (t-PA), the so-called *clotbuster*, which is most effective if administered intravenously within the first 3 h after a stroke. However, both necrosis and apoptosis contribute to the damaging effects of ischemic stroke. Within minutes, the dramatically reduced blood flow to the ischemic core induces irrecoverable necrotic death. This necrotic core, however, is surrounded by the *ischemic penumbra* (the peri-infarct zone), a region of tissue that is less severely injured by the precipitating event. The ischemic penumbra may constitute over half the total lesion volume. Although it is functionally silenced by reduced blood flow, it maintains metabolic integrity temporarily. Many potentially recoverable neurons in this zone undergo apoptosis over the course of hours or days post stroke.[112] If these apoptotic processes could be mitigated during this critical period, then neurons in the region could be salvaged.[112] Targeting and preventing apoptosis in the ischemic penumbra is a rational therapeutic goal for limiting the volume of neural tissue irreparably damaged by a clinical stroke. Unfortunately, no such therapies currently exist.

10.14.1 Caspase-3 and the Intrinsic and Extrinsic Apoptotic Pathways

Both initiator and executioner caspases have been shown to play a major role in the regulation of apoptosis. Several studies have shown that caspase-3 is of particular importance as a key mediator of both the intrinsic and extrinsic apoptotic pathways in animal models of ischemic stroke (Figure 10.12).[112]

The intrinsic and extrinsic apoptotic pathways have many features in common and both lead to caspase-3-dependent cell death. For both pathways, this requires the assemblage of the apoptosome following cleavage of *Bcl-2 interacting domain* (Bid) to truncated Bid (tBid). At the mitochondrial membrane, tBid interacts with the *Bcl-2-associated death promoter* (Bad) and *bcl-2-like protein 4* (Bax), which leads to the opening of the *mitochondrial transition pores* (MTP) and the release of cytochrome C (Cyt C). *B cell lymphoma-2* (Bcl-2) itself is

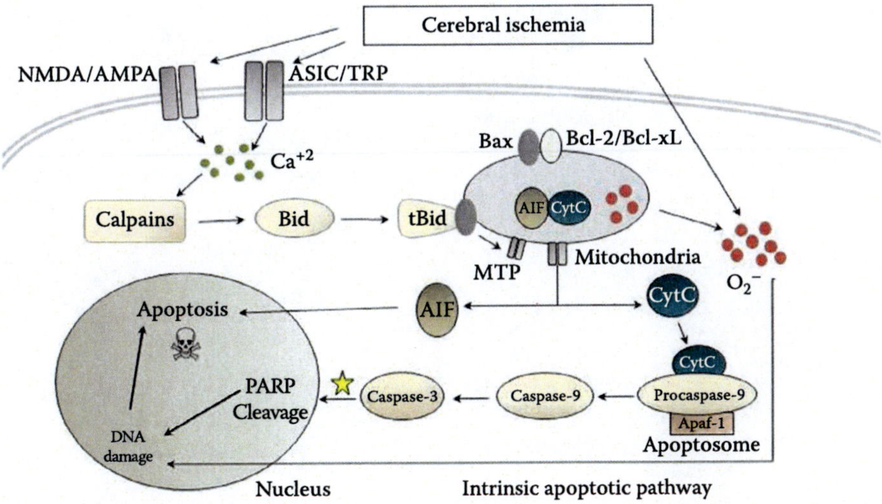

FIGURE 10.12 In the intrinsic apoptotic pathway, cerebral ischemia elevates cytosolic calcium levels through the stimulation AMPA and NMDA receptors by glutamate. Increased intracellular calcium activates calpains and mediates cleavage of Bid to truncated tBid and the subsequent formation of the apoptosome and ultimately to the activation of caspase-3. Activated caspase-3 cleaves nuclear DNA repair enzymes such as *poly (ADP-ribose) polymerase* (PARP), which leads to nuclear DNA damage and apoptosis. The intrinsic apoptotic pathway can also proceed without the involvement of caspase-3.

an important antiapoptotic protein and is thus classified as an oncogene.[112] Once released into the cytosol, Cyt C binds to *apoptotic protein-activating factor-1* (APAF-1) and pro-caspase-9 to form the apoptosome, which activates caspase-9, and subsequently, caspase-3. Activated caspase-3 cleaves nuclear DNA repair enzymes such as *poly (ADP-ribose) polymerase* (PARP), which leads to nuclear DNA damage and apoptosis.[112]

In the intrinsic apoptotic pathway, cerebral ischemia elevates cytosolic calcium levels through the stimulation of α-amino-3-hydroxy-5-methyl-4-isoxazolepropionic acid (AMPA) and *N*-methyl-D-aspartate (NMDA) receptors by glutamate. Increased intracellular calcium activates calpains and mediates cleavage of Bid to truncated tBid, as described above. However, when the MTPs are opened, either Cyt C or *apoptosis-inducing factor* (AIF) is released. The Cyt C pathway proceeds as described, while AIF translocates rapidly to the nucleus and induces caspase-independent cell death.

The extrinsic pathway involves the binding of FasL to the *cell surface death receptor* FasR, which triggers the recruitment of *Fas-associated death domain protein* (FADD). FADD binds to and cleaves procaspase-8, which activates caspase-3 either by direct cleavage or by mitochondrial-dependent mechanisms and the assemblage of the apoptosome, as described above (Figure 10.13).[112]

In essence, caspase-3 catalyzes the final step in programmed cell death, which makes it an attractive target for intervention strategies.

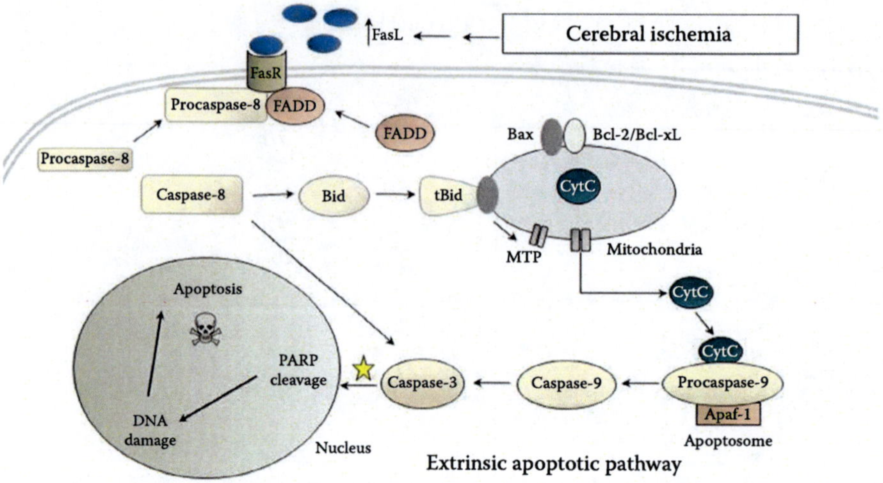

FIGURE 10.13 The extrinsic apoptotic pathway is activated by the binding of the extracellular FasL to FasR, which triggers the recruitment of the FADD. FADD binds to procaspase-8 to create a *death-inducing signaling complex* (DISC), which in turn activates caspase-8. Caspase-8 either directly cleaves caspase-3 to its active form or mediates the formation of the apoptosome and the subsequent caspase-9 activation of caspase-3.[112]

10.14.2 Animal Studies of Caspase-3 Upregulation Following Induced Strokes

Several animal studies have demonstrated the upregulation of caspase-3 following induced strokes. Caspase-3 has been detected in mouse brain during early reperfusion following a 2 h middle cerebral artery (MCA) occlusion,[112] while Asahi and his coworkers demonstrated that caspase-3 mRNA was upregulated in rat brains within 1 h of focal ischemia.[113] It is important to note that caspase-3 has also been shown to be upregulated in ischemic human brain tissue,[114] and that following an ischemic stroke, both pharmacological inhibition[115] and genetic disruption of caspases[116–118] have shown a neuroprotective effect. Caspase-3 activation has been shown to occur up to 9 h after brief MCA occlusion.[119] Caspase inhibitors injected up to 9 h after reperfusion mitigated ischemic damage in a murine model. These findings support two key concepts:

1. There is an extended treatment window for caspase-3 inhibition post stroke.
2. Caspase-3 inhibition is an appropriate therapeutic target for poststroke treatment.

Of course, maintaining high levels of caspase-3 is important in killing cancer cells through these same apoptotic biochemical cascades. An effective cancer therapy might involve the upregulation of caspase-3 through inhibition of XIAP. Indeed, we have recently begun looking for XIAP inhibitors but are at an early point in that research. However, the upregulation of caspase-3 following a cerebral stroke contributes significantly to the delayed neuronal death that occurs following a stroke. If apoptotic processes could be mitigated during this critical period, then neurons in the ischemic penumbra could be salvaged.[112]

10.15 OTHER NEUROPATHOLOGIES ASSOCIATED WITH THE UPREGULATION OF CASPASE-3

Caspase-3 has also been implicated in the development of Alzheimer's disease by cleaving Tau proteins, making them more likely to form neurofibrillary tangles. Alzheimer's disease is a progressive neurodegenerative disorder characterized by accelerated neuronal cell death, leading to dementia.[120] The characteristic pathology includes extracellular amyloid plaques and intraneuronal fibrillar structures, including neurofibrillary tangles (NFTs). Amyloid plaques are formed by the extracellular deposition of proteolytic fragments of amyloid β, whereas the fibrillar structures are composed of the microtubule-associated protein tau assembled into polymeric filaments.[121]

Gamblin and colleagues demonstrated that the proteolytic cleavage of protein tau at Asp421 by caspase-3, caspase-7, and caspase-8 generated a truncated tau protein that was more prone to filament assembly *in vitro*.[122] Proteolysis of tau at Asp421 removes 20 amino acids from its C terminus, a domain the authors had previously shown to have inhibited tau filament assembly *in vitro*.[122,123] The authors proposed that this inhibitory effect is due to tau's C terminal tail folding back on the microtubule-binding repeat region and disrupting its polymerization-promoting effects.[123] This specific proteolysis also occurs in cortical neurons treated with amyloid β and in the hallmark neurofibrillar pathologies of Alzheimer's disease.[122]

There has been some research into the efficacy of caspase-3 inhibitors as potential treatments for Alzheimer's disease.[124] Minocycline is a second-generation tetracycline that has been shown to be neuroprotective in mouse models of Huntington's disease, amyotrophic lateral sclerosis, Parkinson's disease, and multiple sclerosis.[125,126] This neuroprotective effect appears to be associated with its ability to directly inhibit the release of cytochrome c and prevent the activation of caspase-3.[127,128] Minocycline was tested in an animal model of Alzheimer's disease and was shown to slow neuronal cell death and improve cognitive impairment in both mouse and rat models.[129]

10.16 RESULTS OF STUDIES IN THE STIERLE RESEARCH LABORATORY

10.16.1 Small-Molecule Inhibitors That Target MMP-3 and Caspase-1

Our studies have supported the concept that mechanisms associated with inflammation and EMT are interconnected. Some of the compounds isolated as MMP-3 inhibitors actually mitigated the production of proinflammatory cytokines in the IIA[130–132] as effectively as caspase-1 inhibitors. At the same time, caspase-1 inhibitors blocked cell invasion and migration. So at the cellular level, small-molecule inhibitors of enzymes that are associated with EMT (MMP-3) inhibit inflammation, while small-molecule inhibitors of inflammation-associated enzymes (caspase-1) also affect pathways associated with EMT. In the past, EMT and inflammation were studied as discreet phenomena, but more recent studies have highlighted their synergistic relationship.

It is important to point out that many of these molecules could have multiple protein targets and our approach has an inherent bias. That is undeniable. But we continue to find that bioactivity in our biochemical assays is validated in cell-based assays. We have found within our studies that individual compounds may inhibit only one enzyme, two enzymes, or all three enzymes with varying degrees of potency. Whether or not there are other targets will be addressed in the future.

10.16.2 Results of Cell-Based Assays

10.16.2.1 Induced Inflammasome Assay and Caspase-1

Although specific MMP-3 and caspase-1 inhibition assays guide compound isolation in our laboratory, these biochemical *in vitro* assays must be validated by orthogonal cell-based *in vitro* assays. In 2009, immunologist/toxicologist Girtsman introduced the *induced inflammasome assay* (IIA) to validate the caspase-1 inhibition assay. The IIA is used to assess the ability of crude extracts, column fractions, or pure compounds to block production of proinflammatory cytokines in *human monocytic leukemia cells* (THP-1 cells), a cellular system that is analogous to TAMs.[44,130–132] TAMs represent the predominant population of inflammatory cells in solid tumors. THP-1 cells are differentiated into macrophages by the phorbol ester PMA 24 h prior to experimentation.[130–132] When exposed to titanium nanowires and bacterial LPS, differentiated THP-1 cells produce large numbers of inflammasomes and proinflammatory cytokines, including pro-IL-1β and IL-18.

The inflammasome binds to and activates caspase-1, which in turn cleaves the cytokines to produce their active forms, IL-1β and IL-18. In carefully controlled experiments, induced THP-1 cells were exposed to test compounds at concentrations of 100, 10, and 1 μM, and the concentrations of IL-1β, IL-18, and TNF-α post exposure were determined.[44,130–132] Experiments are conducted in 96-well plates for 24 h in 37°C water-jacketed CO_2 incubators. Quantitation of IL-1β is determined using *Human IL-1β DuoSet* from R&D Systems, and ELISA assays are performed according to the manufacturer's protocol. The plates are read at 490 nm. In repeated assays, the IL-1β mitigation potential of the various compounds isolated from these microbes has been measured.[130–132]

Test compounds included MMP-3 inhibitors, caspase-1 inhibitors, and MMP-3/caspase-1 inhibitors (Figure 10.14). As expected, all caspase-1 inhibitors isolated from BPL microbes were active in the assay. Caspase-1 is required to activate proinflammatory cytokines, so caspase-1 inhibition should effectively prevent cytokine production in a cellular system—if the compound can cross cell membranes. Compounds that were dual enzyme inhibitors were also active in the assay. Several compounds that were not caspase-1 inhibitors (in the enzyme assay), however, also mitigated proinflammatory cytokine production in the IIA. Unlike an enzyme assay, in a cellular system, signaling crosstalk is possible, so MMP-3 inhibitors could affect proinflammatory cytokine production through a *non-caspase-mediated* pathway (Figure 10.15).

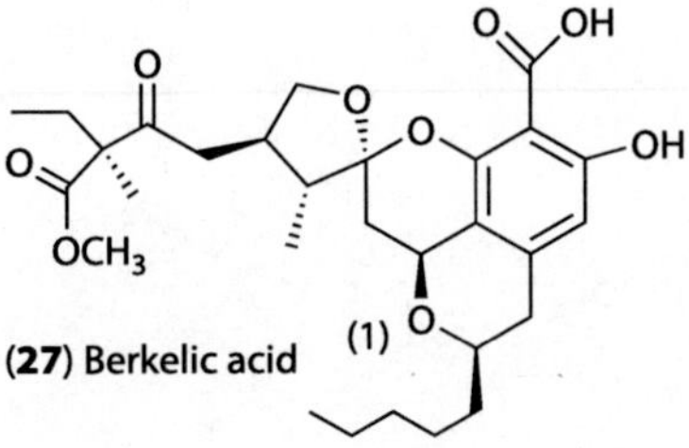

(**27**) Berkelic acid

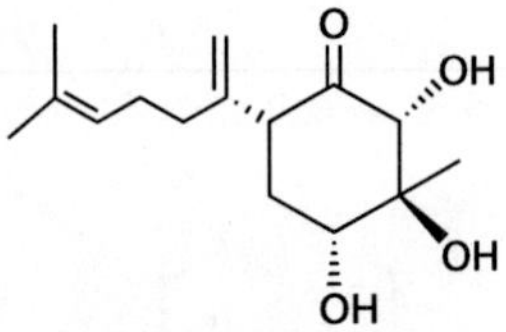

(**28**) Berkebisabolane A

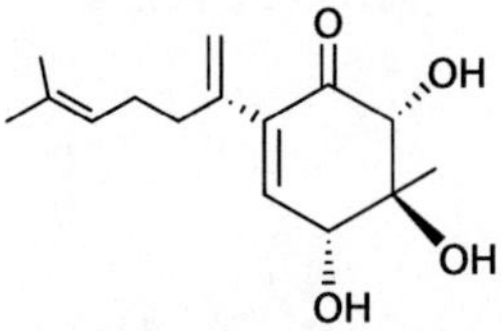

(**29**) Berkebisabolane B

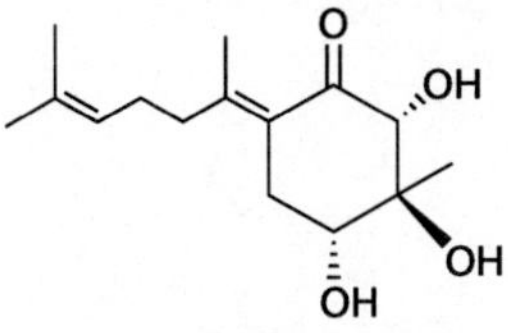

(**30**) Berkebisabolane C

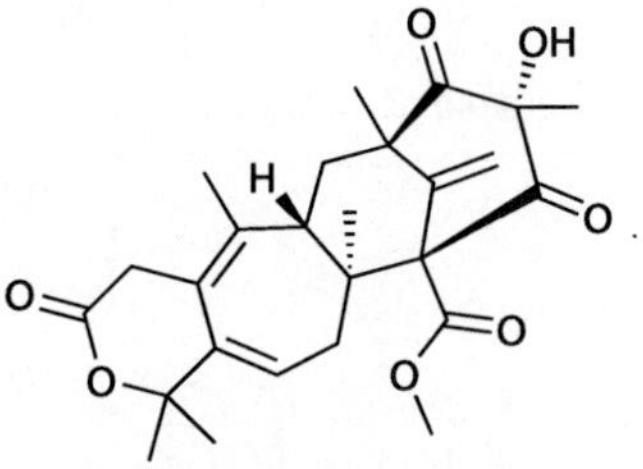

(**31**) Berkeleydione

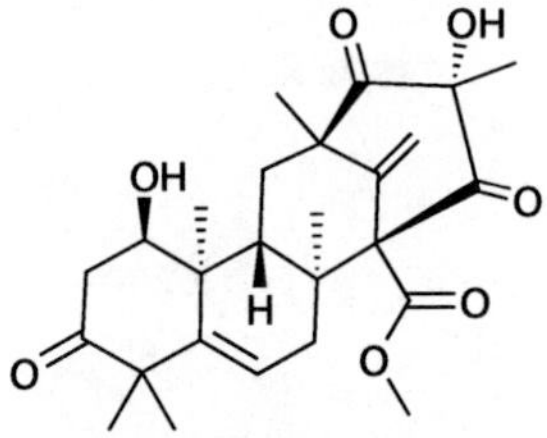

(**32**) Berkeleytrione

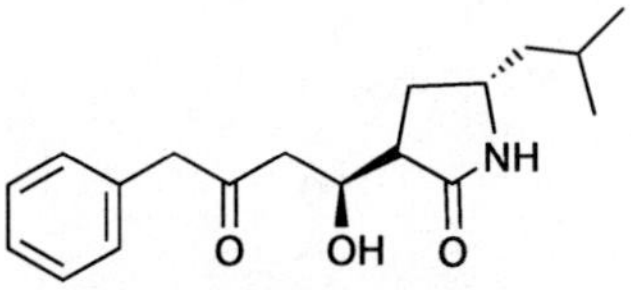

(**33**) Berkeleyamide A

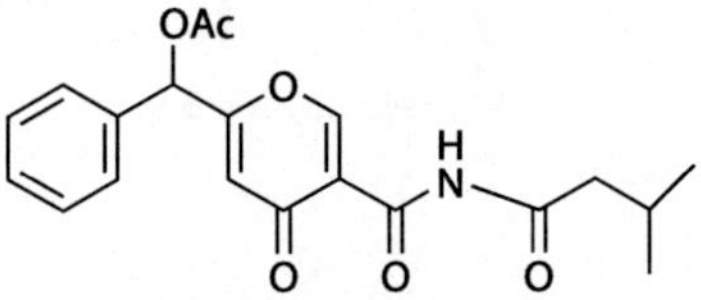

(**34**) Berkeleyamide B

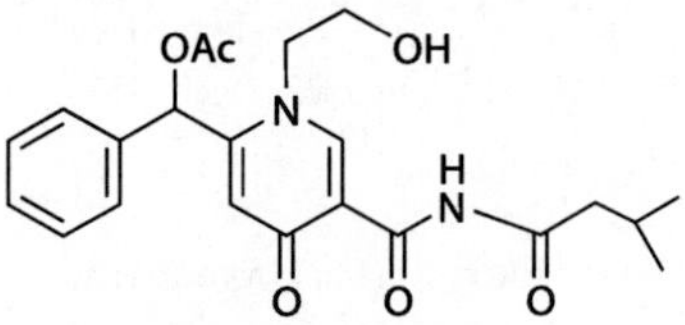

(**35**) Berkeleyamide C

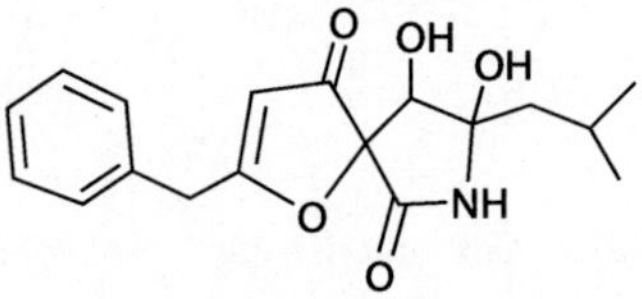

(**36**) Berkeleyamide D

FIGURE 10.14 MMP-3 and caspase-1 inhibitors isolated from Berkeley Pit Lake extremophiles. *(Continued)*

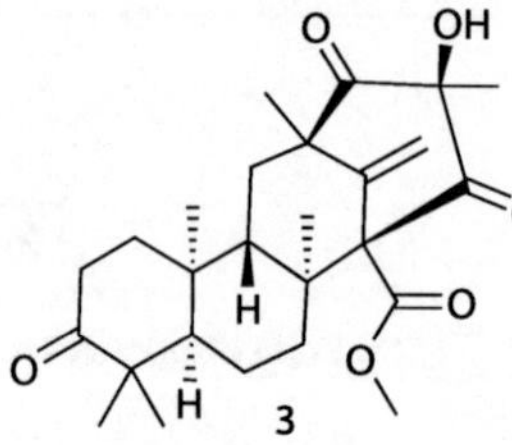

(**37**) Preaustinoid A

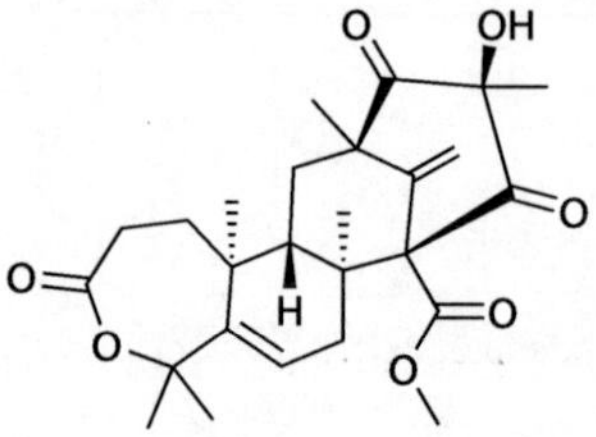

(**38**) Preaustinoid A1

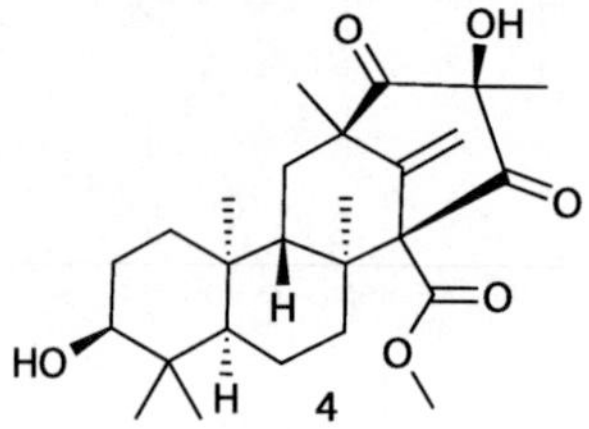

(**39**) Berkeleyone A

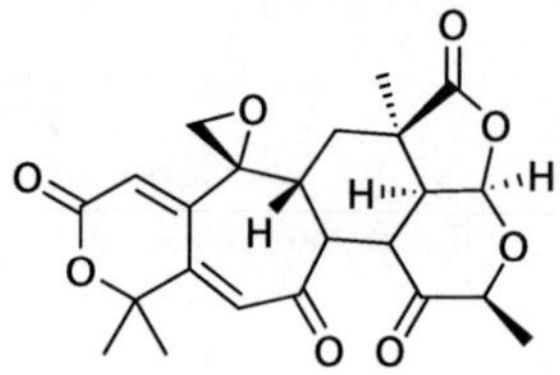

(**40**) Berkeleyacetal C

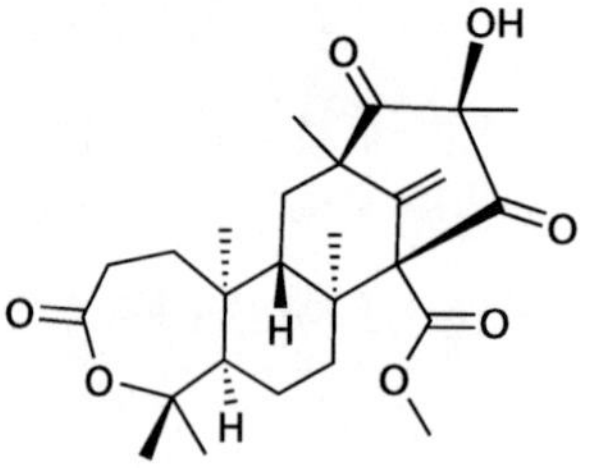

(**41**) Berkeleyone B

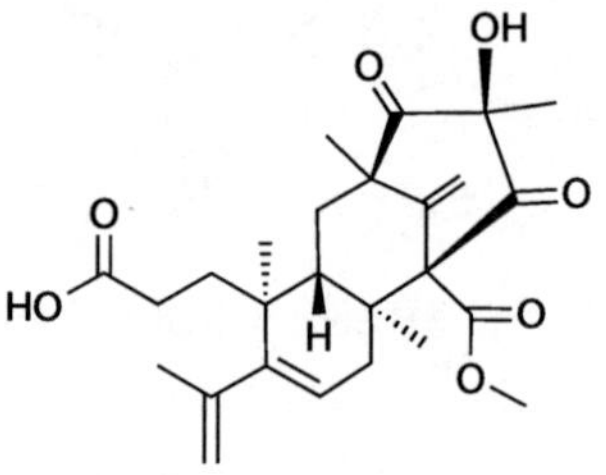

(**42**) Berkeleyone C

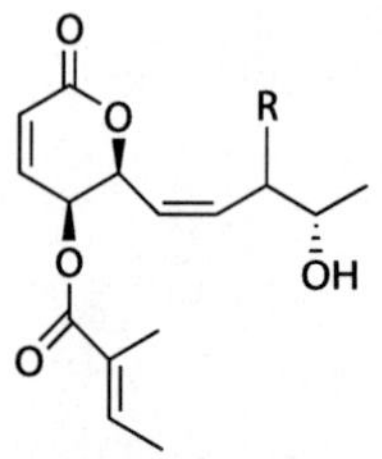

(**43**) R = O phomopsolide A
(**44**) R = OH phomopsolide B

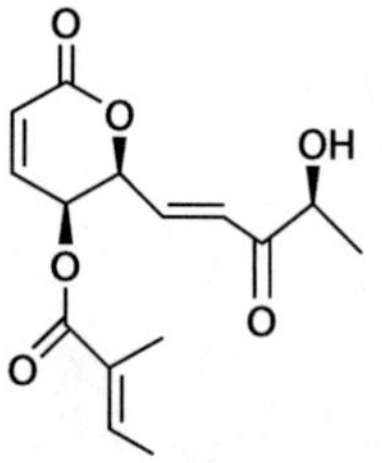

(**45**) Phomopsolide C

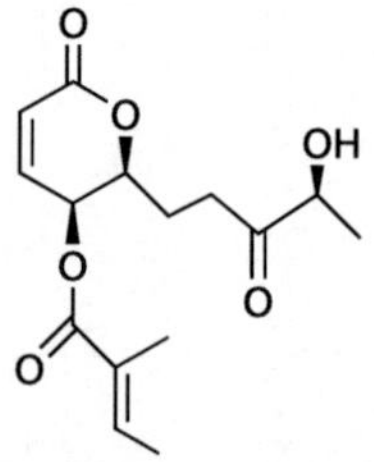

(**46**) Phomopsolide E

FIGURE 10.14 (*Continued*) MMP-3 and caspase-1 inhibitors isolated from Berkeley Pit Lake extremophiles.

(*Continued*)

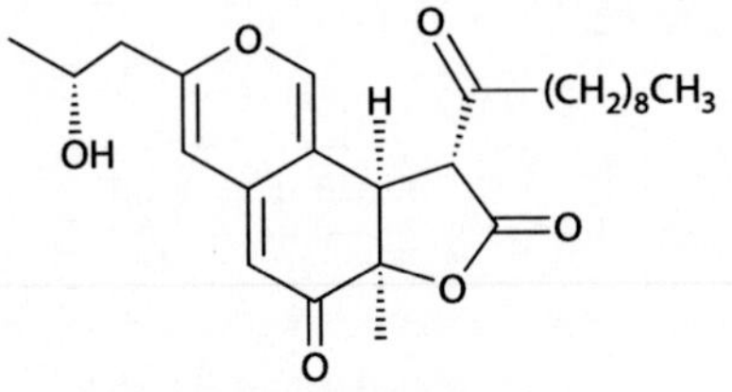

(**47**) Berkchaetoazaphilone A

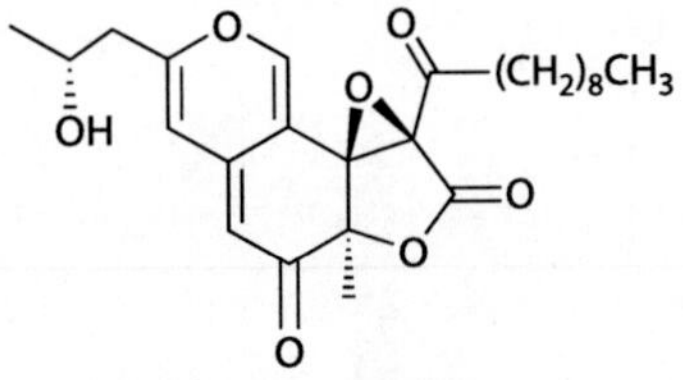

(**48**) Berkchaetoazaphilone B

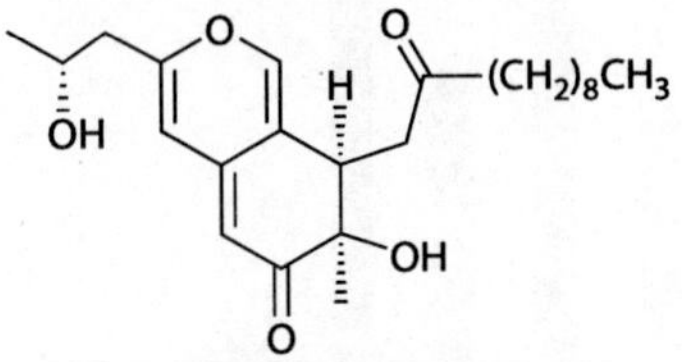

(**49**) Berkchaetoazaphilone C

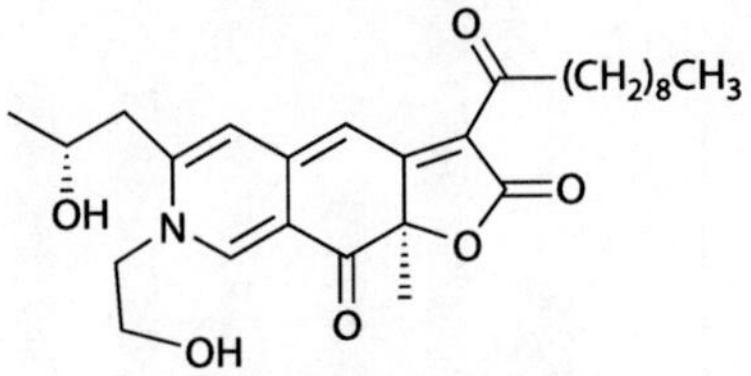

(**50**) Berkchaetorubramine

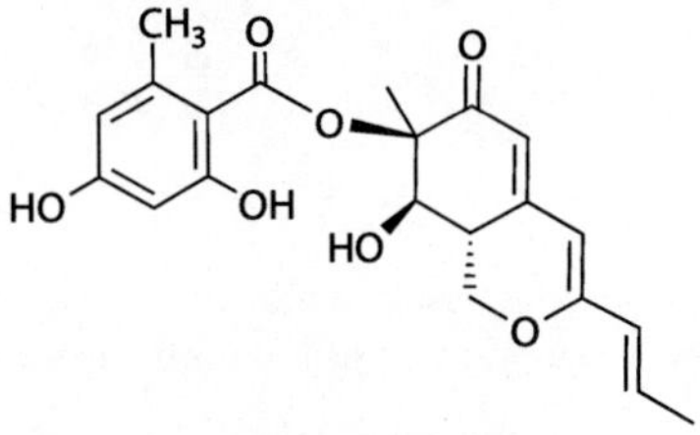

(**51**) Berkazaphilone B

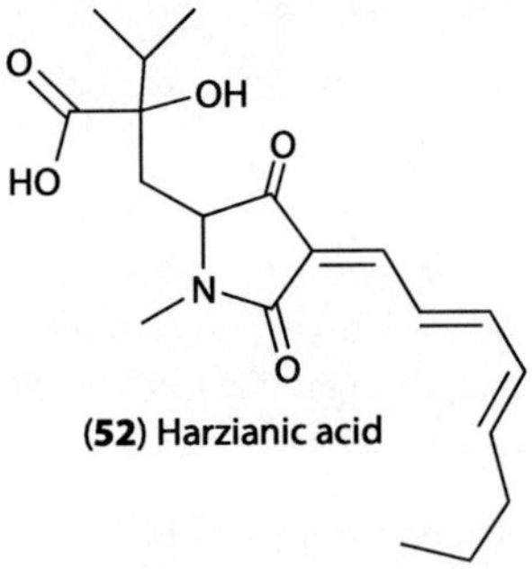

(**52**) Harzianic acid

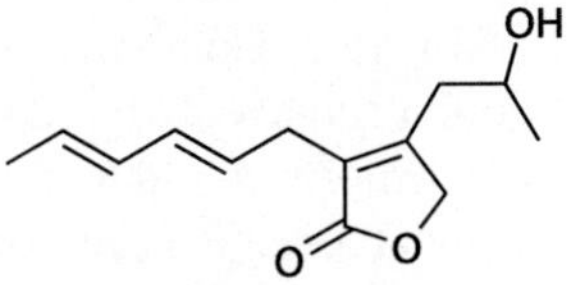

(**53**) Harzianolide

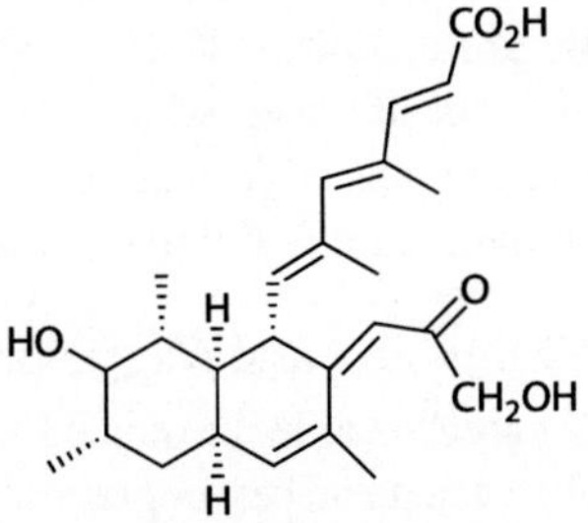

(**54**) 8-2B

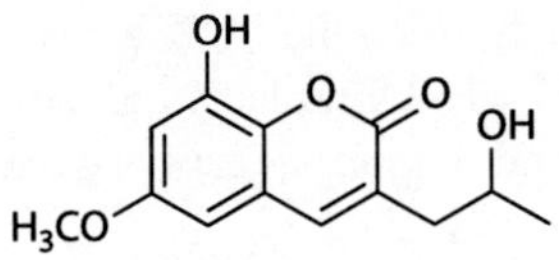

(**55**) BPL66

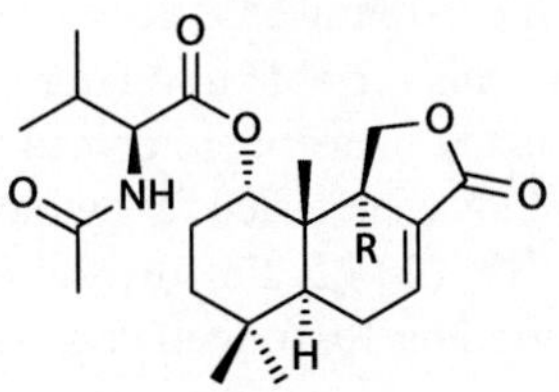

(**56**) berkedrimane A R = H
(**57**) berkedrimane B R = OH

FIGURE 10.14 (*Continued*) MMP-3 and caspase-1 inhibitors isolated from Berkeley Pit Lake extremophiles.

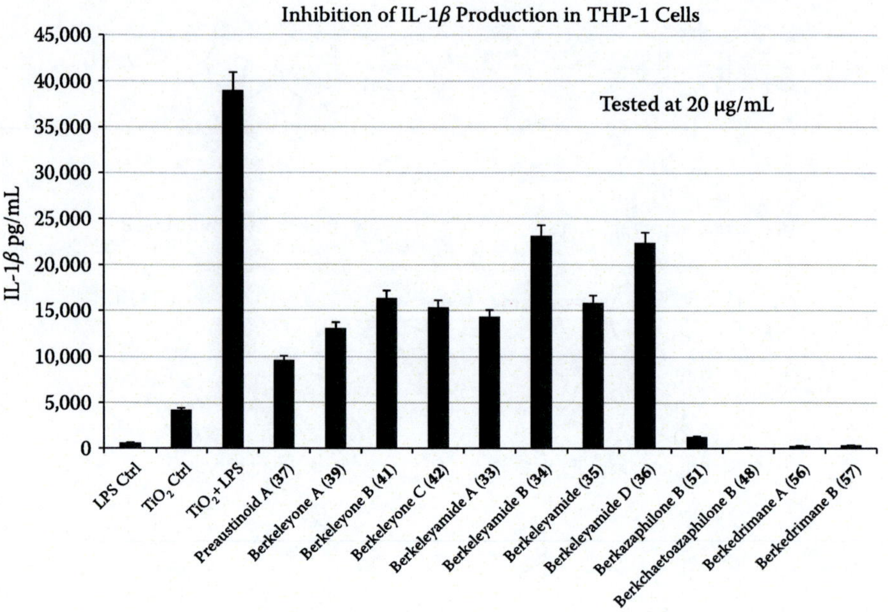

FIGURE 10.15 Results of the induced inflammasome assay demonstrate that MMP-3 and caspase-1 inhibitors can effectively mitigate the production of proinflammatory cytokines in differentiated macrophage-like THP-1 cells.[70]

The most active compound tested to date was berkchaetoazaphilone B (**48**).[133] At a test concentration of 10 μM, it completely inhibited the production of IL-6 and IL-33, and mitigated the production of TNF-α and IL-1β by 95%. Compound **47**, which differs only in the absence of the epoxide, showed similar activity at 100 μM, but no effect at 10 μM. Berkazaphilones B (**51**) and C and berkedrimanes A and B (**56** and **57**) were also effective inhibitors and maintained activity even at lower concentrations.[132]

10.16.2.2 Cell Invasion and Migration Assays and MMP-3

Murdock scholar Lily Apedaile introduced two new cell-based assays into the laboratory this summer: cell invasion and cell migration. These assays were run in Corning Fluorblok 96 multiwell insert plates, which allowed for the plates to be read by a fluorometric plate reader without having to remove the upper well inserts. In the Corning Fluorblok 96 multiwell insert plates, cells are able to migrate to the lower chamber through 8 μM pores. The cell invasion assay plates were coated with Matrigel to simulate the extracellular matrix. All compounds were tested for cytotoxicity using the MTT assay, which uses the tetrazolium dye MTT (3-(4,5-dimethylthiazol-2-yl)-2,5-diphenyltetrazolium bromide). Compound concentrations for the cell invasion and migrations assays were several orders of magnitude below any observed cytotoxicity levels.[134]

Two metastatic human cancer cell lines were selected: MDA-MB-231, which is an aggressive, triple-negative breast cancer (TNBC) that highly expresses MMP-3, and A549, which is a human lung adenocarcinoma. We ran kinetics studies on 13 natural products (compounds were either produced by BPL fungi or by an endophytic fungus of *Sequoia sempervirens*) isolated in the laboratory at a concentration of 125 μM. The potent MMP-3 inhibitor NNGH (*N*-Isobutyl-*N*-(4-methoxy-phenylsulfonyl) glycyl hydroxamic acid) was run at a concentration of 0.65 μM. The cells were starved of fetal bovine serum (FBS), stained with CellTracker Green, and then loaded onto the plate with the compounds. The bottom wells of the plate contained medium with 10% FBS as a chemoattractant. The cells are allowed to incubate at 37°C and 5% CO_2, and readings were taken at 0, 1, 2, 4, 6, 10, and 24 h time points. Compounds with *DNA* designation were compounds from other projects.[134]

Phomopsolide E (**46**) is the most potent inhibitor of cell migration (Figure 10.16), while berkchaetoazaphilone B (**48**) and berkazaphilone B (**51**) inhibit cell invasion

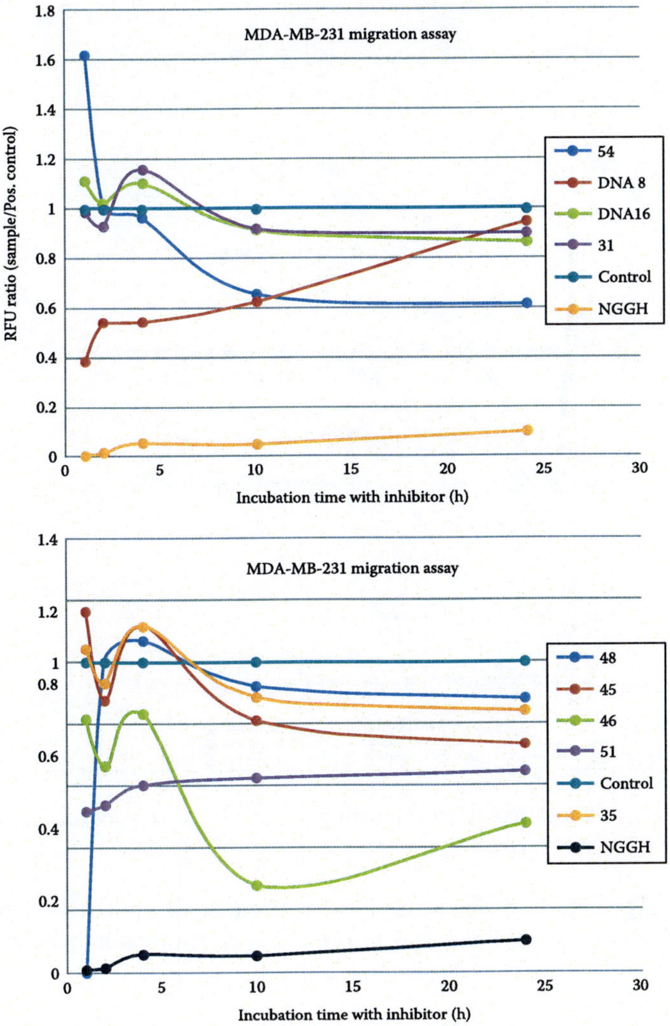

FIGURE 10.16 Cell migration assay comparing the activity of compounds from BPL fungi and the known inhibitor NNGH.

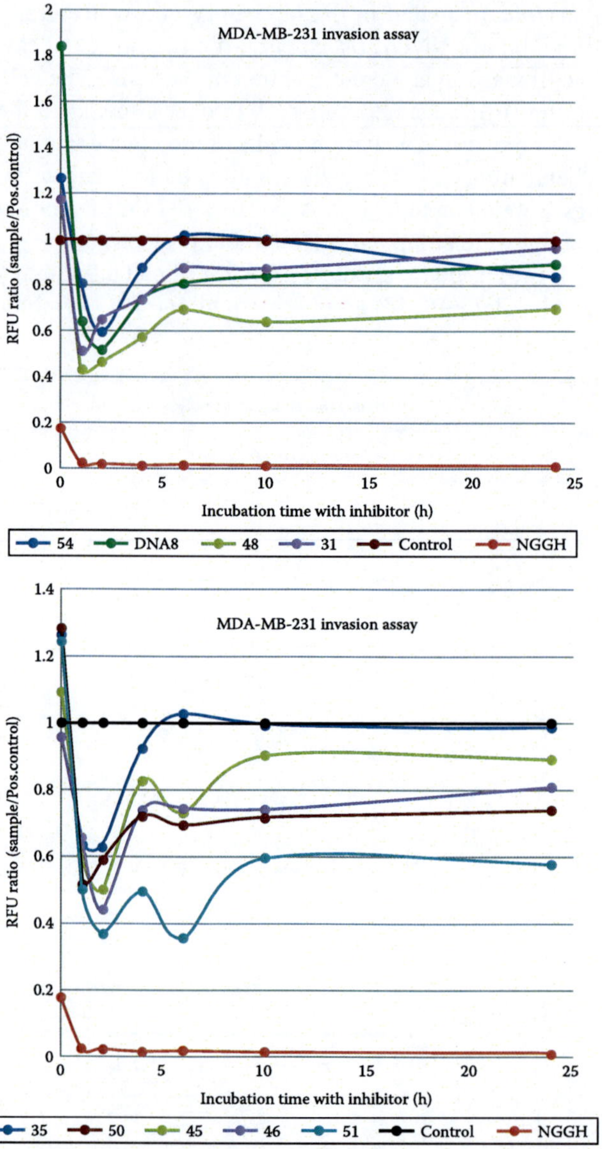

FIGURE 10.17 Cell invasion assay comparing the activity of compounds from BPL fungi and the known inhibitor NNGH.

most effectively (Figure 10.17). All of these compounds exhibit low μM activity against specific cancer cell lines. It is also notable that compounds **48** and **51** were effective inhibitors of cytokine production in the IIA.

10.16.2.3 Oxygen-Glucose Deprivation Reperfusion Assay and Caspase-3

Colleague Darrell Jackson developed a cell-based assay using a mouse N2a neuroblastoma cell line to simulate severe stroke-like conditions. In this assay, the cells are subjected to oxygen-glucose deprivation (OGD) for 60 min. After OGD (60 min), the cells underwent a 4 h reperfusion period in the presence of three different test compounds. Three different caspase-3 inhibitors were selected for this cell-based experiment: BPL66 (**55**), sequoiatone B (**58**), and thiomarinol (**59**). These compounds were selected because they were all potent inhibitors of caspase-3 but differed in their potency toward caspase-1: BPL66 (**55**) inhibited both caspases equally, sequoiatone B (**58**) was a moderate inhibitor of caspase-1, and thiomarinol (**59**) was a weak inhibitor of caspase-1. Sequoiatone B and thiomarinol were not isolated from BPL microbes but were included in this assay as part of a larger search for caspase-3 inhibitors.

Cells were also treated with staurosporine, a strong inducer of caspase-3 expression as a (−) control and with the potent peptidyl inhibitor DEVD-CHO as a (+) control. Following this 60/4 treatment, the cells were analyzed using a caspase-3 inhibition assay (ApoAlert™ Caspase-3 Fluorescent Assay Kit). The kit is designed for fluorescent detection of caspase-3 protease activity in mammalian cells. It uses the substrate conjugate DEVD-AFC (Ac-Asp-Glu-Val-Asp-AFC (AFC = 7-Amino-4-trifluoromethylcoumarin)), which forms a fluorescent product, 7-amino-4-trifluoromethyl coumarin (AFC), when cleaved by caspase-3 in N2a cells. Inhibition of caspase-3 diminishes the production of the fluorescent product. Each of the three compounds completely inhibited caspase-3 when tested at 40 μM. The experiment was repeated with BPL66 at a concentration of 20 μM, and again, caspase-3 was completely inhibited. These preliminary results showed promising trends (Figure 10.18). They demonstrated that all three of these compounds could not only inhibit caspase-3 in a cell-free biochemical assay, but also cross the cell membrane and inhibit caspase-3 *in situ*.

10.16.2.4 Overview of the MMP-3–ROS–Caspase-1 Cycle

In this chapter, we have highlighted some of the key points of the inflammation/EMT cycle, what we have christened the MMP-3–ROS–caspase-1 cycle.

MMP-3 directly promotes EMT and inflammation through the following mechanisms:

- Direct cleavage of E-cadherin[86]
- Activation of MMP-9, which cleaves the basal lamina[92]
- Induced expression of Rac1b, which leads to the formation of ROS and subsequent upregulation of Snail,[86] which suppresses E-cadherin production
- Formation of ROS, which stimulates the assemblage of the NLRP3 inflammasome, which in turn activates caspase-1[54,58]

Caspase-1 promotes inflammation and EMT through the following mechanisms:

- Caspase-1 activates the proinflammatory cytokines IL-1β and IL-18, which leads to inflammation.[54,58]
- Inflammation promotes the expression of MMP-3 and the cycle continues to EMT.[99–102]

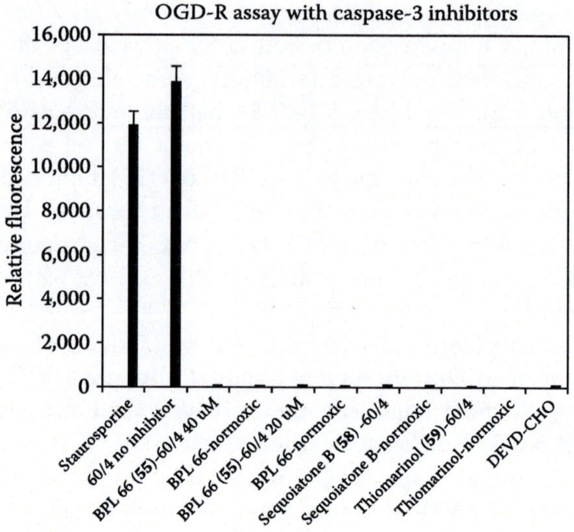

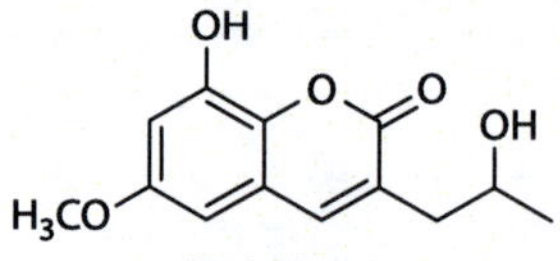

(**55**) BPL66

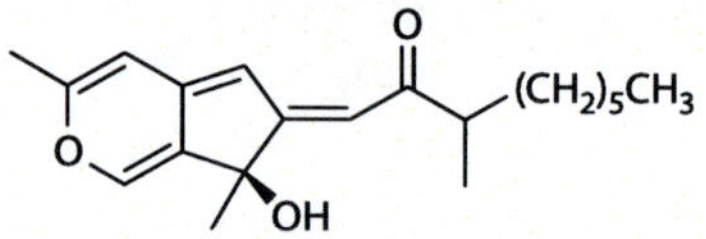

(**58**) Sequoiatone B

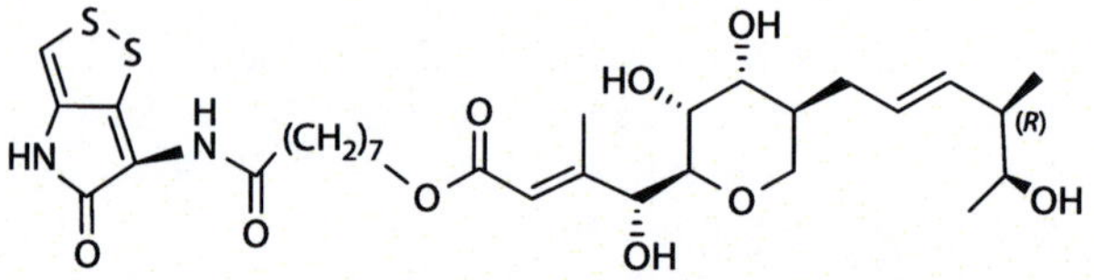

(**59**) Thiomarinol

FIGURE 10.18 Caspase-3 inhibition in murine N2a cells after 60 min of OGD and 4 h reperfusion.

10.17 SELECTED SECONDARY METABOLITES OF BPL FUNGAL EXTREMOPHILES

We have published three review articles that discuss many of the active compounds isolated to date, the BPL fungus from which they were derived, and their biological activities.[40,130,131] Only a fraction of the BPL microbes have been thoroughly studied: a *Pithomyces* sp., a *Penicillium* sp., a deep-water isolate of *P. rubrum*, *P. solitum*, *P. clavigerum*, *Pleurostomophora* sp., *Oidiodendrum tenuissimum*,

Trichoderma virens, and the actinomycete *Streptomyces griseoplanus*. Many of the compounds reported were isolated from iterative fermentations of four different fungi: an alga-associated *Penicillium* sp., *P. solitum*, *P. clavigerum*, and *P. rubrum*. We have long observed that a single organism can produce a diverse array of secondary metabolites if the culture conditions or duration of fermentation is varied even slightly, which has indeed been the case with BPL fungi.

One of our most prolific fungi is the deep-water isolate *Penicillium rubrum*. In culture, it produces berkeleydione (**31**) and berkeleytrione (**32**),[41] the berkeleyacetals,[42] the berkeleyamides,[43] the berkeleyones,[44] and the berkazaphilones.[133]

The two- and three-dimensional (3D) structures of our collection of natural product inhibitors of caspase-1, caspase-3, and MMP-3 have been carefully determined through spectroscopic and x-ray crystallographic methods. Each of these molecules was isolated from complex organic extracts of microbial fermentations using enzyme inhibition assays to direct each step of the chromatographic separations. Each pure compound was tested to establish IC_{50} values for enzyme inhibition. Active compounds were then evaluated in orthogonal cell line assays intended to validate the biochemical assays. These assays include the IIA to assess anti-inflammatory potential, cancer cell line screening at the NCI Developmental Therapeutics Program (NCI-DTP) and the Memorial Sloan Kettering Cancer Center (MSKCC), and OGD studies of neuroblastoma cells to assess the downregulation of caspase-3. We have recently incorporated cell invasion and cell migration assays into the protocol, and are currently testing compounds against three metastatic cancer lines.

10.17.1 Enzyme Inhibitors and Inactive Analogs Produced by Berkeley Pit Microbes

We have previously reported the isolation and characterization of several compounds that were isolated by enzyme inhibition–guided purification from crude fungal extract to pure compound.[40–44,130–133] For every fungal extract studied, ^{1}H-NMR spectra are obtained for both active and inactive column fractions, and once active compounds are fully characterized, the NMR data are scrutinized for compounds with similar chemotypes. These compounds are also isolated, characterized, and tested for activity. Closely related but inactive natural analogs provide important SARs that facilitate the development of semisynthetic analogs with greater potency and/or selectivity.

10.17.2 X-Ray Crystallographic Studies of Inhibitor–Protein Interactions

At this point in the research, we are assessing the interaction of each inhibitor with its target enzyme and will highlight a few of these compounds that are currently being examined as potential scaffolds for analog development. Although molecular modeling is a very useful tool in the exploration of ligand–protein interaction, it does not generate actual data. One of the true gold standards for assessing this interaction is X-ray analysis of the crystalline ligand–protein complex. High-quality protein crystals are required to study the coordination of various ligands with a target protein at the atomic level to facilitate structure-aided drug design.

Colleague T.C. Mou designed two N-terminal hexahistidine-tagged constructs of MMP-3 that encompassed either the catalytic domain alone (residues 100–272), or both the catalytic domain and its C-terminal HPEX domain (residues 100–477). MMP-3 was overexpressed in the *Escherichia coli* strain BL21(DE3)-pLys. The cells were harvested, lysed, and cleared by ultracentrifugation. The supernatant was subjected to iterative chromatographic separations and pure MMP-3 protein (catalytic domain only) was collected. Crystallization trays were set up by the hanging-drop and sitting-drop vapor diffusion methods at 4°C. Rod-shaped crystals were observed after a few weeks and growth was continued for 4 weeks before harvest. For ligand-bound MMP-3 structures, MMP-3 crystals were soaked in 1–10 mM ligands for 2–5 days before harvest. Crystals were passed through a cryoprotectant with 20% glycerol in reservoir solution before being flash-frozen in liquid nitrogen.

The X-ray datasets of MMP-3 catalytic domain crystals with or without ligand soaking were collected to 1.35–1.9 Å resolutions at Stanford Synchotron Radiation Lightsource (SSRL) 14-1 and 7-1 beamlines. Diffraction data were indexed, integrated, and scaled with the program HKL2000.[135] The structure was solved by molecular replacement using the ligand-free structure of MMP-3 (PDB code: 1CQR)[136] as a search model for phasing the data. A clear solution comprising two molecules in one asymmetric unit was obtained. The structure was further subjected to multiple rounds of refinements in PHENIX[137] and manually inspected electron density fitting using the program COOT[138] before the final structure of the MMP-3 structures were modeled. The crystals were soaked in a solution of berkchaetoazaphilone B (**48**) for several days and then reexamined. The ligand-bound crystal exhibited a positional shift of Tyr223 typical of inhibitor binding (as shown in PDB code: 1B3D).[136] Unfortunately, ligand density was not observed in the active site.

This same methodology is also being pursued for both caspase-1 and caspase-3. At this point, the enzymes have been successfully expressed in *E. coli* cultures and are now being purified and crystallized. Although analog development can proceed without 3D crystallographic studies using traditional medicinal chemistry methodologies, the data provided by these studies would greatly enhance our efforts. Once a target is clearly defined, molecular modeling can be used to assess analogs within the target site.

10.18 MMP-3 AND CASPASE-1 INHIBITORS

10.18.1 Berkelic Acid and Berkebisabolanes from a Green Alga–Associated *Penicillium* sp.

Several different families with related scaffolds have been isolated using MMP-3 or caspase-1 inhibition assays (Figure 10.14). These include berkelic acid (**27**), which was isolated from the $CHCl_3$ extract of a *Chlorella mutabilis*–associated *Penicillium* sp. grown in PDB (still culture, 21 days).[139] In our hands, it inhibited both MMP-3 (1.87 μM) and caspase-1 (98 μM) in the assay systems, and the NCI-DTP determined that it selectively inhibited OVCAR-3 (91 nM).

The stereoconfiguration of berkelic acid generated some disagreement, however. When Furstner initiated his synthesis of berkelic acid, he found that the nuclear Overhauser effect data they observed for their precursor molecules were inconsistent with our data.[140] He also indicated that the originally proposed relative stereoconfiguration did not represent a thermodynamic minimum because of a key syn-periplanar interaction between the C-25 methyl substituent and the C-16 methylene group. Consequently, he inverted the configuration of our proposed structure at both C-18 and C-19, and presented a revised structure of berkelic acid. All subsequent syntheses of berkelic acid in other laboratories conformed to this revised structure.[140] Unfortunately, none of these synthetic molecules were active in either enzyme or cell line assays.

Total synthesis has not simplified the berkelic acid story. In NCI-DTP's hands, the natural product was selective and active, yet the synthetic products are neither. It is possible that a minor contaminant of the natural product was actually responsible for the observed activity. This is always a potential problem with compound purification—either from natural or from synthetic sources. It is also possible that the actual structure of berkelic acid is the less thermodynamically stable molecule we originally proposed. Unfortunately, despite repeated attempts to repeat the fermentation protocol that yielded berkelic acid, we have not been able to detect any trace of berkelic acid in extracts of *Penicillium* sp.

The same isolate of *Penicillium* that produced berkelic acid also produced berkebisabolanes A–C (**28–30**), a series of sesquiterpenes when grown in PDBH+.[141] All three berkebisabolanes showed MIC_{50} values in the 30 μM range against caspase-1 and in the 300 nM range against MMP-3.

10.18.2 Meroterpenes from a Deep-Water Isolate of *Penicillium rubrum*

The $CHCl_3$ extract of a *P. rubrum* isolate collected from a depth of 275 m and grown as a 21-day still culture in PDBH+ (pH 2.7) inhibited both MMP-3 and caspase-1. Assay-guided fractionation yielded two different families of meroterpenes: the berkeleyones, including berkeleydione (**31**) and berkeleytrione (**32**),[41] berkeleyones A–C (**39**, **41**, **42**)[44] and the known compounds preaustinoids A (**37**) and A1 (**38**),[44] and berkeleyacetals (e.g., **40**) (Figure 10.14).[42]

Berkeleydione (**31**) and berkeleytrione (**32**)[41] effectively inhibited both MMP-3 and caspase-1 in the micromolar range. Berkeleydione (**31**) was tested in the NCI's antitumor screen against 60 human cell lines. It showed selective activity toward non-small-cell lung cancer NCI-H460, with a $\log_{10}$ GI_{50} of 398 nM. This extreme selectivity is noteworthy in a natural product that has not been derivatized or tailored toward a particular cancer type.

Berkeleyacetal C (**40**) inhibited both MMP-3 and caspase-1 in the micromolar range, while berkeleyacetals A and B inhibited these enzymes in the millimolar range.[42] Berkeleyacetal C was accepted in the NCI-DTP human cell line assay and was tested in a single dose–response assay. It inhibited the growth of non-small-cell lung cancer NCI H460 (IC_{50} = 10 μM) as well as all of the leukemia cell lines (Table 10.1).[42] It was also active against A2058 and MES-SA, with IC_{50} values of 8.6 and 6.4 μM, respectively.

TABLE 10.1
Berkeley Pit Microbes with MMP-3 Inhibition Data and Activity against Human Cancer Cell Lines

	Compounds	MMP-3	Cancer Cell Line	IC_{50}/GI_{50}
27	Berkelic acid	2 µM	OVCAR-3	91 nM
28	Berkebisabolane A	280 nM	NT	
29	Berkebisabolane B	315 nM	NT	
30	Berkebisabolane C	340 nM	NT	
31	Berkeleydione	22 µM	NCI 460	398 nM
32	Berkeleytrione	180 µM	NA	
40	Berkeleyacetal C	25 µM	NCI 460	10 µM
	Berkeleyacetal C		A2058	8.6 µM
	Berkeleyacetal C		MES-SA	6.4 µM
43	Phomopsolide A	100 µM	HeLaS3	5.0 µM
	Phomopsolide A		UM-UC-3	4.8 µM
	Phomopsolide A		A2058	2.53 µM
	Phomopsolide A		MES-SA	2.16 µM
44	Phomopsolide B	>200 µM	UM-UC-3	6.1 µM
	Phomopsolide B		A2058	5.53 µM
45	Phomopsolide C	100 µM	HeLaS3	3.6 µM
	Phomopsolide C		UM-UC-3	1.2 µM
	Phomopsolide C		Y79	1.4 µM
	Phomopsolide C		A2058	6.1 µM
	Phomopsolide C		MES-SA	4.74 µM
46	Phomopsolide E	>200 µM	HeLaS3	7.3 µM
	Phomopsolide E		UM-UC-3	6.6 µM
	Phomopsolide E		A2058	2.91 µM
	Phomopsolide E		MES-SA	2.86 µM
47	Berkchaetoazaphilone A	150 µM	NA	
48	Berkchaetoazaphilone B	6 µM	Y79	1.1 µM
	Berkchaetoazaphilone B		SR, MOLT-4	10 µM
	Berkchaetoazaphilone B		CCRF-CEM	10 µM
	Berkchaetoazaphilone B		RPMI-8226	10 µM
	Berkchaetoazaphilone B		LOX IMIV	10 µM

Human Cancer Cell Line

	NCI-DTP		MSKCC
OVCAR-3	Ovarian adenocarcinoma	HeLaS3	Cervical adenocarcinoma
NCI-460	Non-small-cell (NSC) lung cancer	UM-UC-3	Bladder carcinoma
CCRF-CEM	Leukemia cell line	Y79	Retinoblastoma
RPMI-8226	Leukemia cell line		**Eisai**
MOLT-4, SR	Leukemia cell line	A2058	Metastatic skin cancer
LOX IMVI	Melanoma	MES-SA	Uterine sarcoma

10.18.3 Azaphilones from *Pleurostomophora* sp.

The three novel berkchaetoazaphilones A–C (**47–49**) as well as berkchaetorubramine (**50**) were isolated from *Pleurostomophora* sp (Figure 10.14). Compound **47** is a moderately active MMP-3 inhibitor, while B (**48**) is a more potent inhibitor of MMP-3, and has demonstrated micromolar activity against human retinoblastoma, leukemia, and melanoma cell lines (Table 10.1).[136]

10.18.4 Phomopsolides from an Alga-Associated *Penicillium* sp.

Although the isolation of the phomopsolides was guided by MMP-3 inhibition assays, they are only moderate inhibitors. However, they have exhibited micromolar activity against several human cancer cell lines when tested by colleague Hakim Djaballah at the MSKCC. Phomopsolides A (**43**) and C (**45**) are E–Z stereoisomers, and E (**46**) is the dihydro-analog of **45**, but all three compounds demonstrated comparable cytotoxicity/antiproliferative activity, as shown in Table 10.1. Clearly, this activity is not based solely on their abilities to act as moderate MMP-3 inhibitors. In the cell migration assay (Figure 10.17), phomopsolide E (**46**) was the most potent inhibitor of cell migration (A was not available for testing). It was also a more potent inhibitor of cell invasion than C (**45**) (Figure 10.18), even though both compounds were similarly antiproliferative against A2058, a highly metastatic skin cancer.

10.19 USE OF SAR INFORMATION TO DIRECT THE SYNTHESIS OF MORE POTENT ANALOGS

The production of secondary metabolites by higher plants, fungi, bacteria, or any other source generally offers first-order SARs. Most of the compounds produced by these microbes were produced as part of a series of closely related analogs. Enzyme inhibition assays were used to guide the isolation of bioactive metabolites. After the bioactive compounds were isolated and the structures were determined, *NMR-guided* isolation was used to purify compounds with similar structural characteristics that were inactive (or significantly less active). These compounds constitute a series, and different members of that series differ in their bioactivity profiles. The SAR evaluation can provide insights into specific structural features that are responsible for the observed activity. They can also assist in the synthesis of new analogs with enhanced biological activity.

Molecular computational methods will also facilitate these studies. Each enzyme inhibitor is modeled in the *catalytic domain* of well-defined crystal structures of the target enzyme. These carefully refined models are hypothesis generators that indicate (a) ligand structural features that confer high inhibition potency and/or selectivity, (b) protein–ligand interaction sites, and (c) ligand modifications that enhance interaction within these sites. These data will help us design and synthesize analogs that will be tested for enhanced enzyme inhibitory activities. They will then be assessed in the appropriate cell-based assays. Results from these functional assays will directly test SAR hypotheses generated by computational and molecular modeling efforts, allowing for a refinement of the models and guiding further iterations of organic synthesis of novel compounds.

10.20 CONCLUSIONS AND FUTURE DIRECTIONS

Although our research coupling inflammation and EMT pathways and determining the ability of caspase-3 inhibitors to minimize poststroke brain damage is at an early stage, these studies have already demonstrated that the microbes of BPL are a valuable source of new and interesting secondary metabolites. The identification of drug-like molecules and therapeutic leads often begins with chemical or biological screening of compound libraries. Natural product libraries isolated from bacteria, fungi, higher plants, and assorted marine invertebrates represent a vast and diverse source of lead compounds with diverse biological activities. Translating promising natural product leads into therapeutically relevant molecules can be impeded by their scarcity, complexity, and nonoptimal pharmacokinetic characteristics. This collection of BPL extremophile metabolites will provide the basic scaffolds for the synthesis of new analogs for further study and development.

It is not often that scientists have the opportunity to explore such a unique environment. Based on our previous studies, *bioprospecting* in the Berkeley Pit will continue to yield interesting new chemistry that will provide novel insights into how two important life processes, inflammation and EMT, interact. Some of these insights could lead to the development of new drugs that could mitigate inflammation and the array of pathologies caused by chronic inflammation. They could also lead to drugs that could block metastatic processes and mitigate the mortality associated with metastatic diseases. Or they could lead to drugs that could protect stroke patients from the unnecessary collateral damage associated with upregulated caspase-3.

A collection of small molecules to probe these connections will be generated, which will serve as pharmacophores for analogs with enhanced potency. They will provide (a) intriguing targets for chemical synthesis, (b) excellent probes for the specific pathways they inhibit, and (c) potential pharmaceutical agents. We have not proven our speculation that the low pH, high metal concentration, and oxidizing environment of BPL have actually selected for the production of inhibitors of caspases and metalloproteinases. Indeed, we have not even begun to test this hypothesis directly. However, as for the secondary metabolites and their microbial producers, they could be the richest products ever mined from *the richest hill on earth.*

REFERENCES

1. Gaynor, M.; Mankin, A. S. Macrolide antibiotics: Binding site, mechanism of action, resistance. *Curr. Top. Med. Chem.* **2003**, *3*, 949–961.
2. Hofle, G.; Bedorf, N.; Steinmetz, H.; Schomburg, D.; Gerth, K.; Reichenbach, H. Epothilone A and B—Novel 16-membered macrolides with cytotoxic activity: Isolation, crystal structure, and conformation in solution. *Angew. Chem.* **1996**, *35*, 1567–1569.
3. Maiese, W. M.; Lechevauert, M. P.; Lechevalier, H. A.; Korshalla, J. Calicheamicins, a novel family of antitumor antibiotics: Taxonomy, fermentation and biological properties. *J. Antibiot.* **1989**, *42*, 558–563.
4. Stierle, A.; Strobel, G.; Stierle, D. Taxol and taxane production by *Taxomyces andreanae* an endophytic fungus of Pacific yew. *Science* **1993**, *260*, 214–216.

5. Zhao, J.; Zhou, L.; Wang, J.; Shan, T.; Zhong, L.; Liu, X.; Gao, X. Endophytic fungi for producing bioactive compounds originally from their host plants. In *Current Research, Technology and Education Topics in Applied Microbiology and Microbial Biotechnology*, Mendez-Vilas, A. (Ed.), Vol. 1, Formatex Research Center, Badajoz, Spain, **2012**, pp. 567–576.
6. Chandra, S. Endophytic fungi: Novel sources of anticancer lead molecules. *Appl. Microbiol. Biotechnol.* **2012**, *95*, 47–59.
7. Kharwar, R. N.; Mishra, A.; Gond, S. K.; Stierle, A.; Stierle, D. Anticancer compounds derived from fungal endophytes: Their importance and future challenges. *Nat. Prod. Rep.* **2011**, *28*(7), 1208–1228.
8. Stierle, D.; Stierle, A. Bioactive compounds from four endophytic *Penicillium* sp. isolated from the northwest Pacific yew tree. In *Bioactive Natural Products*, Atta-Ur-Rahman (Ed.), Vol. 24, Elsevier Science Publishers, Amsterdam, the Netherlands, **2000**, pp. 933–978.
9. Montana Bureau of Mines and Geology. http://www.mbmg.mtech.edu/env-berkeley.html, accessed August 15, 2011.
10. Duaime, T. E. Long term changes in the limnology and geochemistry of the Berkeley Pit Lake, Butte Montana. *Mine Water Environ.* **2006**, *25*, 76–85.
11. Nordstrom, D. K.; Southam, G. Geomicrobiology of sulfide mineral oxidation. In *Reviews in Mineralogy*, Banfield, J. F.; Nealson, K. H. (Eds.), Vol. 35, Mineralogical Society of America, Washington, DC, **1997**, pp. 361–390.
12. Crutzen, P. J.; Stoermer, E. F. The "Anthropocene". *Global Change Newslett.* **1997**, *41*, 12–13.
13. Brock, T. D.; Freeze, H. *Thermus aquaticus* gen. n. and sp. n., a nonsporulating extreme thermophile. *J. Bacteriol.* **1969**, *98*, 289–297.
14. Mitman, G. G. Algal bioremediation of the Berkeley Pit. *J. Phycol.* **2000**, *36*, 49.
15. Adams, D. **1995**. Did toxic stew cook the goose? https://www.hcn.org/issues/49/1520, accessed March 10, 2015.
16. MacElroy, R. D. Some comments on the evolution of extremophiles. *Biosystems* **1974**, *6*, 74–75.
17. Barnes, K. A. **2013**. Extreme life. https://www.kevinabarnes.com/extreme-life, accessed March 10, 2015.
18. Giddings, L. A.; Newman, D. Bioactive compounds from terrestrial extremophiles. In *Springer Briefs in Microbiology*, Tiquia-Arashiro, S. M.; Mormile, M. (Eds.), Springer, New York, **2015**, pp. 2–4.
19. Saiki, R. K.; Gelfand, D. H.; Stoffel, S.; Scharf, S. J.; Higuchi, R.; Horn, G. T.; Mullis, K. B.; Erlich, H. A. Primer-directed enzymatic amplification of DNA with a thermo-stable DNA polymerase. *Science* **1988**, *239*(4839), 487–491.
20. Mullis, K. B. The unusual origin of the polymerase chain reaction. *Sci. Am.* **1990**, *262*, 56–61.
21. Pace, N. R. A molecular view of microbial diversity and the biosphere. *Science* **1997**, *276*, 734–740.
22. Kashefi, K.; Lovley, D. R. Extending the upper temperature limit for life. *Science* **2003**, *301*, 934.
23. Stetter, K. O. Extremophiles and their adaptation to hot environments. *FEBS Lett.* **1999**, *452*, 22–25.
24. Huber, H.; Hohn, M. J.; Rachel, R.; Fuchs, T.; Wimmer, V. C.; Stetter, K. O. A new phylum of Archaea represented by a nanosized hyperthermophilic symbiont. *Nature* **2002**, *417*, 63–67.
25. Pettit, R. K. Culturability and secondary metabolite diversity of extreme microbes: Expanding contribution of deep sea and deep-sea vent microbes to natural product discovery. *Mar. Biotechnol.* **2011**, *13*, 1–11.

26. Thornburg, C. C.; Zabriskie, M.; McPhail, K. L. Deep-sea hydrothermal vents: Potential hot spots for natural products discovery? *J. Nat. Prod.* **2010**, *73*, 489–499.
27. Andrianasolo, E. H.; Haramaty, L.; Rosario-Passapera, R.; Bidle, K.; White, E.; Vetriani, C.; Falkowski, P.; Lutz, R. Ammonificins A and B, hydroxyethylamine chroman derivatives from a cultured marine hydrothermal vent bacterium *Thermovibrio ammonificans*. *J. Nat. Prod.* **2009**, *72*, 1216–1219.
28. Andrianasolo, E. H.; Haramaty, L.; Rosario-Passapera, R.; Vetriani, C.; Falkowski, P.; White, E.; Lutz, R. Ammonificins C and D, hydroxyethylamine chromene derivatives from a cultured marine hydrothermal vent bacterium *Thermovibrio ammonificans*. *Mar. Drugs* **2012**, *10*, 2300–2311.
29. Homann, V. V.; Sandy, M.; Tincu, A. J.; Templeton, A. S.; Tebo, B. M.; Butler, A. Loihichelins A-F, a suite of amphiphilic siderophores produced by the marine bacterium *Halomonas* LOB-5. *J. Nat. Prod.* **2009**, *72*, 884–888.
30. Boone, D. R.; Liu, Y.; Zhao, Z. J.; Balkwill, D. L.; Drake, G. R.; Stevens, T. O.; Aldrich, H. C. *Bacillus infernus* sp. *nov*., an Fe (III)- and Mn (IV)-reducing anaerobe from the deep terrestrial subsurface. *Int. J. Syst. Evol. Microbiol.* **1995**, *45*, 441–448.
31. Cavicchioli, R.; Thomas, T. Extremophiles. In *Encyclopaedia of Microbiology*, Lederberg, J. (Ed.), Vol. 2, Academic Press Inc, San Diego, CA, **2000**, pp. 317–337.
32. Roadcap, G.; Bethke, C. M.; Sanford, R. A.; Pardinas, J.; Qusheng, J. Microbial community found thriving in very alkaline (pH 12–13) groundwater. *Geol. Soc. Am. Abstr.* **2003**, *35*, 379.
33. Park, H. B.; Kim, Y. J.; Lee, J. K.; Lee, K. R.; Kwon, H. C. Spirobacillenes A and B, unusual spirocyclopentenones from *Lysinibacillus fusiformis* $KMCOO_3$. *Org. Lett.* **2012**, *14*, 5002–5005.
34. He, J.; Roemer, E.; Lange, C. et al. Structure, derivatization, and antitumor activity of new griseusins from *Nocardiopsis* sp. *J. Med. Chem.* **2007**, *50*, 5168–5175.
35. Ding, Z. G.; Li, M. G.; Zhao, J. Y.; Ren, J.; Huang, R.; Xie, M. J.; Cui, X. L.; Zhu, H. J.; Wen, M. L. Naphthospironone A: An unprecedented and highly functionalized polycyclic metabolite from an alkaline mine waste extremophile. *Chem. Eur. J.* **2010**, *16*, 3902–3905.
36. Ding, Z. G.; Zhao, J. Y.; Li, M. G.; Huang, R.; Li, Q. M.; Cui, X. L.; Zhu, H. J.; Wen, M. L. Griseusins F and G, spiro-naphthoquinones from a tin mine tailings-derived alkalophilic *Nocardiopsis* species. *J. Nat. Prod.* **2012**, *75*, 1994–1998.
37. Wang, X.; Shaaban, K. A.; Elshahawi, S. I. et al. Frenolicins C-G, pyranonaphthoquinones from *Streptomyces* sp. RM-4-15. *J. Nat. Prod.* **2013**, *76*, 1441–1447.
38. Sánchez-Andrea, I.; Stams, A. J.; Amils, R.; Sanz, J. L. Enrichment and isolation of acidophilic sulfate-reducing bacteria from Tinto River sediments. *Environ. Microbiol. Rep.* **2013**, *5*, 672–678.
39. Sen, A.; Johnson, B. Acidophilic sulphate-reducing bacteria: Candidates for bioremediation of acid mine drainage. *Process Metall.* **1999**, *9*, 709–718.
40. Stierle, A. A.; Stierle, D. B. Bioprospecting in the Berkeley Pit: Bioactive metabolites from acid mine waste extremophiles. In *Bioactive Natural Products*, Atta-Ur-Rahman (Ed.), Vol. 32, Elsevier Science, Amsterdam, the Netherlands, **2005**, pp. 1123–1175.
41. Stierle, D. B.; Stierle, A. A.; Hobbs, J. D.; Stokken, J.; Clardy, J. Berkeleydione and berkeleytrione, new bioactive metabolites from an acid mine organism. *Org. Lett.* **2004**, *6*, 1049–1052.
42. Stierle, D. B.; Stierle, A. A.; Patacini, B. The berkeleyacetals, three meroterpenes from a deep water acid mine waste *Penicillium*. *J. Nat. Prod.* **2007**, *70*, 1820–1823.
43. Stierle, A. A.; Stierle, D. B.; Patacini, B. The berkeleyamides: Four new amides from *Penicillium rubrum*, a deep water acid mine waste fungus. *J. Nat. Prod.* **2008**, *71*, 856–860.

44. Stierle, D.; Stierle, A.; Patacini, B.; McIntyre, K.; Girtsman, T.; Bolstad, E. Berkeleyones and related meroterpenes from a deep water acid mine waste fungus that inhibit the production of interleukin 1-β from induced inflammasomes. *J. Nat. Prod.* **2011**, *74*, 2273–2277.
45. Visagie, C. M.; Houbraken, J.; Frisvad, J. C.; Hong, S. B.; Klaassen C. H. W.; Perrone, G.; Seifert, K. A.; Varga, J.; Yaguchi, T.; Samson, R. A. Identification and nomenclature of the genus *Penicillium*. *Stud. Mycol.* **2014**, *78*, 343–371.
46. Liu, K. L.; Porras-Alfredo, A.; Kiske, C. R.; Eichorst, S. A.; Xie, G. Accurate, rapid taxonomic classification of fungal large-subunit rRNA genes. *Appl. Environ. Microbiol.* **2012**, *78*, 1523–1533.
47. Ambrose, C.; Christapher, V.; Bhore, S. J. Endophytic bacteria as a source of novel antibiotics: An overview. *Pharmacogn. Rev.* **2013**, *7*, 11–16.
48. Prazeres dos Santos, I.; Cláudio Nascimento da Silva, L.; Vanusa da Silva, M.; Magali de Araújo, J.; da Silva Cavalcanti, M.; Lucia de Menezes Lima, V. Antibacterial activity of endophytic fungi from leaves of *Indigofera suffruticosa* Miller (Fabaceae). *Front. Microbiol.* **2015**, *6*, 350–356.
49. Vieira, M. L.; Johann, S.; Hughes, F. M.; Rosa, C. A.; Rosa, L. H. The diversity and antimicrobial activity of endophytic fungi associated with medicinal plant *Baccharis trimera* (Asteraceae) from the Brazilian savannah. *Can. J. Microbiol.* **2014**, *12*, 847–856.
50. Mousa, W. K.; Raizada, M. N. The diversity of anti-microbial secondary metabolites produced by fungal endophytes: An interdisciplinary perspective. *Front. Microbiol.* **2013**, *4*, 65.
51. Kato, Y.; Ozawa, S.; Miyamoto, C.; Maehata, Y.; Baba, Y. Acidic extracellular micro environment and cancer. *Cancer Cell Int.* **2013**, *13*, 89–96.
52. Imlay, J. A.; Chin, S. M.; Linn, S. Toxic DNA damage by hydrogen peroxide through the Fenton reaction *in vivo* and *in vitro*. *Science* **1988**, *240*, 640–642.
53. Liou, G.-Y.; Storz, P. Reactive oxygen species in cancer. *Free Radic. Res.* **2010**, *44*, 1–20.
54. Mittal, M.; Siddiqui, M. R.; Tran, K.; Reddy, S. P.; Malik, A. B. Reactive oxygen species in inflammation and tissue injury. *Antioxid. Redox Signal* **2014**, *20*, 1126–1167.
55. Warburg, O.; Posener, K.; Negelein, E. Über den stoffwechsel der tumoren (On metabolism of tumors). *Biochem. Z.* **1924**, *152*, 319–344.
56. Sanders, T. J.; McCarthy, N. E.; Giles, E. M.; Davidson, K. L.; Haltalli, M. L.; Hazell, S.; Lindsay, J. O.; Stagg, A. J. Increased production of retinoic acid by intestinal macrophages contributes to their inflammatory phenotype in patients with Crohn's disease. *Gastroenterology* **2014**, *146*, 1278–1288.
57. Smith, D. G.; Guillemin, G. J.; Pemberton, L.; Kerr, S.; Nath, A.; Smythe, G. A.; Brew, B. J. Quinolinic acid is produced by macrophages stimulated by platelet activating factor Nef and Tat. *J. Neurovirol.* **2001**, *7*, 56–60.
58. Davis, B. K.; Ting, J. P.-Y. NLRP3 has a sweet tooth. *Nat. Immunol.* **2010**, *11*, 105–106.
59. Ghayur, T.; Banerjee, S.; Hugunin, M. et al. Caspase-1 processes IFN-gamma-inducing factor and regulates LPS-induced IFN-gamma production. *Nature* **1997**, *386*, 619–623.
60. Brown, P. D. Matrix metalloproteinase inhibitors: A novel class of anticancer agents. *Adv. Enzyme Regul.* **1995**, *35*, 293–301.
61. Nagase, H. The biological role of zinc and zinc metalloproteases. In *Zinc Metalloproteases in Health and Disease*, Hooper, N. M. (Ed.), Taylor & Francis, London, U.K., **1996**, pp. 153–204.
62. Coussens, L. M.; Werb, Z. Matrix metalloproteinases and the development of cancer. *Chem. Biol.* **1996**, *3*, 895–904.
63. Stetler-Steveson, W. G.; Hewitt, R.; Corcoran, M. Matrix metalloproteinases and tumour invasion: From correlation and causality to the clinic. *Semin. Cancer Biol.* **1996**, *7*, 147–154.
64. Coussens, L. M.; Fingleton, B.; Matrisian, L. M. Matrix metalloproteinase inhibitors and cancer: Trials and tribulations. *Science* **2002**, *295*, 2387–2392.
65. Zucker, S.; Cao, J.; Chen, W.-T. Critical appraisal of the use of matrix metallo-proteinase inhibitors in cancer treatment. *Oncogene* **2000**, *19*, 6642–6650.

66. Hasty, K. A.; Reife, R. A.; Kang, A. H.; Stuart, J. M. The role of stromelysin in the cartilage destruction that accompanies inflammatory arthritis. *Arthritis Rheum.* **1990**, *33*, 388–397.
67. Docherty, A. J.; O'Connell, J.; Crabbe, T.; Angal, S.; Murphy, G. The matrix metalloproteinases and their natural inhibitors: Prospects for treating degenerative tissue diseases. *Trends Biotechnol.* **1992**, *10*, 200–207.
68. Tsiatsiani, L.; Breusegem, F.; Gallois, P.; Zaviolov, A.; Lam, E.; Bozhkov, P. V. Metacaspases. *Cell Death Differ.* **2011**, *18*, 1279–1288.
69. Carmona-Gutierrez, D.; Frohlich, K. U.; Kroemer, G.; Madeo, F. Metacaspases are caspases. Doubt no more. *Cell Death Differ.* **2010**, *17*, 377–378.
70. Stierle, A. A.; Stierle, D. B. Secondary metabolites of mine waste acidophilic fungi. In *Topics in Biodiversity and Conservation: Bioprospecting*, Paterson, R.; Lima, N. (Eds.), Vol. 16, Springer Nature, Chambord, Switzerland, **2017**, 214–243.
71. Uren, A. G.; O'Rouke, K.; Aravind, L.; Pisabarro, M. T.; Seshagiri, S. Paracaspases and Metacaspases: Two ancient families of caspase-like proteins, one of which plays a key role in MALT lymphoma. *Mol. Cell* **2000**, *6*, 961–967.
72. McIlwain, D. R.; Berger, T.; Mak, T. W. Caspase functions in cell death and disease. *Cold Spring Harb. Perspect. Biol.* **2013**, *5*, a008656.
73. Clark, R. S. B.; Kochanek, P. M.; Watkin, S. C. et al. Caspase-3 mediated neuronal death after traumatic brain injury in rats. *J. Neurochem.* **2000**, *74*, 740–753.
74. Lu, H.; Ouyang, W.; Huang, C. Inflammation, a key event in cancer development. *Mol. Cancer Res.* **2006**, *4*, 221–223.
75. Orlikova, B.; Legrand, N.; Panning, J.; Dicato, M.; Diederich, M. Anti-inflammatory and anticancer drugs from nature. *Cancer Treat. Res.* **2014**, *159*, 123–143.
76. Liggetta, J. L.; Zhang, X.; Eling, T. E.; Baeka, S. J. Anti-tumor activity of non-steroidal anti-inflammatory drugs: Cyclooxygenase-independent targets. *Cancer Lett.* **2014**, *346*, 217–224.
77. Garcia-Albeniz, X.; Chan, A. T. Aspirin for the prevention of colorectal cancer. *Best Pract. Res. Clin. Gastroenterol.* **2011**, *25*, 461–472.
78. Understanding autoimmune diseases, NIH Publication Number 98-4273. May **1998**. http://purl.access.gpo.gov/GPO/LPS15824, accessed October 15, 2016.
79. Franchi, L.; Eigenbrod, T.; Muñoz-Planillo, R.; Nuñez, G. The inflammasome: A caspase-1-activation platform that regulates immune responses and disease pathogenesis. *Nat. Immunol.* **2009**, *10*, 241–256.
80. Sternlicht, M. D.; Lochter, A.; Sympson, C. J.; Huey, B.; Rougier, J.-P.; Gray, J. W.; Pinkel, D.; Bissell, M. J. The stromal proteinase MMP3/stromelysin-1 promotes mammary carcinogenesis. *Cell* **1999**, *98*, 137–146.
81. Granot, T.; Milhas, D.; Carpentier, S.; Dagan, A.; Segui, B.; Gatt, S.; Levade, T. Caspase-dependent and -independent cell death of Jurkat human leukemia cells induced by novel synthetic ceramide analogs. *Leukemia* **2006**, *20*, 392–399.
82. Okamoto, M.; Liu, W.; Luo, Y.; Tanaka, A.; Cai, X.; Norris, D. A.; Dinarello, C.; Fujita, M. Constitutively active inflammasome in human melanoma cells mediating autoinflammation via caspase-1 processing and secretion of interleukin-1β. *J. Biol. Chem.* **2010**, *285*, 6477–6488.
83. Paugh, B. S.; Bryan, L.; Paugh, S. W. et al. Interleukin-1 regulates the expression of sphingosine kinase-1 in glioblastoma cells. *J. Biol. Chem.* **2009**, *284*, 3408–3417.
84. Schlosser, S.; Gansauge, F.; Ramadani, M.; Beger, H.-G.; Gansauge, S. Inhibition of caspase-1 induces cell death in pancreatic carcinoma cells and potentially modulates expression levels of bcl-2 family proteins. *FEBS Lett.* **2001**, *491*, 104–108.
85. Muerkoster, S. S.; Lust, J.; Arlt, A.; Hasler, R.;Witt, M.; Sebens, T.; Schreiber, S.; Folsch, U. R.; Schafer, H. Acquired chemoresistance in pancreatic carcinoma cells: Induced secretion of IL-1β and NO lead to inactivation of caspases. *Oncogene* **2006**, *25*, 3973–3981.

86. Radisky, D. C.; Levy, D. D.; Littlepage, L. E. et al. Rac1b and reactive oxygen species mediate MMP-3-induced EMT and genomic instability. *Nature* **2005**, *436*, 123–127.
87. Coffelt, S. B.; de Visser, K. E. Cancer: Inflammation lights the way to metastasis. *Nature* **2014**, *507*, 48–49.
88. Wu, Y.; Zhou, B. Inflammation: A driving force speeds cancer metastasis. *Cell Cycle* **2009**, *8*, 3267–3273.
89. American Cancer Society. *Cancer Facts & Figures*, American Cancer Society, Atlanta, GA, **2014**, pp. 17–18.
90. Mannello, F.; Tonti, G.; Papa, S. Matrix metalloproteinase inhibitors as anticancer therapeutics. *Curr. Cancer Drug Targets* **2005**, *5*, 285–298.
91. Correia, A. L.; Mori, H.; Chen, E. I.; Schmitt, F. C.; Bissell, M. J. The hemopexin domain of MMP3 is responsible for mammary epithelial invasion and morphogenesis through extracellular interaction with HSP90b. *Gene Dev.* **2013**, *27*, 805–817.
92. Vandooren, J.; Van den Steen, P. E.; Opdenakker, G. Biochemistry and molecular biology of gelatinase B or matrix metalloproteinase-9 (MMP-9): The next decade. *Crit. Rev. Biochem. Mol. Biol.* **2013**, *48*, 222–272.
93. Nussinov, R.; Tsai, C.-J. Allostery in disease and in drug discovery. *Cell* **2013**, *153*, 293–305.
94. Jordan, P.; Brazao, R.; Boavida, M. G.; Gespach, C.; Chastre, E. Cloning of a novel human Rac1b splice variant with increased expression in colorectal tumors. *Oncogene* **1999**, *18*, 6835–6839.
95. Schnelzer, A.; Prechtel, D.; Knaus, U.; Dehne, K.; Gerhard, M.; Graeff, H.; Harbeck, N.; Schmitt, M.; Lengyel, E. Rac1 in human breast cancer: Overexpression, mutation analysis, and characterization of a new isoform, Rac1b. *Oncogene* **2000**, *19*, 3013–3020.
96. Mehner, C.; Miller, E.; Khauv, D. et al. Tumor cell-derived MMP3 orchestrates Rac1b and tissue alterations that promote pancreatic adenocarcinoma. *Mol. Cancer Res.* **2014**, *12*, 1430–1439.
97. Kessenbrock, K.; Dijkgraaf, G. J. P.; Lawson, D. A.; Littlepage, L. E.; Shahi, P.; Pieper, U.; Werb, Z. A role for matrix metalloproteinases in regulating mammary stem cell function via the Wnt signaling pathway. *Cell Stem Cell* **2013**, *13*, 300–313.
98. Dufour, A.; Sampson, N. S.; Li, J. et al. Small-molecule anticancer compounds selectively target the hemopexin domain of matrix metalloproteinase-9. *Cancer Res.* **2011**, *71*, 4977–4988.
99. Kelley, M. J.; Rose, A. Y.; Song, K.; Chen, Y.; Bradley, J. M.; Rookhuizen, D.; Acott, T. S. Synergism of TNF and IL-1 in the induction of matrix metalloproteinase-3 in trabecular meshwork. *Invest. Ophthalmol. Vis. Sci.* **2007**, *48*, 2634–2643.
100. Tsuzaki, M.; Guyton, G.; Garrett, W. et al. IL-1β induces COX2, MMP-1, -3 and -13, ADAMTS-4, IL-1β and IL-6 in human tendon cells. *J. Orthopaed. Res.* **2003**, *21*, 256–264.
101. Kim, H. G.; Shrestha, B.; Lim, S. Y. et al. Cordycepin inhibits lipopolysaccharide-induced inflammation by the suppression of NF-κB through Akt and p38 inhibition in RAW 264.7 macrophage cells. *Eur. J. Pharmacol.* **2006**, *545*, 192–199.
102. Noh, E. M.; Kim, J. S.; Hur, H. et al. Cordycepin inhibits IL-1b-induced MMP-1 and MMP-3 expression in rheumatoid arthritis synovial fibroblasts. *Rheumatology* **2009**, *48*, 45–48.
103. Steenport, M.; Khan, K. M.; Du, B.; Barnhard, S. E.; Dannenberg, A. J.; Falcone, D. J. Matrix metalloproteinase (MMP)-1 and MMP-3 induce macrophage MMP-9: Evidence for the role of TNF-alpha and cyclooxygenase-2. *J. Immunol.* **2009**, *183*, 8119–8127.
104. Harijith, A.; Ebenezer, D. L.; Natarajan, V. Reactive oxygen species at the crossroads of inflammasome and inflammation. *Front. Physiol.* **2014**, *5*(352), 1–11.
105. Martinon, F. Signaling by ROS drives inflammasome activation. *Eur. J. Immunol.* **2010**, *40*, 616–619.

106. Bratton, S. B.; Lewis, J.; Butterworth, M.; Duckett, C. S.; Cohen, G. M. XIAP inhibition of caspase-3 preserves its association with the Apaf-1 apoptosome and prevents CD95- and Bax-induced apoptosis. *Cell Death Differ.* **2002**, *9*, 881–892.
107. University of Washington, Institute for Health Metrics and Evaluation. GBD Compare: Global Burden of Disease data visualizations. Global, deaths, both sexes, all ages, 2010. http://vizhub.healthdata.org/gbd-compare/, accessed July 31, 2014.
108. Mackay, J.; Mensah, G.A. The atlas of heart disease and stroke. World Health Organization and Center for Disease Control and Prevention. Available from: http://www.who.int/cardiovascular_diseases/resources/atlas/en, accessed February 14, 2013.
109. Global report: UNAID report on the Global AIDS epidemic, **2013**. http://www.unaids.org/sites/default/files/media_asset/JC2571_AIDS_by_the_numbers_en_1.pdf, accessed June 10, 2016.
110. World Health Organization Global Tuberculosis Report, **2014**. http://apps.who.int/iris/bitstream/10665/137094/1/9789241564809_eng.pdf, accessed June 1, 2016.
111. World Health Organization World Malaria report. http://www.who.int/malaria/media/world_malaria_report_2014/en/, accessed June 1, 2016.
112. Broughton, B. R. S.; Reutens, D. C.; Sobey, C. G. Apoptotic mechanisms after cerebral ischemia. *Stroke* **2009**, *40*, e331–e339.
113. Asahi, M.; Hoshimaru, M.; Uemura, Y.; Tokime, T.; Kojima, M.; Ohtsuka, T.; Matsuura, N.; Aoki, T.; Shibahara, K.; Kikuchi, H. Expression of interleukin-1 beta converting enzyme gene family and bcl-2 gene family in the rat brain following permanent occlusion of the middle cerebral artery. *J. Cereb. Blood Flow Metab.* **1997**, *17*, 11–18.
114. Rami, A.; Sims, J.; Botez, G.; Winckler, J. Spatial resolution of phospholipid scramblase 1 (PLSCR1), caspase-3 activation and DNA-fragmentation in the human hippocampus after cerebral ischemia. *Neurochem. Int.* **2003**, *43*, 79–87.
115. Le, D. A.; Wu, Y.; Huang, Z. et al. Caspase activation and neuroprotection in caspase-3-deficient mice after in vivo cerebral ischemia and in vitro oxygen glucose deprivation. *Proc. Natl. Acad. Sci. USA* **2002**, *99*, 15188–15193.
116. Endres, M.; Namura, S.; Shimizu-Sasamata, M.; Waeber, C.; Zhang, L.; Gomez-Isla, T.; Hyman, B. T.; Moskowitz, M. A. Attenuation of delayed neuronal death after mild focal ischemia in mice by inhibition of the caspase family. *J. Cereb. Blood Flow. Metab.* **1998**, *18*, 238–247.
117. Ma, J.; Endres, M.; Moskowitz, M. A. Synergistic effects of caspase inhibitors and MK-801 in brain injury after transient focal cerebral ischaemia in mice. *Br. J. Pharmacol.* **1998**, *124*, 756–762.
118. Mouw, G.; Zechel, J. L.; Zhou, Y.; Lust, W. D.; Selman, W. R.; Ratcheson, R. A. Caspase-9 inhibition after focal cerebral ischemia improves outcome following reversible focal ischemia. *Metab. Brain Dis.* **2002**, *17*, 143–151.
119. Fink, K.; Zhu, J.; Namura, S.; Shimizu-Sasamata, M.; Endres, M.; Ma, J.; Dalkara, T.; Yuan J.; Moskowitz, M. A. Prolonged therapeutic window for ischemic brain damage caused by delayed caspase activation. *J. Cereb. Blood Flow. Metab.* **1998**, *18*, 1071–1076.
120. Yuan, J.; Yankner, B. A. Apoptosis in the nervous system. *Nature* **2000**, *407*, 802–809.
121. Trojanowski, J. Q.; Schmidt, M. L.; Shin, R. W.; Bramblett, G. T.; Rao, D.; Lee, V. M. Altered tau and neurofilament proteins in neuro-degenerative diseases: Diagnostic implications for Alzheimer's disease and Lewy body dementias. *Brain Pathol.* **1993**, *3*, 45–54.
122. Gamblin, T. C.; Chen, F.; Zambrano, A. et al. Caspase cleavage of tau: Linking amyloid and neurofibrillary tangles in Alzheimer's disease. *Proc. Natl. Acad. Sci. USA* **2003**, *100*, 10032–10037.

123. Abraha, A.; Ghoshal, N.; Gamblin, T. C.; Cryns, V.; Berry, R. W.; Kuret, J.; Binder, L. I. C-terminal inhibition of tau assembly in vitro and in Alzheimer's disease. *J. Cell Sci.* **2000**, *113*, 3737–3745.
124. Rohn, T. T.; Head, E. Caspases as therapeutic targets in Alzheimer's disease: Is it time to "cut" to the chase? *Int. J. Clin. Exp. Pathol.* **2009**, *2*, 108–118.
125. Hersch, S.; Fink, K.; Vonsattel, J. P.; Friedlander, R. M. Minocycline is protective in a mouse model of Huntington's disease. *Ann. Neurol.* **2003**, *54*, 841–843.
126. Zhang, W.; Narayanan, M.; Friedlander, R. M. Additive neuroprotective effects of minocycline with creatine in a mouse model of ALS. *Ann. Neurol.* **2003**, *53*, 267–270.
127. Zhu, S.; Stavrovskaya, I. G.; Drozda, M. et al. Minocycline inhibits cytochrome c release and delays progression of amyotrophic lateral sclerosis in mice. *Nature* **2002**, *417*, 74–78.
128. Wang, X.; Zhu, S.; Drozda, M.; Zhang, W.; Stavrovskaya, I. G.; Cattaneo, E.; Ferrante, R. J.; Kristal, B. S.; Friedlander, R. M. Minocycline inhibits caspase-independent and -dependent mitochondrial cell death pathways in models of Huntington's disease. *Proc. Natl. Acad. Sci. USA.* **2003**, *100*, 10483–10487.
129. Choi, Y.; Kim, H. S.; Shin, K. Y. et al. Minocycline attenuates neuronal cell death and improves cognitive impairment in Alzheimer's disease models. *Neuropsychopharmacology* **2007**, *32*, 2393–2404.
130. Stierle, A.; Stierle D. Bioprospecting in the Berkeley Pit: The use of signal transduction enzyme inhibition assays to isolate bioactive secondary metabolites from the extremophilic fungi of an Acid Mine Waste Lake. In *Bioactive Natural Products*, Atta-Ur-Rahman (Ed.), Vol. 39, Elsevier Science, Amsterdam, the Netherlands, **2013**, pp. 1–47.
131. Stierle, A. A.; Stierle, D. B. Bioactive secondary metabolites of acid mine waste extremophiles. *Nat. Prod. Commun.* **2014**, *9*, 1037–1044.
132. Stierle, A.; Stierle, D.; Girtsman, T. Caspase-1 inhibitors from an extremophilic fungus that target specific leukemia cell lines. *J. Nat. Prod.* **2012**, *75*, 344–350.
133. Stierle, A. A.; Stierle, D. B.; Girtsman, T.; Antczak, C.; Djaballah, H. Azaphilones from the acid mine waste extremophile *Pleurostomophora* sp. *J. Nat. Prod.* **2015**, *78*, 2917–2923.
134. Apedaile, L.; Stierle, A.; Stierle, D. Cell invasion and migration assays of MMP-3 inhibitors. Manuscript in preparation.
135. Otwinowski, Z.; Minor, M. Processing of X-ray diffraction data collected in oscillation mode. *Methods Enzymol.* **1999**, *276A*, 307–326.
136. Chen, L.; Rydel, T. J.; Gu, F.; Dunaway, C. M.; Pikul, S.; Dunham, K. M.; Barnett, B. L. Crystal structure of the stromelysin catalytic domain at 2.0 A resolution: Inhibitor-induced conformational changes. *J. Mol. Biol.* **1999**, *293*, 545–557.
137. Adams, P. D.; Afonine, P. V.; Bunkóczi, G. et al. PHENIX: A comprehensive Python-based system for macromolecular structure solution. *Acta Crystallogr. Sect. D Biol. Crystallogr.* **2012**, *66*, 213–221.
138. Emsley, P.; Lohkamp, B.; Scott, W. G.; Cowtan, K. Features and development of Coot. *Acta Crystallogr. Sect. D Biol. Crystallogr.* **2012**, *66*, 486–501.
139. Stierle, A. A.; Stierle, D. B.; Kelly, K. Berkelic acid, a novel spiroketal with highly specific anti-tumor activity from an acid-mine waste fungal extremophile. *J. Org. Chem.* **2006**, *71*, 5357–5360.
140. Buchgraber, P.; Snaddon, T. N.;Wirtz, C.; Mynott, R.; Goddard, R.; Furstner, A. A concise synthesis of berkelic acid inspired by combining the natural products spicifernin and pulvilloric acid. *Angew. Chem. Int. Ed.* **2008**, *47*, 8450–8454.
141. Stierle, A.; Stierle, D.; Kemp, K. Novel sesquiterpenoid matrix metalloproteinase-3 inhibitors from an acid mine waste extremophile. *J. Nat. Prod.* **2004**, *67*, 1392–1395.

11 Deep-Sea Hydrothermal Vent Organisms as Sources of Natural Products

Kerry L. McPhail, Eric H. Andrianasolo, David A. Gallegos, and Richard A. Lutz

CONTENTS

11.1 INTRODUCTION

The template macromolecules of *conspicuous/discernible* life support conserved patterns of primary metabolic processes for which molecular and cellular biological research continues to reveal inimitably complex networks of cell signaling pathways. Growing *molecular* understanding of these pathways elucidates mechanisms of biological development, disease, and ageing, and guides the development of strategies for prevention and treatment of diseases, efficient agriculture, and bioinspired new materials. Coupled to primary metabolism is the context-dependent secondary metabolism responsible for the diverse array of small organic molecules (natural products) that are recognized to have evolved for optimal interactions with biological macromolecules,[1] to facilitate interaction of organisms with their environment. The target affinity and specificity of structurally diverse natural products have been reviewed to illustrate their enduring and expanding role not only as drug leads and templates for combinatorial library development, but also as dynamic molecular probes for understanding biological mechanisms.[2]

For the most part, the natural products described and utilized in current chemical biological research are derived from aerobic life supported by photosynthetic primary metabolism in the superficial terrestrial and shallow marine biomes. With technological advances, the ecosystems considered as viable sources of natural products have expanded to macroorganism-inhabited (including gut) microbiomes, the deep ocean, and deep in the earth's crust. These expansive ecosystems generally comprise limited or no light and oxygen, and may experience other physicochemical parameters, or sharp gradients thereof, outside of the range traditionally considered to be life sustaining, such as pressure, temperature, pH, salt, and high metal concentrations. The small-molecule chemistry of extreme-tolerant and extremophilic organisms from a variety of habitats, reported through 2008, has been reviewed in detail by Wilson and Brimble,[3] organized by chemical class into the categories of high and low temperatures, high pressure, high salt, and high and low pH. This review appears to be the first to systematically group small-molecule compounds according to specific environmental parameters, and relatively little can yet be surmised as far as cause and effect, particularly when a combination of these conditions exist.

Since light is required for photosynthesis, the discovery of abundant macroscopic life in its absence in deep-sea hydrothermal vents[4] presented an alternative paradigm for primary production based on chemical redox reactions. In this process of chemosynthesis, the oxidation of reduced compounds such as HS^-, H_2S, S^0, CH_4, H_2, and NH_4^+ or Fe(II)- and Mn(II)-containing minerals provides energy for the synthesis of useable organic carbon from inorganic sources such as CO_2 and CH_4.[5,6] Fixation of CO_2 by chemoautolithotrophic bacteria is identical to the (light-independent) Calvin–Benson–Bessham (CBB) cycle used by photosynthetic plants. While aerobic microbes use O_2 as the electron acceptor during the energy-yielding chemosynthetic reaction, anaerobic hydrothermal microorganisms use CO_2, Fe^{3+}, NO_3^{2-}, or organic compounds (to oxidize H_2).[7,8] Habitats ranging from whale and wood falls in the deeper ocean to some shallow-water coastal sediments, cold seeps, organic-rich mud flats, and even sewage outfalls, predominantly characterized by high sulfide concentrations and the presence of free-living macroorganisms with reduced digestive systems,[9] have subsequently also been identified as chemosynthetic ecosystems, with variable influence from photosynthetic primary production. More recently, it has been realized that high densities of biomass in the deep sea are not restricted to hydrothermal vents or whale and wood falls. An assemblage of megabenthos with by far the highest peak biomass reported in the deep sea outside of vent communities was found in a seamount 2–2.5 km deep off the southeast coast of Australia near the Sub-Antarctic Zone in an area characterized by high rates of surface productivity and carbon export to the deep ocean.[10] The environmental conditions and taxa in the assemblage are widely distributed around the southern mid-latitudes, implying that the high-biomass assemblage is also widespread.

The deep (dark) ocean may be defined as depths beyond the euphotic zone (upper 200–300 m) and also the continental shelf, where the sea bottom receives less than 1% of organic matter from traditional photosynthetic primary production.[11] Thus, it could be expected that metabolomes of marine ecosystems below the euphotic zone are shaped by chemosynthesis. Beyond the limits of light penetration, *deep* has also been defined according to hydrostatic pressure, which increases approximately 10 MPa (~100 atm) per kilometer in the water column. Water depths of 1000 m or more are arbitrarily assigned as the deep biosphere, which accounts for 88% of the volume of the oceans (average depth of 3800 m = 38 MPa). The effects of high hydrostatic pressures on the primary biomolecules of life in the deep ocean were reviewed in 2015 by Jebbar et al.,[12] while Skropeta and coworkers [13,14] have reviewed the natural products reported from organisms collected at depths beyond those accessible by (self-contained underwater breathing apparatus) CUBA (50 to >5000 m). The striking difference in obligate hydrothermal vent fauna above and below 200 m (660 ft) reported by Tarasov et al.[15] was used to delimit our previous review focused on *deep-sea* hydrothermal vents as potential hot spots for natural products investigations.[16] The latter 2010 review considered the geological setting and geochemical nature of deep-sea vents that impact the biogeography of vent organisms, chemosynthesis, and the known biological and metabolic diversity of eukaryotes and prokaryotes at vent sites, as well as the handful of small-molecule natural products isolated directly from deep-sea vent organisms. Deep-sea hydrothermal vent environments are variously characterized by the presence of extreme hydrostatic pressures, high temperatures (5°C–400°C), very low pH (as low as 2.6), H_2S (up to 100 mM), heavy metals (Co, Cd, Pb, Sr, Ba, etc.), radionuclides, and toxic polycyclic aromatic hydrocarbons. The capacity of deep-sea organisms to utilize, detoxify, or degrade toxic compounds has led to the current intense interest on the part numerous diverse corporations engaged in natural products discovery and bioremediation.

In illustration of the premise that "microbes are the principal custodians of the environment," Orcutt et al.[17] reviewed habitats and microbial processes in the dark ocean, including known and speculative metabolic types that ultimately delineate global biogeochemical cycles. Consideration of different metabolisms highlights that identification of compounds as *natural products* is somewhat subjective. In extreme environments, where different metabolic types and even new types of organisms (e.g., Archaea) have been characterized, a loose definition of natural products as small organic molecules produced by a subset of organisms (to the best of current knowledge) is appropriate. Overall, investigation of complex and dynamic extreme environments, such as deep-sea hydrothermal vents, warrants an inventory of small-molecule chemistry that may ultimately be assigned as part of primary or secondary metabolism, in tandem with comparisons to well-established natural products sources and considerations of the biological environments within which small molecules may be used as dynamic molecular probes. This chapter reviews the handful of deep-sea vent-derived natural products reported to date and discusses the ongoing research to link phylogenetic diversity of organisms to the observed physiological and metabolic diversity in deep-sea hydrothermal vents, and relevance to many biological contexts. In so doing, parallels may be drawn between the natural products biological source and the biological context in which they may find utility.

11.1.1 Distribution and Geology of Vents

Exploration of deep-sea hydrothermal vent systems has been dictated largely by proximity and/or accessibility to countries involved in deep-sea hydrothermal vent research, and has been focused in the eastern Pacific and the northcentral Atlantic. Thus, although deep-sea hydrothermal vent sites extend from several hundred to several million square meters around mid-ocean ridges, along active tectonic plate margins, and around active submarine volcanoes or seamounts located in the center of tectonic plates, they are largely unexplored and new discoveries continue to be reported. For example, an extensive, long-lived black smoker vent field has even been discovered[18] at the northern end of the *ultraslow-spreading* Arctic Mid-Ocean Ridge (AMOR). The InterRidge Global Database of Active Submarine Hydrothermal Vent Fields is a list of documented hydrothermal vent fields maintained by InterRidge at the Institut de Physique du Globe de Paris (Figure 11.1).

Where rates of seafloor spreading are more rapid, as on the East Pacific Rise, eruptive disturbances are frequent enough that individual chimneys or diffuse-flow areas in a vent field may be present for less than 20 years. Alternately, in areas where volcanism is less frequent, such as the Main Endeavor Field (Juan de Fuca Ridge), some active mounds are thought to be more than 200 years old.[6]

A range of different rock types, including basalt, peridotite, and felsic rocks, can host active deep-sea hydrothermal vents and influence the chemistry of the emitted vent fluids.[19] As introduced in our 2010 review of deep-sea vent biodiversity and natural products potential,[16] hydrothermal fluid emissions from vents vary in their constituents, mineralogy, and temperature, resulting in different rates and morphologies of deposited formations,[20] which may be classified as black smokers (hot), white smokers (intermediate), beehives, flanges, or complex sulfides, for example. The complex mineralization processes ongoing in even mature sulfide structures result in convoluted internal plumbing that may create diffuse warm-water flows at temperatures and mineral fluxes suitable for the growth of macroorganisms. In the focused hot flows of black smokers, it is common to measure temperature changes from 350°C to 10°C over a distance of just a few centimeters under high hydrostatic pressures. All of these parameters in combination influence the composition and metabolism of vent communities. Additionally, a surprisingly significant influence of atmospheric forcing on deep-ocean dynamics was reported in 2011.[21] Surface-generated mesoscale eddies are proposed to play a role in spreading hydrothermal vent efflux in the deep ocean and disperse vent larvae hundreds of kilometers between isolated and ephemeral communities around the northern East Pacific Rise. Since these eddies are formed seasonally and are sensitive to phenomena such as El Niño, they have the potential to introduce seasonal and atmospheric variations into the deep ocean.[21]

11.1.2 Diversity and Biogeography of Vent Fauna

Despite dispersal of larval organisms over vast distances by deep-ocean currents, many hydrothermal vent fields exhibit a surprisingly high degree of endemism, with distinctive ecosystems presenting unique combinations of habitats.[6] The Southern Ocean, with its high deep-sea species diversity, is recognized as the center of origin for the global deep-sea fauna and is a putative gateway connecting hydrothermal vents in different oceans. Explorations of the East Scotia Ridge (ESR) in the Southern Ocean revealed that

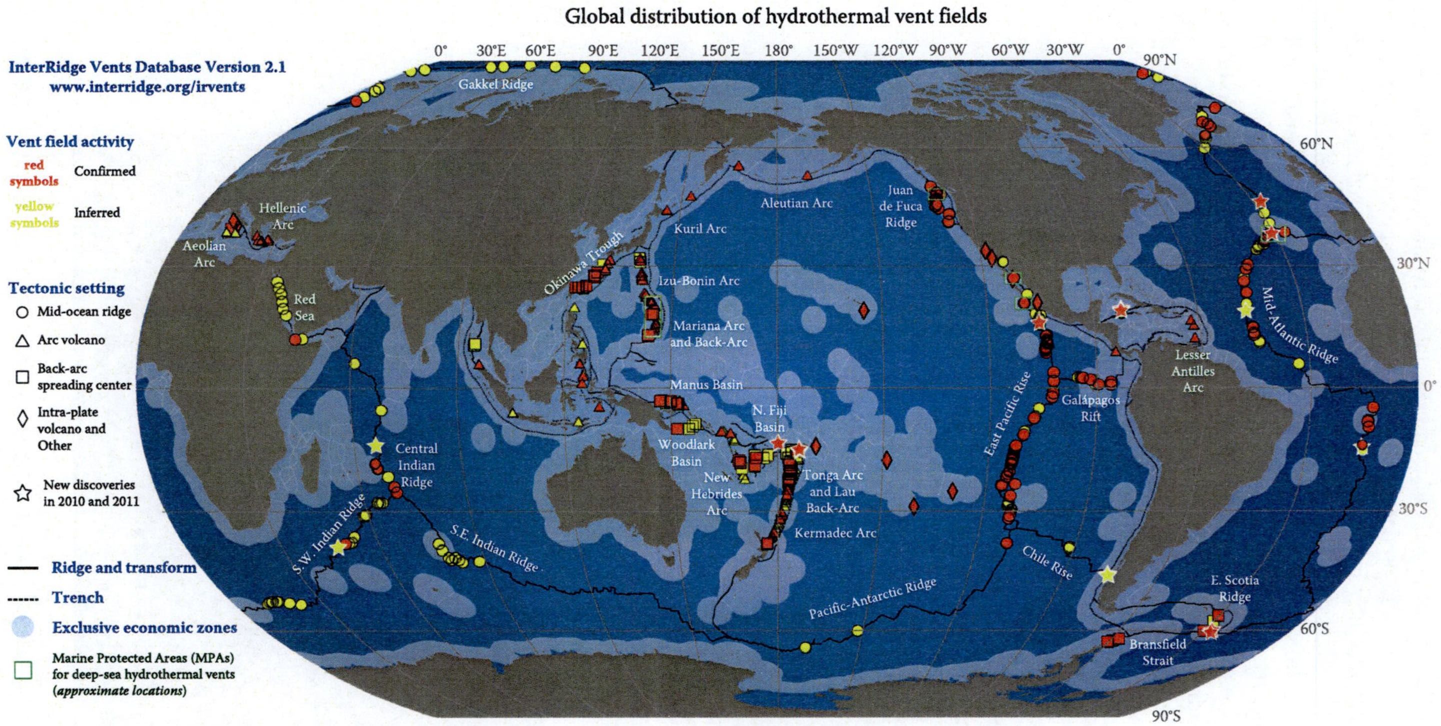

FIGURE 11.1 Global map of hydrothermal vents identified to date. (Data compiled in InterRidge's vents database; http://vents-data.interridge.org/maps; Beaulieu, S.E., InterRidge Global Database of active submarine hydrothermal vent fields: Prepared for Inter-Ridge, Version 3.4, World Wide Web electronic publication, Version 3.2, 2015, http://vents-data.interridge.org, Accessed December 13, 2016.[157])

macrofauna such as polychaete worms, bathymodiolid mussels, and alvinocaridid shrimp are absent from the ESR vents, which instead host stalked barnacles, limpets, peltospiroid gastropods, anemones, and predatory sea stars, as well as a new type of yeti crab.[22] Mature hydrothermal vent sites in the eastern Pacific typically host scattered clusters of tubeworms cloaked in white microbial mats (*Beggiatoa*) and provide substrate for other polychaetes such as palm worms and scale worms, together with limpets, snails, and the occasional hydrothermal vent shrimp, zoarcid fish, mussels, and crabs (Figure 11.2). In contrast, vent sites along the Mid-Atlantic Ridge are characterized by shrimp (*Rimicaris*) swarming on the sides of chimneys near high flows that lack the tubeworms and polychaetes found in the Pacific vent communities[6] (Figures 11.3 and 11.4). In both oceans, beds of hydrothermal vent mussels (*Bathymodiolus* spp.) are found in areas of diffuse venting (Figure 11.5). Numerous genera reported in Atlantic vents are shared with the

FIGURE 11.2 Tubeworms, crabs, and a zoarcid fish in a deep-sea hydrothermal vent at 9° 50′ N along the East Pacific Rise.

FIGURE 11.3 Shrimp (*Rimicaris*) swarming on the sides of chimneys in the TAG hydrothermal vent field along the Mid-Atlantic Ridge.

FIGURE 11.4 Close-up of shrimp (*Rimicaris*) on the side of a chimney in the TAG hydrothermal vent field along the Mid-Atlantic Ridge.

FIGURE 11.5 A bed of mussels (*Bathymodiolus thermophilus*), together with a zoarcid fish, inhabiting a deep-sea hydrothermal vent at 9° 50′ N along the East Pacific Rise.

Pacific vent fauna. In the Indian Ocean, deep-sea vent communities are different enough to constitute a separate biogeographic province from either the Atlantic or the Pacific Ocean, although Indian Ocean communities do share vent macrofauna common to both. The vent fauna associated with the recently explored AMOR is distinct from the fauna along the Mid-Atlantic Ridge to the south, and is proposed to arise by local specialization and migration of fauna from cold seeps and the Pacific.[18]

11.1.3 Phylogenetic Diversity of Vent Microorganisms

A high diversity of largely undescribed bacteria and archaea,[16] as well as fungi,[23] exists in deep-sea hydrothermal vent systems, where environmental gradients support innumerable ecological niches. The diversity of archaea and bacteria from hydrothermal versus nonhydrothermal (cold) deep-ocean environments, based on phylogenetic sequence data through 2010, is reproduced in Figure 11.6. This wealth of data has been collected and reported on the phylogenetic diversity of microbial communities

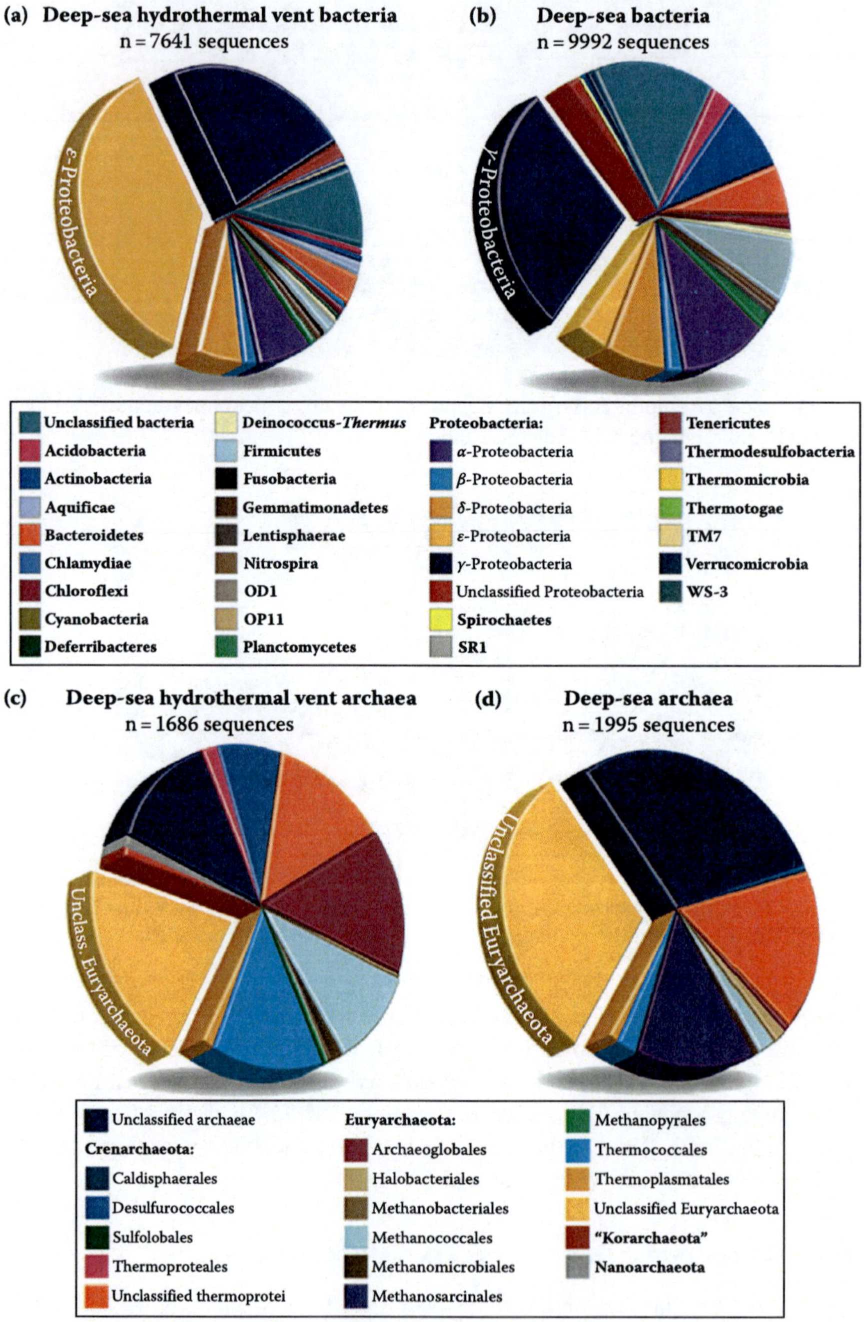

FIGURE 11.6 Relative abundances in the deep sea of taxa of bacteria in **(a)** hydrothermal vent samples and **(b)** cold water column and sediment samples, and archaea in **(c)** hydrothermal vent samples and **(d)** cold water column and sediment samples, determined by 16S rRNA gene sequence analysis. (Reproduced from Thornburg, C.C. et al., *J. Nat. Prod.*, 73, 489, 2010.)

in specific vent chimneys or fields, but the interfield variability among these communities is relatively understudied due to the logistics of broad sampling and sequencing efforts. To address this gap in knowledge,[24] we used barcoded pyrosequencing of the variable region 4 (V4) of the 16S rRNA gene to characterize the archaeal and bacterial communities of over 30 hydrothermal deposit samples from six vent fields located along the Eastern Lau Spreading Center. Distinct communities at certain sites were attributed to the unique geochemistry of specific vent fluids resulting from active degassing of a subsurface magma chamber. Otherwise, the results showed that hydrothermal vent deposits in back-arc basins hosted microbial communities that were taxonomically similar to those from mid-ocean ridge systems.[24] Subsequently, for comparison across three different oceans, pyrosequencing of 16S rRNA gene was used to characterize the bacterial communities of the venting sulfide, seawater, and tubeworm trophosome from the East Pacific Rise, South Atlantic Ridge, and Southwest Indian Ridge.[25] Differences in bacterial diversity were observed between samples from the three oceanic regions, with bacterial communities of the East Pacific Rise sulfide chimneys being most diverse, although Proteobacteria, Actinobacteria, and Bacteroidetes were the predominant phyla in all vents.

While it is well recognized that the majority of distinct microbial lineages in vent communities occur in very low abundance, the ecological role and distribution of these rare versus abundant lineages are not well understood. Anderson et al.[26] used 16S rRNA tag sequencing to describe the biogeography and microbial community structure of both rare and abundant archaea and bacteria in hydrothermal vent systems, with results suggesting that while most archaeal and bacterial lineages in vents are rare and display a highly regional distribution, a small percentage of mostly archaeal lineages are successful at widespread dispersal and colonization.

Extensive description of the phylogenetic diversity of vent microorganisms is now being followed by investigation of the adaptation and metabolic potential of the microbial communities using comparative genomic analyses of metagenomes from a variety of microenvironments, as has been done for samples from a black smoker chimney in the Mothra hydrothermal vent field at the Juan de Fuca Ridge.[27] Metagenomes highly enriched in genes for mismatch repair and homologous recombination suggest that these microbial communities have evolved extensive DNA repair systems to cope with the extreme conditions that have potential deleterious effects on genomes. Enrichment of genes for chemotaxis and flagellar assembly in chimney metagenomes also reflects adaptation to the highly dynamic conditions present within the chimney walls, while a high proportion of transposases implies that horizontal gene transfer may be occurring routinely in the deep-sea vent chimney biosphere.[27]

11.2 SMALL-MOLECULE CHEMISTRY AND REPORTED BIOLOGICAL ROLE OR ACTIVITY

Marine bacteria collected and cultured from deep-sea cold sediments have yielded a diversity of novel natural products since the first characterization of an obligate marine actinomycete species,[28] and the subsequent isolation of many new taxa of obligate bacterial halophiles over the last 20 years. Seminal research by Fenical and Jensen established the value of marine sediment-derived actinomycete bacteria as a source of drug discovery leads,[29] and unusual new metabolites continue to be isolated from sediments collected by mud missiles, trawling, push cores, or customized samplers for submersibles.

11.2.1 Primary-Type Metabolites

There are relatively few small-molecule compounds reported directly from deep-sea hydrothermal vent collections and microbial cultures, and those published may be compared to products from terrestrial or shallow-water thermophilic and also piezophilic (barophilic) cold-water microorganisms.

Unusual glycerol ether lipids were reported from anaerobic cultures of deep-sea vent archaea as early as 1983.[30] These important membrane components are substitutes for the more thermally and chemically labile esterified fatty acids found in terrestrial and shallow-water organisms. Such stabilized small molecules may find specific applications in chemical biology, just as thermally stable enzymes from deep-sea vent organisms have become important tools in molecular and cellular biology. Early characterization of archaeal membrane components was performed in order to identify source-specific biomarkers of microbial organic matter in deep-sea hydrothermal vents, and the previously unknown macrocyclic glycerol diether **1** (Figure 11.7) comprised 95% of the polar lipid extract of the methanogen *Methanococcus jannaschii* (cultured under H_2, CO_2, 85°C), with open chain **2** as a minor constituent.[31] Isoprenoid hydrocarbons and alkylglycerol ether–derived polar

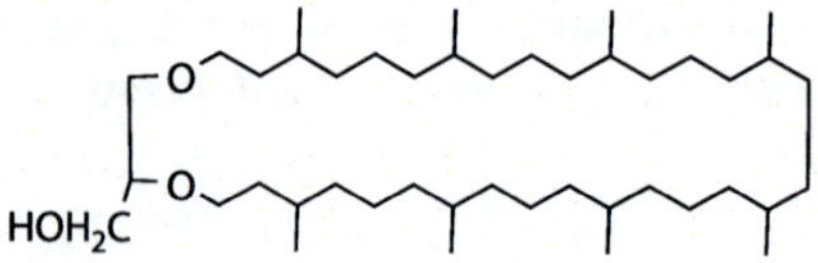

(1)

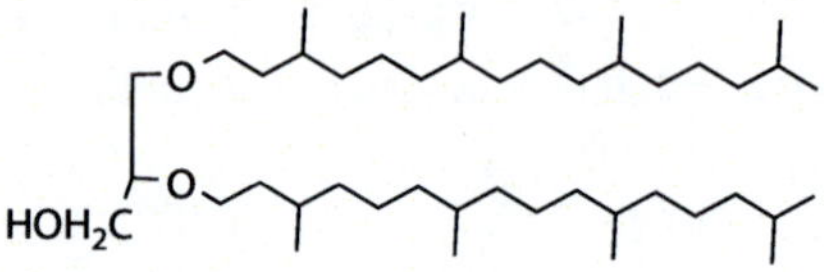

(2)

Polysulfide core:

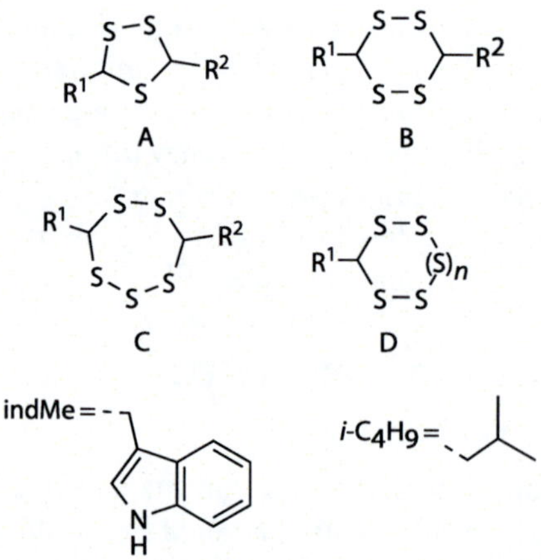

Compound	Core	R1	R2	n
Lenthionine (3)	C	H	H	---
(4)	A	CH3	CH3	---
(5)	A	C2H5	CH3	---
(6)	A	i-C4H9	H	---
(7)	A	i-C4H9	CH3	---
(8)	A	C2H5	i-C4H9	---
(9)	A	i-C4H9	i-C3H7	---
(10)	A	i-C4H9	i-C4H9	---
(11)	A	Benzyl	CH3	---
(12)	A	Benzyl	i-C4H9	---
(13)	A	IndMe	i-C4H9	---
(14)	B	i-C4H9	CH3	---
(15)	B	i-C4H9	i-C4H9	---
(16)	B	IndMe	i-C4H9	---
(17)	C	i-C4H9	H	---
(18)	C	i-C4H9	CH3	---
(19)	C	i-C4H9	i-C3H7	---
(20)	C	i-C4H9	i-C4H9	---
(21)	C	Benzyl	i-C4H9	---
(22)	C	IndMe	i-C4H9	---
(23)	D	i-C4H9	---	1
(24)	D	CH3	---	2
(25)	D	i-C4H9	---	2
(26)	D	i-C4H9	---	3

FIGURE 11.7 Potential primary or shunt metabolites from deep-sea hydrothermal vent organisms.

lipids are the two main classes of archaeal lipids, and are reviewed in more depth by Skropeta[13] and Wilson and Brimble.[3] Interestingly, the lipids of thermophilic archaea are not only distinguished from those of other organisms in containing isoprenoid rather than straight hydrocarbon chains and ether linkages to glycerol, but the glycerol found in archaea also has the reverse stereochemistry.

Wilson and Brimble[3] also reviewed other classes of metabolites reported from deep-sea vent hyperthermophilic microorganisms, including tRNA nucleosides, polyamines, and cyclic polysulfides, for which research is ongoing to understand their biological roles and relevance to human health. Cyclic polysulfides with varying core motifs have been reported from sulfur-metabolizing archaea of the genus *Thermococcus* (Figure 11.7, **3–26**). The first examples of these molecules were reported in 1993,[32] and included lenthionine (1,2,3,5,6-pentathiepane, **3**), reported initially from shitake mushrooms.[33] Detection by gas chromatograph–mass spectrometry (GC-MS) of seven cyclic polysulfides[34] from cultures of shallow-water *Cytophaga* marine bacteria from the North Sea led to structure confirmation by syntheses, and investigation of the conformational flexibility and equilibria, of these five-, six-, seven-, and eight-membered cyclic polysulfides (dithia-, trithia-, tetrathia-cycloalkanes). The structural flexibility and potential difficulty in the assignment of polysulfide metabolites are illustrated by the proposed existence of the tunicate-derived lissoclinotoxin A as a heptasulfane,[35] in addition to the trisulfane and pentasulfane structures assigned previously. The mass spectrum for this *o*-benzopolysulfane shows mass peaks for tri-, penta-, and heptasulfane molecules in the (equilibrated) natural products sample. In an ecological context, cyclic polysulfides are antifungal and proposed to serve a defensive role in producing organisms such as plants and red algae. They may also be formed at high temperatures.[36] In deep-sea hydrothermal vents, where energy production is associated with elemental sulfur reduction to H_2S by *Desulfuromonas* bacteria and archaea, or with H_2S oxidation to elemental sulfur by *Beggiatoa* bacteria, it seems that cyclic polysulfides are shunt metabolites.

The biological role of H_2S in human health and the importance of organic polysulfides, as a source of readily reducible sulfane sulfur and thus slow-release H_2S delivery systems, are reviewed by Pluth et al.[37] H_2S-synthesizing enzymes are differentially expressed in the neuronal, cardiovascular, immune, and endocrine systems, indicating the importance and possible therapeutic potential of H_2S-releasing sulfanes, especially related to cardiovascular disease. A cyclic polysulfide motif, capable of forming reversible covalent bonds with an active-site cysteine, has also been identified as the pharmacophore of new STEP (STriatal-Enriched protein tyrosine Phosphatase) enzyme inhibitors. STEP is a neuron-specific phosphatase that is overactive in several neuropsychiatric and neurodegenerative disorders, including Alzheimer's disease.[38]

The critical cellular roles of polyamines range from involvement in cellular translation in eukaryotes and archaea to bacterial biofilm formation and specialized roles in natural products biosynthesis. While there is conservation of aminobutyl-containing spermidine (or homospermidine) in eukaryotes and archaea, for posttranslational modification of translation factor eIF5A and aIF5A, respectively, there is no known conserved role for any polyamine across bacteria and a corresponding increased variety of bacterial polyamines. The distribution and biosynthesis of polyamines in eukaryotes, archaea, and bacteria are reviewed by Michael.[39] Hyperthermophilic bacteria and archaea are reported to have distinct long-chain polyamines. In a survey of eight *Thermus* bacterial species growing at 70°C, triamines and tetramines were ubiquitously distributed, while linear pentamines and hexamines, and branched tetra- and penta-amines were present in the hyperthermophilic species, but not in the moderate thermophiles.[40] Similarly, archaeal hyperthermophiles are reported to produce numerous different polyamines and a range of longer-chain polyamines.[39,41] It has been shown that these long-chain polyamines are essential for protein synthesis and stabilize nucleic acids at high temperatures,[41] as do the tRNA-derived unusual nucleosides reviewed by Wilson and Brimble.[3]

11.2.2 Secondary-Type Metabolites

As far as secondary metabolite natural products are concerned, the updated 2014 review by Skropeta and Wei[14] added 188 new marine natural products from deep-sea organisms to the 400 metabolites reviewed in their 2008 comprehensive review, encompassing 50 m to bathyal (200–4000 m), abyssal (4000–6500 m), and hadal (>6500 m) ocean depths. Of the 22 deep-sea species yielding new natural products since 2008 that were collected from greater than 200 m, 15 were bacteria or fungi, and 13 of these were from depths greater than 1000 m (Table 3 in Skropeta and Wei[14]). It is noteworthy that some deep-sea samples collected by dredging or submersible appear to have been collected from zones of documented hydrothermal vent fields. For example, south of the Aguni Knoll in the Okinawa Trough is the collection site (1800 m deep) of the crinoid echinoderm *Proisocrinus rubberimus*, which may well be present in proximal areas of diffuse venting. This scarlet-red organism apparently yielded the first tri- and tetra-brominated anthraquinones from a natural source, proisocrinins A–F (**27–32**; Figure 11.8).[42]

What may be considered the first confirmed deep-sea vent natural products were reported in 2009. Brominated hydroxyethylamine chromans, ammonificins A and B (**33** and **34**), were characterized from the chemolithotrophic, anaerobic bacterium *Thermovibrio ammonificans*,[43] which was isolated from the walls of an active deep-sea hydrothermal vent chimney on the East Pacific Rise and cultured at 75°C (pH 5.5, 2% w/v NaCl) in the presence of H_2 and CO_2, with nitrate or sulfur as the electron acceptor. While chroman derivatives are well known, elaboration of the chroman core with hydroxylethylamine and phenol in **33**, or brominated phenol in **34**, is unique.

The observed antimicrobial and apoptosis-inducing biological activity of the parent extract was not matched by the pure compounds **33** and **34**. Further work was undertaken to reisolate and purify minor inseparable compounds from *T. ammonificans* fractions, resulting in a subsequent report of chromene-containing ammonificins C (**35**) and D (**36**).[44] These unsaturated analogs of **33** and **34** showed low micromolar proapoptotic activity, illustrating the importance of the double bond for the biological activity.

The loihichelins A–F (**37–42**)[45] were isolated from a cultured *Halomonas* strain, and belong to the ubiquitous group of amphiphilic siderophores, which all chelate Fe(III) via the bidentate coordination of two hydroxamate groups and an α-hydroxyaspartate residue.[46] In addition to a role in the acquisition of iron as a trace nutrient, it is suggested that they may be required for energy generation by *Halomonas* during the metabolism of reduced Fe(II).[45]

While new actinomycete bacteria are reported to be isolated and cultured from hydrothermal vent fluids, as reviewed in the subsequent discussion on laboratory cultivation of deep-sea vent-derived organisms, new biologically active natural products from these more recent collections have yet to appear in the literature.

The most recent report of a microbial natural product from a (shallow) hydrothermal vent appears to be that of aspergstressin (**43**).[47] This new meroterpenoid is derived from a marine strain of *Aspergillus* fungus isolated from the digestive

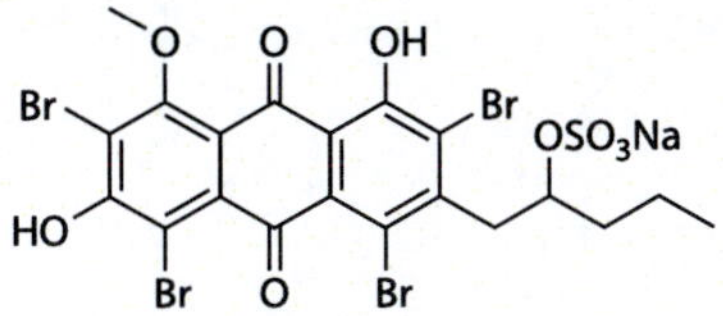

Proisocrinin A (**27**)

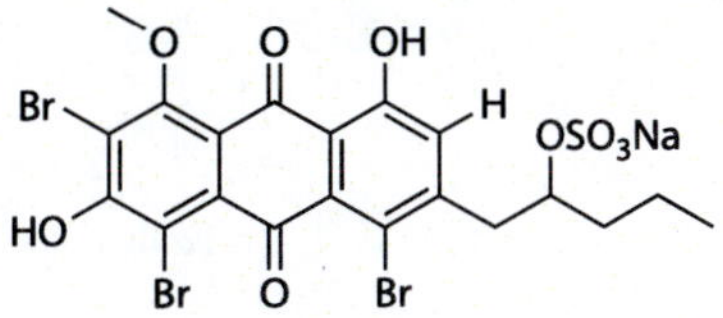

Proisocrinin B (**28**)

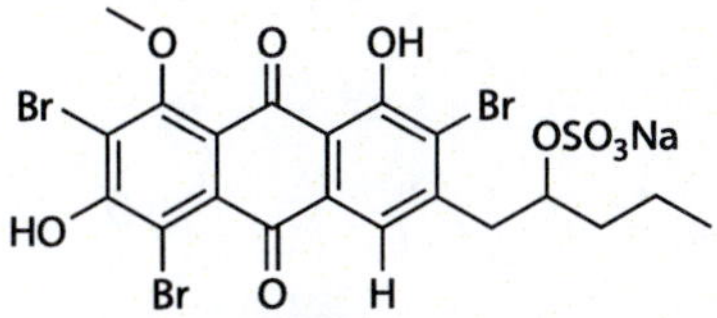

Proisocrinin C (**29**)

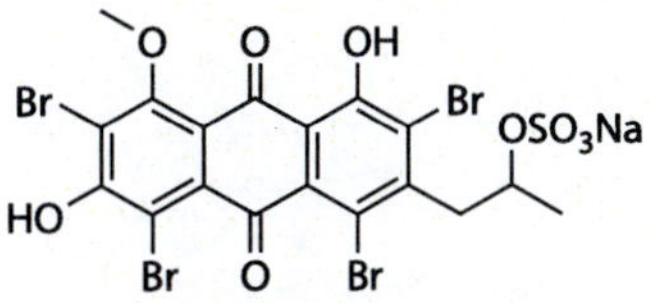

Proisocrinin D (**30**)

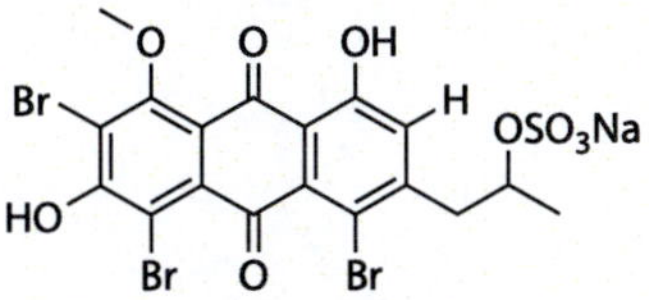

Proisocrinin E (**31**)

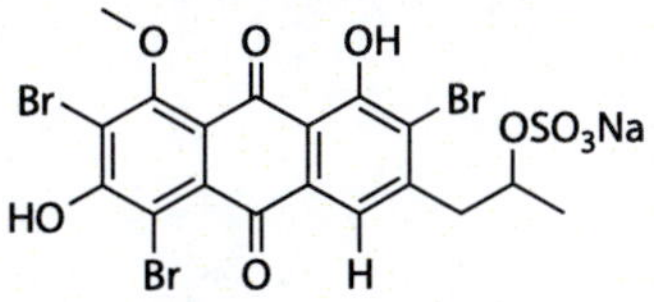

Proisocrinin F (**32**)

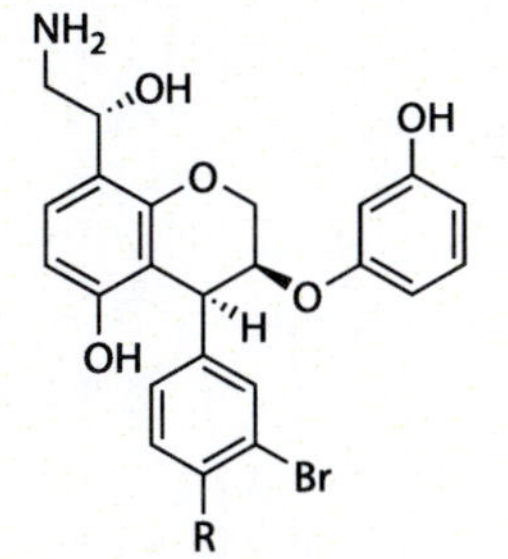

Ammonificin A (**33**) R = OH
Ammonificin B (**34**) R = Br

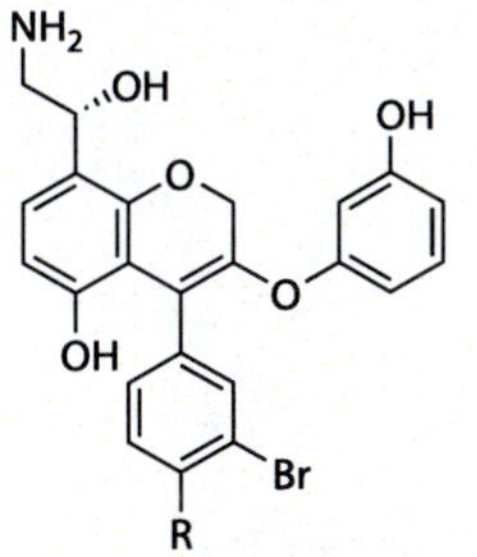

Ammonificin C (**35**) R = OH
Ammonificin D (**36**) R = Br

FIGURE 11.8 Secondary-type metabolites from deep-sea hydrothermal vent organisms. *(Continued)*

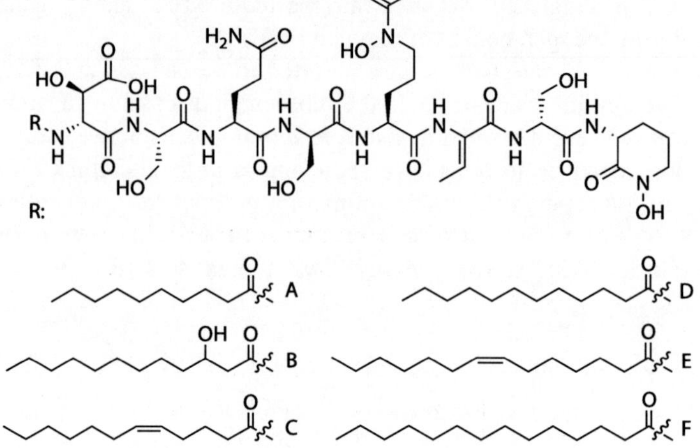

Loihichelins A (**37**)-F (**42**)

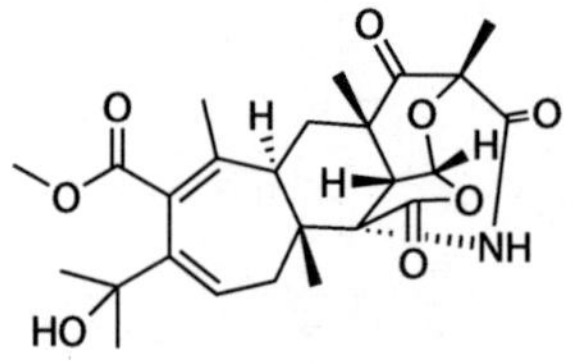

Aspergstressin (**43**)

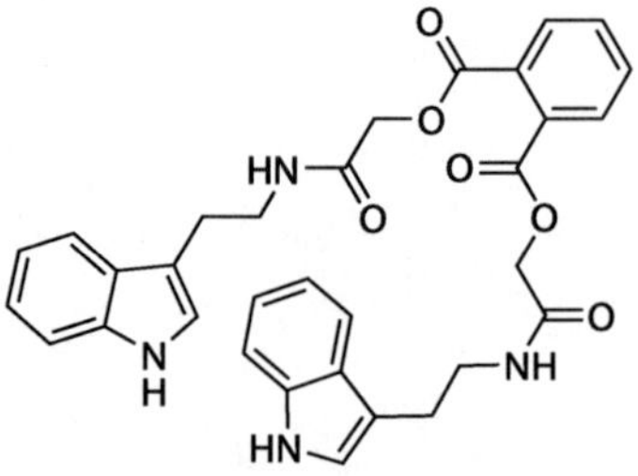

Ditryptophenaline (**44**)

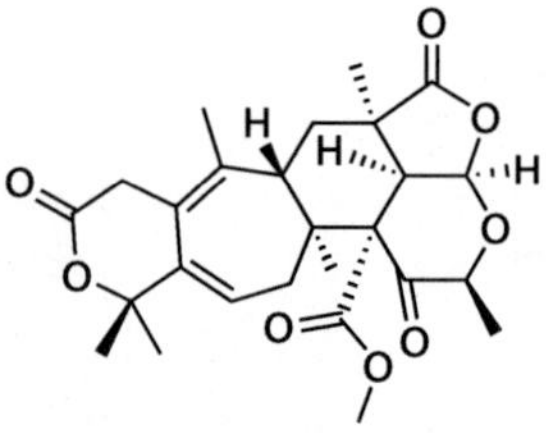

Berkeleyacetal A (**45**)

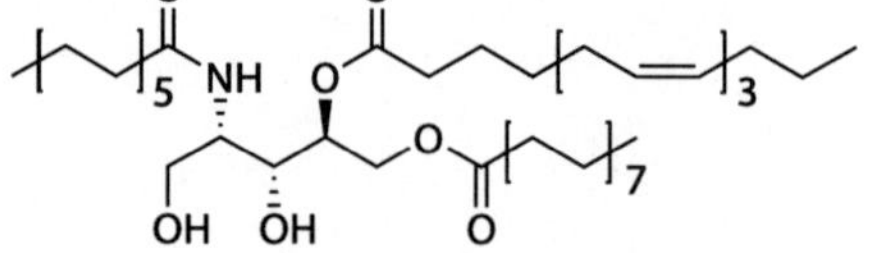

Bathymodiolamide A (**46**)

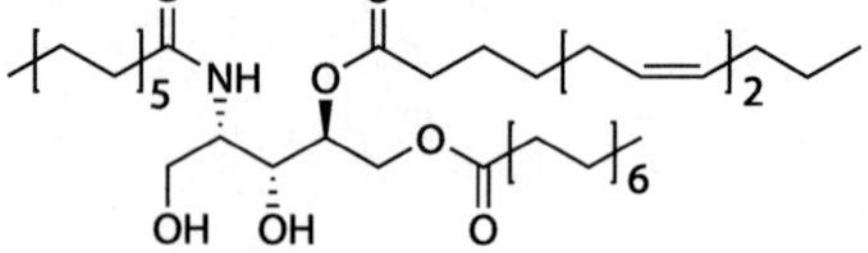

Bathymodiolamide B (**47**)

FIGURE 11.8 (*Continued*) Secondary-type metabolites from deep-sea hydrothermal vent organisms.

gland of the crab *Xenograpsus testudinatus*, which was collected near Kueishantao Island off the northeastern coast of Taiwan, at the southern end of the Okinawa Trough. Although no collection depth is reported, this area comprises around 30 shallow hydrothermal vents, rich in heavy metals, at 10–30 m depths. This compound warrants mention as an example of a natural product that is not produced under optimal growth conditions for the source organism. The *Aspergillus* strain was cultivated in potato dextrose broth (PDB) and tested for its susceptibility to cadmium, zinc, cobalt, and nickel. The induction of stress metabolites upon exposure to cobalt led to growth experiments using six different cobalt ion concentrations that showed that the normally barely detectable levels of aspergstressin (**43**) increased to optimal production at 6 mM cobalt, in parallel with significantly decreased levels of the regular major product (under metal-free culture conditions). The latter metabolite was identified as known compound ditryptophenaline (**44**), originally isolated from a mangrove endophytic fungus.[48] Aspergstressin (**43**) is structurally related to berkeleyacetal A (**45**) produced by a deep-water (270 m = 855 ft) isolate of *Penicillium* fungus collected from acidic (pH 2.5), metal sulfate–rich water of Berkeley Pit Lake.[49]

Bivalve mollusks are the only documented invertebrate natural products sources collected from deeper bathyal depths around hydrothermal vents since 2008, for which active pure compounds have been isolated and characterized. The variety of bivalve-derived sterols reported prior to 2008 from deep-sea environments has been reviewed by Skropeta.[13] These were isolated from the hydrothermal vent mussel *Bathymodiolus septemdierum* collected at 1244 m in the Myojin Knoll at the Izu–Bonin Island Arc, and the cold-seep clam *Calyptogena soyoae* collected at 1100 m in Sagami Bay, Japan. More recently, new ceramide derivatives bathymodiolides A and B (**46** and **47**) were reported from a vent mussel (*Bathymodiolus azoricus*) from the Lucky Strike vent field along the northern region of the Mid-Atlantic Ridge. In their 2011 paper, Andrianasolo et al. refer to the vent mussel species from which the compounds were isolated as *Bathymodiolus thermophilus*; it was subsequently determined that this vent mussel species was *Bathymodiolus azoricus*. Thirty specimens collected by submersible from the Lucky Strike vent site (1733 m depth) were dissected, and samples of adductor muscle, gill, and mantle tissues were frozen separately for subsequent analyses. The methanol extract of gill tissue was active in a screen for proapoptotic activity, which was shown to be caused by bathymodiolides A and B (**46** and **47**). The development of the high-throughput cell-based ApopScreen assay used, as well as the isolation and characterization of the ammonificins and bathymodiolides, was reviewed in 2012.[50]

Although the component pure compounds responsible for the activity were not isolated and characterized, significant antimalarial activity was detected in the organic extract of the hydrothermal vent shrimp *Mirocaris fortunata*, collected near the Azores Islands (North Atlantic Ocean). Of the samples screened from a variety of habitats, this vent shrimp extract showed both potent growth inhibition and a good selectivity index in the screen for activity against the Dd2 (chloroquine-resistant) and 3D7 (chloroquine-sensitive) strains of *Plasmodium falciparum*.[51]

11.3 VENT MICROBIAL METABOLIC TYPES

11.3.1 Influence of Vent Geochemistry on Metabolic Types of Microbes

It has been demonstrated experimentally that key redox reactions involving iron, sulfur, and hydrogen are energetically favorable for microbial metabolism at hydrothermal conditions (100°C) relevant to life because they remain at disequilibrium even in a heterogeneous system.[52] This lab-based experimental and theoretical modeling was coupled with field observations of both microbial diversity and geochemical heterogeneity for distinct intrachimney microbial populations from the East Pacific Rise in an attempt to translate observed genetic diversity into physiological diversity.

As reviewed by Orcutt et al.,[17] there can be a sharp transition from anaerobic microbial processes within high-temperature vent fluids to aerobic processes due to rapid mixing with associated diffuse venting and surrounding cold water. Advances in high-throughput metagenomics technologies are beginning to reveal the exceptional metabolic and physiological diversity of hydrothermal microbial communities, which appear to be undergoing rapid dynamic succession and adaptation in response to the steep temperature and chemical gradients across vent chimneys.[53] In general, methanogenesis, based on H_2 oxidation coupled to CO_2 or SO_4 reduction, appears to predominate in high-temperature (above 45°C) vent fluids associated with basalt, while S_2, metal, and CH_4 oxidation are more prevalent in cooler hydrothermal vent plumes.[54–56] A recent *Trends in Microbiology* spotlight discussion[57] highlights that CH_4 metabolism is phylogenetically more widespread than previously recognized and summarizes key aspects of research on methanogenesis, and the associated reductive acetyl-CoA pathway (Wood–Ljungdahl pathway). The latter is among the oldest metabolic pathways on earth and, as a linear carbon fixation pathway that operates without energetically costly intermediates, may occur abiotically in a thermodynamically favorable environment where its iron–sulfur(–nickel) chemical catalysts are present. Sulfur cycling processes are less important in vents hosted in ultramafic (high magnesium and iron) rock, where H_2 and CH_4 metabolism dominates.[58] A recently characterized microbial metabolism, thought to be a globally significant sink for oceanic nitrogen,[59] is anaerobic ammonium oxidation (anammox) in which NH_4^+ is oxidized to N_2 gas, with NO_2^- as the electron acceptor.[60]

Reconstruction of the metabolic pathways from the microbial metagenome of a black smoker vent sample in the Mothra hydrothermal vent field (Juan De Fuca Ridge)[27] revealed that the microbial community in the wall of this sulfide chimney was mainly fueled by sulfur oxidation, putatively coupled to nitrate reduction to perform inorganic carbon fixation through the CBB cycle. Based on the genomic organization of key genes of the carbon fixation and sulfur oxidation pathways, both obligate and facultative autotrophs appeared to contribute to biomass production.[27]

In a study designed to examine the link between lithoautotrophic primary production (inorganic compounds used as electron donors to fix CO_2) and the use of the resulting organic compounds by organotrophic microorganisms (both as a source of carbon for biosynthesis and electrons for energy production), a biofilm from the surface of a black smoker chimney was investigated with respect to structure and metabolic capabilities.[61] Imaging revealed rod-shaped organotrophic Bacteroidetes growing as ectobionts on the polysaccharide sheaths of filamentous lithoautotrophic *Sulfurovum* (Epsilonproteobacteria). Additionally, recycling of organic matter in the biofilm was indicated by in situ expression of acetyl-CoA synthetase by the *Sulfurovum* bacterium.

Around a single vent chimney, genes detected may include those involved in methanogenesis, aerobic and anaerobic CH_4 oxidation, nitrification, denitrification, sulfate reduction, degradation of complex carbon substrates, and metal resistance. Intriguingly, there is also increasing evidence in hydrothermal vents for nonsolar phototrophic metabolism. Since the 1989 discovery of vent shrimp that possess novel photoreceptors[62] and subsequent evidence that high-temperature vent fluids generate infrared and visible light from black body radiation at the seafloor,[63] microbial metagenomic sequencing continues to indicate the presence of microorganisms capable of using photoradiation for energy. GSB1 is an anaerobic phototrophic sulfur-oxidizing bacterium[64] that is phylogenetically related to green sulfur bacteria and uses bacteriochlorophyll c and the carotenoid chlorobactene for *photosynthesis*. Notably, bacteriochlorophylls have found application in photodynamic therapy (PDT), given that they absorb longer wavelengths of light, not absorbed by chlorophylls in plants or Cyanobacteria, to effectively generate cytotoxic reactive oxygen species (ROS) at wavelengths that penetrate tissues deeply. Padeliporfin (WST11) is a third-generation water-soluble palladium bacteriochlorophyll that was developed as a photo-sensitive agent for vascular-targeted PDT.[65] It has very recently shown efficacy in a phase III clinical trial across 47 European university centers and community hospitals for low-risk prostate cancer patients.[66]

Recently, comparison of microbial metabolic functions inside a newly formed chimney and the outer section of a mature chimney used GeoChip microarray–based, high-throughput metagenomics to reveal that only about 1% of the microbial functional genes detected were common to both *internal* and *external* chimney populations.[67] The internal chimney population showed a predominance of genes encoding RuBisCO, the protein central to light-dependent carbon fixation, for which the sequences were most closely related to those in photosynthetic cyanobacteria and green sulfur bacteria. By comparison, the external chimney population was dominated by genes for the light-independent CBB cycle. Very few functional genes were detected in the young internal chimney population compared with the mature external chimney population, indicating that the composition of the microbial community changes rapidly during chimney growth. In agreement with metagenomics-based studies, consideration of the geochemistry of vent fluids from vents hosted in different types of rocks and the energetics of possible catabolic and anabolic reactions using numerical models led to the conclusion that variation in microbial catabolic strategies (and therefore phylotypes) may be far more diverse in some rock types than others, and anabolic processes may *produce* rather than use energy.[19] Variation in the geochemical composition of hydrothermal fluids is also believed to dictate community structure and productivity of chemolithotrophic archaea.[68]

In a detailed study of metabolic constraints correlated to the distribution of hyperthermophiles, and thus their biogeochemical impact in deep-sea hydrothermal vent systems, the relative abundances of methanogens, autotrophic iron(III) oxide reducers, and heterotrophic sulfur reducers within samples from the Endeavour Segment and Axial Volcano in the northeastern Pacific Ocean were determined.[69] These abundances were correlated to environmental conditions such as reduction potential, organic carbon, and hydrogen availability. Laboratory cultivation of hyperthermophilic methanogen species of *Methanocaldococcus* demonstrated the hydrogen-dependent growth kinetics for the three strains investigated, which could be correlated with collection site field microbiology and fluid geochemistry data. Thus, these methanogens were predicted to be hydrogen limited, and potentially rely on hydrogen produced by heterotrophs, whereas models and kinetic experiments indicated that iron(III) oxide reducers were not hydrogen limited under the same conditions.[69]

Microbes that are symbionts of vent macrofauna clustered in diffuse, warm-water flows with temperatures of 10°C–40°C and even up to 60°C[6] may exploit much broader redox zones than free-living microbes.[70] For example, thiotrophic bacteria in the gill tissues of large *Calyptogena magnifica* clams or the trophosomes of *Riftia pachyptila* tubeworms may exploit redox zones over 1 m, while those living in the gill chamber of *Rimicaris exoculata* shrimp are transported between high-temperature anoxic vent water and cold-temperature oxic seawater. *Bathymodiolus* mussels exploit a wide range of marine environments[71] since various species are able to use one of more gases, including H_2S, CH_4, and H_2.[72]

11.3.2 Influence of Hydrostatic Pressure on Hydrothermal Vent Metabolisms

Hydrostatic pressures in the ocean rise to greater than 1000 atm in the deep trenches (10 m water = 1 atm). Consequently, genetic, biochemical, and physiological processes of deep-sea inhabitants, including pressure-sensitive gene regulation, structure and function of proteins and other cellular components, and metabolism and physiology, have been studied intensively, and research through 2007 on these piezophiles (barophiles) has been reviewed by Skropeta.[13] Given the effect of high pressures on the conformational shape of proteins[12] and membranes, it is suggested that the secondary metabolites resulting from metabolic processes at high pressures may differ from those produced at atmospheric pressure. In particular, the question arises as to whether exposure and adaptation of deep-sea organisms to the extreme conditions of the deep sea present the potential for the discovery of inhibitors or activators of stress-related pathways that may play a key role in a number of diseases, including cancer and other chronic diseases.

Transcriptomics was used to profile changes in gene expression in response to pressure for two hyperthermophilic archaeal species, the piezophilic *Thermococcus barophilus* strain MP and the piezo-sensitive *Thermococcus kodakarensis* strain KOD1. A total of 378 genes were differentially expressed in *T. barophilus* cells grown at 0.1 MPa (1 atm), 40 MPa (395 atm), and 70 MPa (690 atm), whereas 141 genes were differentially regulated in *T. kodakarensis* cells grown at 0.1 and 25 MPa (247 atm).[73] The impact of hydrostatic pressure on amino acid metabolism in cultured *Thermococcus barophilus* was further investigated. Amino acid metabolism changed from a requirement for 3 to 17 amino acids at atmospheric pressure and

40 MPa, respectively.[74] It was suggested that energetic constraints imposed by high hydrostatic pressure make polypeptides and chitin a preferable carbon resource to de novo amino acid synthesis at high pressures, despite relatively low energy yields upon fermentation of these polymers.

The first obligate piezophilic hyperthermophilic microorganism, *Pyrococcus yayanosii* CH1, has been discovered most recently[75] from the Ashadze hydrothermal vent (4100 m depth). Multiomics analyses showed a loss of aromatic amino acid biosynthesis pathways in the *P. yayanosii* genome and constitutive high expression of energy metabolism compared with nonobligate piezophilic *Pyrococcus* sp. Differential proteomics and transcriptomics analyses identified key hydrostatic pressure–responsive genes involved in translation, chemotaxis, and energy metabolism (hydrogenases and formate metabolism), and clustered regularly interspaced short palindromic repeats (CRISPR) sequences associated with cellular apoptosis susceptibility (cas) proteins.

Besides archaea and bacteria, recent investigations of the composition of deep-sea microbial communities have shown the presence of fungi dominating the micro-eukaryotic population. There is no universal genetic marker for piezoresistance in fungi; however, morphological and physiological responses of fungi to high hydrostatic pressures could be examined using a collection of deep-sea yeasts to determine limits of pressure tolerance and metabolic activity.[76] Targeted yeasts originally isolated from deep-sea hydrothermal vents were all able to grow under high hydrostatic pressure, with some being better adapted than others, and some piezo-sensitive yeast strains displayed a unique and complex morphological switch induced by increasing hydrostatic pressure. The remarkable taxonomic-based dichotomic response to pressure, with piezo-sensitive ascomycetes and piezo-tolerant basidiomycetes, was unexpected.

11.3.3 Microbial Symbionts of Deep-Sea Vent Organisms

The chemolithoautrophic epi- and endosymbionts living on or within various specialized tissues and structures of hydrothermal vent invertebrates have elicited broad interest. A variety of evidence indicates that nearly all invertebrates associated with hydrothermal vents acquire most, if not all, of their needs for fixed carbon and nitrogen from their symbionts, while others may also require a substantial harvest of free-living microorganisms.[77,78] For example, suspension-feeding hydrothermal vent clams (Bivalvia: Vesicomyidae) and mussels (Bivalvia: Mytilidae) harbor chemoautotrophic microorganisms associated with their gills, which supply approximately 45% of fixed carbon to their host.[8,77] Invertebrates from six metazoan phyla and ciliate protists from hydrothermal vents are known to host chemosynthetic bacteria either episymbiotically on the exterior of the host (e.g., alvinellid tubeworms, rimicarid shrimp) or endosymbiotically (e.g., vestimentiferan tubeworms, solemyid clams) within cytoplasmic vacuoles of specialized host cells called bacteriocytes. The biology, metabolism, and link between symbiont physiology and vent geochemistry are reviewed by Stewart et al.[70]

Symbionts of vestimentiferan tubeworms have not been cultured successfully in the laboratory, but have been surveyed by metagenome sequencing.[79] Nucleotide sequences from the small-subunit rRNA gene suggest that intracellular symbionts of the eastern Pacific vent tubeworms *Oasisia alvinae*, *Riftia pachyptila*, *Tevnia jerichonana*, and *Ridgeia piscesae* all belong to the same phylotype of γ-proteobacteria, *Candidatus* Endoriftia persephone.

Mussels of the deep-sea vent genus *Bathymodiolus* are the most abundant and widespread members of vent and seep communities. They differ from tidal-zone mussels in the morphology of their gills, labial palps, and alimentary tract, as well as in having chemoautotrophic bacteria in the gills, since they have reduced filter-feeding activity and rely instead on symbiotic association with CH_4-oxidizing and/or sulfur-oxidizing bacteria.[80] Bathymodiolin mussels have also evolved interactions with a higher diversity of bacterial lineages than other bivalve groups found in deep-sea hydrothermal vents, cold seeps, and organic falls. It is remarkable that certain *Bathymodiolus* mussels support both chemoautotrophs and methanotrophs. Phylogenetic analyses of bacteria 16S rRNA-encoding gene sequences were performed for comparison of bacteria associated with *Bathymodiolus mauritanicus* and *Idas*-like specimens from three sites in the Northeast Atlantic (two mud volcanoes in the Gulf of Cadiz and one seamount of the Gorringe Bank).[81] *Bathymodiolus mauritanicus* hosts a dual symbiosis dominated by two phylotypes of CH_4-oxidizing bacteria and a less abundant phylotype of a sulfur-oxidizing bacterium, which was also the dominant phylotype in a sympatric population of *Idas-like* mussels at the Darwin mud volcano. This was the first report of a bacterial phylotype shared between two different deep-sea mussels from divergent clades. Since the description of *Bathymodiolus* in 1985, other genera specific for vent and seep taxa have been established, including *Gigantidas*, *Tamu*, and *Vulcanidas*.[82]

The various modes for transmission of symbionts between invertebrate hosts and to their offspring, as well as between hydrothermal vents and vent fields, and implications for the evolution and demography of chemosynthetic symbionts are reviewed by Vrijenhoek.[83] Until recently, it was doubted that viable symbionts could be released to augment environmental populations of hydrothermal vent tubeworms because the adult worms lack obvious openings and their symbionts were largely regarded as terminally differentiated. Klose et al.[84] show that tubeworm symbionts rapidly escape their hosts upon death and recruit to surfaces where they proliferate. It is proposed that the release of symbionts may enable adaptations that evolve within host individuals to spread within host populations and further to new environments.

It is proposed that the ability to use hydrogen as an energy source is widespread in hydrothermal vent symbioses where hydrogen is abundant. It has been shown that the symbionts of the *Bathymodiolus* mussels from the Mid-Atlantic Ridge use hydrogen for primary production, and that the symbionts of *Bathymodiolus* mussels from Pacific vents have *hupL*, the key gene for hydrogen oxidation, as do symbionts of the tubeworm *Riftia pachyptila* and the shrimp *Rimicaris exoculata*.[72] In contrast to deep-sea vent bivalves, adult vestimentiferan tubeworms of the family *Siboglinidae* (formerly included in phyla Pogonophora and Vestimentifera)[85,86] lack a digestive tract entirely and derive their nutrition solely from culturing sulfur-oxidizing bacteria within a specialized organ known as the trophosome. This organ accounts for approximately 16% of the animal's wet weight and consists primarily of symbiont-containing lobes (bacteriocytes), crystals of elemental sulfur, and blood vessels. Bacterial densities between 10^9 and 10^{11} cells per gram of wet tissue have been observed within the trophosome of the giant vestimentiferan tubeworm *Riftia pachyptila*, which is capable of growing up to 2.0 m in length. Although numerous early reports indicate dominance by only a single bacterial phylotype within trophosomes of *R. pachyptila* and other vestimentiferans, more recent molecular evidence suggests that a more diverse community may colonize these and other structures within the trunk of these tubeworms.[87–89] A highly diverse assemblage of episymbionts were identified from cuticular secretions extending from the dorsal integument of the extremely thermotolerant polychaetous annelid *Alvinella pompejana*.[90] Other epibiont communities include microbes *farmed* on dense aggregations of shrimp in Mid-Atlantic Ridge hydrothermal vents. The shrimp compete for space near warm, sulfide-rich water emissions to support their *crop* of microorganisms.[91] Microorganisms are also associated with the scale-shaped, iron sulfide–containing sclerites of a newly described scaly snail, *Crysomallon squamiferum*, from hydrothermal vents in the Indian Ocean.[92] Without these chemosynthetic microorganisms, the rapid growth rates required to prosper and reach reproductive maturity in such an extreme and dynamic environment would not be possible.[6,77,93] In addition to providing critical food resources to their hosts, epi- and endosymbiotic bacteria may also produce chemical metabolites capable of deterring predation within these communities.[94]

Deep-sea sponges found at hydrothermal vent sites have also been investigated for their associated microbial communities by bacterial 16S rRNA gene sequencing.[95] *Characella* sp. (Pachastrellidae) was collected from a hydrothermal vent site (686 m depth) in the Sumisu Caldera, Ogasawara Island chain, Japan, while *Pachastrella* sp. (Pachastrellidae) and an unidentified Poecilosclerida sponge were collected from an oil seep (572 m depth) in the Gulf of Mexico. Bacterial gene sequences detected were most similar (>99% identity) to those of the γ-proteobacterial thioautotrophic symbionts of *Bathymodiolus* vent mussels. This was the first report of a likely stable association between sponges and thioautotrophic bacteria, although associations between sponges and methanotrophs have been reported previously.[96]

Recently, the microbial community associated with the shrimp *Alvinocaris longirostris* from hydrothermal fields of the Okinawa Trough was investigated using high-throughput sequencing and clone library construction and analysis to compare community structures and metabolic profiles of microbes associated with the shrimp's gill and gut.[97] Fourteen taxa were detected in the gill and gut communities, of which 11 taxa were shared by both tissues, although the community structures were significantly different between tissues: for example, Firmicutes were abundant only in gut tissue. In addition, genes (cbbM and aclB) encoding the key enzymes of CBB and reductive tricarboxylic acid cycles were both present and significantly more abundant in the gill than in the gut, providing the first evidence that at least two carbon fixation pathways are present in both the gill and gut communities.

11.4 CULTIVATION OF DEEP-SEA VENT ORGANISMS

The number of new marine natural products reported annually in the literature continues to increase, particularly for those produced by marine microbes[98]: 677 in 2016 versus 493 in 2013. Not unexpectedly, the majority of the microbial compounds reviewed by Blunt et al. were obtained from laboratory cultures of the source microorganism.

Despite the very small percentage of microorganisms estimated to be culturable by traditional methods, a majority of pharmaceutical candidates in development are microbial natural products, and actinomycete soil bacteria (order Actinomycetales) are responsible for over 50% of the microbial antibiotics discovered to date, most of which originate from the *Streptomyces* and *Micromonospora* genera.[99] Only 4% of the bacteria that have been successfully cultured are extremophiles.[100] While microbes from extreme environments have been cultured from deep-sea hydrothermal vent environments under standard conditions,[101] many of these organisms likely fall into the extreme-tolerant category, and the natural products reported from these sources may not necessarily be new or novel. Nevertheless, the ability to isolate new organisms, even under suboptimal growth conditions for production of natural products relevant to the deep-sea vent environment, allows genome sequencing and assessment of the potential for natural products biosynthesis.

Access to hydrothermal vent samples typically occurs through research collaboration with such institutions or through national culture collections where samples are deposited. For example, Japan Agency for Marine-Earth Science and Technology (JAMSTEC) promotes collaborative efforts with industry through their Cooperative Research Project for Extremophiles program.[102] Similarly, the Brittany Microbe Culture Collection (BMCC) allows academic and industrial access to over 1300 microorganisms isolated from deep-sea hydrothermal vents by French Research Institute for Exploitation of the Sea (IFREMER). In the United States, the American Type Culture Collection (ATCC) publicly lists hydrothermal vent microorganisms available for purchase, albeit a small collection with only 24 microorganisms currently listed.[103] Undoubtedly, ongoing culture-independent molecular phylogenetic surveys of microbial deep-vent habitats, such as chimney structures, sediments, vent emissions, and chemosynthetic macrofauna, provide information about the ecophysiological requirements of uncultivated microbial lineages in deep-sea hydrothermal vent environments for use in approximating culture conditions for future cultivation efforts.[104]

11.4.1 New Strains Isolated in Culture

Success in culturing new hydrothermal vent microorganisms through 2010 is reviewed by Pettit.[101] Reports of subsequent isolation in culture and characterization of new archaea and bacteria through 2016 are summarized in Tables 11.1 and 11.2.

11.4.2 New Technologies for Culturing

The isolation and cultivation of microorganisms remains the central contemporary challenge for microbial natural products research related to undiscovered, newly detected or characterized, and also known natural products. Only a small fraction of the total microbial diversity in nature has been grown in the laboratory, and apparently only a fraction of the biogenetic potential of currently cultured microorganisms is translated to phenotypic products. While DNA and RNA sequencing of environmental samples provides a wealth of information independent of the ability to culture organisms, new gene and pathway functions cannot be inferred from pure sequence data. Approaches to culturing those microorganisms currently unculturable in the

TABLE 11.1
New Archaea Cultured from Deep-Sea Hydrothermal Environments (2010–2016)

Source Vent Field/ Ridge[a]	Depth (m)	Sample Description	Essential Culture Conditions?	Type Strain Identifier	Phylogenetic Identity	Reference
12° 58.4′ N 44° 51.8′ W, MAR	4100	Black smoker sample	98°C, 3.5% NaCl, pH 7.5–8.0, 52 MPa, anaerobe	$CH1^T$ (=JCM 16557)	*Pyrococcus yayanosii* sp. nov.	Birrien et al.[105]
7° 25′ 24″ S 107° 47′ 66″ W, EPR	2700	Hydrothermal chimney sample	80°C, 2% NaCl, pH 7.0, anaerobic	Bio-pl-0405IT2^T (=CSUR P577^T=JCM 16307^T)	*Thermococcus prieurii* sp. nov.	Gorlas et al.[106]
12° 50.342′ N 103° 56.903′ W, EPR	2633	Hydrothermal chimney sample	87.5°C, 2% NaCl, pH 7.0, anaerobic	30-1^T (=CNCM 4275=JCM 19601)	*Thermococcus nautili* sp. nov.	Gorlas et al.[107]
47° 57′ 00″ N 129° 06′ 00″ W, JFR	2200	Polychaete worm *Paralvinella*	82°C, 3.23% NaCl, pH 8.0, obligate anaerobe	ES1^T (=DSM 27261^T= KACC 17923^T)	*Thermococcus paralvinellae* sp. nov.	Hensley et al.[108]
44° 30′ 00″ N 130° 30′ 00″ W, JFR	2350	Polychaete worm *Paralvinella*	88°C, obligate anaerobe	CL1^T (=DSM 27260^T= KACC 17922^T)	*Thermococcus. cleftensis* sp. nov.	Hensley et al.[108]
45° 55′ N 129° 59′ W, JFR	1520	Hydrothermal vent fluid	88°C, obligate anaerobe	JH146^T (=DSM 27223^T= KACC 18232^T)	*Methanocaldococcus bathoardescens*	Stewart et al.[109]

[a] *Ridge abbreviations:* MAR, Mid-Atlantic Ridge; JFR, Juan de Fuca Ridge; EPR, East Pacific Rise.

TABLE 11.2
New Bacteria Cultured from Deep-Sea Hydrothermal Environments (2010–2016)

Source Vent Field/ Ridge[a]	Depth (m)	Sample Description	Essential Culture Conditions?	Type Strain Identifier	Phylogenetic Identity	Reference
9° 50′ N 104° 17′ W EPR	2513	Hydrothermal fluids	35°C, 2.5% NaCl, pH 7.5, aerobic, ASW medium	EPR92T (5DSM 232095JCM 16666T)	*Parvibaculum hydrocarboniclasticum* sp. nov.	Rosario-Passapera et al.[110]
21° 28′ 09″ N 144° 04′ 22″ E Nikko Seamount, Mariana Arc	456	Deep-sea hydrothermal vent chimney	37°C, 2.5% NaCl, pH 6.0, obligate anaerobe, MMJHS medium	496Chim^T (=DSM 22050^T=JCM 15747^T=NBRC 105224^T)	*Thiofractor thiocaminus* gen. nov., sp. nov.	Makita et al.[111]
20° 45.8′ S 176° 11.5′ W Eastern Lau Spreading Center	2150	Deep-sea hydrothermal vent chimney	50°C, 2.0%–2.5% NaCl, pH 6.5, anaerobic	S3R1^T (=DSM 24185^T=VKM B-2672^T)	*Deferrisoma camini* gen. nov., sp. nov.	Slobodkina et al.[112]
22° 16.25′ S 176° 54.17′ W Eastern Lau Spreading Center	1925	Deep-sea hydrothermal vent deposit	57°C–60°C, 3% NaCl, pH 5.5–5.7, anaerobic	cd-1655R^T (=DSM 25116^T=OCM 1212^T)	*Mesoaciditoga lauensis* gen. nov., sp. nov.	Reysenbach et al.[113]
22° 10.82′ S 176° 36.09′ W Eastern Lau Spreading Center	1910	Deep-sea hydrothermal vent chimney	74°C, 1.5%–3.5% NaCl, pH 7.0, anaerobic	S95^T (=DSM 24515^T=VKM B-2683^T)	*Thermosulfurimonas dismutans* gen. nov., sp. nov.	Slobodkin et al.[114]

(Continued)

TABLE 11.2 (*Continued*)
New Bacteria Cultured from Deep-Sea Hydrothermal Environments (2010–2016)

Source Vent Field/ Ridge[a]	Depth (m)	Sample Description	Essential Culture Conditions?	Type Strain Identifier	Phylogenetic Identity	Reference
24° 51′ N, 123° 50′ E Hatoma Knoll, OT	1470	Deep-sea hydrothermal vent chimney	55°C, 2.5% NaCl, pH 6.6, obligate anaerobe	AC55^T (=JCM 17643^T=DSM 24660^T=NBRC 107904^T)	*Thermotomaculum hydrothermale* gen. nov., sp. nov.	Izumi et al.[115]
22° 10.82′ S 176° 36.09′ W Eastern Lau Spreading Center	1910	Deep-sea hydrothermal vent chimney	61°C, 1.8%–2.7% NaCl, pH 6.8, anaerobic	S69^T (=DSM 25762^T=VKM B-2760^T)	*Dissulfuribacter thermophilus* gen. nov., sp. nov.	Slobodkin et al.[116]
12° 48′ N 103° 56′ W EPR	2621	Tube of tubeworm *Alvinella pompejana*	75°C, 3% NaCl, pH 6.0, anaerobic	HB-8^T (=DSM 24425^T=JCM 17384^T)	*Phorcysia thermohydrogeniphila* gen. nov., sp. nov.	Pérez-Rodríguez et al.[117]
37° 17.5240′ N, 32° 16.5085′ W, MAR	1624	Deep-sea hydrothermal vent deposit	60°C–65°C, 1%–6% NaCl, pH 4.5–5.0, obligate anaerobe	Mar08-272r^T (=DSM 24585^T=OCM 985^T)	*Hippea jasoniae* sp. nov.	Flores et al.[118]
9° 46.5003′ N, 104° 16.8100′ W, EPR	2520	Deep-sea hydrothermal vent deposit	60°C, 1%–5% NaCl, pH 4.5–5.0, obligate anaerobe	EP5-r^T (=DSM 24586^T=OCM 986^T)	*Hippea alviniae* sp. nov.	Flores et al.[118]
9° 50′ N, 104° 17′ W, EPR	2520	Water surrounding sulfide chimneys	Up to 50°C, up to 10% NaCl, anaerobe	EX25	*Vibrio antiquaries*	Hasan et al.[119]
47° 57′ N 129° 05′ W, JFR	1762	Above the hot water plume of a deep-sea hydrothermal vent	30°C, 2%–6% NaCl, pH 7.0, aerobe	MCS 33^T (=LMG 27140^T=CCUG 62981^T)	*Glycocaulis abyssi* gen. nov., sp. nov.	Abraham et al.[120]

(*Continued*)

TABLE 11.2 (*Continued*)
New Bacteria Cultured from Deep-Sea Hydrothermal Environments (2010–2016)

Source Vent Field/Ridge[a]	Depth (m)	Sample Description	Essential Culture Conditions?	Type Strain Identifier	Phylogenetic Identity	Reference
Southwest Indian Ridge	Not reported	Deep-sea sediment of a hydrothermal vent field	35°C, 0%–1% NaCl, pH 7.0, aerobic	9-2^T (=CGMCC 1.12749^T=JCM 19653^T=MCCC 1K00278^T)	*Pontibacter amylolyticus* sp. nov.	Oren et al.[121]
3° 06′ 15″ S 102° 33′ 16″ W, EPR	Not reported	Deep-sea sediment of a hydrothermal vent field	30°C–35°C, 0%–1% NaCl, pH 7.5. aerobic	22DY15^T (=JCM 19489^T=DSM 27767^T=CGMCC 1.12416^T=MCCC 1K00276^T)	*Brevirhabdus pacifica* gen. nov., sp. nov.	Wu et al.[122]
102.6° W 3.1° S, EPR	2891	Deep-sea hydrothermal sulfide deposit	60°C–62°C, 2.3% NaCl, pH 7.0, 20 MPa, anaerobic	DY22613^T (=JCM 19466^T=DSM 28033^T=MCCC 1A06456^T)	*Anoxybacter fermentans* gen. nov., sp. nov.	Zeng et al.[123]
9° 49′ N, 104° 17′ W, EPR	2500	Black smoker chimney	55°C–60°C, 2%–3% NaCl, pH 5.5–6.0, anaerobic	TB-6^T (=DSM 27783^T=JCM 19563^T)	*Cetia pacifica* gen. nov., sp. nov.	Grosche et al.[124]

[a] *Ridge abbreviations:* MAR, Mid-Atlantic Ridge; JFR, Juan de Fuca Ridge; EPR, East Pacific Rise; OT, Okinawa Trough; Mariana, Mariana Back-Arc Basin.

laboratory are reviewed by Stewart,[125] and include simulating the natural environmental conditions through media adjustments, coculturing two or more microbial species, and/or including a host organism. The natural products potential of one or a handful of strains may be investigated by methodical modification of standard (petri plate or conical flask) culture conditions of media, salinity, pH, temperature, oxygen, and so on. However, the latter OSMAC (one strain, many compounds) approach[126,127] is cumbersome and impractical for a large-scale screening program. Bioreactors allow careful control of culture (including aerobic versus anaerobic) conditions, and numerous fermentation approaches using customized bioreactors have been patented. In addition to liquid culture at varying scales that include microfluidic bioreactors,[128,129] packed-column bioreactors allow solid media such as soil to be used.[130] The concept of arrays of microbioreactors to quantitatively measure even thousands of cellular phenotypes at once has led to phenotype microarrays, which comprise a phenotyping technology that is compatible with high-throughput molecular genetics.[131] This phenotypic testing (*phenomics*) uses living cells and thus enables testing of gene function and improving genome annotation, even under conditions where cellular respiration but not growth occurs. Phenotypic microarrays have also been used successfully for isolating and growing unusual anaerobic microorganisms[132]; for example, in systems biology research on sulfate-reducing anaerobic delta proteobacterium *Desulfovibrio vulgaris* Hildenborough.[133]

11.5 DEVELOPMENT OF OTHER BIOPRODUCTS

With regard to deep-sea environments, biotechnological interests have focused mainly on the development of new enzymes and exopolysaccharides (EPS) to improve agriculture, biotechnology, cosmetics, pharmaceutical, and waste management processes. Energy production has also been a recent focus of deep-sea vent research, including microbial hydrogen production as a sustainable fuel source.[134]

Enzymes from thermophiles play an important role in everyday practices such as research, and industrial and manufacturing processes. One of the best-known examples is the thermostable DNA polymerase, Taq polymerase[135,136], from the thermophile *Thermus aquaticus*, which was isolated from thermal pools in Yellowstone National Park. The use of this enzyme not only simplified polymerase chain reaction applications for their more effective use in molecular research, but it is also estimated to have brought royalties of nearly US$2 billion to Roche in the licensing of this product.[137] The recognized potential utility of thermophiles and their bioproducts is evident from the increasing genome-sequencing efforts, yielding over 120 complete genome sequences of hyperthermophiles in the Aquificae, Thermotogae, Crenarchaeota, and Euryarchaeota. Enzymes from thermophiles and hyperthermophiles have been reviewed,[138,139] and those specifically from deep-sea vent microbes have been summarized by Pettit.[101]

Most recently,[140] Urbieta et al. summarize major current applications (*white biotechnology*) of thermophiles and thermozymes from a wide range of habitats, and review the application and role of gene sequencing in understanding the biodiversity, genomes, transcriptomes, metagenomes, and single-cell sequencing of thermophiles.

Numerous biotechnology companies are actively involved in product development and licensing of derivatives from thermophilic and hyperthermophilic hydrothermal vent organisms. Five of these companies, Verenium (now BASF), New England Biolabs Inc., Invitrogen Corporation, Sederma, and California Tan, have commercialized products derived from deep-sea and hydrothermal vent organisms.[141,142] Examples of products obtained from the reassembly of genes encoding unique enzyme sequences from microorganisms collected from hydrothermal vents include Fuelzyme™-LF, an α-amylase enzyme for the cost-effective production of ethanol from corn; AccuPrime™ DNA Polymerase, which provides a high-yield PCR amplification of GC-rich targets;[143] and Pyrolase™ 160 Enzyme, a broad-spectrum β-glycosidase used in industry to reduce the viscosity of guar polymers.[144] The company New England BioLabs Inc. has also developed the DNA Polymerase Deep $Vent_R$™, which was purified from a strain of *Escherichia coli* that carries the Deep $Vent_R$ DNA Polymerase gene from a deep-sea hydrothermal vent *Pyrococcus* sp.[145] Current research is also ongoing concerning CO_2 fixation and assimilation in the deep-sea environment for the possible biotechnological application of enzymes for CO_2 capture to reduce greenhouse gases in the atmosphere.[146]

In addition to characterizing the diversity of thermal esterases and pullulanases produced by hyperthermophilic archaea and bacteria from several hydrothermal vents in the Pacific Ocean,[147,148] researchers at IFREMER and elsewhere have also been interested in EPS from these microbes, with potentially novel and unusual characteristics and functional activities under extreme conditions. EPS biosynthesis is a common protective mechanism found in extremophiles, and EPS make up a substantial component of the extracellular polymers surrounding most microbial cells in extreme environments. Three main EPS producers have been identified from either hydrothermal vent fluid or as endosymbionts of polychaetes, which may provide clinical applications in the area of cancer therapy, cardiovascular disease, and bone healing.[149,150] Most notable was the first *Vibrio* isolate from a deep-sea hydrothermal vent. *Vibrio diabolicus* was isolated from a Pompei worm (polychaete *Alvinella pompejana*) tube collected from a hydrothermal vent on the East Pacific Rise.[151] This bacterium produces an EPS, termed HE 800, which is characterized as a linear, high-molecular-weight (800 kDa) EPS with a repetitive tetrasaccharide sequence of *N*-acetyl glucosamine and *N*-acetyl galactosamine in equal amounts. This EPS has shown remarkable results in bone regeneration studies in the parietal bone of male rats. HE 800–treated rats had 95.9% (±6.2) bone healing (n = 20), while control and collagen-treated animals demonstrated only 14.5% (±16.0) and 17.8% (±18.1) bone healing, respectively.[152,153] Chemical characterization of EPS from piezophiles has been reviewed by Wilson and Bramble,[3] while potential biotechnological advantages and commercial value of EPS from extremophiles are discussed by Nicolaus et al.[154] The likely ecological roles in responses to environmental stress, in recognition processes and cell–cell interactions, and in adherence of biofilms to surfaces are reviewed by Le Costaouëc et al.[155]

11.6 CONCLUDING REMARKS

As eloquently stated by Smanski et al. with respect to microorganisms[156]: "[T]he ecological and evolutionary pressures that drive the non-uniform distribution of NP biosynthesis provide a rational framework for the targeted isolation of strains enriched in new NP potential." In a broad sense, it cannot be assumed that sampling a subset of biological diversity will provide a coherent, sufficiently complete spectrum of biologically relevant chemical diversity. Despite conservation of the template biomolecules of life, knowledge of the patterns and processes of diverse metabolic strategies, and flexibility, in energy production and biosynthesis, remains far from comprehensive, especially in the vast depths of the unexplored oceans. The challenges of accessing, isolating, and culturing micro- and macroorganisms, and characterizing new organic molecules require ongoing technological innovations and interdisciplinary collaborative research, which is well justified given the tremendous contributions of natural products to chemical and biological research, and to global health, to date.

ACKNOWLEDGMENTS

Financial support is gratefully acknowledged for D.A.G. from the Oregon Sea Grant under award number NA10OAR4170059 (project number R/BT-52) from the National Oceanic and Atmospheric Administration's National Sea Grant College Program, U.S. Department of Commerce, and by appropriations made by the Oregon State legislature; support for R.A.L. and E.H.A. was provided by the New Jersey Agricultural Experiment Station and the School of Environmental and Biological Sciences, Rutgers University.

REFERENCES

1. Hong, J. Role of natural product diversity in chemical biology. *Curr. Opin. Chem. Biol.* **2011**, 15, 350–354.
2. Dixon, N.; Wong, L. S.; Geerlings, T. H.; Micklefield, J. Cellular targets of natural products. *Nat. Prod. Rep.* **2007**, 24, 1197–1432.
3. Wilson, Z. E.; Brimble, M. A. Molecules derived from the extremes of life. *Nat. Prod. Rep.* **2009**, 26, 44–71.
4. Lonsdale, P. Clustering of suspension-feeding macrobenthos near abyssal hydrothermal vents at oceanic spreading centers. *Deep-Sea Res.* **1977**, 24, 857–863.
5. Jannasch, H. W.; Wirsen, C. O. Morphological survey of microbial mats near deep-sea thermal vents. *Appl. Environ. Microbiol.* **1981**, 41, 528–538.
6. Van Dover, C. L. *The Ecology of Deep-Sea Hydrothermal Vents*. Princeton University Press: Princeton, NJ, **2000**; p. 425.
7. Kelley, D. S.; Baross, J. A.; Delaney, J. R. Volcanoes, fluids, and life at mid-ocean ridge spreading centers. *Annu. Rev. Earth Planet. Sci.* **2002**, 30, 385–491.
8. Schmidt, C.; Vuillemin, R.; Le Gall, C.; Gaill, F.; Le Bris, N. Geochemical energy sources for microbial primary production in the environment of hydrothermal vent shrimps. *Mar. Chem.* **2008**, 108, 18–31.
9. Dubilier, N.; Bergin, C.; Lott, C. Symbiotic diversity in marine animals: The art of harnessing chemosynthesis. *Nat. Rev. Microbiol.* **2008**, 6, 725–740.
10. Thresher, R. E.; Adkins, J.; Fallon, S. J.; Gowlett-Holmes, K.; Althaus, F.; Williams, A. Extraordinarily high biomass benthic community on Southern Ocean seamounts. *Sci. Rep.* **2011**, 1, 119–119.
11. Thiel, M. Cindy Lee Van Dover: The ecology of deep-sea hydrothermal vents. *Helgol. Mar. Res.* **2002**, 55, 308–309.

12. Jebbar, M.; Franzetti, B.; Girard, E.; Oger, P. Microbial diversity and adaptation to high hydrostatic pressure in deep-sea hydrothermal vents prokaryotes. *Extremophiles* **2015**, 19, 721–740.
13. Skropeta, D. Deep-sea natural products. *Nat. Prod. Rep.* **2008**, 25, 1131–1166.
14. Skropeta, D.; Wei, L. Recent advances in deep-sea natural products. *Nat. Prod. Rep.* **2014**, 31, 999–1025.
15. Tarasov, V. G.; Gebruk, A. V.; Mironov, A. N.; Moskalev, L. I. Deep-sea and shallow-water hydrothermal vent communities: Two different phenomena? *Chem. Geol.* **2005**, 224, 5–39.
16. Thornburg, C. C.; Zabriskie, T. M.; McPhail, K. L. Deep-sea hydrothermal vents: Potential hot spots for natural products discovery? *J. Nat. Prod.* **2010**, 73, 489–499.
17. Orcutt, B. N.; Sylvan, J. B.; Knab, N. J.; Edwards, K. J. Microbial ecology of the dark ocean above, at, and below the seafloor. *Microbiol. Mol. Biol. Rev.* **2011**, 75, 361–422.
18. Pedersen, R. B.; Rapp, H. T.; Thorseth, I. H.; Lilley, M. D.; Barriga, F. J. A. S.; Baumberger, T.; Flesland, K.; Fonseca, R.; Früh-green, G. L.; Jorgensen, S. L. Discovery of a black smoker vent field and vent fauna at the Arctic Mid-Ocean Ridge. *Nat. Commun.* **2010**, 1, 126–126.
19. Amend, J. P.; McCollom, T. M.; Hentscher, M.; Bach, W. Catabolic and anabolic energy for chemolithoautotrophs in deep-sea hydrothermal systems hosted in different rock types. *Geochim. Cosmochim. Acta* **2011**, 75, 5736–5748.
20. Haymon, R. M. Growth history of hydrothermal black smoker chimneys. *Nature* **1983**, 301, 695–698.
21. Adams, D. K.; McGillicuddy Jr., D. J.; Zamudio, L.; Thurnherr, A. M.; Liang, X.; Rouxel, O.; German, C. R.; Mullineaux, L. S. Surface-generated mesoscale eddies. *Science* **2011**, 332, 580–583.
22. Rogers, A. D.; Tyler, P. A.; Connelly, D. P.; Copley, J. T.; James, R.; Larter, R. D.; Linse, K. et al. The discovery of new deep-sea hydrothermal vent communities in the Southern Ocean and implications for biogeography. *PLoS Biol.* **2012**, 10, e1001234.
23. Manohar, C. S.; Raghukumar, C. Fungal diversity from various marine habitats deduced through culture-independent studies. *FEMS Microbiol. Lett.* **2013**, 341, 69–78.
24. Flores, G. E.; Wagner, I. D.; Liu, Y.; Reysenbach, A.-L. Distribution, abundance, and diversity patterns of the thermoacidophilic "deep-sea hydrothermal vent euryarchaeota 2". *Front. Microbiol.* **2012**, 3, 47.
25. He, T.; Zhang, X. Characterization of bacterial communities in deep-sea hydrothermal vents from three oceanic regions. *Mar. Biotechnol.* **2016**, 18, 232–241.
26. Anderson, R. E.; Sogin, M. L.; Baross, J. A. Biogeography and ecology of the rare and abundant microbial lineages in deep-sea hydrothermal vents. *FEMS Microbiol. Ecol.* **2015**, 91, 1–11.
27. Xie, W.; Wang, F.; Guo, L.; Chen, Z.; Sievert, S. M.; Meng, J.; Huang, G. et al. Comparative metagenomics of microbial communities inhabiting deep-sea hydrothermal vent chimneys with contrasting chemistries. *ISME J.* **2011**, 5, 414–426.
28. Helmke, E.; Weyland, H. *Rhodococcus marinonascens* sp. nov., an Actinomycete from the sea. *Int. J. Syst. Bacteriol.* **1984**, 34, 127–138.
29. Fenical, W.; Jensen, P. R. Developing a new resource for drug discovery: Marine actinomycete bacteria. *Nat. Chem. Biol.* **2006**, 2, 666–673.
30. Comita, P. B.; Gagosian, R. B. Membrane lipid from deep-sea hydrothermal vent methanogen: A new macrocyclic glycerol diether. *Science* **1983**, 222, 1329–1331.
31. Comita, P. B.; Gagosian, R. B.; Pang, H.; Costello, C. E. Structural elucidation of a unique macrocyclic membrane lipid from a new, extremely thermophilic, deep-sea hydrothermal vent Archaebacterium, *Methanococcus jannaschii*, *J. Biol. Chem.* **1984**, 259, 15234–15241.

32. Ritzau, M.; Keller, M.; Wessels, P.; Stetter, K. O.; Zeeck, A. New cyclic polysulfides from hyperthermophilic archaea of the genus *Thermococcus*. *Liebigs Ann. Chem.* **1993**, 8, 871–876.
33. Morita, K.; Kobayashi, S. Isolation, structure, and synthesis of lenthionine and its analogs. *Chem. Pharm. Bull.* **1967**, 15, 988–993.
34. Sobik, P.; Grunenberg, J.; Böröczky, K.; Laatsch, H.; Wagner-Döbler, I.; Schulz, S. Identification, synthesis, and conformation of tri- and tetrathiacycloalkanes from marine bacteria. *J. Org. Chem.* **2007**, 72, 3776–3782.
35. Aebisher, D.; Brzostowska, E. M.; Sawwan, N.; Ovalle, R.; Greer, A. Implications for the existence of a heptasulfur linkage in natural *o*-benzopolysulfanes. *J. Nat. Prod.* **2007**, 70, 1492–1494.
36. Waring, R. H. Sulfur-sulfur compounds. In *Biological Interactions of Sulfur Compounds*, Mitchell, S. C., Ed. Taylor & Francis Ltd: London, U.K., **2006**; pp. 145–173.
37. Pluth, M. D.; Bailey, T. S.; Hammers, M. D.; Hartle, M. D.; Henthorn, H. A.; Steiger, A. K. Natural products containing hydrogen sulfide releasing moieties. *Synlett* **2015**, 26, 2633–2643.
38. Xu, J.; Chatterjee, M.; Baguley, T. D.; Brouillette, J.; Kurup, P.; Ghosh, D.; Kanyo, J. et al. Inhibitor of the tyrosine phosphatase STEP reverses cognitive deficits in a mouse model of Alzheimer's disease. *PLoS Biol.* **2014**, 12, e1001923.
39. Michael, A. J. Polyamines in eukaryotes, bacteria, and archaea. *J. Biol. Chem.* **2016**, 291, 14896–14903.
40. Hamana, K.; Niitsu, M.; Samejima, K.; Matsuzaki, S. Polyamine distributions in thermophilic eubacteria belonging to *Thermus* and *Acidothermus*. *J. Biochem.* **1991**, 109, 444–449.
41. Oshima, T. Unique polyamines produced by an extreme thermophile, *Thermus thermophilus*. *Amino Acids* **2007**, 33, 367–372.
42. Wolkenstein, K.; Schoefberger, W.; Müller, N.; Oji, T. Proisocrinins A-F, brominated anthraquinone pigments from the stalked crinoid *Proisocrinus ruberrimus*. *J. Nat. Prod.* **2009**, 72, 2036–2039.
43. Andrianasolo, E. H.; Haramaty, L.; Rosario-Passapera, R.; Bidle, K.; White, E.; Vetriani, C.; Falkowski, P.; Lutz, R. Ammonificins A and B, hydroxyethylamine chroman derivatives from a cultured marine hydrothermal vent bacterium, *Thermovibrio ammonificans*. *J. Nat. Prod.* **2009**, 72, 1216–1219.
44. Andrianasolo, E. H.; Haramaty, L.; Rosario-Passapera, R.; Vetriani, C.; Falkowski, P.; White, E.; Lutz, R. Ammonificins C and D, hydroxyethylamine chromene derivatives from a cultured marine hydrothermal vent bacterium, *Thermovibrio ammonificans*. *Mar. Drugs* **2012**, 10, 2300–2311.
45. Homann, V. V.; Sandy, M.; Tincu, J. A.; Templeton, A. S.; Tebo, B. M.; Butler, A. Loihichelins A-F, a suite of amphiphilic siderophores produced by the marine bacterium *Halomonas* LOB-5. *J. Nat. Prod.* **2009**, 72, 884–888.
46. Vraspir, J. M.; Butler, A. Chemistry of marine ligands and siderophores. *Annu. Rev. Mar. Sci.* **2009**, 1, 43–63.
47. Ding, C.; Wu, X.; Auckloo, B.; Chen, C.-T.; Ye, Y.; Wang, K.; Wu, B. An unusual stress metabolite from a hydrothermal vent fungus *Aspergillus* sp. WU 243 induced by cobalt. *Molecules* **2016**, 21, 105.
48. Yang, J. X.; Qiu, S. X.; She, Z. G. Metabolites of mangrove endophytic fungus Gx-3a from South China Sea. *Technol. Dev. Chem. Ind.* **2013**, 2, 168–170.
49. Stierle, D. B.; Stierle, A. A.; Patacini, B. The berkeleyacetals, three mereterpenes from a deep water acid mine waste penicillium. *J. Nat. Prod.* **2007**, 70, 1820–1823.
50. Andrianasolo, E. H.; Lutz, R.; Falkowski, P. Deep-sea hydrothermal vents as a new source of drug discovery. In *Bioactive Natural Products*, 1st edn., Atta-ur-Rahman, Ed. Elsevier: Amsterdam, the Netherlands, **2012**; Vol. 36, pp. 43–66.

51. Lino, S. P. P.; Machado, M.; Lopes, D.; Rosario, V. D.; Santos, R. S.; Colaco, A. Evaluation of antimalarial properties from Azores deep-sea invertebrate extracts: A first contribution. *bioRxiv* 030007, doi: https://doi.org/10.1101/030007.
52. Houghton, J. L.; Seyfried Jr, W. E. An experimental and theoretical approach to determining linkages between geochemical variability and microbial biodiversity in seafloor hydrothermal chimneys. *Geobiology* **2010**, 8, 457–470.
53. Wang, F.; Zhou, H.; Meng, J. GeoChip-based analysis of metabolic diversity of microbial communities at the Juan de Fuca Ridge hydrothermal vent. *Proc. Natl. Acad. Sci. USA* **2009**, 106, 4840–4845.
54. McCollom, T. M.; Shock, E. L. Geochemical constraints on chemolithoautotrophic metabolism by microorganisms in seafloor hydrothermal systems. *Geochim. Cosmochim. Acta* **1997**, 61, 4375–4391.
55. McCollom, T. M. Geochemical constraints on primary productivity in submarine hydrothermal vent plumes. *Deep Sea Res.* **2000**, 47, 85–101.
56. Tivey, M. K. Environmental conditions within active seafloor vent structures: Sensitivity to vent fluid composition and fluid flow. *Geophys. Monogr. Ser.* **2004**, 144, 137–152.
57. Lever, M. A. A new era of methanogenesis research. *Trends Microbiol.* **2016**, 24, 84–86.
58. McCollom, T. M. Geochemical constraints on sources of metabolic energy for chemolithoautotrophy in ultramafic-hosted deep-sea hydrothermal systems. *Astrobiology* **2007**, 7, 933–950.
59. Dalsgaard, T.; Canfield, D. E.; Petersen, J.; Thamdrup, B.; Acuna-Gonzalez, J. N_2 production by the anammox reaction in the anoxic water column of Golfo Dulce, Costa Rica. *Nature* **2003**, 422, 606–608.
60. Oshiki, M.; Satoh, H.; Okabe, S. Ecology and physiology of anaerobic ammonium oxidizing bacteria. *Environ. Microbiol.* **2016**, 18, 2784–2796.
61. Stokke, R.; Dahle, H.; Roalkvam, I.; Wissuwa, J.; Daae, F. L.; Tooming-Klunderud, A.; Thorseth, I. H.; Pedersen, R. B.; Steen, I. H. Functional interactions among filamentous Epsilonproteobacteria and Bacteroidetes in a deep-sea hydrothermal vent biofilm. *Environ. Microbiol.* **2015**, 17, 4063–4077.
62. Van Dover, C. L.; Szuts, E. Z.; Chamberlain, S. C.; Cann, J. R. A novel eye in 'eyeless' shrimp from hydrothermal vents of the Mid-Atlantic Ridge. *Nature* **1989**, 337, 458–460.
63. Van Dover, C. L.; Reynolds, G. T.; Chave, A. D.; Tyson, J. A. Light at deep-sea hydrothermal vents. *Geophys. Res. Lett.* **1996**, 23, 2049–2052.
64. Beatty, J. T.; Overmann, J.; Lince, M. T.; Manske, A. K.; Lang, A. S.; Blankenship, R. E.; Van Dover, C. L.; Martinson, T. A.; Plumley, F. G. An obligately photosynthetic bacterial anaerobe from a deep-sea hydrothermal vent. *Proc. Natl. Acad. Sci. USA* **2005**, 102, 9306–9310.
65. Mazor, O.; Brandis, A.; Plaks, V.; Neumark, E.; Rosenbach-Belkin, V.; Salomon, Y.; Scherz, A. WST11, a novel water-soluble bacteriochlorophyll derivative; cellular uptake, pharmacokinetics, biodistribution and vascular-targeted photodynamic activity using melanoma tumors as a model. *Photochem. Photobiol.* **2005**, 81, 342–351.
66. Azzouzi, A.-R.; Vincendeau, S.; Barret, E.; Cicco, A.; Kleinclauss, F.; van der Poel, H. G.; Stief, C. G. et al. Padeliporfin vascular-targeted photodynamic therapy versus active surveillance in men with low-risk prostate cancer (CLIN1001 PCM301): An open-label, phase 3, randomised controlled trial. *Lancet Oncol.* **2016**, 18(2), 181–191.
67. Wang, F.; Zhou, H.; Meng, J. GeoChip-based analysis of metabolic diversity of microbial communities at the Juan de Fuca Ridge hydrothermal vent. *Proc. Natl. Acad. Sci. USA* **2009**, 106, 4840–4845.
68. Takai, K.; Nakamura, K. Archaeal diversity and community development in deep-sea hydrothermal vents. *Curr. Opin. Microbiol.* **2011**, 14, 282–291.
69. Ver Eecke, H. C. Growth kinetics and constraints related to metabolic diversity and abundances of hyperthermophiles in deep-sea hydrothermal vents. Open Access dissertation, University of Massachusetts, Amherst, MA, **2011**.

70. Stewart, F. J.; Newton, I. L. G.; Cavanaugh, C. M. Chemosynthetic endosymbioses: Adaptations to oxic-anoxic interfaces. *Trends Microbiol.* **2005**, 13, 439–448.
71. Johnson, S. B.; Won, Y.-J.; Harvey, J. B. J.; Vrijenhoek, R. C. A hybrid zone between *Bathymodiolus* mussel lineages from eastern Pacific hydrothermal vents. *BMC Evol. Biol.* **2013**, 13, 21(1–18).
72. Petersen, J. M.; Zielinski, F. U.; Pape, T.; Seifert, R.; Moraru, C.; Amann, R.; Hourdez, S. et al. Hydrogen is an energy source for hydrothermal vent symbioses. *Nature* **2011**, 476, 176–180.
73. Vannier, P.; Michoud, G. G.; Oger, P.; Marteinsson, V. P. R.; Jebbar, M. Genome expression of *Thermococcus barophilus* and *Thermococcus kodakarensis* in response to different hydrostatic pressure conditions. *Res. Microbiol.* **2015**, 166 717–725.
74. Cario, A.; Lormieres, F.; Xiang, X.; Oger, P. High hydrostatic pressure increases amino acid requirements in the piezo-hyperthermophilic archaeon *Thermococcus barophilus*. *Res. Microbiol.* **2015**, 166, 710–716.
75. Michoud, G.; Jebbar, M. High hydrostatic pressure adaptive strategies in an obligate piezophile *Pyrococcus yayanosii*. *Sci. Rep.* **2016**, 6, 27289.
76. Burgaud, G.; Hu, N. T. M.; Arzur, D.; Coton, M.; Perrier-Cornet, J. M.; Jebbar, M.; Barbier, G. Effects of hydrostatic pressure on yeasts isolated from deep-sea hydrothermal vents. *Res. Microbiol.* **2015**, 166, 700–709.
77. Van Dover, C. L.; German, C. R.; Speer, K. G.; Parson, L. M.; Vrijenhoek, R. C. Evolution and biogeography of deep-sea vent and seep invertebrates. *Science* **2002**, 295, 1253–1257.
78. Woese, C. R. A new biology for a new century. *Microbiol. Mol. Biol. Rev.* **2004**, 68, 173–186.
79. Perez, M.; Juniper, S. K. Insights into symbiont population structure among three vestimentiferan tubeworm host species at Eastern Pacific Spreading Centers. *Appl. Environ. Microbiol.* **2016**, 82, 5197–5205.
80. Duperron, S.; Lorion, J.; Samadi, S.; Gros, O.; Gaill, F. Symbioses between deep-sea mussels (Mytilidae: Bathymodiolinae) and chemosynthetic bacteria: Diversity, function and evolution. *C. R. Biol.* **2009**, 332, 298–310.
81. Rodrigues, C. F.; Cunha, M. R.; Génio, L.; Duperron, S. A complex picture of associations between two host mussels and symbiotic bacteria in the northeast Atlantic. *Naturwissenschaften* **2013**, 100, 21–31.
82. Génio, L.; Kiel, S.; Cunha, M. R.; Grahame, J.; Little, C. T. S. Shell microstructures of mussels (Bivalvia: Mytilidae: Bathymodiolinae) from deep-sea chemosynthetic sites: Do they have a phylogenetic significance?. *Deep-Sea Res.* **2012**, 64, 86–103.
83. Vrijenhoek, R. C. Genetics and evolution of deep-sea chemosynthetic bacteria and their invertebrate hosts. In *The Vent and Seep Biota*, Kiel, S., Ed. Springer: Dodrecht, the Netherlands, **2010**; pp. 16–42.
84. Klose, J.; Polz, M. F.; Wagner, M.; Schimak, M. P.; Gollner, S.; Bright, M. Endosymbionts escape dead hydrothermal vent tubeworms to enrich the free-living population. *Proc. Natl. Acad. Sci. USA* **2015**, 112, 11300–11305.
85. Childress, J. J.; Fisher, C. R. The biology of hydrothermal vent animals: Physiology, biochemistry, and autotrophic symbiosis. In *Oceanography and Marine Biology: An Annual Review*, Barnes, M.; Ansell, A. D.; Gibson, R. N., Eds. UCL Press Limited: London, U.K., **1992**; Vol. 30, pp. 337–441.
86. Desbruyères, D.; Segonzac, M.; Bright, M. *Handbook of Deep-Sea Hydrothermal Vent Fauna*. Denisia 18: Linz, Austria, **2006**; p. 544.
87. Chao, L. S. L.; Davis, R. E.; Moyer, C. L. Characterization of bacterial community structure in vestimentiferan tubeworm *Ridgeia piscesae* trophosomes. *Mar. Ecol.* **2007**, 28, 72–85.

88. Harmer, T. L.; Rotjan, R. D.; Nussbaumer, A. D.; Bright, M.; Ng, A. W.; DeChaine, E. G.; Cavanaugh, C. M. Free-living tubeworm endosymbionts found at deep-sea vents. *Appl. Environ. Microbiol.* **2008**, 74, 3895–3898
89. Lopez-Garcia, P.; Gaill, F.; Moreira, D. Wide bacterial diversity associated with tubes of the vent worm *Riftia pachyptila*. *Environ. Microbiol.* **2002**, 4, 204–215.
90. Haddad, A.; Camacho, F.; Durand, P.; Cary, S. C. Phylogenetic characterization of the epibiotic bacteria associated with the hydrothermal vent polychaete *Alvinella pompejana*. *Appl. Environ. Microbiol.* **1995**, 61, 1679–1687.
91. Polz, M. F.; Robinson, J. J.; Cavanaugh, C. M.; Dover, C. L. V. Trophic ecology of massive shrimp aggregations at a Mid-Atlantic Ridge hydrothermal vent site. *Limnol. Oceanogr.* **1998**, 43, 1631–1638.
92. Goffredi, S. K.; Waren, A.; Orphan, V. J.; Van Dover, C. L.; Vrijenhoek, R. C. Novel forms of structural integration between microbes and a hydrothermal vent gastropod from the Indian Ocean. *Appl. Environ. Microbiol.* **2004**, 70, 3082–3090.
93. Santelli, C. M.; Orcutt, B. N.; Banning, E.; Bach, W.; Moyer, C. L.; Sogin, M. L.; Staudigel, H.; Edwards, K. J. Abundance and diversity of microbial life in ocean crust. *Nature* **2008**, 453, 653–656.
94. Kicklighter, C. E.; Fisher, C. R.; Hay, M. E. Chemical defense of hydrothermal vent and hydrocarbon seep organisms: A preliminary assessment using shallow-water consumers. *Mar. Ecol. Prog. Ser.* **2004**, 275, 11–19.
95. Nishijima, M.; Lindsay, D. J.; Hata, J.; Nakamura, A.; Kasai, H.; Ise, Y.; Fisher, C. R.; Fujiwara, Y.; Kawato, M.; Maruyama, T. Association of thioautotrophic bacteria with deep-sea sponges. *Mar. Biotechnol.* **2010**, 12, 253–260.
96. Petersen, J. M.; Dubilier, N. Methanotrophic symbioses in marine invertebrates. *Environ. Microbiol. Rep.* **2009**, 1, 319–335.
97. Sun, Q. L.; Zeng, Z. G.; Chen, S.; Sun, L. First comparative analysis of the community structures and carbon metabolic pathways of the bacteria associated with *Alvinocaris longirostris* in a hydrothermal vent of Okinawa Trough. *PLoS One* **2016**, 11(4), e0154359.
98. Blunt, J. W.; Copp, B. R.; Keyzers, R. A.; Munro, M. H. G.; Prinsep, M. R. Marine natural products. *Nat. Prod. Rep.* **2016**, 33, 382–431.
99. Subramani, R.; Aalbersberg, W. Marine actinomycetes: An ongoing source of novel bioactive metabolites. *Microbiol. Res.* **2012**, 167, 571–580.
100. Ferrer, M.; Golyshina, O.; Beloqui, A.; Golyshin, P. N. Mining enzymes from extreme environments. *Curr. Opin. Microbiol.* **2007**, 10, 207–214.
101. Pettit, R. K. Culturability and secondary metabolite diversity of extreme microbes: Expanding contribution of deep sea and deep-sea vent microbes to natural product discovery. *Mar. Biotechnol.* **2011**, 13, 1–11.
102. JAMSTEC. Cooperative Research Project for extremophiles. http://www.jamstec.go.jp/jamstec-j//XBR/bv/en/menubiov.html (December 12, 2016).
103. ATCC. ATCC bacteriology collection, search terms: Deep sea hydrothermal vent. The Global BioResource Center, Manassas, VA. https://www.atcc.org/Search_Results.aspx?dsNav=Ntk:PrimarySearch%7cdeep+sea+hydrothermal+vent%7c3%7c,Ny:True,Ro:0,N:1000552&searchTerms=deep+sea+hydrothermal+vent&redir=1 (July 14, 2017).
104. Nunoura, T.; Oida, H.; Nakaseama, M.; Kosaka, A.; Ohkubo, S. B.; Kikuchi, T.; Kazama, H. et al. Archaeal diversity and distribution along thermal and geochemical gradients in hydrothermal sediments at the yonaguni knoll IV hydrothermal field in the southern Okinawa Trough. *Appl. Environ. Microbiol.* **2010**, 76, 1198–1211.
105. Birrien, J.-L.; Zeng, X.; Jebbar, M.; Cambon-Bonavita, M.-A.; Querellou, J.; Oger, P.; Bienvenu, N.; Xiao, X.; Prieur, D. *Pyrococcus yayanosii* sp. nov., an obligate piezophilic hyperthermophilic archaeon isolated from a deep-sea hydrothermal vent. *Int. J. Syst. Evol. Microbiol.* **2011**, 61, 2827–2831.

106. Gorlas, A.; Alain, K.; Bienvenu, N.; Geslin, C. *Thermococcus prieurii* sp. nov., a hyperthermophilic archaeon isolated from a deep-sea hydrothermal vent. *Int. J. Syst. Evol. Microbiol.* **2013**, 63, 2920–2926.
107. Gorlas, A.; Croce, O.; Oberto, J.; Gauliard, E.; Forterre, P.; Marguet, E. *Thermococcus nautili* sp. nov., a hyperthermophilic archaeon isolated from a hydrothermal deep-sea vent. *Int. J. Syst. Evol. Microbiol.* **2014**, 64, 1802–1810.
108. Hensley, S. A.; Jung, J. H.; Park, C. S.; Holden, J. F. *Thermococcus paralvinellae* sp. nov. and *Thermococcus cleftensis* sp. nov. of hyperthermophilic heterotrophs from deep-sea hydrothermal vents. *Int. J. Syst. Evol. Microbiol.* **2014**, 64, 3655–3659.
109. Stewart, L. C.; Jung, J. H.; Kim, Y. T.; Kwon, S. W.; Park, C. S.; Holden, J. F. *Methanocaldococcus bathoardescens* sp. nov., a hyperthermophilic methanogen isolated from a volcanically active deep-sea hydrothermal vent. *Int. J. Syst. Evol. Microbiol.* **2015**, 65, 1280–1283.
110. Rosario-Passapera, R.; Keddis, R.; Wong, R.; Lutz, R. A.; Starovoytov, V.; Vetriani, C. *Parvibaculum hydrocarboniclasticum* sp. nov., a mesophilic, alkane-oxidizing alphaproteobacterium isolated from a deep-sea hydrothermal vent on the East Pacific Rise. *Int. J. Syst. Evol. Microbiol.* **2012**, 62, 2921–2926.
111. Makita, H.; Nakagawa, S.; Miyazaki, M.; Nakamura, K. I.; Inagaki, F.; Takai, K. *Thiofractor thiocaminus* gen. nov., sp. nov., a novel hydrogen-oxidizing, sulfur-reducing epsilonproteobacterium isolated from a deep-sea hydrothermal vent chimney in the Nikko Seamount weld of the northern Mariana Arc. *Arch. Microbiol.* **2012**, 194, 785–794.
112. Slobodkina, G. B.; Reysenbach, A. L.; Panteleeva, A. N.; Kostrikina, N. A.; Wagner, I. D.; Bonch-Osmolovskaya, E. A.; Slobodkin, A. I. *Deferrisoma camini* gen. nov., sp. nov., a moderately thermophilic, dissimilatory iron(III)-reducing bacterium from a deep-sea hydrothermal vent that forms a distinct phylogenetic branch in the Deltaproteobacteria. *Int. J. Syst. Evol. Microbiol.* **2012**, 62, 2463–2468.
113. Reysenbach, A. L.; Liu, Y.; Lindgren, A. R.; Wagner, I. D.; Sislak, C. D.; Mets, A.; Schouten, S. *Mesoaciditoga lauensis* gen. nov., sp. nov., a moderately thermoacidophilic member of the order Thermotogales from a deep-sea hydrothermal vent. *Int. J. Syst. Evol. Microbiol.* **2013**, 63, 4724–4729.
114. Slobodkin, A. I.; Reysenbach, A. L.; Slobodkina, G. B.; Baslerov, R. V.; Kostrikina, N. A.; Wagner, I. D.; Bonch-Osmolovskaya, E. A. *Thermosulfurimonas dismutans* gen. nov., sp. nov., an extremely thermophilic sulfur-disproportionating bacterium from a deep-sea hydrothermal vent. *Int. J. Syst. Evol. Microbiol.* **2012**, 62, 2565–2571.
115. Izumi, H.; Nunoura, T.; Miyazaki, M.; Mino, S.; Toki, T.; Takai, K.; Sako, Y.; Sawabe, T.; Nakagawa, S. *Thermotomaculum hydrothermale* gen. nov., sp. nov., a novel heterotrophic thermophile within the phylum Acidobacteria from a deep-sea hydrothermal vent chimney in the Southern Okinawa Trough. *Extremophiles* **2012**, 16, 245–253.
116. Slobodkin, A. I.; Reysenbach, A. L.; Slobodkina, G. B.; Kolganova, T. V.; Kostrikina, N. A.; Bonch-Osmolovskaya, E. A. *Dissulfuribacter thermophilus* gen. nov., sp. nov., a thermophilic, autotrophic, sulfur-disproportionating, deeply branching deltaproteobacterium from a deep-sea hydrothermal vent. *Int. J. Syst. Evol. Microbiol.* **2013**, 63, 1967–1971.
117. Pérez-Rodríguez, I.; Grosche, A.; Massenburg, L.; Starovoytov, V.; Lutz, R. A.; Vetriani, C. *Phorcysia thermohydrogeniphila* gen. nov., sp. nov., a thermophilic, chemolithoautotrophic, nitrateammonifying bacterium from a deep-sea hydrothermal vent. *Int. J. Syst. Evol. Microbiol.* **2012**, 62, 2388–2394.
118. Flores, G. E.; Hunter, R. C.; Liu, Y.; Mets, A.; Schouten, S.; Reysenbach, A. L. *Hippea jasoniae* sp. nov. and *Hippea alviniae* sp. nov., thermoacidophilic members of the class Deltaproteobacteria isolated from deep-sea hydrothermal vent deposits. *Int. J. Syst. Evol. Microbiol.* **2012**, 62, 1252–1258.
119. Hasan, N. A.; Grim, C. J.; Grim, C. J.; Lipp, E. K.; Lipp, E. K.; Rivera, I. N. G.; Rivera, I. N. G. et al. Deep-sea hydrothermal vent bacteria related to human pathogenic *Vibrio* species. *Proc. Natl. Acad. Sci. USA* **2015**, 112(21), E2813–E2819.

120. Abraham, W. R.; Lünsdorf, H.; Vancanney, M.; Smit, J. Cauliform bacteria lacking phospholipids from an abyssal hydrothermal vent: Proposal of *Glycocaulis abyssi* gen. nov., sp. nov., belonging to the family Hyphomonadaceae. *Int. J. Syst. Evol. Microbiol.* **2013**, 63, 2207–2215.
121. Oren, A.; Zhou, P.; Jian, S.-L.; Liu, Z.-S.; Xu, X.-W.; Wang, C.-S.; Wu, Y.-H. *Pontibacter amylolyticus* sp. nov., isolated from a deep-sea sediment hydrothermal vent field. *Int. J. Syst. Evol. Microbiol.* **2016**, 66, 1760–1767.
122. Wu, Y.-H.; Xu, L.; Zhou, P.; Wang, C.-S.; Oren, A.; Xu, X.-W. *Brevirhabdus pacifica* gen. nov., sp. nov., isolated from deep-sea sediment in a hydrothermal vent field. *Int. J. Syst. Evol. Microbiol.* **2015**, 65, 3645–3651.
123. Zeng, X.; Zhang, Z.; Li, X.; Zhang, X.; Cao, J.; Jebbar, M.; Alain, K.; Shao, Z. *Anoxybacter fermentans* gen. nov., sp. nov., a piezophilic, thermophilic, anaerobic, fermentative bacterium isolated from a deep-sea hydrothermal vent. *Int. J. Syst. Evol. Microbiol.* **2015**, 65, 710–715.
124. Grosche, A.; Sekaran, H.; Pérez-Rodríguez, I.; Starovoytov, V.; Vetriani, C. *Cetia pacifica* gen. nov., sp. nov., a chemolithoautotrophic, thermophilic, nitrate-ammonifying bacterium from a deep-sea hydrothermal vent. *Int. J. Syst. Evol. Microbiol.* **2015**, 65, 1144–1150.
125. Stewart, E. J. Growing unculturable bacteria. *J. Bacteriol.* **2012**, 194, 4151–4160.
126. Bode, H. B.; Bethe, B.; Hofs, R.; Zeeck, A. Big effects from small changes: Possible ways to explore nature's chemical diversity. *ChemBioChem* **2002**, 3, 619–627.
127. Scherlach, K.; Hertweck, C. Triggering cryptic natural product biosynthesis in microorganisms. *Org. Biomol. Chem.* **2009**, 7, 1753–1760.
128. Connon, S. A.; Giovannoni, S. J. High-throughput methods for culturing microorganisms in very-low-nutrient media yield diverse new marine isolates. *Appl. Environ. Microbiol.* **2002**, 68, 3878–3885.
129. Au, S. H.; Shih, S. C. C.; Wheeler, A. R. Integrated microbioreactor for culture and analysis of bacteria, algae and yeast. *Biomed. Microdevices* **2011**, 13, 41–50.
130. Wcry, N.; Gerike, U.; Sharman, A.; Chaudhuri, J. B.; Hough, D. W.; Danson, M. J. Use of a packed-column bioreactor for isolation of diverse protease-producing bacteria from Antarctic soil. *Appl. Environ. Microbiol.* **2003**, 69, 1457–1464.
131. Bochner, B. R. *Phenomics and Phenotype Microarrays: Applications Complementing Metagenomics*. Wiley-Blackwell: Hoboken, NJ, **2011**; pp. 533–540.
132. Borglin, S.; Joyner, D.; DeAngelis, K. M.; Khudyakov, J.; D'haeseleer, P.; Joachimiak, M. P.; Hazen, T. Application of phenotypic microarrays to environmental microbiology. *Curr. Opin. Biotechnol.* **2012**, 23, 41–48.
133. Borglin, S.; Joyner, D.; Jacobsen, J.; Mukhopadhyay, A.; Hazen, T. C. Overcoming the anaerobic hurdle in phenotypic microarrays: Generation and visualization of growth curve data for *Desulfovibrio vulgaris* Hildenborough. *J. Microbiol. Methods* **2009**, 76, 159–168.
134. Jiang, L.; Long, C.; Wu, X.; Xu, H.; Shao, Z.; Long, M. Optimization of thermophilic fermentative hydrogen production by the newly isolated *Caloranaerobacter azorensis* H53214 from deep-sea hydrothermal vent environment. *Int. J. Hydrog. Energy* **2014**, 39, 14154–14160.
135. Gelfand, D. H.; Stoffel, S.; Lawyer, F. C.; Saiki, R. K. Purified thermostable enzyme. U.S. Patent 4889818, December 26, **1989**.
136. Innis, M. A.; Myambo, K. B.; Gelfand, D. H.; Brow, M. A. D. Methods for DNA sequencing with *Thermus aquaticus* DNA polymerase. U.S. Patent 5075216, December 24, **1991**.
137. Fore, J.; Wiechers, I. R.; Cook-Deegan, R.; Mullis, K.; Rabinow, P.; Daniell, E.; Mullis, K. et al. The effects of business practices, licensing, and intellectual property on development and dissemination of the polymerase chain reaction: Case study. *J. Biomed. Discov. Collab.* **2006**, 1, 7.
138. Elleuche, S.; Schäfers, C.; Blank, S.; Schröder, C.; Antranikian, G. Exploration of extremophiles for high temperature biotechnological processes. *Curr. Opin. Microbiol.* **2015**, 25, 113–119.

139. Elleuche, S.; Schröder, C.; Sahm, K.; Antranikian, G. Extremozymes—biocatalysts with unique properties from extremophilic microorganisms. *Curr. Opin. Biotechnol.* **2014**, 29, 116–123.
140. Urbieta, M. S.; Donati, E. R.; Chan, K. G.; Shahar, S.; Sin, L. L.; Goh, K. M. Thermophiles in the genomic era: Biodiversity, science, and applications. *Biotechnol. Adv.* **2015**, 33, 633–647.
141. Arico, S.; Salpin, C. *Bioprospecting of Genetic Resources in the Deep Seabed: Scientific, Legal and Policy Aspects*. United Nations University Institute of Advanced Studies: Tokyo, Japan, **2005**; p. 72.
142. Leary, D. Bioprospecting and the genetic resources of hydrothermal vents on the high seas. What is the existing legal position? Where are we heading and what are our options? In *Deep Sea 2003: Conference on the Governance and Management of Deep-Sea Fisheries*, Shotton, R., Ed. FAO: Queenstown, New Zealand, **2003**; pp. 455–487.
143. Product website: http://www.thermofisher.com/order/catalog/product/12337016 (accessed June 6, 2017).
144. Product website: https://www.basf.com/en/products-and-industries/general-business-topics/enzymes/markets.html (accessed June 6, 2017).
145. Kadar, E.; Costa, V.; Santos, R. S. Distribution of micro-essential (Fe, Cu, Zn) and toxic (Hg) metals in tissues of two nutritionally distinct hydrothermal shrimps. *Sci. Total Environ.* **2006**, 358, 143–150.
146. Minic, Z.; Thongbam, P. D. The biological deep sea hydrothermal vent as a model to study carbon dioxide capturing enzymes. *Mar. Drugs* **2011**, 9, 719–738.
147. Cornec, L.; Robineau, J.; Rolland, J. L.; Dietrich, J.; Barbier, G. Thermostable esterases screened on hyperthermophilic archaeal and bacterial strains isolated from deep-sea hydrothermal vents: Characterization of esterase activity of a hyperthermophilic archaeum, *Pyrococcus abyssi*. *J. Mar. Biotechnol.* **1998**, 6, 104–110.
148. Gantelet, H.; Ladrat, C.; Godfroy, A.; Barbier, G.; Duchiron, F. Characteristics of pullulanases from extremely thermophilic archaea isolated from deep-sea hydrothermal vents. *Biotechnol. Lett.* **1998**, 20, 819–823.
149. Guezennec, J. Deep-sea hydrothermal vents: A new source of innovative bacterial exopolysaccharides of biotechnological interest? *J. Ind. Microbiol. Biotechnol.* **2002**, 29, 204–208.
150. Nichols, C. A. M.; Guezennec, J.; Bowman, J. P. Bacterial exopolysaccharides from extreme marine environments with special consideration of the southern ocean, sea ice, and deep-sea hydrothermal vents: A review. *Mar. Biotechnol.* **2005**, 7, 253–271.
151. Raguenes, G.; Christen, R.; Guezennec, J.; Pignet, P.; Barbier, G. *Vibrio diabolicus* sp. nov., a new polysaccharide-secreting organism isolated from a deep-sea vent polychaete annelid, *Alvinella pompejana*. *Int. J. Syst. Bacteriol.* **1997**, 47, 989–995.
152. Zanchetta, P.; Lagarde, N.; Guezennec, J. A new bone-healing material: A hyaluronic acid-like bacterial exopolysaccharide. *Calcif. Tissue Int.* **2003**, 72, 74–79.
153. Zanchetta, P.; Lagarde, N.; Guezennec, J. Systemic effects on bone healing of a new hyaluronic acid-like bacterial exopolysaccharide. *Calc. Tissue Intl.* **2003**, 73, 232–236.
154. Nicolaus, B.; Kambourova, M.; Toksoy Oner, E. Exopolysaccharides from extremophiles: From fundamentals to biotechnology. *Environ. Technol.* **2010**, 31, 1145–1158.
155. Le Costaouëc, T.; Cérantola, S.; Ropartz, D.; Ratiskol, J.; Sinquin, C.; Colliec-Jouault, S.; Boisset, C. Structural data on a bacterial exopolysaccharide produced by a deep-sea *Alteromonas macleodii* strain. *Carbohydr. Polym.* **2012**, 90, 49–59.
156. Smanski, M. J.; Schlatter, D. C.; Kinkel, L. L. Leveraging ecological theory to guide natural product discovery. *J. Ind. Microbiol. Biotechnol.* **2016**, 43, 115–128.
157. Beaulieu, S. E. InterRidge Global Database of active submarine hydrothermal vent fields: Prepared for Inter-Ridge, Version 3.4. World Wide Web electronic publication. Version 3.2, **2015**. http://vents-data.interridge.org (December 13, 2016).

12 Cone Snail Venom Peptides and Future Biomedical Applications of Natural Products

Baldomero M. Olivera, Helena Safavi-Hemami, Shrinivasan Raghuraman, and Russell W. Teichert

CONTENTS

12.1 INTRODUCTION

12.1.1 General Biological Framework for Natural Products

A major theme of this chapter is that natural products research is at a critical juncture. To explain our reasoning, we present an overview of natural products, followed by insights, broadly applicable to all natural products research, gained from studying a specific class of natural products, the peptidic components of cone snail venoms.

Natural products are compounds produced by living organisms that were previously referred to as *secondary metabolites*. Such compounds are distinguished from primary metabolites, the intermediates and products of cellular metabolic pathways or endogenous physiological processes of organisms,[1] which were the major focus of early biochemical research. There has been a gradual replacement of the term *secondary metabolites* with *natural products*. We will use *natural products*, rather than *secondary metabolites*, throughout this chapter. Unfortunately, the term *secondary metabolites* suggested secondary importance and these compounds were initially largely treated as such by the scientific community. One of the early hypotheses about the roles of secondary metabolites was that they were merely waste products used by organisms to eliminate excess nitrogen.

A guiding principle is that each natural product evolved for a specific biological purpose, but this notion has not had much impact on most prior research into natural products. Medicinal chemists have traditionally explored natural products as a source of chemically diverse molecules for drug discovery. Consequently, while remarkable progress has been made in the technology for the chemical characterization of natural products, there has been relatively little effort made to investigate their true biological roles. Natural products, unlike primary metabolites, are not broadly distributed among organisms. The biosynthesis of each natural product is carried out by a taxonomically restricted spectrum of organisms, in some cases perhaps by only a single species.[2] We suggest that an individual natural product is evolved by an organism (the *producer*) to mediate a specific chemical interaction that perturbs the physiology of another organism (the *target*) or small groups of organisms. If this were true, then we could expect that over eons of time, natural selection should have honed each of these compounds to interact optimally with a specific molecular target, causing the desired physiological effect on the target species. The homologs of such molecular targets in humans have potential as therapeutic drug targets. In this chapter, we present evidence from research on *Conus* venom peptides, demonstrating that biomedical applications emerge naturally from basic research that uncovers the mechanisms that underlie the bioactivity of a natural product.[3] We suggest that elucidating the native biological roles of natural products translates basic research into therapeutic drug discovery more efficiently. This translational research requires an interdisciplinary approach at the interface of chemistry and biology.

12.1.2 *Conus* Venom Peptides as Natural Products

The bioactive peptides found in the venoms of cone snails (*conopeptides, conotoxins*) are used by this large, biodiverse lineage of predatory marine snails (~800 species) to capture prey, defend against predators, or deter competitors.[4] This field of study has progressed into a distinctly different trajectory from studies of other natural products, primarily because venoms are injected directly into the targeted organism.[5] In contrast to most natural products research, the biochemical characterization of venom components has always been accompanied by a biological characterization, with the broad goal of elucidating the molecular mechanisms responsible for the observed effects of a venom. As injectable natural products, the peptide components of cone snail venoms have some unique features, but they also share other important characteristics with noninjectable natural products that are delivered to the target organisms through diverse routes.

12.1.3 Case Study of Chemical Interactions: The Skunk

The skunk illustrates a familiar chemical interaction between animals. As a deterrent to potential predators, the skunk's odorous secretions are volatile natural products that mediate this chemical interaction. Identifying the secreted chemicals only scratches the surface. At a more mechanistic level, we would like to identify and characterize the receptors in man and other potential predators that receive the aversive signals. We would also like to know whether these volatile skunk compounds act synergistically: if more than one odorant receptor is activated, is a much more powerful aversive response elicited? What is the spectrum of predators targeted by such compounds? Do all potential predators have homologous receptors? Could the diversity of volatile compounds reflect multiple predators with different odorant receptors and different neuronal circuitry? These are just a few of the types of biological questions associated with this specific chemical interaction that should be addressed.

There are also another set of significant questions regarding the producer organism (the skunk). How much do skunk secretions vary? Does this cocktail of odorants change in a systematic way depending on the ecology or latitude and longitude? How did the system evolve? Do species closely related to the skunk have some homologous mechanisms? How are the natural products stored, and how does the skunk control when to use them? Thus, the chemical characterization of the skunk's odorant compounds is clearly only a first step in addressing the depth of questions that can now be elucidated in a systems biology context. We have presented the skunk as a familiar example of natural products mediating chemical interactions as it illustrates how a broad conceptual framework raises a set of questions that can be framed for all chemical interactions between organisms.

12.1.4 Chemical Neuroethology Defined: A Division of Systems Biology

Although the biological interactions between cone snails and their specific prey, predators, and competitors are less familiar than is the case for skunk and man, the scientific questions are the same. Indeed, for a few cone snail species, there may be more molecular mechanisms elucidated than are presently available for the interaction between the skunk and its potential predators. In both cases however, natural products are central to eliciting a change in the behavior of the targeted animal.

Neuroethology is a field of study that seeks to understand how animal behavior is controlled by the nervous system. We propose the term *chemical neuroethology* to refer to a specific branch of neuroethology that seeks to understand how one organism can modify the behavior of a target animal through chemical compounds that it (the former) produces. The compound needs to be effectively delivered (secrete, inject, spray, etc.) to the nervous system of the targeted animal for purposes advantageous to the producer (e.g., predation, defense, competitive interactions, social interactions, etc.). *When a natural product is evolved in one species to alter the behavior of another species, the comprehensive investigation of both the chemistry and the associated biology requires a new research paradigm that we refer to as chemical neuroethology*. Chemical neuroethology may seem to overlap with the field of chemical ecology, which also studies the chemicals that mediate interactions between organisms, but unlike chemical ecology, chemical neuroethology focuses on the neurobiological mechanisms through which a natural product alters the behavior of the target. In this chapter, we hope to demonstrate that one application of chemical neuroethology is to provide a general platform that can potentially accelerate natural products drug discovery.

12.1.5 Drug Development: Why Natural Products Have Largely Been Abandoned (and Why They Should Not Be)

Almost all large pharmaceutical companies were once dominated by a search for bioactive natural products from cultures of microorganisms.[1] The modern pharmaceutical industry was largely established to discover novel antibiotics, which they pursued by screening compounds secreted by diverse microorganisms for the requisite bioactivity. Medicinal chemists could then alter a natural product to give it an improved pharmacological profile (e.g., increased potency or specificity, among other desirable characteristics that would make the compound a better drug).

This approach to drug discovery was gradually replaced as microbial cultures became less productive. An alternative approach was adopted, based on two scientific advances: (1) Molecular genetics enabled heterologous expression of protein drug targets in immortalized cell lines that could be mass produced for high-throughput screening; (2) Large synthetic chemical libraries were constructed by combinatorial chemistry for high-throughput screening against heterologously expressed drug targets. This industrialization of drug discovery research has been an intellectually seductive approach, but in practice, it has proven to be very costly and relatively unproductive, which may explain, in part, why many pharmaceutical companies are again turning to collaborations with academic research labs and small biotech companies for early-stage drug discovery.[6] Two major problems with the high-throughput screening approach are the following: (1) many of the drug candidates identified have proven to have intolerable off-target side effects and (2) many others fail for lack of therapeutic efficacy,[7] presumably because drug targets were insufficiently validated and the biology of the disease was insufficiently understood. Too often, these problems have only been discovered in late-stage human clinical trials,[7] rendering the current process of drug discovery unsustainably expensive and risky. Typically, natural products are not included in chemical libraries for high-throughput screening because they are often relatively complex molecules that are difficult to synthesize and only a small amount of each natural product is available from natural sources.

Instead of trying to fit natural products into the conventional screening paradigm, we suggest that an entirely different approach be pursued, based on a different set of scientific principles and disciplines. We believe that in the future, many drug leads will come from a *small-shop approach*, with specialized teams of chemists and biologists deriving medicinal compounds that mediate chemical interactions between organisms, rather than from high-throughput screening. Through this approach, investigators should discover structurally novel and diverse compounds that have evolved to target signaling proteins in other organisms with high potency and selectivity. Such compounds will find applications as therapeutic drugs and for identifying and validating novel therapeutic drug targets. These novel compounds and their targets could not be discovered by the present paradigm of molecular genetics, synthetic chemical libraries, and high-throughput screening.

12.2 *CONUS GEOGRAPHUS*, VENOM PEPTIDES, AND BEHAVIOR

12.2.1 Correlating Behavior and Chemistry

In this and the next section, we discuss specific examples of chemical interactions between fish-hunting cone snail species and their prey. Several species will be discussed, with emphasis on *Conus geographus* (Figure 12.1), whose venom components have been extensively characterized chemically and biologically, in regard to their various roles in facilitating prey capture, defense against predators, and deterrence of competitors. Efforts to understand how this piscivorous species captures fish

FIGURE 12.1 Left panel, the shell of *Conus geographus*. Top panels, a cartoon showing how *C. geographus* is believed to hunt in the wild. It approaches a school of small fish hiding in reef crevices at night and releases components of its venom that disable the fish by making them hypoactive and insensitive to water movement. This increases the probability that the snail can successfully engulf the entire school. It then injects venom into each fish, causing an irreversible block of neuromuscular transmission. Bottom panels, right, aquarium observations of *C. geographus* engulfing a fish. The snail detects fish by a chemosensory mechanism using its siphon and begins to open its rostrum (*false mouth*) **(a, b)**. The snail releases some venom components into the water toward the targeted fish **(c)**. The fish is debilitated sufficiently, so the snail approaches it closely and engulfs it **(d)**. The fish is then believed to be injected with paralytic toxins, and in about an hour, the scales and bones of the fish will be regurgitated along with the hypodermic needle-like harpoon that the snail used to inject venom.

led to the characterization of many different bioactive venom compounds and their protein targets, which all turned out to be ion channels and receptors (Table 12.1). Understanding the mechanisms that underlie the bioactivity of each of these compounds provided insights into how the snail targets the motor and sensory circuitry of potential fish prey, as well as its energy metabolism. These natural products in the venom are essential elements in the strategy *C. geographus* has evolved for prey capture.

A most remarkable and unanticipated consequence of these studies is that four of the first seven *C. geographus* venom peptides that were comprehensively characterized have led to significant biomedical applications. Two of these native venom compounds have reached human clinical trials, and one is in preclinical development as a drug lead. The fourth compound has become a diagnostic agent, as well

TABLE 12.1
List of Conotoxins Characterized from *Conus geographus*

	Peptide	Molecular Target	Peptide Sequence	Reference
Nirvana cabal	Conantokin-G	NMDA receptor (NR2B)	GEγγLQγNQγLIRγKSN*	[11]
	Contulakin-G	Neurotensin receptor	ZSEEGGSNAT+KKPYIL	[12]
	Con-Ins G1	Insulin receptor	GVVγHCCHRPCSNAEFKKYC*-TFDTOKHRCSGSγITNSYMDLCYR	[13]
	σ-GVIIIA	5-HT_3 receptors	GCTRTCGGOKCTGTCTCTNSSKCGCRYNVHPSGwGCGCACS*	[22]
Motor cabal	ω-GVIA	$Ca_V2.2$	CKSOGSSCSOTSYNCCRSCNOYTKRCY*	[17]
	ω-GVIIA	Ca_V	CKSOGTOCSRGMRDCCTSCLLYSNKCRRY	[133]
	ω-GVIIB	Ca_V	CKSOGTOCSRGMRDCCTSCLSYSNKCRRY	[133]
	μ-GIIIA	$Na_V1.4$	RDCCTOOKKCKDRQCKOQRCCA*	[134]
	μ-GIIIB	$Na_V1.4$	RDCCTOORKCKDRRCKOMKCCA*	[15]
	μ-GIIIC	Na_V	RDCCTOOKKCKDRRCKOLKCCA*	[15]
	α-GI	nAChR (muscle subtype)	ECCNPACGRHYSC*	[135]
	α-GIA	nAChR (muscle subtype)	ECCNPACGRHYSCGK	[135]
	α-GII	nAChR (muscle subtype)	ECCHPACGKHFSC*	[135]
Miscellaneous	Scratcher peptide	N.D.	KFLSGGFKγIVCHRYCAKGIAKEFCNCPD*	[136]
	ω-GVIC	N.D.	CKSOGSSCSOTSYNCCRSCNOYTKRC	[133]
	Conotoxin GS	Na_V	ACSGRGSRCOOQCCMGLRCGGNPQKCIGAHγDV	[137]
	α-GIC	nAChR (α3β2, α3β4, α4β2)	GCCSHPACAGNNQHIC*	[138]
	α-GID	nAChR (α7, α3β2)	IRDγCCSNPACRVNNOHVC	[139]
	μO§-GVIIJ	Na_V complexes (w/o β2/β4)	GwCGDOGATCGKLRLYCCSGFCDC§YTKTCKDKSSA	[42]
	αS-GVIIIB	nAChR (α9α10)	SGSTCTCFTSTNCQGSCECLSPPGCYCSNNGIRQRGCSCTCPGT*	[57]
	Conopressin G	Vasopressin receptor	CFIRNCPKG*	[23]

Note: Special characters that represent post-translational modifications (*, γ, O, +, w, §) of peptide sequences are explained in Table 12.7.

as a widely used pharmacological research tool, and was key to the development of a Food and Drug Administration (FDA)–approved drug. These four peptides will be discussed in detail in Section 12.4, but a brief overview is presented here. One of these peptides, ω-conotoxin GVIA, was the first selective blocker of the N-type calcium (Ca) channel, $Ca_V2.2$,[8] and has become an important research tool for the neuroscience community. It was critical for establishing $Ca_V2.2$ as a validated drug target for intractable pain. A closely related *Conus* venom peptide that also selectively blocks $Ca_V2.2$ later became an FDA-approved drug (ziconotide, Prialt®) for neuropathic pain.[9] Indeed, $Ca_V2.2$ is one of the nonopioid pain targets identified in the last 50 years that resulted in the development of a novel FDA-approved drug. Radiolabeled ω-conotoxin GVIA is a diagnostic agent for an autoimmune disease, Lambert–Eaton syndrome.[10] A second *C. geographus* venom peptide, conantokin-G,[11] is a selective blocker of *N*-methyl-D-aspartate (NMDA) receptors. This peptide reached Phase I clinical trials for intractable epilepsy. A third peptide, contulakin-G, a homolog of neurotensin,[12] reached Phase I clinical trials for neuropathic pain. The fourth peptide is a small, fast-acting form of insulin, called Con-Ins G1, which may find clinical application for diabetes.[13] This surprisingly high fraction of drug leads from studying a single chemical interaction between one cone snail species and its fish prey is evidence for the utility of chemical neuroethology for the future of natural products–based drug discovery.

C. geographus is arguably the animal with the most intensively studied diverse repertoire of natural products (i.e., bioactive venom peptides) whose molecular neurobiological mechanisms have been elucidated. It is infamous for having caused the greatest number of human fatalities of any molluscan species; in the absence of medical intervention, it is estimated that 70% of human stinging cases are fatal.[4] The biology of *C. geographus* in the wild has not been extensively characterized, but a combination of aquarium and field observations suggest that it is a piscivorous species that specializes in hunting schools of fish. Most cone snails are primarily nocturnal, so it is likely that *C. geographus* goes foraging at night, when schools of small fish hide in reef crevices. A variety of evidence suggests that *C. geographus* attempts to capture as many fish in a school as possible, in contrast to most fish-hunting cone snail species that target one fish at a time (see Section 12.3). The postulated prey capture strategy of *C. geographus* is illustrated in the cartoon shown in Figure 12.1.

Characteristically, in aquaria, *C. geographus* is able to almost immediately detect the presence of a fish and begins moving systematically toward the fish, primarily using a chemosensory mechanism. As it approaches the fish, it opens its rostrum (Figure 12.1) and apparently releases components of its venom into the water. If there is a single fish, it can shape its rostrum like the barrel of a gun to focus the release of venom components close to the fish. The fish begin swimming in an abnormal manner and seem to become disoriented. The cone snail then begins to engulf as many fish as possible. The engulfed fish are stung individually using a hollow radular tooth as the injection device. When *C. geographus* is fed fish sequentially in an aquarium, each time a fish is presented, the snail opens its rostrum and previously engulfed fish can be seen totally paralyzed deep within the rostrum.

As described in the preceding paragraph, when *C. geographus* is foraging, there is a precapture stage to render the prey vulnerable to being engulfed by its large rostrum. Part of the strategy is to inhibit the sensory circuitry of the fish, which prevents them from swimming away as the cone snail approaches. In addition, the snail is believed to release a specialized insulin that makes the fish hypoglycemic,[13] and therefore lethargic and unable to escape. The individual components that underlie this prey capture strategy will be described below (Section 12.2.3).

The roles of several *Conus* venom peptides can be correlated to the behavior described; two biological endpoints are achieved to capture prey. The venom peptides released into the water cause fish to become lethargic, hypoglycemic, and increasingly unaware of the approaching predator, unable to detect movement in the water. The sensory circuitry of the fish is jammed, particularly the lateral line system, comprising hair cells that sensitively alert fish to any movement in the water. We have adopted the term *cabal* for venom peptides that act together in a concerted fashion to achieve a specific physiological endpoint.[5] The *C. geographus* venom peptides released into the water to debilitate a school of fish are known as the *nirvana cabal*.[2] Nirvana cabal peptides facilitate engulfing the fish by rendering them semisedated and lethargic. Once engulfed, each fish is injected with paralytic toxins that cause a total inhibition of the neuromuscular circuitry. The engulfed fish are thus irreversibly paralyzed, allowing the snail to continue foraging and engulfing more fish. The second group of peptides that blocks neuromuscular transmission is known as the *motor cabal*.[5] We discuss the individual venom components of the two cabals below.

12.2.2 Motor Cabal of *Conus geographus*

With regard to their biological roles, as well as the molecular mechanisms that underlie their bioactivity, the venom peptides that are best understood are the components of the motor cabal of *C. geographus* venom.[5] The overall physiological strategy is to use a combination of venom peptide *drugs*, with each individual venom peptide targeted to a specific, critical signaling component of the neuromuscular circuitry of a fish. The detailed physiological mechanisms involve the (1) inhibition of neurotransmitter release from the presynaptic terminus, (2) block of the postsynaptic nicotinic acetylcholine receptor (nAChR) that is activated by the neurotransmitter released from the presynaptic terminus, and (3) direct inhibition of action potentials at the muscle membrane. Any one of these venom peptides is sufficient to block neuromuscular transmission, yet all are expressed at high levels, ensuring the rapid and irreversible block of neuromuscular transmission in the envenomated fish. This was the first experimental evidence for the use of a combination drug strategy by cone snails.

The first motor cabal peptide to be characterized was α-conotoxin GI, which binds competitively to the nAChR at the postsynaptic terminus.[14] α-Conotoxin GI is 13 amino acids in length and selectively binds to only one of the two acetylcholine-binding sites (between the $\alpha 1$ and δ subunits of the muscle nAChR), which is sufficient to prevent channel opening. Several variants (α-conotoxins GII and GIA) that are clearly homologs were subsequently characterized (Table 12.1).

In addition to blocking nicotinic receptors, other conotoxins also help to block action potentials in muscle by blocking voltage-gated sodium (Na) channels expressed in muscle membranes. Three μ-conotoxins were identified from *C. geographus* venom (μ-conotoxins GIIIA, GIIIB, and GIIIC; Table 12.1)[15] that were shown to selectively block the voltage-gated Na channel subtype in skeletal muscle ($Na_V1.4$); their discovery was an early indication that there was more than one voltage-gated Na channel subtype (this was before the cloning of the different voltage-gated Na channel subtypes). Because of the universality of the Hodgkin–Huxley formulation of action potential generation, most neuroscientists at the time had thought that there was only one type of voltage-gated Na channel. (Many decades later, it is amusing to note that in the original paper describing the μ-conotoxins, the reviewer objected to any suggestion that there might be different Na channel subtypes; only a vague statement was allowed to be incorporated in this manuscript.[15]) μ-Conotoxins are pore blockers and compete for binding with tetrodotoxin and saxitoxin. However, because of their larger size, they occupy a larger volume of the pore vestibule, not necessarily overlapping completely with the tetrodotoxin-binding site, as demonstrated by Yoshikami and coworkers.[15] Thus, although in the conventional formulation of pharmacological sites on the voltage-gated Na channel, μ-conotoxins are considered a site-1 ligand, the binding sites of μ-conotoxins and saxitoxin/tetrodotoxin are overlapping but non-identical.[16]

The discovery that ω-conotoxins from *C. geographus* venom targeted presynaptic Ca channels first demonstrated that *Conus* venom peptides defined novel signaling complexes in the nervous system. Two peptides that both inhibited synaptic transmission but have divergent pharmacological properties, ω-conotoxins GVIA and GVIIA,[2,17] were initially characterized. ω-Conotoxin GVIA is a remarkably selective ligand for the $Ca_V2.2$ voltage-gated Ca channel,[8] which is the major Ca channel subtype expressed presynaptically in the fish (but not mammalian) neuromuscular junctions. The very high selectivity of ω-conotoxin GVIA for a specific subtype of voltage-gated Ca channels facilitated the definition of Ca channel subtypes.[18] Indeed, the biochemical characterization of $Ca_V2.2$ was initially described as the purification of the ω-conotoxin receptor[19]. ω-Conotoxin GVIA is a pore blocker, with extremely high potency for Ca_V 2.2, from all vertebrate species examined so far. In contrast, the subtype selectivity of ω-conotoxin GVIIA has not been elucidated; it is possible that it targets a subtype of voltage-gated Ca channels at the presynaptic terminus of fish that is significantly divergent in its pharmacological properties from the spectrum of mammalian Ca channel subtypes.

12.2.3 Nirvana Cabal

All of the conopeptides discussed in the previous section are injected by the snail to cause paralysis of engulfed fish prey. Success in engulfing prey is dependent on a different set of peptides that are believed to be released by the snail into the water and are presumably taken up through the gills. These peptides are known as the nirvana cabal,[2] and the physiological endpoint is a loss of sensory acuity, probably targeted primarily to the hair cell circuitry of the lateral line, accompanied by a low-energy, hypoglycemic state. The effects of the nirvana cabal (originally named because the fish seemed to be in a quiescent, sedated state reminiscent of people in an opium den) are not as well characterized in terms of their endogenous mechanisms (i.e., on fish prey) as are the *Conus* venom peptides targeted to the neuromuscular circuitry. Nevertheless, nirvana cabal peptides have proven to be unique pharmacological agents, with very high biomedical potential.

The first peptide of the nirvana cabal to be characterized was conantokin-G, first isolated by Craig Clark, an undergraduate at the University of Utah, who found a *C. geographus* venom fraction that put mice to sleep, which he designated *the sleeper peptide*.[11] After purification and characterization (by another University of Utah undergraduate, J. Michael McIntosh), the peptide proved to be a landmark biochemically, because it demonstrated for the first time that a specific post-translational modification, the γ-carboxylation of glutamate, occurred outside the vertebrate blood clotting cascade (there were four γ-carboxyglutamate residues in the *C. geographus* sleeper peptide).[20] This peptide was designated conantokin-G (*antokin* is the Filipino word for sleepy). A notable feature of the symptomatology induced by this peptide is that only neonatal mice went to sleep following intracranial injection of the peptide. Mice have a strong righting reflex that caused saline-injected controls to quickly roll over to their stomachs when turned to their backs after injection of the peptide (even if they fell asleep). However, mice injected with conantokin-G would fall asleep and remain asleep after being rolled over to their backs, eventually recovering from the effects of the peptide after a few hours. By about 3 weeks of age, sleep symptoms were no longer observed following intracranial injection.[11] It was later demonstrated that the peptide inhibited NMDA receptors that contained an NR2B subunit. As will be discussed in Section 12.4, conantokin-G has been shown to have therapeutic potential as an antiepileptic.

A second unusual peptide, first identified by another Utah undergraduate, David Griffin, was originally called *the sluggish peptide*, based on the phenotype observed upon intracranial injection into mice.[12,21] The peptide was characterized by Anthony Craig, then at the Salk Institute; the amino acid sequence demonstrated clear homology to neurotensin, a well-characterized vertebrate neuropeptide,[12,21] but unlike the native neurotensin, the peptide was blocked for Edman sequencing at the N-terminus with pyroglutamate and contained a glycosylated threonine residue. As is detailed in Section 12.4, this peptide has remarkable analgesic properties, and has been through Phase I human clinical trials. As far as is known, neurotensin is strictly a vertebrate neuropeptide; that *C. geographus* evolved a peptide targeted to neurotensin receptors is therefore remarkable. It appears to have evolved *de novo*, without its evolutionary origin based on an endogenous neuropeptide.

A final *Conus* venom peptide candidate of the nirvana cabal targets the serotonin 5-HT_3 receptor. This peptide is unusually large, with 10 cysteine residues (five disulfide linkages), and contains an unusual post-translational modification, a bromotryptophan residue.[22] So far, there is only one characterization of this peptide reported in the literature. These peptides all cause hypoactivity in the neuronal circuitry when injected intracranially into mice. It is likely that these target homologous receptors present in the lateral line circuitry of fish, but this remains to be rigorously demonstrated.

A most remarkable natural product evolved by *C. geographus* is Con-Ins G1, the smallest functional insulin ever discovered.[13] This venom component, expressed at high levels, has been demonstrated to cause hypoactivity when released into water containing zebrafish, and has striking sequence similarity to endogenous fish insulins (but far less sequence similarity to the endogenous cone snail insulin).[13] It has been hypothesized that this peptide has evolved to be an extremely-quick-acting insulin; unlike mammalian insulins secreted by the β-cells of the pancreas that are hexamers, Cons-Ins G1 is a monomer, and therefore can act immediately upon being taken up in the bloodstream of a fish. Another unique feature of Con-Ins GI is the presence of post-translationally modified amino acids (γ-carboxyglutamate, C-terminal amidation and hydroxyproline).

Thus, together, the characterized components of the nirvana cabal are suggested to make a fish insensitive to movement in the water because its hair-cell system is jammed by these components in *C. geographus* venom. In addition, the fish become lethargic and unable to escape because blood glucose levels are lowered by Con-Ins G1. The uptake of nirvana cabal peptides may be facilitated by venom components such as conopressin G, which causes constriction of major blood vessels, which could in turn accelerate the uptake of venom peptides by the capillary bed.[23]

12.3 TASER-AND-TETHER STRATEGY OF FISH-HUNTING CONE SNAILS

12.3.1 Overview

C. geographus, the subject of the case study described above, is unusual in its prey capture strategy because it targets schools of fish (only one other species uses this strategy, the closely related *Conus tulipa*; see Figure 12.3a). Many fish-hunting cone snail species target one prey at a time, using a hunting strategy called *taser-and-tether* (Figure 12.2). This strategy is used by cone snail species in at least four different phylogenetic lineages (*Chelyconus*, *Pionoconus*, *Textilia*, and *Gastridium*; see Figure 12.3b).[24] The molecular mechanisms underlying this prey capture strategy overlap with those of the *net-hunting strategy* of *C. geographus* in the presence of a motor cabal. However, there is no nirvana cabal equivalent; instead, a different set of toxins comprising the *lightning-strike cabal*[2] (to be described below) are used to effect an almost-instantaneous immobilization of the envenomated prey, an essential and characteristic feature of the *taser-and-tether* strategy.

Almost all of the *Conus* species that employ the taser-and-tether strategy exhibit a behaviorally similar pattern after they detect the presence of fish prey.[2] They begin to extend a highly distensible proboscis (Figure 12.2). In essence, the proboscis is used as a line, and when they (cone snails) are within striking distance of the prey, they eject a harpoon-like tooth with sufficient force to pierce the skin of the fish. All cone snails that use the taser-and-tether strategy have a highly specialized radular tooth for injecting their venom. It is shaped like a harpoon and is highly barbed, thereby serving as both a hypodermic needle for injecting venom and a harpoon that mechanically tethers the fish. Some examples of these harpoons are shown in Figure 12.4, contrasting with the needle-like tooth of *C. geographus*, which is not barbed, more like a hypodermic needle. The radular tooth is propelled from the

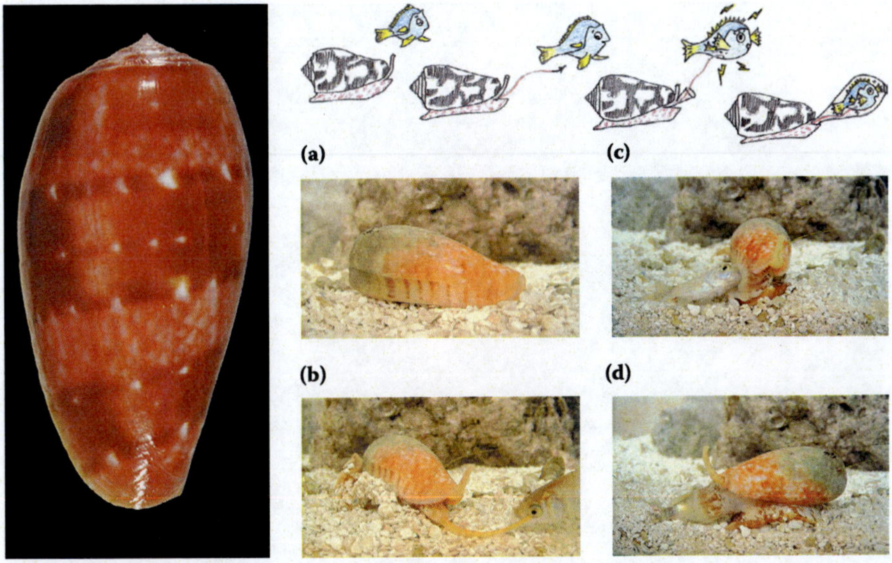

FIGURE 12.2 An example of the taser-and-tether strategy: *Conus bullatus*, the bubble cone. A shell of *C. bullatus* is shown in the left panel, followed by a cartoon illustrating how the snail captures its prey. Right panels show an aquarium prey capture sequence. In **(a)**, the snail is quiescent, but once it detects a fish **(b)**, it extends its proboscis toward the fish and harpoons it, injecting venom and tethering the fish, which is then drawn into the rostrum of the snail by retraction of the proboscis **(c)**. In **(d)**, the fish is almost totally engulfed, and predigestion will take place within the rostrum of the snail. *C. bullatus* is one of perhaps over 30 species that use the taser-and-tether strategy, species distributed in four different subgenera. *C. bullatus* belongs to the subgenus *Textilia*, but the taser-and-tether strategy has also been observed in all species in the subgenera *Pionoconus* and *Chelyconus*, and in some (but not all) species in the subgenus *Gastridium*.

proboscis to pierce the skin of the fish. Venom is injected through the hollow barbed tooth by which the fish is tethered and held firmly by circular muscles at the tip of the proboscis. The harpooned fish, which is almost instantaneously immobilized, is then engulfed by the rostrum of the snail simply by retracting the proboscis. Predigestion of the fish takes place in the rostrum, and the scales and bones of the partially digested fish are regurgitated.

It is likely that the taser-and-tether strategy is a more ancient strategy than the net-hunting strategy employed by *C. geographus*. An interesting biogeographic feature of the lineages of cone snails that employ taser-and-tether is that three of them are exclusively Indo-Pacific, but one is phylogenetically more distant and biogeographically isolated (*Chelyconus*), found only in the New World. The present molecular evidence suggests that the taser-and-tether strategy evolved from worm-hunting cone snails twice: once generating the New World lineage (*Chelyconus*) and separately in the Indo-Pacific, to give rise to three distinct lineages, *Pionoconus*, *Textilia*, and *Gastridium*, whereas the net strategy of *C. geographus* subsequently evolved within the subgenus *Gastridium*.

FIGURE 12.3 Shells of cone snails that prey on fish, other snails, and worms. **(a, b)** Piscivorous *Conus* species. **(a)** Two fish-hunting species known to use a net-hunting strategy: left, *C. geographus*; right, *C. tulipa*. Both belong to the subgenus *Gastridium*. **(b)** Fish-hunting species known to use the taser-and-tether strategy. Top row, left, *C. bullatus* (subgenus *Textilia*), *C. catus*, *C. striatus*, and *C. magus* (subgenus *Pionoconus*). Lower row, *C. consors* (*Pionoconus*), *C. purpurascens*, and *C. ermineus* (subgenus *Chelyconus*). **(c)** Molluscivorous (snail-hunting) *Conus* species. From left to right, *C. marmoreus*, *C. textile*, *C. ammiralis*, *C. bandanus*, *C. aulicus*, *C. gloriamaris*, *C. victoriae*, and *C. canonicus*. **(d)** Vermivorous species. Top row, *C. chiangi*, *C. floridulus*, *C. litteratus*, *C. tessulatus*, and *C. pulicarius*. Bottom row, *C. imperialis*, *C. andremenezi*, *C. tribblei*, *C. coralinus*, and *C. leopardus*. The shells are illustrated the same size in each group, but there are large differences in the actual size of the individual shells.

12.3.2 Lightning-Strike Cabal

A signature of venoms from fish-hunting cone snails that employ a taser-and-tether strategy is a set of excitatory components that are known collectively as *the lightning-strike cabal*.[2] Two types of venom peptides are central to eliciting these effects: peptides that delay inactivation of voltage-gated Na channels and a second group that block voltage-gated potassium (K) channels.[25]

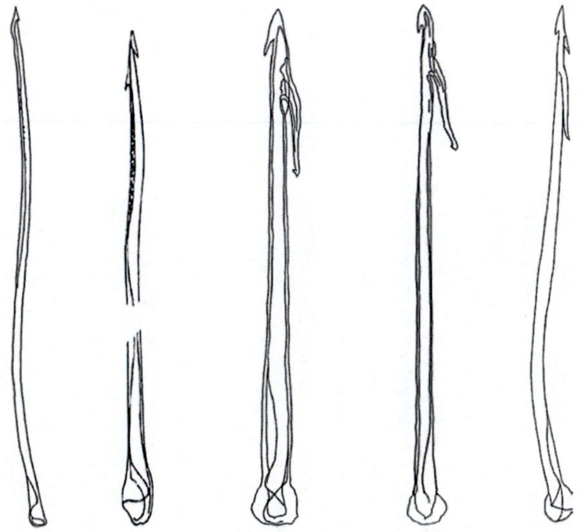

FIGURE 12.4 Comparison of radula teeth of fish-hunting cone snails that use a net-hunting (from left to right: *C. geographus* and *C. tulipa*) and taser-and-tether hunting strategy (*C. bullatus*, *C. magus*, and *C. ermineus*). (Reproduced from Tucker, J.K. and Tenorio, M.J., *Systematic Classification of Recent and Fossil Conoidean Gastropods*, ConchBooks, Hackenheim, Germany, 2009.[131])

The peptides that delay inactivation of voltage-gated Na channels comprise a highly conserved group of hydrophobic peptides with three disulfide bonds, known as δ-conotoxins.[25] Delaying inactivation of Na channels results in a greater and long-lasting membrane depolarization with each channel-opening event. δ-Conotoxins from fish-hunting cone snails (Table 12.2), always have six cysteine residues, arranged in an inhibitory cysteine knot (ICK) motif. The inhibition of Na channel inactivation by δ-conotoxins inhibits a major mechanism for repolarizing an axon after an action potential. Action potentials are also terminated by the opening of voltage-gated K channels.[25] Taser-and-tether snail venoms also contain voltage-gated K channel blockers, which act synergistically with δ-conotoxins to elicit a massive and prolonged depolarization of neuronal membranes.

In contrast to the δ-conotoxins that are highly conserved in all four lineages of taser-and-tether fish hunters, there is no conservation of conopeptide families that inhibit K channels. In the New World fish-hunting clade, *Chelyconus*, κ-conotoxins, which are O superfamily peptides, serve as K channel inhibitors (Table 12.2).[25] The K channel inhibitors in *Pionoconus*, the major fish-hunting clade in the Indo-Pacific, are the conkunitizins,[26] genetically and structurally unrelated to κ-conotoxins. The synergy obtained by inhibiting Na channel inactivation, while also blocking K channels, elicits a massive and prolonged depolarization in the affected axons, generating tonic firing of anterograde and retrograde action potentials, resulting in the nearly instant, tetanic immobilization observed in the envenomated fish.

TABLE 12.2
Conotoxin Cabals from *Conus* Species Other than *geographus*

	Peptide	*Conus* Species	Molecular Target	Peptide Sequence	Reference
Lightning-strike cabal	δ-SVIE	*C. striatus*	$Na_V1.4$	DGCSSGGTFCGIHOGLCCSEFCFLWCITFID	[43]
	δ-PVIA	*C. purpurascens*	$Na_V1.2$, 1.4, 1.7	EACYAOGTFCGIKOGLCCSEFCLPGVCFG*	[44]
	δ-MVIA	*C. magus*	Putative Na_V	DGCYNAGTFCGIROGLCCSEFCFLWCITFVDS*	[43]
	Conkunitzin S1	*C. striatus*	Shaker K channel	Long peptide (60 amino acids)	[50]
	Con-ikot-ikot	*C. striatus*	AMPA receptors	Long peptide (27 amino acids)	[27]
	κA-SIVA	*C. striatus*	Putative K_V blocker	ZKSLVPS⁺VITTCCGYDOGTMCOOCRCTNSC*	[21]
	κ-PVIIA	*C. purpurascens*	K_V Shaker	CRIONQKCFQHLDDCCSRKCNRFNKCV*	[25]
	κA-PIVE	*C. purpurascens*	Putative K_V blocker	DCCGVKLEMCHPCLCDNSCKNYGK*	[26]
	κA-MIVA	*C. magus*	Putative K_V blocker	AOγLVVT⁺AT⁺TNCCGYNOMTICOOCMCTYSCOOKRKO*	[140]
Motor cabal	α-SI	*C. striatus*	nAChR ($\alpha1\beta1\gamma\delta$, $\alpha1\beta1\delta\varepsilon$)	ICCNPACGPKYSC*	[141]
	α-SIA	*C. striatus*	nAChR ($\alpha1\beta1\gamma\delta$, $\alpha1\beta1\delta\varepsilon$)	YCCHPACGKNFDC*	[142]
	α-PIB	*C. purpurascens*	nAChR ($\alpha1\beta1\gamma\delta$, $\alpha1\beta1\delta\varepsilon$)	ZSOGCCWNPACVKNRC*	[143]
	αA-OIVA	*C. obscurus*	nAChR ($\alpha1\beta1\gamma\delta > \alpha1\beta1\delta\varepsilon$)	CCGVONAACHOCVCKNTC*	[32]
	αA-PIVA	*C. purpurascens*	nAChR ($\alpha1\beta1\gamma\delta$, $\alpha1\beta1\delta\varepsilon$)	GCCGSYONAACHOCSCKDROSYCGQ*	[144]
	α-MI	*C. magus*	nAChR ($\alpha1\beta1\gamma\delta$, $\alpha1\beta1\delta\varepsilon$)	GRCCHPACGKNYSC*	[145]
	α-MIC	*C. magus*	nAChR ($\alpha1\beta1\gamma\delta$, $\alpha1\beta1\delta\varepsilon$)	CCHPACGKNYSC*	[146]
	α-MII	*C. magus*	nAChR ($\alpha6\beta2 > \alpha3\beta2 > \alpha6\beta4$)	GCCSNPVCHLEHSNLC*	[147]
	α-BuIA	*C. bullatus*	nAChRs	GCCSTPPCAVLYC*	[148]
	μ-BuIIIA	*C. bullatus*	$Na_V1.2$, 1.4	VTDRCCKGKRECGRWCRDHSRCC*	[149]
	μ-BuIIIB	*C. bullatus*	$Na_V1.2$, 1.4	VGERCCKNGKRGCGRWCRDHSRCC*	[149]
	μ-PIIIA	*C. purpurascens*	$Na_V1.1$, 1.2, 1.4, 1.6	ZRLCCGFOKSCRSRQCKOHRCC*	[150]

(*Continued*)

TABLE 12.2 (*Continued*)
Conotoxin Cabals from *Conus* Species Other than *geographus*

Peptide	*Conus* Species	Molecular Target	Peptide Sequence	Reference
μ-SIIIA	*C. striatus*	TTX-R	ZNCCNGGCSSKWCRDHARCC*	[151]
μ-SIIIB	*C. striatus*	$Na_V1.2$, 1.7	ZNCCNGGCSSKWCKGHARCC*	[151]
μ-MIIIA	*C. magus*	$Na_V1.2$, 1.4	ZGCCNVPNGCSGRWCRDHAQCC*	[152]
ω-SVIA	*C. striatus*	$Ca_V2.2 > Ca_V2.1$	CRSSGSOCGVTSICCGRCYRGKCT*	[153]
ω-SVIB	*C. striatus*	$Ca_V2.1 > Ca_V2.2$	CKLKGQSCRKTSYDCCSGSCGRSGKC*	[153]
ω-MVIIA	*C. magus*	$Ca_V2.2 > Ca_V2.1$	CKGKGAKCSRLMYDCCTGSCRSGKC*	[52]
ω-MVIIB	*C. magus*	Ca_V2	CKGKGASCHRTSYDCCTGSCNRGKC*	[52]
ω-MVIIC	*C. magus*	$Ca_V2.1 > Ca_V2.2$	CKGKGAPCRKTMYDCCSGSCGRRGKC*	[154]
ω-MVIID	*C. magus*	$Ca_V2.2 >> Ca_V2.1$	CQGRGASCRKTMYNCCSGSCNRGRC*	[155]

Note: Special characters that represent post-translational modifications of peptide sequences are explained in Table 12.7.

In order to effectively cause the entire animal to become tetanically immobilized, synaptic connections need to faithfully transmit the hyperexcitability elicited at the venom injection site with undiminished intensity. In at least some of the taser-and-tether cone snails, synapses that may decrease the excitability are specifically targeted. These are primarily glutamatergic (known to be present in the lateral line system of the fish). The depolarization elicited by δ- and κ-conopeptides would release a massive amount of neurotransmitter. However, to ensure that this results in a massive depolarization of the postsynaptic circuitry, the snails have evolved a class of peptides called con-ikot-ikot peptides that powerfully inhibit desensitization of α-amino-3-hydroxy-5-methyl-4-isoxazolepropionic acid (AMPA) receptors.[27] Only one of these peptides has been thoroughly characterized, from *C. striatus* (in the Indo-Pacific subgenus *Pionoconus*), and preliminary evidence for their presence has been obtained for other species of *Pionoconus*. Thus, at the venom injection site, axons are massively depolarized, generating a storm of action potentials. A minimal cocktail of δ-conotoxins that inhibit Na channel inactivation, κ-type conotoxins that block voltage-gated K channels, and finally, con-ikot-ikot peptides that prevent the desensitization of AMPA receptors once they are activated by glutamate, together form a core triad that underlies an almost instantaneous tetanic immobilization of fish after a strike by the diverse piscivorous cone snail species that employ the taser-and-tether strategy.

12.3.3 Motor Cabal of Taser-And-Tether Cone Snail Species

The composition of the motor cabal found in the venoms of taser-and-tether species is analogous to the motor cabal of *C. geographus*. The μ-conotoxins that inhibit the action potential on the muscle membrane are found in most (if not all) of the cone snails that immobilize their prey through a taser-and-tether strategy. This family is highly conserved across the different fish-hunting lineages.[28] μ-Conotoxins have a characteristic arrangement of cysteine residues (CC...C...C...CC) and genetically belong to the M superfamily.[29] μ-Conotoxins are highly positively charged peptides that potently block the pore of the voltage-gated Na channel on the plasmalemma of muscle.

In addition to Na channel blockers, all the taser-and-tether species also express conopeptides that inhibit the postsynaptic nAChRs. However, in contrast to μ-conotoxins, there is lineage specificity and structural divergence in these peptides. In the Indo-Pacific *Pionoconus* (*C. magus* and *C. striatus*), the nAChR blockers are peptides that belong to the A superfamily, with four cysteine residues and two disulfide bonds, as shown in Table 12.2. These are the well-known α-conotoxins, and they are homologous to α-conotoxin G1 of *C. geographus*. However, in *Chelyconus* (*C. purpurascens*), the major nAChR antagonists also belong to the A superfamily, but have six cysteine residues. These peptides are known as αA-conotoxins.[30,31] Examples of these peptides are shown in Table 12.2. In the one species of *Gastridium* that is known to use a taser-and-tether strategy, *C. obscurus*, αA-conotoxins are also used, but these differ in structure from the αA-conotoxins of the New World *Conus* species (the former are known as *the short αA-conotoxins*).[31–33] Finally, in *C. bullatus* from the *Textilia* clade, the α-conotoxin family has also been recruited, but the *Textilia* peptides have a characteristically different spacing from the α-conotoxins found in the *Pionoconus* clade. All of these are illustrated in Table 12.2.

There may be a more sophisticated targeting of the postsynaptic circuitry in taser-and-tether species than there is in *C. geographus*. It is known that *C. obscurus* has two nAChR antagonists, one of which is targeted to the fetal muscle subtype, which contains $\alpha 1$, $\beta 1$, γ, and δ subunits. In mammals, the γ subunit is replaced by an ε subunit during late gestation or shortly after birth. However, the γ subunit is expressed even in muscles of adult fish.[31,33] Furthermore, in *Chelyconus*, there are not only αA-conotoxins, but also more conventional α-conotoxins in at least some species, such as *C. ermineus*,[30] as well as a third group of nAChR antagonists, ψ-conotoxins, which are noncompetitive antagonists.[34–36]

12.4 MAJOR CHARACTERIZED MOLECULAR TARGETS OF *CONUS* VENOM PEPTIDES

12.4.1 Overview

Conopeptides that have a common molecular target (e.g., nAChRs, voltage-gated Na channels), as well as a shared structural framework (largely determined by the pattern of disulfide bonds), are grouped together into conopeptide families, each designated by a Greek letter. A summary of conopeptide families is given in Table 12.3. The known molecular targets of conopeptides fall into four broad classes, voltage-gated ion channels, ligand-gated ion channels (or ionotropic receptors), G-protein-coupled receptors (GPCRs or metabotropic receptors), and other targets. The most extensively characterized conopeptides target ion channels, particularly those that are critical to neuromuscular transmission or electrical signaling. There are four families of ion channels that are likely targeted by venom components of most

TABLE 12.3
Pharmacological Families of Conopeptides[42]

Family	Molecular Target	Representative Peptide
α (alpha)	nAChRs	α-GI
γ (gamma)	Neuronal pacemaker cation currents	γ-PnVIIA
δ (delta)	Voltage-gated Na channel	δ-TxVIA
ε (epsilon)	GPCRs	ε-TxVA
ι (iota)	Voltage-gated Na channels	ι-RXIA
κ (kappa)	Voltage-gated K channels	κ-PVIIA
μ (mu)	Voltage-gated Na channels	μ-GIIIA
ρ (rho)	$\alpha 1$ adrenoreceptors	ρ-TIA
σ (sigma)	Serotonin-gated ion channel	σ-GVIIIA
τ (tao)	Somatostatin receptor	τ-CnVA
χ (chi)	Norepinephrine transporter	χ-MrIA
ω (omega)	Voltage-gated Ca channels	ω-GVIA

(and possibly all) cone snails: voltage-gated Na, Ca, and K channels, and nAChRs. We discuss conopeptides that target the three voltage-gated ion channels first, followed by those that target nAChRs and other ligand-gated ion channels.

12.4.2 Conopeptides Targeted to Voltage-Gated Na Channels

This topic has recently been reviewed.[37] Conopeptides that modulate Na channel activity can be generally divided into those that block the pore of the channel (pore blockers) and those that act on voltage sensors (gating modifiers). The gating modifiers include *Conus* peptides that prevent activation of the voltage-gated Na channel, inhibit inactivation, or shift activation to more negative potentials. *Conus* peptides that act through the last two mechanisms cause a greater activity of the targeted ion channel, thereby increasing the influx of sodium into the cell, whereas blockers of gating activation inhibit electrical activity in the targeted circuitry. The pore blockers also inhibit electrical activity in the targeted circuitry.

12.4.2.1 Pore Blockers of Voltage-Gated Na Channels: μ-Conotoxins

As described in Sections 12.2 and 12.3, μ-conotoxins are well-characterized components of fish-hunting cone snail venoms and are important components of the motor cabal that inhibits neuromuscular transmission.[5] Specific examples are shown in Tables 12.1 and 12.2 and in the Appendix. Although most μ-conotoxins have a high affinity for the muscle subtype of Na channels ($Na_V1.4$), consistent with their physiological role in the motor cabal, different μ-conotoxins target neuronal subtypes of voltage-gated Na channels with vastly different potencies, as shown in Table 12.4. These differences in subtype selectivity have been used by Yoshikami and coworkers to identify functional expression of different Na channel subtypes in native neurons.[38]

12.4.2.2 μO-Conotoxins: *Conus* Peptides That Inhibit Activation

A second general class of *Conus* peptides targeted to voltage-gated Na channels that inhibit their activity is the μO-conotoxins; they fundamentally differ in their mechanism by preventing activation of the targeted voltage-gated Na channel. Two species of cone snails are known to have such peptides in their venom, *C. marmoreus*, a snail-hunting cone (this produces μO-conotoxin MrVIA[39,40] and MrVIB[39,41]), and *C. geographus* (which has an unusual μO-conotoxin, μO-conotoxin GVIIJ[42]). The latter peptide is covalently tethered to the targeted voltage-gated Na channel through a disulfide linkage.[42] Both of these *Conus* peptides have six cysteine residues in an ICK motif, but μO-conotoxin GVIIJ has a seventh cysteine residue, which is cysteinylated (i.e., disulfide bonded to a free Cys residue). The Cys amino acid serves as the leading group, and a cysteine residue on the voltage-gated Na channel forms a disulfide bond with Cys 23 of μO-GVIIJ. Because this Cys residue on the channel is also the site for covalently binding β2 or β4 subunits, μO-GVIIJ has the unusual property of being selective for voltage-gated Na channels that lack β2 and β4 subunits. If these subunits are present, the Cys residue to which the *Conus* peptide is covalently attached is unavailable, rendering the channel resistant to inhibition by this peptide. In contrast, μO-conotoxins MrVIA and MrVIB are extremely hydrophobic

TABLE 12.4
***μ*-Conopeptides That Discriminate between Different Sodium Channel Subtypes**

Na_V	SmIIIA	KIIIA	SIIIA	CnIIIA	MIIIA	GIIIA	PIIIA	SxIIIA	BuIIIA	BuIIIB	TIIIA
1.1	0.0038	0.29	11*	14.2*	22.6*	0.26	0.12*	0.37*	0.35	0.36	0.9*
1.2	0.0013	0.005	0.05	0.25	0.45	17.8*	0.62*	1*	0.012	0.013	0.045
1.3	0.035	8*	11*	11*	7.7	>100*	3.2*	>100*	0.35	0.2	7.9*
1.4	0.00022	0.09*	0.13*	0.27*	0.33	0.019*	0.036*	0.007*	0.012	0.0036	0.005
1.5	1.3*	287*	251*	7.4*	>100*	>100*	>100*	>100*	13.8*	9*	>100*
1.6	0.16*	0.24*	0.76*	7.1*	21.6*	0.68*	0.1*	0.57*	4.4*	1.8*	25*
1.7	1.3*	0.29	65*	>100*	97*	>100*	>100*	>100*	>100*	>100*	>100*
1.8	>100*	>100*	>100*	>100*	>100*	>100*	>100*	>100*	>100*	>100*	>100*

Source: Modified from Wilson, M.J. et al., *Proc. Natl. Acad. Sci. USA*, 108, 10302, 2011.

Notes: Na_V1.x clones were from rat except Na_V1.6, which was from mouse. >100 means that a concentration of 100 μM produced a <10% block.

IC_{50} value; other values are Kd, all in μM.

peptides that prevent voltage sensors from activating the channel. As is characteristic of ligands that bind to voltage sensors, a prepulse of an extremely positive membrane potential dissociates the peptide and the inhibition by these peptides is reversed. The sequences of these μO-conotoxins are shown in the Appendix.

12.4.2.3 Conopeptides Targeted to Voltage-Gated Na Channels That Increase Excitability

There are two classes of conopeptides targeted to voltage-gated Na channels that increase neuronal excitability: δ-conotoxins,[43,44] which inhibit channel inactivation, and ι-conotoxins,[45,46] which shift channel activation to a more negative membrane potential. The known peptides that have been directly characterized are shown in Table 12.2 and in the Appendix.

The δ-conotoxin family is a large, widely distributed family of extremely hydrophobic peptides with six cysteines in an ICK structure. These peptides play a key role in *the lightning-strike cabal* of fish-hunting *Conus* (see Section 12.3). They act by stabilizing the voltage sensor in an open state of the channel, delaying fast inactivation of the Na channel.[25]

ι-conotoxins differ from δ-conotoxins structurally and mechanistically.[45,46] ι-conotoxins have eight cysteine residues and belong to a different gene superfamily (the I1 superfamily, instead of the O superfamily to which all δ-conotoxins belong; see the next section). ι-conotoxins have not been extensively characterized; however, one distinguishing feature is that the native peptides have a post-translationally modified D-amino acid. The preliminary work that has been done to compare analogs with and without the D-amino acid modification suggest that the post-translational modification is important for activity. So far, ι-conotoxins have only been characterized from a single subgenus of fish-hunting cone snails, *Phasmoconus*, in contrast to the very wide phylogenetic distribution of δ-conotoxins. Sequences of ι-conotoxins are shown in the Appendix.

12.4.3 Conopeptides Targeted to Voltage-Gated K Channels

It is likely that every *Conus* venom has components that are targeted to K channels. As outlined in Section 12.3, these K channel blockers play a central role in the lightning-strike cabal of several lineages of fish-hunting cone snail species. Thus, they are key components in the toxin combination that elicits the very rapid immobilization of fish prey, which is essential for the slow-moving cone snails to catch fish that could otherwise swim away. A major problem in investigating this ubiquitous class of *Conus* venom components is that the identification of native molecular targets is very challenging. In contrast to voltage-gated Na channels and Ca channels, in which the pore-forming subunit is a single gene product with four domains, K channels comprise four different subunits, which may form channels with homomeric or heteromeric combinations of subunits. Each K channel gene encodes a single K channel subunit. In mammals, there are ~70 genes that encode K channel subunits; approximately 40 of those genes encode subunits of voltage-gated K channels.[47] The large number of K channel genes and the potential for heteromeric combinations creates a vast K channel complexity that has not been addressed in molecular neuroscience. At the present time, there is no facile way to identify whether a physiologically relevant K channel that functions in a specific cell or circuitry is a homomeric combination or a heteromeric combination with two, three, or four different types of subunits. It is necessary in these heteromeric combinations to identify which subunits are present. For example, if a particular venom peptide were targeted to a native K channel that contains three different K channel subunits, it would be very difficult to identify the different subunits by presently available methods. For this reason, the characterization of *Conus* peptides targeted to voltage-gated K channels is at a relatively immature stage, compared with other conopeptide families. It seems likely that many conopeptides evolved to specifically target heteromeric K channels whose subunit compositions are presently unknown.

Unlike the known Na channel–targeted conopeptides that fall into discrete peptide families, and bind to different sites on voltage-gated Na channels, some conopeptides that target K channels may bind to identical sites of the same K channel subtype, even though they belong to different gene superfamilies. This complicates the prediction of which peptide in a venom is likely to be a K channel antagonist. For each lineage of cone snail, this has to be determined independently. In contrast, for voltage-gated Na channels, conopeptides that inhibit channel inactivation are all likely to belong to the δ-conotoxin family (O1 superfamily), which are all structurally, genetically, and functionally related. Thus, we will review K channel–targeted *Conus* peptides based on the lineage of cone snails from which they are derived.

Chelyconus: The first conopeptides that target K channels were characterized from *C. purpurascens*. Together with its sister species, *C. ermineus*, *C. purpurascens* peptides belong to the O1-gene superfamily (see Section 12.6) and have been called κ-conotoxins. The first κ-conotoxin characterized, which was the basis for elucidating the components of the lightning-strike cabal, was κ-conotoxin PVIIA.[25] When examined for activity on homomeric K channel complexes, κ-conotoxin PVIIA has a preference for mammalian $K_V1.2$ and also inhibits the *Drosophila* shaker channel. However, the precise heteromeric combination targeted by this peptide is unknown at the present time. *K*-conotoxin PVIIA has some potential biomedically relevant activity as a cardioprotective compound that prevents the cell death that occurs after reperfusion (see Section 12.5).

Phasomoconus: This is a large group of fish-hunting cone snails. Their biology is not well understood because there are few records of how they actually capture prey.[24] The morphology of their radular tooth, used to inject venom, suggests that they do not tether their prey, and therefore the true function of their K channel–targeted peptides is largely speculative. After venom injection, however, in the one case where it has been observed, the envenomated fish clearly went through a hyperexcitable state and was in fact immobilized in what appeared to be a tetanic-like state. Thus, the K channel–targeted conopeptides from this group may have a role analogous to those in the lineages that have a taser-and-tether strategy. There are two well-characterized peptides from the *Phasmoconus* species *C. radiatus*. These are κM-conotoxin RIIIJ[48] (κM-RIIIJ) and κM-conotoxin RIIIK (κM-RIIIK).[49] They are related to each other in being relatively small, and are members of the M-gene superfamily. Thus, they are genetically unrelated to the *Chelyconus* K channel–targeted peptides described above. κM-RIIIJ and κM-RIIIK were both found to preferentially target the $K_V1.2$ homomer when compared with other K_V1 subfamily homomeric complexes, albeit with relatively low (micromolar) affinity. Recently, it has been found that certain heteromeric combinations have a much higher affinity for κM-conotoxin RIIIJ (H. Terlau and coworkers, personal communication). Thus, this is the first *Conus* venom peptide for which the true targeted heteromeric combination will be elucidated. This work has also revealed that κM-conotoxins RIIIJ and RIIIK do not overlap in their high-affinity targets, and therefore, presumably, both target heteromeric complexes containing a $K_V1.2$ subunit, but differ in the other K channel subunits that make up the functional complex. These peptides are therefore an important model system for elucidating how venom peptides from cone snails target specific heteromeric K channel complexes in a selective and potent manner.

Pionoconus: A large (60-residue) kunitz-domain polypeptide was discovered from the venom of *C. striatus*, from the *Pionoconus* clade.[50] Similar in structure to the dendrotoxins from mamba snake venoms, which are also kunitz-domain polypeptides that block voltage-gated K channels, the available evidence indicates that conkunitzin peptides evolved to target voltage-gated K channels. Conkunitzin S1 blocks $K_V1.7$ channels,[51] although it may have a higher-affinity heteromeric K channel target.

12.4.4 Conopeptides Targeted to Voltage-Gated Ca Channels

ω-conotoxins were among the first conopeptides to find important applications in neuroscience research and in biomedical applications. ω-conotoxins GVIA[17] (from *C. geographus*) and MVIIA[52] (from *C. magus*) were the first pharmacological agents discovered for selectively blocking N-type Ca currents, which were later shown to be mediated by $Ca_V2.2$ channels. This discovery complemented the discoveries of other compounds that selectively blocked P/Q-type and L-type Ca currents. ω-conotoxin MVIIA eventually became an FDA-approved drug for neuropathic pain, sold under the generic name ziconotide (trade name Prialt). Recently, an O1 superfamily of glycine-rich peptides from *Virgiconus* clade was identified that potentially targets Ca channels.[53]

12.4.5 Conopeptides Targeted to Nicotinic Acetylcholine Receptors

Conopeptides that have evolved in cone snail venoms to inhibit nAChRs are more homogeneous in their genetic origins than conopeptides that target other major signaling molecules characterized so far. Over a wide range of different cone snail species, nAChR-targeted conopeptides belong to the A superfamily, with most having two disulfide bonds.

α-conotoxins from fish-hunting cone snails are among the best characterized of all *Conus* peptides (see Sections 12.2 and 12.3). There is a general motif in the amino acid sequence (...CCXXXXCXXXXXXXC..., where C = cysteine and X = any amino acid) that is phylogenetically very broadly distributed across multiple subgenera of *Conus*. These conopeptides are generally referred to as the α4/7-conotoxins because there are four and seven amino acids, respectively, between cysteine residues in the primary amino acid sequence. Different lineages however have altered this basic motif. In the fish-hunting cone snail lineages that use a taser-and-tether strategy in the Indo-Pacific region, most species have a shortened version known as the α3/5 motif (...CCXXXCXXXXXC...). Thus, well-known fish-hunting species such as *C. geographus*, *C. magus*, and *C. striatus* all have the shortened α3/5-conotoxins, as shown in Table 12.2 and in the Appendix. As already summarized in Section 12.3, species in *Chelyconus* and some *Gastridium* species have an additional disulfide linkage, and although these peptides also genetically belong to the A superfamily, they are recognized as a separate pharmacological family (αA-conotoxins[30,31]).

There is a vast series of nAChR-targeted conopeptides belonging to the α-conotoxin family from both snail-hunting and worm-hunting *Conus*. Those that have been characterized are summarized in Table 12.2 and in the Appendix. Many of these share a considerable sequence similarity and have a similar spectrum of selectivity for different nAChR subtypes.

A very distinctive group of α-conotoxins has evolved in certain species of worm-hunting cone snails that belong to the subgenus *Stephanoconus*, which are known to eat amphinomid polychaetes (*fireworms*). These peptides are apparently targeted to a specific subclass of nAChRs, those that contain only α subunits. In mammals, this is a relatively minor group of the total spectrum of nAChRs, and only two have been functionally characterized, the homomeric $\alpha7$ nAChR and the heteromeric $\alpha9\alpha10$ nAChR. There are reports that the spectrum of nAChRs containing $\alpha7$, $\alpha9$, and $\alpha10$ subunits is more complex, but this remains a subject of continuing research.[54] In certain invertebrate lineages (such as *Caenorhabditis elegans*), there is a much greater diversity of α subunits that can assemble into all α-complexes. In *Caenorhabditis elegans*, there are at least 32 different nAChR subunits, of which 22 are α subunits.[55] It is likely for this reason that worm-hunting *Conus* species have been a rich source of venom peptides that target nAChRs containing only α subunits.

Michael McIntosh and coworkers have systematically developed conopeptides and analogs of native conopeptides that are selective for particular nAChR subtypes. A few of the highly selective peptides for nAChR subtypes are those shown in Table 12.5. These peptides, among others, provide a useful pharmacological toolkit for identifying which nAChR plays a role in a specific circuitry or signaling pathway.

12.4.6 Other Families of *Conus* Venom Peptides Targeted to Nicotinic Acetylcholine Receptors

In addition to α-conotoxins, which share a common genetic origin (A superfamily), there are additional families of conotoxins that block nAChRs, including ψ-conotoxins,[34–36] which belong to the M superfamily. Examples of ψ-conotoxins are shown in the Appendix. Unlike α-conotoxins, which compete with acetylcholine for binding, ψ-conotoxins are noncompetitive inhibitors of nAChRs and may block the pore of the nAChR channel. These peptides have not been extensively characterized, but have been found in the venoms of fish-hunting cone snails.

In addition to α- and ψ-conotoxins, we previously identified a novel nAChR antagonist that blocks a variety of both muscle and neuronal subtypes. This peptide, known as αS-conotoxin RVIIIA,[56] belongs to a different gene superfamily, the S superfamily, which is characterized by having 10 cysteine residues (five disulfide bonds). A second homologous peptide from a different *Conus* species is highly selective for the $\alpha9\alpha10$ nAChR.[57]

TABLE 12.5
α-Conopeptides That Discriminate between Different nAChR Subtypes (Rat)

nAChR Subtype	α2β2	α2β4	α3β2	α3β4	α4β2	α4β4	α6β2	α6β4	α7	α9
α2β2	—	—	—	—	—	—	—	—	—	—
α2β4	—	—	—	—	—	—	—	—	—	—
α3β2	PeIA##	PeIA##	—	PeIA##	PeIA##	PeIA##	—	PeIA##	PeIA##	PeIA##
α3β4	TxID	TxID	TxID	—	TxID	TxID	TxID	—	TxID	TxID
α4β2	—	—	—	—	—	PeIA##	—	—	—	PeIA##
α4β4	—	—	—	—	—	—	—	—	—	—
α6β2	PeIA#	PeIA#	PeIA#	PeIA#	PeIA#	PeIA#	—	TxIB	PeIA#	PeIA#
α6β4	TxID	TxID	TxID	PeIA#	TxID	TxID	TxID	—	TxID	TxID
α7	ArIB#	ArIB#	ArIB#	ArIB#	ArIB#	ArIB#	ArIB#	ArIB#	—	ArIB#
α9	RgIA	RgIA	RgIA	RgIA	RgIA	RgIA	RgIA	RgIA	RgIA	—

References 65,156–159

PeIA# PeIA[A7V,S9H,V10A,N11R,E14A]

PeIA## PeIA[S9H,V10A,E14N]

ArIB# ArIB[V11L,V16D]

Note: Each conopeptide in the table blocks the corresponding nAChR subtype shown in the same row to the left with significantly greater potency than the corresponding nAChR subtype shown in the same column at the top.

12.4.7 Conopeptides Targeted to Glutamate Receptors

At the present time, there are two groups of conopeptides known to target different types of glutamate receptors. The first of these to be discovered were the conantokins[11] (see Section 12.2), which characteristically lack any disulfide bonds, but have multiple residues of the post-translationally modified amino acid γ-carboxyglutamate. All members of the conantokin family target NMDA receptors and have no effects on other classes of glutamate receptors. Several conantokins have high selectivity for specific NMDA receptor subtypes; the best-characterized are targeted to NMDA receptors with NR2B subunits.[58–60] Conantokins have been found in two lineages of fish-hunting cone snails, the *Gastridium* lineage (i.e., *C. tulipa* and *C. geographus*) and the *Asprella* lineage. In the former case, the physiological role of conantokins is known; these are part of the Nirvana cabal that the snail releases to debilitate potential fish prey. There is little information about the biology of any species in *Asprella*, since these occur in relatively deep waters. However, at least some *Asprella* species have a greater diversity of conantokins than are found in any other cone snail.[60]

Many peptides that belong to the same gene superfamily as the conantokins do not have any activity on NMDA receptors. The targets of these peptides have not yet been determined. Unlike the conantokins, which have Gly–Glu or Gly–Asp residues at their N-terminus, many closely related peptides (e.g., with many γ-carboxyglutamate residues and other sequence similarities) that do not share this N-terminal motif do not block NMDA receptors and are not considered part of the conantokin family.

A second group of glutamate receptor–targeted peptides are the con-ikot-ikot peptides, which were first characterized from the fish-hunting *Conus* species *C. striatus*.[27] The physiological role appears to be to facilitate the lightning-strike cabal, by inhibiting the desensitization of postsynaptic AMPA receptors found in the lateral line circuitry of fish. Transcriptomic methods have revealed that members of the con-ikot-ikot family are widely distributed, but the biological activities of these homologs have not been directly assessed. Examples of con-ikot-ikot peptides and conantokin peptides are shown in the Appendix.

12.4.8 Other Conopeptide Targets

In previous sections, we have described many well-characterized conotoxins that target voltage-gated ion channels (Na, K, and Ca channels) and ligand-gated ion channels (such as nAChRs and glutamate receptors). A less explored family of peptides with promising therapeutic potential includes peptides targeting monoamine transporters and GPCRs. These peptides are listed under miscellaneous targets (see the Appendix). Some of these peptides, such as ρ-contoxins (ρ-TIA targets the a1 adrenoreceptor), conopressins (conopressin G and conopressin T target vasopressin receptors), and σ-GVIIIA (targets 5 HT_3 receptors) provide avenues to explore similar peptides from closely related species for generating a pharmacological tool box of target selective conotoxins. Notably, χ-MrIA (targets norepinephrine transporters) and contulakin-G (targets neurotensin receptors) have reached human clinical trials for the treatment of neuropathic pain and are discussed in detail in Section 12.5.

12.5 BIOMEDICALLY SIGNIFICANT *CONUS* VENOM PEPTIDES: THERAPEUTIC, DIAGNOSTIC, AND BASIC RESEARCH APPLICATIONS

A summary of biomedically significant peptides is provided in Table 12.6.

12.5.1 ω-Conotoxins, ω-Conotoxin GVIA, and ω-Conotoxin MVIIA

The most significant biomedical impact of *Conus* venom peptides so far has resulted from the discovery of ω-conotoxins and the realization that they target a specific voltage-gated Ca channel subtype that is present at many presynaptic termini. The most extensively characterized of these peptides is ω-conotoxin GVIA (ω-GVIA),[2,8,17] which became a standard research tool. The widespread use of this tool led to the extensive characterization of its molecular target, the N-type Ca channel ($Ca_V2.2$). In fact, radiolabeled ω-GVIA was the reagent that made the purification of $Ca_V2.2$ possible. In addition to its use as a research tool, it is also used as a diagnostic tool. In Lambert–Eaton syndrome, an autoimmune disease against voltage-gated Ca channels, radiolabeled ω-GVIA is used to differentiate the presynaptic targeting of the antibodies from clinically similar syndromes such as myasthenia gravis.[61,62]

Ultimately, ω-conotoxin MVIIA (ω-MVIIA), although less well characterized than ω-GVIA, is the peptide that ultimately became a therapeutic drug for intractable pain (generic name, ziconotide; trade name, Prialt). This peptide was used in a number of basic research studies, including some carried out on neurotransmitter release by George Miljanich, who initiated the development of ω-MVIIA as a commercial drug.[9] Miljanich and J. Ramachandran, then in a small biotech company called Neurex, discovered that the $Ca_V2.2$ channel was present in the dorsal horn in layers where afferent pain fibers synapse with spinal neurons, which led to the systematic exploration of ω-MVIIA for the treatment of pain. It is now an approved drug for intractable pain that is delivered intrathecally to the spinal cord. It is notable that native ω-MVIIA, as synthesized by *C. magus*, is chemically identical to the commercial drug (ziconotide).

12.5.2 α-Conotoxins Vc1.1 and Rg1A

α-Conotoxin Vc1.1 (α-Vc1.1),[63,64] from a snail-hunting *Conus* species, *C. victoriae*, and α-conotoxin RgIA (α-RgIA),[65] from a worm-hunting species, *C. regius*, have emerged as biomedically relevant peptides. The original impetus for investigating the clinical possibilities for α-conotoxins came from Bruce Livett, then at the University of Melbourne. At the time, the conventional wisdom was that stimulating the activity of nAChRs could alleviate pain. However, Livett believed that inhibiting selected nAChRs might produce analgesia. His laboratory isolated α-conotoxin Vc1.1 from *C. victoriae*[63,64] and demonstrated that the peptide had efficacy in animal models of neuropathic pain after nerve injury. The effects were remarkable because the analgesic symptomatology persisted far beyond the life of the peptide, suggesting a disease-altering mechanism.[64] Ultimately, the peptide was developed by a biotech company in Australia and completed a Phase I human clinical trial for safety, but was not sufficiently efficacious in its initial Phase II trial and was ultimately abandoned.[66]

TABLE 12.6
Conopeptides with Recognized Biomedical Potential

Peptide	Species	Peptide Sequence	Molecular Target	Condition	Stage in Development	Biotech Partner
α-Vc1.1	*C. victoriae*	GCCSDPRCNYDHPEIC*	*α9α10*, *α6β2*	Neuropathic pain	Preclinical	Metabolic, Inc.
ω-CVID	*C. catus*	CKSKGAKCSKLMYDCCSGSCSGTVGRC*	$Ca_V2.2$	Neuropathic pain	Phase II	Amrad, Inc.
ω-MVIIA	*C. magus*	CKGKGAKCSRLMYDCCTGSCRSGKC*	$Ca_V2.2$	Cancer pain	Phase III	Elan
χ-MrIA	*C. marmoreus*	NGVCCGYKLCHOC	NE Transporter	Neuropathic pain	Preclinical	Xenome, Inc.
Contulakin-G	*C. geographus*	ZSEEGGSNAT⁺KKPYIL	Neurotensin receptor	Chronic pain	Phase II	Cognetix, Inc.
Conantokin-G	*C. geographus*	GEγγLQγNQγLIRγKSN*	NMDA NR2B	Epilepsy	Preclinical	Cognetix, Inc.
α-RgIA	*C. regius*	GCCSDPRCRYRCR	*α9α10* nAChR	Nerve injury pain	Preclinical	Kineta, Inc.
κ-PVIIA	*C. purpurascens*	CRIONQKCFQHLDDCCSRKCNRFNKCV*	K_V1 subfamily	Cardioprotection	Preclinical	Cognetix, Inc.
κM-RIIIK	*C. radiatus*	LOSCCSLNLRLCOVOACKRNOCCT*	K_V1 subfamily	Cardioprotection	Preclinical	

Note: Special characters that represent post-translational modifications of peptide sequences are explained in Table 12.7.

Abbreviation: NE, Norepinephrine.

At the time that α-Vc1.1was in development as a therapeutic drug, the molecular target of the peptide had not been identified. McIntosh and coworkers showed that α-Vc1.1 had highest affinity for $\alpha9\alpha10$ nAChRs.[66] A comparison of the affinity of α-Vc1.1 for the $\alpha9\alpha10$ nAChR in rodents and in humans provided a potential explanation for the lack of efficacy in Phase II human clinical trials.[67] There was a big difference in the potency of the peptide, with two orders of magnitude decrease in potency when the human $\alpha9\alpha10$ nAChR was compared with the homologous rodent nAChR.[67] This could potentially explain the disappointing results in the Phase II clinical trials with α-Vc1.1.

However, α-RgIA, from *C. regius*, had an even higher selectivity for the $\alpha9\alpha10$ nAChR,[65] although it was similarly less potent in the human versus the rodent nAChR.[67] McIntosh and coworkers have since developed a variety of derivatives of α-RgIA that have very high potency for the human $\alpha9\alpha10$ nAChR subtype, while retaining its high selectivity over all other tested nAChR subtypes (unpublished data). These studies have revealed the probable mechanism for the therapeutic effects originally observed by Livett and coworkers.

The inhibition of $\alpha9\alpha10$ nAChR after nerve injury has a clear effect on neuroinflammation.[64,66] In the chronic constriction injury model, which is a nerve injury model of neuropathic pain, lymphocytes and macrophages accumulate at the site of injury, but α-RgIA was shown to decrease significantly the accumulation of immune cells at the site of nerve injury. Thus, α-conotoxin RgIA is a potent antineuroinflammatory compound.[68,69] McIntosh and coworkers also used another α-conotoxin to selectively block the closely related $\alpha7$ nAChR subtype,[70] which did not produce any measurable effects. This peptide, when tested in the same neuropathic pain model after nerve injury (the CCI rat model), caused an increase in the inflammatory response, instead of inhibiting the inflammatory response, which was observed after administration of α-RgIA (unpublished data). The degeneration of dorsal root ganglion (DRG) neurons and infiltration of immune cells into the DRG and spinal cord were all inhibited by α-RgIA.[69] These results suggest that α-RgIA has a disease-altering effect on nerve injury that leads to neuropathic pain.

It should be emphasized that the precise molecular mechanism by which this occurs is not known. The degeneration of DRG cells and infiltration of immune cells into the DRG, and in addition, infiltration into the spinal cord, were all inhibited by α-RgIA.[69]

The anti-inflammatory and analgesic mechanisms mediated by α-RgIA have not been completely vetted. There is evidence for the presence of $\alpha9\alpha10$ nAChR in immune cells,[54,71] suggesting that nerve injury leads to signaling via $\alpha9\alpha10$ nAChRs, although the downstream effects of activating $\alpha9\alpha10$ nAChRs are not yet understood. One group of researchers has suggested that the analgesic effects of α-RgIA are mediated by activation of gamma-aminobutyric acid B ($GABA_B$) receptors, rather than by inhibition of $\alpha9\alpha10$ nAChRs.[72] However, most of the analgesic and anti-inflammatory effects of α-RgIA are abolished in $\alpha9$ knockout mice, and these mice are resistant to the development of neuropathic pain after nerve injury (unpublished data).

As discussed in Section 12.4, it is likely that α-RgIA is used by *C. regius* as a paralytic toxin for its prey, which are amphinomid polychaetes (*fireworms*). As far as we are aware, the precise nAChR present at the neuromuscular junction of the gastropod mollusc prey of *C. victoriae* has not yet been elucidated. In mammalian systems, only $\alpha7$, $\alpha9$, and $\alpha10$ subunits form nAChRs without any β or other non-α subunits. However, α subunits of nAChRs that putatively form ion channels composed of only α subunits are much expanded in their diversity in most invertebrate systems.

12.5.3 Contulakin-G

As described in Section 12.2, one component of the nirvana cabal of *C. geographus* is a GPCR-targeted peptide, contulakin-G, which is homologous to the neuropeptide neurotensin.[12] Interestingly, there are no known neurotensin-like peptides outside of vertebrate systems. Thus, the origins of contulakin-G and how it evolved in *C. geographus* are problematic. A transcriptomic analysis of the precursor peptide that contains contulakin-G shows that it belongs to the C superfamily, although other C superfamily peptides appear to be completely unrelated to contulakin-G, both structurally and functionally. The only link is a highly conserved signal sequence shared by all C superfamily peptides.

Contulakin-G (also known as CGX-1160, as an investigational new drug) has reached Phase I human clinical trials as an analgesic drug for spinal cord injury patients.[73] The preclinical data demonstrated efficacy in a wide variety of animal models of pain,[74] with a broad therapeutic index, in contrast to other pain medications, such as opioids or ziconotide. However, the mechanism by which contulakin-G produces analgesia is unclear. When compared with neurotensin activity on the two known neurotensin receptors, contulakin-G was significantly less potent than neurotensin. However, it was a much more effective analgesic than neurotensin in animal pain models,[75] raising questions about its analgesic mechanism.

There are a few possibilities that may explain the greater analgesic activity of contulakin-G than that of neurotensin. First, contulakin-G produces less receptor desensitization or internalization than neurotensin, which retains a higher density of neurotensin receptors in the plasma membrane that can be activated repeatedly.[75] A second possibility that is increasingly explored in the GPCR field is that some agonists of the same receptor can trigger different downstream signal transduction pathways (biased signaling). A third possibility is that contulakin-G is more metabolically stable than neurotensin.[75] Contulakin-G has two post-translational modifications, pyroglutamate at the N-terminus and a glycosylated threonine residue,[12,21] which may contribute to increased stability.

12.5.4 Conantokins

As described above, conantokin-G is a 17-amino acid-long peptide with an unusual post-translational modification, gamma-carboxyglutamate.[11] Conantokin-G potently and selectively blocks the NR1/NR2B receptor subtype of NMDA receptors.[58,59] It has exhibited therapeutic potential for multiple disease states, including pain, epilepsy, and neuroprotection.[76,77] In multiple animal models of epilepsy, the effective therapeutic dose (ED_{50}) of conantokin-G was significantly lower than doses that cause toxicity (TD_{50}). These models include the Frings audiogenic seizure mouse model, and maximal electroshock (MES) and subcutaneous pentylenetetrazol (scPTZ) tests.[78] Conantokin-G (also known as CGX-1007, as an investigational new drug) previously reached a Phase I human clinical trial for the treatment of intractable epilepsy. Additionally, this peptide has demonstrated efficacy in mouse pain models, including partial sciatic nerve ligation, the formalin test, and the complete Freund's adjuvant model of allodynia.[76] The peptide was first found to be neuroprotective because of its ability to inhibit glutamate-induced neurotoxicity in cerebellar granule neurons.[79] It was later found to be neuroprotective in a rat model of focal ischemia, at doses that did not cause side effects or toxicity.[77,80] In addition to conantokin-G, a related peptide, known as conantokin-T, has shown therapeutic efficacy in models of chronic pain.[76] Another conantokin peptide, conantokin-R, was effective at suppressing epileptic seizures in animal models.[81]

12.5.5 Con-Insulin

The discovery of Con-Ins G1 in the venom of *C. geographus* represented the first and, thus far, sole example of the use of insulin for prey capture.[13] Con-Ins G1 is the smallest insulin reported from any natural source. Critically, it lacks the canonical C-terminal segment of the insulin B chain, which, in human insulin, plays a key role in mediating the assembly of the storage form of the hormone[82] and in engaging the primary binding site on the surface of the insulin receptor.[83,84] As explained next, this structural feature of Con-Ins G1 provides a unique opportunity for the development of novel, ultra-fast-acting insulin analogs for the treatment of diabetes.

In healthy humans, insulin is stored in pancreatic β-cells as a hexamer consisting of three insulin dimers held together by two central zinc ions, the insulin monomer itself consisting of an A and a B chain, cross-linked by two disulfide bridges and with a third disulfide bridge within the A chain.[85] Insulin hexamer-to-monomer conversion is crucial to its bioavailability and can lead to a delay in glucose control following injection into diabetic patients. Thus, insulin administration often involves a combination of a rapid-acting mealtime insulin and a longer-acting basal insulin.[86] Rapid-acting insulin analogs contain amino acid substitutions in the C-terminus of the B chain that result in reduced rates of self-association but do not completely abolish oligomerization. Thus, even the best rapid-acting insulin analogs still require 15–30 min to effectively lower blood glucose levels.[87] This can result in severe hyperglycemia and diabetic complications. Attempts to merely remove the C-terminus of the B chain in order to abolish self-association have resulted in near-complete loss of activity. For example, *des*-octapeptide insulin, a monomeric analog, preserves less than 0.1% bioactivity.[88] In contrast, Con-Ins G1 is monomeric and lacks any equivalent of the C-terminus of the human B chain.[89] Despite these features, Con-Ins G1 potently binds to the human insulin receptor and activates the receptor signaling pathway.[89] Thus, Con-Ins G1 represents a naturally occurring, monomeric mimetic of human insulin. Preclinical investigations into developing this peptide as an ultra-fast-acting insulin analog for the treatment of diabetes are underway. In addition to its therapeutic potential, Con-Ins G1 has provided a unique tool for investigating the minimum binding site of insulin at the human insulin receptor. It appears that two tyrosines in the B chain of Con-Ins G1 (B15 and B20) engage with the receptor to accommodate for the loss of the B chain C-terminus.[89] Introducing these residues into *des*-octapeptide insulin and/or other B chain–truncated versions of human insulin may result in the recovery of activity in the absence of oligomerization. All currently available fast-acting insulin analogs were designed nearly two decades ago and the discovery of Con-Ins G1 represents an exciting new avenue for the design of novel and truly rapid-acting insulin therapeutics.

12.5.6 χ-Conotoxins

Chi-conotoxins (χ-conotoxins) belong to the T superfamily that are expressed in a wide range of *Conus* species but, unlike the other members of the T superfamily, they differ in their sequences and cysteine arrangement.[90] The cysteine patterns in χ-conotoxins resemble previously described α-conotoxins with a different disulfide-bonding framework, as exemplified by a well-characterized χ-conotoxin MrIA (χ-MrIA).[91] χ-MrIA is a 13-amino-acid-long peptide isolated from *C. marmoreus* that targets norepinephrine transporters (NETs) without affecting other monoamine transporters such as dopamine and serotonin.[92] The binding studies on human NETs implied that they compete with other tricyclic antidepressants but not with norepinephrine.[93] Intrathecal administration of χ-MrIA was shown to have antinociceptive properties in a mouse hot-plate assay.[91] This peptide (also known as Xen2174 as an investigational new drug) has reached human clinical trials for the treatment of neuropathic pain.[94]

12.6 *Conus* Venom Peptides: Unusual Gene Products

The following section provides an overview of conotoxin diversity and classification, and highlights unusual characteristics of *Conus* venom components that distinguish them from other animal venoms. Conotoxins are unusually short,[95] a characteristic that is only shared with spider venom peptides.[96] Their sequences diversify more rapidly than most other animal venoms,[97] and their mature-peptide products carry a high density and diversity of post-translational modifications not observed in any other animal venom.[98] It is also more straightforward to glean evolutionary patterns in cone snail venoms because the phylogenetic relationships between the ~800 species in the family Conidae are relatively well understood.[99,100]

12.6.1 Conotoxin Classification and Genetic Diversity

Almost all conotoxins identified to date share a characteristic precursor organization containing an N-terminal signal sequence that directs the peptides to the endoplasmic reticulum, an intermediate propeptide region, which has been suggested to assist in conotoxin biosynthesis,[101–103] followed by the mature toxin at the C-terminus, which is proteolytically cleaved from the propeptide during toxin maturation (Figure 12.5). Conotoxins that belong to the same gene superfamily share a highly conserved signal sequence, whereas the toxin region is hypervariable, with the exception of a conserved arrangement of cysteine residues (in most cases).

Signal Sequence | Propeptide | Toxin

A-superfamily signal sequence

```
GI       MGMRMMFTVFLLVVLATTVVSFPS-ERASDGRDDTAKDEGSDMDKLVEKK-ECC-NPACGRH--YSCGR
Lp1.2    MGMRMMFTVFLLVVLATTVVSFTS-DRAFDGRNAAASDKASDLISLAVR--GCCSHPACSVNNPYFCGGKR
Mr1.7    MGMRMMFTMFLLVVLATTVVSFTS-NRAFRRRNAVA--KASDLIALNARRPECCTHPACHVSHPELCG
Tx1.2    MGMRMMFTVFLLVVLATTVVSFTSGRRTFHGRNAAA--KASGLVSLTDRRPECCSHPACNVDHPEICR
Im1.8    MGMRMMFTVFLLVVLATAVLPVTL-DRASDGRNAAANAKTARLIAPFIRD-YCCPRGPCMV----WCG
Bu1.1    MGMRMMFTVFLLVVLATTVVSFST-DDESDGSNEEPSADQTARSSMNRAP-GCCNNPACVKH---RCG
MIVA     MGMRMMFTVFLLVVLATTVVSIPS-DRASDGRNAVVHERAPELVVTATT--NCCGYNPMTICPPCMCTYSCPPKRKPGRRND
```

M-superfamily signal sequence

```
GIIIB    MMSKLGVLLTICLLLFPLTALPMDGDEPANRPVERMQDNISSEQY----PLFEK--RRDCCTPPRK----CKDRRCKPMKCCAGR
Ca3-TP3  MMSKLGVLLITCLLLFPLTAVPLDGDQHADRPAERLQDDISSENH----PFFDP--VKRCCNAGF-----CR-FGCTP--CCY
Mr3.3    MMSKLGVLLTICLLLFPLTAVPLDGDQPADRPAERLQDDISSEKQITNTPILDS--GRECCGSFA-----CR-FGCVP--CCV
Lt3.6    MMSKLGVLLTICLLLFPLTALPMDGDQPVDRPAERMQGKISSEQH----PMFDP--IEGCCTQS------CT--TCFP--CCLI
S3-TS01  MMSKLGVLLTICLLLFPLTAVPLDGDQPLDRHAERMHDGISPKRH----PWFDP--VKRCCKVQ------CE--SCTS--CC
BuIIIA   MMSKLGVLLTICLLLFPLFALPQDGDQPADRPAERMQDDISSEQN----SLLEKRVTDRCCKGKRECGRWCR--DHSR--CCGRR
```

FIGURE 12.5 Characteristic conotoxin precursor organization with a conserved N-terminal signal sequence, a propeptide region with intermediate mutation rates, and a hypervariable mature-toxin region at the C-terminus. Members of the A and M superfamilies are shown.

Propeptide sequences typically exhibit intermediate mutation rates. This striking feature of contrasting mutation rates for adjacent regions was noted in the first study that reported the cloning of conotoxin genes[104] and has been consistently reported since.[105–107] The molecular mechanism behind this unusual sequence juxtaposition remains unknown.

In the early days of *Conus* venom research, conotoxins were typically classified by their pharmacological profile and/or conserved cysteine arrangement. This was feasible because most newly identified toxin sequences were indeed pharmacologically characterized. Nowadays, next-generation nucleotide and protein sequencing techniques are generating large libraries of novel sequences that cannot be functionally characterized at the same pace as the accelerated rate of discovery. Thus, novel conotoxin sequences are predominantly classified based on their conserved N-terminal signal sequences (Figure 12.5 shows examples of sequences belonging to the A and M superfamilies). The cysteine arrangement, defined as a characteristic pattern of cysteine residues, can be readily deduced from the primary amino acid sequence but does not contribute to superfamily classification because conotoxins sharing the same cysteine arrangement can belong to different gene families. To date, more than 53 gene superfamilies (Figure 12.6) and 26 cysteine arrangements have been described in *Conus*. Some gene superfamilies appear to be ubiquitously found in all *Conus* species (e.g., the O1 and M superfamilies), while others are restricted to a subset of species (e.g., the B superfamily, which is mostly expressed in some fish-hunting cone snails). Figure 12.6 provides an overview of all conotoxin superfamilies identified to date, with their number and frequencies. By far, the most studied superfamilies are the A superfamily, which comprises the majority of α-conotoxins characterized to date, and the O1 superfamily, which contains the majority of δ and ω conotoxins identified to date. A comprehensive review of the diverse superfamilies found in *Conus* was recently provided.[108] However, it should be noted that several additional superfamilies have since been discovered and additional ones will be discovered over the next decade.

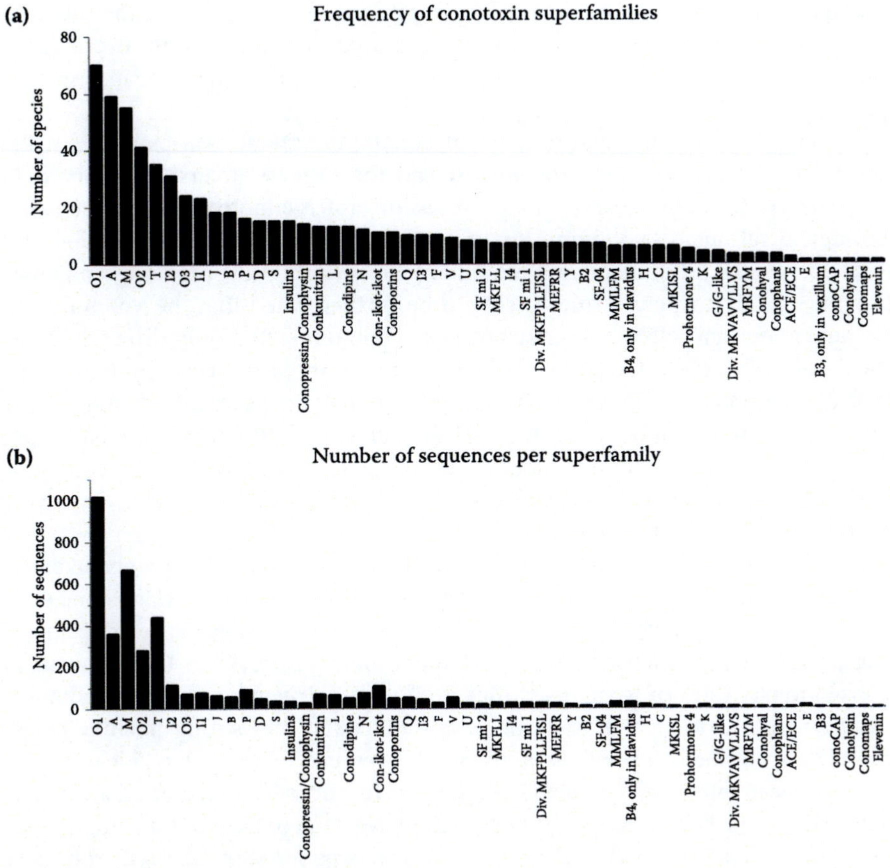

FIGURE 12.6 Diversity of gene superfamilies. **(a)** Number of cone snail species for which a particular superfamilies has been described. Several superfamilies, such as the O1, A, and M superfamilies, are almost ubiquitously expressed in all *Conus* species. Others, such as the B superfamily and insulins are less widely distributed. Several superfamilies have only been described in a single species to date, including conolysin, conomaps, and elevenin. **(b)** Total number of distinct conotoxin sequences described to date. With more than 1000 sequences, the O1 superfamily represents the most diverse gene superfamily. This superfamily comprises a diverse set of pharmacological families, including δ and ω conotoxins. Numbers were calculated from the Conoserver database[132] and recently published transcriptomes.[109]

There is ongoing debate in the literature regarding the number of conotoxins found in a single species of cone snails. This is because deep-sequencing technologies, which are required to comprehensively sequence all conotoxins expressed in the venom gland, have high sequencing error rates that have not been appropriately taken into account by some researchers in the field. This has led to vast overestimates of conotoxin diversity, such as 3303 conotoxin sequences recently reported in *C. episcopatus*. We believe that these estimates are a result of inadequate processing of next-generation sequencing datasets, as suggested by others,[109] and that the true diversity lies between 100 and 500 different conotoxins, as reported in several independent studies that used more stringent data processing techniques of transcriptome sequencing data.[109–113]

It is emerging that the total number of conotoxins greatly varies between the different *Conus* species, with the lowest and the highest diversity observed in *C. imperialis* (68 conotoxins) and *C. sponsalis* (401 conotoxins), respectively. A biological rationale explaining this divergence has not yet been provided, but species-to-species differences in biotic interactions (with prey, predator, and competitor) are likely to play a role. It should be further noted that the venom of an individual cone snail often comprises pairs of conotoxins that only differ by a few amino acids. These are likely to be allelic variants inherited from parent snails. Because some conotoxins are highly polymorphic (i.e., have a high number of allelic variants within a population),[114,115] the number of conotoxin sequences per species increases when venom samples are pooled from more than one individual. This should be taken into account when discussing the diversity of conotoxin sequences found in *Conus* venoms.

Despite the discrepancies reported for individual *Conus* venoms, agreement exists on the rapid diversification of conotoxin sequences upon speciation. Each cone snail species expresses its own repertoire of conotoxins, with almost no overlap in the primary amino acid sequence between different *Conus* species.[110,112] These unusually fast evolutionary rates of toxin diversification [97,111] are driven by at least two known molecular mechanisms: high rates of gene duplication and strong positive selection.[105,106] As outlined in previous sections, the venoms of several species of cone snails have been intensively studied over the past few decades (a list of all characterized peptides is provided in the Appendix) and have been of tremendous importance for biomedical research and as drug leads. These peptides were derived from only 5% of extant species and many *Conus* venoms remain completely uncharacterized. Technological advances in next-generation transcriptome and proteome sequencing technologies have already transformed the field of conopeptide discovery and novel conotoxin sequences are being discovered at rates never seen before.[109–111,113,116] Pharmacological testing of these sequences is likely to lead to novel discoveries and the generation of new classes of therapeutics at a scale that cannot be predicted at this stage.

12.6.2 Diversity of Post-Translational Modifications Observed in *Conus*

In addition to their high evolutionary rates, another unusual feature of conotoxins is their high diversity and density of post-translational modifications. To date, 16 different modifications have been reported for conotoxins (Table 12.7 provides an overview of these). Many of these, such as disulfide bond formations, C-terminal amidation, and γ-carboxylation, are commonly found across diverse cone snail venoms, while others are more infrequent (e.g., L-to-D epimerization and glycosylation). Modifications are known to affect conotoxin structure, potency, and target selectivity, and represent a key mechanism of venom diversification.[117]

TABLE 12.7
Post-translational Modifications of Conopeptides

Modification	First Reported for	Molecular Target	Sequence	Species	Reference	Reported Number[a]
Disulfide bond formation	α-GI	nAChR, muscle	ECCNPACGRHYSC*	*C. geographus*	[160]	>2000
Amidation of the C-terminus	α-GI	nAChR, muscle	ECCNPACGRHYSC*	*C. geographus*	[160]	175
Hydroxylation of						
Proline	μ-GIIIA	$Na_V1.4$	RDCCTOOKKCKDRQCKOQRCCA	*C. geographus*	[161]	110
Lysine	DeXIIIA	N.D.	DCOTSCOTTCANGwECC hyLGYOCVNhyLACSGCTH*	*C. delessertii*	[162]	1
Valine	Conophan–gld–V	N.D.	APANShyVWS	*C. gladiator*	[123]	2
γ-Carboxylation of glutamate	Conantokin-G	NMDA-R NR2B	GEγγLQγNQγLIRgKSN*	*C. geographus*	[20]	58
Bromination of tryptophan	Bromocontryphan	N.D.	GCO(D-w)EPWC*	*C. radiatus*	[121]	19
Cyclization of N-terminal	Bromoheptapeptide	N.D.	ZCGQAwC*	*C. imperialis*	[119]	16
Sulfation of tyrosine	α-EpI	nAChR	GCCSDPRCNMNNPDY (SO4) C	*C. episcopatus*	[163]	6
L-to-D Epimerization of						
Tryptophan	Bromocontryphan	N.D.	GCO(D-w)EPWC*	*C. radiatus*	[121]	8
Leucine	Leu–Contryphan–P	N.D.	GCV(D-L)LPWC	*C. purpurascens*	[164]	4
Phenylalanine	RXIA	N.D.	GOSFCKADEKOCEYHADCC NCCLSGIC AOSTNWILPGCSTSSF(D-F)KI	*C. radiatus*	[46]	4
Valine	Conophan-mus-V	N.D.	SOANS(D-V)WS	*C. mus*	[123]	2
O-Glycosylation of						
Serine	κA-SIVA	K channels	ZKSLVPS+VITTCCGYDOGT MCOOCRCTNSC*	*C. striatus*	[21]	2
Threonine	Contulakin-G	Neurotensin-R	ZSEEGGSNAT+KKPYIL	*C. geographus*	[12]	3
S-Cysteinylation	μO§-GVIIJ	$Na_V1.2$	GWCGDOGATCGKLRLYCCSG FCDSCC§ YTKTCKDKSSA	*C. geographus*	[42]	1

[a] Reported number (far-right column) is the number of conopeptides in Conoserver that are known to contain each post-translational modification.[132] The number of peptides containing disulfide bonds was predicted based on the presence of cysteines in all conotoxins available in Conoserver (http://www.conoserver.org/). All other modifications cannot be predicted based on sequence information alone and only those confirmed by mass spectrometry or Edman sequencing are listed. Underlined amino acids in the sequences represent examples of the post-translational modifications indicated in the far-left column.

Despite being common in *Conus*, many of these modifications have been rarely reported in other biological systems. For instance, the discovery of γ-carboxylation of glutamate in *C. geographus* venom by a vitamin K–dependent carboxylase was surprising, because at the time, this modification was believed to be a unique feature of the mammalian blood clotting cascade.[20,118] Similarly, bromination of tryptophan in *C. imperialis* venom was a significant finding. First described in cone snails,[119] it was subsequently shown to be present in the mammalian nervous system.[120] The role of bromination remains elusive, and the enzyme, most likely a haloperoxidase, has not been identified in eukaryotes, including *Conus* and humans.

The sheer density of modifications reported for some conotoxins is also astounding. For instance, in bromocontryphan-R, a peptide isolated from *C. radiatus* that elicits a stiff-tail syndrome in mice, five of eight amino acids are modified: L-to-D epimerization of tryptophan 4, hydroxylation of proline 6, bromination of tryptophan 7, disulfide bond formation between cysteine 2 and 8, and C-terminal amidation of cysteine 8.[121] Similarly, TxVA, a 13-amino acid-long peptide from *C. textile* that causes hyperactivity and spasticity in mice, contains nine modified residues: two γ-carboxyglutamates, a bromotryptophan, an *O*-glycosylated threonine, a 4-hydroxyproline residue, and four cysteines forming two disulfide bonds.[122] The high density of modifications is further exemplified by the finding that some amino acid alterations can occur sequentially at the same residue. For instance, in conophans isolated from *C. gladiator* and *C. mus*, the already unusual L-to-D epimerizations of valines are accompanied by hydroxylations of the same residue.[123]

12.6.3 *Conus* Venom Insulins: Insights into How Natural Product Evolution Refines Structure and Function

From a conventional viewpoint, insulins would not generally be regarded as natural products. However, the venom insulins of cone snails do represent an extreme outlier class of natural products, and partly because they are so unusual, their characterization provides some insights into how natural products evolve, to be able to mediate chemical interactions between organisms. Because cone snails that express an insulin in their venom gland also express at least one conventional signaling insulin in their neuroendocrine cells, one can directly compare the evolutionary trajectory of the natural product to its endogenous counterpart. This feature is not unique to cone snail insulins, as many components of other animal venoms evolved from conventional *housekeeping proteins*,[124] but insulin is one of the best functionally and structurally characterized proteins and has served a central role in the advancement of peptide chemistry, pharmacology, cell signaling, and structural biology.[125] This vast knowledge of a single molecule offers the unique opportunity to efficiently trace the structural and functional adaptations of a compound that evolved from an endogenous protein into a natural product. An additional unique opportunity arises from the fact that several lineages of cone snails experienced a major shift in prey preference from invertebrates (worms) to vertebrates (fish), providing a unique window into monitoring evolutionary adaptations of a natural product as it changes its target from invertebrates to vertebrates.

In mammals, insulin is produced and released by the endocrine β-cells of the pancreas, where its primary role is the regulation of blood glucose homeostasis.[85] By contrast, invertebrate insulins are more variable and can serve more diverse functions, including regulation of hemolymph glucose levels, neuronal signaling, memory, reproduction, and growth.[126,127] In molluscs, including cone snails, insulins are produced in endocrine cells associated with the gastrointestinal tract and neuroendocrine cells of the central nervous system.[126–128] Molluscan insulins differ structurally from the vertebrate hormone by having longer B chains and one additional cysteine in each chain that presumably forms an interchain disulfide bond.[127] A recent analysis of the venom glands and neuroendocrine tissues of diverse cone snail species showed that the conventional signaling insulin shares the characteristic structural features of other molluscan insulins (longer B chains and an additional interchain disulfide) but that venom insulins expressed in the venom glands closely resembled insulins expressed in the prey organisms. Venom insulins sequenced from snail hunters are similar to snail insulins and venom insulins expressed in fish hunters share the same cysteine framework and chain lengths with fish insulins.[129] This finding highlights that venom insulins, particularly those found in fish hunters, not only greatly diverged from their own ancestral signaling insulin but also mimic the endogenous insulins of their vertebrate prey to efficiently target the prey's insulin receptor. An additional structural feature is that, unlike the endogenous fish insulin, venom insulins are post-translationally modified. These modifications are known to be of functional importance; they improve binding to the human insulin receptor and enhance downstream insulin signaling in mammalian cells.[89] Additionally, as discussed in Section 12.5.5, because venom insulins must act rapidly, the snail has evolved venom insulins that circulate as monomers (and not hexamers, as do almost all vertebrate insulins). Furthermore, venom insulins presumably have biochemical characteristics that facilitate rapid uptake into the circulation of the targeted fish prey, structural features that are not yet completely understood. Thus, cone snail venom insulins evolved structural features suitable for their specialized function.

On a molecular level, this has been achieved by a duplication of the insulin gene in the genome of cone snails or an early ancestor. One gene copy continued to encode the hormonal regulator of the endogenous energy metabolism of the cone snail, but the other copy was ultimately co-opted for expression in the venom gland, and was therefore used for biotic interactions of each individual cone snail species. Thus, the insulin genes of cone snails, while having originated from an ancient duplication, have been subjected for tens of millions of years to very different types of natural selection. The endogenous insulin is subject to stringent purifying selection,[129] to interact optimally with the insulin receptor of the cone snail. In contrast, the insulin genes expressed in the venom are subject to intense diversifying selection, tracing the adaptive radiation of cone snails. In particular, whenever there was a major shift in prey (e.g., from worms to fish), then a powerful selection of the venom insulin gene resulted in accelerated, episodic evolution so that the gene product could interact optimally with a very different insulin receptor (e.g., from the worm insulin receptor to the vertebrate insulin receptor found in teleost fish). It is the juxtaposition of the evolutionary trajectories of the endogenous insulin versus venom insulins of cone snails that provides a unique window into natural product evolution.

The evolutionary insights gained from studying cone snail insulins are likely to apply to many other natural products for which the molecular targets have not yet been identified or not comprehensively characterized.

12.7 PERSPECTIVES

In the first section of this chapter, we discussed natural products using a somewhat idiosyncratic perspective. We return to this framework, which was based on insights gleaned from cone snail venom components, described in Sections 12.2 through 12.6 above. In this section, we extend these insights generally to all natural products.

From a purely chemical perspective, components found in the venom of any given species of cone snails comprise a library of approximately 100–500 distinct natural products, mostly disulfide-rich, post-translationally modified peptides. Our studies on fish-hunting cone snails revealed that from a biological perspective, this complex mixture is designed to optimally achieve specific physiological endpoints in targeted animals, such as the fish prey of *C. geographus*. For *C. geographus*, two distinct physiological endpoints were defined: one group of venom components (*the nirvana cabal*[2]) act together on a potential prey to target sensory circuitry and energy metabolism, ultimately allowing the snail to capture multiple fish. Another set of venom components (*the motor cabal*[5]) is injected to cause an irreversible block of neuromuscular transmission, resulting in total paralysis of the fish.

The venom gland of cone snails is just one example of many different types of secretory organs attached to the foregut in the largest group of predatory marine snails called neogastropods (cone snails are only one of the multiple lineages within this major clade of gastropod molluscs). There is evidence that other lineages of neogastropods similarly debilitate their potential prey (such as the colubrarids, which anesthetize the fish from which they will suck blood).[130] Thus, the secretions of glands that cause these effects are, like the venom gland of cone snails, presumably targeted to another animal and have evolved to achieve specific physiological endpoints (e.g., local anesthesia at the site from which the colubrarid snail is going to suck blood). Thus, specialized neogastropod glands likely secrete a complex cocktail of diverse natural products, and some of these allow the predatory neogastropod snail to achieve a set of specific physiological endpoints that enhance prey capture. Our insight into the complexity of the neuroethological mechanisms involved in the interaction between venomous cone snails and their fish prey (detailed in Sections 12.2 and 12.3) suggests that the strategy evolved by other neogastropods that do not inject their secretions may be even more sophisticated (since unlike injected venom peptides, these would need to cross permeability barriers to access their molecular targets).

The characterization of cone snail venom components has demonstrated that knowing the underlying biological role of an individual venom component greatly facilitates identifying its potential biomedical applications. This is illustrated by the high frequency of biomedical applications from the subset of *C. geographus* venom components whose underlying molecular mechanisms have been elucidated. It is desirable to have both a comprehensive chemical characterization of a natural product and an insight into the molecular mechanisms of the biological activity for which that compound evolved. There are fundamentally two strategies to achieve this. The more conventional approach (*chemistry first*) is to isolate an individual natural product, followed by chemical characterization, synthesis, and/or expression. The pure compound is then used in various assays to try to identify its biological activity. This is in effect what the pharmaceutical industry does. Purified or synthetic natural products are stockpiled and tested for activity on disease-relevant molecular targets. This process requires substantial amounts of material for multiple screening campaigns over time.

The framework that we propose is encompassed by the new discipline that we refer to as *chemical neuroethology* (see Section 12.1). A whole organism biological context for the natural product is obtained first, if possible. Thus, a specialized gland of a neogastropod snail may secrete a mixture of natural products, and if the effects of the secretion on a potential prey, predator, or competitor of the species that evolved these secretions can be elucidated, then a limited set of mechanistic possibilities can be systematically explored. The entire glandular secretion can be tested on potential targets, and if biological effects are observed, the contribution of individual components in the mixture can be assessed and the individual compound responsible for a specific bioactivity purified. Only then is a comprehensive chemical characterization carried out (a *biology first* approach).

By starting with the probable physiological endpoints that a multicomponent secretion achieves, the specialized glandular secretions of carnivorous neogastropod molluscs that contain a diversity of compounds with unique mechanisms of action can be understood from a more holistic perspective. Using this approach, we can extend the biological mechanisms used by cone snails to the wider set of carnivorous neogastropod snails, comprising large lineages of nonvenomous predators. These generalizations apply to all chemical interactions in the marine environment, not just those of neogastropod molluscs. By first seeking to understand the true physiological endpoints and diverse mechanisms of novel natural products, we are enabled to more effectively pursue biomedical applications for these unique medicinal compounds.

ACKNOWLEDGMENTS

This work was supported by the National Institutes of Health Grant GM 048677 (to B.M.O.), an International Outgoing Fellowship Grant from the European Commission (CONBIOS 330486; to H.S-H.), and the Esther Fujimoto Memorial Fellowship (to S.R.).

12A APPENDIX: LIST OF CONOTOXINS GROUPED BY MOLECULAR TARGETS

Voltage-Gated Na Channels					
Peptide Family	**Peptides**	**Species**	**Subtype Targeted**	**Peptide Sequence**	**References**
μ-Conotoxins	μ-BuIIIA	*C. bullatus*	$Na_V1.2$, 1.4	VTDRCCKGKRECGRWCRDHSRCC*	[134]
	μ-BuIIIB	*C. bullatus*	$Na_V1.4$, 1.2	VGERCCKNGKRGCGRWCRDHSRCC*	[134]
	μ-CIIIA	*C. catus*	Na_V	GRCCEGPNGCSSRWCKDHARCC*	[152]
	μ-CnIIIA	*C. consors*	$Na_V1.2$, 1.4	GRCCDVPNACSGRWCRDHAQCC*	[134,152]
	μ-CnIIIB	*C. consors*	TTX-R	ZGCCGEPNLCFTRWCRNNARCCRQQ	[152]
	μ-GIIIA	*C. geographus*	$Na_V1.4$, 1.1	RDCCTOOKKCKDRQCKOQRCCA*	[15]
	μ-GIIIB	*C. geographus*	$Na_V1.4$	RDCCTOORKCKDRRCKOMKCCA*	[15]
	μ-GIIIC	*C. geographus*	Na_V	RDCCTOOKKCKDRRCKOLKCCA*	[15]
	μ-KIIIA	*C. kinoshitai*	$Na_V1.2$, 1.4, 1.6, 1.7	CCNCSSKWCRDHSRCC*	[165]
	μ-MIIIA	*C. magus*	$Na_V1.4$, 1.2	ZGCCNVPNGCSGRWCRDHAQCC*	[152]
	μ-PIIIA	*C. purpurascens*	$Na_V1.4$, 1.6, 1.1, 1.2	ZRLCCGFOKSCRSRQCKOHRCC*	[150]
	μ-SIIIA	*C. striatus*	$Na_V1.2$, 1.4	ZNCCNGGCSSKWCRDHARCC*	[166]
	μ-SIIIB	*C. striatus*	$Na_V1.2$, 1.7	ZNCCNGGCSSKWCKGHARCC*	[151]
	μ-SmIIIA	*C. stercusmuscarum*	$Na_V1.4$, 1.2, 1.1	ZRCCNGRRGCSSRWCRDHSRCC	[38,167]
	μ-SxIIIA	*C. striolatus*	$Na_V1.4$	RCCTGKKGSCSGRACKNLKCCA*	[168]
	μ-TIIIA	*C. tulipa*	$Na_V1.4$, 1.2	RHGCCKGOKGCSSRECROQHCC*	[38,169]
μO-Conotoxins	μO-MrVIA	*C. marmoreus*	$Na_V1.2$, 1.4, 1.7	ACRKKWEYCIVPIIGFIYCCPGLICGPFVCV	[39,40]
	μO-MrVIB	*C. marmoreus*	TTX-R	ACSKKWEYCIVPILGFVYCCPGLICGPFVCV	[39,41]
	μO§-GVIIJ	*C. geographus*	$Na_V1.2(\beta1, \beta3)$	GwCGDOGATCGKLRLYCCSGFCDC§YTKTCKDKSSA	[42]
	μO-MfVIA	*C. magnificus*	$Na_V1.4$, 1.8	RDCQEKWEYCIVPILGFVYCCPGLICGPFVCV	[170]

(Continued)

Voltage-Gated Na Channels					
Peptide Family	**Peptides**	**Species**	**Subtype Targeted**	**Peptide Sequence**	**References**
δ-Conotoxins	δ-PVIA	*C. purpurascens*	$Na_V1.2$, 1.4, 1.7	EACYAOGTFCGIKOGLCCSEFCLPGVCFG*	[44]
	δ-SVIE	*C. striatus*	$Na_V1.4$	DGCSSGGTFCGIHOGLCCSEFCFLWCITFID	[171]
	δ-TxVIA	*C. textile*	Na_V	WCKQSGEMCNLLDQNCCDGYCIVLVCT	[172,173]
	δ-SuVIA	*C. suturatus*	$Na_V1.3$, 1.4, 1.6, 1.7	CAGIGSFCGLPGLVDCCSDRCFIVCLP	[174]
	δ-TsVIA	*C. tessulatus*	Na_V	CAAFGSFCGLPGLVDCCSGRCFIVCLL	[175]
	δ-EVIA	*C. ermineus*	$Na_V1.2$, 1.3, 1.6	DDCIKOYGFCSLPILKNGLCCSGACVGVCADL*	[176]
ι-Conotoxins	ι-RXIA	*C. radiatus*	Na_V	GOSFCKADEKOCEYHADCCNCCLSGICAOSTNWILPGCSTSSFfKI	[45,46]
	ι-LtIIIA	*C. literratus*	TTX-S	DγCCγOQWCDGACDCCS	[177]
Miscellaneous	μ-PnIVA	*C. pennaceus*	Na_V	CCKYGWTCLLGCSPCGC	[178]
	μ-PnIVB	*C. pennaceus*	Na_V	CCKYGWTCWLGCSPCGC	[178]
	δ-Am2766	*C. amadis*	Na_V	CKQAGESCDIFSQNCCVGTCAFICIE*	[179]
	δ-GmVIA	*C. gloriamaris*	Na_V	VKPCRKEGQLCDPIFQNCCRGWNCVLFCV	[180]
	δ-ErVIA	*C. eburneus*	Na_V	CAGIGSFCGLPGLVDCCSGRCFIVCLP	[175]
	μ-LtVd	*C. litteratus*	TTX-S	DCCPAKLLCCNP	[181]
	Conotoxin GS	*C. geographus*	Na_V	ACSGRGSRCOOQCCMGLRCGRGNPQKCIGAHγDV	[137]

Voltage-Gated Ca Channels				
Peptide	**Species**	**Molecular Target**	**Sequence**	**References**
ω-CVIA	*C. catus*	$Ca_V2.2$, $Ca_V2.1$	CKSTGASCRRTSYDCCTGSCRSGRC*	[182]
ω-CVIB	*C. catus*	$Ca_V2.1$, $Ca_V2.2$	CKGKGASCRKTMYDCCRGSCRSGRC*	[182]
ω-CVIC	*C. catus*	$Ca_V2.1$, $Ca_V2.2$	CKGKGQSCSKLMYDCCTGSCSRRGKC*	[182]
ω-CVID	*C. catus*	$Ca_V2.2$	CKSKGAKCSKLMYDCCSGSCSGTVGRC*	[182]
ω-CVIE-2	*C. catus*	$Ca_V2.2$	CKGKGASCRRTSYDCCTGSCRSGRC*	[183]
ω-CVIF	*C. catus*	$Ca_V2.2$	CKGKGASCRRTSYDCCTGSCRLGRC*	[183]
ω-FVIA	*C. fulmen*	$Ca_V2.2$^^	CKGTGKSCSRIAYNCCTGSCRSGKC*	[184]
ω-GVIA	*C. geographus*	$Ca_V2.2$	CKSOGSSCSOTSYNCCRSCNOYTKRCY*	[17]
ω-GVIIA	*C. geographus*	$Ca_V2.2$	CKSOGTOCSRGMRDCCTSCLLYSNKCRRY	[133]
ω-GVIIB	*C. geographus*	$Ca_V2.2$	CKSOGTOCSRGMRDCCTSCLSYSNKCRRY	[133]
ω-MVIIA	*C. magus*	$Ca_V2.2$	CKGKGAKCSRLMYDCCTGSCRSGKC*	[133]
ω-MVIIB	*C. magus*	$Ca_V2.2$	CKGKGASCHRTSYDCCTGSCNRGKC*	[52]
ω-MVIIC	*C. magus*	$Ca_V2.2$	CKGKGAPCRKTMYDCCSGSCGRRGKC*	[154]
ω-SVIA	*C. striatus*	$Ca_V2.2$, $Ca_V2.1$	CRSSGSOCGVTSICCGRCYRGKCT*	[153]
ω-SVIB	*C. striatus*	$Ca_V2.2$, $Ca_V2.1$	CKLKGQSCRKTSYDCCSGSCGRSGKC*	[153]
Miscellaneous				
ω-Cl16a	*C. californicus*	Ca_V	NCPAGCRSQGCCM	[185,186]
ω-MVIID	*C. magus*	Ca_V	CQGRGASCRKTMYNCCSGSCNRGRC*	[154]
ω-PnVIA	*C. pennaceus*	Ca_V	GCLEVDYFCGIPFANNGLCCSGNCVFVCTPQ	[187]
ω-PnVIB	*C. pennaceus*	Ca_V	DDDCEPPGNFCGMIKIGPPCCSGWCFFACA	[187]

Voltage-Gated K Channels				
Peptide	**Species**	**Molecular Target**	**Sequence**	**References**
κ-BtX	*C. betulinus*	Calcium-activated BK channel	CRAγGTYCγNDSQCCLNγCCWGGCGHOCRHP*	[188]
κ-MIVA	*C. magus*	Putative K_V blocker	AOγLVVT$^+$AT$^+$TNCCGYNOMTICOOCMCTYSCOOKRKO*	[140]
κ-PIVE	*C. purpurascens*	Putative K_V blocker	DCCGVKLEMCHPCLCDNSCKNYGK*	[26]
κ-PlXIVA	*C. planorbis*	K_V1.6	FPRPRICNLACRAGIGHKYPFCHCR*	[189]
κ-PVIIA	*C. purpurascens*	Shaker K channel	CRIONQKCFQHLDDCCSRKCNRFNKCV*	[25,190]
κM-RIIIJ	*C. radiatus*	K_V1.2	LOOCCTOOKKHCOAOACKYKOCCKS	[48]
κM-RIIIK	*C. radiatus*	Kv blocker	LOSCCSLNLRLCOVOACKRNOCCT*	[49]
κA-SIVA	*C. striatus*	Putative K_V blocker	ZKSLVPS$^+$VITTCCGYDOGTMCOOCRCTNSC*	[21]
κ-ViTx	*C. virgo*	K_V1.1, K_V1.3	SRCFPPGIYCTPYLPCCWGICCGTCRNVCHLRI	[191]
κ-CPY-Fe1	*C. ferrugineus*	K_V1.6^^	GTYLYPFSYYRLWRYFTRFLHKQPYYYVHI	[192]
κ-CPY-Pl1	*C. planorbis*	K_V1.6	ARFLHPFQYYTLYRYLTRFLHRYPIYYIRY	[192]
κ-SrXIA	*C. spurius*	K_V1.2, K_V1.6	CRTEGMSCγγNQQCCWRSCCRGECEAPCRFGP	[193]
Conkunitzin-S1	*C. striatus*	Shaker K channel	Long peptide (60 amino acids)	[50]

Nicotinic Acetylcholine Receptors (nAChRs)

	Peptide	Species	Molecular Target	Sequence	References
α-Conotoxin	α-Ac1.1a	*C. achatinus*	$\alpha1\beta1\gamma\delta$, $\alpha1\beta1\delta\varepsilon$	NGRCCHPACGKHFNC*	[194]
	α-Ac1.1b	*C. achatinus*	$\alpha1\beta1\gamma\delta$, $\alpha1\beta1\delta\varepsilon$	NGRCCHPACGKHFSC*	[194]
	α-AnIA	*C. anemone*	$\alpha3\beta2$^^	CCSHPACAANNQDY(SO4)C*	[195]
	α-AnIB	*C. anemone*	$\alpha3\beta2$, $\alpha7$	GGCCSHPACAANNQDY(SO4)C*	[195]
	α-ArIA	*C. arenatus*	$\alpha7$, $\alpha3\beta2$	IRDECCSNPACRVNNPHVCRRR	[158]
	α-ArIB	*C. arenatus*	$\alpha7$, $\alpha3\beta2$, $\alpha6\alpha3\beta2\beta3$	DECCSNPACRVNNPHVCRRR	[158]
	α-AuIA	*C. aulicus*	$\alpha3\beta4$^^	GCCSYPPCFATNSDYC*	[196]
	α-AuIB	*C. aulicus*	$\alpha3\beta4$, $\alpha6\beta4$	GCCSYPPCFATNPDC*	[196]
	α-AuIC	*C. aulicus*	$\alpha3\beta4$	GCCSYPPCFATNSGYC*	[196]
	α-BnIA	*C. bandanus*	$\alpha7$^^	GCCSHPACSVNNPDIC*	[140,197]
	α-BuIA	*C. bullatus*	$\alpha6\alpha3\beta2 > \alpha6\alpha3\beta4 = \alpha3\beta2$	GCCSTPPCAVLYC*	[148]
	α-CnIA	*C. consors*	$\alpha1\beta1\gamma\delta$, $\alpha1\beta1\delta\varepsilon$	GRCCHPACGKYYSC*	[198]
	α-EI	*C. ermenius*	$\alpha1\beta1\gamma\delta$, $\alpha1\beta1\delta\varepsilon$	RDOCCYHPTCNMSNPQIC*	[199]
	α-EIIA	*C. ermenius*	$\alpha1\beta1\gamma\delta$, $\alpha1\beta1\delta\varepsilon$	ZTOGCCWNPACVKNRC*	[200]
	αA-EIVA	*C. ermenius*	$\alpha1\beta1\gamma\delta$, $\alpha1\beta1\delta\varepsilon$	GCCGPYONAACHOCGCKVGROOYCDROSGG*	[30]
	αA-EIVB	*C. ermenius*	$\alpha1\beta1\gamma\delta$, $\alpha1\beta1\delta\varepsilon$	GCCGKYONAACHOCGCTVGROOYCDROSGG*	[30]
	α-EpI	*C. episcopatus*	$\alpha3$*, $\alpha7$	GCCSDPRCNMNNPDY(SO4)C*	[163]
	α-GeXXA	*C. generalis*	nAChRs	PCQSVRPGRVWGKCCLTRLCSTMCCARADCTCVYH TWRGHGCSCVM	[201]
	α-GI	*C. geographus*	$\alpha1\beta1\gamma\delta$, $\alpha1\beta1\delta\varepsilon$	ECCNPACGRHYSC*	[135]
	α-GIA	*C. geographus*	$\alpha1\beta1\gamma\delta$, $\alpha1\beta1\delta\varepsilon$	ECCNPACGRHYSCGK	[135]
	α-GIC	*C. geographus*	$\alpha3\beta2 > \alpha3\beta4$	GCCSHPACAGNNQHIC*	[138]
	α-GID	*C. geographus*	$\alpha7 \sim \alpha3\beta2 > \alpha3\beta4$	IRDγCCSNPACRVNNOHVC	[139]
	α-GII	*C. geographus*	$\alpha1\beta1\gamma\delta$, $\alpha1\beta1\delta\varepsilon$	ECCHPACGKHFSC*	[135]
	α-ImI	*C. imperialis*	$\alpha3\beta2 > \alpha7$	GCCSDPRCAWRC*	[202]

(*Continued*)

Nicotinic Acetylcholine Receptors (nAChRs)				
Peptide	**Species**	**Molecular Target**	**Sequence**	**References**
α-ImII	*C. imperialis*	α7	ACCSDRRCRWRC*	[70]
α-Lp1.1	*C. leopardus*	α3β2, α6β2	GCCARAACAGIHQELC*	[203]
α-Lp1.4	*C. leopardus*	α1β1γδ, α6/α3β2	GCCSHPACSGNHQELCD*	[204]
α-LsIA	*C. limpusi*	α7, α3β2	SGCCSNPACRVNNPNIC*	[205]
α-LtIA	*C. litteratus*	α3β2^^	GCCARAACAGIHQELC*	[206]
α-LvIA	*C. lividus*	α3β2 > α6β2	RGCCSHPACNVDHPEIC*	[207]
α-MI	*C. magus*	α1β1γδ, α1β1δε	GRCCHPACGKNYSC*	[135,208]
α-MIC	*C. magus*	α1β1γδ, α1β1δε	CCHPACGKNYSC*	[146]
α-MII	*C. magus*	α6β2 > α6β4	GCCSNPVCHLEHSNLC*	[147]
α-Mr1.1	*C. marmoreus*	nAChRs	GCCSHPACSVNNPDIC*	[204]
α-MrIC	*C. marmoreus*	α7 agonist	PECCTHPACHVSNPELC*	[209]
αA-OIVA	*C. obscurus*	α1β1γδ > α1β1δε	CCGVONAACHOCVCKNTC*	[31,32]
αA-OIVB	*C. obscurus*	α1β1γδ > α1β1δε	CCGVONAACPOCVCNKTCG*	[31,33]
α-OmIA	*C. omaria*	α3β2 > α7 > α6β2	GCCSHPACNVNNPHICG*	[210]
α-PeIA	*C. pergrandis*	α9α10 = α3β2 = α6β2 > α3β4 > α7	GCCSHPACSVNHPELC*	[211]
αA-PeIVA	*C. pergrandis*	α1β1γδ, α1β1δε	CCGVONAACHOCVCTGKC	[31]
α-PIA	*C. purpurascens*	α6β2 > α6β4 ~ α3β2	RDPCCSNPVCTVHNPQIC*	[212]
α-PIB	*C. purpurascens*	α1β1γδ, α1β1δε	ZSOGCCWNPACVKNRC*	[143]
α-PIVA	*C. purpurascens*	α1β1γδ, α1β1δε	GCCGSYONAACHOCSCKDROSYCGQ*	[144]
α-PnIA	*C. pennaceus*	α7 > α3β4	GCCSLPPCAANNPDY(SO4)C*	[213]
α-Qc1.2	*C. quercinus*	α3β2, α3β4	QCCANPPCKHVNC*	[214]
α-RgIA	*C. regius*	α9α10	GCCSDPRCRYRCR	[65]
α-RgIB	*C. regius*	α3β4^^	TWEECCKNPGCRNNHVDRCRGQV	[215]
α-SI	*C. striatus*	α1β1γδ, α1β1δε	ICCNPACGPKYSC*	[153]

(Continued)

	Peptide	Species	Molecular Target	Sequence	References
	Nicotinic Acetylcholine Receptors (nAChRs)				
	α-SIA	*C. striatus*	$\alpha1\beta1\gamma\delta$, $\alpha1\beta1\delta\varepsilon$	YCCHPACGKNFDC*	[142]
	α-SII	*C. striatus*	$\alpha1\beta1\gamma\delta$, $\alpha1\beta1\delta\varepsilon$	GCCCNPACGPNYGCGTSCS	[153]
	α-SrIA	*C. spurius*	$\alpha1\beta1\gamma\delta$, $\alpha4\beta2$ (potentiation)	RTCCSROTCRMγYPγLCG*	[143]
	α-SrIB	*C. spurius*	$\alpha1\beta1\gamma\delta$, $\alpha4\beta2$ (potentiation)	RTCCSROTCRMEYPγLCG*	[143]
	α-TiIA	*C. tinianus*	nAChRs	GGCCSHPACQNNPDY(SO4)C*	[216]
	α-TxIA	*C. textile*	AChBP ~ $\alpha3\beta2 > \alpha7$	GCCSROOCIANNPDLC*	[217]
	α-TxIB	*C. textile*	$\alpha6/\alpha3\beta2\beta3$	GCCSDPPCRNKHPDLC*	[218]
	α-TxID	*C. textile*	$\alpha3\beta4 > \alpha6\beta4 >> \alpha3\beta2$	GCCSHPVCSAMSPIC*	[159]
	α-TxVC	*C. textile*	$\alpha4\beta2 > \alpha3\beta2$	KPCCSIHDNSCCGL	[219]
	α-VcIA	*C. victoriae*	$\alpha9\alpha10$	GCCSDORCNYDHPγIC*	[63,64]
	α-Vc1.2	*C. victoriae*	$\alpha3\beta2 > \alpha7$	GCCSNPACMVNNPQIC	[220]
	α-VxXXB	*C. vexilium*	$\alpha7 > \alpha3\beta2 > \alpha4\beta2$	DDγSγCIINTRDSPWGRCCRTRMCGSMCCPRNGCTCV YHWRRGHGCSCPG	[221]
αC-Conotoxins	αC-PrXA	*C. parius*	$\alpha1\beta1\gamma\delta$, $\alpha1\beta1\delta\varepsilon$	TYGIYDAKPOFSCAGLRGGCVLPONLROKFKE*	[222]
αD-Conotoxins	αD-VxXIIA	*C. vexillium*	$\alpha7$, $\alpha3\beta2$, $\alpha4\beta2$	DVQDCQVSTOGSKWGRCCLNRVCGPMCCPASHCYC VYHRGRGHGCSG*	[221]
αO-Conotoxins	αO-GeXIVA	*C. generalis*	$\alpha9\alpha10$	TCRSSGRYCRSPYDRRRRYCRRITDACV	[223]
αS-Conotoxins	αS-GVIIIB	*C. geographus*	$\alpha9\alpha10$	SGSTCTCFTSTNCQGSCECLSPPGCYCSNNGIRQRGC SCTCPGT*	[57]
	αS-RVIIIA	*C. radiatus*	nAChRs	KCNFDKCKGTGVYNCGγSCSCγGLHSCR CTYNIGSMKSGCACICTYY	[56]
ψ-Conotoxins	ψ-PIIIE	*C. purpurascens*	$\alpha1\beta1\gamma\delta$, $\alpha1\beta1\delta\varepsilon$	HOOCCLYGKCRRYOGCSSASCCQR*	[36]
	ψ-PIIIF	*C. purpurascens*	$\alpha1\beta1\gamma\delta$, $\alpha1\beta1\delta\varepsilon$	GOOCCLYGSCROFOGCYNALCCRK*	[35]
	ψ-PrIIIE	*C. parius*	$\alpha1\beta1\gamma\delta$, $\alpha1\beta1\delta\varepsilon$	AARCCTYHGSCLKEKCRRKYCC*	[34]

Glutamate Receptors				
Peptide	**Species**	**Molecular Target**	**Sequence**	**Reference**
Conantokin-Br	*C. brettinghami*	NR2B,NR2D,NR2A	GDγγYSKFIγRERγAGRLDLSKFP	[224]
Conantokin-G	*C. geographus*	NR2B	GEγγLQγNQγLIRγKSN*	[11]
Conantokin-P	*C. purpurascens*	NR2B	GEγγHSKYQγCLRγIRVNKVQQγC	[225]
Conantokin-Pr1	*C. parius*	NR2B	GEDγYAγGIRγYQLIHGKI	[58]
Conantokin-Pr2	*C. parius*	NR2B	DEPγYAγAIRγYQLKYGKI	[58]
Conantokin-Pr3	*C. parius*	NR2B	GEPγVAKWAγGLRγKAASN*	[58]
Conantokin-R	*C. radiatus*	NR2B,NR2A	GEγγVAKMAAγLARγNIAKGCKVNCYP	[81]
Conantokin-RlA	*C. rolani*	NR2B,NR2D	ADγγYLKFIγEQRKQGKLDPTKFP	[226]
Conantokin-RlB	*C. rolani*	NR2B	GEγγLAγKAOγFARγLAN*	[60]
Miscellaneous				
Conantokin-E	*C. ermenius*	N.D.	GEγγHSKYQγCLRγIRVNNVQQγC	[225]
Conantokin-L	*C. lynceus*	NMDAR	GEγγVAKMAAγLARγDAVN*	[227]
Conantokin-T	*C. tulipa*	NMDAR	GEγγYQKMLγNLRγAEVKKNA*	[228]
Conantokin-Bk-A	*C. bocki*	NMDAR	GDγγYSγFIγRERγLVSSKIPR	[229]
ConantokinBk-B	*C. bocki*	NMDAR	GEγγYSgAI*	[229]
Con-ikot-ikot	*C. striatus*	AMPA receptors	Long peptide (27 amino acids)	[27]

Miscellaneous Targets				
Peptide	**Species**	**Molecular Target**	**Sequence**	**Reference**
χ-MrIA	*C. marmoreus*	Norepinephrine transporter	NGVCCGYKLCHOC	[230]
σ-GVIIIA	*C. geographus*	5-HT_3 receptor	GCTRTCGGOKCTGTCTCTNSSKCGCRYNVHPSGwGCGCACS*	[22]
τ-LiC32	*C. lividus*	SST_3 receptor	LWQNTWCCRDHLRCC*	[231]
ρ-TIA	*C. tulipa*	α1 adrenoreceptor	FNWRCCLIPACRRNHKKFC*	[230]
γ-AsVIIA	*C. austini*	N.D.	TCKQKGEGCSLDVγCCSSSCKPGGPLFDFDC	[232]
γ-PnVIIA	*C. pennaceus*	Neuronal pacemaker cation currents	DCTSWFGRCTVNSγCCSNSCDQTYCγLYAFOS	[233]
ε-TxVA	*C. textile*	Putative GPCRs	γCCγDGwCCT+AAO	[90]
Conopressin G	*C. geographus*	Vasopressin receptor	CFIRNCPKG*	[23]
Conopressin T	*C. tulipa*	Vasopressin receptor	CYIQNCLRV*	[234]
Contryphan Am	*C. amadis*	Calcium channel	GCOwDPWC*	[235]
Contryphan M	*C. marmoreus*	Calcium channels	NγSγCPwHPWC*	[236]
Contryphan Vn	*C. ventricosus*	Calcium-dependent K channel	GDCPwKPWC*	[237]
Contulakin-G	*C. geographus*	Neurotensin receptor	ZSEEGGSNAT+KKPYIL	[12]
Con-Ins G1	*C. geographus*	Insulin receptor	GVVγHCCHRPCSNAEFKKYC*-TFDTOKHRCSGSγITNSYMDLCYR	[13]

Note: Special characters that represent post-translational modifications of peptide sequences are explained in Table 12.7.

^^ Indicates that the peptides were not tested on other subtypes of targets.

REFERENCES

1. Katz, L.; Baltz, R. H. Natural product discovery: Past, present, and future. *J. Ind. Microbiol. Biotechnol.* **2016**, 43, 155–176.
2. Olivera, B. M. *Conus* venom peptide: Reflections from the biology of clades and species. *Annu. Rev. Ecol. Syst.* **2002**, 33, 25–47.
3. Han, T. S.; Teichert, R. W.; Olivera, B. M.; Bulaj, G. *Conus* venoms: A rich source of peptide-based therapeutics. *Curr. Pharm. Des.* **2008**, 14, 2462–2479.
4. Terlau, H.; Olivera, B. M. *Conus* venoms: A rich source of novel ion channel-targeted peptides. *Physiol. Rev.* **2004**, 84, 41–68.
5. Olivera, B. M.; Just Lecture, E. E. *Conus* venom peptides, receptor and ion channel targets, and drug design: 50 million years of neuropharmacology. *Mol. Biol. Cell* **1997**, 8, 2101–2109.
6. Vallance, P. Industry-academic relationship in a new era of drug discovery. *J. Clin. Oncol.* **2016**, 34, 3570–3575.
7. Hwang, T. J.; Carpenter, D.; Lauffenburger, J. C.; Wang, B.; Franklin, J. M.; Kesselheim, A. S. Failure of investigational drugs in late-stage clinical development and publication of trial results. *JAMA Intern. Med.* **2016**, 176, 1826–1833.
8. Reynolds, I. J.; Wagner, J. A.; Snyder, S. H.; Thayer, S. A.; Olivera, B. M.; Miller, R. J. Brain voltage-sensitive calcium channel subtypes differentiated by omega-conotoxin fraction GVIA. *Proc. Natl. Acad. Sci. USA* **1986**, 83, 8804–8807.
9. Miljanich, G. P. Ziconotide: Neuronal calcium channel blocker for treating severe chronic pain. *Curr. Med. Chem.* **2004**, 11, 3029–3040.
10. Oger, J.; Frykman, H. An update on laboratory diagnosis in myasthenia gravis. *Clin. Chim. Acta* **2015**, 449, 43–48.
11. Olivera, B. M.; McIntosh, J. M.; Clark, C.; Middlemas, D.; Gray, W. R.; Cruz, L. J. A sleep-inducing peptide from *Conus geographus* venom. *Toxicon* **1985**, 23, 277–282.
12. Craig, A. G.; Norberg, T.; Griffin, D.; Hoeger, C.; Akhtar, M.; Schmidt, K.; Low, W. et al. Contulakin-G, an O-glycosylated invertebrate neurotensin. *J. Biol. Chem.* **1999**, 274, 13752–13759.
13. Safavi-Hemami, H.; Gajewiak, J.; Karanth, S.; Robinson, S. D.; Ueberheide, B.; Douglass, A. D.; Schlegel, A. et al. Specialized insulin is used for chemical warfare by fish-hunting cone snails. *Proc. Natl. Acad. Sci. USA* **2015**, 112, 1743–1748.
14. Gray, W. R.; Luque, A.; Olivera, B. M.; Barrett, J.; Cruz, L. J. Peptide toxins from *Conus geographus* venom. *J. Biol. Chem.* **1981**, 256, 4734–4740.
15. Cruz, L. J.; Gray, W. R.; Olivera, B. M.; Zeikus, R. D.; Kerr, L.; Yoshikami, D.; Moczydlowski, E. *Conus geographus* toxins that discriminate between neuronal and muscle sodium channels. *J. Biol. Chem.* **1985**, 260, 9280–9288.
16. French, R. J.; Yoshikami, D.; Sheets, M. F.; Olivera, B. M. The tetrodotoxin receptor of voltage-gated sodium channels—Perspectives from interactions with micro-conotoxins. *Mar. Drugs* **2010**, 8, 2153–2161.
17. Olivera, B. M.; McIntosh, J. M.; Cruz, L. J.; Luque, F. A.; Gray, W. R. Purification and sequence of a presynaptic peptide toxin from *Conus geographus* venom. *Biochemist* **1984**, 23, 5087–5090.
18. Cruz, L. J.; Johnson, D. S.; Olivera, B. M. Characterization of the omega-conotoxin target. Evidence for tissue-specific heterogeneity in calcium channel types. *Biochemist* **1987**, 26, 820–824.
19. Witcher, D. R.; De Waard, M.; Campbell, K. P. Characterization of the purified N-type Ca^{2+} channel and the cation sensitivity of omega-conotoxin GVIA binding. *Neuropharmacology* **1993**, 32, 1127–1139.

20. McIntosh, J. M.; Olivera, B. M.; Cruz, L. J.; Gray, W. R. γ-Carboxyglutamate in a neuroactive toxin. *J. Biol. Chem.* **1984**, 259, 14343–14346.
21. Craig, A. G.; Zafaralla, G.; Cruz, L. J.; Santos, A. D.; Hillyard, D. R.; Dykert, J.; Rivier, J. E. et al. An O-glycosylated neuroexcitatory *Conus* peptide. *Biochemist* **1998**, 37, 16019–16025.
22. England, L. J.; Imperial, J.; Jacobsen, R.; Craig, A. G.; Gulyas, J.; Akhtar, M.; Rivier, J.; Julius, D.; Olivera, B. M. Inactivation of a serotonin-gated ion channel by a polypeptide toxin from marine snails. *Science* **1998**, 281, 575–578.
23. Cruz, L. J.; de Santos, V.; Zafaralla, G. C.; Ramilo, C. A.; Zeikus, R.; Gray, W. R.; Olivera, B. M. Invertebrate vasopressin/oxytocin homologs. Characterization of peptides from *Conus geographus* and *Conus straitus* venoms. *J. Biol. Chem.* **1987**, 262, 15821–15824.
24. Olivera, B. M.; Seger, J.; Horvath, M. P.; Fedosov, A. E. Prey-capture strategies of fish-hunting cone snails: Behavior, neurobiology and evolution. *Brain Behav. Evol.* **2015**, 86, 58–74.
25. Terlau, H.; Shon, K. J.; Grilley, M.; Stocker, M.; Stuhmer, W.; Olivera, B. M. Strategy for rapid immobilization of prey by a fish-hunting marine snail. *Nature* **1996**, 381, 148–151.
26. Teichert, R. W.; Jacobsen, R.; Terlau, H.; Yoshikami, D.; Olivera, B. M. Discovery and characterization of the short kappaA-conotoxins: A novel subfamily of excitatory conotoxins. *Toxicon* **2007**, 49, 318–328.
27. Walker, C. S.; Jensen, S.; Ellison, M.; Matta, J. A.; Lee, W. Y.; Imperial, J. S.; Duclos, N. et al. A novel *Conus* snail polypeptide causes excitotoxicity by blocking desensitization of AMPA receptors. *Curr. Biol.* **2009**, 19, 900–908.
28. Jacob, R. B.; McDougal, O. M. The M-superfamily of conotoxins: A review. *Cell. Mol. Life Sci.* **2010**, 67, 17–27.
29. Corpuz, G. P.; Jacobsen, R. B.; Jimenez, E. C.; Watkins, M.; Walker, C.; Colledge, C.; Garrett, J. E. et al. Definition of the M-conotoxin superfamily: Characterization of novel peptides from molluscivorous *Conus* venoms. *Biochemist* **2005**, 44, 8176–8186.
30. Jacobsen, R.; Yoshikami, D.; Ellison, M.; Martinez, J.; Gray, W. R.; Cartier, G. E.; Shon, K. J. et al. Differential targeting of nicotinic acetylcholine receptors by novel alphaA-conotoxins. *J. Biol. Chem.* **1997**, 272, 22531–22537.
31. Teichert, R. W.; Lopez-Vera, E.; Gulyas, J.; Watkins, M.; Rivier, J.; Olivera, B. M. Definition and characterization of the short alphaA-conotoxins: A single residue determines dissociation kinetics from the fetal muscle nicotinic acetylcholine receptor. *Biochemist* **2006**, 45, 1304–1312.
32. Teichert, R. W.; Rivier, J.; Dykert, J.; Cervini, L.; Gulyas, J.; Bulaj, G.; Ellison, M.; Olivera, B. M. AlphaA-Conotoxin OIVA defines a new alphaA-conotoxin subfamily of nicotinic acetylcholine receptor inhibitors. *Toxicon* **2004**, 44, 207–214.
33. Teichert, R. W.; Rivier, J.; Torres, J.; Dykert, J.; Miller, C.; Olivera, B. M. A uniquely selective inhibitor of the mammalian fetal neuromuscular nicotinic acetylcholine receptor. *J. Neurosci.* **2005**, 25, 732–7366.
34. Lluisma, A. O.; Lopez-Vera, E.; Bulaj, G.; Watkins, M.; Olivera, B. M. Characterization of a novel psi-conotoxin from *Conus parius* Reeve. *Toxicon* **2008**, 51, 174–180.
35. Van Wagoner, R. M.; Jacobsen, R. B.; Olivera, B. M.; Ireland, C. M. Characterization and three-dimensional structure determination of psi-conotoxin Piiif, a novel noncompetitive antagonist of nicotinic acetylcholine receptors. *Biochemist* **2003**, 42, 6353–6362.
36. Shon, K. J.; Grilley, M.; Jacobsen, R.; Cartier, G. E.; Hopkins, C.; Gray, W. R.; Watkins, M. et al. A noncompetitive peptide inhibitor of the nicotinic acetylcholine receptor from *Conus purpurascens* venom. *Biochemist* **1997**, 36, 9581–9587.

37. Green, B. R.; Olivera, B. M. Venom peptides from cone snails: Pharmacological probes for voltage-gated sodium channels. *Curr. Top. Membr.* **2016**, 78, 65–86.
38. Zhang, M.-M.; Wilson, M. J.; Azam, L.; Gajewiak, J.; Rivier, J. E.; Bulaj, G.; Olivera, B. M.; Yoshikami, D. Co-expression of Na(V)beta subunits alters the kinetics of inhibition of voltage-gated sodium channels by pore-blocking mu-conotoxins. *Br. J. Pharmacol.* **2013**, 168, 1597–1610.
39. McIntosh, J. M.; Hasson, A.; Spira, M. E.; Gray, W. R.; Li, W.; Marsh, M.; Hillyard, D. R.; Olivera, B. M. A new family of conotoxins that blocks voltage-gated sodium channels. *J. Biol. Chem.* **1995**, 270, 16796–16802.
40. Safo, P.; Rosenbaum, T.; Shcherbatko, A.; Choi, D. Y.; Han, E.; Toledo-Aral, J. J.; Olivera, B. M.; Brehm, P.; Mandel, G. Distinction among neuronal subtypes of voltage-activated sodium channels by mu-conotoxin PIIIA. *J. Neurosci.* **2000**, 20, 76–80.
41. de Araujo, A. D.; Callaghan, B.; Nevin, S. T.; Daly, N. L.; Craik, D. J.; Moretta, M.; Hopping, G.; Christie, M. J.; Adams, D. J.; Alewood, P. F. Total synthesis of the analgesic conotoxin MrVIB through selenocysteine-assisted folding. *Angew. Chem. Int. Ed. Engl.* **2011**, 50, 6527–6529.
42. Gajewiak, J.; Azam, L.; Imperial, J.; Walewska, A.; Green, B. R.; Bandyopadhyay, P. K.; Raghuraman, S. et al. A disulfide tether stabilizes the block of sodium channels by the conotoxin muO section sign-GVIIJ. *Proc. Natl. Acad. Sci. USA* **2014**, 111, 2758–2763.
43. Bulaj, G.; DeLa Cruz, R.; Azimi-Zonooz, A.; West, P.; Watkins, M.; Yoshikami, D.; Olivera, B. M. d-Conotoxin structure/function through a cladistic analysis. *Biochemist* **2001**, 40, 13201–13208.
44. Shon, K. J.; Grilley, M. M.; Marsh, M.; Yoshikami, D.; Hall, A. R.; Kurz, B.; Gray, W. R.; Imperial, J. S.; Hillyard, D. R.; Olivera, B. M. Purification, characterization, synthesis, and cloning of the lockjaw peptide from *Conus purpurascens* venom. *Biochemist* **1995**, 34, 4913–4918.
45. Buczek, O.; Wei, D.; Babon, J. J.; Yang, X.; Fiedler, B.; Chen, P.; Yoshikami, D.; Olivera, B. M.; Bulaj, G.; Norton, R. S. Structure and sodium channel activity of an excitatory I1-superfamily conotoxin. *Biochemist* **2007**, 46, 9929–9940.
46. Jimenez, E. C.; Shetty, R. P.; Lirazan, M.; Rivier, J.; Walker, C.; Abogadie, F. C.; Yoshikami, D.; Cruz, L. J.; Olivera, B. M. Novel excitatory *Conus* peptides define a new conotoxin superfamily. *J. Neurochem.* **2003**, 85, 610–621.
47. Gutman, G. A.; Chandy, K. G.; Grissmer, S.; Lazdunski, M.; McKinnon, D.; Pardo, L. A.; Robertson, G. A. et al. International Union of Pharmacology. LIII. Nomenclature and molecular relationships of voltage-gated potassium channels. *Pharmacol. Rev.* **2005**, 57, 473–508.
48. Chen, P.; Dendorfer, A.; Finol-Urdaneta, R. K.; Terlau, H.; Olivera, B. M. Biochemical characterization of kappaM-RIIIJ, a Kv1.2 channel blocker: Evaluation of cardioprotective effects of kappaM-conotoxins. *J. Biol. Chem.* **2010**, 285, 14882–14889.
49. Ferber, M.; Sporning, A.; Jeserich, G.; DeLaCruz, R.; Watkins, M.; Olivera, B. M.; Terlau, H. A novel *Conus* peptide ligand for K^+ channels. *J. Biol. Chem.* **2003**, 278, 2177–2183.
50. Bayrhuber, M.; Vijayan, V.; Ferber, M.; Graf, R.; Korukottu, J.; Imperial, J.; Garrett, J. E. et al. Conkunitzin-S1 is the first member of a new Kunitz-type neurotoxin family: Structural and functional characterization. *J. Biol. Chem.* **2005**, 280, 23766–23770.
51. Finol-Urdaneta, R. K.; Remedi, M. S.; Raasch, W.; Becker, S.; Clark, R. B.; Struver, N.; Pavlov, E.; Nichols, C. G.; French, R. J.; Terlau, H. Block of Kv1.7 potassium currents increases glucose-stimulated insulin secretion. *EMBO Mol. Med.*. **2012**, 4, 424–434.
52. Olivera, B. M.; Cruz, L. J.; de Santos, V.; LeCheminant, G.; Griffin, D.; Zeikus, R.; McIntosh, J. M. et al. Neuronal Ca channel antagonists. Discrimination between Ca channel subtypes using w-conotoxin from *Conus magus* venom. *Biochemist* **1987**, 26, 2086–2090.

53. Espino, S. S.; Dilanyan, T.; Imperial, J. S.; Aguilar, M. B.; Teichert, R. W.; Bandyopadhyay, P.; Olivera, B. M. Glycine-rich conotoxins from the *Virgiconus* clade. *Toxicon* **2016**, 113, 11–17.
54. Hecker, A.; Kullmar, M.; Wilker, S.; Richter, K.; Zakrzewicz, A.; Atanasova, S.; Mathes, V. et al. Phosphocholine-modified macromolecules and canonical nicotinic agonists inhibit ATP-induced IL-1beta release. *J. Immunol.* **2015**, 195, 2325–2334.
55. Holden-Dye, L.; Joyner, M.; O'Connor, V.; Walker, R. J. Nicotinic acetylcholine receptors: A comparison of the nAChRs of *Caenorhabditis elegans* and parasitic nematodes. *Parasitol. Int.* **2013**, 62, 606–615.
56. Teichert, R. W.; Jimenez, E. C.; Olivera, B. M. Alpha S-conotoxin RVIIIA: A structurally unique conotoxin that broadly targets nicotinic acetylcholine receptors. *Biochemist* **2005**, 44, 7897–7902.
57. Christensen, S. B.; Bandyopadhyay, P. K.; Olivera, B. M.; McIntosh, J. M. alphaS-conotoxin GVIIIB potently and selectively blocks alpha9alpha10 nicotinic acetylcholine receptors. *Biochem. Pharmacol.* **2015**, 96, 349–356.
58. Teichert, R. W.; Jimenez, E. C.; Twede, V.; Watkins, M.; Hollmann, M.; Bulaj, G.; Olivera, B. M. Novel conantokins from *Conus parius* venom are specific antagonists of *N*-methyl-D-aspartate receptors. *J. Biol. Chem.* **2007**, 282, 36905–36913.
59. Donevan, S. D.; McCabe, R. T. Conantokin-G is an NR2B-selective competitive antagonist of *N*-methyl-D-aspartate receptors. *Mol. Pharmacol.* **2000**, 58, 614–623.
60. Gowd, K. H.; Han, T. S.; Twede, V.; Gajewiak, J.; Smith, M. D.; Watkins, M.; Platt, R. J. et al. Conantokins derived from the *Asprella* clade impart conRl-B, an *N*-methyl D-aspartate receptor antagonist with a unique selectivity profile for NR2B subunits. *Biochemist* **2012**, 51, 4685–4692.
61. Lang, B.; Waterman, S.; Pinto, A.; Jones, D.; Moss, F.; Boot, J.; Brust, P. et al. The role of autoantibodies in Lambert-Eaton myasthenic syndrome. *Ann. N. Y. Acad. Sci.* **1998**, 841, 596–605.
62. Motomura, M.; Johnston, L.; Lang, B.; Vincent, A.; Newsom-Davis, J. An improved diagnostic assay for Lambert-Eaton myasthenic syndrome. *J. Neurol. Neurosurg. Psychiatry* **1995**, 58, 85–87.
63. Nevin, S. T.; Clark, R. J.; Klimis, H.; Christie, M. J.; Craik, D. J.; Adams, D. J. Are alpha9alpha10 nicotinic acetylcholine receptors a pain target for alpha-conotoxins? *Mol. Pharmacol.* **2007**, 72, 1406–1410.
64. Satkunanathan, N.; Livett, B.; Gayler, K.; Sandall, D.; Down, J.; Khalil, Z. Alpha-conotoxin Vc1.1 alleviates neuropathic pain and accelerates functional recovery of injured neurones. *Brain Res.* **2005**, 1059, 149–158.
65. Ellison, M.; Haberlandt, C.; Gomez-Casati, M. E.; Watkins, M.; Elgoyhen, A. B.; McIntosh, J. M.; Olivera, B. M. Alpha-RgIA: A novel conotoxin that specifically and potently blocks the alpha9alpha10 nAChR. *Biochemist* **2006**, 45, 1511–1517.
66. McIntosh, J. M.; Absalom, N.; Chebib, M.; Elgoyhen, A. B.; Vincler, M. Alpha9 nicotinic acetylcholine receptors and the treatment of pain. *Biochem. Pharmacol.* **2009**, 78, 693–702.
67. Azam, L.; McIntosh, J. M. Molecular basis for the differential sensitivity of rat and human alpha9alpha10 nAChRs to alpha-conotoxin RgIA. *J. Neurochem.* **2012**, 122, 1137–1144.
68. Vincler, M.; Wittenauer, S.; Parker, R.; Ellison, M.; Olivera, B. M.; McIntosh, J. M. Molecular mechanism for analgesia involving specific antagonism of alpha9alpha10 nicotinic acetylcholine receptors. *Proc. Natl. Acad. Sci. USA* **2006**, 103, 17880–17884.
69. Di Cesare Mannelli, L.; Cinci, L.; Micheli, L.; Zanardelli, M.; Pacini, A.; McIntosh, J. M.; Ghelardini, C. alpha-conotoxin RgIA protects against the development of nerve injury-induced chronic pain and prevents both neuronal and glial derangement. *Pain* **2014**, 155, 1986–1995.

70. Ellison, M.; Gao, F.; Wang, H.-L.; Sine, S. M.; McIntosh, J. M.; Olivera, B. M. Alpha-conotoxins ImI and ImII target distinct regions of the human alpha7 nicotinic acetylcholine receptor and distinguish human nicotinic receptor subtypes. *Biochemist* **2004**, 43, 16019–16026.
71. Richter, K.; Mathes, V.; Fronius, M.; Althaus, M.; Hecker, A.; Krasteva-Christ, G.; Padberg, W. et al. Phosphocholine—An agonist of metabotropic but not of ionotropic functions of alpha9-containing nicotinic acetylcholine receptors. *Sci. Rep.* **2016**, 6, 28660.
72. Callaghan, B.; Haythornthwaite, A.; Berecki, G.; Clark, R. J.; Craik, D. J.; Adams, D. J. Analgesic alpha-conotoxins Vc1.1 and Rg1A inhibit N-type calcium channels in rat sensory neurons via GABAB receptor activation. *J. Neurosci.*. **2008**, 28, 10943–10951.
73. Sang, C. N.; Barnabe, K. J.; Kern, S. E. Phase IA clinical trial evaluating the tolerability, pharmacokinetics, and analgesic efficacy of an intrathecally administered neurotensin A analogue in central neuropathic pain following spinal cord injury. *Clin. Pharmacol. Drug. Dev.* **2016**, 5, 250–258.
74. Allen, J. W.; Hofer, K.; McCumber, D.; Wagstaff, J. D.; Layer, R. T.; McCabe, R. T.; Yaksh, T. L. An assessment of the antinociceptive efficacy of intrathecal and epidural contulakin-G in rats and dogs. *Anesth. Analg.* **2007**, 104, 1505–1513, table of contents.
75. Lee, H. K.; Zhang, L.; Smith, M. D.; Walewska, A.; Vellore, N. A.; Baron, R.; McIntosh, J. M.; White, H. S.; Olivera, B. M.; Bulaj, G. A marine analgesic peptide, Contulakin-G, and neurotensin are distinct agonists for neurotensin receptors: Uncovering structural determinants of desensitization properties. *Front. Pharmacol.* **2015**, 6, 11.
76. Malmberg, A. B.; Gilbert, H.; McCabe, R. T.; Basbaum, A. I. Powerful antinociceptive effects of the cone snail venom-derived subtype-selective NMDA receptor antagonists conantokins G and T. *Pain* **2003**, 101, 109–116.
77. Williams, A. J.; Dave, J. R.; Phillips, J. B.; Lin, Y.; McCabe, R. T.; Tortella, F. C. Neuroprotective efficacy and therapeutic window of the high-affinity *N*-methyl-D-aspartate antagonist conantokin-G: *In vitro* (primary cerebellar neurons) and *in vivo* (rat model of transient focal brain ischemia) studies. *J. Pharmacol. Exp. Ther.* **2000**, 294, 378–386.
78. Layer, R. T.; Wagstaff, J. D.; White, H. S. Conantokins: Peptide antagonists of NMDA receptors. *Curr. Med. Chem.* **2004**, 11, 3073–3084.
79. Skolnick, P.; Zhou, L.-M.; Chandler, P.; Nashed, N. T.; Pennington, M.; Maccecchini, M.-L. Conantokin-G and its Analogs: Novel probes of the NMDA receptor associated polyamine site. In *Pharmacology and Toxicology: Basic and Clinical Aspects: Direct and Allosteric Control of Glutamate Receptors*, Palfreyman, M. G.; Reynolds, I. J.; Skolnick, P., Eds. CRC Press: Boca Raton, FL, **1994**, 155–165.
80. Williams, A. J.; Ling, G.; McCabe, R. T.; Tortella, F. C. Intrathecal CGX-1007 is neuroprotective in a rat model of focal cerebral ischemia. *NeuroReport* **2002**, 13, 821–824.
81. White, H. S.; McCabe, R. T.; Armstrong, H.; Donevan, S.; Cruz, L. J.; Abogadie, F. C.; Torres, J. et al. *In vitro* and *in vivo* characterization of conantokin-R, a selective NMDA antagonist isolated from the venom of the fish-hunting snail *Conus radiatus*. *J. Pharmacol. Exp. Therap.* **2000**, 292, 425–432.
82. Dodson, G.; Steiner, D. The role of assembly in insulin's biosynthesis. *Curr. Opin. Struct. Biol.* **1998**, 8, 189–194.
83. Menting, J. G.; Whittaker, J.; Margetts, M. B.; Whittaker, L. J.; Kong, G. K.; Smith, B. J.; Watson, C. J. et al. How insulin engages its primary binding site on the insulin receptor. *Nature* **2013**, 493, 241–245.
84. Menting, J. G.; Yang, Y.; Chan, S. J.; Phillips, N. B.; Smith, B. J.; Whittaker, J.; Wickramasinghe, N. P. et al. Protective hinge in insulin opens to enable its receptor engagement. *Proc. Natl. Acad. Sci. USA* **2014**, 111, E3395–E3404.
85. Adams, M. J.; Blundell, T. L.; Dodson, E. J.; Dodson, G. G.; Vijayan, M.; Baker, E. N.; Harding, M. M.; Hodkin, D. C.; Rimmer, B.; Sheat, S. Structure of rhombohedral 2 zinc insulin crystals. *Nature* **1969**, 224, 491–495.

86. Owens, D. R. New horizons-alternative routes for insulin therapy. *Nat. Rev. Drug Discov.* **2002**, 1, 529–540.
87. Noble, S. L.; Johnston, E.; Walton, B. Insulin lispro: A fast-acting insulin analog. *Am. Fam. Physician* **1998**, 57, 279–286, 289–292.
88. Bao, S. J.; Xie, D. L.; Zhang, J. P.; Chang, W. R.; Liang, D. C. Crystal structure of desheptapeptide(B24-B30)insulin at 1.6 A resolution: Implications for receptor binding. *Proc. Natl. Acad. Sci. USA* **1997**, 94, 2975–2980.
89. Menting, J. G.; Gajewiak, J.; MacRaild, C. A.; Chou, D. H.; Disotuar, M. M.; Smith, N. A.; Miller, C. et al. A minimized human insulin-receptor-binding motif revealed in a *Conus geographus* venom insulin. *Nat. Struct. Mol. Biol.* **2016**, 23, 916–920.
90. Walker, C. S.; Steel, D.; Jacobsen, R. B.; Lirazan, M. B.; Cruz, L. J.; Hooper, D.; Shetty, R. et al. The T-superfamily of conotoxins. *J. Biol. Chem.* **1999**, 274, 30664–30671.
91. McIntosh, J. M.; Corpuz, G. O.; Layer, R. T.; Garrett, J. E.; Wagstaff, J. D.; Bulaj, G.; Vyazovkina, A.; Yoshikami, D.; Cruz, L. J.; Olivera, B. M. Isolation and characterization of a novel *Conus* peptide with apparent antinociceptive activity. *J. Biol. Chem.* **2000**, 275, 32391–32397.
92. Sharpe, I. A.; Palant, E.; Schroeder, C. I.; Kaye, D. M.; Adams, D. J.; Alewood, P. F.; Lewis, R. J. Inhibition of the norepinephrine transporter by the venom peptide chi-MrIA. Site of action, Na^+ dependence, and structure-activity relationship. *J. Biol. Chem.* **2011**, 278, 40317–40323.
93. Paczkowski, F. A.; Sharpe, I. A.; Dutertre, S.; Lewis, R. J. chi-Conotoxin and tricyclic antidepressant interactions at the norepinephrine transporter define a new transporter model. *J. Biol. Chem.* **2007**, 282, 17837–17844.
94. Carstens, B. B.; Clark, R. J.; Daly, N. L.; Harvey, P. J.; Kaas, Q.; Craik, D. J. Engineering of conotoxins for the treatment of pain. *Curr. Pharm. Des.* **2011**, 17, 4242–4253.
95. Cruz, L. J.; Gray, W. R.; Yoshikami, D.; Olivera, B. M. *Conus* venoms—A rich source of neuroactive peptides. *J. Toxicol. Toxin Rev.* **1985**, 4, 107–132.
96. Teichert, R. W.; Olivera, B. M.; McIntosh, J. M.; Bulaj, G.; Horvath, M. P. The molecular diversity of conoidean venom peptides and their targets: From basic research to therapeutic applications. In *Venoms to Drugs: Venom as a Source for the Development of Human Therapeutics*, King, G. F., Ed. Royal Society of Chemistry: Cambridge, U.K., **2015**; pp. 163–203.
97. Sunagar, K.; Moran, Y. The rise and fall of an evolutionary innovation: Contrasting strategies of venom evolution in ancient and young animals. *PLoS Genet.* **2015**, 11, e1005596.
98. Buczek, O.; Bulaj, G.; Olivera, B. M. Conotoxins and the posttranslational modification of secreted gene products. *Cell. Mol. Life Sci.* **2005**, 62, 3067–3079.
99. Puillandre, N.; Bouchet, P.; Duda, T. F., Jr.; Kauferstein, S.; Kohn, A. J.; Olivera, B. M.; Watkins, M.; Meyer, C. Molecular phylogeny and evolution of the cone snails (Gastropoda, Conoidea). *Mol. Phylogenet. Evol.* **2014**, 78, 290–303.
100. Puillandre, N.; Duda, T. F.; Meyer, C.; Olivera, B. M.; Bouchet, P. One, four or 100 genera? A new classification of the cone snails. *J. Molluscan Stud.* **2015**, 81, 1–23.
101. Bandyopadhyay, P. K.; Colledge, C. J.; Walker, C. S.; Zhou, L.-M.; Hillyard, D. R.; Olivera, B. M. Conantokin-G precursor and its role in g-carboxylation by a vitamin K-dependent carboxylase from a *Conus* snail. *J. Biol. Chem.* **1998**, 273, 5447–5450.
102. Buczek, O.; Olivera, B. M.; Bulaj, G. Propeptide does not act as an intramolecular chaperone but facilitates protein disulfide isomerase-assisted folding of a conotoxin precursor. *Biochemist* **2004**, 43, 1093–1101.
103. Conticello, S. G.; Kowalsman, N. D.; Jacobsen, C.; Yudkovsky, G.; Sato, K.; Elazar, Z.; Munck Petersen, C.; Aronheim, A.; Fainzilber, M. The prodomain of a secreted hydrophobic mini-protein facilitates its export from the endoplasmic reticulum by hitchhiking on sorting receptors. *J. Biol. Chem.* **2003**, 278, 26311–26314.

104. Woodward, S. R.; Cruz, L. J.; Olivera, B. M.; Hillyard, D. R. Constant and hypervariable regions in conotoxin propeptides. *EMBO J.* **1990**, 9, 1015–1020.
105. Chang, D.; Duda, T. F. J. Extensive and continuous duplication facilitates rapid evolution and diversification of gene families. *Mol. Biol. Evol.* **2012**, 29, 2019–2029.
106. Duda, T. F.; Palumbi, S. R. Evolutionary diversification of multigene families: allelic selection of toxins in predatory cone snails. *Mol. Biol. Evol.* **2000**, 17, 1286–1293.
107. Olivera, B. M.; Walker, C.; Cartier, G. E.; Hooper, D.; Santos, A. D.; Schoenfeld, R.; Shetty, R.; Watkins, M.; Bandyopadhyay, P.; Hillyard, D. R. Speciation of cone snails and interspecific hyperdivergence of their venom peptides potential: Evolutionary significance of introns. In *Molecular Strategies in Biological Evolution*, Caporale, L. H., Ed. New York Academy of Science: New York, **1999**; Vol. 870, pp. 223–237.
108. Robinson, S. D.; Norton, R. S. Conotoxin gene superfamilies. *Mar. Drugs* **2014**, 12, 6058–6101.
109. Phuong, M. A.; Mahardika, G. N.; Alfaro, M. E. Dietary breadth is positively correlated with venom complexity in cone snails. *BMC Genomics* **2016**, 17, 401.
110. Barghi, N.; Concepcion, G. P.; Olivera, B. M.; Lluisma, A. O. Comparison of the venom peptides and their expression in closely related *Conus* species: Insights into adaptive post-speciation evolution of *Conus* exogenomes. *Genome Biol. Evol.* **2015**, 7, 1797–1814.
111. Hu, H.; Bandyopadhyay, P. K.; Olivera, B. M.; Yandell, M. Elucidation of the molecular envenomation strategy of the cone snail *Conus geographus* through transcriptome sequencing of its venom duct. *BMC Genomics* **2012**, 13, 1–12.
112. Li, Q.; Robinson, S. D.; Barghi, N.; Lu, A.; Fedosov, A. E.; Lluisma, A.; Bandyopadhyay, P.; Olivera, B.; Yandell, M.; Safavi-Hemami, H. Divergence of venom exogene repertoire in *Turriconus*. Submitted for publication, **2016**.
113. Robinson, S. D.; Safavi-Hemami, H.; McIntosh, L. D.; Purcell, A. W.; Norton, R. S.; Papenfuss, A. T. Diversity of conotoxin gene superfamilies in the venomous snail, *Conus victoriae*. *PLoS One* **2014**, 9, e87648.
114. Chang, D.; Olenzek, A. M.; Duda, T. F., Jr. Effects of geographical heterogeneity in species interactions on the evolution of venom genes. *Proc. Biol. Sci.* **2015**, 282, 20141984.
115. Duda, T. F., Jr.; Chang, D.; Lewis, B. D.; Lee, T. Geographic variation in venom allelic composition and diets of the widespread predatory marine gastropod *Conus ebraeus*. *PLoS One* **2009**, 4, e6245.
116. Hu, H.; Bandyopadhyay, P. K.; Olivera, B. M.; Yandell, M. Characterization of the *Conus bullatus* genome and its venom-duct transcriptome. *BMC Genomics* **2011**, 12, 60.
117. Olivera, B. M.; Safavi-Hemami, H.; Horvarth, M. P.; Teichert, R. W. Conopeptides, marine natural products from venoms: Biomedical applications and future research applications. In *Marine Biomedicine: From Beach to Bedside*, Baker, B. J., Ed. CRC Press: Boca Raton, FL, **2015**.
118. Stanley, T. B.; Stafford, D. W.; Olivera, B. M.; Bandyopadhyay, P. K. Identification of a vitamin K-dependent carboxylase in the venom duct of a *Conus* snail. *FEBS Lett.* **1997**, 407, 85–88.
119. Craig, A. G.; Jimenez, E. C.; Dykert, J.; Nielsen, D. B.; Gulyas, J.; Abogadie, F. C.; Porter, J. et al. A novel post-translational modification involving bromination of tryptophan. *J. Biol. Chem.* **1997**, 272, 4689–4698.
120. Fujii, R.; Yoshida, H.; Fukusumi, S.; Habata, Y.; Hosoya, M.; Kawamata, Y.; Yano, T. et al. Identification of a neuropeptide modified with bromine as an endogenous ligand for GPR7. *J. Biol. Chem.* **2002**, 277, 34010–34016.
121. Jimenez, E. C.; Craig, A. G.; Watkins, M.; Hillyard, D. R.; Gray, W. R.; Gulyas, J.; Rivier, J.; Cruz, L. J.; Olivera, B. M. Bromocontryphan: Post-translational bromination of tryptophan. *Biochemist* **1997**, 36, 989–994.

122. Rigby, A. C.; Lucas-Meunier, E.; Kalume, D. E.; Czerwiec, E.; Hambe, B.; Dahlqvist, I.; Fossier, P. et al. A conotoxin from *Conus textile* with unusual posttranslational modifications reduces presynaptic Ca^{2+} influx. *Proc. Natl. Acad. Sci. USA* **1999**, 96, 5758–5763.
123. Pisarewicz, K.; Mora, D.; Pflueger, F. C.; Fields, G. B.; Mari, F. Polypeptide chains containing D-gamma-hydroxyvaline. *J. Am. Chem. Soc.* **2005**, 127, 6207–6215.
124. Fry, B. G.; Roelants, K.; Champagne, D. E.; Scheib, H.; Tyndall, J. D.; King, G. F.; Nevalainen, T. J. et al. The toxicogenomic multiverse: Convergent recruitment of proteins into animal venoms. *Annu. Rev. Genomics Hum. Genet.* **2009**, 10, 483–511.
125. Mayer, J. P.; Zhang, F.; DiMarchi, R. D. Insulin structure and function. *J. Pept. Sci.* **2007**, 88, 687–713.
126. Ebberink, R. H. M.; Smit, A. B.; Van Minnen, J. The insulin family: Evolution of structure and function in vertebrates and invertebrates. *Biol. Bull.* **1989**, 177, 176–182.
127. Smit, A. B.; van Kesteren, R. E.; Li, K. W.; Van Minnen, J.; Spijker, S.; Van Heerikhuizen, H.; Geraerts, W. P. Towards understanding the role of insulin in the brain: Lessons from insulin-related signaling systems in the invertebrate brain. *Prog. Neurobiol.* **1998**, 54, 35–54.
128. Floyd, P. D.; Li, L.; Rubakhin, S. S.; Sweedler, J. V.; Horn, C. C.; Kupfermann, I.; Alexeeva, V. Y. et al. Insulin prohormone processing, distribution, and relation to metabolism in *Aplysia californica*. *J. Neurosci.* **1999**, 19, 7732–7741.
129. Safavi-Hemami, H.; Lu, A.; Li, Q.; Fedosov, A. E.; Biggs, J.; Showers Corneli, P.; Seger, J.; Yandell, M.; Olivera, B. M. Venom insulins of cone snails diversify rapidly and track prey taxa. *Mol. Biol. Evol.* **2016**, 33(11), 2924–2934.
130. Modica, M. V.; Lombardo, F.; Franchini, P.; Oliverio, M. The venomous cocktail of the vampire snail *Colubraria reticulata* (Mollusca, Gastropoda). *BMC Genomics* **2015**, 16, 1–21.
131. Tucker, J. K.; Tenorio, M. J. *Systematic Classification of Recent and Fossil Conoidean Gastropods*. ConchBooks: Hackenheim, Germany, **2009**.
132. Kaas, Q.; Yu, R.; Jin, A. H.; Dutertre, S.; Craik, D. J. ConoServer: Updated content, knowledge, and discovery tools in the conopeptide database. *Nucleic Acids Res.* **2012**, 40, D325–D330.
133. Olivera, B. M.; Gray, W. R.; Zeikus, R.; McIntosh, J. M.; Varga, J.; Rivier, J.; de Santos, V.; Cruz, L. J. Peptide neurotoxins from fish-hunting cone snails. *Science* **1985**, 230, 1338–1343.
134. Wilson, M. J.; Yoshikami, D.; Azam, L.; Gajewiak, J.; Olivera, B. M.; Bulaj, G.; Zhang, M.-M. mu-Conotoxins that differentially block sodium channels NaV1.1 through 1.8 identify those responsible for action potentials in sciatic nerve. *Proc. Natl. Acad. Sci. USA* **2011**, 108, 10302–10307.
135. Gray, W. R.; Rivier, J. E.; Galyean, R.; Cruz, L. J.; Olivera, B. M. Conotoxin MI: Disulfide bonding and conformational states. *J. Biol. Chem.* **1983**, 258, 12247–12251.
136. Olivera, B. M.; Rivier, J.; Clark, C.; Ramilo, C. A.; Corpuz, G. P.; Abogadie, F. C.; Mena, E. E.; Woodward, S. R.; Hillyard, D. R.; Cruz, L. J. Diversity of *Conus* neuropeptides. *Science* **1990**, 249, 257–263.
137. Yanagawa, Y.; Abe, T.; Satake, M.; Odani, S.; Suzuki, J.; Ishikawa, K. A novel sodium channel inhibitor from *Conus geographus*: Purification, structure, and pharmacological properties. *Biochemist* **1988**, 27, 6256–6262.
138. McIntosh, J. M.; Dowell, C.; Watkins, M.; Garrett, J. E.; Yoshikami, D.; Olivera, B. M. a-Conotoxin GIC from *Conus geographus*, a novel peptide antagonist of nAChRs. *J. Biol. Chem.* **2002**, 277, 33610–33615.
139. Nicke, A.; Loughnan, M. L.; Millard, E. L.; Alewood, P. F.; Adams, D. J.; Daly, N. L.; Craik, D. J.; Lewis, R. J. Isolation, structure, and activity of GID, a novel alpha 4/7-conotoxin with an extended N-terminal sequence. *J. Biol. Chem.* **2003**, 278, 3137–3144.

140. Santos, A. D.; McIntosh, J. M.; Hillyard, D. R.; Cruz, L. J.; Olivera, B. M. The A-superfamily of conotoxins: Structural and functional divergence. *J. Biol. Chem.* **2004**, 279, 17596–17606.
141. Zafaralla, G. C.; Ramilo, C.; Gray, W. R.; Karlstrom, R.; Olivera, B. M.; Cruz, L. J. Phylogenetic specificity of cholinergic ligands: Alpha-conotoxin SI. *Biochemist* **1988**, 27, 7102–7105.
142. Myers, R. A.; Zafaralla, G. C.; Gray, W. R.; Abbott, J.; Cruz, L. J.; Olivera, B. M. a-Conotoxins, small peptide probes of nicotinic acetylcholine receptors. *Biochemist* **1991**, 30, 9370–9377.
143. Lopez-Vera, E.; Aguilar, M. B.; Schiavon, E.; Marinzi, C.; Ortiz, E.; Restano Cassulini, R.; Batista, C. V. F. Novel alpha-conotoxins from *Conus spurius* and the alpha-conotoxin EI share high-affinity potentiation and low-affinity inhibition of nicotinic acetylcholine receptors. *FEBS J.* **2007**, 274, 3972–3985.
144. Hopkins, C.; Grilley, M.; Miller, C.; Shon, K. J.; Cruz, L. J.; Gray, W. R.; Dykert, J.; Rivier, J.; Yoshikami, D.; Olivera, B. M. A new family of *Conus* peptides targeted to the nicotinic acetylcholine receptor. *J. Biol. Chem.* **1995**, 270, 22361–22367.
145. McIntosh, M.; Cruz, L. J.; Hunkapiller, M. W.; Gray, W. R.; Olivera, B. M. Isolation and structure of a peptide toxin from the marine snail *Conus magus*. *Arch. Biochem. Biophys.* **1982**, 218, 329–334.
146. Kapono, C. A.; Thapa, P.; Cabalteja, C. C.; Guendisch, D.; Collier, A. C.; Bingham, J.-P. Conotoxin truncation as a post-translational modification to increase the pharmacological diversity within the milked venom of *Conus magus*. *Toxicon* **2013**, 70, 170–178.
147. Cartier, G. E.; Yoshikami, D.; Gray, W. R.; Luo, S.; Olivera, B. M.; McIntosh, J. M. A new a-conotoxin which targets a3b2 nicotinic acetylcholine receptors. *J. Biol. Chem.* **1996**, 271, 7522–7528.
148. Azam, L.; Dowell, C.; Watkins, M.; Stitzel, J. A.; Olivera, B. M.; McIntosh, J. M. Alpha-conotoxin BuIA, a novel peptide from *Conus bullatus*, distinguishes among neuronal nicotinic acetylcholine receptors. *J. Biol. Chem.* **2005**, 280, 80–87.
149. Holford, M.; Zhang, M.-M.; Gowd, K. H.; Azam, L.; Green, B. R.; Watkins, M.; Ownby, J.-P.; Yoshikami, D.; Bulaj, G.; Olivera, B. M. Pruning nature: Biodiversity-derived discovery of novel sodium channel blocking conotoxins from *Conus bullatus*. *Toxicon* **2009**, 53, 90–98.
150. Shon, K. J.; Olivera, B. M.; Watkins, M.; Jacobsen, R. B.; Gray, W. R.; Floresca, C. Z.; Cruz, L. J. et al. mu-Conotoxin PIIIA, a new peptide for discriminating among tetrodotoxin-sensitive Na channel subtypes. *J. Neurosci.* **1998**, 18, 4473–4481.
151. Schroeder, C. I.; Ekberg, J.; Nielsen, K. J.; Adams, D.; Loughnan, M. L.; Thomas, L.; Adams, D. J.; Alewood, P. F.; Lewis, R. J. Neuronally micro-conotoxins from *Conus striatus* utilize an alpha-helical motif to target mammalian sodium channels. *J. Biol. Chem.* **2008**, 283, 21621–21628.
152. Zhang, M.-M.; Fiedler, B.; Green, B. R.; Catlin, P.; Watkins, M.; Garrett, J. E.; Smith, B. J.; Yoshikami, D.; Olivera, B. M.; Bulaj, G. Structural and functional diversities among mu-conotoxins targeting TTX-resistant sodium channels. *Biochemist* **2006**, 45, 3723–3732.
153. Ramilo, C. A.; Zafaralla, G. C.; Nadasdi, L.; Hammerland, L. G.; Yoshikami, D.; Gray, W. R.; Kristipati, R.; Ramachandran, J.; Miljanich, G.; Olivera, B. M. Novel alpha- and omega-conotoxins from *Conus striatus* venom. *Biochemist* **1992**, 31, 9919–9926.
154. Hillyard, D. R.; Monje, V. D.; Mintz, I. M.; Bean, B. P.; Nadasdi, L.; Ramachandran, J.; Miljanich, G.; Azimi-Zoonooz, A.; McIntosh, J. M.; Cruz, L. J. A new *Conus* peptide ligand for mammalian presynaptic Ca^{2+} channels. *Neuron* **1992**, 9, 69–77.
155. Monje, V. D.; Haack, J. A.; Naisbitt, S. R.; Miljanich, G.; Ramachandran, J.; Nasdasdi, L.; Olivera, B. M.; Hillyard, D. R.; Gray, W. R. A new *Conus* peptide ligand for Ca channel subtypes. *Neuropharmacology* **1993**, 32, 1141–1149.

156. Hone, A. J.; Scadden, M.; Gajewiak, J.; Christensen, S.; Lindstrom, J.; McIntosh, J. M. alpha-Conotoxin PeIA[S9H,V10A,E14N] potently and selectively blocks alpha6beta-2beta3 versus alpha6beta4 nicotinic acetylcholine receptors. *Mol. Pharmacol.* **2012**, 82, 972–982.
157. Hone, A. J.; Ruiz, M.; Scadden, M. A. L.; Christensen, S.; Gajewiak, J.; Azam, L.; McIntosh, J. M. Positional scanning mutagenesis of alpha-conotoxin PeIA identifies critical residues that confer potency and selectivity for alpha6/alpha3beta2beta3 and alpha3beta2 nicotinic acetylcholine receptors. *J. Biol. Chem.* **2013**, 288, 25428–25439.
158. Whiteaker, P.; Christensen, S.; Yoshikami, D.; Dowell, C.; Watkins, M.; Gulyas, J.; Rivier, J.; Olivera, B. M.; McIntosh, J. M. Discovery, synthesis, and structure activity of a highly selective alpha7 nicotinic acetylcholine receptor antagonist. *Biochemist* **2007**, 46, 6628–6638.
159. Luo, S.; Zhangsun, D.; Zhu, X.; Wu, Y.; Hu, Y.; Christensen, S.; Harvey, P. J.; Akcan, M.; Craik, D. J.; McIntosh, J. M. Characterization of a novel alpha-conotoxin TxID from *Conus textile* that potently blocks rat alpha3beta4 nicotinic acetylcholine receptors. *J. Med. Chem.* **2013**, 56, 9655–9663.
160. Cruz, L. J.; Gray, W. R.; Olivera, B. M. Purification and properties of a myotoxin from *Conus geographus* venom. *Arch. Biochem. Biophys.* **1978**, 190, 539–548.
161. Stone, B. L.; Gray, W. R. Occurrence of hydroxyproline in a toxin from the marine snail *Conus geographus*. *Arch. Biochem. Biophys.* **1982**, 216, 765–767.
162. Aguilar, M. B.; Lopez-Vera, E.; Ortiz, E.; Becerril, B.; Possani, L. D.; Olivera, B. M.; Heimer de la Coteranovel, E. P. A novel conotoxin from *Conus delessertii* with post-translationally modified lysine residues. *Biochemist* **2005**, 44, 11130–11136.
163. Loughnan, M.; Bond, T.; Atkins, A.; Cuevas, J.; Adams, D. J.; Broxton, N. M.; Livett, B. G. et al. a-Conotoxin EpI, a novel sulfated peptide from *Conus episcopatus* that selectively targets neuronal nicotinic acetylcholine receptors. *J. Biol. Chem.* **1998**, 273, 15667–15674.
164. Jacobsen, R. B.; Jimenez, E. C.; De la Cruz, R. G.; Gray, W. R.; Cruz, L. J.; Olivera, B. M. A novel D-leucine-containing *Conus* peptide: Diverse conformational dynamics in the contryphan family. *J. Pept. Res.* **1999**, 54, 93–99.
165. Zhang, M.-M.; Green, B. R.; Catlin, P.; Fiedler, B.; Azam, L.; Chadwick, A.; Terlau, H. et al. Structure/function characterization of micro-conotoxin KIIIA, an analgesic, nearly irreversible blocker of mammalian neuronal sodium channels. *J. Biol. Chem.* **2007**, 282, 30699–30706.
166. Bulaj, G.; West, P. J.; Garrett, J. E.; Watkins, M.; Zhang, M.-M.; Norton, R. S.; Smith, B. J.; Yoshikami, D.; Olivera, B. M. Novel conotoxins from *Conus striatus* and *Conus kinoshitai* selectively block TTX-resistant sodium channels. *Biochemist* **2005**, 44, 7259–7265.
167. West, P. J.; Bulaj, G.; Garrett, J. E.; Olivera, B. M.; Yoshikami, D. Mu-conotoxin SmIIIA, a potent inhibitor of tetrodotoxin-resistant sodium channels in amphibian sympathetic and sensory neurons. *Biochemist* **2002**, 41, 15388–15393.
168. Walewska, A.; Skalicky, J. J.; Davis, D. R.; Zhang, M.-M.; Lopez-Vera, E.; Watkins, M.; Han, T. S.; Yoshikami, D.; Olivera, B. M.; Bulaj, G. NMR-based mapping of disulfide bridges in cysteine-rich peptides: Application to the mu-conotoxin SxIIIA. *J. Am. Chem. Soc.* **2008**, 130, 14280–14286.
169. Lewis, R. J.; Schroeder, C. I.; Ekberg, J.; Nielsen, K. J.; Loughnan, M.; Thomas, L.; Adams, D. A.; Drinkwater, R.; Adams, D. J.; Alewood, P. F. Isolation and structure-activity of mu-conotoxin TIIIA, a potent inhibitor of tetrodotoxin-sensitive voltage-gated sodium channels. *Mol. Pharmacol.* **2007**, 71, 676–685.
170. Vetter, I.; Dekan, Z.; Knapp, O.; Adams, D. J.; Alewood, P. F.; Lewis, R. J. Isolation, characterization and total regioselective synthesis of the novel muO-conotoxin MfVIA from *Conus magnificus* that targets voltage-gated sodium channels. *Biochem. Pharmacol.* **2012**, 84, 540–548.

171. Leipold, E.; Hansel, A.; Olivera, B. M.; Terlau, H.; Heinemann, S. H. Molecular interaction of delta-conotoxins with voltage-gated sodium channels. *FEBS Lett.* **2005**, 579, 3881–3884.
172. Hillyard, D. R.; Olivera, B. M.; Woodward, S.; Corpuz, G. P.; Gray, W. R.; Ramilo, C. A.; Cruz, L. J. A molluscivorous *Conus* toxin: Conserved frameworks in conotoxins. *Biochemist* **1989**, 28, 358–361.
173. Kohno, T.; Sasaki, T.; Kobayashi, K.; Fainzilber, M.; Sato, K. Three-dimensional solution structure of the sodium channel agonist/antagonist delta-conotoxin TxVIA. *J. Biol. Chem.* **2002**, 277, 36387–36391.
174. Jin, A.-H.; Israel, M. R.; Inserra, M. C.; Smith, J. J.; Lewis, R. J.; Alewood, P. F.; Vetter, I.; Dutertre, S. delta-Conotoxin SuVIA suggests an evolutionary link between ancestral predator defence and the origin of fish-hunting behaviour in carnivorous cone snails. *Proc. Biol. Sci.* **2015**, 282, 20150817.
175. Aman, J. W.; Imperial, J. S.; Ueberheide, B.; Zhang, M.-M.; Aguilar, M.; Taylor, D.; Watkins, M. et al. Insights into the origins of fish hunting in venomous cone snails from studies of *Conus tessulatus*. *Proc. Natl. Acad. Sci. USA* **2015**, 112, 5087–5092.
176. Barbier, J.; Lamthanh, H.; Le Gall, F.; Favreau, P.; Benoit, E.; Chen, H.; Gilles, N. et al. A delta-conotoxin from *Conus ermineus* venom inhibits inactivation in vertebrate neuronal Na^+ channels but not in skeletal and cardiac muscles. *J. Biol. Chem.* **2004**, 279, 4680–4685.
177. Wang, L.; Liu, J.; Pi, C.; Zeng, X.; Zhou, M.; Jiang, X.; Chen, S.; Ren, Z.; Xu, A. Identification of a novel M-superfamily conotoxin with the ability to enhance tetrodotoxin sensitive sodium currents. *Arch. Toxicol.* **2009**, 83, 925–932.
178. Fainzilber, M.; Nakamura, T.; Gaathon, A.; Lodder, J. C.; Kits, K. S.; Burlingame, A. L.; Zlotkin, E. A new cysteine framework in sodium channel blocking conotoxins. *Biochemist* **1995**, 34, 8649–8656.
179. Sudarslal, S.; Majumdar, S.; Ramasamy, P.; Dhawan, R.; Pal, P. P.; Ramaswami, M.; Lala, A. K. et al. Sodium channel modulating activity in a delta-conotoxin from an Indian marine snail. *FEBS Lett.* **2003**, 553, 209–212.
180. Shon, K. J.; Hasson, A.; Spira, M. E.; Cruz, L. J.; Gray, W. R.; Olivera, B. M. Delta-conotoxin GmVIA, a novel peptide from the venom of *Conus gloriamaris*. *Biochemist* **1994**, 33, 11420–11425.
181. Liu, J.; Wu, Q.; Pi, C.; Zhao, Y.; Zhou, M.; Wang, L.; Chen, S.; Xu, A. Isolation and characterization of a T-superfamily conotoxin from *Conus litteratus* with targeting tetrodotoxin-sensitive sodium channels. *Peptides* **2007**, 28, 2313–2319.
182. Lewis, R. J.; Nielsen, K. J.; Craik, D. J.; Loughnan, M. L.; Adams, D. A.; Sharpe, I. A.; Luchian, T. et al. Novel omega-conotoxins from *Conus catus* discriminate among neuronal calcium channel subtypes. *J. Biol. Chem.* **2000**, 275, 35335–35344.
183. Berecki, G.; Motin, L.; Haythornthwaite, A.; Vink, S.; Bansal, P.; Drinkwater, R.; Wang, C. I. et al. Analgesic (omega)-conotoxins CVIE and CVIF selectively and voltage-dependently block recombinant and native N-type calcium channels. *Mol. Pharmacol.* **2010**, 77, 139–148.
184. Lee, S.; Kim, Y.; Back, S. K.; Choi, H.-W.; Lee, J. Y.; Jung, H. H.; Ryu, J. H. et al. Analgesic effect of highly reversible omega-conotoxin FVIA on N type Ca^{2+} channels. *Mol. Pain* **2010**, 6, 97.
185. Bernaldez, J.; Lopez, O.; Licea, A.; Salceda, E.; Arellano, R. O.; Vega, R.; Soto, E. Electrophysiological characterization of a novel small peptide from the venom of *Conus californicus* that targets voltage-gated neuronal Ca^{2+} channels. *Toxicon* **2011**, 57, 60–67.
186. Biggs, J. S.; Watkins, M.; Puillandre, N.; Ownby, J.-P.; Lopez-Vera, E.; Christensen, S.; Moreno, K. J. et al. Evolution of *Conus* peptide toxins: Analysis of *Conus californicus* Reeve, 1844. *Mol. Phylogenet. Evol.* **2010**, 56, 1–12.

187. Kits, K. S.; Lodder, J. C.; van der Schors, R. C.; Li, K. W.; Geraerts, W. P.; Fainzilber, M. Novel omega-conotoxins block dihydropyridine-insensitive high voltage-activated calcium channels in molluscan neurons. *J. Neurochem.* **1996**, 67, 2155–2163.
188. Fan, C.-X.; Chen, X.-K.; Zhang, C.; Wang, L.-X.; Duan, K.-L.; He, L.-L.; Cao, Y. et al. A novel conotoxin from *Conus betulinus*, kappa-BtX, unique in cysteine pattern and in function as a specific BK channel modulator. *J. Biol. Chem.* **2003**, 278, 12624–12633.
189. Imperial, J. S.; Bansal, P. S.; Alewood, P. F.; Daly, N. L.; Craik, D. J.; Sporning, A.; Terlau, H.; Lopez-Vera, E.; Bandyopadhyay, P. K.; Olivera, B. M. A novel conotoxin inhibitor of Kv1.6 channel and nAChR subtypes defines a new superfamily of conotoxins. *Biochemist.* **2006**, 45, 8331–8340.
190. Jacobsen, R. B.; Koch, E. D.; Lange-Malecki, B.; Stocker, M.; Verhey, J.; Van Wagoner, R. M.; Vyazovkina, A.; Olivera, B. M.; Terlau, H. Single amino acid substitutions in kappa-conotoxin PVIIA disrupt interaction with the shaker K+ channel. *J. Biol. Chem.* **2000**, 275, 24639–24644.
191. Kauferstein, S.; Huys, I.; Lamthanh, H.; Stocklin, R.; Sotto, F.; Menez, A.; Tytgat, J.; Mebs, D. A novel conotoxin inhibiting vertebrate voltage-sensitive potassium channels. *Toxicon* **2003**, 42, 43–52.
192. Imperial, J. S.; Chen, P.; Sporning, A.; Terlau, H.; Daly, N. L.; Craik, D. J.; Alewood, P. F.; Olivera, B. M. Tyrosine-rich conopeptides affect voltage-gated K^+ channels. *J. Biol. Chem.* **2008**, 283, 23026–23032.
193. Aguilar, M. B.; Perez-Reyes, L. I.; Lopez, Z.; de la Cotera, E. P. H.; Falcon, A.; Ayala, C.; Galvan, M.; Salvador, C.; Escobar, L. I. Peptide sr11a from *Conus spurius* is a novel peptide blocker for Kv1 potassium channels. *Peptides* **2010**, 31, 1287–1291.
194. Liu, L.; Chew, G.; Hawrot, E.; Chi, C.; Wang, C. Two potent alpha3/5 conotoxins from piscivorous *Conus achatinus*. *Acta Biochim. Biophys. Sin. Shanghai* **2007**, 39, 438–444.
195. Loughnan, M. L.; Nicke, A.; Jones, A.; Adams, D. J.; Alewood, P. F.; Lewis, R. J. Chemical and functional identification and characterization of novel sulfated alpha-conotoxins from the cone snail *Conus anemone*. *J. Med. Chem.* **2004**, 47, 1234–1241.
196. Luo, S.; Kulak, J. M.; Cartier, G. E.; Jacobsen, R. B.; Yoshikami, D.; Olivera, B. M.; McIntosh, J. M. alpha-conotoxin AuIB selectively blocks alpha3 beta4 nicotinic acetylcholine receptors and nicotine-evoked norepinephrine release. *J. Neurosci.* **1998**, 18, 8571–8579.
197. Nguyen, B.; Le Caer, J.-P.; Araoz, R.; Thai, R.; Lamthanh, H.; Benoit, E.; Molgo, J. Isolation, purification and functional characterization of alpha-BnIA from *Conus bandanus* venom. *Toxicon* **2014**, 91, 155–163.
198. Favreau, P.; Krimm, I.; Le Gall, F.; Bobenrieth, M. J.; Lamthanh, H.; Bouet, F.; Servent, D. et al. Biochemical characterization and nuclear magnetic resonance structure of novel alpha-conotoxins isolated from the venom of *Conus consors*. *Biochemist* **1999**, 38, 6317–6326.
199. Martinez, J. S.; Olivera, B. M.; Gray, W. R.; Craig, A. G.; Groebe, D. R.; Abramson, S. N.; McIntosh, J. M. alpha-Conotoxin EI, a new nicotinic acetylcholine receptor antagonist with novel selectivity. *Biochemist* **1995**, 34, 14519–14526.
200. Quinton, L.; Servent, D.; Girard, E.; Molgo, J.; Le Caer, J.-P.; Malosse, C.; Haidar, E. A.; Lecoq, A.; Gilles, N.; Chamot-Rooke, J. Identification and functional characterization of a novel alpha-conotoxin (EIIA) from *Conus ermineus*. *Anal. Bioanal. Chem.* **2013**, 405, 5341–5351.
201. Xu, S.; Zhang, T.; Kompella, S. N.; Yan, M.; Lu, A.; Wang, Y.; Shao, X. et al. Conotoxin alphaD-GeXXA utilizes a novel strategy to antagonize nicotinic acetylcholine receptors. *Sci. Rep.* **2015**, 5, 14261.
202. McIntosh, J. M.; Yoshikami, D.; Mahe, E.; Nielsen, D. B.; Rivier, J. E.; Gray, W. R.; Olivera, B. M. A nicotinic acetylcholine receptor ligand of unique specificity, alpha-conotoxin ImI. *J. Biol. Chem.* **1994**, 269, 16733–16739.

203. Peng, C.; Han, Y.; Sanders, T.; Chew, G.; Liu, J.; Hawrot, E.; Chi, C.; Wang, C. alpha4/7-conotoxin Lp1.1 is a novel antagonist of neuronal nicotinic acetylcholine receptors. *Peptides* **2008**, 29, 1700–1707.
204. Peng, C.; Chen, W.; Sanders, T.; Chew, G.; Liu, J.; Hawrot, E.; Chi, C. Chemical synthesis and characterization of two alpha4/7-conotoxins. *Acta Biochim. Biophys. Sin. Shanghai* **2010**, 42, 745–753.
205. Inserra, M. C.; Kompella, S. N.; Vetter, I.; Brust, A.; Daly, N. L.; Cuny, H.; Craik, D. J.; Alewood, P. F.; Adams, D. J.; Lewis, R. J. Isolation and characterization of alpha-conotoxin LsIA with potent activity at nicotinic acetylcholine receptors. *Biochem. Pharmacol.* **2013**, 86, 791–799.
206. Luo, S.; Akondi, K. B.; Zhangsun, D.; Wu, Y.; Zhu, X.; Hu, Y.; Christensen, S. et al. Atypical alpha-conotoxin LtIA from *Conus litteratus* targets a novel microsite of the alpha3beta2 nicotinic receptor. *J. Biol. Chem.* **2010**, 285, 12355–12366.
207. Luo, S.; Zhangsun, D.; Schroeder, C. I.; Zhu, X.; Hu, Y.; Wu, Y.; Weltzin, M. M. et al. A novel alpha4/7-conotoxin LvIA from *Conus lividus* that selectively blocks alpha3beta2 vs. alpha6/alpha3beta2beta3 nicotinic acetylcholine receptors. *FASEB J.* **2014**, 28, 1842–1853.
208. Jacobsen, R. B.; DelaCruz, R. G.; Grose, J. H.; McIntosh, J. M.; Yoshikami, D.; Olivera, B. M. Critical residues influence the affinity and selectivity of alpha-conotoxin MI for nicotinic acetylcholine receptors. *Biochemist* **1999**, 38, 13310–13315.
209. Mueller, A.; Starobova, H.; Inserra, M. C.; Jin, A.-H.; Deuis, J. R.; Dutertre, S.; Lewis, R. J.; Alewood, P. F.; Daly, N. L.; Vetter, I. alpha-Conotoxin MrIC is a biased agonist at alpha7 nicotinic acetylcholine receptors. *Biochem. Pharmacol.* **2015**, 94, 155–163.
210. Talley, T. T.; Olivera, B. M.; Han, K.-H.; Christensen, S. B.; Dowell, C.; Tsigelny, I.; Ho, K.-Y.; Taylor, P.; McIntosh, J. M. Alpha-conotoxin OmIA is a potent ligand for the acetylcholine-binding protein as well as alpha3beta2 and alpha7 nicotinic acetylcholine receptors. *J. Biol. Chem.* **2006**, 281, 24678–24686.
211. McIntosh, J. M.; Plazas, P. V.; Watkins, M.; Gomez-Casati, M. E.; Olivera, B. M.; Elgoyhen, A. B. A novel alpha-conotoxin, PeIA, cloned from *Conus pergrandis*, discriminates between rat alpha9alpha10 and alpha7 nicotinic cholinergic receptors. *Int. J. Biol. Chem.* **2005**, 280, 30107–30112.
212. Dowell, C.; Olivera, B. M.; Garrett, J. E.; Staheli, S. T.; Watkins, M.; Kuryatov, A.; Yoshikami, D.; Lindstrom, J. M.; McIntosh, J. M. Alpha-conotoxin PIA is selective for alpha6 subunit-containing nicotinic acetylcholine receptors. *J. Neurosci.* **2003**, 23, 8445–8452.
213. Fainzilber, M.; Hasson, A.; Oren, R.; Burlingame, A. L.; Gordon, D.; Spira, M. E.; Zlotkin, E. New mollusc-specific alpha-conotoxins block *Aplysia* neuronal acetylcholine receptors. *Biochemist* **1994**, 33, 9523–9529.
214. Peng, C.; Chen, W.; Han, Y.; Sanders, T.; Chew, G.; Liu, J.; Hawrot, E.; Chi, C.; Wang, C. Characterization of a novel alpha4/4-conotoxin, Qc1.2, from vermivorous *Conus quercinus*. *Acta Biochim. Biophys. Sin. Shanghai* **2009**, 41, 858–864.
215. Braga, M. C. V.; Nery, A. A.; Ulrich, H.; Konno, K.; Sciani, J. M.; Pimenta, D. C. alpha-RgIB: A novel antagonist peptide of neuronal acetylcholine receptor isolated from *Conus regius* venom. *Int. J. Pept.* **2013**, 2013, 543028.
216. Kauferstein, S.; Porth, C.; Kendel, Y.; Wunder, C.; Nicke, A.; Kordis, D.; Favreau, P.; Koua, D.; Stocklin, R.; Mebs, D. Venomic study on cone snails (*Conus* spp.) from South Africa. *Toxicon* **2011**, 57, 28–34.
217. Dutertre, S.; Ulens, C.; Buttner, R.; Fish, A.; van Elk, R.; Kendel, Y.; Hopping, G. et al. AChBP-targeted alpha-conotoxin correlates distinct binding orientations with nAChR subtype selectivity. *EMBO J.* **2007**, 26, 3858–3867.
218. Luo, S.; Zhangsun, D.; Wu, Y.; Zhu, X.; Hu, Y.; McIntyre, M.; Christensen, S.; Akcan, M.; Craik, D. J.; McIntosh, J. M. Characterization of a novel alpha-conotoxin from *Conus textile* that selectively targets alpha6/alpha3beta2beta3 nicotinic acetylcholine receptors. *J. Biol. Chem.* **2013**, 288, 894–902.

219. Wang, S.; Du, T.; Liu, Z.; Wang, S.; Wu, Y.; Ding, J.; Jiang, L.; Dai, Q. Characterization of a T-superfamily conotoxin TxVC from *Conus* textile that selectively targets neuronal nAChR subtypes. *Biochem. Biophys. Res. Commun.* **2014**, 454, 151–156.
220. Safavi-Hemami, H.; Siero, W. A.; Kuang, Z.; Williamson, N. A.; Karas, J. A.; Page, L. R.; MacMillan, D. et al. Embryonic toxin expression in the cone snail *Conus victoriae*: Primed to kill or divergent function? *J. Biol. Chem.* **2011**, 286, 22546–22557.
221. Loughnan, M.; Nicke, A.; Jones, A.; Schroeder, C. I.; Nevin, S. T.; Adams, D. J.; Alewood, P. F.; Lewis, R. J. Identification of a novel class of nicotinic receptor antagonists: Dimeric conotoxins VxXIIA, VxXIIB, and VxXIIC from *Conus vexillum*. *J. Biol. Chem.* **2006**, 281, 24745–24755.
222. Jimenez, E. C.; Olivera, B. M.; Teichert, R. W. AlphaC-conotoxin PrXA: A new family of nicotinic acetylcholine receptor antagonists. *Biochemist* **2007**, 46, 8717–8724.
223. Luo, S.; Zhangsun, D.; Harvey, P. J.; Kaas, Q.; Wu, Y.; Zhu, X.; Hu, Y. et al. Cloning, synthesis, and characterization of alphaO-conotoxin GeXIVA, a potent alpha9alpha10 nicotinic acetylcholine receptor antagonist. *Proc. Natl. Acad. Sci. USA* **2015**, 112, E4026–E4035.
224. Twede, V. D.; Teichert, R. W.; Walker, C. S.; Gruszczynski, P.; Kazmierkiewicz, R.; Bulaj, G.; Olivera, B. M. Conantokin-Br from *Conus brettinghami* and selectivity determinants for the NR2D subunit of the NMDA receptor. *Biochemist* **2009**, 48, 4063–4073.
225. Gowd, K. H.; Twede, V.; Watkins, M.; Krishnan, K. S.; Teichert, R. W.; Bulaj, G.; Olivera, B. M. Conantokin-P, an unusual conantokin with a long disulfide loop. *Toxicon* **2008**, 52, 203–213.
226. Gowd, K. H.; Watkins, M.; Twede, V. D.; Bulaj, G. W.; Olivera, B. M. Characterization of conantokin Rl-A: Molecular phylogeny as structure/function study. *J. Pept. Sci.* **2010**, 16, 375–382.
227. Jimenez, E. C.; Donevan, S.; Walker, C.; Zhou, L.-M.; Nielsen, J.; Cruz, L. J.; Armstrong, H.; White, H. S.; Olivera, B. M. Conantokin-L, a new NMDA receptor antagonist: Determinants for anticonvulsant potency. *Epilepsy Res.* **2002**, 51, 73–80.
228. Haack, J. A.; Rivier, J.; Parks, T. N.; Mena, E. E.; Cruz, L. J.; Olivera, B. M. Conantokin-T: A gamma-carboxyglutamate containing peptide with *N*-methyl-D-aspartate antagonist activity. *J. Biol. Chem.* **1990**, 265, 6025–6029.
229. Platt, R. J.; Curtice, K. J.; Twede, V. D.; Watkins, M.; Gruszczynski, P.; Bulaj, G.; Horvath, M. P.; Olivera, B. M. From molecular phylogeny towards differentiating pharmacology for NMDA receptor subtypes. *Toxicon* **2014**, 81, 67–79.
230. Sharpe, I. A.; Gehrmann, J.; Loughnan, M. L.; Thomas, L.; Adams, D. A.; Atkins, A.; Palant, E. et al. Two new classes of conopeptides inhibit the alpha1-adrenoceptor and noradrenaline transporter. *Nat. Neurosci.* **2001**, 4, 902–907.
231. Petrel, C.; Hocking, H. G.; Reynaud, M.; Upert, G.; Favreau, P.; Biass, D.; Paolini-Bertrand, M. et al. Identification, structural and pharmacological characterization of tau-CnVA, a conopeptide that selectively interacts with somatostatin sst3 receptor. *Biochem. Pharmacol.* **2013**, 85, 1663–1671.
232. Zugasti-Cruz, A.; Maillo, M.; Lopez-Vera, E.; Falcon, A.; Heimer de la Cotera, E. P.; Olivera, B. M.; Aguilar, M. B. Amino acid sequence and biological activity of a gamma-conotoxin-like peptide from the worm-hunting snail *Conus austini*. *Peptides* **2006**, 27, 506–511.
233. Fainzilber, M.; Nakamura, T.; Lodder, J. C.; Zlotkin, E.; Kits, K. S.; Burlingame, A. L. gamma-Conotoxin-PnVIIA, a gamma-carboxyglutamate-containing peptide agonist of neuronal pacemaker cation currents. *Biochemist* **1998**, 37, 1470–1477.
234. Dutertre, S.; Croker, D.; Daly, N. L.; Andersson, A.; Muttenthaler, M.; Lumsden, N. G.; Craik, D. J.; Alewood, P. F.; Guillon, G.; Lewis, R. J. Conopressin-T from *Conus tulipa* reveals an antagonist switch in vasopressin-like peptides. *J. Biol. Chem.* **2008**, 283, 7100–7108.

235. Sabareesh, V.; Gowd, K. H.; Ramasamy, P.; Sudarslal, S.; Krishnan, K. S.; Sikdar, S. K.; Balaram, P. Characterization of contryphans from *Conus loroisii* and *Conus amadis* that target calcium channels. *Peptides* **2006**, 27, 2647–2654.
236. Hansson, K.; Ma, X.; Eliasson, L.; Czerwiec, E.; Furie, B.; Furie, B. C.; Rorsman, P.; Stenflo, J. The first gamma-carboxyglutamic acid-containing contryphan: A selective L-type calcium ion channel blocker isolated from the venom of *Conus marmoreus*. *J. Biol. Chem.* **2004**, 279, 32453–32463.
237. Massilia, G. R.; Eliseo, T.; Grolleau, F.; Lapied, B.; Barbier, J.; Bournaud, R.; Molgo, J. et al. Contryphan-Vn: A modulator of Ca^{2+}-dependent K^{+} channels. *Biochem. Biophys. Res. Commun.* **2003**, 303, 238–246.

13 Naturally Occurring Disulfide-Rich Cyclic Peptides from Plants and Animals

Synthesis and Biosynthesis

Simon J. de Veer and David J. Craik

CONTENTS

13.1 INTRODUCTION

Plants and animals produce a range of disulfide-rich peptides for defense against predators or prey capture. An example of the former are the plant cyclotides[1] and an example of the latter are the conotoxins, described in Chapter 12. In the current chapter, we focus on a subset of disulfide-rich peptides that have the distinguishing feature of a head-to-tail cyclic peptide backbone. In general, cyclization significantly stabilizes peptides, and the combination of disulfide bonds *and* a cyclic backbone leads to peptides of exceptional stability. Peptides with these features have been the focus of studies over the last two decades, both in our laboratory[2–7] and in other laboratories.[8–11]

In this chapter, we first provide a background on the discovery, structures, and bioactivities of the various classes of disulfide-rich cyclic peptides and then focus on their synthesis and biosynthesis. We note that many of these cyclic peptides occur naturally in plants or animals, but that one group in which we have a particular interest, the cyclic conotoxins, are nonnatural substances. We consider them here along with naturally occurring cyclic peptides because the two groups share topological features and have some similar properties, including high stability. Indeed, naturally occurring cyclic plant peptides provided the inspiration for the development of synthetic cyclic conotoxins.

Since an extensive history of conotoxins and their potential applications is given in Chapter 12 and in other recent reviews,[12–15] we refer readers to those articles for background reading on these marine-derived disulfide-rich peptides. Similarly, we note that the scope of this chapter does not cover several classes of naturally cyclic peptides that do not have disulfide bonds. These include the orbitides[16] from plants and the cyclic bacteriocins[17] from bacteria, which have been reviewed elsewhere. In addition, we will not consider synthetic cyclic templates that lack disulfide bonds.

Figure 13.1 shows prototypic examples from the various classes of disulfide-rich cyclic peptides that we will describe in this chapter and illustrates their size range (10–50 amino acids) and disulfide content (one to four disulfide bonds). The simplest example is a class of seed-derived cyclic peptides with a single disulfide bond, and the most complex is a cyclic chlorotoxin with four disulfide bonds. In this section, we introduce each class of cyclic peptide, and in Section 13.2, we describe chemical and biological approaches that have been used to synthesize them. In Section 13.3, we provide selected applications of synthetic or biosynthetic approaches to exploit these molecules in drug design.

13.1.1 Sunflower Trypsin Inhibitor-1 and Other PawS-Derived Peptides

The prototypic member of this family of cyclic peptides is sunflower trypsin inhibitor-1 (SFTI-1), whose name describes both its origin (from sunflower seeds) and its function (a trypsin inhibitor). This naturally occurring 14-amino acid peptide was originally deduced to be cyclic based on unexpected electron density between its presumed termini in its crystal structure in complex with trypsin.[18] Subsequent solid-phase peptide synthesis followed by nuclear magnetic resonance (NMR) structure determination confirmed the cyclic nature of SFTI-1[19] and demonstrated the molecule to have identical solution and solid-state structures, as would be expected for a highly constrained cyclic peptide.

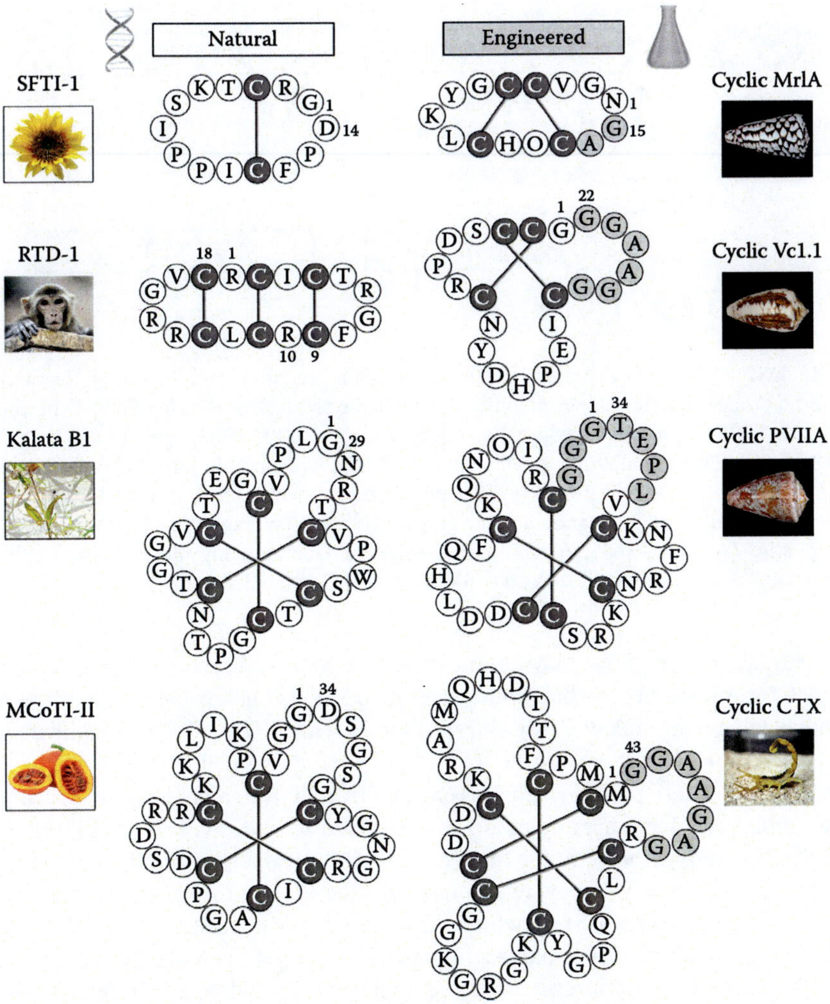

FIGURE 13.1 Disulfide-rich peptides found in plants and animals that are naturally cyclic or have been cyclized using an engineered linker. Naturally occurring cyclic peptides are shown in the left column and range from the single disulfide-bridged peptide, SFTI-1, to cyclotides (kalata B1 and MCoTI-II) that contain a cystine knot formed by three disulfide bonds. Engineered cyclic peptides are shown in the right column and include the conotoxins, MrIA, Vc1.1, and PVIIA, together with the scorpion toxin, chlorotoxin (CTX). Sequences are shown in single-letter amino acid code and graphics below each peptide name depict the source organism: SFTI-1, *Helianthus annuus* (sunflower); RTD-1, *Macacca mulata* (rhesus macaque); kalata B1, *Oldenlandia affinis*; MCoTI-II, *Momordica cochinchinensis*; MrIA, *Conus marmoreus*; Vc1.1, *Conus victoriae*; PVIIA, *Conus purpurascens*; CTX, *Leiurus quinquestriatus*. Cysteine residues are shaded dark gray, disulfide bonds are represented by solid lines, and engineered linker segments (right panels) are shaded light gray. The cyclization point for each peptide is indicated by numbering the first and last residues of the naturally occurring (or synthetic) precursor peptide. Note that RTD-1 is formed by the ligation of two precursor peptides (1–9 and 10–18). *Abbreviations*: CTX, chlorotoxin; MCoTI-II, *Momordica cochinchinensis* trypsin inhibitor-II; O, hydroxyproline; RTD-1, rhesus theta defensin-1; SFTI-1, sunflower trypsin inhibitor-1.

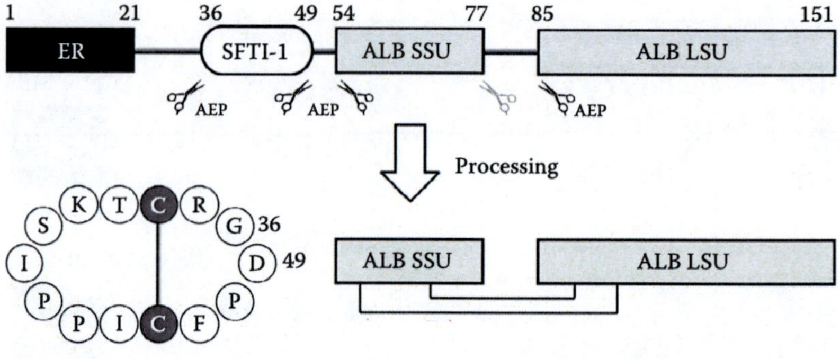

FIGURE 13.2 SFTI-1 is expressed within an albumin precursor protein and is liberated by proteolytic processing. The upper graphic illustrates the arrangement of the PawS1 (preproalbumin with SFTI-1) protein, showing the endoplasmic reticulum (ER) signal (residues 1–21), SFTI-1 domain (36–49), albumin small subunit (54–77), and albumin large subunit (85–151). PawS1 undergoes post-translational processing to liberate cyclic SFTI-1, as well as the small and large albumin subunits, as shown in the graphics below. Asparaginyl endopeptidase (AEP) cleavage sites (Asn35, Asp49, Asn53, and Asn84) are labeled using black scissors and the cleavage site for an additional protease (following Lys77) is labeled using gray scissors.

Investigations into the biosynthetic origin of SFTI-1 revealed that the coding sequence for the mature cyclic peptide is embedded within a proalbumin precursor, which has been named PawS1 for *p*repro*a*lbumin *w*ith *S*FTI-*1*.[20] It now appears that SFTI-1 is just one of a large number of similarly sized cyclic peptides found in sunflower plants.[21] These peptides are also excised from within a proalbumin precursor, and, as such, they have been named *P*awS-*d*erived *p*eptides (PDPs).[21] Figure 13.2 illustrates the biosynthetic route that leads to the production of SFTI-1, and, by inference, other recently discovered members of the PDP family of cyclic peptides.[22] The evolutionary mechanism by which these *stowaway* peptides have infiltrated an upstream region of existing proalbumin genes has not yet been elucidated.

SFTI-1 is particularly interesting from a pharmaceutical perspective because of its stability and small size. Based on its small size, and hence amenability to rapid and efficient solid-phase peptide synthesis, it has been the subject of a large number of chemical engineering studies, particularly for the development of protease inhibitors (described further in Section 13.3.1). Such applications are congruent with its natural function as a trypsin inhibitor. A selection of naturally occurring and chemically reengineered sequences of SFTI-1 and related cyclic peptides are summarized in Figure 13.3 to highlight the breadth of these applications.

13.1.2 Two-Disulfide Cyclic Conotoxins

The two-disulfide cyclic peptides that we will cover here are the cyclic conotoxins. So far, no naturally occurring cyclic conotoxins have been reported, but several groups have made synthetic cyclic conotoxin analogues, with the aim of improving the biopharmaceutical properties of natural (acyclic) conotoxins.[2,23] Consideration of the disulfide bonding pattern plays a very important role in these studies; for a

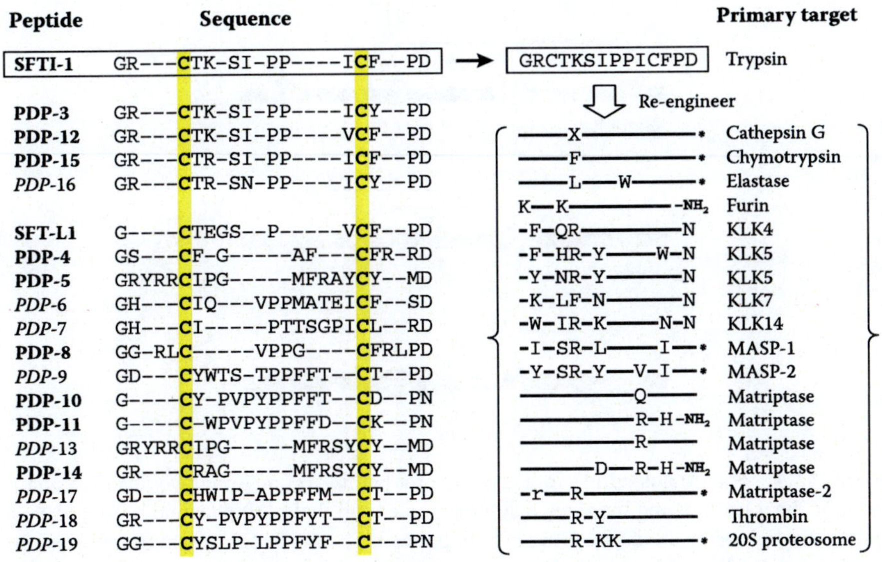

FIGURE 13.3 Naturally occurring SFTI-1 and PDPs and their engineered variants. PDPs found in plants (Heliantheae and Millerieae tribes) are shown in the left panel, with the prototypic peptide, SFTI-1, enclosed by a rectangle. PDPs in bold font have been identified in seed extracts, and those in italics have been predicted from gene sequencing. Engineered SFTI-1 variants are shown in the right panel, with the primary target for each inhibitor indicated in the far-right column. Amino acid substitutions are shown in single-letter amino acid code, with D-amino acids in lower case and X representing 4-guanidyl-L-phenylalanine. Acyclic peptides with a C-terminal carboxylate are marked by an asterisk (*) and acyclic peptides truncated after residue 12 are indicated by showing the peptide's C-terminal amide.

peptide with two disulfide bonds, there are three possible disulfide connectivities, which are generically referred to as the globular, ribbon, or beads forms, corresponding to disulfide connectivities (I–III, II–IV), (I–IV, II–III), or (I–II, III–IV), respectively, as illustrated in Figure 13.4. Examples of natural (i.e., acyclic) conotoxins with each of these connectivities have been made synthetically,[24] but in the cyclic forms, only the globular and ribbon connectivities have been studied.

The most extensively studied peptide in the class of two-disulfide conotoxins is cyclo-Vc1.1, which has the globular disulfide connectivity and was developed as an orally active peptide with potential for treating neuropathic pain.[2] The motivation for this work was the finding that a synthetic linear Vc1.1 peptide based on a cDNA sequence derived from the venom duct of *Conus victoriae* had activity in pain assays.[25] As illustrated in Figure 13.1, cyclo-Vc1.1 comprises the amino acid sequence of linear Vc1.1 with the termini joined together with a linker of six amino acids. In a rat chronic constriction injury model of neuropathic pain, this cyclic conotoxin is more than 100-fold more potent than gabapentin, the *gold standard* drug used clinically for the treatment of neuropathic pain in humans.[2] A mixture of Ala and Gly residues was used in the linker, with the rationale being that these small nonpolar amino acids would not introduce new functionality into the parent

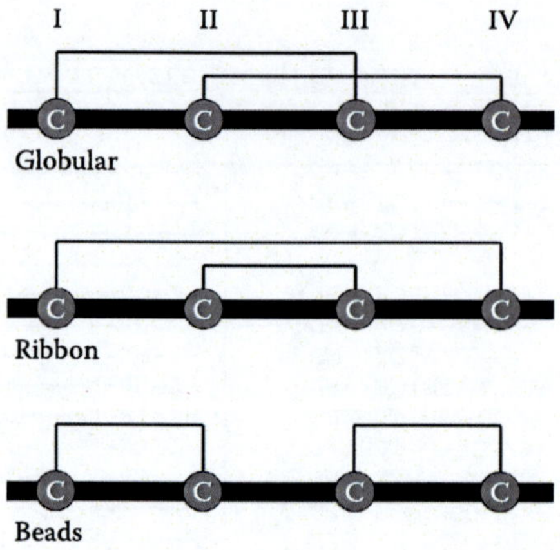

FIGURE 13.4 Potential disulfide connectivities for peptides containing two disulfide bonds. Schematic diagram showing the three disulfide isomers (globular, ribbon, and beads) that can be produced following oxidative folding of peptides that contain four cysteine residues; for example, two-disulfide conotoxins. Cysteine residues are labeled using Roman numerals and can be connected as follows: I–III and II–IV (globular), I–IV and II–III (ribbon), or I–II, III–IV (beads).

peptide—a principle that had been earlier developed for the model conotoxin, MII.[26] A five-amino acid linker was also trialed to achieve cyclization, but this variant did not maintain biological activity.

Cyclic MrIA, also illustrated in Figure 13.1, is another example of a synthetic cyclic conotoxin, in this case having the ribbon connectivity of its two disulfide bonds.[27] The parent linear version of this peptide has potent activity in blocking the noradrenaline transporter and was thus identified as having potential use in the treatment of pain, albeit via a different mechanism from Vc1.1.[28] An N-terminally modified derivative of this peptide designed to have increased stability over the native peptide reached phase II clinical trials[29] but was ultimately withdrawn because of safety concerns. The cyclic form is also very stable but has not progressed into clinical trials.

Other examples of cyclic two-disulfide conotoxins include cyclo-RgIA,[30] cyclo-IMI,[31] and cyclo-AuIB.[32] These cyclic peptides have been extensively reviewed recently,[33] so we will not cover them in detail here. Interestingly, in the case of cyclo-ImI, the folding outcome (i.e., disulfide connectivity) is influenced by the length of the linker used to achieve cyclization.[31] This finding demonstrates that cyclization is not always an innocuous change with respect to overall structure; consequently, careful design of the size and nature of the linker is required so as not to reduce the desired biological activity of the parent linear conotoxin.

Overall, the approach of cyclizing conotoxins has proved to be useful in increasing the stability of native conotoxins and, in one case, in endowing the molecule with oral activity, which is a significant breakthrough for a peptide-based drug lead. As will be seen later in this chapter, cyclization is indeed a relatively generalizable strategy that can be used to improve the biopharmaceutical properties of many peptides. We now turn to studies of naturally occurring three-disulfide cyclic peptides.

13.1.3 Theta-Defensins

Theta-defensins (θ-defensins) are antimicrobial peptides originally discovered in rhesus monkey leukocytes.[34] The prototypic θ-defensin, RTD-1, comprises 18 amino acids, including six cysteine residues forming three disulfide bonds in a laddered connectivity, as shown in Figure 13.1. Intriguingly, despite their small size, θ-defensins are the product of *two* genes, with each gene encoding a precursor protein that contributes just nine amino acids to the mature cyclic peptide. This unusual mechanism appears to have arisen as a result of the evolutionary truncation of an ancestral α-defensin gene, as illustrated in Figure 13.5. Incorporation of a stop codon midway through the ancestral α-defensin gene leads to a peptide containing three,

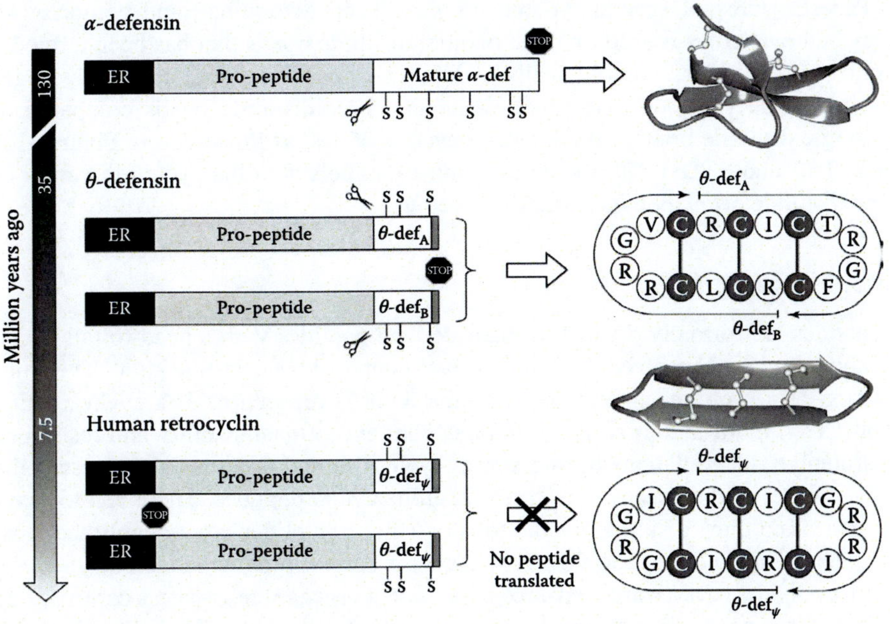

FIGURE 13.5 Evolution of θ-defensins by duplication of a truncated α-defensin gene. The ancestral α-defensin gene is thought to have evolved before placental mammals diverged from marsupials (~130 million years ago). α-defensins are expressed as preproproteins and undergo proteolytic processing to release the cationic, mature α-defensin (29–35 amino acids) from its anionic propeptide (~60 amino acids). Introduction of a premature stop codon within the mature α-defensin domain led to the emergence of θ-defensin genes in Old World monkeys (~35 million years ago). θ-defensins are the product of two peptide precursors (shown here as θ-def$_A$ and θ-def$_B$) that each contain three cysteine residues and are connected to generate an 18-residue, backbone-cyclized peptide. Currently, the mechanism by which θ-defensins are cyclized and the enzymes involved remain unknown. In humans, θ-defensin orthologues (retrocyclins) are transcribed but do not produce functional θ-defensin peptides due to the introduction of an additional premature stop codon in the ER signal (~7.5 million years ago). In each schematic diagram, cysteine residues located in the α-defensin or θ-defensin domain are indicated by showing the thiol sulfur atom (–s), and the structures illustrated are human neutrophil peptide-1 (α-defensin) and rhesus theta-defensin-1 (θ-defensin).

rather than the usual six, cysteine residues, which then pairs with another similar peptide to produce the θ-defensin. The mechanism of the double head-to-tail ligation to form the mature cyclic peptide is still unknown.

Despite their presence in some of our evolutionary cousins, including macaques, baboons, and bonobos, humans do not produce θ-defensin peptides; even though our genome encodes θ-defensin-like sequences, they are present as pseudogenes that contain a premature stop codon upstream of the propeptide segment, which prevents translation. Nonetheless, the chemical synthesis of some of these sequences led to molecules termed retrocyclins (Figure 13.5), which have potent anti-HIV activity.[35] It seems ironic that humans have evolutionally lost the ability to express these natural antiviral peptides, but perhaps synthetic chemistry will come to our aid as synthetic retrocyclins have been proposed as having potential for use as topical microbicides.[36]

The structures of several θ-defensins have been determined and comprise an elongated oval cross-braced by three parallel disulfide bonds that has been termed a cyclic cystine ladder.[37] Topologically, θ-defensins can be regarded as slightly larger cousins of the PawS-derived peptides, having three disulfide bonds compared to the single disulfide bond of the PawS-derived peptides, as illustrated in Figure 13.1 for SFTI-1 and RTD-1. They also have some parallels with the cyclotide family of three-disulfide cyclic peptides described below.

13.1.4 Cyclotides

Cyclotides are formally defined as plant-derived peptides with a head-to-tail cyclic backbone and six conserved cysteine residues connected in a cystine knot.[1] The combination of a cyclic backbone and a cystine knot is referred to as a cyclic cystine knot (CCK) motif and provides cyclotides with exceptional stability and resistance to proteolytic degradation. In one sense, cyclotides and θ-defensins represent the two extremes of the spectrum of three-disulfide cyclic peptides. There are 15 ways of connecting three disulfide bonds; topologically, cyclotides are arguably the most complex (being knotted) and θ-defensins are the simplest (being laddered).

All cyclotides have a conserved three-dimensional structure comprising a core made up of the cystine knot motif and a small associated β-sheet, with six backbone *loops* between successive cysteine residues effectively protruding from that core. Most structural studies on cyclotides have been performed using NMR spectroscopy, following the NMR structure determination of kalata B1 reported in 1995.[38] There are two main reasons for this: first, cyclotides give very high quality and well-dispersed NMR spectra and are particularly amenable to NMR structure determination; second, they are quite difficult to crystallize, as is the case for conotoxins and other small disulfide-rich peptides. The latter problem has recently been overcome with the use of racemic crystallography to determine the structures of kalata B1 and a range of other disulfide-rich peptides.[39] This new approach to structure determination of recalcitrant peptides requires the synthesis of the all-D (mirror image) form of the native peptide, which is then used to make a racemic mixture with the native (all-L) peptide. It works because symmetry considerations favor crystallization of racemic mixtures relative to pure enantiomers. We note that this approach provides a nice example of the power of modern chemical peptide synthesis methods, as production of all-D peptides is not possible using recombinant synthesis.

Cyclotides do not occur in every plant and, at the time of writing, have been discovered in 58 species from five major plant families, including the Violaceae, Rubiaceae, Cucurbitaceae, Fabaceae, and Solanaceae.[40] Of these plant families, cyclotides appear to be ubiquitous in species within the Violaceae,[11] but are sparsely distributed in the other families. An individual plant species may contain anywhere from a few cyclotides to several hundred cyclotides, with different suites of cyclotides found in different plant tissues (e.g., flowers, leaves, stems, and roots).[41] More than 300 cyclotide sequences have been characterized to date and are reported in CyBase,[42] but it is estimated that the family may comprise more than 50,000 members.[43] Cyclotides have been classified into three subfamilies, termed Möbius, bracelet, or trypsin inhibitors, based on structural and/or functional characteristics. Figure 13.1 shows an example of a prototypic cyclotide from the Möbius subfamily, kalata B1, and an example of a cyclotide from the trypsin inhibitor subfamily, MCoTI-II. These peptides have been two of the most extensively studied cyclotides.

The natural function of cyclotides is as host defense agents, principally against insect pests,[44] but they also are active against nematodes[45] and mollusks.[46] Naturally occurring cyclotides also have a range of other activities that have been discovered in various screening programs, including anti-HIV,[47,48] antimicrobial,[49] and antibarnacle[50] activities. Some cyclotides have mild hemolytic activity.[51] In general, it appears that these various activities derive from the ability of cyclotides to interact with and disrupt biological membranes.[52,53] The cyclic backbone appears to be important here as certain biological activities are lost in synthetic acyclic analogues of cyclotides. This is the case, for example, for the hemolytic activity[54] and anti-HIV[55] activities of kalata B1.

The *in planta* biosynthesis of cyclotides involves the ribosomal production of cyclotide precursor proteins that are subsequently processed by asparaginyl endopeptidase (AEP) enzymes to excise and cyclize the mature peptide products.[10,56–58] As described in Section 13.3, this *in planta* cyclization of cyclotides is remarkably efficient and, generally, no traces of linear precursors of cyclotides are found in cyclotide-producer plants. An exception is the *acyclotide* sequences that are found in some Poaceae plants, including maize and other grasses, which have similar sequences to cyclotides but lack a cyclic backbone. The first evidence for these linear cyclotide analogues was a cyclotide-like transcript identified in a study examining how maize responds to fungal infection.[59] This transcript contained a putative cyclotide domain, signposted by six conserved cysteine residues with characteristic spacing, but the C-terminal processing site was absent, indicating that it could not be cyclized by AEP. A broader search of nucleotide sequences from Poaceae plants revealed 22 additional acyclotide transcripts, predominantly in maize and wheat, but also in barley and rice.[60]

The first acyclotide peptide (as opposed to nucleic acid sequence) extracted from plants was violacin A. This peptide was isolated from the cyclotide-producing plant, *Viola odorata*, and does not undergo cyclization due to a point mutation that introduces a premature stop codon immediately before the C-terminal Asn residue.[61] Another example of a family of acyclotides, named the panitides, was discovered recently in *Panicum laxum* extracts and served to confirm the presence of linear cyclotide peptides in Poaceae plants.[62] Multiple acyclotides encoded by a single gene have also been identified in *Momordica angiosantha*, an African plant from the Cucurbitaceae family.[63]

A wide range of synthetic acyclotides have been made to explore structure–activity relationships and, in particular, to define the role of the cyclic backbone. As noted above, some interesting effects on activities are seen, and in general, linearization of the cyclic backbone results in a loss of activity despite the fact that the three-dimensional structures of these acyclic derivatives are typically not very different from the cyclic counterparts. The structural similarity suggests that the cyclic backbone is not necessarily a driving force in the folding of cyclotides, but by contrast, cyclic derivatives tend to be more stable than their acyclic counterparts. Cyclization may thus have evolved as a mechanism to improve both stability and bioactivity of peptides without diverging too far from existing structures.

13.1.5 Cyclic Three-Disulfide Conotoxins

The topology of many three-disulfide conotoxins from animals is identical to that of acyclotides from plants. By corollary, the topology of cyclic three-disulfide conotoxins is identical to that of cyclotides, as illustrated in Figure 13.6. Indeed, as noted earlier, it is this topological similarity, combined with the remarkable stability of cyclotides, that led us to propose the artificial synthesis of cyclic conotoxins to improve their biopharmaceutical properties.[64] The prototypic example for this work was for the 25-amino acid conotoxin MVIIA, which had earlier been developed as an analgesic (known as ziconotide, or Prialt) for neuropathic pain.[65,66]

Despite the topological similarity to cyclotides, in general, the synthetic production of three-disulfide cyclic conotoxins has proved to be relatively difficult compared with the facile synthesis of two-disulfide cyclic conotoxins, but has been achieved for MVIIA[23,64] and PVIIA[67] as well as for the P-superfamily conotoxins gm9a and bru9a.[5] The latter two peptides display the highest sequence homology with cyclotides of any known conotoxins to date. In the case of PVIIA, cyclization was achieved using the enzyme sortase A to link the termini of a slightly modified PVIIA precursor. This approach of using enzymes to achieve peptide cyclization is currently an active area of investigation in the literature, and we return to it later in this chapter.

13.1.6 Cyclic Four-Disulfide Chlorotoxins

Chlorotoxin, a cystine-stabilized α/β peptide isolated from the venom of the death stalker scorpion (*Leiurus quinquestriatus*), is the most complex disulfide-rich peptide cyclized to date. Chlorotoxin has attracted significant interest due to its ability to preferentially bind to cancer cells[68,69] and has been investigated for several applications, including tumor imaging and the targeted delivery of agents for chemotherapy or radiotherapy. For example, labeling chlorotoxin with the fluorescent tag Cy5.5 led to the development of *Tumor Paint*,[69] an imaging agent that is currently being trialed for glioma patients undergoing brain surgery. During the development of Tumor Paint, a cyclic version of chlorotoxin was designed by inserting a seven-residue linker between the peptide's termini.[70] Compared with wild-type chlorotoxin, the cyclic variant showed improved stability in human serum and favored selective labeling with Cy5.5 at just one of the three Lys residues present in chlorotoxin. This property is beneficial for reproducible production and subsequent characterization,

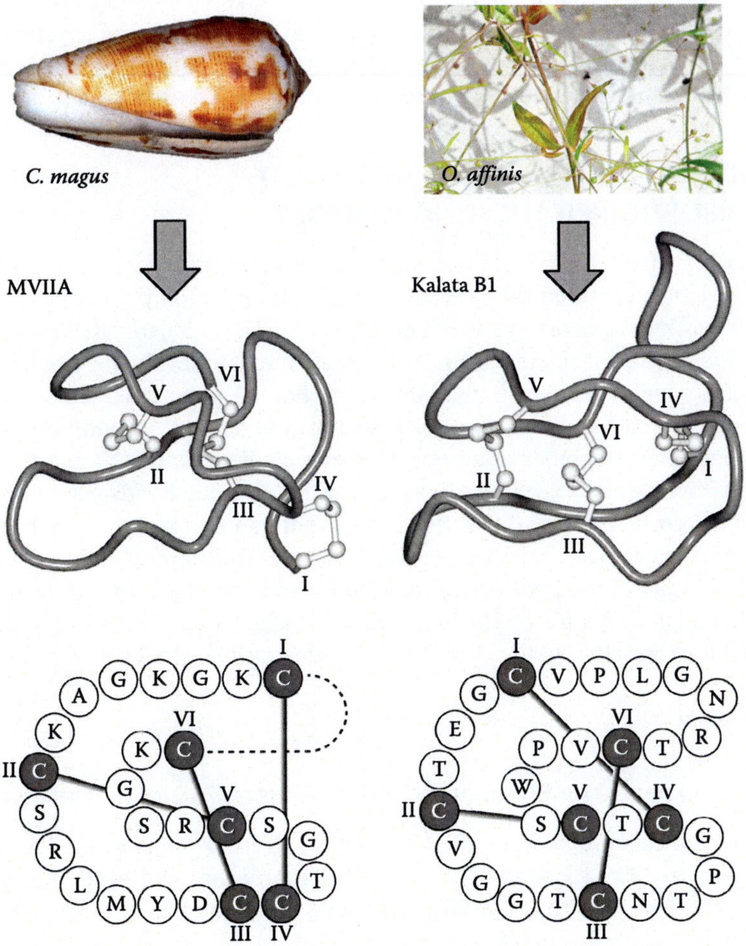

FIGURE 13.6 Structural similarity between the conotoxin MVIIA and the cyclotide kalata B1. MVIIA is a 25-amino acid peptide produced by the cone snail, *Conus magus*, which contains an inhibitor cystine knot (ICK) motif. Kalata B1 is a 29-amino acid cyclic peptide produced by the plant *Oldenlandia affinis*, which contains a cyclic cystine knot (CCK) motif. Both peptides have a similar fold, as revealed by NMR spectroscopy (for clarity, one structure from the ensemble of NMR structures is shown [PDB IDs: 1mvi and 1nb1]). Disulfide bonds are shown in ball-and-stick representation and cysteine residues are labeled using Roman numerals to illustrate the I–IV, II–V, III–VI disulfide connectivity. Schematic diagrams show the sequence of each peptide (viewed from the top of each structure, as indicated by the gray arrow), with cysteine residues highlighted by gray shading. The dashed line connecting the N- and C-termini in MVIIA represents an engineered linker that could be introduced to produce a cyclic conotoxin that resembles a cyclotide.

potentially providing an easier pathway to regulatory approval of a homogeneous product, compared with randomly conjugated chlorotoxin, with mono-, di-, and trilabeling of the dye on the three Lys residues. This example demonstrates that cyclization can sometimes have unexpected benefits. The cyclic and acyclic chlorotoxin–Cy5.5 conjugates had similar serum half-lives and equivalent tumor-binding activity when administered intravenously to mice.[70]

13.2 APPROACHES TO THE SYNTHESIS OF DISULFIDE-RICH CYCLIC PEPTIDES

The discovery of increasing numbers of cyclic peptides in nature, as described in Section 13.1, has captured the interest of chemists and biologists alike. Initially, very little was known about cyclic peptide biosynthesis, and early studies relied on isolating the peptide of interest from its natural source or developing new methods for producing cyclic peptides in the laboratory. For the latter case, although linear peptides and proteins are relatively straightforward to produce, disulfide-rich cyclic peptides presented a number of unique synthetic challenges: the N- and C-termini must be efficiently and selectively ligated, and the disulfide bonds need to be connected in the correct order so that the proper fold can be obtained. Over the past decade, these challenges have been solved in several different ways, giving rise to a diverse range of methods for synthesizing cyclic peptides. In this section, we explore the approaches for producing cyclic peptides using chemical synthesis or recombinant expression, with an emphasis on the various strategies for achieving backbone cyclization, specifically those that link the N- and C-termini by forming a peptide bond.

13.2.1 Strategies for Producing Cyclic Peptides: Synthetic Chemistry

Chemical synthesis provided the first breakthrough in attempts to produce cyclic peptides *in vitro*. The foundation for chemical approaches had been laid decades earlier by the development of solid-phase peptide synthesis,[71] enabling the stepwise assembly of linear peptides on porous resin. In this method, synthesis starts with the C-terminal residue of the target sequence, which is typically coupled to the resin as an *N*-protected amino acid. After washing the resin with solvent to remove excess reagents, the *N*-protecting group (*tert*-butoxycarbonyl [Boc] or 9-fluorenylmethoxycarbonyl [Fmoc]) is selectively removed using acid (Boc) or base (Fmoc) to unmask the amino terminus. Subsequent residues are added via successive coupling and deprotection reactions that involve incubating the resin with a solution of the next *N*-protected amino acid and chemical activators (coupling), followed by removal of the *N*-protecting group (deprotection). The main advantage of this method is its versatility, as virtually any sequence combination, including proteinogenic and nonproteinogenic amino acids, can be assembled from a single set of starting reagents. Once the linear precursor has been synthesized, the cyclic backbone can be formed by several different methods, including using coupling reagents, native chemical ligation (NCL), or specialized enzymes, as described in the following sections.

13.2.1.1 Cyclization Using Chemical Coupling Reagents

This approach for cyclizing synthetic peptides proceeds by chemically activating the peptide C-terminus using coupling reagents. In essence, this type of reaction parallels a routine coupling step during solid-phase peptide synthesis, with the exception that the cyclization reaction involves the peptide C-terminus rather than an amino acid in solution. Of the cyclic peptides described in this chapter, one of the most challenging to cyclize using conventional coupling reagents was kalata B1. Nonetheless, reasonable yields of cyclic kalata B1 were achieved by first folding the linear peptide to form the cystine knot, followed by a rapid cyclization reaction using 2-(1*H*-benzotriazol-1-yl)-1,1,3,3-tetramethyluronium (HBTU).[54] A similar approach has been used to synthesize RTD-1, using either Fmoc[34] or Boc chemistry,[72] with oxidative folding carried out first to form the cystine ladder, followed by backbone cyclization using different coupling reagents. Synthesis of SFTI-1 was found to be more straightforward and backbone cyclization could be carried out without first forming the peptide's intramolecular disulfide bond. Several combinations of coupling reagents have been used to cyclize SFTI-1, including methods conducted on resin[19] or in solution.[73,74] Although initial strategies for cyclization using coupling reagents were successful, peptides that required the removal of side chain–protecting groups prior to cyclization were prone to lower yields due to undesired side reactions.

Recently, the need to deprotect and fold the acyclic precursor peptide before cyclization has been overcome by using more efficient coupling reagents. A robust method for producing diverse cyclic peptides, including kalata B1 and MCoTI-II, was recently described, and involved Fmoc synthesis, followed by cyclization using 1-[bis(dimethylamino)-methylene]-1*H*-1,2,3-triazolo[4,5-b]pyridinium 3-oxid hexafluorophosphate (HATU) as the coupling reagent.[75] This strategy has also been used to synthesize 62 different cyclic hexapeptides, including peptides containing up to four N-methylated residues,[76] further demonstrating the efficiency of HATU as a cyclization reagent. Microwave-assisted synthesis methods have also been developed, and allow rapid cyclization of SFTI-1 and engineered variants using benzotriazole-1-yl-oxytripyrrolidinophosphonium hexafluorophosphate (PyBOP) and 1-hydroxy-7-azabenzotriazole (HOAt).[77] Indeed, these new strategies provide solutions to several of the challenges initially encountered when using coupling reagents for backbone cyclization. Concerns regarding racemization (epimerization) of the C-terminal residue during the cyclization reaction can be minimized by selecting a suitable point of cyclization, for example, Gly7-Gly8 in kalata B1,[54] or using new classes of coupling reagents.[75] Additionally, lower yields associated with the removal of side chain–protecting groups prior to cyclization[54] can be overcome by HATU-mediated cyclization, which has been applied to a diverse range of side chain–protected cyclic peptides.[75]

13.2.1.2 Native Chemical Ligation

Native chemical ligation (NCL) was originally introduced as an approach for ligating two peptide segments, the first having a C-terminal thioester and the second having an N-terminal cysteine residue.[78] The C-terminal thioester undergoes nucleophilic attack by the thiol group of the cysteine residue to link the two segments via a covalent bond (transthioesterification), and when this reaction involves the N-terminal

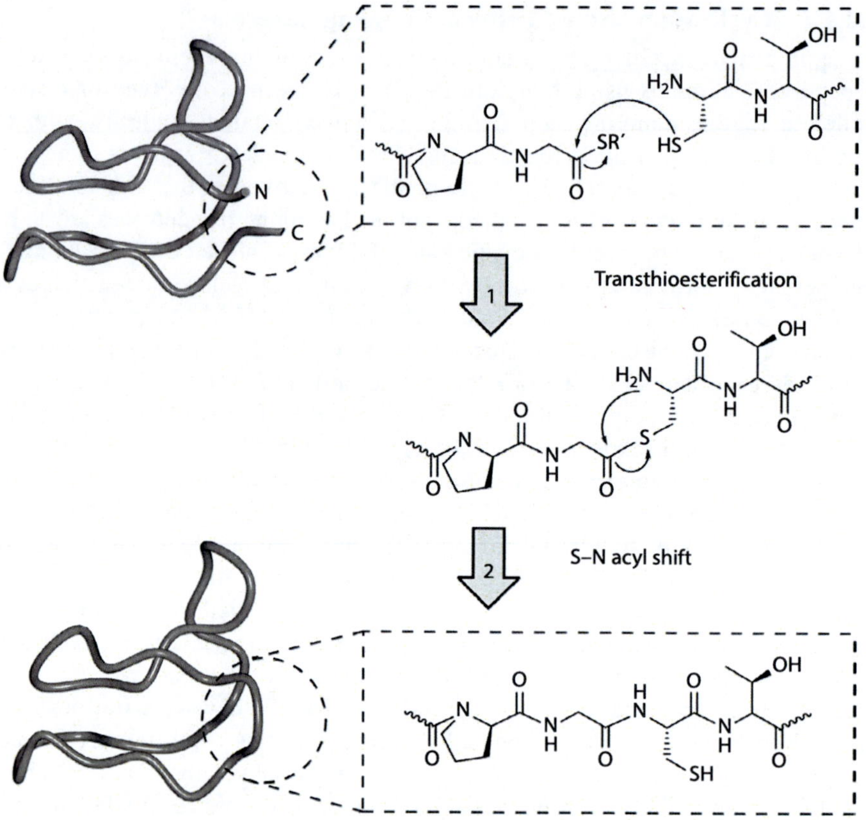

FIGURE 13.7 Cyclization of kalata B1 using intramolecular NCL. Reaction scheme showing the key steps in NCL that facilitate backbone cyclization of kalata B1. The cyclization point is between Gly18 and Cys19, as described in the first studies reporting the chemical synthesis of kalata B1. In the first step (1), nucleophilic attack by the thiol group (Cys19) on the C-terminal thioester (Gly18) generates a covalent bond between the N- and C-terminal residues (transthioesterification). In the second step (2), spontaneous rearrangement via an irreversible S–N acyl shift yields a native peptide bond at the point of cyclization.

cysteine residue, a spontaneous and irreversible S–N acyl shift occurs to generate a peptide bond (Figure 13.7). NCL can be adapted for backbone cyclization by placing the N-terminal cysteine residue and the C-terminal thioester in the same peptide chain. Indeed, disulfide-rich peptides have several inherent advantages for NCL-mediated cyclization. They are amenable to cyclization at more than one site by virtue of their continuous backbone and multiple cysteine residues. Additionally, internal cysteine residues have been suggested to promote efficient cyclization via successive (reversible) thiol–thiolactone exchanges, termed the thia zip mechanism,[79] although this concept has so far not been proven with direct experimental evidence.

NCL has been successfully applied to cyclic peptides containing up to four disulfide bonds. Initially, peptides intended to be cyclized by NCL could only be synthesized using Boc chemistry, as the C-terminal linker used to generate the thioester was susceptible to cleavage by piperidine (used in Fmoc synthesis during each N-deprotection cycle). Using similar approaches, kalata B1 was synthesized in two independent studies using Boc chemistry and NCL cyclization, followed by formation of the cystine knot using either one-step[54] or two-step[49] oxidative folding strategies. Compared with using coupling reagents (Section 13.2.1.1), NCL allows backbone cyclization under much milder conditions (phosphate buffer with a reducing agent, tris(2-carboxyethyl)phosphine) and provides higher yields of kalata B1 than strategies based on postfolding cyclization using standard coupling reagents.[54] SFTI-1[80] and RTD-1[81] have also been produced using a Boc/NCL strategy, with cyclization and oxidative folding achieved in a single step using ammonium bicarbonate as the buffer system. Additionally, cyclic variants of several animal toxins have been successfully synthesized using Boc chemistry followed by NCL, including the examples discussed earlier of chlorotoxin[70] and a range of conotoxins: α-MII,[26] χ-MrIA,[27] α-ImI,[31] α-Vc1.1,[2] as well as the P-superfamily conotoxins gm9a and bru9a.[5] Although Boc synthesis generally provides higher yields and purity than Fmoc chemistry, it requires the use of extremely hazardous chemicals, including hydrofluoric acid, and is not suitable for peptides containing acid-sensitive functional groups.

Some of the drawbacks of Boc synthesis prompted the development of Fmoc-compatible strategies for peptide thioester synthesis. In one approach, resin functionalized with a safety-catch sulfonamide linker was used to assemble the linear peptide.[82] The desired peptide thioester was subsequently generated by activating the linker using iodoacetonitrile, followed by cleavage using ethyl 3-mercaptopropionate, and this scheme has been successfully used to synthesize MCoTI-II[82] and RTD-1.[83] A C-terminal thioester can also be generated in solution, for example, by microwave-assisted synthesis using PyBOP and *p*-acetamidothiophenol, and this approach has been applied to synthesize kalata B1 and MCoTI-II.[84] Another strategy involves the production of thioester surrogates. Here, the peptide is assembled on resin that has been functionalized with a linker, for example, 2-(butyl-amino)ethanethiol, and is converted to the desired peptide thioester in solution via a tandem N–S and S–S acyl shift by incubation with sodium 2-mercaptoethanesulfonate (MESNa) in 0.1 M sodium phosphate (pH 3) at 40°C.[23] An SFTI-1 precursor produced by this method was efficiently cyclized under typical NCL reaction conditions.[23] Very recently, a strategy for converting a C-terminal cysteine residue into a thioester was reported, and involved incubating the peptide with MESNa in 0.1 M sodium phosphate (pH 5.8) at 55°C.[85] This method was used to synthesize cyclic SFTI-1 from linear precursors that were produced either by chemical synthesis or by expression in *Escherichia coli* as a SFTI–thioredoxin fusion protein.[85] However, the N–S acyl shift involving the C-terminal cysteine appears to be influenced by the adjacent sequence, being most efficient at Gly–Cys or His–Cys motifs, which presents a potential drawback.

NCL remains a widely used approach for producing cyclic peptides. Indeed, the absolute requirement for an N-terminal cysteine residue has not proven to be overly problematic, particularly as larger disulfide-rich peptides that might seem more challenging to cyclize typically contain a higher number of cysteine residues. Moreover, in cases where there is no convenient ligation site, a cysteine residue can be substituted in place of alanine to facilitate cyclization via NCL, after which the cysteine residue can be desulfurized to yield the desired alanine residue.[86] Overall, NCL offers a versatile method for backbone cyclization that has high chemoselectivity, and thus can be carried out on fully deprotected peptides without the risk of side reactions.

13.2.1.3 Chemoenzymatic Cyclization

Synthetic peptides can also be cyclized using enzymes—a process that mimics the natural biosynthetic pathway of many cyclic peptides to some extent. The key benefit of this approach is that efficient cyclization does not require reactive chemicals or modified termini, but is guided by consensus recognition sequences that are specific for particular enzyme classes. Initially, chemoenzymatic cyclization was explored using the serine protease inhibitors SFTI-1 and MCoTI-II, as these peptides have the intrinsic ability to be religated by their target proteases. Both SFTI-1 and MCoTI-II are standard mechanism (or Laskowski) inhibitors, a large and structurally diverse family of reversible protease inhibitors that have a shared mode of action involving cleavage and resynthesis of the inhibitor's reactive site bond.[87] Consequently, although serine proteases, including trypsin, are typically associated with cleaving substrates, Laskowski inhibitors represent an exception where trypsin can act to synthesize a peptide bond.[88]

Trypsin-mediated cyclization of SFTI-1 was performed using a peptide precursor that was produced by Fmoc synthesis (with the backbone open between Ser6 and Lys5; Figure 13.8) and folded to form the intramolecular disulfide bond.[89] Incubating acyclic SFTI-1[6,5] with trypsin (1:1 ratio) resulted in ligation of the Lys5–Ser6 peptide bond, with the reaction reaching an equilibrium of 9:1 in favor of cyclic SFTI-1.[89] Enzymatic cyclization of MCoTI-II was also performed on a folded, acyclic peptide that was synthesized using Fmoc chemistry.[90] Here, trypsin immobilized on Sepharose beads was used for the cyclization reaction to avoid contamination with trypsin autodigestion peptides, and ligation of the Lys-Ile(P1-P1′) peptide bond was observed within 15 min. Additionally, substituting Lys with Phe allowed efficient cyclization using chymotrypsin.[90] An important consideration for trypsin-mediated cyclization is that the reaction product is a high-affinity inhibitor of the cyclizing enzyme, and thus, the reaction is often performed using equimolar concentrations of enzyme and peptide. Additionally, the method is essentially limited to cyclic peptides that act as reversible serine protease inhibitors.

More recently, new approaches have been developed that allow the application of chemoenzymatic cyclization to a broader range of peptides. The transpeptidase sortase A is expressed in many Gram-positive bacteria and has been investigated for various protein engineering applications, including peptide head-to-tail cyclization.[91] In bacteria, sortase A is primarily involved in attaching various surface proteins to the bacterial cell wall.[92] This process involves recognition of a five-residue sequence, LP(X)T↓G, at the C-terminus of a target protein that is cleaved by sortase A, generating a covalent enzyme–substrate complex.[93] Subsequent nucleophilic attack by a second peptide segment containing multiple N-terminal Gly residues results in cleavage of the enzyme–substrate complex and ligation of the two fragments. This mechanism can be adapted to cyclize peptides and proteins by engineering the target peptide sequence to contain the required N-terminal poly-Gly and C-terminal LP(X)TG motifs.[91] Using this strategy, linear kalata B1 and SFTI-1 were cyclized *in vitro* using recombinant *Staphylococcus aureus* sortase A,[94] and a cyclic conotoxin, PVIIA, was recently produced using an engineered sortase A variant, SrtA5°.[67] This reaction has

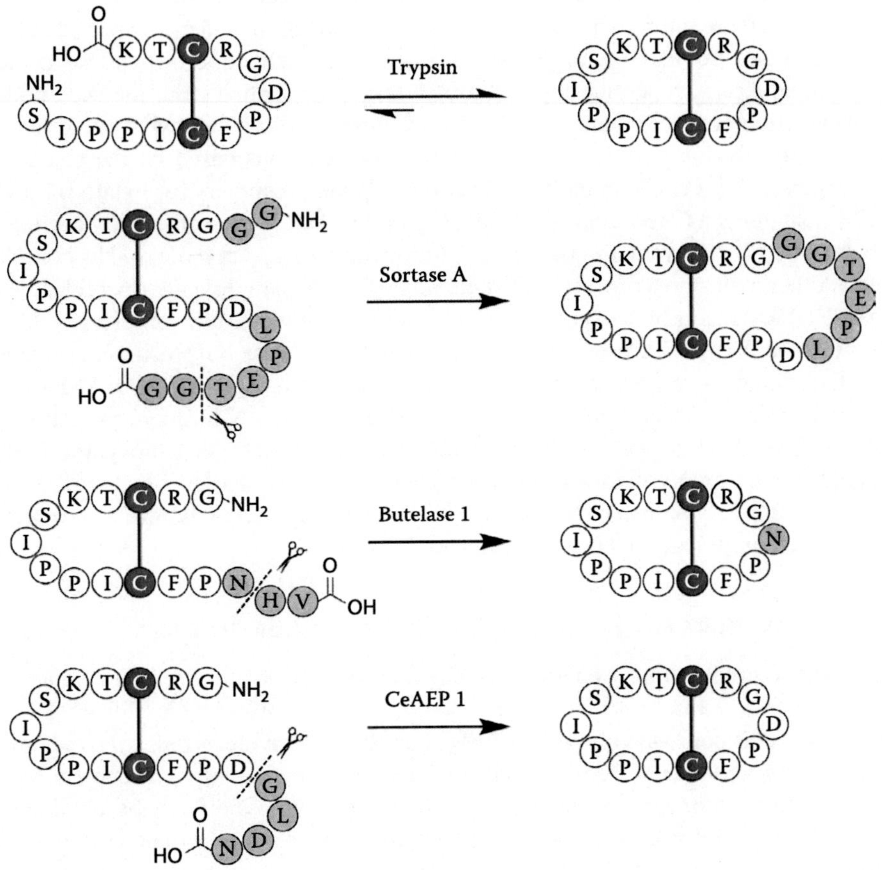

FIGURE 13.8 Chemoenzymatic cyclization of SFTI-1 using different proteases and synthetic precursors. The acyclic SFTI-1 precursor required for each enzyme (trypsin, sortase A, butelase 1, or *Canavalia ensiformis* asparaginyl endopeptidase 1 [CeAEP1]) is shown on the left and the resulting cyclic product is shown on the right. The N- and C-termini are labeled for each precursor peptide, any modified residues are shaded light gray, and protease cleavage sites are indicated using scissors. Trypsin cyclizes SFTI-1 between Lys5 and Ser6 (the reactive site bond of cyclic SFTI-1). This reaction is part of SFTI-1's intrinsic mode of action, whereby target proteases cleave and then religate the Lys–Ser bond. Cyclization of SFTI-1 by sortase A requires an engineered precursor and involves cleavage after the Thr residue in the sortase recognition site (LP[X]TG), followed by a nucleophilic attack involving the N-terminal poly-Gly motif. Note that cyclization using sortase A leaves an additional six residues (LPETGG) in the cyclic product (shaded gray). Butelase 1 cyclizes SFTI-1 between Gly1 and Asn14 (substituted in place of Asp14) following cleavage of the C-terminal dipeptide His-Val. CeAEP1 cyclizes SFTI-1 in a manner that resembles SFTI-1 cyclization *in planta* and involves cleavage of the C-terminal motif, Gly–Leu–Asp–Asn, followed by backbone ligation between Gly1 and Asp14.

the advantage of high specificity, but also has the potential drawback that the sortase recognition sequence is retained in the cyclic product (Figure 13.8).

A significant breakthrough in chemoenzymatic cyclization, and cyclic peptide biosynthesis in general, came with the isolation and characterization of butelase 1, the first reported Asn/Asp peptide ligase.[10] Butelase 1 was purified from the seed pods of the cyclotide-producer plant butterfly pea (*Clitoria ternatea*) and was found to be capable of catalyzing both ligation and cyclization reactions based on the recognition sequence: N↓HV. Chemically synthesized peptide precursors for kalata B1 and SFTI-1 containing a C-terminal HV motif (Figure 13.8) were each efficiently cyclized by butelase 1 (k_{cat}/K_M values ~ 10^4 M^{-1} s^{-1}) following cleavage at the Asn–His bond.[10]

In another major breakthrough, *Oldenlandia affinis* asparaginyl endopeptidase 1b (*Oa*AEP1_b) was recombinantly expressed in *E. coli* and shown to have efficient cyclase activity.[57] *Oa*AEP1_b processed and cyclized a synthetic kalata B1 precursor containing a C-terminal seven-residue propeptide (cleavage sequence: TRN↓GLPSLAA), but was unable to process kalata B1 that also contained the N-terminal propeptide, LQLK.[57] This finding was consistent with earlier hypotheses that biosynthesis of kalata B1 in *O. affinis* involves several processing enzymes (described further in Section 13.2.2.3),[56] and that N-terminal truncation must occur before C-terminal transpeptidation to lead to the cyclic product.

13.2.2 Strategies for Producing Cyclic Peptides: Biosynthesis

In addition to production by chemical or chemoenzymatic synthesis, cyclic peptides are amenable to expression in bacteria, yeast, or plants using recombinant DNA technology. This approach allows the target peptide to be synthesized, cyclized, and folded *in vivo*, and thus overcomes some of the challenges that might be encountered using chemical synthesis. Additionally, recombinant expression can be combined with molecular biology techniques routinely used to generate sequence diversity, enabling the production of libraries of cyclic peptides for screening against a wide range of targets. Although it is relatively straightforward to express linear peptides and proteins in a range of different organisms, producing cyclic peptides is a more complex task. Despite this complication, several methods for expressing cyclic peptides have been developed by adapting biosynthetic strategies from nature, including harnessing the cellular machinery of organisms that produce endogenous cyclic peptides (plants, for example) or using specialized protein domains, such as engineered inteins, that can compensate for the absence of cyclizing enzymes.

13.2.2.1 Expressed Protein Ligation

The first method introduced for cyclizing recombinant peptides was expressed protein ligation.[95] This strategy derives from protein splicing, a self-catalyzed, post-translational processing mechanism first identified in yeast, whereby an internal segment (intein) is excised from within a protein and the two flanking segments (exteins) are subsequently ligated.[96,97] In expressed protein ligation, the sequence of the target cyclic peptide is arranged to start at a cysteine residue and is expressed as a fusion protein with an N-terminal methionine residue (required for translation) and a C-terminal intein segment (Figure 13.9). Cyclization of the peptide–intein fusion

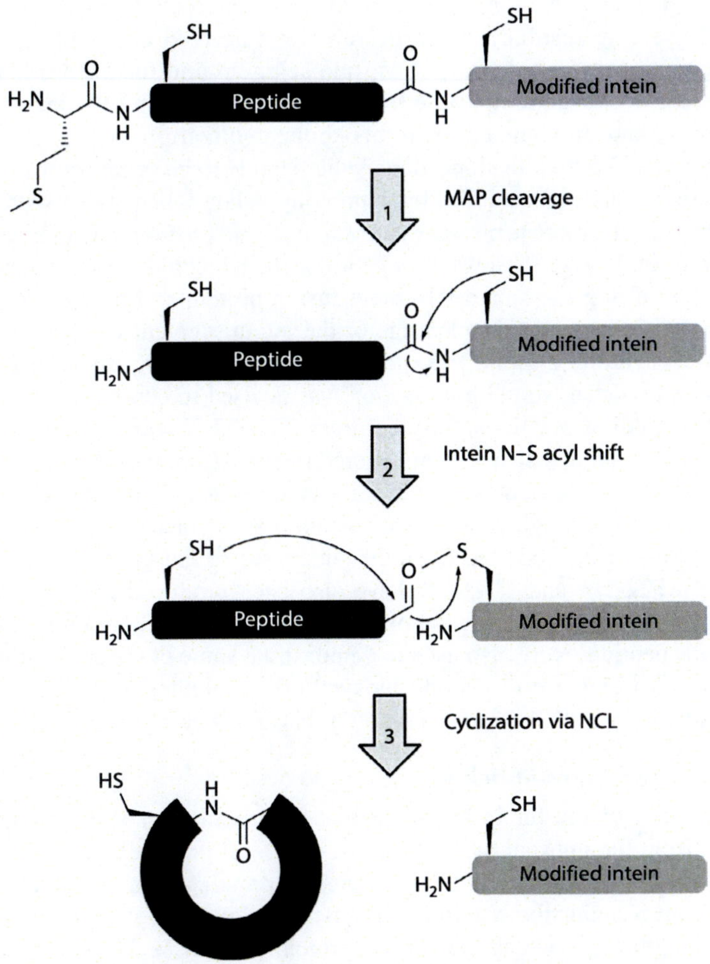

FIGURE 13.9 Biosynthesis of cyclic peptides using expressed protein ligation. The upper graphic shows the typical arrangement of a fusion protein that can be cyclized using expressed protein ligation, with the target peptide sequence flanked by a Met residue at the N-terminus and an engineered intein segment at the C-terminus. The key steps in the cyclization mechanism are shown below. First, the N-terminal Met residue is cleaved by Met aminopeptidase (MAP) (1) to expose the first residue of the target peptide (cysteine) at the new N-terminus. Second, the modified intein undergoes a spontaneous, reversible N–S acyl shift (2) to generate a thioester bond that links the peptide and intein segments. Third, the thioester bond is subjected to nucleophilic attack by the thiol group of the N-terminal cysteine residue, leading to cleavage of the intein segment and head-to-tail cyclization of the target peptide (3). This final step involves transthioesterication followed by an irreversible S–N acyl shift, as illustrated in Figure 13.7 for native chemical ligation.

protein occurs in a manner that resembles NCL. First, the fusion protein is processed by Met aminopeptidase to remove the N-terminal methionine residue and expose the obligatory cysteine residue for NCL. Next, the engineered intein (also containing an N-terminal cysteine residue) undergoes a reversible N–S acyl shift to generate a thioester bond between the target peptide segment and the C-terminal intein. This precursor is subsequently able to undergo NCL, leading to cleavage of the intein segment and, in some cases, to backbone cyclization.

Kalata B1 was the first disulfide-rich cyclic peptide to be produced by expressed protein ligation. This was achieved by expressing kalata B1 in-frame with an engineered variant of the *Saccharomyces cerevisiae* (*S. cerevisiae*) vacuolar membrane ATPase intein in *E. coli*.[98] However, cyclic kalata B1 could only be successfully obtained by purifying the kalata B1–intein fusion protein and then performing the cyclization reaction *in vitro*, as cleavage of the intein segment *in vivo* did not lead to efficient cyclization. Despite this, peptides cyclized *in vitro* underwent efficient folding, and a two-step, single-pot reaction was devised to combine the cyclization and oxidative folding reactions.[98] By contrast, MCoTI-II was able to be cyclized and folded *in vivo* using a different engineered intein (*Mycobacterium xenopi* gyrase A), with purification carried out by affinity capture using immobilized trypsin.[99] Additionally, using an engineered *E. coli* strain that is amenable to disulfide bond formation (Origami 2, DE3) led to further increases in the yield of cyclic, folded MCoTI-II.[99] SFTI-1[100] and RTD-1[101] have also been produced by intein-mediated protein ligation, with cyclization and oxidative folding carried out both *in vitro* and *in vivo*. These peptides were also used to demonstrate some of the applications made available by bacterial cyclic peptide biosynthesis, including genetically encoded cyclic peptide libraries[100] and incorporation of NMR-active isotopes.[101]

13.2.2.2 Protein Trans-Splicing

Another iteration of the intein-based method is protein trans-splicing. This strategy derives from the unusual biosynthesis of the DnaE protein in *Synechocystis* sp. PCC6803, which involves ligation of two separate proteins via a split intein.[102] Here, the first protein contains the N-terminal segment of DnaE followed by a 123-amino acid intein fragment, whereas the second protein contains a 36-amino acid intein fragment followed by the C-terminal DnaE segment. Association of the two intein fragments results in the formation of a functional intein unit, which undergoes self-processing and leads to the excision of the intein domain followed by ligation of the two DnaE extein segments.[102] Split inteins can be adapted for backbone cyclization by expressing a single protein construct, where the target peptide sequence is inserted between the N-terminal (I_N) and C-terminal (I_C) intein subunits. However, to allow the split intein to self-associate, the intein subunits are arranged in reverse order (I_C–target–I_N), which allows cyclization of the target peptide when the intein domain is excised.[103]

Protein trans-splicing has been used to produce several disulfide-rich cyclic peptides, including MCoTI-I.[104] This peptide was expressed in *E. coli* Origami 2 (DE3) cells using the *Nostoc puntiforme* PCC73102 (*Npu*) DnaE split intein, allowing both the cyclization and folding steps to be performed *in vivo*. Additionally, an engineered MCoTI-I variant was expressed where Asp14 was replaced with *p*-azidophenylalanine (encoded by an amber stop codon), enabling the cyclic peptide to be labeled with a fluorescent tag using copper-free click chemistry.[104] Using a similar approach, cyclic SFTI-1 was expressed in *E. coli*,[105] with the reported yield being more than 10-fold higher compared to conventional expressed protein ligation. Very recently, cyclic MCoTI-I was expressed in *S. cerevisiae* via protein trans-splicing based on the *Npu* DnaE split intein.[106] Moreover, the authors demonstrated that protein trans-splicing could be used to express engineered cyclic peptides *in vivo* by producing a variant of MCoTI-I that was able to inhibit cytotoxicity induced by co-overexpression of α-synuclein.[106] This development highlights the great potential of intein-based strategies, particularly protein trans-splicing, for generating genetically encoded libraries of cyclic peptides that can be screened for novel biological activities.

13.2.2.3 In Planta Cyclization

Using plants to produce naturally occurring and engineered cyclic peptides has a number of exciting applications, including large-scale biosynthesis of pharmaceutical and agricultural commodities. Accordingly, there is growing interest in characterizing the endogenous processing pathways for cyclic peptide biosynthesis in plants and the requirements for efficient cyclization. In *O. affinis*, kalata B1 is produced as a precursor protein, Oak1, and is liberated following proteolytic cleavage at N-terminal and C-terminal processing sites.[44] Transforming *Oak1* cDNA into two plant species that do not typically express cyclic peptides, *Arabidopsis thaliana* and *Nicotiana tabacum*, yielded a modest amount of cyclic kalata B1, indicating that these plants are capable of producing cyclic peptides, although linear, misprocessed peptides were more abundant.[56] Additionally, examining the processing and cyclization of several engineered *Oak1* constructs demonstrated the important role of four residues flanking the C-terminal processing site: Asn29 and the tripeptide segment Gly30–Leu31–Pro32.[56] These findings were confirmed in a later study that studied mutations at the N- and C-terminal processing sites and identified a series of substitutions that could be made without impairing the production of cyclic kalata B1, including an Arg28-to-Ala substitution that provided improved yields of the cyclic product.[107]

Like kalata B1, SFTI-1 is excised from within a larger precursor protein, in this case PawS1 in sunflower (*Helianthus annuus*) seeds. Expression of PawS1 in *Arabidopsis thaliana* under the control of the seed-targeting *OLEOSIN* promoter successfully yielded cyclic, oxidized SFTI-1, although acyclic, oxidized SFTI-1 was more abundant.[20] Additionally, transforming the *OLEOSIN–PawS1* construct into *Arabidopsis* lines that carry mutations in all four *aep* genes demonstrated that one or more AEPs were essential for SFTI-1 processing and cyclization. In a following study, sunflower seed extract was used as the enzyme source for cyclizing a synthetic SFTI-1 precursor that contained a C-terminal propeptide (GLDN).[108] This experiment demonstrated that endogenous AEPs in sunflower seeds are capable of cyclizing SFTI-1, but, interestingly, recombinant *Helianthus annuus* AEP1 (HaAEP1), the AEP transcript most abundantly expressed in sunflower seeds, was unable to cleave or cyclize the SFTI-GLDN precursor, indicating the involvement of a different AEP.[108]

A different strategy that plants use to produce cyclic trypsin inhibitors was revealed by a study on the biosynthesis of MCoTI-II.[3] Here, MCoTI-II was shown to be expressed as a multidomain precursor protein, TIPTOP, that contains at least four MCoTI domains that appear to encode cyclic peptides and terminates with an acyclic trypsin inhibitor domain.[3] As with SFTI-1, TIPTOP processing *in planta* was examined by expressing an *OLEOSIN–TIPTOP* construct in wild-type and *aep*-null *Arabidopsis* lines. Each trypsin inhibitor domain was correctly processed and cyclized in wild-type *Arabidopsis*, whereas in *aep* null plants, only the acyclic trypsin inhibitor domain (MCoTI-V) was released from the TIPTOP precursor and none of the cyclic MCoTI domains were processed.[3]

Identifying residues in the cyclic peptide domain and flanking regions that are critical for correct processing and cyclization has provided a more complete understanding of how naturally occurring cyclic peptides are produced in plants. These insights are also important for informing engineering studies, particularly where the eventual aim is to express the modified cyclic peptide in plants, as they reveal which residues are not tolerant to substitution and must be preserved in engineered variants. Strategies have also been developed for expressing cyclic peptides using suspension cell cultures, including *O. affinis* cultures for producing kalata B1,[109] and photobioreactors for large-scale production, with reported yields of kalata B1 reaching 21 mg/day.[110] These techniques offer a viable alternative to cultivation of transgenic plants, and may allow simpler processing to extract and purify the desired cyclic peptide.

13.3 CYCLIC PEPTIDES AS ENGINEERING TEMPLATES FOR DESIGNING NEW CHEMICAL TOOLS

The inherent diversity in structure and function displayed by naturally occurring cyclic peptides makes them excellent templates for designing a variety of new chemical tools. In the past few years, the number and range of chemical tools that have been engineered from naturally occurring cyclic peptides have expanded rapidly, with most embracing one of two engineering philosophies: to harness the existing biological activity of the template cyclic peptide and redirect it to a new target, or to equip the cyclic peptide with a novel biological function by introducing a new functional epitope. Both concepts have been successfully applied, yielding new molecules for specific protein targets as well as compound libraries that can be produced by chemical or biological synthesis.

13.3.1 Engineered Protease Inhibitors

A common objective in natural product–based engineering studies is to harness the built-in activity of a given molecule and redirect it to a related target. For cyclic peptides, this concept has been widely applied to serine protease inhibitors (Figure 13.10), with SFTI-1 being the most prevalent design template. In these studies, several different inhibitors for matriptase, a protease implicated in various cancers, have been produced by substituting various residues of SFTI-1 with proteinogenic or nonproteinogenic amino acids.[111–113] Inhibitors for another cancer-related protease, kallikrein-related peptidase 4 (KLK4), were developed using a substrate-guided approach that involved substituting the P4, P2, and P1 residues of SFTI-1 based on the sequence of an optimal synthetic peptide substrate.[114] Subsequent optimization

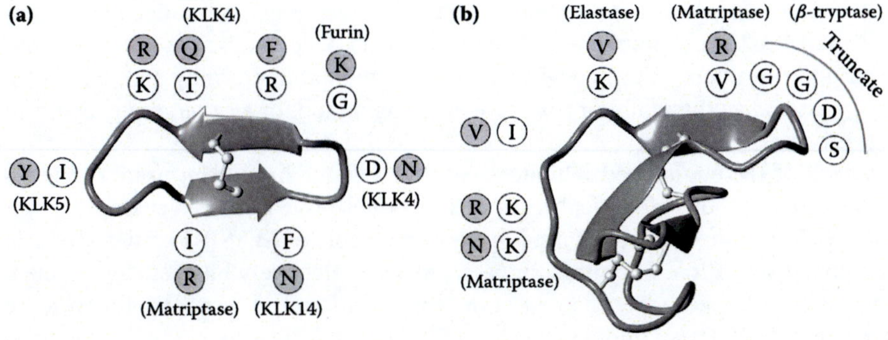

FIGURE 13.10 Engineered inhibitors for diverse protease targets based on SFTI-1 and MCoTI-II. Examples of amino acid substitutions reported in different engineering studies are displayed on structures of SFTI-1 **(a)** and MCoTI-II **(b)**. Substituted residues are shown in gray-shaded circles, and the corresponding residues in the naturally occurring inhibitor are shown in an unshaded circle at their position in the structure. The protease target for each substitution is listed within brackets.

of the inhibitor's intramolecular hydrogen bond network by screening further substitutions *in silico* yielded a second-generation inhibitor that displayed subnanomolar affinity for KLK4.[115] The substrate-guided approach has also been combined with an inhibitor-based library screen for optimizing the P2′ residue to engineer inhibitors for KLK proteases implicated in skin diseases, including KLK5,[116] KLK7,[117] and KLK14.[118] Using phage display, SFTI-based inhibitors were developed for mannose-binding lectin-associated serine protease (MASP)-1 and MASP-2, and shown to selectively block complement activation via the lectin pathway.[119] Recent engineered inhibitors for the 20S proteasome,[120] matriptase-2,[121] and furin[122] have also been described, highlighting the versatility of SFTI-1 as a design template.

MCoTI-II has also been used as a design template for engineering serine protease inhibitors. A potent β-tryptase inhibitor was developed by deleting a four-residue segment (SDGG) from loop 6 of MCoTI-II.[123] Additionally, replacing the P1 Lys residue with Val yielded an effective inhibitor of leukocyte elastase.[123] A second inhibitor for β-tryptase was designed by producing an MCoTI-II/EETI-II chimera. Here, the N-terminal segment of MCoTI-II was merged with the C-terminal segment of EETI-II, and a cyclic variant that showed high affinity for β-tryptase (K_i = 1 nM) was engineered by adding a sequence containing the tripeptide motif, KKV, adjacent to the P4 Val residue.[124] MCoTI-II has also been used to design inhibitors for matriptase. In one study, point mutations were screened in chemically synthesized inhibitor variants, which identified a P4 Val-to-Arg substitution that improved the inhibitor's activity against matriptase by 10-fold (K_i = 290 pM).[113] A second high-affinity matriptase inhibitor (K_i = 830 pM) was produced using yeast surface display.[125] Here, libraries of MCoTI-II variants were generated by fully randomizing the P1′–P4′ residues (loop 1) and partially randomizing a series of residues in loops 2, 5, and 6, with the lead variant showing a 43-fold selectivity over trypsin.[125] Very recently, a selective Factor XIIa inhibitor was developed by substituting five residues in MCoTI-II based on data from substrate screening, inhibitor screening, and molecular modeling.[126] As MCoTI-II is appreciably larger than SFTI-1, it has the advantage of containing additional loops that can be used to optimize inhibitor affinity and selectivity, or to functionalize the inhibitor for use as an imaging tool or as a targeting agent for drug delivery.

Several of the engineered inhibitors described above have been used as chemical tools to study the biological roles of individual proteases. For example, engineered SFTI variants have been used to explore the role of MASP-1 and MASP-2 in complement activation. Inhibitors for each protease selectively blocked complement activation via the lectin pathway *in vitro*, with inhibition of MASP-1 found to be more effective.[119] These findings were confirmed in a following study, where engineered inhibitors with higher selectivity were used to demonstrate that MASP-1 activates MASP-2 and to subsequently define the contribution of each protease toward activating complement components C2 and C4.[127] An engineered SFTI variant has also been used to study the contribution of KLK4 to multicellular aggregation and chemoresistance in ovarian cancer cells. Here, inhibition of KLK4 was shown to reduce cellular aggregation in three-dimensional cultures and increase the sensitivity of ovarian cancer cells to paclitaxel treatment.[128] Additionally, SFTI-based inhibitors for KLK5, KLK7, and KLK14 have been used to explore the direct contribution of each protease to corneocyte shedding (desquamation) in human stratum corneum. In an *ex vivo* desquamation assay, inhibition of KLK7 was shown to almost completely prevent corneocyte shedding, demonstrating that KLK7 activity is nonredundant in this particular assay, whereas inhibition of KLK5 partially blocked corneocyte shedding.[117] As engineered protease inhibitors become more widely used for exploring biological pathways, it is important that these molecules show sufficient selectivity, not just potency, in order to correctly attribute any effects that result from inhibitor treatment to the target protease.

13.3.2 Ultrastable Scaffolds for Displaying Bioactive Peptide Sequences

Cyclic peptides have also been used in engineering studies as ultrastable scaffolds for displaying bioactive peptide epitopes. Whereas linear peptides tend to be degraded by proteases and are generally thought to have limited potential for use in biological assays or as therapeutics, in many cases, this drawback can be overcome by grafting the peptide's bioactive motif into a cyclic peptide scaffold, as illustrated in Figure 13.11. This strategy has led to a number of exciting developments in recent years, and a series of cyclic peptides have been produced that have applications in treatment of a wide range of diseases, including cancer, chronic pain, neurodegeneration, and obesity.

Grafted cyclic peptides have been designed for several protein targets of interest in cancer, including targets that reside intracellularly. As an example of the latter type of target, the activity of the tumor suppressive transcription factor, p53, is negatively regulated by two proteins, Hdm2 and HdmX.[129] Accordingly, blocking the interaction between Hdm2 or HdmX and p53 has emerged as a potential strategy for inducing cell cycle arrest and apoptosis in tumor cells. Recently, cyclic peptides that bind to Hdm2 and HdmX with high affinity were produced by grafting a 15-residue

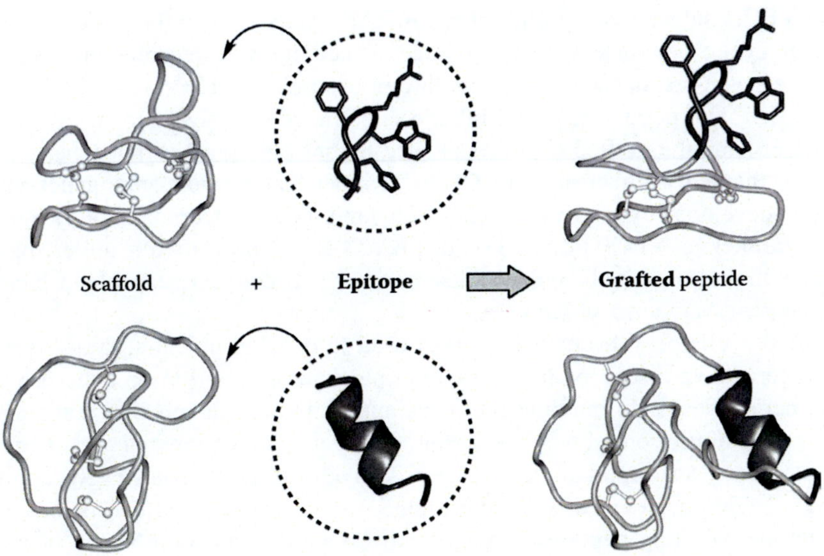

FIGURE 13.11 Grafting strategy for inserting a bioactive epitope into a cyclic peptide scaffold. The cyclic peptide scaffold (light gray) is shown on the left and the bioactive epitope is encircled by a dashed line (black). In each case, the grafted peptide was designed by inserting the sequence of the bioactive epitope into one of the exposed loops of the cyclic peptide scaffold (as indicated by the arrow), and the resulting scaffold–epitope chimera (right panels) was subsequently produced by chemical synthesis or recombinant expression. Examples shown are the melanocortin receptor binding sequence (HFRW) grafted into kalata B1 (upper panels) and a helical Hdm2/HdmX-binding peptide grafted into MCoTI-I (lower panels).

helical peptide into loop 6 of MCoTI-I that targets the p53-binding domain of Hdm2 and HdmX.[9] These peptides were successfully produced by chemical synthesis and expressed protein ligation, and were found to decrease the viability of Hdm2/HdmX-expressing cancer cell lines and suppress the growth of tumor xenografts *in vivo*.[9] Cyclic peptides have also been used to target the BCR-ABL tyrosine kinase, a key intracellular target in chronic myeloid leukemia. In a recent study, the sequence of a linear ABL kinase peptide substrate (Abltide) was grafted into loop 1 and/or loop 6 of MCoTI-II, with the double-grafted peptide showing the highest inhibitory activity.[130] In another study, MCoTI-II was used to engineer cyclic peptide antagonists for SET, a protein that blocks the tumor suppressive activity of protein phosphatase 2A. This was achieved by grafting peptide sequences derived from apolipoprotein E, known as COG peptides, into loop 6 of MCoTI-II, which yielded stable, cell-penetrating peptides that induced cell death in a SET-overexpressing cancer cell line.[131]

Protein–protein interactions that modulate angiogenesis have also been the focus of engineering studies using cyclic peptides. Cyclic peptides that block the interaction between vascular endothelial growth factor receptor-2 (VEGF-R2) and VEGF were developed by grafting the hexapeptide sequence, RRKRRR, into various loops of kalata B1.[132] The most effective VEGF-R2 antagonist was produced by grafting the poly-Arg epitope into loop 3, and the resulting kalata B1 variant was shown to inhibit VEGF-induced cell proliferation in BAF3 cells (IC_{50} = 12 μM). A second series of cyclic peptides with antiangiogenic activity was recently produced by grafting a seven-residue sequence from thrombospondin-1 (TSP-1) into SFTI-1 and MCoTI-II.[133] This peptide sequence inhibits endothelial cell migration via interaction with CD36, and the grafted peptides showed higher biological activity than TSP-1 as well as markedly higher stability. Cyclic peptides that promote angiogenesis have also been developed by grafting epitopes from laminin $\alpha 1$, osteopontin, or a synthetic peptide mimicking VEGF into SFTI-1 or MCoTI-II.[134] The grafted peptides showed markedly higher stability than their linear counterparts and were capable of promoting angiogenesis *in vitro* and *in vivo*.

Grafted cyclic peptides have also been used to develop potential drug leads for neurodegenerative disorders, including multiple sclerosis, Parkinson's disease, and Alzheimer's disease. In multiple sclerosis, autoimmune inflammation leads to the deterioration of the central nervous system due to demyelination and axon damage. In this process, an important self-antigen is myelin oligodendrocyte glycoprotein (MOG),[135] with the segment spanning Met35–Lys55 representing the major bioactive epitope. Grafting peptides derived from MOG_{35-55} into loop 5 or 6 of kalata B1 yielded a series of engineered variants that protected mice from central nervous system inflammation and axonal damage to varying degrees in an animal model of experimental autoimmune encephalomyelitis.[4] In Parkinson's disease, accumulation of misfolded α-synuclein leads to the formation of neurotoxic aggregates (Lewy bodies) in the brain. Recently, cyclic peptides that disrupt α-synuclein aggregation were designed by grafting an eight-residue epitope into loop 6 of MCoTI-I.[106] The resulting engineered variant was expressed in *S. cerevisiae* using a split intein trans-splicing approach and shown to block α-synuclein-mediated cytotoxicity in a yeast synucleopathy model.[106] Aggregation of tau, a microtubule-associated protein, is also a major feature of neurodegenerative disorders, including Alzheimer's disease.[136] A six-residue segment based on residues 306–311 of tau, Ac–VQIVYK–NH_2, has the ability to form fibrils reminiscent of full-length tau, and grafting this sequence into SFTI-1 yielded a series of engineered peptides that interrupted the formation of hexapeptide tau fibrils *in vitro*.[137]

Yet more applications that have been investigated using grafted cyclic peptides include chronic pain, obesity, and HIV infection. For example, blocking the activation of the bradykinin B_1 receptor has been identified as a potential therapeutic strategy for treating chronic pain.[138] To develop B_1 receptor antagonists that show high *in vivo* stability, epitopes derived from kallidin or kinestatin were grafted into loop 6 of kalata B1.[139] These peptides selectively inhibited the activation of the B_1 receptor *in vitro* and effectively relieved pain in an animal model following intraperitoneal injection or oral delivery. In the case of obesity, the melanocortin-4 receptor (MC4R) has emerged as a key modulator of energy homeostasis.[140] MC4R activation has been shown to decrease food intake and increase energy expenditure, prompting efforts to design selective MC4R agonists that can potentially be used to treat obesity. Cyclic peptide MC4R agonists were produced by grafting the conserved MCR-binding sequence (HFRW) into loop 6 of kalata B1, yielding a potent agonist that displayed a more than 100-fold selectivity for MC4R.[141] Finally, in the case of HIV infections, the chemokine receptor CXCR4 is a promising therapeutic target due to its role in mediating virus entry into host T cells.[142] CXCR4 antagonists were designed by grafting the bioactive epitope from CVX15, a linear CXCR4 antagonist derived from horseshoe crab polyphemusin, into loop 6 of MCoTI-I. The most effective grafted variant showed higher activity than CVX15 and potently blocked HIV-1 replication in human T lymphocyte MT-4 cells.[8]

13.3.3 Cyclic Peptide Libraries for Developing New Chemical Tools

In addition to acting as scaffolds for stabilizing bioactive peptide sequences, cyclic peptides have been used as templates for producing sequence-diverse compound libraries that can be screened against a range of targets. Protease inhibitor libraries based on SFTI-1 have been produced by synthetic chemistry that can be used to profile the specificity of different serine proteases at key residues on the SFTI-binding loop. This approach has been used to examine the inhibition of trypsin, chymotrypsin, and elastase with acyclic SFTI-1 variants carrying substitutions at the P1 residue (Lys5) or P4′ residue (Pro9).[143] More recently, an inhibitor library based on an engineered SFTI variant was used to characterize the P2′ specificity of 13 serine proteases, including trypsin, chymotrypsin, matriptase, four kallikrein-related peptidases, and several proteases from the coagulation cascade.[118]

Bacterial display has also been used to produce and screen cyclic peptide–based libraries. This approach allows for the generation of sequence diversity on a much greater scale than chemical synthesis, making it possible to screen complete binding motifs rather than single residues. Bacterial display was first applied to design novel peptides based on kalata B1 that bind to the serine protease, thrombin.[144] Here, acyclic kalata B1 (open between Gly7 and Gly8) was fused to the engineered display protein, eCPX, and used as a structural template for generating a large library of variants (6×10^9 individual transformants), where seven residues in loop 6 were randomized. Thrombin-binding peptides were selected via affinity maturation using fluorescence-activated cell sorting (FACS), and after three rounds of sorting, kalata B1 variants with K_D values in the nanomolar range were identified.[144] This technique was expanded in a following study that developed kalata B1–based antagonists for neuropilin-1 (NRP1), a receptor that is implicated in several cancers. After screening NRP1 against the loop 6–focused kalata B1 library, a second library was constructed to scan substitutions at seven further residues distributed across loops 1, 3, and 5.[145] This library was preincubated with trypsin to deplete variants that were susceptible to proteolytic degradation and then screened against a lower concentration of NRP1 to favor selection of peptides that showed high affinity and stability. Several kalata B1 variants were identified that showed K_D values ranging from 30 to 60 nM, and cyclic and acyclic forms of the lead peptide were able to inhibit the migration of human umbilical vein endothelial cells by blocking the interaction between NRP1 and VEGF.[145] However, when compared with kalata B1, these peptides were less resistant to degradation by a series of proteases, including urokinase plasminogen activator, matrix metalloproteinase-9, and matriptase.

A second library-based strategy that makes use of genetically encoded sequence diversity is the random nonstandard peptide integrated discovery (RaPID) system.[146] This method provides a platform for cell-free translation of cyclic peptides from a

DNA template by incorporating several novel technologies, including protein synthesis using recombinant elements (PURE) system[147] and customized flexizymes to allow loading of tRNA with nonproteinogenic amino acids to facilitate codon reprogramming.[148] Backbone cyclization is achieved by including a Cys–Pro–glycolic acid motif at the C-terminus, which self-rearranges to form a diketopiperadine thioester and subsequently reacts with the amino terminus to ligate the two termini.[149] Cyclic RTD-1 and SFTI-1 were successfully produced using cell-free translation, as well as engineered variants that contained N-methylated amino acids.[149] Additionally, a cyclic peptide library was produced where several residues were randomized and screened against trypsin via several rounds of *in vitro* display and polymerase chain reaction (PCR)–based deconvolution.[149] The RaPID system contains a further innovation whereby each peptide is covalently linked to its mRNA template via puromycin, allowing for the selection of effective peptide variants by affinity maturation using successive rounds of cell-free transcription, translation, and PCR amplification.[146] This approach has been used to produce macrocyclic peptides containing several *N*-methylated amino acids that bind to the ubiquitin ligase, E6AP, with subnanomolar affinity and inhibit polyubiquitination of several E6AP target proteins, including p53.[146] In a further study, codon reprogramming was used to introduce a chemical warhead (ε-*N*-trifluoroacetyl lysine) into a macrocyclic peptide library to generate potent inhibitors of the human deacetylase, sirtuin 2.[150] Collectively, these examples demonstrate the significant potential of genetically encoded cyclic peptide libraries and suggest an exciting future for using engineered cyclic peptides to target certain receptors or enzymes that might lie outside the scope of conventional small molecules or large biologics.

13.4 OVERVIEW AND PERSPECTIVES

Ribosomally produced cyclic disulfide-rich cyclic peptides were virtually unknown two decades ago. Now, these unique peptides form the basis of a maturing field in which they are increasingly being used as research tools and drug design scaffolds. Their rapid rise in the chemical biology arena has largely been driven by their ability to effectively occupy the space between small organic molecules (<500 Da) and larger proteins (>10 kDa) and, in doing so, adopt several desirable traits from each class of molecules. These attributes include their intermediate size, exceptional stability, the capacity for high binding affinity and selectivity, and in some cases, the ability to cross membranes or oral activity. The recognition that backbone cyclization often improves the stability of small disulfide-rich peptides has proven particularly important and has prompted efforts to design cyclic variants of naturally occurring linear peptides, including conotoxins and chlorotoxin. Additionally, this observation has sparked numerous engineering studies that have used cyclic peptides as scaffolds to display bioactive peptide sequences and, thus, overcome the limited metabolic stability typical of linear peptides.

In many respects, the significant progress in characterizing and engineering disulfide-rich cyclic peptides has been made possible by the availability of robust methods for producing them by chemical synthesis or recombinant expression. Now that this groundwork has been laid, the field is well positioned to explore some of the fundamental questions about cyclic peptides that remain incompletely answered at present. One such question is: How many different biosynthetic strategies exist in nature for producing cyclic peptides? Already, biosynthetic mechanisms have been described that involve head-to-tail cyclization of a single peptide domain (kalata B1, SFTI-1, and MCoTI-II) or head-to-tail ligation of two peptide domains encoded by different genes (RTD-1). Additionally, several enzymes have been identified that are capable of generating cyclic peptides from their respective precursor proteins, with the most thoroughly characterized being AEPs (which cleave after an Asn or Asp residue) in plants. However, specifically which AEPs are required for the processing of different cyclic peptides *in vivo* remains unknown at present. Moreover, processing of Oak1 and TIPTOP also involves an additional protease that cleaves between Lys and Gly or Lys and Gln, respectively, and processing of RTD-1 precursors involves cleavage at Leu–Arg, Cys–Arg, and Cys–Gln sites. These observations highlight that, even for those cyclic peptides that are relatively well characterized, there is still much to explore with respect to cyclic peptide biosynthesis, let alone cyclic peptides that have received limited attention or are yet to be discovered.

Grasping how different cyclic peptides are produced in their respective host organisms is not only important for understanding their biology and evolution, but also essential for designing effective strategies for the biosynthesis of engineered cyclic peptides. Although naturally occurring cyclic peptides have a number of promising applications, their existing bioactivities can be improved, redirected, or diversified by modifying the template molecule. In the short amount of time that has elapsed since the first engineered cyclic peptides were reported, a number of very exciting lead molecules have emerged that have potential applications in treatment of cancer, chronic pain, and neurodegenerative disorders. A key obstacle that must be overcome before these peptides could see use in patients, aside from the conventional hurdles that face all drug development programs and clinical trials, relates to devising approaches for efficient, large-scale production. An attractive strategy is to use plants and harness their specialized biosynthetic machinery to express engineered cyclic peptides. This approach has several clear advantages over chemical synthesis, including lower cost, ease to scale up, and avoiding the use of hazardous and toxic chemicals, and may allow simpler extraction and purification. However, it is conceivable that certain modifications made during the engineering process could impair expression or cyclization *in planta*; thus, it may be useful to consider any restrictions imposed by the host organism, such as sequence requirements, as early as practical, rather than focusing solely on target affinity and selectivity.

Some of the other key questions that need to be addressed are: How many cyclic peptides remain to be discovered? Why have some organisms adopted cyclic peptides and others have not? Can we develop improved methods of detecting and discovering cyclic peptides? One particularly intriguing question is: Why did our ancestors apparently lose the ability to make cyclic retrocyclin, a potent anti-HIV agent, while Old World monkeys such as rhesus macaque, baboon, and gorillas retain the ability to make related θ-defensins? Does this suggest that there might be some disadvantages in maintaining the infrastructure to produce cyclic peptides? There is no doubt that the field of cyclic peptides will continue to grow, and we encourage new students to enter this field to address some of these intriguing questions.

ACKNOWLEDGMENTS

Work in our laboratory on disulfide-rich cyclic peptides is funded by a grant from the Australian Research Council (ARC; DP150100443). David Craik is an ARC Australian Laureate Fellow (FL150100146). We also thank the Simon Axelson Foundation for funding conotoxin studies.

REFERENCES

1. Craik, D. J.; Daly, N. L.; Bond, T.; Waine, C. Plant cyclotides: A unique family of cyclic and knotted proteins that defines the cyclic cystine knot structural motif. *J. Mol. Biol.* **1999**, *294*, 1327–1336.
2. Clark, R. J.; Jensen, J.; Nevin, S. T.; Callaghan, B. P.; Adams, D. J.; Craik, D. J. The engineering of an orally active conotoxin for the treatment of neuropathic pain. *Angew. Chem. Int. Ed. Engl.* **2010**, *49*, 6545–6548.
3. Mylne, J. S.; Chan, L. Y.; Chanson, A. H.; Daly, N. L.; Schaefer, H.; Bailey, T. L.; Nguyencong, P.; Cascales, L.; Craik, D. J. Cyclic peptides arising by evolutionary parallelism via asparaginyl-endopeptidase-mediated biosynthesis. *Plant Cell* **2012**, *24*, 2765–2778.
4. Wang, C. K.; Gruber, C. W.; Čemažar, M.; Siatskas, C.; Tagore, P.; Payne, N.; Sun, G.; Wang, S.; Bernard, C. C.; Craik, D. J. Molecular grafting onto a stable framework yields novel cyclic peptides for the treatment of multiple sclerosis. *ACS Chem. Biol.* **2014**, *9*, 156–163.
5. Akcan, M.; Clark, R. J.; Daly, N. L.; Conibear, A. C.; de Faoite, A.; Heghinian, M. D.; Sahil, T.; Adams, D. J.; Mari, F.; Craik, D. J. Transforming conotoxins into cyclotides: Backbone cyclization of P-superfamily conotoxins. *Biopolymers* **2015**, *104*, 682–692.
6. Poth, A. G.; Colgrave, M. L.; Philip, R.; Kerenga, B.; Daly, N. L.; Anderson, M. A.; Craik, D. J. Discovery of cyclotides in the Fabaceae plant family provides new insights into the cyclization, evolution, and distribution of circular proteins. *ACS Chem. Biol.* **2011**, *6*, 345–355.
7. Craik, D. J. Joseph Rudinger memorial lecture: Discovery and applications of cyclotides. *J. Pept. Sci.* **2013**, *19*, 393–407.
8. Aboye, T. L.; Ha, H.; Majumder, S.; Christ, F.; Debyser, Z.; Shekhtman, A.; Neamati, N.; Camarero, J. A. Design of a novel cyclotide-based CXCR4 antagonist with anti-human immunodeficiency virus (HIV)-1 activity. *J. Med. Chem.* **2012**, *55*, 10729–10734.
9. Ji, Y.; Majumder, S.; Millard, M.; Borra, R.; Bi, T.; Elnagar, A. Y.; Neamati, N.; Shekhtman, A.; Camarero, J. A. *In vivo* activation of the p53 tumor suppressor pathway by an engineered cyclotide. *J. Am. Chem. Soc.* **2013**, *135*, 11623–11633.
10. Nguyen, G. K.; Wang, S.; Qiu, Y.; Hemu, X.; Lian, Y.; Tam, J. P. Butelase 1 is an Asx-specific ligase enabling peptide macrocyclization and synthesis. *Nat. Chem. Biol.* **2014**, *10*, 732–738.
11. Burman, R.; Yeshak, M. Y.; Larsson, S.; Craik, D. J.; Rosengren, K. J.; Göransson, U. Distribution of circular proteins in plants: Large-scale mapping of cyclotides in the Violaceae. *Front. Plant Sci.* **2015**, *6*, 855.
12. Kalia, J.; Milescu, M.; Salvatierra, J.; Wagner, J.; Klint, J. K.; King, G. F.; Olivera, B. M.; Bosmans, F. From foe to friend: Using animal toxins to investigate ion channel function. *J. Mol. Biol.* **2015**, *427*, 158–175.
13. Akondi, K. B.; Muttenthaler, M.; Dutertre, S.; Kaas, Q.; Craik, D. J.; Lewis, R. J.; Alewood, P. F. Discovery, synthesis, and structure-activity relationships of conotoxins. *Chem. Rev.* **2014**, *114*, 5815–5847.
14. Halai, R.; Craik, D. J. Conotoxins: Natural product drug leads. *Nat. Prod. Rep.* **2009**, *26*, 526–536.

15. Olivera, B. M.; Showers Corneli, P.; Watkins, M.; Fedosov, A. Biodiversity of cone snails and other venomous marine gastropods: Evolutionary success through neuropharmacology. *Annu. Rev. Anim. Biosci.* **2014**, *2*, 487–513.
16. Shim, Y. Y.; Gui, B.; Arnison, P. G.; Wang, Y.; Reaney, M. J. T. Flaxseed (*Linum usitatissimum* L.) bioactive compounds and peptide nomenclature: A review. *Trends Food Sci. Technol.* **2014**, *38*, 5–20.
17. Montalban-Lopez, M.; Sanchez-Hidalgo, M.; Cebrian, R.; Maqueda, M. Discovering the bacterial circular proteins: Bacteriocins, cyanobactins, and pilins. *J. Biol. Chem.* **2012**, *287*, 27007–27013.
18. Luckett, S.; Garcia, R. S.; Barker, J. J.; Konarev, A. V.; Shewry, P. R.; Clarke, A. R.; Brady, R. L. High-resolution structure of a potent, cyclic proteinase inhibitor from sunflower seeds. *J. Mol. Biol.* **1999**, *290*, 525–533.
19. Korsinczky, M. L.; Schirra, H. J.; Rosengren, K. J.; West, J.; Condie, B. A.; Otvos, L.; Anderson, M. A.; Craik, D. J. Solution structures by ^{1}H NMR of the novel cyclic trypsin inhibitor SFTI-1 from sunflower seeds and an acyclic permutant. *J. Mol. Biol.* **2001**, *311*, 579–591.
20. Mylne, J. S.; Colgrave, M. L.; Daly, N. L.; Chanson, A. H.; Elliott, A. G.; McCallum, E. J.; Jones, A.; Craik, D. J. Albumins and their processing machinery are hijacked for cyclic peptides in sunflower. *Nat. Chem. Biol.* **2011**, *7*, 257–259.
21. Elliott, A. G.; Delay, C.; Liu, H.; Phua, Z.; Rosengren, K. J.; Benfield, A. H.; Panero, J. L. et al. Evolutionary origins of a bioactive peptide buried within Preproalbumin. *Plant Cell* **2014**, *26*, 981–995.
22. Elliott, A. G.; Franke, B.; Armstrong, D. A.; Craik, D. J.; Mylne, J. S.; Rosengren, K. J. Natural structural diversity within a conserved cyclic peptide scaffold. *Amino Acids* **2017**, *49*, 103–116.
23. Hemu, X.; Taichi, M.; Qiu, Y.; Liu, D. X.; Tam, J. P. Biomimetic synthesis of cyclic peptides using novel thioester surrogates. *Biopolymers* **2013**, *100*, 492–501.
24. Gehrmann, J.; Alewood, P. F.; Craik, D. J. Structure determination of the three disulfide bond isomers of alpha-conotoxin GI: A model for the role of disulfide bonds in structural stability. *J. Mol. Biol.* **1998**, *278*, 401–415.
25. Satkunanathan, N.; Livett, B.; Gayler, K.; Sandall, D.; Down, J.; Khalil, Z. Alpha-conotoxin Vc1.1 alleviates neuropathic pain and accelerates functional recovery of injured neurones. *Brain Res.* **2005**, *1059*, 149–158.
26. Clark, R. J.; Fischer, H.; Dempster, L.; Daly, N. L.; Rosengren, K. J.; Nevin, S. T.; Meunier, F. A.; Adams, D. J.; Craik, D. J. Engineering stable peptide toxins by means of backbone cyclization: Stabilization of the alpha-conotoxin MII. *Proc. Natl. Acad. Sci. USA* **2005**, *102*, 13767–13772.
27. Lovelace, E. S.; Armishaw, C. J.; Colgrave, M. L.; Wahlstrom, M. E.; Alewood, P. F.; Daly, N. L.; Craik, D. J. Cyclic MrIA: A stable and potent cyclic conotoxin with a novel topological fold that targets the norepinephrine transporter. *J. Med. Chem.* **2006**, *49*, 6561–6568.
28. Sharpe, I. A.; Gehrmann, J.; Loughnan, M. L.; Thomas, L.; Adams, D. A.; Atkins, A.; Palant, E. et al. Two new classes of conopeptides inhibit the alpha1-adrenoceptor and noradrenaline transporter. *Nat. Neurosci.* **2001**, *4*, 902–907.
29. Brust, A.; Palant, E.; Croker, D. E.; Colless, B.; Drinkwater, R.; Patterson, B.; Schroeder, C. I. et al. chi-Conopeptide pharmacophore development: Toward a novel class of norepinephrine transporter inhibitor (Xen2174) for pain. *J. Med. Chem.* **2009**, *52*, 6991–7002.
30. Halai, R.; Callaghan, B.; Daly, N. L.; Clark, R. J.; Adams, D. J.; Craik, D. J. Effects of cyclization on stability, structure, and activity of alpha-conotoxin RgIA at the alpha9alpha10 nicotinic acetylcholine receptor and GABA(B) receptor. *J. Med. Chem.* **2011**, *54*, 6984–6992.

31. Armishaw, C. J.; Dutton, J. L.; Craik, D. J.; Alewood, P. F. Establishing regiocontrol of disulfide bond isomers of alpha-conotoxin ImI via the synthesis of N-to-C cyclic analogs. *Biopolymers* **2010**, *94*, 307–313.
32. Lovelace, E. S.; Gunasekera, S.; Alvarmo, C.; Clark, R. J.; Nevin, S. T.; Grishin, A. A.; Adams, D. J.; Craik, D. J.; Daly, N. L. Stabilization of alpha-conotoxin AuIB: Influences of disulfide connectivity and backbone cyclization. *Antioxid. Redox Signal.* **2011**, *14*, 87–95.
33. Akcan, M.; Craik, D. J. Engineering venom peptides to improve their stability and bioavailability. In *Venoms to Drugs: Venom as a Source for the Development of Human Therapeutics*, King, G. F. Ed. Royal Society of Chemistry, Cambridge, U.K., **2015**, pp. 275–289.
34. Tang, Y. Q.; Yuan, J.; Osapay, G.; Osapay, K.; Tran, D.; Miller, C. J.; Ouellette, A. J.; Selsted, M. E. A cyclic antimicrobial peptide produced in primate leukocytes by the ligation of two truncated alpha-defensins. *Science* **1999**, *286*, 498–502.
35. Daly, N. L.; Chen, Y. K.; Rosengren, K. J.; Marx, U. C.; Phillips, M. L.; Waring, A. J.; Wang, W.; Lehrer, R. I.; Craik, D. J. Retrocyclin-2: Structural analysis of a potent anti-HIV theta-defensin. *Biochemistry* **2007**, *46*, 9920–9928.
36. Penberthy, W. T.; Chari, S.; Cole, A. L.; Cole, A. M. Retrocyclins and their activity against HIV-1. *Cell. Mol. Life Sci.* **2011**, *68*, 2231–2242.
37. Conibear, A. C.; Craik, D. J. The chemistry and biology of theta defensins. *Angew. Chem. Int. Ed. Engl.* **2014**, *53*, 10612–10623.
38. Saether, O.; Craik, D. J.; Campbell, I. D.; Sletten, K.; Juul, J.; Norman, D. G. Elucidation of the primary and three-dimensional structure of the uterotonic polypeptide kalata B1. *Biochemistry* **1995**, *34*, 4147–4158.
39. Wang, C. K.; King, G. J.; Northfield, S. E.; Ojeda, P. G.; Craik, D. J. Racemic and quasi-racemic X-ray structures of cyclic disulfide-rich peptide drug scaffolds. *Angew. Chem. Int. Ed. Engl.* **2014**, *53*, 11236–11241.
40. Craik, D. J. Overview on the discovery and applications of cyclotides. In *Advances in Botanical Research: Plant Cyclotides*, Craik, D. J. Ed. Academic Press, London, U.K., **2015**, pp. 1–13.
41. Gilding, E. K.; Jackson, M. A.; Poth, A. G.; Henriques, S. T.; Prentis, P. J.; Mahatmanto, T.; Craik, D. J. Gene coevolution and regulation lock cyclic plant defence peptides to their targets. *New Phytol.* **2016**, *210*, 717–730.
42. Mulvenna, J. P.; Wang, C.; Craik, D. J. CyBase: A database of cyclic protein sequence and structure. *Nucleic Acids Res.* **2006**, *34*, D192–D194.
43. Gruber, C. W.; Elliott, A. G.; Ireland, D. C.; Delprete, P. G.; Dessein, S.; Göransson, U.; Trabi, M. et al. Distribution and evolution of circular miniproteins in flowering plants. *Plant Cell* **2008**, *20*, 2471–2483.
44. Jennings, C.; West, J.; Waine, C.; Craik, D.; Anderson, M. Biosynthesis and insecticidal properties of plant cyclotides: The cyclic knotted proteins from *Oldenlandia affinis*. *Proc. Natl. Acad. Sci. USA* **2001**, *98*, 10614–10619.
45. Colgrave, M. L.; Kotze, A. C.; Huang, Y. H.; O'Grady, J.; Simonsen, S. M.; Craik, D. J. Cyclotides: Natural, circular plant peptides that possess significant activity against gastrointestinal nematode parasites of sheep. *Biochemistry* **2008**, *47*, 5581–5589.
46. Plan, M. R.; Saska, I.; Cagauan, A. G.; Craik, D. J. Backbone cyclised peptides from plants show molluscicidal activity against the rice pest *Pomacea canaliculata* (golden apple snail). *J. Agric. Food Chem.* **2008**, *56*, 5237–5241.
47. Daly, N. L.; Koltay, A.; Gustafson, K. R.; Boyd, M. R.; Casas-Finet, J. R.; Craik, D. J. Solution structure by NMR of circulin A: A macrocyclic knotted peptide having anti-HIV activity. *J. Mol. Biol.* **1999**, *285*, 333–345.
48. Gustafson, K. R.; Sowder II, R. C.; Henderson, L. E.; Parsons, I. C.; Kashman, Y.; Cardellina II, J. H.; McMahon, J. B.; Buckheit Jr., R. W.; Pannell, L. K.; Boyd, M. R. Circulins A and B. Novel human immunodeficiency virus (HIV)-inhibitory macrocyclic peptides from the tropical tree *Chassalia parvifolia*. *J. Am. Chem. Soc.* **1994**, *116*, 9337–9338.

49. Tam, J. P.; Lu, Y. A.; Yang, J. L.; Chiu, K. W. An unusual structural motif of antimicrobial peptides containing end-to-end macrocycle and cystine-knot disulfides. *Proc. Natl. Acad. Sci. USA* **1999**, *96*, 8913–8918.
50. Göransson, U.; Sjogren, M.; Svangard, E.; Claeson, P.; Bohlin, L. Reversible antifouling effect of the cyclotide cycloviolacin O2 against barnacles. *J. Nat. Prod.* **2004**, *67*, 1287–1290.
51. Barry, D. G.; Daly, N. L.; Clark, R. J.; Sando, L.; Craik, D. J. Linearization of a naturally occurring circular protein maintains structure but eliminates hemolytic activity. *Biochemistry* **2003**, *42*, 6688–6695.
52. Huang, Y. H.; Colgrave, M. L.; Daly, N. L.; Keleshian, A.; Martinac, B.; Craik, D. J. The biological activity of the prototypic cyclotide kalata B1 is modulated by the formation of multimeric pores. *J. Biol. Chem.* **2009**, *284*, 20699–20707.
53. Göransson, U.; Herrmann, A.; Burman, R.; Haugaard-Jonsson, L. M.; Rosengren, K. J. The conserved Glu in the cyclotide cycloviolacin O2 has a key structural role. *ChemBioChem* **2009**, *10*, 2354–2360.
54. Daly, N. L.; Love, S.; Alewood, P. F.; Craik, D. J. Chemical synthesis and folding pathways of large cyclic polypeptides: Studies of the cystine knot polypeptide kalata B1. *Biochemistry* **1999**, *38*, 10606–10614.
55. Daly, N. L.; Gustafson, K. R.; Craik, D. J. The role of the cyclic peptide backbone in the anti-HIV activity of the cyclotide kalata B1. *FEBS Lett.* **2004**, *574*, 69–72.
56. Gillon, A. D.; Saska, I.; Jennings, C. V.; Guarino, R. F.; Craik, D. J.; Anderson, M. A. Biosynthesis of circular proteins in plants. *Plant J.* **2008**, *53*, 505–515.
57. Harris, K. S.; Durek, T.; Kaas, Q.; Poth, A. G.; Gilding, E. K.; Conlan, B. F.; Saska, I. et al. Efficient backbone cyclization of linear peptides by a recombinant asparaginyl endopeptidase. *Nat. Commun.* **2015**, *6*, 10199.
58. Saska, I.; Gillon, A. D.; Hatsugai, N.; Dietzgen, R. G.; Hara-Nishimura, I.; Anderson, M. A.; Craik, D. J. An asparaginyl endopeptidase mediates *in vivo* protein backbone cyclization. *J. Biol. Chem.* **2007**, *282*, 29721–29728.
59. Basse, C. W. Dissecting defense-related and developmental transcriptional responses of maize during *Ustilago maydis* infection and subsequent tumor formation. *Plant Physiol.* **2005**, *138*, 1774–1784.
60. Mulvenna, J. P.; Mylne, J. S.; Bharathi, R.; Burton, R. A.; Shirley, N. J.; Fincher, G. B.; Anderson, M. A.; Craik, D. J. Discovery of cyclotide-like protein sequences in graminaceous crop plants: Ancestral precursors of circular proteins? *Plant Cell* **2006**, *18*, 2134–2144.
61. Ireland, D. C.; Colgrave, M. L.; Nguyencong, P.; Daly, N. L.; Craik, D. J. Discovery and characterization of a linear cyclotide from *Viola odorata*: Implications for the processing of circular proteins. *J. Mol. Biol.* **2006**, *357*, 1522–1535.
62. Nguyen, G. K.; Lian, Y.; Pang, E. W.; Nguyen, P. Q.; Tran, T. D.; Tam, J. P. Discovery of linear cyclotides in monocot plant *Panicum laxum* of Poaceae family provides new insights into evolution and distribution of cyclotides in plants. *J. Biol. Chem.* **2013**, *288*, 3370–3380.
63. Mahatmanto, T.; Mylne, J. S.; Poth, A. G.; Swedberg, J. E.; Kaas, Q.; Schaefer, H.; Craik, D. J. The evolution of Momordica cyclic peptides. *Mol. Biol. Evol.* **2015**, *32*, 392–405.
64. Craik, D. J.; Daly, N. L.; Nielsen, K. J. Cyclised conotoxin peptides. Patent WO/2000/015654. **2000**.
65. Jones, R. M.; Cartier, G. E.; McIntosh, J. M.; Bulaj, G.; Farrar, V. E.; Olivera, B. M. and Composition and therapeutic utility of conotoxins from genus *Conus*. Patent status 1996–2000. *Expert Opin. Ther. Pat.* **2001**, *11*, 603–623.
66. Miljanich, G. P. Ziconotide: Neuronal calcium channel blocker for treating severe chronic pain. *Curr. Med. Chem.* **2004**, *11*, 3029–3040.

67. Kwon, S.; Bosmans, F.; Kaas, Q.; Cheneval, O.; Conibear, A. C.; Rosengren, K. J.; Wang, C. K.; Schroeder, C. I.; Craik, D. J. Efficient enzymatic cyclization of an inhibitory cystine knot-containing peptide. *Biotechnol. Bioeng.* **2016**, *113*, 2202–2212.
68. Soroceanu, L.; Gillespie, Y.; Khazaeli, M. B.; Sontheimer, H. Use of chlorotoxin for targeting of primary brain tumors. *Cancer Res.* **1998**, *58*, 4871–4879.
69. Veiseh, M.; Gabikian, P.; Bahrami, S. B.; Veiseh, O.; Zhang, M.; Hackman, R. C.; Ravanpay, A. C. et al. Tumor paint: A chlorotoxin:Cy5.5 bioconjugate for intraoperative visualization of cancer foci. *Cancer Res.* **2007**, *67*, 6882–6888.
70. Akcan, M.; Stroud, M. R.; Hansen, S. J.; Clark, R. J.; Daly, N. L.; Craik, D. J.; Olson, J. M. Chemical re-engineering of chlorotoxin improves bioconjugation properties for tumor imaging and targeted therapy. *J. Med. Chem.* **2011**, *54*, 782–787.
71. Merrifield, R. B. Solid phase peptide synthesis. I. The synthesis of a tetrapeptide. *J. Am. Chem. Soc.* **1963**, *85*, 2149–2154.
72. Trabi, M.; Schirra, H. J.; Craik, D. J. Three-dimensional structure of RTD-1, a cyclic antimicrobial defensin from Rhesus macaque leukocytes. *Biochemistry* **2001**, *40*, 4211–4221.
73. Long, Y. Q.; Lee, S. L.; Lin, C. Y.; Enyedy, I. J.; Wang, S.; Li, P.; Dickson, R. B.; Roller, P. P. Synthesis and evaluation of the sunflower derived trypsin inhibitor as a potent inhibitor of the type II transmembrane serine protease, matriptase. *Bioorg. Med. Chem. Lett.* **2001**, *11*, 2515–2519.
74. Zablotna, E.; Kazmierczak, K.; Jaskiewicz, A.; Stawikowski, M.; Kupryszewski, G.; Rolka, K. Chemical synthesis and kinetic study of the smallest naturally occurring trypsin inhibitor SFTI-1 isolated from sunflower seeds and its analogues. *Biochem. Biophys. Res. Commun.* **2002**, *292*, 855–859.
75. Cheneval, O.; Schroeder, C. I.; Durek, T.; Walsh, P.; Huang, Y. H.; Liras, S.; Price, D. A.; Craik, D. J. Fmoc-based synthesis of disulfide-rich cyclic peptides. *J. Org. Chem.* **2014**, *79*, 5538–5544.
76. Wang, C. K.; Northfield, S. E.; Swedberg, J. E.; Colless, B.; Chaousis, S.; Price, D. A.; Liras, S.; Craik, D. J. Exploring experimental and computational markers of cyclic peptides: Charting islands of permeability. *Eur. J. Med. Chem.* **2015**, *97*, 202–213.
77. de Veer, S. J.; Swedberg, J. E.; Akcan, M.; Rosengren, K. J.; Brattsand, M.; Craik, D. J.; Harris, J. M. Engineered protease inhibitors based on sunflower trypsin inhibitor-1 (SFTI-1) provide insights into the role of sequence and conformation in Laskowski mechanism inhibition. *Biochem. J.* **2015**, *469*, 243–253.
78. Dawson, P. E.; Muir, T. W.; Clark-Lewis, I.; Kent, S. B. Synthesis of proteins by native chemical ligation. *Science* **1994**, *266*, 776–779.
79. Tam, J. P.; Lu, Y. A. A biomimetic strategy in the synthesis and fragmentation of cyclic protein. *Protein Sci.* **1998**, *7*, 1583–1592.
80. Daly, N. L.; Chen, Y. K.; Foley, F. M.; Bansal, P. S.; Bharathi, R.; Clark, R. J.; Sommerhoff, C. P.; Craik, D. J. The absolute structural requirement for a proline in the P3′-position of Bowman-Birk protease inhibitors is surmounted in the minimized SFTI-1 scaffold. *J. Biol. Chem.* **2006**, *281*, 23668–23675.
81. Conibear, A. C.; Rosengren, K. J.; Harvey, P. J.; Craik, D. J. Structural characterization of the cyclic cystine ladder motif of theta-defensins. *Biochemistry* **2012**, *51*, 9718–9726.
82. Thongyoo, P.; Tate, E. W.; Leatherbarrow, R. J. Total synthesis of the macrocyclic cysteine knot microprotein MCoTI-II. *Chem. Commun.* **2006**, (27) 2848–2850.
83. Aboye, T. L.; Li, Y.; Majumder, S.; Hao, J.; Shekhtman, A.; Camarero, J. A. Efficient one-pot cyclization/folding of rhesus theta-defensin-1 (RTD-1). *Bioorg. Med. Chem. Lett.* **2012**, *22*, 2823–2826.
84. Park, S.; Gunasekera, S.; Aboye, T. L.; Göransson, U. An efficient approach for the total synthesis of cyclotides by microwave assisted Fmoc-SPPS. *Int. J. Pept. Res. Ther.* **2010**, *16*, 167–176.

85. Shariff, L.; Zhu, Y.; Cowper, B.; Di, W. L.; Macmillan, D. Sunflower trypsin inhibitor (SFTI-1) analogues of synthetic and biological origin via N → S acyl transfer: Potential inhibitors of human Kallikrein-5 (KLK5). *Tetrahedron* **2014**, *70*, 7675–7680.
86. Yan, L. Z.; Dawson, P. E. Synthesis of peptides and proteins without cysteine residues by native chemical ligation combined with desulfurization. *J. Am. Chem. Soc.* **2001**, *123*, 526–533.
87. Laskowski, M., Jr.; Kato, I. Protein inhibitors of proteinases. *Annu. Rev. Biochem.* **1980**, *49*, 593–626.
88. Finkenstadt, W. R.; Laskowski, M., Jr. Resynthesis by trypsin of the cleaved peptide bond in modified soybean trypsin inhibitor. *J. Biol. Chem.* **1967**, *242*, 771–773.
89. Marx, U. C.; Korsinczky, M. L.; Schirra, H. J.; Jones, A.; Condie, B.; Otvos, L., Jr.; Craik, D. J. Enzymatic cyclization of a potent Bowman-Birk protease inhibitor, sunflower trypsin inhibitor-1, and solution structure of an acyclic precursor peptide. *J. Biol. Chem.* **2003**, *278*, 21782–21789.
90. Thongyoo, P.; Jaulent, A. M.; Tate, E. W.; Leatherbarrow, R. J. Immobilized protease-assisted synthesis of engineered cysteine-knot microproteins. *ChemBioChem* **2007**, *8*, 1107–1109.
91. Antos, J. M.; Popp, M. W.; Ernst, R.; Chew, G. L.; Spooner, E.; Ploegh, H. L. A straight path to circular proteins. *J. Biol. Chem.* **2009**, *284*, 16028–16036.
92. Mazmanian, S. K.; Liu, G.; Ton-That, H.; Schneewind, O. *Staphylococcus aureus* sortase, an enzyme that anchors surface proteins to the cell wall. *Science* **1999**, *285*, 760–763.
93. Ton-That, H.; Liu, G.; Mazmanian, S. K.; Faull, K. F.; Schneewind, O. Purification and characterization of sortase, the transpeptidase that cleaves surface proteins of *Staphylococcus aureus* at the LPXTG motif. *Proc. Natl. Acad. Sci. USA* **1999**, *96*, 12424–12429.
94. Jia, X.; Kwon, S.; Wang, C. I.; Huang, Y. H.; Chan, L. Y.; Tan, C. C.; Rosengren, K. J.; Mulvenna, J. P.; Schroeder, C. I.; Craik, D. J. Semienzymatic cyclization of disulfide-rich peptides using sortase A. *J. Biol. Chem.* **2014**, *289*, 6627–6638.
95. Camarero, J. A.; Muir, T. W. Biosynthesis of a head-to-tail cyclized protein with improved biological activity. *J. Am. Chem. Soc.* **1999**, *121*, 5597–5598.
96. Kane, P. M.; Yamashiro, C. T.; Wolczyk, D. F.; Neff, N.; Goebl, M.; Stevens, T. H. Protein splicing converts the yeast TFP1 gene product to the 69-kD subunit of the vacuolar H(+)-adenosine triphosphatase. *Science* **1990**, *250*, 651–657.
97. Xu, M. Q.; Perler, F. B. The mechanism of protein splicing and its modulation by mutation. *EMBO J.* **1996**, *15*, 5146–5153.
98. Kimura, R. H.; Tran, A. T.; Camarero, J. A. Biosynthesis of the cyclotide kalata B1 by using protein splicing. *Angew. Chem. Int. Ed. Engl.* **2006**, *45*, 973–976.
99. Camarero, J. A.; Kimura, R. H.; Woo, Y. H.; Shekhtman, A.; Cantor, J. Biosynthesis of a fully functional cyclotide inside living bacterial cells. *ChemBioChem* **2007**, *8*, 1363–1366.
100. Austin, J.; Kimura, R. H.; Woo, Y. H.; Camarero, J. A. *In vivo* biosynthesis of an Ala-scan library based on the cyclic peptide SFTI-1. *Amino Acids* **2010**, *38*, 1313–1322.
101. Gould, A.; Li, Y.; Majumder, S.; Garcia, A. E.; Carlsson, P.; Shekhtman, A.; Camarero, J. A. Recombinant production of rhesus theta-defensin-1 (RTD-1) using a bacterial expression system. *Mol. Biosyst.* **2012**, *8*, 1359–1365.
102. Wu, H.; Hu, Z.; Liu, X. Q. Protein trans-splicing by a split intein encoded in a split DnaE gene of *Synechocystis* sp. PCC6803. *Proc. Natl. Acad. Sci. USA* **1998**, *95*, 9226–9231.
103. Scott, C. P.; Abel-Santos, E.; Wall, M.; Wahnon, D. C.; Benkovic, S. J. Production of cyclic peptides and proteins *in vivo*. *Proc. Natl. Acad. Sci. USA* **1999**, *96*, 13638–13643.

104. Jagadish, K.; Borra, R.; Lacey, V.; Majumder, S.; Shekhtman, A.; Wang, L.; Camarero, J. A. Expression of fluorescent cyclotides using protein trans-splicing for easy monitoring of cyclotide-protein interactions. *Angew. Chem. Int. Ed. Engl.* **2013**, *52*, 3126–3131.
105. Li, Y.; Aboye, T.; Breindel, L.; Shekhtman, A.; Camarero, J. A. Efficient recombinant expression of SFTI-1 in bacterial cells using intein-mediated protein trans-splicing. *Biopolymers* **2016**, *106*, 818–824.
106. Jagadish, K.; Gould, A.; Borra, R.; Majumder, S.; Mushtaq, Z.; Shekhtman, A.; Camarero, J. A. Recombinant expression and phenotypic screening of a bioactive cyclotide against alpha-synuclein-induced cytotoxicity in baker's yeast. *Angew. Chem. Int. Ed. Engl.* **2015**, *54*, 8390–8394.
107. Conlan, B. F.; Colgrave, M. L.; Gillon, A. D.; Guarino, R.; Craik, D. J.; Anderson, M. A. Insights into processing and cyclization events associated with biosynthesis of the cyclic peptide kalata B1. *J. Biol. Chem.* **2012**, *287*, 28037–28046.
108. Bernath-Levin, K.; Nelson, C.; Elliott, A. G.; Jayasena, A. S.; Millar, A. H.; Craik, D. J.; Mylne, J. S. Peptide macrocyclization by a bifunctional endoprotease. *Chem. Biol.* **2015**, *22*, 571–582.
109. Seydel, P.; Gruber, C. W.; Craik, D. J.; Dornenburg, H. Formation of cyclotides and variations in cyclotide expression in *Oldenlandia affinis* suspension cultures. *Appl. Microbiol. Biotechnol.* **2007**, *77*, 275–284.
110. Seydel, P.; Walter, C.; Dornenburg, H. Scale-up of *Oldenlandia affinis* suspension cultures in photobioreactors for cyclotide production. *Eng. Life Sci.* **2009**, *9*, 219–226.
111. Li, P.; Jiang, S.; Lee, S. L.; Lin, C. Y.; Johnson, M. D.; Dickson, R. B.; Michejda, C. J.; Roller, P. P. Design and synthesis of novel and potent inhibitors of the type II transmembrane serine protease, matriptase, based upon the sunflower trypsin inhibitor-1. *J. Med. Chem.* **2007**, *50*, 5976–5983.
112. Fittler, H.; Avrutina, O.; Glotzbach, B.; Empting, M.; Kolmar, H. Combinatorial tuning of peptidic drug candidates: High-affinity matriptase inhibitors through incremental structure-guided optimization. *Org. Biomol. Chem.* **2013**, *11*, 1848–1857.
113. Quimbar, P.; Malik, U.; Sommerhoff, C. P.; Kaas, Q.; Chan, L. Y.; Huang, Y. H.; Grundhuber, M. et al. High-affinity cyclic peptide matriptase inhibitors. *J. Biol. Chem.* **2013**, *288*, 13885–13896.
114. Swedberg, J. E.; Nigon, L. V.; Reid, J. C.; de Veer, S. J.; Walpole, C. M.; Stephens, C. R.; Walsh, T. P. et al. Substrate-guided design of a potent and selective kallikrein-related peptidase inhibitor for kallikrein 4. *Chem. Biol.* **2009**, *16*, 633–643.
115. Swedberg, J. E.; de Veer, S. J.; Sit, K. C.; Reboul, C. F.; Buckle, A. M.; Harris, J. M. Mastering the canonical loop of serine protease inhibitors: Enhancing potency by optimising the internal hydrogen bond network. *PLoS One* **2011**, *6*, e19302.
116. de Veer, S. J.; Swedberg, J. E.; Brattsand, M.; Clements, J. A.; Harris, J. M. Exploring the active site binding specificity of kallikrein-related peptidase 5 (KLK5) guides the design of new peptide substrates and inhibitors. *Biol. Chem.* **2016**, *397*, 1237–1249.
117. de Veer, S. J.; Furio, L.; Swedberg, J. E.; Munro, C. A.; Brattsand, M.; Clements, J. A.; Hovnanian, A.; Harris, J. M. Selective substrates and inhibitors for kallikrein-related peptidase 7 (KLK7) shed light on KLK proteolytic activity in the stratum corneum. *J. Invest. Dermatol.* **2017**, *137*, 430–439.
118. de Veer, S. J.; Wang, C. K.; Harris, J. M.; Craik, D. J.; Swedberg, J. E. Improving the selectivity of engineered protease inhibitors: Optimizing the P2 prime residue using a versatile cyclic peptide library. *J. Med. Chem.* **2015**, *58*, 8257–8268.
119. Kocsis, A.; Kekesi, K. A.; Szasz, R.; Vegh, B. M.; Balczer, J.; Dobo, J.; Zavodszky, P.; Gal, P.; Pal, G. Selective inhibition of the lectin pathway of complement with phage display selected peptides against mannose-binding lectin-associated serine protease (MASP)-1 and -2: Significant contribution of MASP-1 to lectin pathway activation. *J. Immunol.* **2010**, *185*, 4169–4178.

120. Debowski, D.; Pikula, M.; Lubos, M.; Langa, P.; Trzonkowski, P.; Lesner, A.; Legowska, A.; Rolka, K. Inhibition of human and yeast 20S proteasome by analogues of trypsin inhibitor SFTI-1. *PLoS One* **2014**, *9*, e89465.
121. Gitlin, A.; Debowski, D.; Karna, N.; Legowska, A.; Stirnberg, M.; Gutschow, M.; Rolka, K. Inhibitors of matriptase-2 based on the trypsin inhibitor SFTI-1. *ChemBioChem* **2015**, *16*, 1601–1607.
122. Fittler, H.; Depp, A.; Avrutina, O.; Dahms, S. O.; Than, M. E.; Empting, M.; Kolmar, H. Engineering a constrained peptidic scaffold towards potent and selective furin inhibitors. *ChemBioChem* **2015**, *16*, 2441–2444.
123. Thongyoo, P.; Bonomelli, C.; Leatherbarrow, R. J.; Tate, E. W. Potent inhibitors of beta-tryptase and human leukocyte elastase based on the MCoTI-II scaffold. *J. Med. Chem.* **2009**, *52*, 6197–6200.
124. Sommerhoff, C. P.; Avrutina, O.; Schmoldt, H. U.; Gabrijelcic-Geiger, D.; Diederichsen, U.; Kolmar, H. Engineered cystine knot miniproteins as potent inhibitors of human mast cell tryptase beta. *J. Mol. Biol.* **2010**, *395*, 167–175.
125. Glotzbach, B.; Reinwarth, M.; Weber, N.; Fabritz, S.; Tomaszowski, M.; Fittler, H.; Christmann, A.; Avrutina, O.; Kolmar, H. Combinatorial optimization of cystine-knot peptides towards high-affinity inhibitors of human matriptase-1. *PLoS One* **2013**, *8*, e76956.
126. Swedberg, J. E.; Mahatmanto, T.; Abdul Ghani, H.; de Veer, S. J.; Schroeder, C. I.; Harris, J. M.; Craik, D. J. Substrate-guided design of selective FXIIa inhibitors based on the plant-derived *Momordica cochinchinensis* trypsin inhibitor-II (MCoTI-II) scaffold. *J. Med. Chem.* **2016**, *59*, 7287–7292.
127. Heja, D.; Kocsis, A.; Dobo, J.; Szilagyi, K.; Szasz, R.; Zavodszky, P.; Pal, G.; Gal, P. Revised mechanism of complement lectin-pathway activation revealing the role of serine protease MASP-1 as the exclusive activator of MASP-2. *Proc. Natl. Acad. Sci. USA* **2012**, *109*, 10498–10503.
128. Dong, Y.; Stephens, C.; Walpole, C.; Swedberg, J. E.; Boyle, G. M.; Parsons, P. G.; McGuckin, M. A.; Harris, J. M.; Clements, J. A. Paclitaxel resistance and multicellular spheroid formation are induced by kallikrein-related peptidase 4 in serous ovarian cancer cells in an ascites mimicking microenvironment. *PLoS One* **2013**, *8*, e57056.
129. Kruse, J. P.; Gu, W. Modes of p53 regulation. *Cell* **2009**, *137*, 609–622.
130. Huang, Y. H.; Henriques, S. T.; Wang, C. K.; Thorstholm, L.; Daly, N. L.; Kaas, Q.; Craik, D. J. Design of substrate-based BCR-ABL kinase inhibitors using the cyclotide scaffold. *Sci. Rep.* **2015**, *5*, 12974.
131. D'Souza, C.; Henriques, S. T.; Wang, C. K.; Cheneval, O.; Chan, L. Y.; Bokil, N. J.; Sweet, M. J.; Craik, D. J. Using the MCoTI-II cyclotide scaffold to design a stable cyclic peptide antagonist of SET, a protein overexpressed in human cancer. *Biochemistry* **2016**, *55*, 396–405.
132. Gunasekera, S.; Foley, F. M.; Clark, R. J.; Sando, L.; Fabri, L. J.; Craik, D. J.; Daly, N. L. Engineering stabilized vascular endothelial growth factor-A antagonists: Synthesis, structural characterization, and bioactivity of grafted analogues of cyclotides. *J. Med. Chem.* **2008**, *51*, 7697–7704.
133. Chan, L. Y.; Craik, D. J.; Daly, N. L. Cyclic thrombospondin-1 mimetics: Grafting of a thrombospondin sequence into circular disulfide-rich frameworks to inhibit endothelial cell migration. *Biosci. Rep.* **2015**, *35*, e00270.
134. Chan, L. Y.; Gunasekera, S.; Henriques, S. T.; Worth, N. F.; Le, S. J.; Clark, R. J.; Campbell, J. H.; Craik, D. J.; Daly, N. L. Engineering pro-angiogenic peptides using stable, disulfide-rich cyclic scaffolds. *Blood* **2011**, *118*, 6709–6717.
135. Kerlero de Rosbo, N.; Milo, R.; Lees, M. B.; Burger, D.; Bernard, C. C.; Ben-Nun, A. Reactivity to myelin antigens in multiple sclerosis. Peripheral blood lymphocytes respond predominantly to myelin oligodendrocyte glycoprotein. *J. Clin. Invest.* **1993**, *92*, 2602–2608.

136. Spires-Jones, T. L.; Hyman, B. T. The intersection of amyloid beta and tau at synapses in Alzheimer's disease. *Neuron* **2014**, *82*, 756–771.
137. Wang, C. K.; Northfield, S. E.; Huang, Y. H.; Ramos, M. C.; Craik, D. J. Inhibition of tau aggregation using a naturally-occurring cyclic peptide scaffold. *Eur. J. Med. Chem.* **2016**, *109*, 342–349.
138. Marceau, F.; Regoli, D. Bradykinin receptor ligands: Therapeutic perspectives. *Nat. Rev. Drug Discov.* **2004**, *3*, 845–852.
139. Wong, C. T.; Rowlands, D. K.; Wong, C. H.; Lo, T. W.; Nguyen, G. K.; Li, H. Y.; Tam, J. P. Orally active peptidic bradykinin B1 receptor antagonists engineered from a cyclotide scaffold for inflammatory pain treatment. *Angew. Chem. Int. Ed. Engl.* **2012**, *51*, 5620–5624.
140. Tao, Y. X. The melanocortin-4 receptor: Physiology, pharmacology, and pathophysiology. *Endocr. Rev.* **2010**, *31*, 506–543.
141. Eliasen, R.; Daly, N. L.; Wulff, B. S.; Andresen, T. L.; Conde-Frieboes, K. W.; Craik, D. J. Design, synthesis, structural and functional characterization of novel melanocortin agonists based on the cyclotide kalata B1. *J. Biol. Chem.* **2012**, *287*, 40493–40501.
142. Feng, Y.; Broder, C. C.; Kennedy, P. E.; Berger, E. A. HIV-1 entry cofactor: Functional cDNA cloning of a seven-transmembrane, G protein-coupled receptor. *Science* **1996**, *272*, 872–877.
143. Zablotna, E.; Jaskiewicz, A.; Legowska, A.; Miecznikowska, H.; Lesner, A.; Rolka, K. Design of serine proteinase inhibitors by combinatorial chemistry using trypsin inhibitor SFTI-1 as a starting structure. *J. Pept. Sci.* **2007**, *13*, 749–755.
144. Getz, J. A.; Rice, J. J.; Daugherty, P. S. Protease-resistant peptide ligands from a knottin scaffold library. *ACS Chem. Biol.* **2011**, *6*, 837–844.
145. Getz, J. A.; Cheneval, O.; Craik, D. J.; Daugherty, P. S. Design of a cyclotide antagonist of neuropilin-1 and -2 that potently inhibits endothelial cell migration. *ACS Chem. Biol.* **2013**, *8*, 1147–1154.
146. Yamagishi, Y.; Shoji, I.; Miyagawa, S.; Kawakami, T.; Katoh, T.; Goto, Y.; Suga, H. Natural product-like macrocyclic *N*-methyl-peptide inhibitors against a ubiquitin ligase uncovered from a ribosome-expressed *de novo* library. *Chem. Biol.* **2011**, *18*, 1562–1570.
147. Shimizu, Y.; Inoue, A.; Tomari, Y.; Suzuki, T.; Yokogawa, T.; Nishikawa, K.; Ueda, T. Cell-free translation reconstituted with purified components. *Nat. Biotechnol.* **2001**, *19*, 751–755.
148. Murakami, H.; Ohta, A.; Ashigai, H.; Suga, H. A highly flexible tRNA acylation method for non-natural polypeptide synthesis. *Nat. Methods* **2006**, *3*, 357–359.
149. Kawakami, T.; Ohta, A.; Ohuchi, M.; Ashigai, H.; Murakami, H.; Suga, H. Diverse backbone-cyclized peptides via codon reprogramming. *Nat. Chem. Biol.* **2009**, *5*, 888–890.
150. Morimoto, J.; Hayashi, Y.; Suga, H. Discovery of macrocyclic peptides armed with a mechanism-based warhead: Isoform-selective inhibition of human deacetylase SIRT2. *Angew. Chem. Int. Ed. Engl.* **2012**, *51*, 3423–3427.

14 Synthesis and Target Identification of Natural Product–Inspired Compound Collections

Luca Laraia and Herbert Waldmann

CONTENTS

14.1 INTRODUCTION

Natural products (NPs) have been a rich source of therapeutics for centuries. To expand the toolkit of compounds to probe biology or drugs, accessing molecules inspired or derived from NPs has attracted great interest in the academic and industrial communities alike. To further expedite this process, the development of new strategies for synthesizing NP-inspired libraries efficiently is of great importance. NPs and their derivatives have fared particularly well when coupled to a phenotypic screening process, that is, one which aims to recapitulate the disease in question without any bias for a given molecular target. However, this introduces the additional challenge of target identification (target ID) and confirmation. While knowing the molecular target of a compound is not required for the approval of a new drug, it can immensely aid the development process and allow the compound to be used as a *tool* to investigate new biology.

In this chapter we outline the principles for selecting privileged scaffolds based on NPs for the synthesis of focused compound libraries. Using case studies we highlight the use of cycloaddition reactions, among others, to rapidly generate complex scaffolds. Biological screening of the compound collections synthesized is described where appropriate before focusing on the target ID approaches available for hit compounds identified through phenotypic screening. Case studies based on our research will be used to illustrate different target ID techniques and their suitability for particular situations. Finally, we will discuss the direction of the field as a whole and which areas are the most likely to see advancements in the coming years.

14.1.1 Chemical Space and Natural Products

Chemical biology aims to use the molecular precision available through chemical techniques to study biological systems.[1] To this end, organic chemistry is challenged in choosing which compounds to synthesize to maximize biological relevance. Chemical space, intended as all the potential small molecules that could conceivably be made, is enormous. Even the so-called *drug-like* space contains upward of 10^{60} molecules, a number far greater than all the atoms present in the world.[2] Therefore, even with all the manpower and materials imaginable, one would barely scratch the surface of all interesting compounds amenable to synthesis. The related attempts made in the development of combinatorial chemistry have led to the insight that the *quality* of chemical libraries is more important than the *quantity* of compounds therein.

In addition to synthetic small molecules, NPs are a rich source of bioactive small molecules.[3,4] This has been the case for centuries, from ancient herbal remedies to currently used therapeutics. In fact, a significant portion of commonly used drugs are either NPs or derivatives thereof.[5] Despite a perceived decreasing interest in screening NPs over the last 20 years, and particularly in the 1990s during the early combichem era, a reemergence has recently been reported.[6] NPs differ significantly in several molecular properties when compared with synthetic molecules. Cheminformatic analyses have shown that NPs contain more oxygen and less nitrogen atoms, more stereogenic centers and fused rings, but less aromatic rings and rotatable bonds.[7–9] This suggests that NPs are rigid, but not flat three-dimensional compounds. The above properties have recently been associated with increased rates of success in clinical development and are frequent arguments in favor of NP-containing and NP-derived compound libraries.

When discussing NPs, it is important to consider the biosynthetic processes that lead to their production. Intermediates in NP synthesis bind a series of enzymes and complexes in stepwise or concerted processes leading up to the final product. Organisms often make NPs for a specific purpose, which requires them to be bioavailable and stable in the species of interest. This implies that they often possess the required physiochemical properties for a candidate drug. NPs have been particularly successful therapeutics in the areas of cancer and infectious diseases, which is most likely the result of their original purpose as a defense mechanism against invading pathogens.

Although the overall number of NPs known is large, NPs can be broken down into families, which originate from the same biosynthetic origins. NPs belonging to the same family often only differ from each other in oxidation patterns and

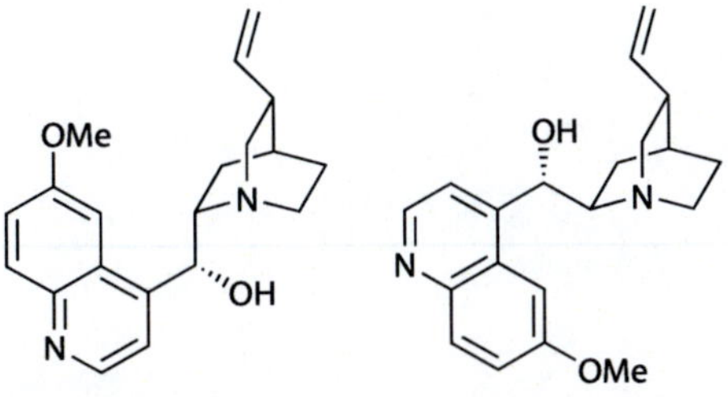

1, Quinine **2**, Quinidine

FIGURE 14.1 Cinchona alkaloids quinine and quinidine.

cyclization strategies. Despite small structural changes, related NPs often display very different bioactivity profiles. For example, the cinchona alkaloids quinine (**1**) and quinidine (**2**) are diastereomers of each other; however, the former is a potent antimalarial, while the latter is a class-I antiarrhythmic drug (Figure 14.1).[10,11]

Because NPs have coevolved with the biosynthetic machinery that is used to make them and display good physiochemical properties and broad and varied bioactivities, they can be considered prevalidated structures. Additionally, it has been suggested that conserved protein-binding motifs are evolutionarily encoded in NPs at the scaffold level, and that selectivity among targets is obtained by varying substituents on the core scaffolds.[12]

14.1.2 Structural Classification of Natural Products (SCONP)

When attempting to tackle the chemical space problem, prevalidated compounds such as NPs are a logical starting point. They occupy biologically relevant chemical space and offer useful starting points to expand into related but as yet unknown areas. One difficulty in this approach is the direction in which to expand and focus the development of novel chemistry for the synthesis of NP-inspired compound collections. To tackle this problem, NPs were first classified using a hierarchical scaffold-based approach (Figure 14.2).[13] In this cheminformatic approach, the molecular scaffolds of all NPs reported in the Dictionary of Natural Products (DNP) were isolated in silico. In this context, a scaffold was described as the core ring systems of the molecule without any substituents or side chains attached. The scaffolds were subsequently deconstructed one ring at a time to reach the final single-ring system. In each step, the smaller scaffold containing less rings is termed the *parent*, whereas the more complex system is termed the *child*. This generates a branch for every deconstructed compound, and if all the branches are combined, a scaffold *tree* is obtained, as depicted in Figure 14.2. To generate scaffold trees, a Java-based program called ScaffoldTreeGenerator has been developed.[14] Another program, ScaffoldHunter, can be used to visualize and navigate several scaffold trees across a compound set, as well as to annotate each member with a property, such as bioactivity. During the construction of a scaffold tree, branches may contain gaps where no compound exists yet. Therefore, virtual scaffolds were introduced to complete trees with gaps in the branching system and highlight opportunities to develop novel chemistry for

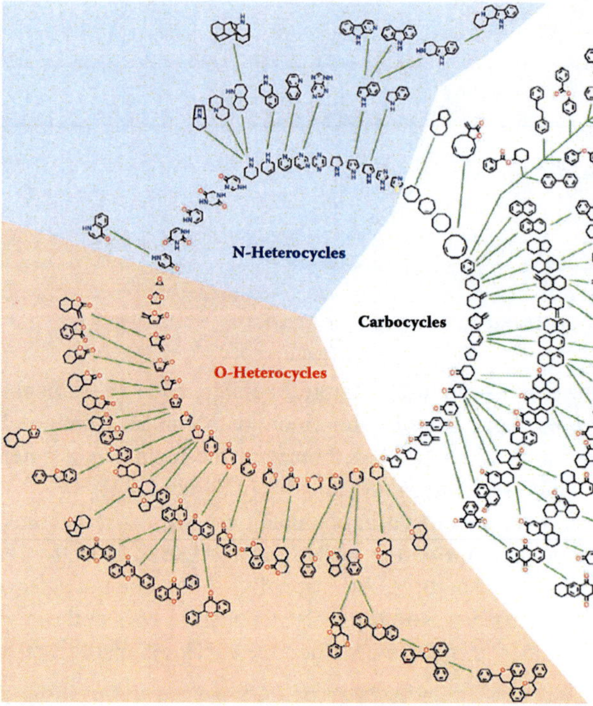

FIGURE 14.2 Natural product scaffold tree created according to the principles of the structural characterization of natural products. (Reproduced with permission from Koch, M.A., Schuffenhauer, A., Scheck, M., Wetzel, S., Casaulta, M., Odermatt, A., Ertl, P., and Waldmann, H., Charting biologically relevant chemical space: A structural classification of natural products (SCONP), *Proc. Natl. Acad. Sci.*, 102, 17272. Copyright 2005 National Academy of Sciences, U.S.A.)

their access. To further enrich the dataset, nonnatural compounds were also included. Both ScaffoldHunter and ScaffoldTreeGenerator are freely available at http://scaffoldhunter.sourceforge.net/.

The anthropological term *brachiation* was adopted to describe the movement along a scaffold tree. In a chemistry-guided approach, one can identify gaps, which could be filled by synthetic chemistry efforts to expand the chemical space around a given scaffold. In a bioactivity-guided approach, one would construct branches based on similar bioactivity profiles. Multiple branches can be constructed if many *parent–child* pairs exist which show similar bioactivity. In the end, the longest branch with the fewest gaps is selected. This approach can be used to identify the minimal essential scaffold required for bioactivity, which could focus the synthesis of a compound library on the most synthetically accessibly scaffold while still retaining the required activity. An example of this approach was the deconstruction of the NP yohimbine (**3**), a Cdc25a phosphatase inhibitor (Figure 14.3).[15] Simplified indolo[2,3-*a*]quinolizidines (**4**) were able to retain the desired activity and were amenable to solid-phase synthesis of a library of over 400 members.

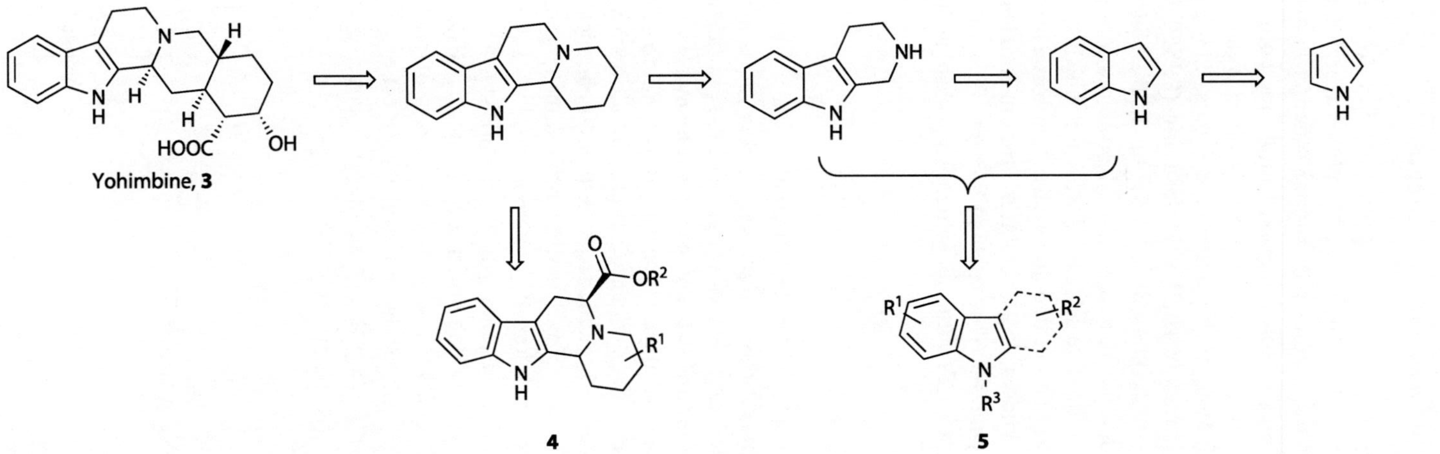

FIGURE 14.3 Brachiation of the natural product Yohimbine (**3**) for the design of new compound libraries containing potent Cdc25a phosphatase inhibitors (**4**, **5**).[15]

14.1.3 Protein Structure Similarity Clustering

In addition to the small-molecule space, protein space is also much smaller than the combination of amino acids would suggest. While a high sequence similarity between proteins is often a predictor of similar function or activity, similarity in three-dimensional structures or folds may also imply the ability to bind similar ligands. This information could be used to identify novel ligands for proteins or predict additional targets for known bioactive compounds. This hypothesis forms the notion behind protein structure similarity clustering (PSSC).[16] Here, proteins are grouped according to structural similarity of folds in their binding sites. This approach was utilized to identify inhibitors of acyl protein thioesterase 1 (APT1), an enzyme known to depalmitoylate H- and N-Ras, for which no small-molecule modulators were known.[17] PSSC revealed dog gastric lipase as a protein with similar subfolds to APT1, and the known lipase inhibitor lipstatin (orlistat) was used as a starting point for library synthesis. This enabled the identification of palmitostatin B (**7**) (Figure 14.4), a potent APT1 inhibitor which was subsequently used to study APT-1 regulation of Ras.

14.1.4 Biology-Oriented Synthesis

Although NPs undoubtedly offer a wealth of potential starting points for drug discovery, there are several limitations to their use.[18] Often, NPs are only produced in small quantities from their organism of origin, and the purification may be long and tedious. Furthermore, the structural complexity of many NPs often prevents the development of a synthetic route to access sufficient material for biological studies, structure–activity relationship (SAR) determination, and ultimately for the use as a therapeutic. Additionally, due to limitations in the natural chemical feedstocks, not all possible variations around an NP are possible through biosynthesis.

To circumvent the aforementioned problems, several strategies to access NP-inspired compound collections have been established, including diversity-oriented synthesis (DOS) and biology-oriented synthesis (BIOS). DOS focuses on maximizing scaffold diversity, to address broad areas of chemical space that have not yet been covered.[19] The synthetic chemistry aspect employs two complementary approaches: applying one set of reagents to a range of starting materials

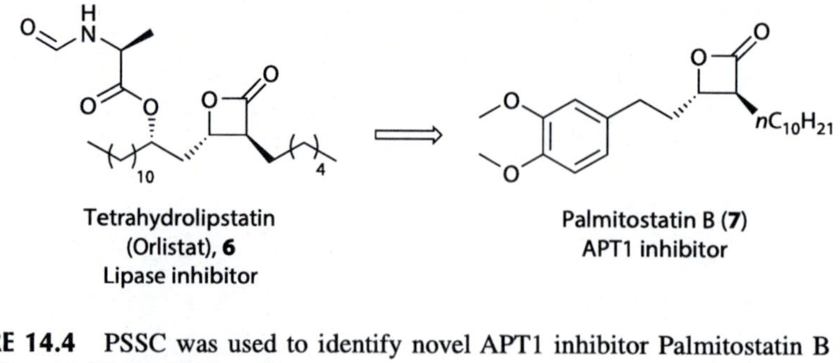

FIGURE 14.4 PSSC was used to identify novel APT1 inhibitor Palmitostatin B (**7**) from known lipase inhibitor Orlistat (**6**).

(substrate-based approach) or exposing the same starting materials to a range of reagents (reagent-based approach). Molecules produced are not required to resemble NPs, but three-dimensional, stereochemically rich compounds are desired. A subclass of this area, known as privileged scaffold (pDOS), has also been developed.[20] This approach focuses on constrained diversity within a biologically validated scaffold. Both of these approaches have been reviewed extensively elsewhere and will not be the focus of this chapter.[21]

BIOS aims to synthesize small-molecule libraries based on NP scaffolds or cores.[22] Using the guiding principles of SCONP and PSSC (individually and in concert) to select synthetic targets enables one to tackle biologically relevant areas of chemical space. The purpose of this approach is to reduce the synthetic complexity of NPs while retaining or even expanding their biological activity.[23] However, it must be noted that while libraries are typically constructed with NPs in mind, the primary goal is biological relevance and hence nonnatural starting points should not be discounted. The synthetic routes should be step-efficient and amenable to SAR exploration: once a hit is identified, medicinal chemistry should no longer be a bottleneck in the discovery efforts. As biomolecules such as proteins or nucleic acids adopt three-dimensional structures and folds, it is of great benefit to generate synthetic routes that are amenable to enantioselective synthesis. Often, the activity within a racemic sample lies entirely with one enantiomer, which must eventually be prepared or purified as such. Counterintuitively, it may be beneficial to prepare the initial compound library as racemic mixtures, which effectively offer twice as many compounds for screening. It is predicted that hit rates for libraries containing prevalidated NP scaffolds should be higher than those prepared by combinatorial chemistry or similar methods, reducing the need for very large libraries and ultra-high-throughput screening apparatus, which is particularly relevant for academic groups that may not have access to such equipment. Experience gained with NP-inspired libraries indeed confirms this notion, with hit rates typically in the range of 0.5%–1.5%. This implies that a library of 200–500 members is sufficient to identify primary hits for further development. While the demands placed on synthetic chemistry are typically higher than on combinatorial libraries, due to the often complex scaffolds present in NPs, the overall library size makes this a manageable endeavor.

The following section will outline recent successful examples of applying BIOS to synthesize NP-inspired compound collections. As extensive reviews on the subject have been published in last years, the focus is on very recent work.[22–26]

14.2 STRATEGIES FOR THE SYNTHESIS OF NATURAL PRODUCT-INSPIRED COLLECTIONS

To efficiently and rapidly synthesize NP-inspired compound collections of a reasonable size, one requires transformations that can add complexity without requiring multistep syntheses. Cycloadditions are some of the most efficient reactions to construct complex NP-inspired molecular architectures and are amenable to enantioselective modifications.[27] Such reactions have successfully been applied to several scaffolds to synthesize biologically active compounds. The tropane scaffold,

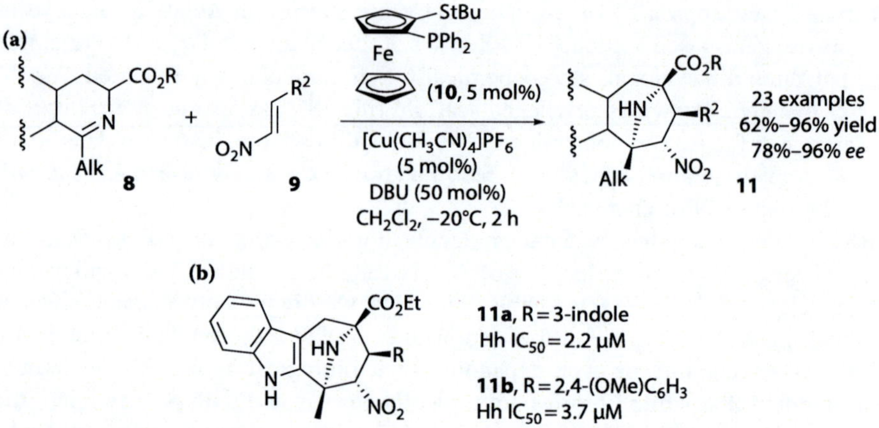

SCHEME 14.1 Synthesis of a tropane-derived compound collection *via* [1, 3]-dipolar cycloaddition reactions: **(a)** synthetic procedure; **(b)** inhibition of the Hedgehog signaling pathway by selected derivatives.

8-azabicyclo[3.2.1]octane, is at the core of over 600 alkaloids and features prominently in a wide variety of pharmaceutical agents. Strategies for the rapid, enantioselective synthesis of tropane-derived compound collections were lacking until recently. A [1, 3]-dipolar cycloaddition strategy was developed to access the tropane scaffold from appropriately functionalized azomethine ylides (**8**) and nitroalkenes (**9**) in one step with high yields and enantiomeric excess (*ee*'s) (Scheme 14.1a).[28] Copper (I) catalysis with ferrocene-derived phosphene ligand (**10**) was responsible for the high enantioselectivities observed.

The synthesized library was tested for its ability to inhibit a range of pathophysiologically relevant signaling pathways. Gratifyingly, several compounds were able to inhibit Hedgehog signaling at low micromolar concentrations (Scheme 14.1b). The identification of the cellular targets of these compounds will hopefully shed light on their unique activity.

Another example of the power of cycloaddition reactions can be observed in recently reported synthesis of an iridoid-inspired compound collection.[29] Iridoids are cyclopentano[c]pyran monoterpene secondary metabolites that are predicted to be the active constituents in many traditional medicines (Scheme 14.2a).[30] Despite their biological relevance, general strategies for the streamlined synthesis of such derivatives were essentially unexplored. To tackle this problem, it was envisaged that these compounds could be made by the kinetic resolution of racemic 2*H*-pyran-3(6*H*)-ones (**16**) *via* an asymmetric [2+3] cycloaddition with appropriate azomethine ylides (**15**) (Scheme 14.2b). This transformation was successfully achieved once again by copper (I) catalysis, coupled with a ferrocene-derived phosphine ligand **10**. The reaction proceeded to deliver a single diastereomer in high yields and with high enantioselectivities.

All in all, a library of 115 compounds was synthesized, which included further modifications on the core scaffold. Several compounds were inhibitors of the Hedgehog and Wnt signaling pathways in the low micromolar range, highlighting

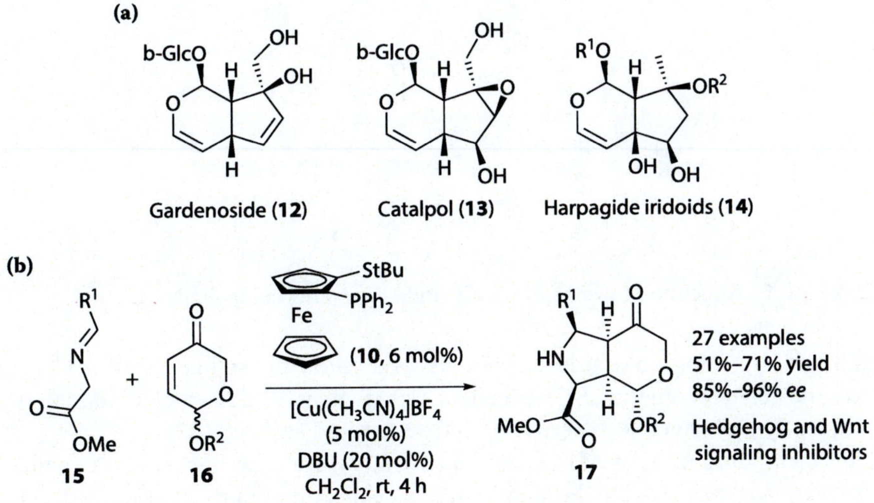

SCHEME 14.2 Synthesis and biological evaluation of an iridoid-inspired compound collection using a kinetic resolution of pyranones *via* a [2+3] cycloaddition of azomethine ylides: **(a)** examples of naturally occurring iridoids; **(b)** synthetic strategy for compound collection synthesis.

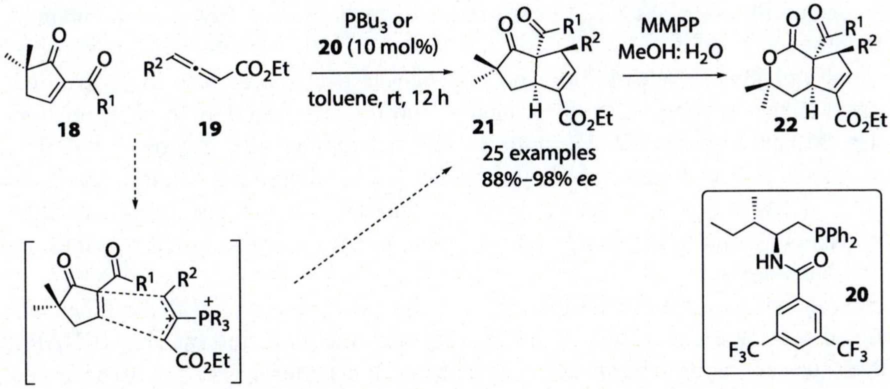

SCHEME 14.3 Synthesis and mechanism of an iridoid-inspired compound collection using phospine-catalyzed [3+2] cycloaddition reactions.

the ability of NP-inspired compounds to modulate distinct biological processes by simple, small, structural changes of the same scaffold.

Another variant of the [3+2] cycloaddition was also used to synthesize iridoid-inspired compound collections (Scheme 14.3).[31] Here, a phosphine-catalyzed formation of an allene-derived zwitterion formed the basis for a cycloaddition with activated unsaturated ketones (**18**) to furnish the desired bicylic ring system. This reaction was initially conducted racemically using tributylphosphine to deliver 25 analogues in moderate-to-excellent yields. An enantioselective variant of this reaction was also

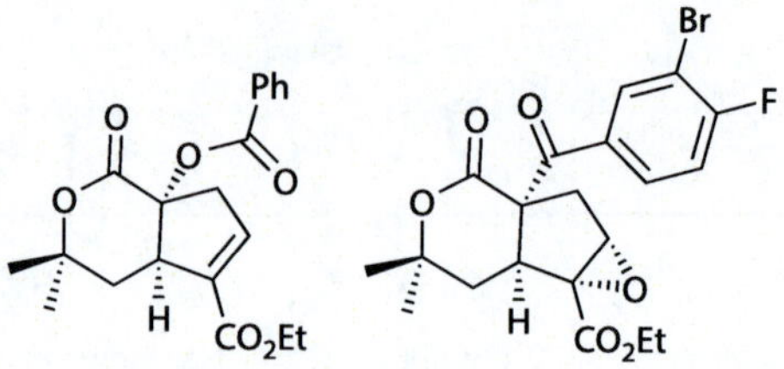

FIGURE 14.5 Structure of two neuritogenic, iridoid-inspired compounds.

established, by using an amino alcohol–derived chiral phosphine catalyst (**20**), to deliver the desired compounds in excellent *ee*'s. Baeyer-Villiger-type oxidation was subsequently performed to access the 5,6 fused iridoid scaffold (**22**).

Following the synthesis of the desired core structures, the library was expanded further, by functionalizing the enone alkene. All compounds synthesized using this approach were screened for their ability to promote neurite outgrowth in primary neuronal cultures of hippocampi from E18/E19 Sprague Dowley rats. Several compounds displayed neuritogenic effects and were subjected to further in-depth phenotypic analysis (Figure 14.5).

Cycloaddition reactions are also powerful tools for merging two different bioactive scaffolds. Both tropanes and pyrrolidines are found ubiquitously in nature, and NPs containing them possess a broad range of bioactivities. However, compounds containing both scaffolds had not been reported until recently. Using a copper (I)-catalyzed [3+2] cycloaddition reaction, it was possible to fuse the pyrrolidine scaffold to the tropane scaffold by using appropriately functionalized azomethine ylides (**26**) and tropanes (**25**) (Scheme 14.4).[32] Conditions involving the chiral phosphine ligand (**27**) were developed, which effected the kinetic resolution racemic tropanes to deliver the desired annulated product **28** as well as the starting material in great yields and *ee*'s. The conditions tolerated a range of substituted tropanes and azomethine ylides.

Subsequently, a one-pot process was developed to directly react the kinetically resolved starting material with a different azomethine ylide and racemic BINAP to deliver two compounds with differing substitution patterns and opposing stereochemistry (Scheme 14.5). This procedure was applied to a range of substrates and all products were obtained in high yields and *ee*'s. Both products were always separable by traditional column chromatography, making this procedure amenable to the synthesis of multiple analogues.

In addition to intermolecular cycloadditions, *intramolecular* variants can be equally efficacious in delivering complex scaffolds in a few synthetic steps. This approach was applied to the synthesis of a compound collection based on the 2,3-pyrrolidino-3,4-piperidine scaffold (**35**; Scheme 14.6).[33] This is found in a number of alkaloids endowed with varying biological activities. For the synthesis of this scaffold, a one-pot Boc deprotection, imine formation, dipolar cycloaddition sequence was developed to generate the desired scaffold in high yield (Scheme 14.6). Using a copper catalyst and a chiral phosphine ligand (**34**)

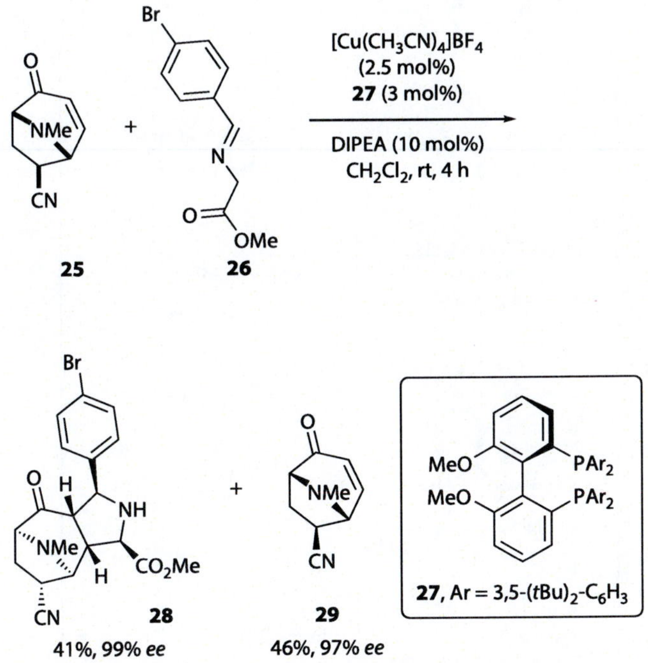

SCHEME 14.4 Fused tropane–pyrrolidine scaffold synthesized by the kinetic resolution of racemic tropanes using asymmetric, copper (I)-catalyzed [3+2] cycloaddition reactions.

rendered the process enantioselective. The resulting products are themselves substrates for azomethine ylide synthesis, and this strategy was employed to perform a second, completely diastereoselective cycloaddition to yield a complex, polycyclic scaffold (**36**) in good yields. This process could also be adapted to be performed in one pot, enabling the rapid synthesis of multiple compounds.

Another interesting scaffold present in biologically relevant compounds is the pyridone. NPs bearing this core had been identified as promoting neurite outgrowth and were thus touted as potential neuroprotective agents (Scheme 14.7a).[34,35] To explore this compound class further, our group developed an expedient synthesis of both a 4-hydroxy-2-pyridone and a 2,4-dimethoxypyridine precursor scaffold (Scheme 14.7b).[36]

From the dibrominated precursor **39**, a regioselective lithium–halogen exchange, followed by trapping with an aldehyde, delivered secondary alcohols (**40**) in good-to-moderate yields. Suzuki couplings allowed the second ring substituent to be installed in excellent yields (**41**). Oxidation of the secondary alcohol furnished the dimethoxypyridines (**42**) in excellent yields, while subsequent demethylation on selected derivatives yielded the final pyridones (**43**). A final library of 59 compounds was submitted for testing in a neurite outgrowth–promoting assay in SH-SY5Y cells. Compound **44** (Scheme 14.7b) was found to potently induce neurite outgrowth in comparison with the DMSO control. This is a particularly significant finding as this compound and related analogues were accessed in just five steps from commercially

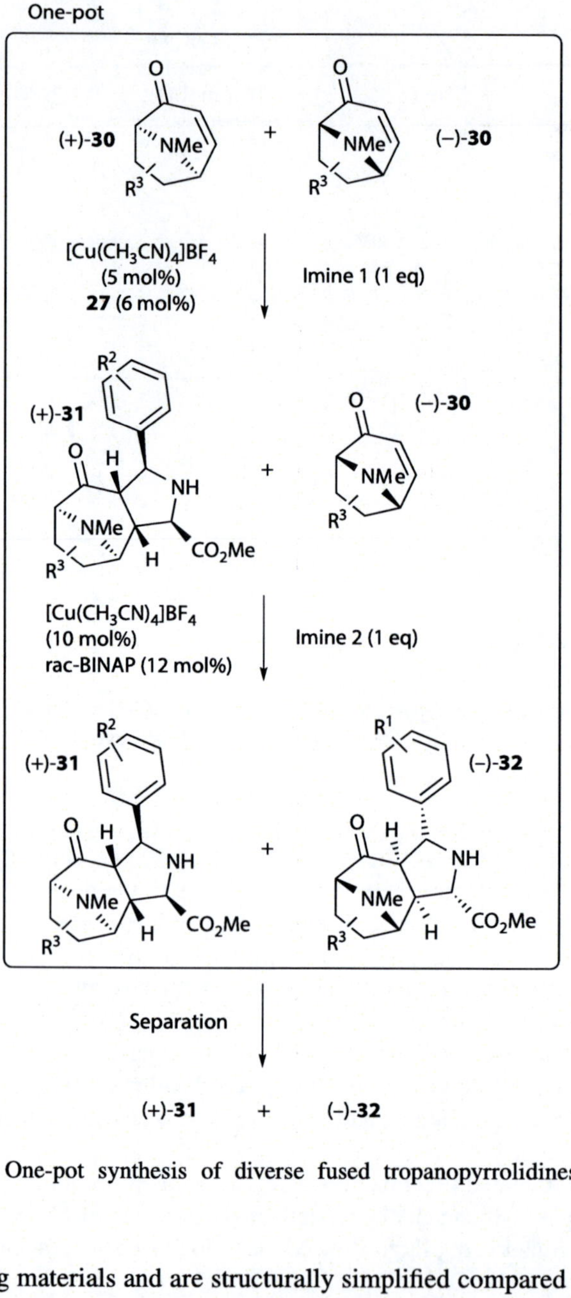

SCHEME 14.5 One-pot synthesis of diverse fused tropanopyrrolidines with opposing stereochemistry.

available starting materials and are structurally simplified compared with the parent NP while still retaining good bioactivity. Compound **44** was subsequently used to identify its protein target and thus the reason for its neuritogenic activity. This and related case studies will be discussed in Section 14.3.

Another example of strategies to access NP-inspired libraries is the recent synthesis of a withanolide-inspired collection.[37] Withanolides belong to a large family of over 300 compounds with a broad range of bioactivities.[38] A vast number of these

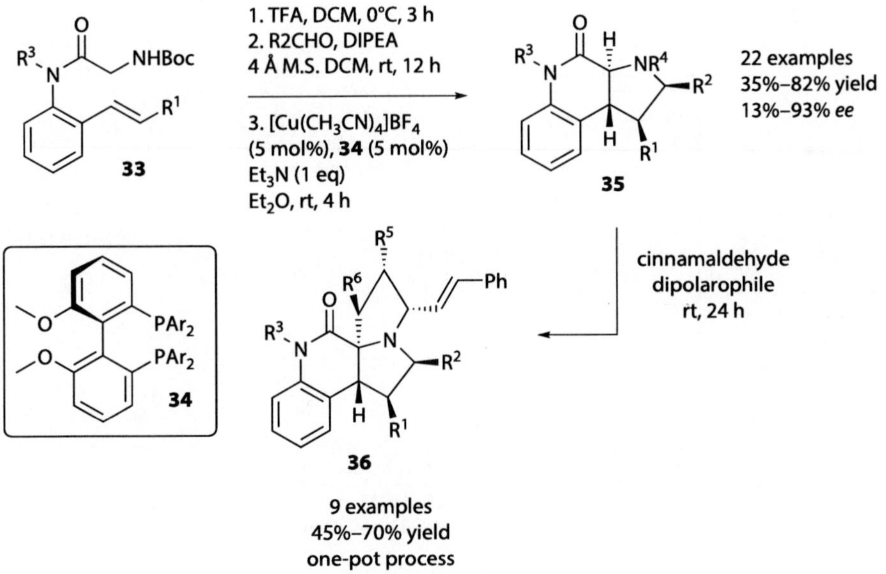

SCHEME 14.6 Enantioselective intramolecular dipolar cycloaddition reactions. Two sequential cycloadditions (the first enantioselective and the second diastereoselective) were performed in one pot to deliver the desired polycyclic scaffold in excellent *ee*'s as a single diastereomer.

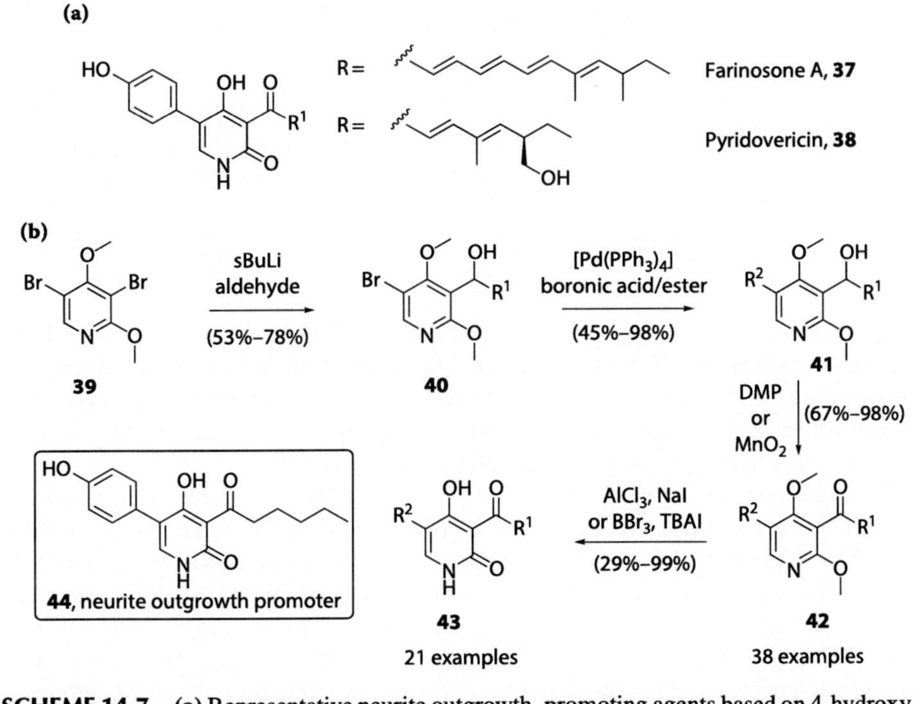

SCHEME 14.7 **(a)** Representative neurite outgrowth–promoting agents based on 4-hydroxy-2-pyridones; **(b)** synthesis of a pyridone- and dimethoxypyridine-based compound collection. A newly discovered neurite outgrowth–promoting compound (**44**) is highlighted.

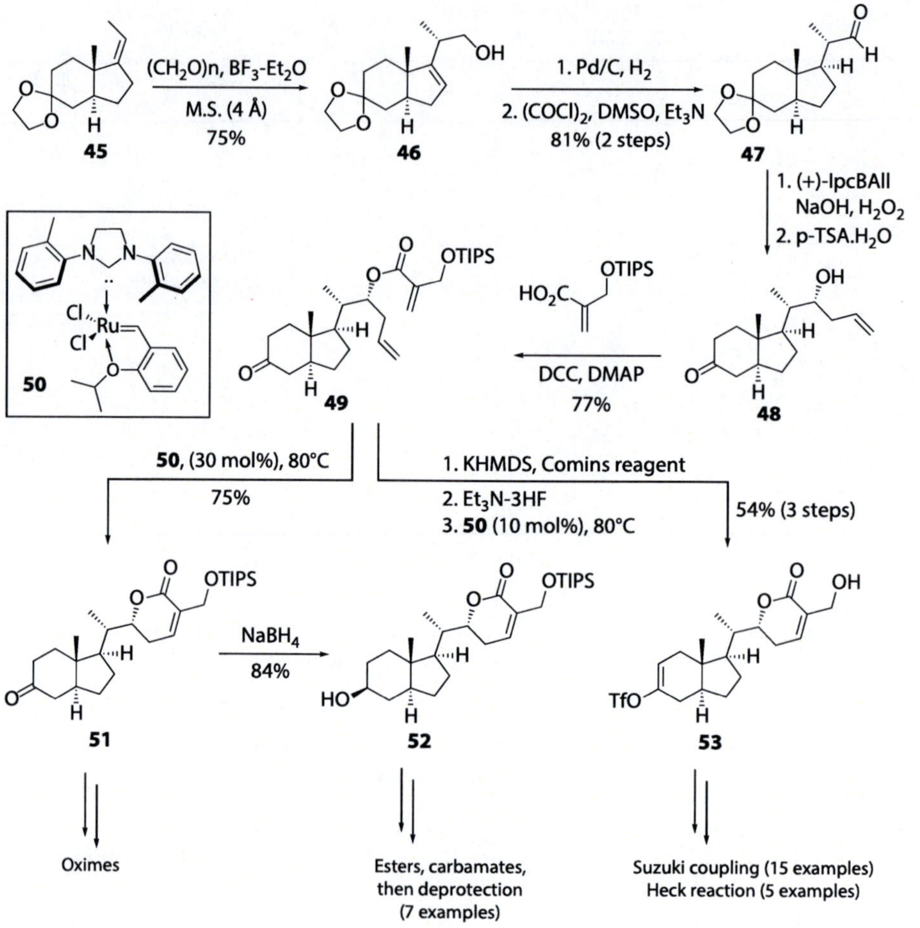

SCHEME 14.8 Synthesis of a withanolide-inspired compound collection.

compounds contain the *trans*-hydrindane-dehydro-∂-lactone scaffold. It was thus reasoned that this core was essential for bioactivity, and the different biological effects could be tailored by varying substitution patterns surrounding the core. As a result of this, a synthetic route was devised to access the parent scaffold with the potential for accessing a range of analogues (Scheme 14.8).

The synthetic route began with (*Z*)-olefin **45**, which had been prepared in large quantities from the Hajos–Parrish ketone in good yields. Synthesis of the allylic alcohol **46**, followed by a diastereoselective hydrogenation and a Swern oxidation, yielded the desired aldehyde **47** in good yields. Asymmetric allylation, followed by deprotection of the acetal, delivered the homoallylic alcohol **48** in good yield over two steps. Ester formation with DCC (N,N′-Dicyclohexylcarbodiimide) and DMAP (4-Dimethylaminopyridine) yielded the precursor (**49**) to the first diversification step. One route involved performing the ring-closing metathesis directly. The resulting product (**51**) contained a ketone that could be converted to an oxime or an alcohol (**52**), which was converted to esters

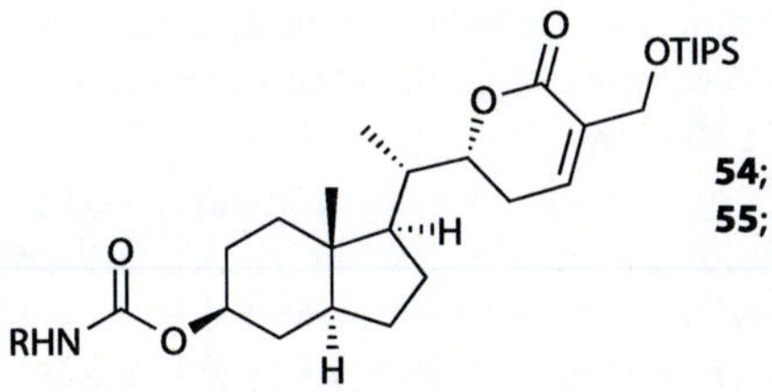

54; R = Ph, IC_{50} = 1.8 μM
55; R = 4-ClC_6H_4, IC_{50} = 2.3 μM

FIGURE 14.6 Hedgehog pathway inhibitors from a withanolide-inspired compound collection.

and carbamates. Alternatively, the enol triflate was synthesized and subsequently subjected to ring-closing metathesis. The resulting compound (**53**) could be subjected to a range of metal-catalyzed reactions to further diversify the core scaffold. All compounds were subsequently screened for the ability to modulate several cellular signaling pathways. Several compounds were found to potently inhibit Hedgehog signaling in C3T/10T1/2 cells (Figure 14.6). Further work enabled the complete characterization of these molecules as Hedgehog signaling inhibitors and the identification of their molecular targets. This will be discussed in more detail in Section 14.3.

14.3 TARGET IDENTIFICATION

It has been widely discussed and acknowledged that NPs and their derivatives have been a rich source of lead compounds and eventually drugs in medicinal research. Key examples include tubulin modulators such as Taxol®, Vinblastine, and Colchicine, which are used for different indications but mostly known for their anticancer activity.[39,40] Other well-studied examples include the immunosuppressants Rapamycin and Cyclosporine, used in patients that have received organ transplants to reduce the rates of rejection, and opiates used to treat pain, including codeine and morphine. Aside from all being NPs, what these compounds have in common is that they were identified through phenotypic screening approaches, without any knowledge of their molecular targets. Phenotypic screens aim to recreate a disease-relevant phenotype in a model organism or cell.[41] The hope is that if a small molecule or therapeutic is found to be active in such a model, it is more likely to succeed in a real-life, clinical setting.

The advantages of the phenotypic approach are that compounds are preselected for their ability to enter the cells and must remain stable enough to exert their activity. As the approach does not require any knowledge of the molecular targets, it is unbiased and not susceptible to a choice of target that has no disease relevance. Recently, it has been reported that a large percentage of *first-in-class* therapeutics, that is, those that display a novel mode of action, were identified *via* phenotypic screens.[42] As a result, there is once again a growing interest in identifying bioactive compounds using this approach.[43] Disadvantages of the phenotypic approach are that the disease model may not accurately recapitulate the disease state, which would result in compounds being inactive in a clinical setting. One of the toughest challenges of the phenotypic approach is that once an interesting compound has been identified, one would typically want to determine the molecular target(s) responsible for the compound's activity.[44] Depending on the phenotype and the structural features of the molecule, this can be a considerable challenge.

Despite the challenges of target ID, phenotypic screening remains a popular choice for screening NP libraries, as a multitude of potential targets are associated with any given phenotype. NPs and their analogues or derivatives are well-positioned to target key cellular processes, as they have often evolved as protective agents or to mediate host–pathogen response and communication. With a returning interest in phenotypic screens, great efforts have been undertaken to enhance possible strategies for target ID and validation.[44–46] Current methods to tackle this problem, as well as case studies relating to the target ID of NP-inspired hit compounds will be discussed in the following section.

14.3.1 Educated Guesses

One of the most rapid methods for target ID is the educated guess. When key proteins are known to influence a particular phenotype, these can directly be assayed for, provided the protein can be readily accessed or produced and activity assays are reported in the literature. A typical example is tubulin, when the phenotype of interest is mitotic arrest. Tubulin plays a central role in all aspects of mitosis, and a large percentage of clinically approved antimitotics target it.[40] Thus, anyone undertaking an antimitotic phenotypic screen should always include a tubulin polymerization assay as a secondary screen in their assay cascade.[47] It is important to note that both stabilizers and inhibitors of polymerization can cause a mitotic arrest.

Another example of a strong correlation between a phenotype and a putative protein target is that between Hedgehog signaling and the membrane-bound receptor Smoothened (Smo).[48] Smo is a G-protein-coupled receptor that is translocated to the primary cilia upon Hedgehog signaling activation resulting from stimulation of the Patched receptor by the Sonic Hedgehog ligand. The NP Cyclopamine (Figure 14.7) has been identified as a Smo inhibitor several years after its cellular effects were first confirmed to be due to Hedgehog pathway inhibition.[49,50] Following this discovery, several Hedgehog inhibitors that bind to Smo have been identified.[51] The most notable of these is Vismodegib (Figure 14.7), a clinically approved antitumor compound that was identified through a phenotypic pathway screen employing a gene reporter assay.[52]

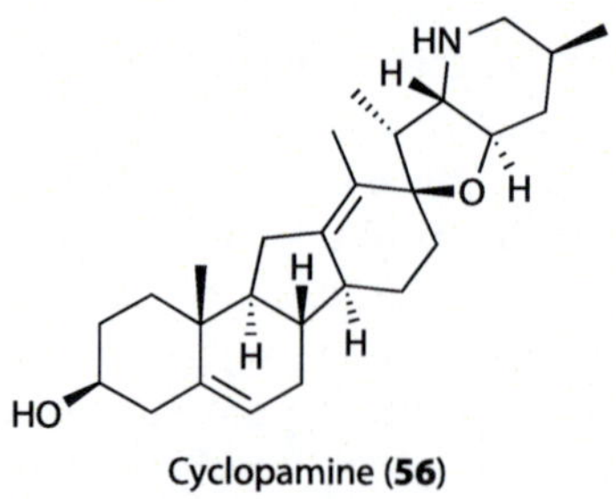

Cyclopamine (**56**)

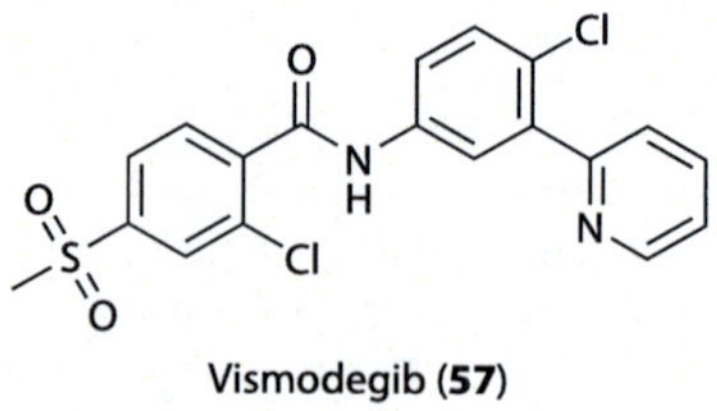

Vismodegib (**57**)

FIGURE 14.7 Structures of the Hedgehog pathway inhibitors that act by binding to Smoothened.

Given the strong link between Hedgehog and Smo, all compounds identified as Hh inhibitors, for example, in reporter gene assay, should subsequently be screened for Smo binding using, for example, a cell-based, fluorescent Cyclopamine competition assay. Fluorescent Cyclopamine binds to Smo in the cell and this can be imaged by fluorescence microscopy.[50] An inhibitor of the interaction gives rise to a marked reduction in fluorescence intensity. This observation was made for the most potent inhibitor based on the withanolide scaffold (Figure 14.6),[37] which shares some structural similarities to Cyclopamine resulting from its steroidal core. Binding to Smo was also confirmed using a radiometric assay using [^3H]-labeled Cyclopamine.

14.3.2 Image Analysis

If the phenotypic change induced by a small molecule can be clearly visualized using fluorescence or normal microscopy, this may offer clues as to the potential macromolecule that is being targeted.[53] This is particularly relevant for antimitotic drug discovery. Mitosis is a process in which multiple large-scale cellular rearrangements take place, and compounds interfering with these often produce very characteristic phenotypes. Tubulin inhibitors cause a complete disruption of the tubulin network, both in mitotic and notably in interphase cells. The latter is not the case in cells treated with different antimitotics. Two novel tubulin inhibitors (**58** and **59**; Figure 14.8) were identified by the Spring group by observing changes in the tubulin network in interphase cells, which were subsequently confirmed by a tubulin polymerization assay.[54,55]

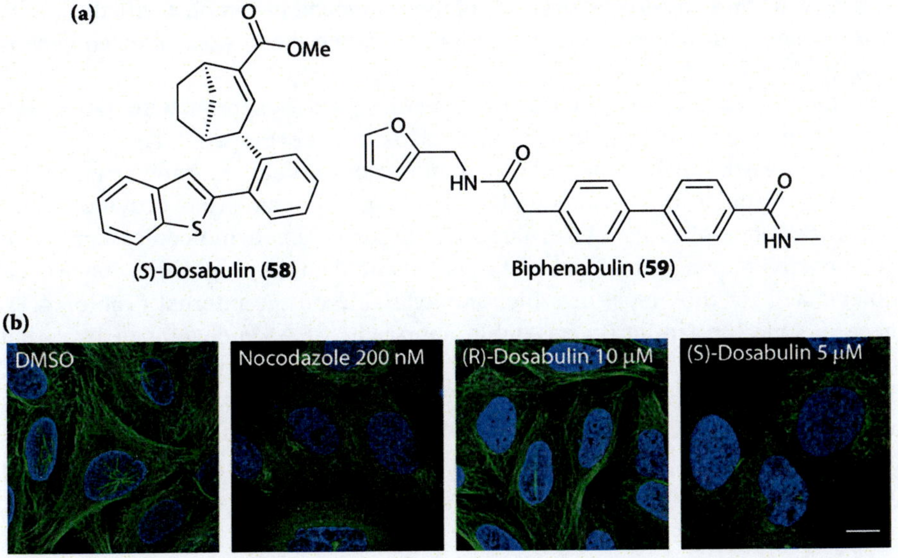

FIGURE 14.8 Tubulin inhibitors identified through phenotypic screening and image-based target identification: **(a)** structures of (*S*)-Dosabulin and Biphenabulin; **(b)** confocal images of cells treated with DMSO, Nocodazole (a tubulin inhibitor control), and both enantiomers of Dosabulin. Circular structures represent nuclei, while filaments represent α-tubulin, scale bar 10 μm. (Reproduced from Ibbeson, B.M. et al., *Nat. Commun.*, 5, 3155, 2014. With permission.)

14.3.3 Structural Similarities and the Computational Approach

Another useful approach for the identification of small-molecule targets is the structural similarity with compounds of known bioactivity and targets, or with a general core that is known to inhibit a particular enzyme class. In the first case, if a compound library has been designed with a particular NP in mind, for which the target(s) are known, there is a good chance that new analogues will bind to the same targets. In the second case, a structural motif may be known to bind a specific enzyme class. For example, aminopyrimidines are known to inhibit several kinases, as they can tightly bind to the hinge region. Kinases are a well-studied enzyme class with many potent and selective inhibitors known and commercially available. Additionally, commercial kinase profiling services are available for relatively modest amounts. If the small molecule of interest displays such a well-recognized motif, then it would be logical to screen said compound against a panel of kinases. Alternatively, kinases may be selected to play a role in the desired phenotype (see Section 14.3.1).

A successful example of this approach is highlighted in the recent discovery of neurite outgrowth–promoting compound (**44**) based on a Militarinone-inspired compound collection (Scheme 14.7b).[36] The hydroxypyridone core was a known hinge-binding motif, and several kinases were known to play a key role in neuritogenesis. Based on this, 67 kinases were tested, and MAP4K4 was the sole one inhibited. siRNA knockdown of MAP4K4 had previously been shown to enhance neurite outgrowth, and hence it was reasoned that a small-molecule inhibitor would behave in the same way. An x-ray crystal structure of the compound in complex with the kinase was obtained, with which the predicted hinge-binding mode was validated experimentally (Figure 14.9).

To rationalize the small-molecule structure-guided approach to target ID, several computational approaches have also been developed.[56] As previously mentioned, when small molecules with annotated targets are known which are structurally similar to a compound of interest, this can often provide a hint toward its potential target. With large databases of small molecules annotated with bioactivity data such as Chembl freely available, many algorithms have been implemented for this, most of which are available on the internet (Table 14.1). These include the similarity ensemble approach (SEA),[57] developed by Brian Shoichet's lab; SwissTargetPrediction,[58] developed by Olivier Michielin's lab; and the SPiDER tool, developed by Gisbert Schneider's lab.[59] Although results can sometimes be inconsistent between different tools, should a significant and consistent result be displayed using all available tools, this would provide a strong testable hypothesis.

14.3.4 Proteomic Approach

For novel compounds or phenotypes for which there is little or no precedent, usually the target cannot be predicted. In these cases, the most widely used target ID method is the affinity-based proteomic approach (Figure 14.10).[60,61] A small molecule of interest is typically immobilized on solid support and then exposed to a

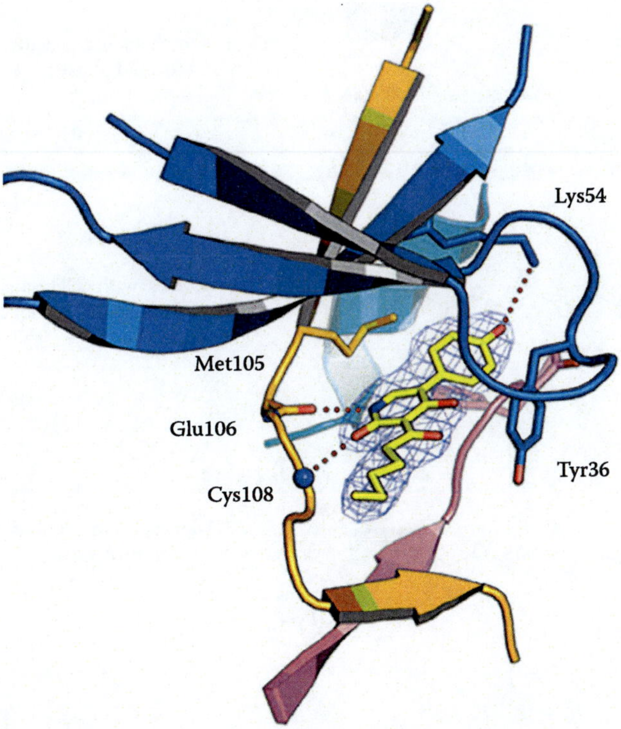

FIGURE 14.9 X-ray crystal structure of **44** in complex with MAP4K4. PDB entry 4RVT. (Reproduced from Schröder, P. et al., *Angew. Chem. Int. Ed.*, 54(42), 12398, 2015. With permission.)

TABLE 14.1
Freely Available Software for the Computational Prediction of Small-Molecule Targets

Name	Producer	Web Address
Similarity Ensemble Approach (SEA)	Kaiser/Shoichet's lab	http://sea.bkslab.org/
SwissTargetPrediction	Michelien's lab	http://www.swisstargetprediction.ch/
ChemProt	TU Denmark and Université Paris Diderot	http://potentia.cbs.dtu.dk/ChemProt/
Superpred	Structural Bioinformatics Group, Charité	http://prediction.charite.de/
TargetHunter	Xie's lab	http://www.cbligand.org/TargetHunter/

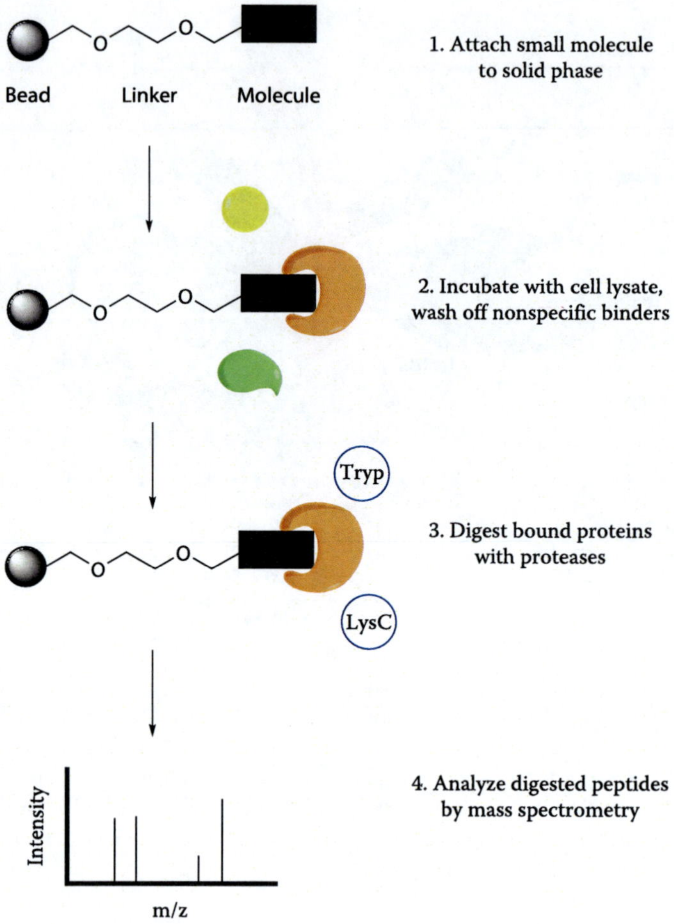

FIGURE 14.10 Schematic representation of affinity-based proteomic profiling for small molecules. A hit compound is covalently linked to a solid phase and incubated with cell lysates. After washing to remove nonspecific binders, the remaining proteins are hydrolyzed and the resulting peptides analyzed by mass spectrometry.

cell lysate. Ideally, the protein target should bind tightly to the immobilized probe. After washing off nonspecific binders, the remaining proteins are hydrolized using proteases such as trypsin or lysyl endopeptidase C (LysC) and the resulting peptides are analyzed by mass spectrometry.

Most of the steps in the proteomic workflow can and have been varied to achieve successful results. A crucial aspect of this approach is the design of a suitable negative probe, that is, a compound that is inactive in the primary screen. This should preferably be closely related to the positive probe, as the goal is to exclude nonspecific binders that bind to similar chemotypes equally. Two examples of successful negative probes are shown in Figure 14.11. For the NP Adenanthin, the exocyclic double bond was required for activity, and thus reducing it made an inactive

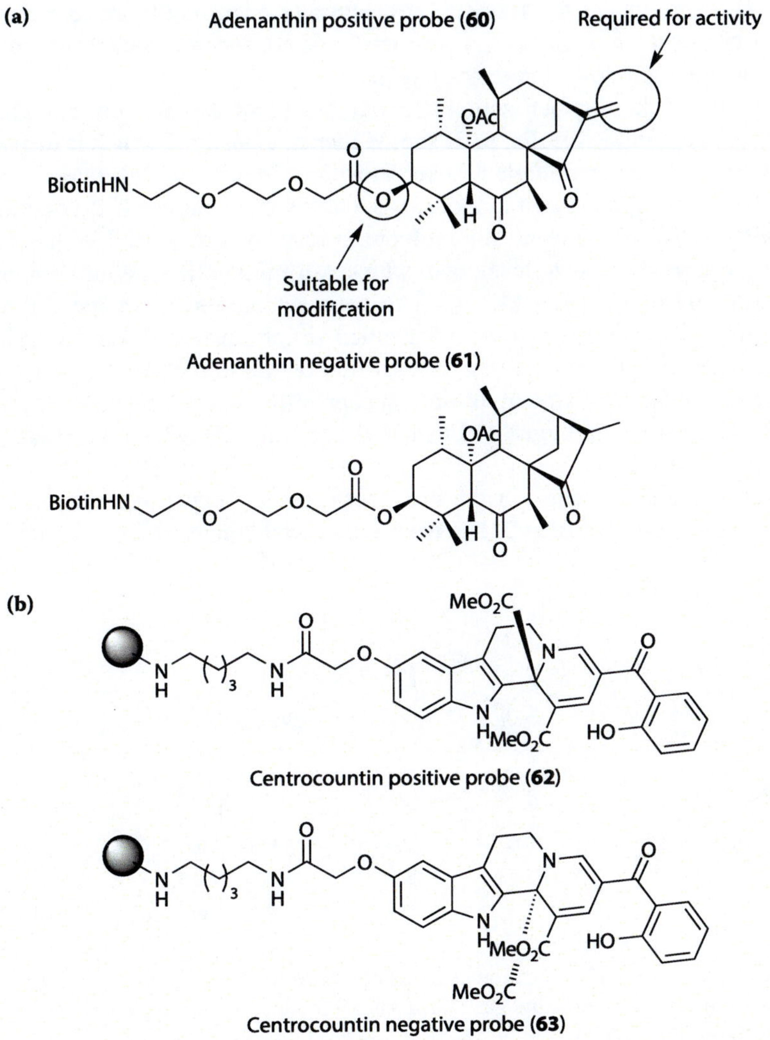

FIGURE 14.11 Positive and negative pull-down probes from two successful proteomic experiments: **(a)** adenanthin-based pull-down probes; **(b)** centrocountin-based pull-down probes.

compound, suitable for a negative probe.[62] For the antimitotic Centrocountins, the activity resided in one single enantiomer. Thus, the other enantiomer could be used as a negative control probe.[63]

Another key parameter that can be optimized is the choice of linker. A frequent choice is a variant of polyethyleneglycol (PEG). There are several prefunctionalized PEG linkers that are either commercially available or synthesized in a few steps to enable a range of chemoselective reactions to be performed on the hit compound of choice. These include amines, carboxylic acids, and azide-containing linkers. An interesting alternative linker contains a polyproline segment.[64] This motif adopts a rod-like structure that is not found in any naturally occurring biomolecules and as such has been shown to have very low levels of nonspecific background binding, enhancing binding to low-abundance targets.

The choice of the immobilization site may be one of the most difficult chemical parameters to optimize. Ideally, SAR should dictate a position, which is dispensable for activity and thus can tolerate a linker. However, in practice, this may not always be possible due to synthetic limitations, particularly if the molecule is complex, like many NPs are. To circumvent this problem, a resin functionalized with a reactive carbene precursor has been developed, which can immobilize the hit compound at several different sites (Figure 14.12).[65] This approach has the advantage of removing what is often a very difficult functionalization to introduce a linker on a hit compound, and one would expect that at least one of the immobilization sites would allow the hit compound to retain its binding capabilities toward the target. It has been successfully used in the target ID of methyl gerfelin (**64**), which was found to be glyoxylase 1 (GLO1).[66]

To improve quantification in the mass spectrometry experiment, stable isotope labeling in cell culture (SILAC) has been introduced (Figure 14.13). This involves

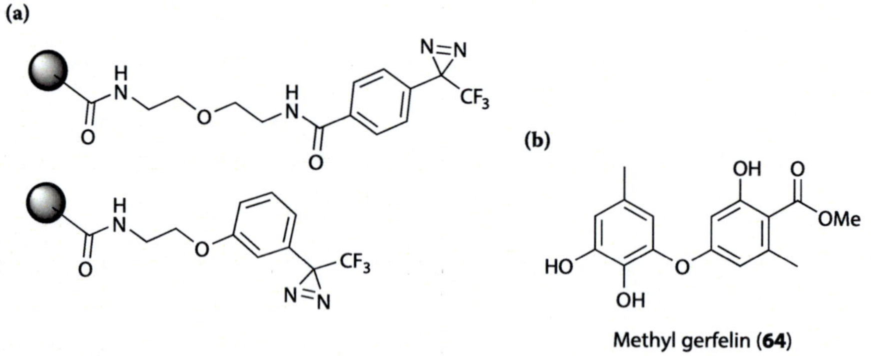

FIGURE 14.12 Target identification using a photochemically activatable resin: **(a)** different resins used; **(b)** structure of methyl gerfelin, a natural product whose target was identified with this approach.

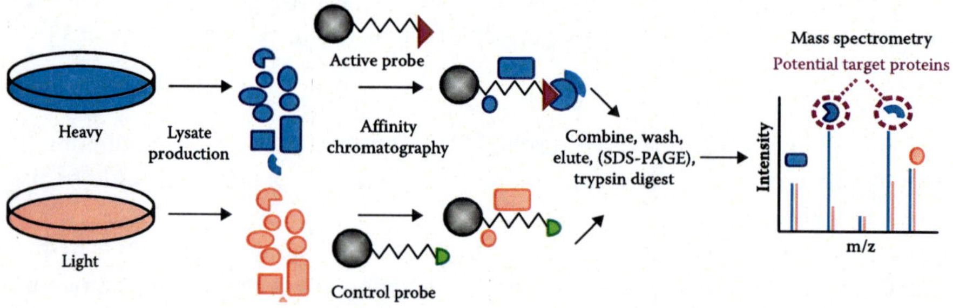

FIGURE 14.13 Workflow of the proteomic experiment using stable isotopes in cell culture (SILAC). (Reproduced from Ziegler, S. et al., *Angew. Chem. Int. Ed.*, 52, 2744, 2013. With permission.)

using two lysates with different isotopic labels on key amino acids (typically lysine and Arginine; however, other options exist). This allows ratios of heavy-to-light lysates to be determined after mixing samples from both positive and negative probe pull-downs. In this setup, it is crucial to perform the experiments using all possible combinations of probes and lysates, that is, positive probe with light lysate combined with negative probe and heavy lysate, as well as positive probe with heavy lysate combined with negative probe and light lysate. Only hits that are identified as such in both instances should be pursued further.

A recent example of the successful use of SILAC-based proteomics was the identification of tubulin and CSE1L as the targets of a tetrahydropyran-based compound collection.[67]

14.4 SUMMARY

NPs remain a rich source of bioactive compounds. To build on the plethora of privileged scaffolds present in nature, NP-derived compound collections allow access to compounds that possess favorable bioactivities but are not produced in nature. Several strategies exist to decide which areas of chemical space to target and how to populate them with synthetic chemistry efforts. These include SCONP, building of the relevant scaffold trees, and PSSC. Once an NP-inspired library has been constructed and screened phenotypically, the relevant targets must be identified. Several approaches for doing this have been outlined, and though further developments are required, with a combination of approaches, it should soon be possible to identify the targets of any bioactive small molecule.

REFERENCES

1. O'Connor, C. J.; Laraia, L.; Spring, D. R. Chemical genetics. *Chem. Soc. Rev.* **2011**, *40*, 4332–4345.
2. Dobson, C. M. Chemical space and biology. *Nature* **2004**, *432*, 824–828.
3. Newman, D. J.; Cragg, G. M. Natural products as sources of new drugs from 1981 to 2014. *J. Nat. Prod.* **2016**, *79*, 629–661.
4. Newman, D. J.; Cragg, G. M. Natural products as sources of new drugs over the 30 years from 1981 to 2010. *J. Nat. Prod.* **2012**, *75*, 311–335.
5. Clardy, J.; Walsh, C. Lessons from natural molecules. *Nature* **2004**, *432*, 829–837.
6. Harvey, A. L.; Edrada-Ebel, R.; Quinn, R. J. The re-emergence of natural products for drug discovery in the genomics era. *Nat. Rev. Drug Discov.* **2015**, *14*, 111–129.
7. Kristina, G.; Gisbert, S. Properties and architecture of drugs and natural products revisited. *Curr. Chem. Biol.* **2007**, *1*, 115–127.
8. Henkel, T.; Brunne, R. M.; Müller, H.; Reichel, F. Statistical investigation into the structural complementarity of natural products and synthetic compounds. *Angew. Chem. Int. Ed.* **1999**, *38*, 643–647.
9. Feher, M.; Schmidt, J. M. Property distributions: Differences between drugs, natural products, and molecules from combinatorial chemistry. *J. Chem. Inf. Comput. Sci.* **2003**, *43*, 218–227.
10. Achan, J.; Talisuna, A. O.; Erhart, A.; Yeka, A.; Tibenderana, J. K.; Baliraine, F. N.; Rosenthal, P. J.; D'Alessandro, U. Quinine, an old anti-malarial drug in a modern world: Role in the treatment of malaria. *Malar. J.* **2011**, *10*, 1–12.

11. Gaughan, C. E.; Lown, B.; Lanigan, J.; Voukydis, P.; Besser, H. W. Acute oral testing for determining antiarrhythmic drug efficacy. *Am. J. Cardiol.* **1976**, *38*, 677–684.
12. van Hattum, H.; Waldmann, H. Biology-oriented synthesis: Harnessing the power of evolution. *J. Am. Chem. Soc.* **2014**, *136*, 11853–11859.
13. Koch, M. A.; Schuffenhauer, A.; Scheck, M.; Wetzel, S.; Casaulta, M.; Odermatt, A.; Ertl, P.; Waldmann, H. Charting biologically relevant chemical space: A structural classification of natural products (SCONP). *Proc. Natl. Acad. Sci.* **2005**, *102*, 17272–17277.
14. Schuffenhauer, A.; Ertl, P.; Roggo, S.; Wetzel, S.; Koch, M. A.; Waldmann, H. The scaffold tree: Visualization of the scaffold universe by hierarchical scaffold classification. *J. Chem. Inf. Model.* **2007**, *47*, 47–58.
15. Nören-Müller, A.; Reis-Corrêa, I.; Prinz, H.; Rosenbaum, C.; Saxena, K.; Schwalbe, H. J.; Vestweber, D. et al. Discovery of protein phosphatase inhibitor classes by biology-oriented synthesis. *Proc. Natl. Acad. Sci.* **2006**, *103*, 10606–10611.
16. Koch, M. A.; Wittenberg, L.-O.; Basu, S.; Jeyaraj, D. A.; Gourzoulidou, E.; Reinecke, K.; Odermatt, A.; Waldmann, H. Compound library development guided by protein structure similarity clustering and natural product structure. *Proc. Natl. Acad. Sci.* **2004**, *101*, 16721–16726.
17. Dekker, F. J.; Rocks, O.; Vartak, N.; Menninger, S.; Hedberg, C.; Balamurugan, R.; Wetzel, S. et al. Small-molecule inhibition of APT1 affects Ras localization and signaling. *Nat. Chem. Biol.* **2010**, *6*, 449–456.
18. Lam, K. S. New aspects of natural products in drug discovery. *Trends Microbiol.* **2007**, *15*, 279–289.
19. Galloway, W. R.; Wilcke, D.; Nie, F.; Hadje-Georgiou, K.; Laraia, L.; Spring, D. R. Diversity-oriented synthesis: Developing new chemical tools to probe and modulate biological systems. In *Concepts and Case Studies in Chemical Biology*, Waldmann, H.; Janning, P., Eds. Wiley-VCH Verlag GmbH & Co. KGaA: Weinheim, Germany, **2014**; pp. 379–390.
20. Kim, J.; Kim, H.; Park, S. B. Privileged structures: Efficient chemical "navigators" toward unexplored biologically relevant chemical spaces. *J. Am. Chem. Soc.* **2014**, *136*, 14629–14638.
21. O'Connor, C. J.; Beckmann, H. S. G.; Spring, D. R. Diversity-oriented synthesis: Producing chemical tools for dissecting biology. *Chem. Soc. Rev.* **2012**, *41*, 4444–4456.
22. Wetzel, S.; Bon, R. S.; Kumar, K.; Waldmann, H. Biology-oriented synthesis. *Angew. Chem. Int. Ed.* **2011**, *50*, 10800–10826.
23. Lachance, H.; Wetzel, S.; Kumar, K.; Waldmann, H. Charting, navigating, and populating natural product chemical space for drug discovery. *J. Med. Chem.* **2012**, *55*, 5989–6001.
24. Kumar, K.; Waldmann, H. Synthesis of natural product inspired compound collections. *Angew. Chem. Int. Ed.* **2009**, *48*, 3224–3242.
25. Kaiser, M.; Wetzel, S.; Kumar, K.; Waldmann, H. Biology-inspired synthesis of compound libraries. *Cell. Mol. Life Sci.* **2008**, *65*, 1186–1201.
26. Breinbauer, R.; Vetter, I. R.; Waldmann, H. From protein domains to drug candidates—Natural products as guiding principles in the design and synthesis of compound libraries. *Angew. Chem. Int. Ed.* **2002**, *41*, 2878–2890.
27. Narayan, R.; Potowski, M.; Jia, Z.-J.; Antonchick, A. P.; Waldmann, H. Catalytic enantioselective 1,3-dipolar cycloadditions of azomethine ylides for biology-oriented synthesis. *Acc. Chem. Res.* **2014**, *47*, 1296–1310.
28. Narayan, R.; Bauer, J. O.; Strohmann, C.; Antonchick, A. P.; Waldmann, H. Catalytic enantioselective synthesis of functionalized tropanes reveals novel inhibitors of hedgehog signaling. *Angew. Chem. Int. Ed.* **2013**, *52*, 12892–12896.

29. Takayama, H.; Jia, Z.-J.; Kremer, L.; Bauer, J. O.; Strohmann, C.; Ziegler, S.; Antonchick, A. P.; Waldmann, H. Discovery of inhibitors of the Wnt and Hedgehog signaling pathways through the catalytic enantioselective synthesis of an iridoid-inspired compound collection. *Angew. Chem. Int. Ed.* **2013**, *52*, 12404–12408.
30. Rosa, T.; Monica, R. L.; Federica, M.; Giancarlo, A. S.; Francesco, M. Biological and pharmacological activities of iridoids: Recent developments. *Mini-Rev. Med. Chem.* **2008**, *8*, 399–420.
31. Dakas, P.-Y.; Parga, J. A.; Höing, S.; Schöler, H. R.; Sterneckert, J.; Kumar, K.; Waldmann, H. Discovery of neuritogenic compound classes inspired by natural products. *Angew. Chem. Int. Ed.* **2013**, *52*, 9576–9581.
32. Xu, H.; Golz, C.; Strohmann, C.; Antonchick, A. P.; Waldmann, H. Enantiodivergent combination of natural product scaffolds enabled by catalytic enantioselective cycloaddition. *Angew. Chem. Int. Ed.* **2016**, *55*, 7761–7765.
33. Vidadala, S. R.; Golz, C.; Strohmann, C.; Daniliuc, C. G.; Waldmann, H. Highly enantioselective intramolecular 1,3-dipolar cycloaddition: A route to piperidino-pyrrolizidines. *Angew. Chem. Int. Ed.* **2015**, *54*, 651–655.
34. Riese, U.; Ziegler, E.; Hamburger, M. Militarinone A induces differentiation in PC12 cells via MAP and Akt kinase signal transduction pathways. *FEBS Lett.* **2004**, *577*, 455–459.
35. Schmid, F.; Jessen, H. J.; Burch, P.; Gademann, K. Truncated militarinone fragments identified by total chemical synthesis induce neurite outgrowth. *Med. Chem. Commun.* **2013**, *4*, 135–139.
36. Schröder, P.; Förster, T.; Kleine, S.; Becker, C.; Richters, A.; Ziegler, S.; Rauh, D.; Kumar, K.; Waldmann, H. Neuritogenic militarinone-inspired 4-hydroxypyridones target the stress pathway kinase MAP4K4. *Angew. Chem. Int. Ed.* **2015**, *54*(42), 12398–12403.
37. Švenda, J.; Sheremet, M.; Kremer, L.; Maier, L.; Bauer, J. O.; Strohmann, C.; Ziegler, S.; Kumar, K.; Waldmann, H. Biology-oriented synthesis of a withanolide-inspired compound collection reveals novel modulators of hedgehog signaling. *Angew. Chem. Int. Ed.* **2015**, *54*, 5596–5602.
38. Liffert, R.; Hoecker, J.; Jana, C. K.; Woods, T. M.; Burch, P.; Jessen, H. J.; Neuburger, M.; Gademann, K. Withanolide A: Synthesis and structural requirements for neurite outgrowth. *Chem. Sci.* **2013**, *4*, 2851–2857.
39. Dumontet, C.; Jordan, M. A. Microtubule-binding agents: A dynamic field of cancer therapeutics. *Nat. Rev. Drug Discov.* **2010**, *9*, 790–803.
40. Jordan, M. A.; Wilson, L. Microtubules as a target for anticancer drugs. *Nat. Rev. Cancer* **2004**, *4*, 253–265.
41. Wagner, B. K.; Schreiber, S. L. The power of sophisticated phenotypic screening and modern mechanism-of-action methods. *Cell Chem. Biol.* **2016**, *23*, 3–9.
42. Swinney, D. C.; Anthony, J. How were new medicines discovered? *Nat. Rev. Drug Discov.* **2011**, *10*, 507–519.
43. Moffat, J. G.; Rudolph, J.; Bailey, D. Phenotypic screening in cancer drug discovery—Past, present and future. *Nat. Rev. Drug Discov.* **2014**, *13*, 588–602.
44. Ziegler, S.; Pries, V.; Hedberg, C.; Waldmann, H. Target identification for small bioactive molecules: Finding the needle in the haystack. *Angew. Chem. Int. Ed.* **2013**, *52*, 2744–2792.
45. Kapoor, S.; Waldmann, H.; Ziegler, S. Novel approaches to map small molecule–target interactions. *Bioorg. Med. Chem.* **2016**, *24*(15), 3232–3245.
46. Schürmann, M.; Janning, P.; Ziegler, S.; Waldmann, H. Small-molecule target engagement in cells. *Cell Chem. Biol.* **2016**, *23*, 435–441.
47. Mayer, T. U.; Kapoor, T. M.; Haggarty, S. J.; King, R. W.; Schreiber, S. L.; Mitchison, T. J. Small molecule inhibitor of mitotic spindle bipolarity identified in a phenotype-based screen. *Science* **1999**, *286*, 971–974.
48. Lin, T. L.; Matsui, W. Hedgehog pathway as a drug target: Smoothened inhibitors in development. *OncoTargets Ther.* **2012**, *5*, 47–58.

49. Taipale, J.; Chen, J. K.; Cooper, M. K.; Wang, B.; Mann, R. K.; Milenkovic, L.; Scott, M. P.; Beachy, P. A. Effects of oncogenic mutations in Smoothened and Patched can be reversed by cyclopamine. *Nature* **2000**, *406*, 1005–1009.
50. Chen, J. K.; Taipale, J.; Cooper, M. K.; Beachy, P. A. Inhibition of Hedgehog signaling by direct binding of cyclopamine to Smoothened. *Genes Dev.* **2002**, *16*, 2743–2748.
51. Chen, J. K. I only have eye for ewe: The discovery of cyclopamine and development of Hedgehog pathway-targeting drugs. *Nat. Prod. Rep.* **2016**, *33*, 595–601.
52. Robarge, K. D.; Brunton, S. A.; Castanedo, G. M.; Cui, Y.; Dina, M. S.; Goldsmith, R.; Gould, S. E. et al. GDC-0449—A potent inhibitor of the hedgehog pathway. *Bioorg. Med. Chem. Lett.* **2009**, *19*, 5576–5581.
53. Fetz, V.; Prochnow, H.; Bronstrup, M.; Sasse, F. Target identification by image analysis. *Nat. Prod. Rep.* **2016**, *33*, 655–667.
54. Laraia, L.; Stokes, J.; Emery, A.; McKenzie, G. J.; Venkitaraman, A. R.; Spring, D. R. High content screening of diverse compound libraries identifies potent modulators of tubulin dynamics. *ACS Med. Chem. Lett.* **2014**, *5*, 598–603.
55. Ibbeson, B. M.; Laraia, L.; Alza, E.; O'Connor, C. J.; Tan, Y. S.; Davies, H. M. L.; McKenzie, G.; Venkitaraman, A. R.; Spring, D. R. Diversity-oriented synthesis as a tool for identifying new modulators of mitosis. *Nat. Commun.* **2014**, *5*, 3155.
56. Cereto-Massagué, A.; Ojeda, M. J.; Valls, C.; Mulero, M.; Pujadas, G.; Garcia-Vallve, S. Tools for in silico target fishing. *Methods* **2015**, *71*, 98–103.
57. Keiser, M. J.; Roth, B. L.; Armbruster, B. N.; Ernsberger, P.; Irwin, J. J.; Shoichet, B. K. Relating protein pharmacology by ligand chemistry. *Nat. Biotechnol.* **2007**, *25*, 197–206.
58. Gfeller, D.; Grosdidier, A.; Wirth, M.; Daina, A.; Michielin, O.; Zoete, V. SwissTargetPrediction: A web server for target prediction of bioactive small molecules. *Nucleic Acids Res.* **2014**, *42*, w32–w38.
59. Reker, D.; Rodrigues, T.; Schneider, P.; Schneider, G. Identifying the macromolecular targets of de novo-designed chemical entities through self-organizing map consensus. *Proc. Natl. Acad. Sci.* **2014**, *111*, 4067–4072.
60. Vendrell-Navarro, G.; Brockmeyer, A.; Waldmann, H.; Janning, P.; Ziegler, S. Identification of the targets of biologically active small molecules using quantitative proteomics. In *Chemical Biology: Methods and Protocols*, Hempel, E. J.; Williams, H. C.; Hong, C. C., Eds. Springer: New York, **2015**; pp. 263–286.
61. Rix, U.; Superti-Furga, G. Target profiling of small molecules by chemical proteomics. *Nat. Chem. Biol.* **2009**, *5*, 616–624.
62. Liu, C.-X.; Yin, Q.-Q.; Zhou, H.-C.; Wu, Y.-L.; Pu, J.-X.; Xia, L.; Liu, W. et al. Adenanthin targets peroxiredoxin I and II to induce differentiation of leukemic cells. *Nat. Chem. Biol.* **2012**, *8*, 486–493.
63. Dückert, H.; Pries, V.; Khedkar, V.; Menninger, S.; Bruss, H.; Bird, A. W.; Maliga, Z. et al. Natural product–inspired cascade synthesis yields modulators of centrosome integrity. *Nat. Chem. Biol.* **2012**, *8*, 179–184.
64. Sato, S.-I.; Kwon, Y.; Kamisuki, S.; Srivastava, N.; Mao, Q.; Kawazoe, Y.; Uesugi, M. Polyproline-rod approach to isolating protein targets of bioactive small molecules: Isolation of a new target of indomethacin. *J. Am. Chem. Soc.* **2007**, *129*, 873–880.
65. Kanoh, N. Photo-cross-linked small-molecule affinity matrix as a tool for target identification of bioactive small molecules. *Nat. Prod. Rep.* **2016**, *33*, 709–718.
66. Kawatani, M.; Okumura, H.; Honda, K.; Kanoh, N.; Muroi, M.; Dohmae, N.; Takami, M. et al. The identification of an osteoclastogenesis inhibitor through the inhibition of glyoxalase I. *Proc. Natl. Acad. Sci. USA.* **2008**, *105*, 11691–11696.
67. Voigt, T.; Gerding-Reimers, C.; Ngoc Tran, T. T.; Bergmann, S.; Lachance, H.; Schölermann, B.; Brockmeyer, A.; Janning, P.; Ziegler, S.; Waldmann, H. A natural product inspired tetrahydropyran collection yields mitosis modulators that synergistically target CSE1L and tubulin. *Angew. Chem. Int. Ed.* **2013**, *52*, 410–414.

15 On the Chemistry and Biology of the Marine Macrolides Zampanolide and Dactylolide

Karl-Heinz Altmann, Simon Glauser, and Tobias Brütsch

CONTENTS

15.1 INTRODUCTION

Natural products are a highly prolific source of new drugs and leads for drug discovery, with a substantial fraction of current prescription medicines being derived from a natural product, either directly or indirectly.[1–3] In addition, at a more basic level, natural products, or related (semi)synthetic derivatives, are indispensable tools for mechanistic investigations in chemical biology.[3,4] The high propensity of natural products to interact with biological macromolecules in general, and with proteins in particular (i.e., potential drug targets), has been suggested to arise from their inherent, inborn interactions with cellular proteins during their own biosynthesis or when exerting a natural biological function. Thus, natural products are thought to occupy biologically relevant chemical space *a priori* and as a consequence represent validated starting points for the identification of ligands that may modulate protein function.[5–7] Perhaps equally important, the occurrence of sp^3 carbon-rich scaffolds in many natural products allows for the fine-tuning of the spatial arrangement of those functional groups that interact with target biomacromolecules.[8] In this context, natural products of marine origin hold a distinct position, in light of their enormous structural diversity and broad range of biological activities. Compared to terrestrial natural products, a higher incidence of significant bioactivity has been noted for marine secondary metabolites. These activities are often associated with a high degree of chemical novelty; that is, many newly discovered bioactive marine natural products display novel types of chemical structures.[5,9] When trying to understand these observations, it needs to be noted that organisms from the kingdoms *Archaea*, *Bacteria*, *Protozoa*, *Chromista*, and *Fungi* are not equipped with any physical means of protection[10]; instead, they defend themselves by releasing secondary metabolites, for example, toxins,[11] into the environment.[12] The same is also true for many algae and lower animals of the phyla *Porifera* and *Cnidaria*. In this context, it has been speculated that the sometimes extraordinary potency of marine secondary metabolites could be traced back to their high dilution in ocean water.[13] It is important to note, however, that the biological activities of marine secondary metabolites are not limited to cytotoxic or antiproliferative effects, but cover a much broader range of pharmaceutically relevant outcomes.

The vast majority of natural product–derived drugs as of today originate from terrestrial organisms.[1,2] However, in light of the above observations, it is not surprising that, more recently, the marine environment has emerged as an attractive, complementary source of bioactive secondary metabolites with potential applications in drug discovery and chemical biology.[5] Marine organisms are the source of an ever-increasing number of new structures, with more than 1100 new compounds described in 2013 alone.[14] The pace of discovery of new marine natural products is driven by advances in a number of different areas that include the discovery of a large number of new marine species every year; the increased efficiency of genome sequencing, which enables the identification of a steadily growing number of biosynthetic gene clusters[5]; an increase in the number of investigations of samples from previously difficult-to-access regions; and, finally and perhaps most importantly, advances in nuclear magnetic resonance (NMR) spectroscopy that allow for structure elucidation with trace amounts of material.[15] At the same time, the elucidation of molecular targets and mechanisms of action of natural products is becoming both faster and more material efficient.[5] Notwithstanding these recent advances, the set of organisms that has been scrutinized so far for bioactive secondary metabolites is still very limited and much remains to be discovered.[14]

Marine secondary metabolites have served as starting points for successful drug development, although the number of marine natural product–derived marketed drugs at this point in time is still limited. Marine natural product–derived drugs include the anticancer drugs trabectidin, which is an unmodified natural product (ecteinascidine-743) that is, however, produced by semisynthesis from a terrestrial natural product (cyanosafracin B)[4]; the truncated halichondrin B derivative eribulin[16]; and the antibody–drug conjugate brentuximab vedotin; the latter contains a derivative of the marine natural product dolastatin 10, monomethyl auristatin,[17] as the active drug payload.[18] The only marine-derived drug that has been approved so far outside of the oncology area is ziconotide, a peptide from a cone snail that is used for the treatment of pain.[19] Many more compounds that are derived from marine natural products are at various stages of clinical development.[20–23]

Within the broader context of natural product–based drug discovery and development, an important distinguishing attribute of trabectidin and eribulin is the fact that their production relies on complex multistep synthetic processes: While trabectidin is produced in 21 chemical steps from the terrestrial natural product cyanosafracin B (which can be obtained by fermentation of *Pseudomonas fluorescens*),[24] the fully synthetic truncated halichondrin B derivative eribulin is obtained in 35 steps (longest linear sequence) from commercially available starting materials.[25] These examples are illuminating, as they reflect a profound change in the perception of the role of chemical synthesis in natural product–based drug discovery and development, which had long been confined to semisynthetic derivatization; in comparison, total synthesis was mostly considered impractical and too costly. The rising importance of the *de novo* chemical synthesis of natural products and natural product analogs in drug-related research is due to (i) the profound advances in synthetic methodology that we have witnessed over the last decades (and that will continue to occur in the future), but it is also owed to the fact that (ii) a growing number of natural products with interesting lead potential can be obtained from the natural source only in low quantities. This is particularly true for, but not limited to, marine secondary metabolites, where detailed biochemical, cellular, or pharmacological studies are often restricted, or even rendered completely impossible, by the insufficient availability of material for broad-based profiling. In many cases, compound (re)supply is highly limited and nonsustainable[4]; not even considering ecological issues, the sheer cost to collect a marine organism as the source of the desired natural product(s) can be a prohibitive factor.[11] While this problem can be overcome for some marine bacteria by large-scale fermentation,[5] more delicate organisms, such as sponges, are difficult to cultivate.[11] As an additional complication, relevant gene clusters may remain silent under breeding conditions, although this problem has been resolved by mixed fermentation, in some cases.[5] As a consequence, comprehensive biological studies with marine secondary metabolites often have to rely on *de novo* chemical synthesis as a means to access sufficient quantities of material.[8,26] Importantly, apart from eribulin, a fully synthetic, highly complex natural product (the marine polyketide discodermolide)[27] and a complex natural product analog (the epothilone analog sagopilone)[28] have entered clinical trials.

However, the value of total synthesis for drug discovery and chemical biology extends far beyond the basic issue of substance supply for biological studies or clinical trials. At the most fundamental level, the availability of stereochemically defined synthetic material can be essential to confirm (or refute) the original structure assignment of a natural product,[29–32] notwithstanding the existence of highly sophisticated modern analytical tools for structure elucidation. From a medicinal chemistry perspective, the chemistry developed in the course of a total synthesis provides the basis for structure–activity relationship (SAR) studies in areas of structural space that are not accessible either through semisynthesis or the genetic manipulation of producer organisms.[33–36] The exploration of structural space around a natural product, however, is essential for understanding the relationship between molecular structure and biological activity, which is required, for example, for the design and synthesis of proper mechanistic probes for chemical biology studies. It may also lead to highly active, but less complex analogs and even to completely new chemotypes for interference with a given pharmacological target.[34,36,37] As an illustrative example, our own work on epothilones has produced hypermodified analogs that may be considered as new structural scaffolds for microtubule inhibition and some of which represent *nonnatural* natural products.[34]

Our own group has a long-standing interest in the total synthesis of bioactive natural products and natural product analogs as an enabling means for extended biochemical and cellular studies and the elucidation of SARs. In this chapter, we will detail some of our recent synthetic work on the marine macrolide zampanolide and discuss how this work has allowed us to address questions of biological relevance and/or has led to structurally simplified templates for further medicinal chemistry research.

15.2 ZAMPANOLIDE AND DACTYLOLIDE: ISOLATION AND STRUCTURE

(−)-Zampanolide (**1**) (Figure 15.1) is a 20-membered macrolide that was first reported in 1996, when Tanaka and Higa described the isolation of this intriguing marine metabolite from the sponge *Fasciospongia rimosa*, collected near Cape Zampa, off the coast of Okinawa.[38] The gross structure of **1**, including the relative stereochemistry of its macrolactone core, was established by means of high-resolution fast-atom bombardment mass spectrometry (HR-FABMS) and extensive two-dimensional NMR experiments; however, the configuration of the C20 stereocenter could not be determined at the time. The absolute configuration of **1** was ascertained only 5 years later by means of total synthesis of its nonnatural (+)-isomer (*ent*-**1**).[39,40] It is noteworthy, however, that no X-ray crystal structure of **1** has been reported to date.

(−)-Zampanolide (**1**) features a highly unsaturated 20-membered macrolactone core structure with an embedded 2,6-*syn*-disubstituted tetrahydropyran (THP) ring, carrying an exocyclic C=C bond. C19 of the macrocycle is connected to a doubly unsaturated amide side chain via a highly unusual hemiaminal linkage; the latter structural motif is only found in a limited number of other bioactive marine metabolites such as the antibiotic echinocandin[41] or the antitumor natural products mycalamide,[42,43] spergualin,[44,45] and upenamide (Figure 15.2).[46]

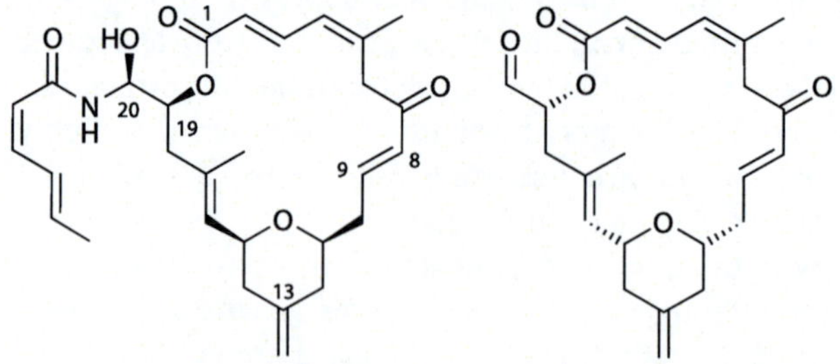

(−)-Zampanolide (**1**) (+)-Dactylolide (**2**)

FIGURE 15.1 Molecular structure of (−)-zampanolide (**1**) and (+)-dactylolide (**2**).

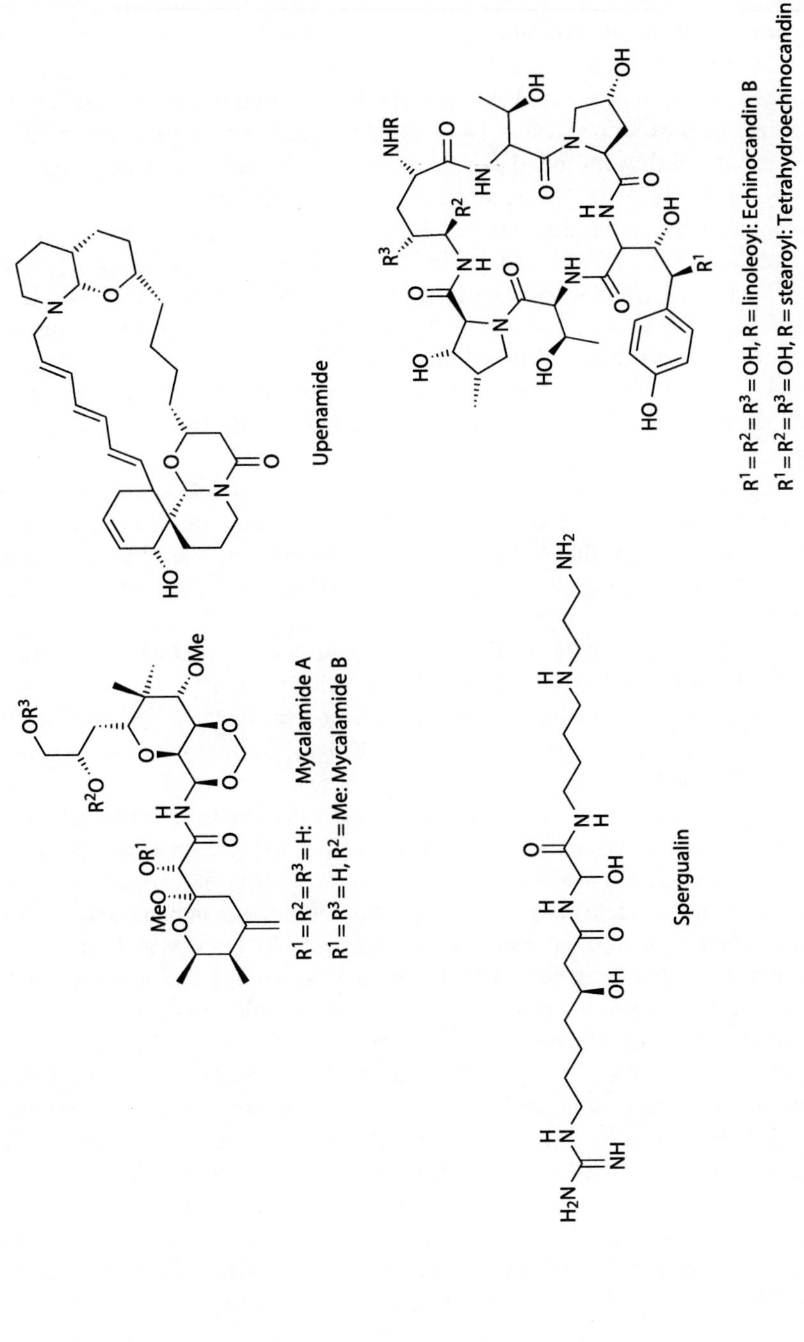

FIGURE 15.2 Molecular structures of natural products incorporating a hemiaminal linkage.

The isolation process delivered 3.9 mg of (−)-zampanolide (**1**) from 480 g of sponge material, alongside of 13.7 mg of the known bioactive marine natural product latrunculin A.[38] In 2009, that is, 13 years after its original discovery, Northcote, Miller, and coworkers reported the reisolation of (−)-zampanolide (**1**) from the Tongan sponge *Cacospongia mycofijiensis*[47] and also confirmed that the compound exhibits potent *in vitro* antiproliferative activity against different cancer cell lines, as had also been found by Tanaka and Higa.[38] (Quite surprisingly, no biological data on **1** had been reported between 1996 and 2009.) Importantly, Northcote, Miller, and coworkers also established that (−)-zampanolide (**1**) was a new microtubule-stabilizing agent and thus inhibited cancer cell proliferation by a mechanism that is analogous to that of the established anticancer drug paclitaxel.

Five years after the initial discovery of (−)-zampanolide (**1**), in 2001, Riccio and coworkers reported the isolation and preliminary biological evaluation of (+)-dactylolide (**2**) (Figure 15.1) from a sponge of the genus *Dactylospongia*, collected off the Vanuatu islands.[48] The structure of **2**, which was obtained as a minor metabolite, was assigned by one-dimensional and two-dimensional NMR spectroscopy and by mass spectrometry (MS); however, the absolute configuration of the compound remained unassigned (in fact, not even the relative configuration at C19 was determined) and, quite surprisingly, no reference is made in the Riccio paper to the previously discovered (−)-zampanolide (**1**). The assignment of the relative and absolute configuration of **2** as shown in Figure 15.1 is based on its first total synthesis by Smith and Safonov in 2001,[39,40] who also showed that the thermolytic degradation of synthetic, nonnatural *ent*-**1** leads to **2**. As had been described for the natural product, the synthetic materials were dextrorotatory, but the absolute values of the specific rotations reported for the natural product and synthetic **2** at similar concentrations in MeOH were substantially different from each other (+30° for natural dactylolide vs. +235° for synthetic **2**). The specific rotations determined for **2** in subsequent total syntheses[49,50] were +134° and +163°, while values between −128° and −258° have been reported for synthetic *ent*-**2**.[51–56] Sanchez and Keck have suggested that these discrepancies could be due to different enolization equilibria in the solutions used to determine specific rotations.[49] More recently, it has even been speculated that the low specific rotation reported by Riccio and coworkers (compared to the absolute values of the specific rotations of synthetic **2** or *ent*-**2**) may indicate that the true absolute configuration of natural dactylolide is in fact that of *ent*-**2**, with the (+) sign of the rotation in Riccio's work being caused by enolization and/or hemiacetal formation.[57] In the absence of additional data for the natural material, the validity of this speculation is impossible to verify; it should be remembered, however, that all specific rotations for **2** or *ent*-**2** have been reported with MeOH as the solvent, and that independent of the variations in absolute values (which may be caused by variations in solvent quality, sample history, and/or other experimental conditions), synthetic **2** (three independent examples) has always been found to be dextrorotatory and *ent*-**2** (seven independent examples) was always levorotatory. Thus, whatever role enolization and hemiacetal formation may play for the ultimately observable specific rotation of samples of **2** or *ent*-**2**, it would be surprising if these effects should have led to a reversal in the sign of the specific rotation just in the case of the natural product. At the same time, it should be noted that *ent*-**2** so far has not been found in nature (unless the compound isolated by Riccio and coworkers was in fact *ent*-**2**),[54] although it would be a plausible biosynthetic precursor for **1**; alternatively, *ent*-**2** could arise from **1** by cleavage of the C–N bond of the hemiaminal moiety.[57] In analogy, the biogenetic precursor of **2** would be *ent*-**1**, which is a compound that (like *ent*-**2**) so far is not known to exist in nature. While these latter considerations lend some plausibility to the hypothesis that natural dactylolide may be *ent*-**2**, it needs to be stressed that the biosynthesis of (–)-zampanolide (**1**) has not been investigated.

One of the questions that regularly arises in discussions about (–)-zampanolide (**1**) is that of chemical stability, as its *N*-acyl hemiaminal substructure is perceived to be susceptible to pH-dependent hydrolysis. While the compound is clearly stable during purification by flash column chromatography, normal-phase high-performance liquid chromatography (HPLC), or reverse-phase HPLC (RP-HPLC), the chemical stability of **1** under different experimental conditions, to the best of our knowledge, has not been explicitly investigated, except that Smith and Safonov have shown that heating *ent*-**1** to 85°C in toluene for 100 min results in its clean and quantitative conversion into **2** and (2*Z*,4*E*)-hexa-2,4-dienamide.[58] As indicated above, an *N*-acyl hemiaminal moiety is also found in a (limited) number of other natural products, which may suggest that certain stabilizing elements are operative in those cases. In particular, Troast and Porco have proposed a hydrogen-bonding network as a plausible stabilizing element in the case of (–)-zampanolide (**1**).[59] Independent of these considerations, however, preliminary studies in our laboratory have shown that significant differences exist between the stability of **1** in different cell culture media that are commonly used in proliferation experiments, with half-lives varying between 10 h and >70 h. Stability was lowest in human plasma, while the compound was essentially stable in HEPES buffer (half-life of ≫72 h). Due to the sensitivity limits of the analytical method used, these experiments were carried out at concentrations of 50 μM **1**, which is more than 1000-fold higher than the IC_{50} values that have been observed for the inhibition of cancer proliferation *in vitro* (*vide infra*). The true significance of the above finding, thus, remains to be established.

15.3 BIOLOGICAL ACTIVITY OF ZAMPANOLIDE AND DACTYLOLIDE

As alluded to in Section 15.2, (–)-zampanolide (**1**) was reported to be a potent cytotoxic agent as part of the original isolation work by Tanaka and Higa, who found the compound to be cytotoxic on the P388, A549, HT29, and MEL28 cell lines, with IC_{50} values between 2 and 10 nM.[38] (No specific information on the IC_{50} values for **1** against the individual cell lines or the exposure time of cells to the compound is available from the paper.) These findings were later confirmed and extended by Northcote, Miller, and coworkers, who reported IC_{50} values for **1** of 4.3 nM against HL60 leukemia cells and 14.3 nM against ovarian carcinoma cells (after 48 h of incubation).[47] Remarkably, essentially identical IC_{50} values were obtained against the ovarian cancer cell line A2780 and its multidrug-resistant, P-glycoprotein (P-gp)-overexpressing variant A2780AD (7.1 and 7.5 nM, respectively), thus demonstrating that **1** is not a substrate for the P-gp efflux pump. Treatment of 1A9 cells with (–)-zampanolide (**1**) led to cell cycle arrest in G2/M and apoptosis induction. At the same time, the compound induced microtubule bundles in interphase cells and multiple asters in dividing cells, and caused a dose-dependent shift of the cellular equilibrium between soluble and polymerized tubulin toward the polymeric state. The assembly-promoting effect on tubulin could also be reproduced biochemically with purified tubulin preparations in an extracellular context. Collectively, the data reported by Northcote, Miller, and coworkers established that (−)-zampanolide (**1**) was a new microtubule-stabilizing agent and thus was able to inhibit cancer cell growth through a taxol-like mechanism of action. (The molecular mechanism of action of **1** will be discussed in detail in Section 15.7.) However, in contrast to taxol, **1** is not a P-gp substrate and, therefore, may be expected to also inhibit the growth of multidrug-resistant tumors that are not responsive to taxol treatment. This latter conclusion has been reconfirmed in a recent study by Cerchietti and coworkers,[60] who demonstrated that (−)-zampanolide (**1**) is a highly potent growth inhibitor of a number of AML and ALL cell lines, including the P-gp-overexpressing, taxol-resistant KG-1a (AML) and CCRF-CEM/VBL (ALL) lines, with GI_{50} values below 1 nM (72 h exposure time). Remarkably, (−)-zampanolide (**1**) was also found to be highly effective in killing leukemic cells from an AML patient with a particularly aggressive disease phenotype (due to an FLT3 mutation) that is associated with poor outcome; the compound was able to kill the tumor bulk as well as the quiescent leukemic stem cell population.

In comparison with (−)-zampanolide (**1**), the antiproliferative activity of natural dactylolide is substantially lower. Riccio and coworkers reported the compound to inhibit the growth of L1210 cells and SKOV-3 cells by 63% and 40%, respectively, at a concentration of 8.3 μM.[48] As these numbers were obtained for a 24 h exposure period, they are not directly comparable with the data summarized above for (−)-zampanolide (**1**); nevertheless, they clearly point to a significantly reduced growth inhibitory potential of *natural* dactylolide vs. (−)-zampanolide (**1**). Unfortunately, to the best of our knowledge, no data on the biological activity of *synthetic* **2** have been reported in the literature, which makes it difficult to put the data from the Riccio paper[48] in context. In contrast to synthetic **2**, data for the inhibition of cancer cell proliferation by synthetic *ent*-**2** are available from a number of studies. In particular, Ding and Jennings have reported an IC_{50} for *ent*-**2** of 4.7 μM against the SKOV-3 cell line, which is similar to the activity described by Riccio for natural dactylolide (*vide supra*).[53] On the other hand, *ent*-**2** showed submicromolar GI_{50} values against 19 out of the 60 cell lines of the NCI 60 cell line panel, which is in line with the activity data reported by our group for A549, MCF-7, HCT-116, and PC-3 cells (IC_{50} values of 301, 247, 210, and 751 nM, respectively).[61] Data in the latter set of cell lines also showed that (−)-zampanolide (**1**) is a 60- to 260-fold more potent cell growth inhibitor than *ent*-**2**. These findings are in agreement with results obtained by Uenishi et al. for the activity of **1** and *ent*-**2** against SKM-1 leukemia and U937 lymphoma cells[54] and by Field et al.[62] for 1A9 ovarian carcinoma cells, and they highlight the crucial importance of the hemiaminal-linked dienamide side chain in (−)-zampanolide (**1**) for highly potent biological activity.

15.4 TOTAL SYNTHESIS OF ZAMPANOLIDE AND DACTYLOLIDE

15.4.1 Previous Syntheses

While the focus of this chapter is on our own synthetic work on (−)-zampanolide (**1**) and (−)-dactylolide (*ent*-**2**),[56,61] it is also appropriate to briefly review some of the work from other groups in this area and to outline the different concepts and some of the key transformations that have come to bear on the synthesis of these macrolides. A comprehensive review on the total synthesis of zampanolide and dactylolide has recently appeared[57] and is outside of the scope of this chapter. The first total synthesis of zampanolide was accomplished in 2001 by Smith and coworkers, who produced the nonnatural (+)-enantiomer *ent*-**1**, thereby establishing the relative and absolute configurations of the natural product with reasonable certainty.[39,40] Four total syntheses of (−)-zampanolide (**1**) have been reported subsequently, including one from our own group (to be discussed in detail below).[51,54,61,63,64] In addition, three total syntheses have been described of (+)-dactylolide (**2**)[49,50,58] and three of (−)-dactylolide (*ent*-**2**), where the compound was not subsequently transformed into **1**.[52,53,55] However, as *ent*-**2** figures as an intermediate in all existing approaches to **1**, the additional syntheses of the former are also formal total syntheses of the latter.

The elaboration of *ent*-**2** into **1** involves what has been termed an *aza-aldol* reaction with (2*Z*,4*E*)-hexandienamide (*Z*,*E*-sorbamide) (**4**), which can either be catalyzed by a proton acid or involves the treatment of the amide with di-isobutylaluminium hydride (DIBALH) and subsequent reaction of the resulting aluminum species with *ent*-**2** (Scheme 15.1).

The reaction shows no *inherent* stereoselectivity and the reported yields of **1** are low to moderate. Thus, Hoye and Hu have reported the DIBALH-promoted reaction to produce a 1:1 mixture of **1** and its C20 epimer (*epi*-**1**), but no yield is provided in their publication either for pure **1** or for the mixture of **1** and *epi*-**1**.[51] Using camphorsulfonic acid (CSA) as a promoter, Tanaka and coworkers obtained **1** in 12% isolated yield, together with 12% of *epi*-**1** and 16% of a bis-acylated aminal product; 35% of *ent*-**2** was recovered unchanged.[54] Ghosh and Chen subsequently showed that the efficiency and selectivity of the acid-promoted aza-aldol reaction between *ent*-**2** and *Z*,*E*-sorbamide (**4**) could be significantly improved by the use of the chiral phosphoric acid (*S*)-3,3′-bis(2,4,6-triisopropylphenyl)-1,1′-binaphthyl-2,2′-diylhydrogenphosphate ((*S*)-TRIP), which furnished **1** with a 2.8:1 selectivity (over *epi*-**1**) and in 51% isolated yield.[63,64] None of the bis-acylated aminal that was the major product under Uenishi's conditions was found in the (*S*)-TRIP-promoted reaction. As the only exception, Smith and coworkers in their synthesis of *ent*-**1** did not employ the above aza-aldol reaction to establish the *N*-acyl hemiaminal subunit; rather, they relied on the acylation of the protected hemiaminal **3** with acid chloride **5** (Scheme 15.1).[39,40] Hemiaminal **3** was obtained from acid **14** by way of Curtius rearrangement and *in situ* trapping of the intermediate isocyanate with 2-trimethylsilyl ethanol (Scheme 15.2).

While the C20 configuration of hemiaminal **3** is epimeric to that of *ent*-**1** (which was unknown at the time and only established as a result of Smith's work), this intermediate was nevertheless found to lead to both *ent*-**1** and C20-*epi*-*ent*-**1**. This was due to significant epimerization at the C20 stereocenter upon final removal of the PMB group with 2,3-dichloro-5,6-dicyano-1,4-benzoquinone (DDQ) in wet dichloromethane (DCM), which yielded a 1.3:1 mixture of *ent*-**1** and its C20*S* epimer. Obviously, this outcome had not been intended; rather, the preinstallment of the C20 stereocenter had in fact been thought to provide access to both C20 epimers selectively, without having to rely on a difficult-to-control addition reaction of an amide to an aldehyde.

As illustrated in Scheme 15.2, key steps in the assembly of the core macrocycle in Smith's synthesis were the construction of the tetrahydropyran (THP) moiety by the

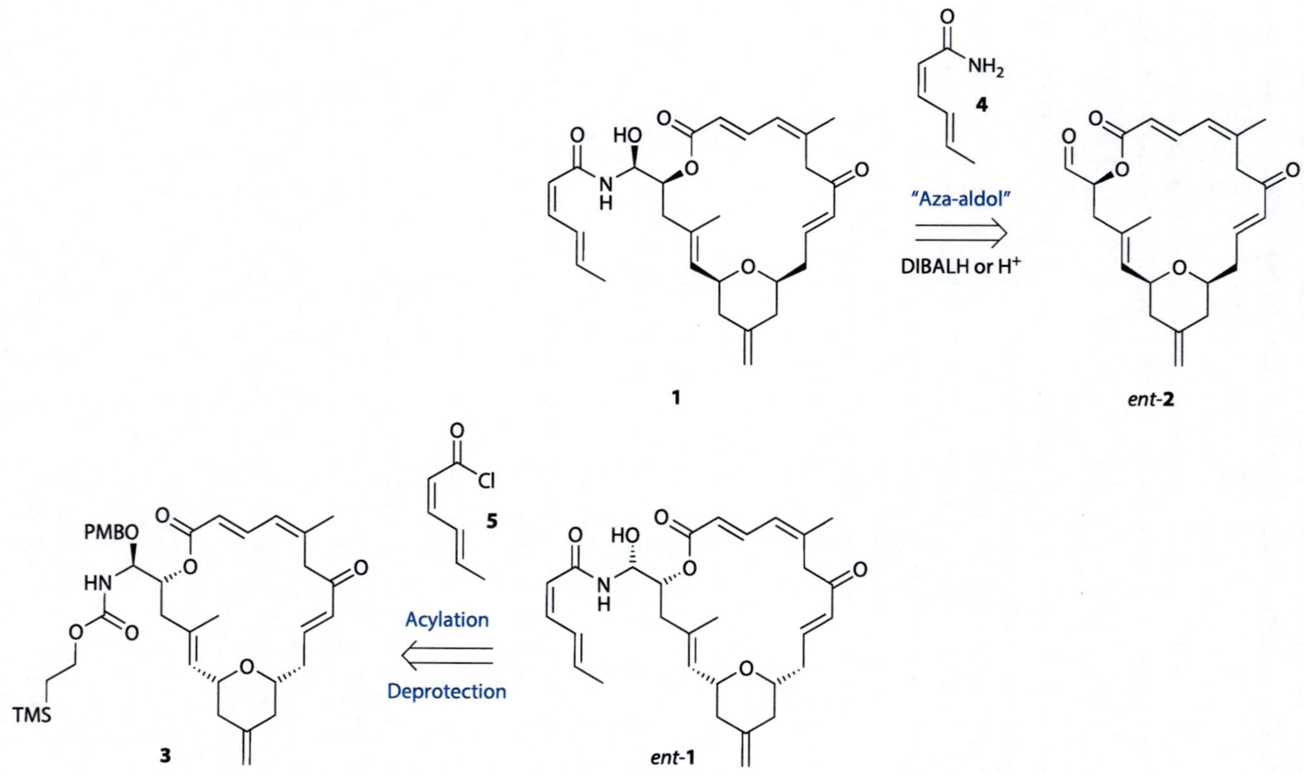

SCHEME 15.1 Alternative strategies for the elaboration of the hemiaminal-linked dienamide side chain in zampanolide. Note that the configuration at C20 of intermediate **3** in Smith's synthesis of (+)-zampanolide (*ent*-**1**) was indeed *S*, as shown. Cleavage of the *p*-methoxybenzyl (PMB) ether in the last step of the synthesis led to partial epimerization at this center and thus to the formation of both C20-*epi*-*ent*-**1** and *ent*-**1**.[40]

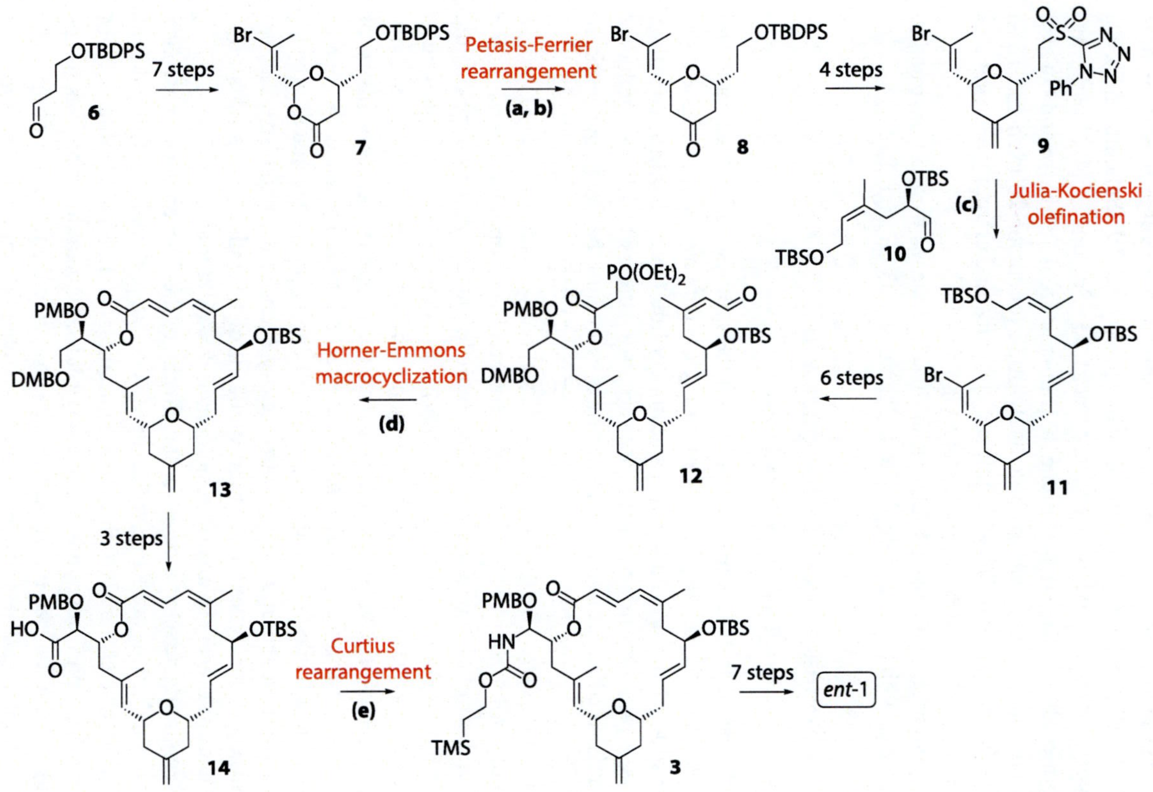

SCHEME 15.2 Key steps in the synthesis of (+)-zampanolide (*ent*-**1**) by Smith and coworkers.[40] Reagents and conditions: *Petasis–Ferrier rearrangement*: **(a)** Cp_2TiMe_2, THF, 65°C, 19 h, 72%. **(b)** Me_2AlCl, CH_2Cl_2, −78°C → 0°C, then $NaHCO_3$, NEt_3, 0°C → RT (room temperature), 59%. *Julia–Kocienski olefination*: **(c)** KHMDS, THF, −78°C, then aldehyde **10**, 88%. *Horner–Wadsworth–Emmons macrocyclization*: **(d)** NaHMDS, THF, 0.006M, −78°C → RT, 72%. *Curtius rearrangement*: **(e)** i. $EtN\textit{i}Pr_2$, *i*-BuOCOCl; ii. NaN_3, H_2O, 0°C; iii. Toluene, heat, 15 min; iv. $TMSCH_2CH_2OH$, heat, 3 h, 66%.

Petasis–Ferrier rearrangement of acetal **7**, the Julia–Kocienski olefination between **9** and **10**, and macrocyclic ring closure through intramolecular Horner–Wadsworth–Emmons (HWE) olefination of ω-diethylphosphono β-keto aldehyde **12**. HWE-based ring closure between C2 and C3 has also been employed by Floreancig and coworkers[50] and by Sanchez and Keck[49] in their syntheses of **2** and *ent*-**2**, respectively. Other approaches to macrocyclic ring closure in the synthesis of **1** include a Ti(IV)-promoted epoxide–acid coupling with epoxy acid **19** (Scheme 15.3),[51] a Kita–Trost-type macrolactonization with seco acid **23** (Scheme 15.4),[54] and ring-closing olefin metathesis (RCM) with diene **30** (Scheme 15.5).[63,64] Hoye and Hu also described an alternative approach to their dactylolide precursor **20** (Scheme 15.3) that features an *intermolecular* Ti(IV)-catalyzed epoxide–acid coupling and ring closure by ring-closing olefin metathesis between C8 and C9.[51]

Different approaches have been followed in these syntheses for the construction of the THP moiety, which are also depicted in Schemes 15.3 through 15.5.

15.4.2 Total Synthesis of (–)-Zampanolide and (–)-Dactylolide by Zurwerra et al.

Given the significant efforts that had been spent on the synthesis of zampanolide and dactylolide even before 2009, the sparsity of biological data available on these compounds at the time of the reisolation of (–)-zampanolide (**1**) by Northcote, Miller, and coworkers is rather surprising. Likewise, no SAR studies had been conducted on these potential lead structures prior to our own entry into the field. Our work on zampanolide was triggered by a number of different considerations: First, in part, the work was in fact driven by the (soft) hypothesis that (–)-zampanolide (**1**) might be a microtubule-stabilizing agent that we had formed based on what we felt was a certain structural similarity with existing microtubule stabilizers (even before the publication of the work of Field et al. in 2009).[47] Second, and more importantly, we were intrigued by the question if the profound difference in biological activity between **1** and **2** was related to the difference in the absolute stereochemistry of the macrolide ring (assuming this difference was real) or to the presence/absence of the hemiaminal-linked side chain (or perhaps both). Third, and very much in keeping with some of the discussion in the introductory section, the chemistry that we planned to develop in the context of the total synthesis of **1** should serve as a platform for the subsequent synthesis of analog structures and SAR studies.

When contemplating novel strategies for the closure of the 20-membered macrolide ring, we recognized that one attractive option for this key step, namely the formation of the double bond between C8 and C9 through an intramolecular HWE reaction, had not been part of any of the previous syntheses of dactylolide/zampanolide. This was somewhat surprising, as HWE-based macrocyclizations involving the formation of the C=C bond in α,β-unsaturated ketone units are well-precedented in natural product synthesis (even if they have not been used extensively) (for examples, see References 65–67). Our synthetic plan for **1**/*ent*-**2** was thus developed around a ring-opening disconnection between C8 and C9 as the key innovative step (Scheme 15.6).

For the elaboration of the hemiaminal-linked sorbamide side chain, we would rely on the previously reported aza-aldol reaction with the aluminate species derived from

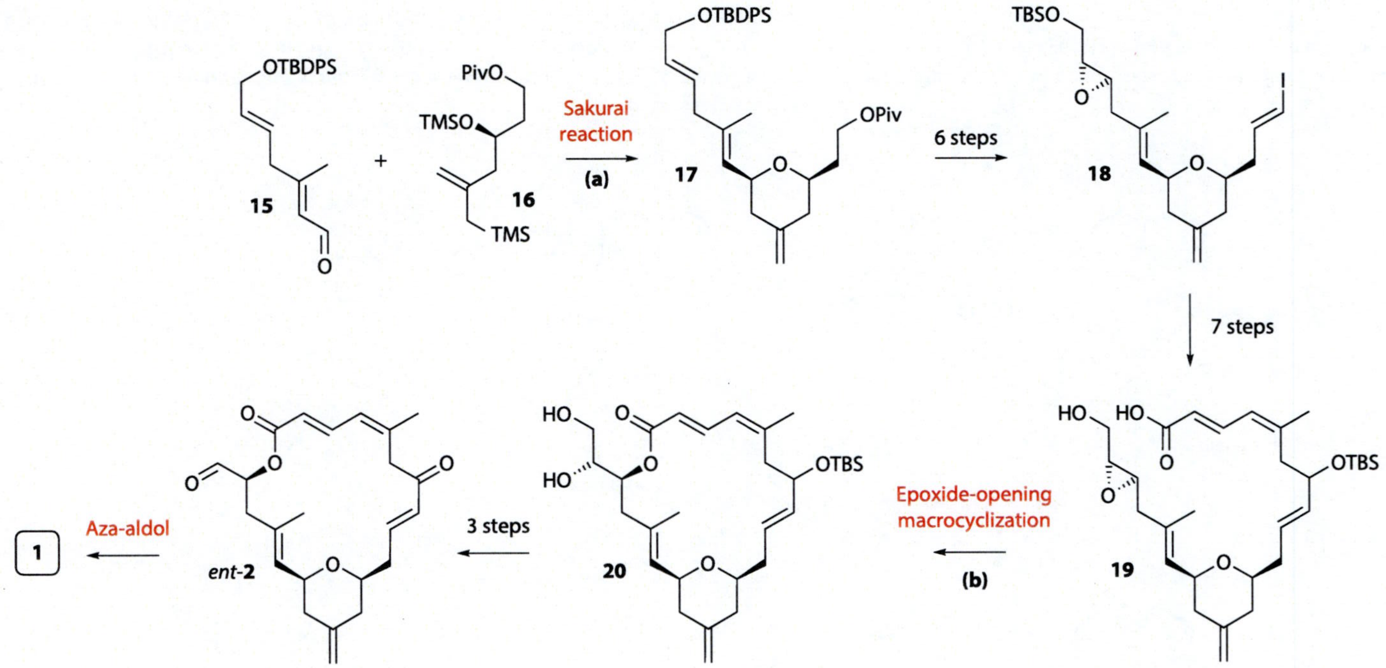

SCHEME 15.3 Key steps in the synthesis of (–)-zampanolide (**1**) by Hoye and Hu.[51] Reagents and conditions: *Sakurai cyclization*: **(a)** CSA (5 mol%), Et_2O, 78%. *Epoxide opening macrocyclization*: **(b)** $Ti(OiPr)_4$, CH_2Cl_2, 75°C, 40% (30% recovered starting material). *Aza-aldol*: **(c)** (*Z*,*E*)-sorbamide, DIBALH, THF, RT.

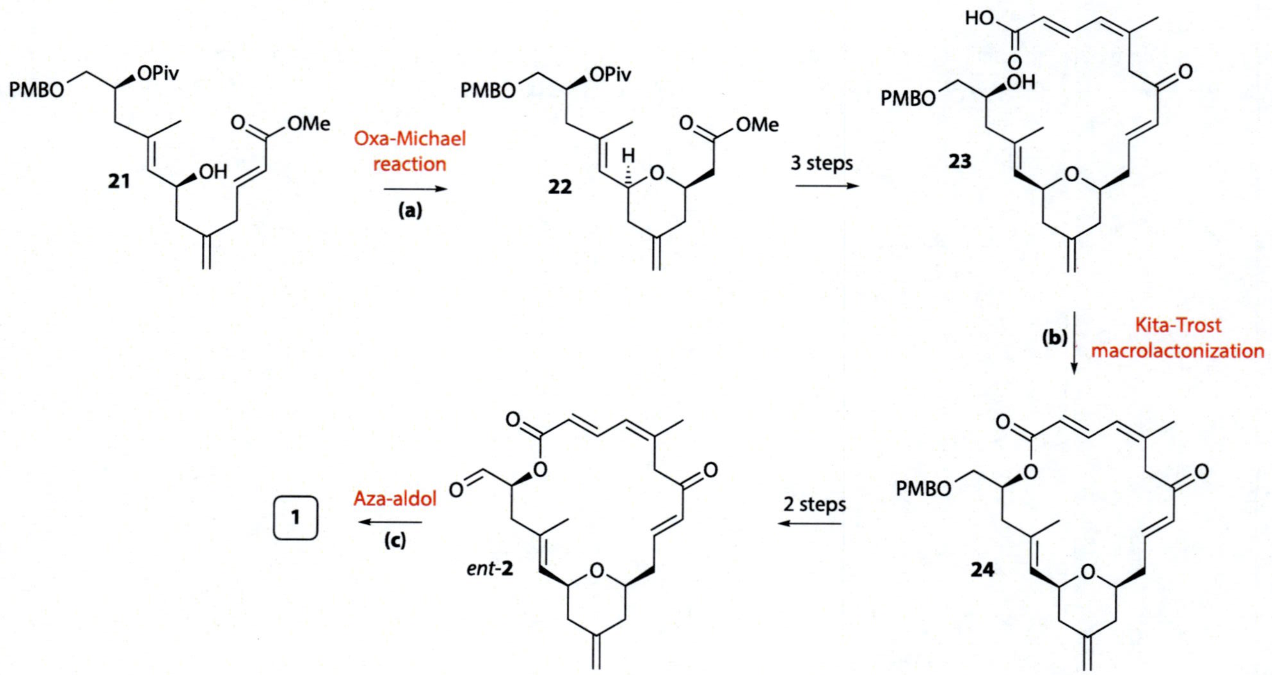

SCHEME 15.4 Key steps in the synthesis of (−)-zampanolide (**1**) by Uenishi and coworkers.[54] Reagents and conditions: *Oxa–Michael addition:* **(a)** LiHMDS (cat.), TMEDA, toluene, 60% *cis*-isomer, 34% *trans*-isomer. *Kita–Trost macrolactonization:* **(b)** ethoxyacetylene[$RuCl_2$(*p*-cymene)]$_2$, acetone, then CSA, toluene, 48%. *Aza-aldol reaction:* **(c) 4**, CSA, CH_2Cl_2, 12% for **1**.

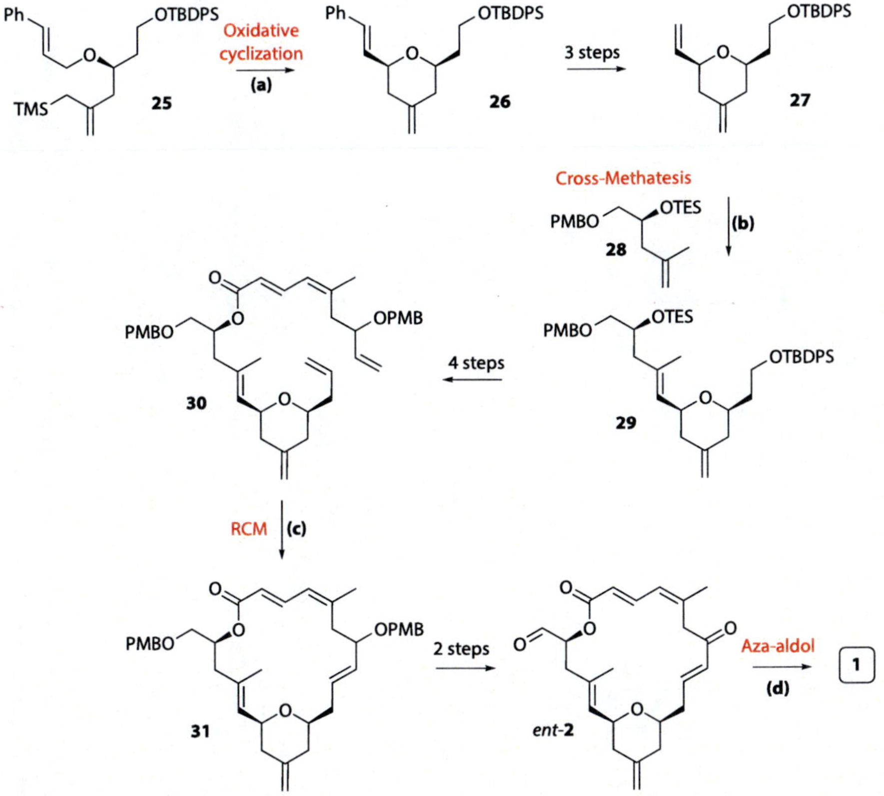

SCHEME 15.5 Key steps in the synthesis of (−)-zampanolide (**1**) by Ghosh and coworkers.[63,64] Reagents and conditions: *Oxidative cyclization*: **(a)** DDQ, PPTS, MeCN, −38°C, 81%. *Cross-metathesis*: **(b)** Grubbs II catalyst, 10 mol%, CH_2Cl_2, refl, *E*/*Z* = 1.7:1, 57% (*E*/*Z* mixture); 23% pure isomer after silyl ether cleavage. *RCM*: **(c)** Grubbs II catalyst, 12 mol%, benzene, 60°C, 65% (after PMB removal). *Aza-aldol*: **(d)** (*S*)-TRIP, 20 mol%, 51% **1**; 18% *epi*-**1**.

(*Z*,*E*)-sorbamide (**4**) by treatment with DIBALH; at the time of initiation of this project, no work on the acid-catalyzed variant of the reaction between *ent*-**2** and **4** had been reported. The ω-dialkylphosphono *β*-keto aldehyde **32** as the crucial precursor for the macrocyclization reaction was to be obtained by esterification of acid **33** with an appropriately protected alcohol **34**, followed by protecting group manipulations and oxidation. The construction of the THP ring was to be achieved through the Prins cyclization of the acetylated acetal **37**; the alkyne moiety of the resulting THP derivative would then be converted into a vinyl iodide, which would be metalated and the metal vinyl species reacted with PMB-protected (*R*)-glycidol (**35**) to produce alcohol **34**.

The synthesis of vinyl iodide **36** departed from D-aspartic acid as the ultimate and readily available precursor (Scheme 15.7). D-Aspartic acid can be converted into epoxide **39** in four simple steps, including the stereoselective retentive conversion into (*R*)-*α*-bromo succinic acid (**38**), reduction of the diacid to the diol, and *in situ* treatment of the diol with NaH and tert-Butylchlorodiphenylsilane (TBDPSCl), which results in epoxide formation and TBDPS protection of the distal hydroxy group.[68] The epoxide was then

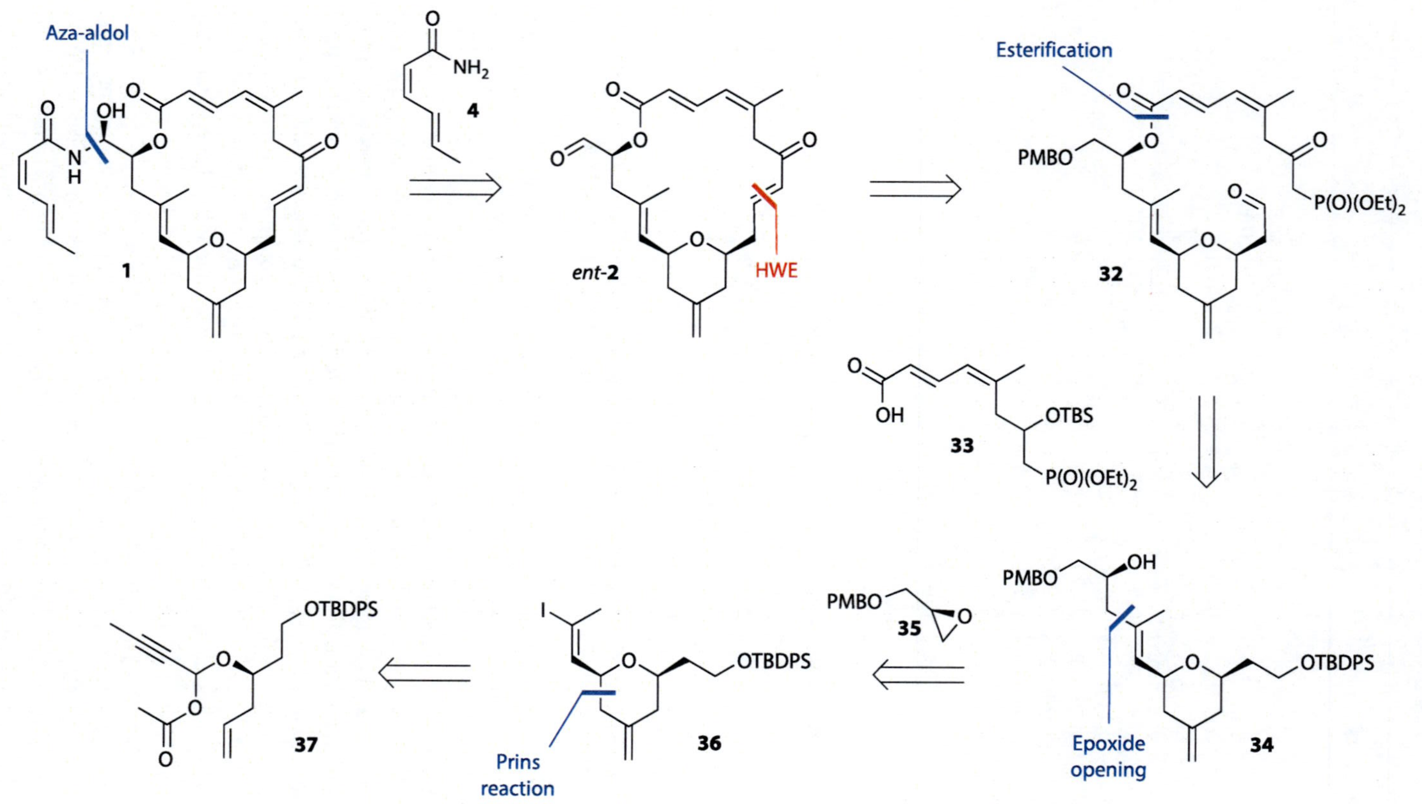

SCHEME 15.6 Retrosynthesis of (–)-zampanolide (**1**). For simplicity, the protecting groups shown are those that were used in the final total synthesis of **1**. At the retrosynthesis stage, a number of other protecting groups/protecting group combinations were also considered that are not shown here.

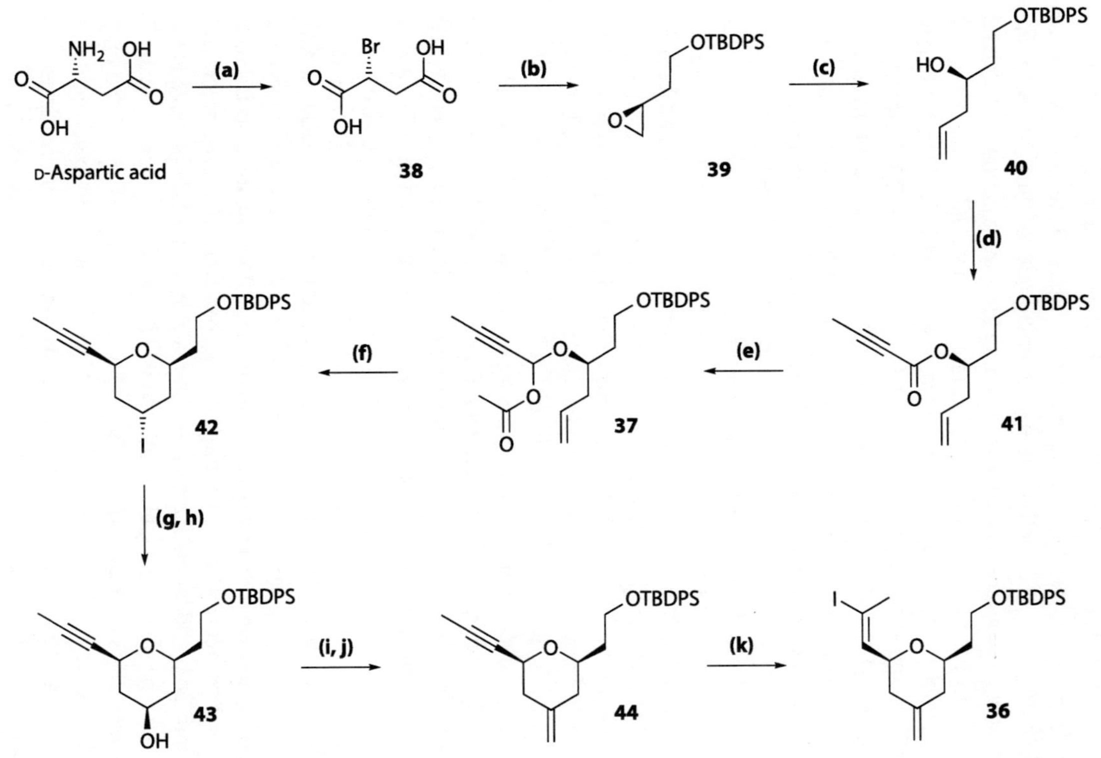

SCHEME 15.7 Reagents and conditions: **(a)** KBr, H_2SO_4, $NaNO_2$, H_2O, 0°C, 90%; **(b)** i. $BH_3 \cdot THF$ or $BH_3 \cdot DMS$, THF, 0°C → RT, 96%, ii. NaH, THF, then TBDPSCl, THF, −10°C, 90%; **(c)** CH_2=CHMgBr, CuI (cat.), THF, −55°C → −30°C, 98%; **(d)** 2-butynoic acid, DCC, DMAP, CH_2Cl_2, 0°C → RT, 85%; **(e)** DIBAL-H, then Ac_2O, pyridine, DMAP, CH_2Cl_2, −78°C, 92%; **(f)** TMSI, 2,6-dimethylpyridine (0.2 equiv), CH_2Cl_2, −19°C, 85%; **(g)** CsOAc, 18-c-6, toluene, 60°C, 4d, 72%; **(h)** K_2CO_3, MeOH/H_2O (20/1), RT; **(i)** DMP, CH_2Cl_2, RT, 85%; **(j)** CH_3Ph_3PBr, *n*-BuLi, THF, 0°C → 50°C, 94%; **(k)** i. Bu_3SnH, *n*-BuLi, CuCN, THF, MeOH, −78°C, ii. NIS, THF, −17°C, 97%.

converted into homoallylic alcohol **40** by regioselective Cu-mediated epoxide opening with CH_2=CHMgBr[69]; **40** was obtained in 98% yield and the reaction could be carried out on multigram scale, thus making the synthesis of this intermediate from D-aspartic acid an attractive alternative to approaches based on asymmetric allylation chemistry.

After esterification of alcohol **40** with 2-butynoic acid,[70] reductive acylation of ester **41** provided the acid-sensitive acetylated acetal **37** as the critical precursor for our projected Prins cyclization. This cyclization was most effectively promoted by trimethylsilyl iodide (TMSI) as the Lewis acid (2.5 equiv),[71] which gave THP derivative **42** as a single isomer in 85% yield. The configuration of the two stereocenters formed in the course of the cyclization reaction was fully controlled by the configuration of the chiral center originating from homoallylic alcohol **40**, leading to the axial iodide **42** as the only observable product. Cyclization of **37** was also possible with TMSBr, leading to the axial bromide corresponding to **42** in 69% yield. Compared with TMSI, however, a significantly larger excess of TMSBr had to be employed (*ca.* 24-fold) and longer reaction times were required to achieve full consumption of starting material. Only mixtures of 2,6-*syn* and 2,6-*anti* isomers were obtained with either CF_3COOH or $SnBr_4$.

With the construction of the THP ring achieved, the iodo substituent at the 4-position needed to be transformed into the exocyclic methylene group present in the natural product. To this end, **42** was converted into the corresponding acetate in yields of about 70% by reaction with CsOAc in the presence of [18]crown-6 at 55°C–60°C; these conditions offered the best compromise between a practical reaction rate and the suppression of elimination side products that were observed at higher temperatures. Thus, substantial amounts of both possible dihydropyran isomers were formed at 90°C in a *ca.* 1:1 ratio; other oxygen nucleophiles investigated, such as $AgOCOCF_3$[72] or $AgClO_4$[73] gave only elimination products, while no conversion was observed with $PhI(OCOCF_3)_2$.[74] Base-mediated hydrolytic cleavage of the acetate moiety gave alcohol **43**, which was then transformed into the desired olefin **44** by oxidation with DMP and subsequent Wittig methylenation of the ensuing ketone (Scheme 15.7); **44** was obtained in 58% overall yield from **42**. Finally, stannylcupration/iodination of **44** with $Bu_3Sn(Bu)CuCNLi_2$[75,76] and iodine (or N-iodo succinimide (NIS)) in THF/CH_2Cl_2 provided the desired vinyl iodide **36** in good yields as a single isomer. In contrast, attempted hydrozirconation with Schwartz reagent[77] followed by treatment with iodine afforded only unchanged starting material. At this point, the synthesis of alcohol building block **34** came to an abrupt halt: In spite of extensive experimentation, no conditions could be identified for the successful conversion of **36** into **34** by lithiation and subsequent reaction with PMB-protected (*R*)-glycidol (**35**). These experiments were all conducted in THF, Et_2O, or mixtures thereof, with $BF_3{\cdot}OEt_2$ as the Lewis acid catalyst for the addition step. In most cases, the only observable product was the iodohydrin derived from **35** by reaction with iodide anion, in addition to reisolated starting material. While these results were highly surprising and are not really understood, in light of the fact that similar transformations are precedented in the literature,[54] they forced us to develop an alternative approach to alcohol **34**, if we did not want to abandon our overall concept of the synthesis of *ent*-**2** and **1**. This alternative approach, obviously, would not include an epoxide opening step with a metalated vinyl species and was centered on the stepwise construction of the trisubstituted C=C bond between C16 and C17, while the formation of the THP ring was still to rely on a segment coupling–based Prins-type cyclization (Scheme 15.8).

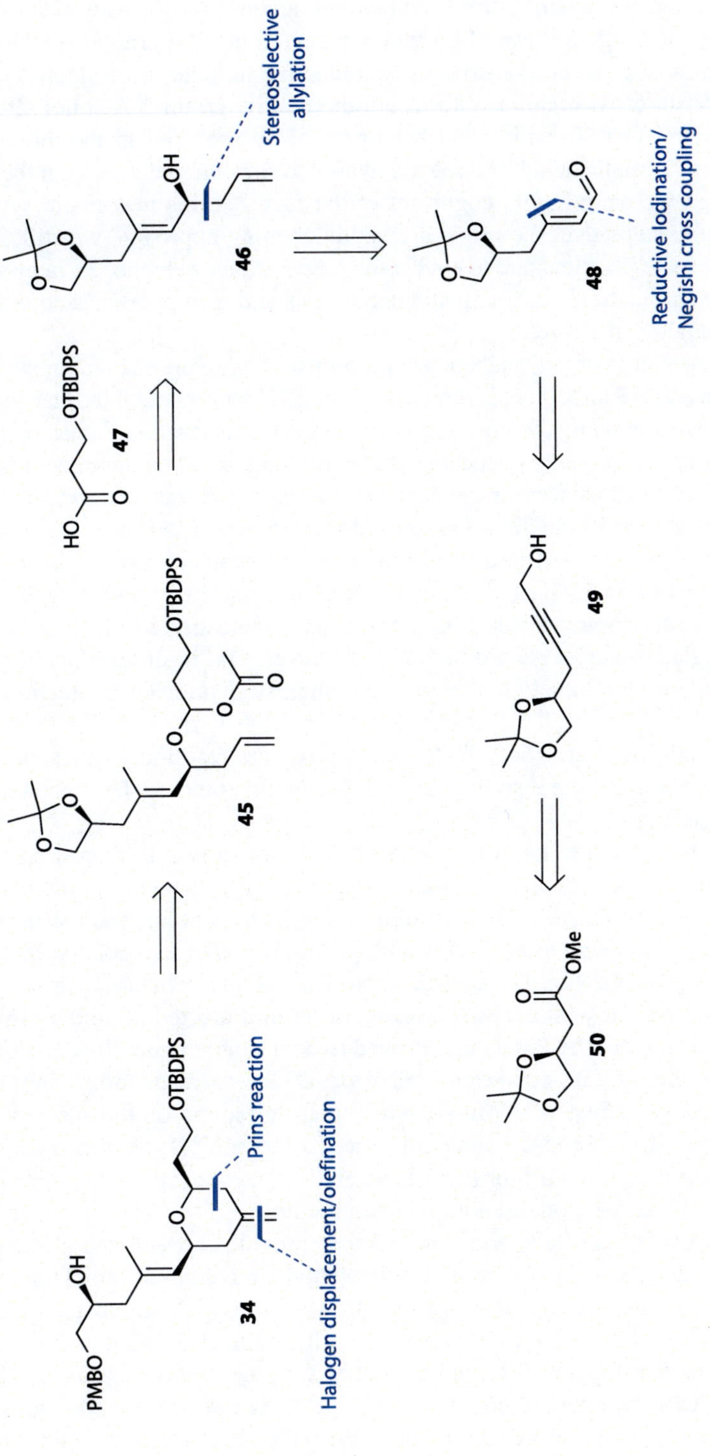

SCHEME 15.8 Alternative retrosynthesis of alcohol building block **34**.

We regarded acylated acetal **45** as a suitable precursor for the cyclization reaction, which would be obtained from homoallylic alcohol **46** *via* esterification with an appropriate acid **47**, followed by reductive acylation. The critical trisubstituted C=C bond in **34** was to be constructed by reductive iodination of alcohol **49** followed by Negishi cross-coupling[78] to install the methyl group.[79] Alcohol **49** would be derived from ester **50**, which had been previously prepared in the literature in three steps from L-malic acid,[80,81] a readily available starting material from the chiral pool. Although we were highly cognizant of the fact that this new linear synthesis comprised more steps than the original epoxide opening approach, we still considered the route to be practical and to enable the preparation of alcohol **34** in quantities that would be sufficient to sustain the total synthesis and to support subsequent SAR studies with analog structures.

The implementation of our synthetic plan commenced with the elaboration of **50** into propargylic alcohol **49** through reduction with DIBALH, conversion of the ensuing aldehyde into the corresponding dibromoolefin in a Corey–Fuchs reaction,[82] treatment of the latter with *n*-BuLi, and finally quenching of the resulting acetylide anion with paraformaldehyde. No difficulties were encountered with these transformations and propargylic alcohol **49** was obtained from **50** in 45% overall yield (Scheme 15.9).

Alkyne **49** was converted into vinyl iodide **51** by means of aluminum-mediated reductive iodination in 75% yield; vinyl iodide **51** underwent smooth Negishi cross-coupling with Me_2Zn and $Pd(dppf)Cl_2$, to furnish trisubstituted olefin **52** in high yield as a single isomer. It should be noted, however, that in order to avoid reduction of the vinyl iodide moiety in the cross-coupling step, transient protection of the allylic alcohol moiety as a TMS ether was required.[79,83] Cross-coupling could also be achieved with Me_2CuLi (64%–77% yield),[84] but the reaction was accompanied by partial reduction and the resulting deiodinated olefin could not be separated from the coupling product.

While this turned out to be of no practical relevance, it is still interesting to note that the success of the reductive iodination step strongly depended on the nature of the protecting groups for the C20/C19-diol moiety (zampanolide numbering), with acetonide protection being essential for high yields. We have also conducted experiments assessing the feasibility of PMB protection of the primary hydroxy group at C20 in combination with a TBS-protected or an unprotected secondary hydroxy group, respectively, but the PMB ether proved to be unstable under the reaction conditions. For a doubly TBS-protected variant of **49**, the corresponding vinyl iodide was obtained only in 34% yield, in spite of literature reports on the successful use of a related bis-TBS ether in the analogous transformation.[79] If the primary hydroxy group was protected as a TIPS or trityl ether and the secondary hydroxy group was left unprotected, the reaction did not go to completion.

The Swern oxidation of allylic alcohol **52** then provided unsaturated aldehyde **48**, which now needed to be transformed into homoallylic alcohol **46**. Initial attempts to establish the C15 stereocenter through the Brown ((–)-DIPCl/allylmagnesium bromide)[85,86] or Keck ((*S*)-BINOL/Ti(O*i*-Pr)$_4$)[87,88] allylation afforded **46** with varying diastereoselectivities (up to 10:1) and in highly variable yields (0%–70%) (Brown allylation) or gave incomplete conversion (Keck allylation; the stereoselectivity of the reaction was not determined in this case). We were then delighted to find that the

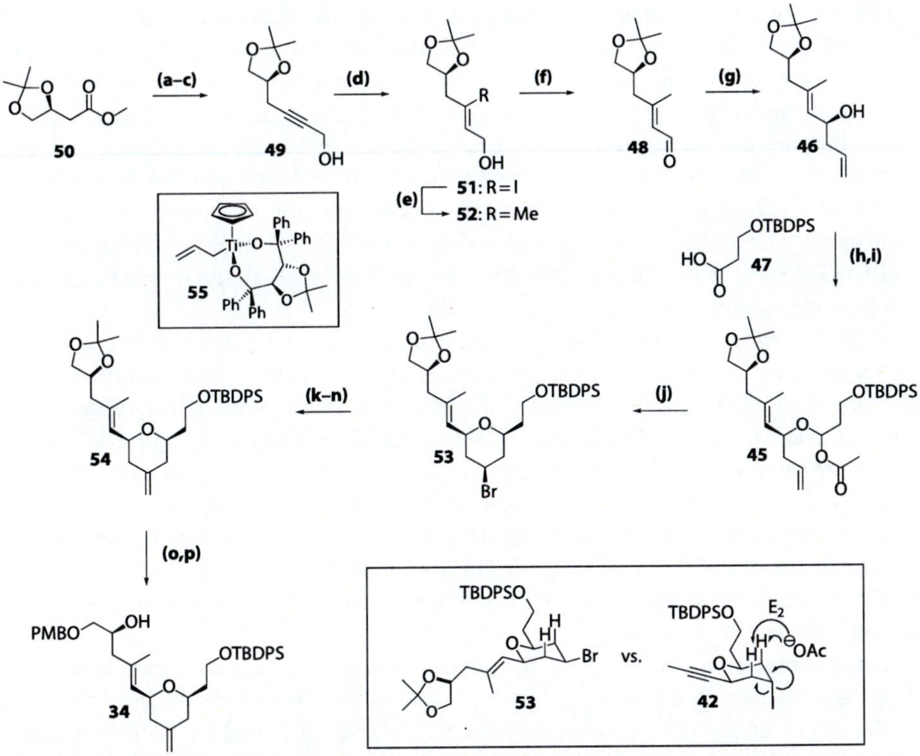

SCHEME 15.9 Reagents and conditions: **(a)** DIBALH, CH_2Cl_2, −70°C, 82%; **(b)** CBr_4, PPh_3, 2,6-dimethylpyridine, THF, 0°C, 86%, **(c)** *n*-BuLi, THF, −78°C, then $(CHO)_n$, 64%; **(d)** $NaAlH_2(OCH_2CH_2OMe)_2$, THF, 0°C → RT, then EtOAc, I_2, THF, −78°C, 75%; **(e)** i. TMSCl, NEt_3, ii. Me_2Zn, Pd(dppf)Cl_2, THF, 80°C, iii. K_2CO_3, MeOH, RT, 81% (three steps); **(f)** $(COCl)_2$, DMSO, NEt_3, CH_2Cl_2, −78°C, 84%; **(g)** **55**, Et_2O, −78°C, 15 min, then NH_4F, −78°C → RT, 80%–97%; **(h)** **47**, EDCI, DMAP, CH_2Cl_2, 0°C, 94%; **(i)** DIBALH then Ac_2O, pyridine, DMAP, CH_2Cl_2, −78°C, 91%; **(j)** i. $SnBr_4$, CH_2Cl_2, −78°C, ii. $Me_2C(OMe)_2$, pTsOH·H_2O, RT, 62% (two steps); **(k)** CsOAc, 18-c-6, toluene, 130°C, 20 h, 88%; **(l)** K_2CO_3, MeOH/H_2O (10/1), RT, **(m)** DMP, CH_2Cl_2, RT, 94% (two steps); **(n)** $MePh_3PBr$, *n*-BuLi, THF, 0°C → 45°C, 92%; **(o)** $CuCl_2$·$2H_2O$, MeOH, 60°C, 84%; **(p)** Bu_2SnO, toluene, Dean-Stark, 140°C, 1.5 d, then PMBCl, TBAI, 120°C, 1.5 h, 57%.

asymmetric allyltitanation of **48** with chiral cyclopentadienyl dialkoxyallyl titanium complex **55** as described by Duthaler, Hafner, and coworkers[89] provided the homoallylic alcohol **46** as a single isomer in yields between 80% and 97% after only 15 min at −78°C; as judged by ^{1}H- and ^{13}C-NMR analysis, no other isomer was formed in the reaction. The bidentate (*R*,*R*)-Taddol ligand could be recovered after NH_4F workup. The allyltitanation protocol was amenable to scale-up and more than 1 g of **46** could be produced in a single run without any erosion in yield and stereoselectivity compared with small-scale preparations.

The stage was now set for the construction of the THP subring, which began with the esterification of secondary alcohol **46** with acid **47** (obtained in three steps from propane-1,3-diol by mono-TBDPS protection followed by a two-step oxidation to

O-TBDPS-3-hydroxy propanal and finally **47**). This was followed by reductive acylation with DIBALH and Ac_2O, to give acyl acetal **45** in good yield (85% from **46**) as a *ca.* 1.7:1 mixture of epimers. Surprisingly, the Prins-type cyclization of **45** was successful only with $SnBr_4$ as the Lewis acid; no THP ring formation was observed with TMSI; instead, the major products formed in the reaction were those that would be expected from an oxonia-Cope rearrangement (based on NMR and thin-layer chromatography [TLC] analysis of crude reaction products after extractive workup and comparison with authentic reference samples).[90] Attempts to induce cyclization with TMSBr, TMSOTf, $BF_3 \cdot OEt_2$/AcOH, TFA, or TFA/$NaOCOCF_3$ did not produce any of the desired product.

THP ring formation with $SnBr_4$ was accompanied by cleavage of the acetonide moiety, which had to be reinstalled after the cyclization step, to furnish the cyclization product in 62% overall yield (Scheme 15.9). Although inconsequential, it is interesting to note that the THP ring in **53** exhibits an all-*syn* configuration of the three substituents at positions 2, 4, and 6 (based on nuclear Overhauser effect [NOE] measurements); that is, the bromo substituent occupies an equatorial position. This is distinctly different from the stereochemical outcome of the TMSI-induced cyclization of **37** (*vide supra*). Similar observations on the stereochemical course of segment coupling–based Prins-type cyclizations have been reported by Rychnovsky and coworkers.[71]

The elaboration of bromide **53** into olefin **54** was based on the same sequence of transformations that had been followed for the conversion of **42** into **44** (i.e., displacement of the bromo substituent with CsOAc, followed by acetate hydrolysis and oxidation of the resulting secondary alcohol, and finally conversion of the ensuing ketone into **54** by Wittig olefination (Scheme 15.9)). Olefin **54** was obtained in 76% overall yield for the four-step sequence from **53**. Compared with iodide **42**, acetate formation from bromide **53** was less problematic and higher yielding, as no side products were formed through elimination. As illustrated in Scheme 15.9, this is a direct consequence of the equatorial orientation of the bromo substituent in **53**, which effectively precludes elimination of HBr by an E_2-type mechanism, due to the gauche arrangement of the bromine atom and the axial hydrogens on carbon atoms 3 and 5 of the THP ring. In contrast, the axial orientation of the iodo substituent in **42** leads to an antiperiplanar arrangement with the axial hydrogens on C3 and C5, which allows for facile elimination of HI (which is what is observed experimentally; *vide supra*). Cleavage of the acetonide moiety in **54** with $CuCl_2 \cdot 2H_2O$[91] followed by regioselective PMB protection of the primary hydroxy group in the resulting diol *via* a cyclic Sn-acetal[92] then concluded the synthesis of secondary alcohol **34**.

In parallel with the work on alcohol building block **34**, we also elaborated a scalable route for the synthesis of the ω-diethylphosphono carboxylic acid building block **33**, which we assumed would be the less challenging task. The synthesis of this acid departed from 2-butynol (**55**), which was first converted into a *Z*-vinyl iodide according to the methodology for the reductive alumination/iodination of propargylic alcohols originally developed by Corey and coworkers.[93] PMB protection of the free hydroxy group by reaction with PMB trichloroacetimidate then provided PMB ether **56** (Scheme 15.10).[94]

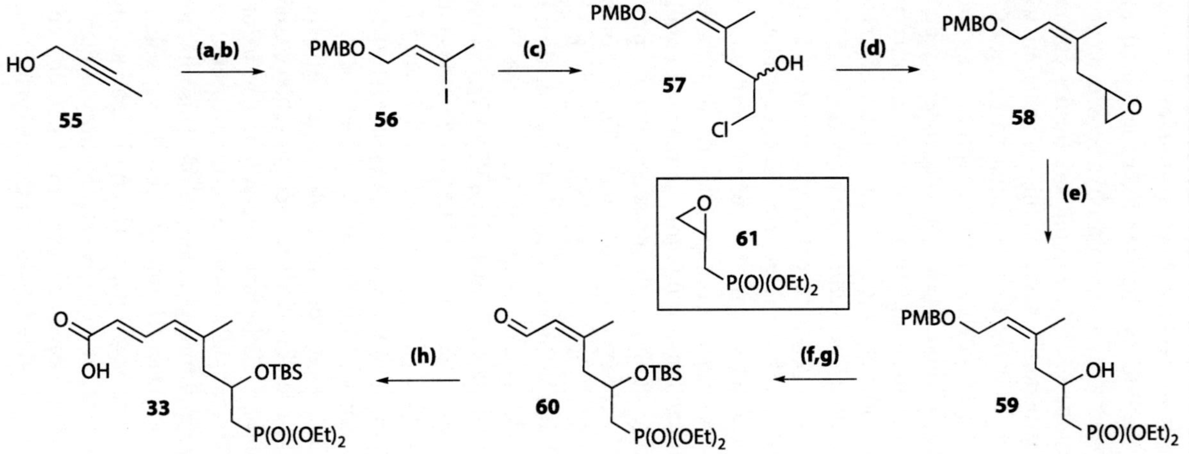

SCHEME 15.10 Reagents and conditions: **(a)** $NaAlH_2(OCH_2CH_2OMe)_2$, Et_2O, 0°C, then EtOAc, then I_2 in THF, −78°C, 88%; **(b)** $PMBO(C{=}N)CCl_3$, PPTS (cat.), CH_2Cl_2/cyclohexane (2/1), RT, 90%; **(c)** *n*-BuLi, epichlorohydrin, $BF_3{\cdot}OEt_2$, toluene, −85°C, 50%–70%; **(d)** KOH, EtOH, 0°C, 89%; **(e)** $HP(O)(OEt)_2$, *n*-BuLi, $BF_3{\cdot}OEt_2$, THF, −78°C, 80%; **(f)** TBSCl, ImH, DMAP, DMF, RT, 84%; **(g)** i. DDQ, CH_2Cl_2/H_2O (20/1), 0°C; ii. $(COCl)_2$, DMSO, NEt_3, CH_2Cl_2, −78°C → RT, 88% (two steps); **(h)** i. $(EtO)_2P(O)COOEt$, *n*-BuLi, THF, 0°C, ii. NaOH, EtOH, 0°C, 94%.

In the next step, **56** was meant to be homologated by reaction with epichlorohydrin, but as for the projected elaboration of vinyl iodide **36** into alcohol **34** (Scheme 15.7), none of the desired product **57** was observed upon formation of the vinyllithium species with *n*-BuLi followed by treatment with racemic epichlorohydrin under Lewis acid catalysis ($BF_3 \cdot OEt_2$) in THF. At this point in time (when we had already embarked on the alternative synthesis of alcohol **34** described above), we became aware of a report on the total synthesis of the natural product yatakemycin by Okano et al., where the opening of (*S*)-epichlorohydrin with an aryllithiuim species is described in toluene as the solvent, without any further comments on other solvent systems possibly investigated.[95] Based on this information, we investigated the use of toluene as the solvent for the transformation of **56** into **57** and, indeed, the epoxide opening reaction proceeded smoothly, reproducibly providing **57** in yields between 60% and 70%. While the reaction was generally clean (according to TLC analysis), it never went to completion, however. As a consequence, the yield for this transformation did not exceed 70%, in spite of extensive optimization efforts, including variations in reaction temperature (−95°C up to RT) or the use of different apolar solvents or solvent mixtures (such as CH_2Cl_2, Et_2O, THF, toluene/cyclohexane, toluene/Et_2O, toluene/hexane, or cyclohexane/hexane). Varying the concentration of **56** in the range between 0.1 and 0.25 M had no significant effect; however, the use of *n*-BuLi in general afforded higher yields of **57** than the use of *t*-BuLi. Optimized conditions for the reaction entailed the use of an excess of epichlorohydrin (3 equiv) in the presence of 1.3 equiv of $BF_3 \cdot OEt_2$, which gave **57** in at least 50% yield even on a 10 g scale. The use of larger amounts of $BF_3 \cdot OEt_2$ did not produce any improvement in yield. Obviously, the observation of a profound solvent effect for this reaction also had a distinct impact on our thinking about the opening of **35** with metalated **36** (cf. Scheme 15.6). This aspect will be addressed below. Treatment of **57** with KOH/EtOH gave the epoxide **58**, which could be opened regioselectively with lithiated diethylphosphite in THF,[96,97] to afford *β*-hydroxy phosphonate **59** in 71% overall yield (from **57**). Attempts to prepare **59** directly by reaction of lithiated **56** with epoxy phosphonate **61** were unsuccessful, either with THF or with toluene as the solvent. Reaction of **59** with TBSCl under standard conditions (imidazole, DMF) was surprisingly slow and needed to be accelerated by the use of DMAP, in order to achieve complete conversion of **59** into the corresponding TBS ether. The subsequent oxidative cleavage of the PMB ether with DDQ afforded a mixture of the expected primary allylic alcohol and the corresponding unsaturated aldehyde **60**. This mixture was submitted to Swern oxidation conditions, which finally allowed the isolation of pure **60** in 79% yield from **59**. A two-step sequence of HWE olefination with the lithium anion of triethyl phosphonoacetate followed by ester hydrolysis then completed the synthesis of acid **33**. The sequence depicted in Scheme 15.10 proved to be suitable for large-scale preparations and provided **33** in multigram quantities. This level of efficiency was indeed critical for the successful completion of the total synthesis of (−)-zampanolide (**1**).

As alluded to above, the successful implementation of the epoxide opening reaction with metalated **56** in toluene, but not in THF, provided a strong impetus to revisit the reaction of the vinyllithium species derived from vinyl iodide **36**

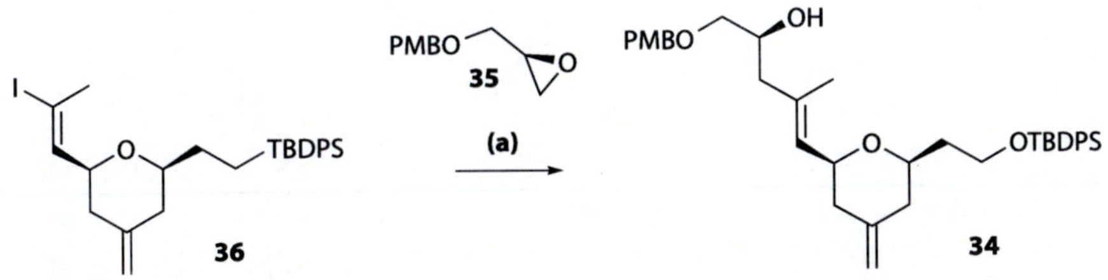

SCHEME 15.11 Reagents and conditions: **(a)** *t*-BuLi, toluene, −78°C, 30 min, then **35** in toluene, $BF_3 \cdot OEt_2$, −85°C → −78°C, 1 h, 61%.

with epoxide **35**, even if a viable route to alcohol **34** had been established in the meantime. Intriguingly, as for vinyl iodide **56** and epichlorohydrin, treatment of **36** with *t*-BuLi in toluene followed by addition of a solution of **35** in toluene and $BF_3 \cdot OEt_2$ produced the desired homoallylic alcohol in 61% yield, thus completing a significantly shorter route to alcohol building block **34** than the one from malic acid (Scheme 15.11).

While we believe that the total synthesis of **1** would have been feasible even if we had had to rely on material supply of **34** through the malic acid route, the D-aspartic acid–based approach depicted in Schemes 15.8 and 15.11 was incomparably more efficient, delivering **34** in 17% yield over 14 steps. In comparison, the route from malic acid, while scalable, comprises 22 steps and gave **34** only in 2.4% overall yield.

With reliable access to acid **33** and alcohol **34** secured, we turned our attention to the assembly of the macrocycle and the subsequent elaboration of the dienamide side chain of zampanolide. In this context, the esterification of **33** with **34** under Yamaguchi conditions proved to be uneventful and gave the desired ester in excellent yield (85%) (Scheme 15.12). Interestingly, only trace amounts of the ester were obtained under standard carbodiimide conditions. Global desilylation with HF·pyridine followed by double oxidation of the resulting free diol to the *β*-keto phosphonate aldehyde **32** then set the stage for the crucial HWE-based macrocyclization reaction. Initial experiments with sodium hexamethyldisilazide (NaHMDS) as the base showed the reaction to be feasible and highly selective for the desired *E* isomer, but yields were highly variable and reaction times were exceedingly long (up to 4 days). These findings were independent of the scale of the reaction (0.007–0.046 mmol). The concentration was kept low in all cases (0.005 M), in order to avoid polymerization of **32**. The deficiencies revealed in those initial experiments could be fully eliminated by the replacement of NaHDMS with $Ba(OH)_2$,[98,99] which not only led to a reduction in the reaction time to 30–60 min, but under optimized conditions also afforded much higher yields (in the range of 80%) reproducibly. In the largest single preparation, 430 mg of **32** were successfully converted into macrocycle **25** in 78% yield. PMB removal under oxidative conditions with DDQ and oxidation with Dess-Martin periodinane (DMP) then completed our new synthesis of (−)-dactylolide (*ent*-**2**) (Scheme 15.12).[56]

As expected, the subsequent introduction of the dienamide side chain proved to be challenging and, unfortunately, low yielding. Following the aza-aldol approach that had been reported previously by Hoye and Hu[51] (and also based on some more

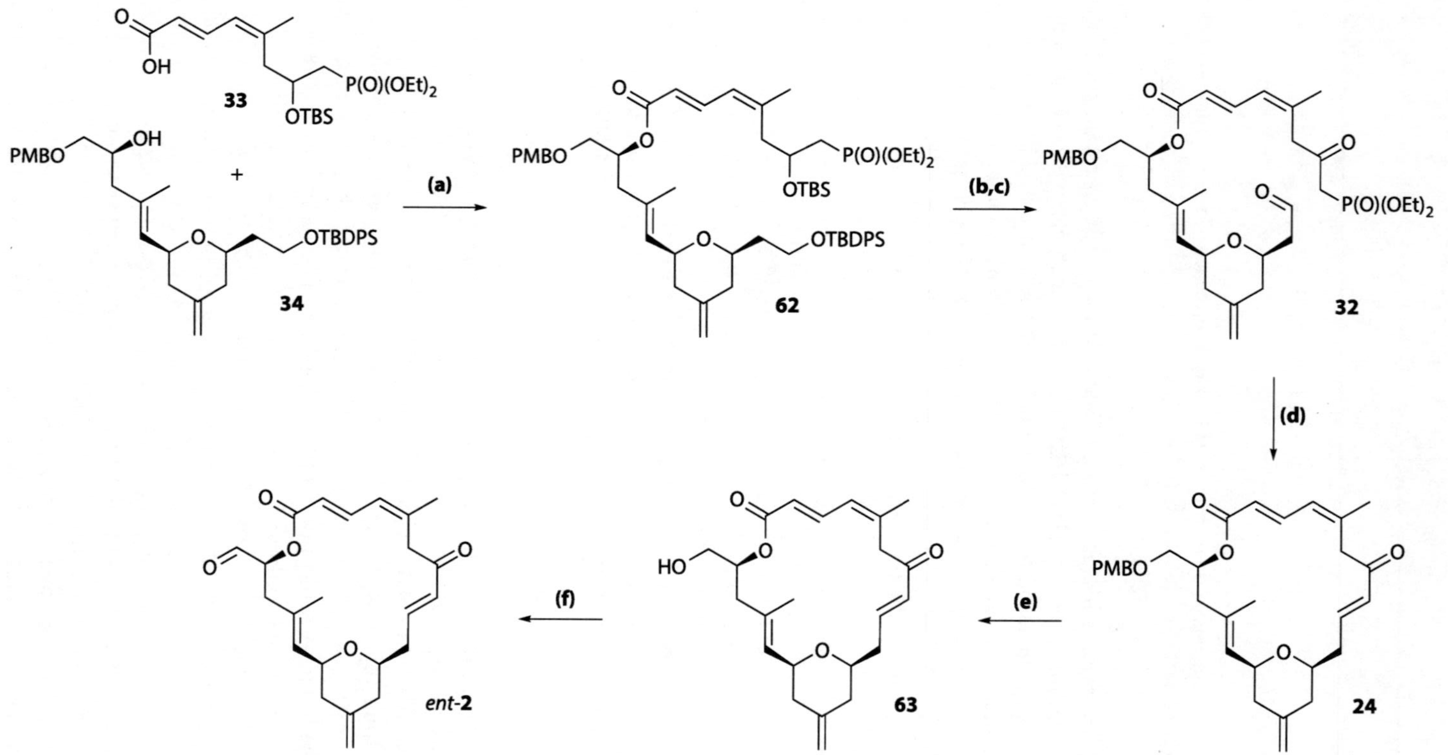

SCHEME 15.12 Reagents and conditions: **(a)** 2,4,6-Trichlorobenzoyl chloride, NEt_3, DMAP, **33**, toluene, RT, 85%; **(b)** HF·py, THF, 0°C → RT, 85%; **(c)** DMP, CH_2Cl_2, RT, 74%; **(d)** $Ba(OH)_2 \cdot 0.8H_2O$, THF/H_2O (40/1), 0°C → RT, 81%; **(e)** DDQ, CH_2Cl_2/H_2O (5/1), RT, 82%; **(f)** DMP, CH_2Cl_2, 78%.

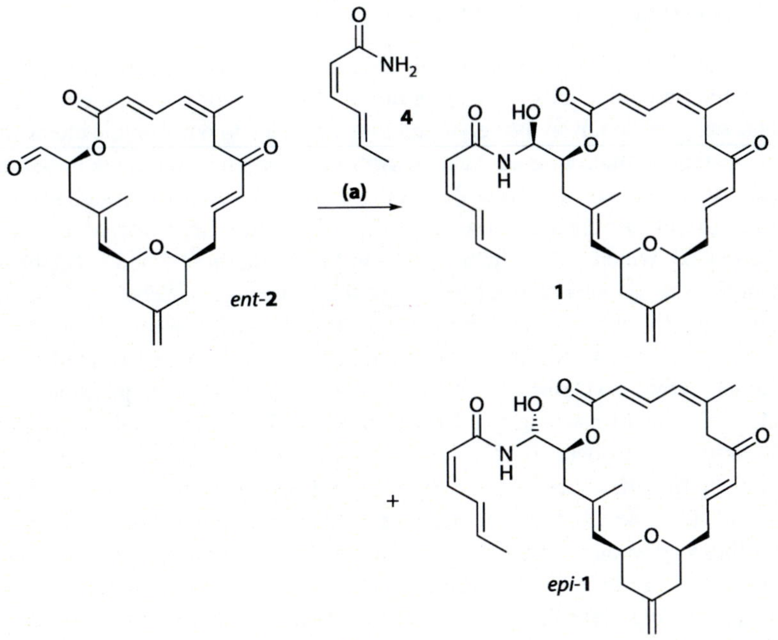

SCHEME 15.13 Reagents and conditions: **(a) 4**, DIBALH, RT, 30 min, then *ent*-**2**, THF, RT, 2.5 h, 18% **1**; 12% *epi*-**1**.

general work on aza-aldol reactions by Bayer and Maier),[100] *ent*-**2** was converted into a 1.1/1 mixture of **1** and its C20 epimer (*epi*-**1**) in 46% yield after flash chromatography on deactivated silica gel (Scheme 15.13). The separation of these isomers was only possible by normal-phase HPLC; the individual isomers, that is, **1** and *epi*-**1**, obtained in this purification step were subsequently submitted to final purification by RP-HPLC. This process delivered analytically pure **1** and *epi*-**1** in 18% and 12% yield, respectively (based on *ent*-**2**). While these final yields clearly leave something to be desired, they have to be ascribed largely to the lack of selectivity in the aza-aldol step and the resulting need for isomer separation by HPLC. Overall, our synthesis proved to be highly reliable, and notwithstanding the modest overall yield in the aza-aldol step, it has allowed us to produce sufficient amounts of material for extensive biological and biochemical profiling of **1**; in addition, this material has been used to prepare a tubulin-bound complex for X-ray crystallographic studies. These aspects of our work will be discussed below.

No fully asymmetric synthesis of **1** has been reported to date. However, as indicated earlier in this chapter, Ghosh and Cheng, after completion of our own work, have shown that the reaction of *ent*-**2** and **4** in the presence of a matched chiral phosphoric acid catalyst ((*S*)-TRIP) proceeds with a *ca*. 3:1 selectivity, thus providing **1** in 51% yield after isomer separation by HPLC.[63,64] We have also applied this methodology in the preparation of one subsequent batch of **1**; in this experiment, **1** and *epi*-**1** were obtained in 21% and 8% yield, respectively (after double HPLC purification on normal phase and reverse phase, which we usually find necessary).

15.5 SYNTHESIS OF ANALOGS

Virtually no SAR data had been available on dactylolide/zampanolide at the outset of our work on these compounds. Thus, one of the objectives of our synthetic studies on dactylolide/zampanolide from the very beginning had been to build on the chemistry that we would develop in the course of the total synthesis work, to explore the importance of individual structural features of **1**/*ent*-**2** for antiproliferative activity (through synthesis of appropriate analogs). At the most basic level, these SAR inquiries involved the investigation of alcohol **63**, which was an intermediate in the synthesis of **1** and *ent*-**2**, of methyl ether **64**, and of the side chain–modified zampanolide analog **66** (Scheme 15.14).

Methyl ether **64** was obtained from **63** by reaction with Meerwein salt in 83% yield (Scheme 15.14). Amide-based zampanolide analog **66** was obtained from *ent*-**2** by Pinnick–Kraus oxidation,[101–103] to give acid **65** in almost quantitative yield, followed by 1-[bis(dimethylamino)methylene]-1H-1,2,3-triazolo[4,5-b]pyridinium 3-oxid hexafluorophosphate (HATU)-mediated coupling with *n*-hexylamine. For reasons unknown, the efficiency of the coupling reaction was low and provided **66** only in 13% yield (from **65**), but no attempts were made to optimize this transformation. Although these compounds will not be discussed here, we note that we have also prepared a series of variants of desTHP-zampanolide (*vide infra*) with amide-based side chains. In no case was the yield of the HATU-mediated coupling reaction >50% (M. Jordi, unpublished results).

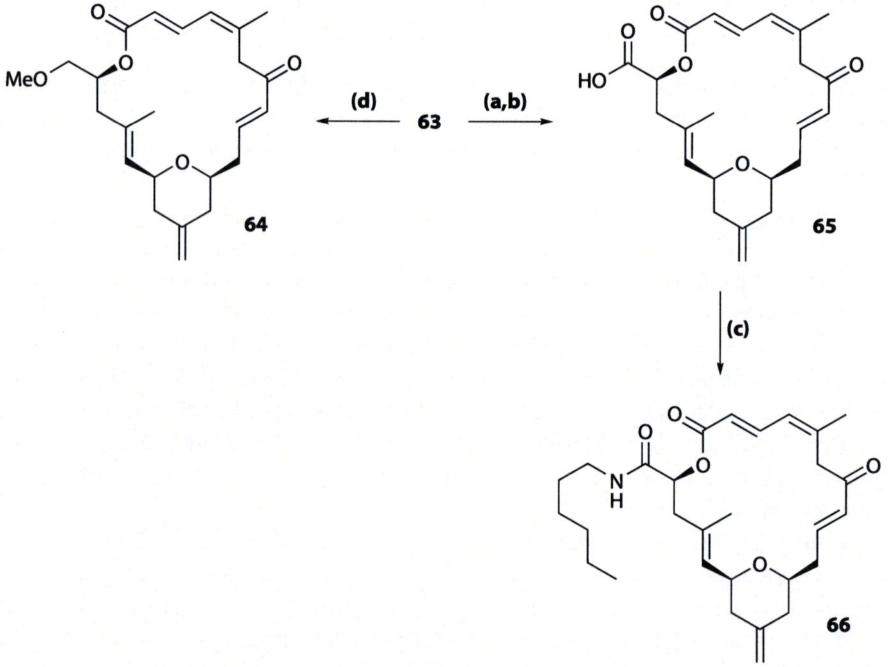

SCHEME 15.14 Reagents and conditions: **(a)** DMP, CH_2Cl_2, 78%; **(b)** $NaClO_2$, $NaH_2PO_4{\cdot}H_2O$, *t*-BuOH/H_2O, 2-methyl-2-butene, RT, 97%; **(c)** *n*-Hexylamine, HATU, DIEA, DMF, RT, 13%; **(d)** Me_3OBF_4, Proton Sponge®, CH_2Cl_2, RT, 83%.

One of the general themes of our analog work on bioactive natural products is the creation of structurally simplified, but hopefully still biologically active analogs that are more readily accessible than the parent natural product. In the case of (−)-zampanolide (**1**), the initial question that we addressed in this context was the significance of the exo-methylene group on the THP ring for biological activity. As discussed above, this double bond in our synthesis of **1** is introduced in a four-step sequence from iodide **42**; the simple reduction of this intermediate in one step would give access to 13-desmethylene dactylolide/zampanolide, and it was to be determined if these simplified variants retained the activity of the parent macrolides *ent*-**2** and **1**, respectively.

Although we were somewhat uncertain about the compatibility of the triple bond in intermediate **42** with radical reduction conditions,[76] in the event, the reduction of **42** with Bu_3SnH/AIBN proceeded in excellent yield to furnish the simplified THP derivative **67** in 88% yield (Scheme 15.15).

Attempts to transform **42** into **67** by iodide/lithium exchange with *t*-BuLi followed by protonolysis resulted in significantly lower yields that did not exceed 35%. As for alkyne **44**, the reductive iodination of **67** was achieved by stannylcupration/iodination with $Bu_3Sn(Bu)CuCNLi_2$[76] and NIS to provide the trisubstituted vinyl iodide **68** in 73% yield. Subsequent lithiation of **68** with *t*-BuLi and reaction of the vinyllithium species with epoxide **35** under the conditions that had proven to be optimal for the conversion of **36** into **34** (toluene, $BF_3{\cdot}OEt_2$) gave homoallylic alcohol **69** in 31% yield. Esterification of **69** with acid **33** followed by simultaneous cleavage of TBS and TBDPS ethers and subsequent oxidation of the free hydroxy groups with DMP gave the cyclization precursor **70** in 43% overall yield (based on **69**); macrocyclization with NaHMDS as the base then furnished the desired protected macrolactone **71** in 49% yield. It should be noted that none of the steps leading from **42** to **71** have been optimized at this point, and we are confident that improvements in yield will be possible for almost every individual transformation. In fact, preliminary work on the epoxide opening step and the macrocyclization reaction (which was carried out before investigation of the $Ba(OH)_2$-mediated macrocyclization of **32**) has already shown this to be the case. In analogy to the synthesis of *ent*-**2**, 13-desmethylene (−)-dactylolide **73** was obtained from **71** by oxidative PMB removal and DMP oxidation of the free alcohol in 55% overall yield.

As will be discussed below, the cellular activity of dactylolide analogs **72** and **73** was very much comparable with that of the corresponding methylenated parent compounds **63** and *ent*-**2**, respectively. These data suggested that not only should we interrogate the significance of the methylene group attached to the THP ring, but perhaps even that of the THP ring as a whole, thus defining desTHP-dactylolide/zampanolide analogs **75** and **74**, respectively, as new target structures for synthesis (Scheme 15.16).

The overall approach that we would follow to access those analogs was based on the same strategic considerations as the synthesis of **1** (Scheme 15.16); in particular, macrocyclic ring closure was to be achieved through intramolecular HWE reaction, which would allow to retain acid **33** as one of the advanced building blocks, while a new synthesis had to be developed for the modified alcohol building block **76**.

The synthesis of alcohol **76** departed from propargylic alcohol **55**, which was converted into vinyl iodide **77** according to literature procedures (Scheme 15.17).[75,76]

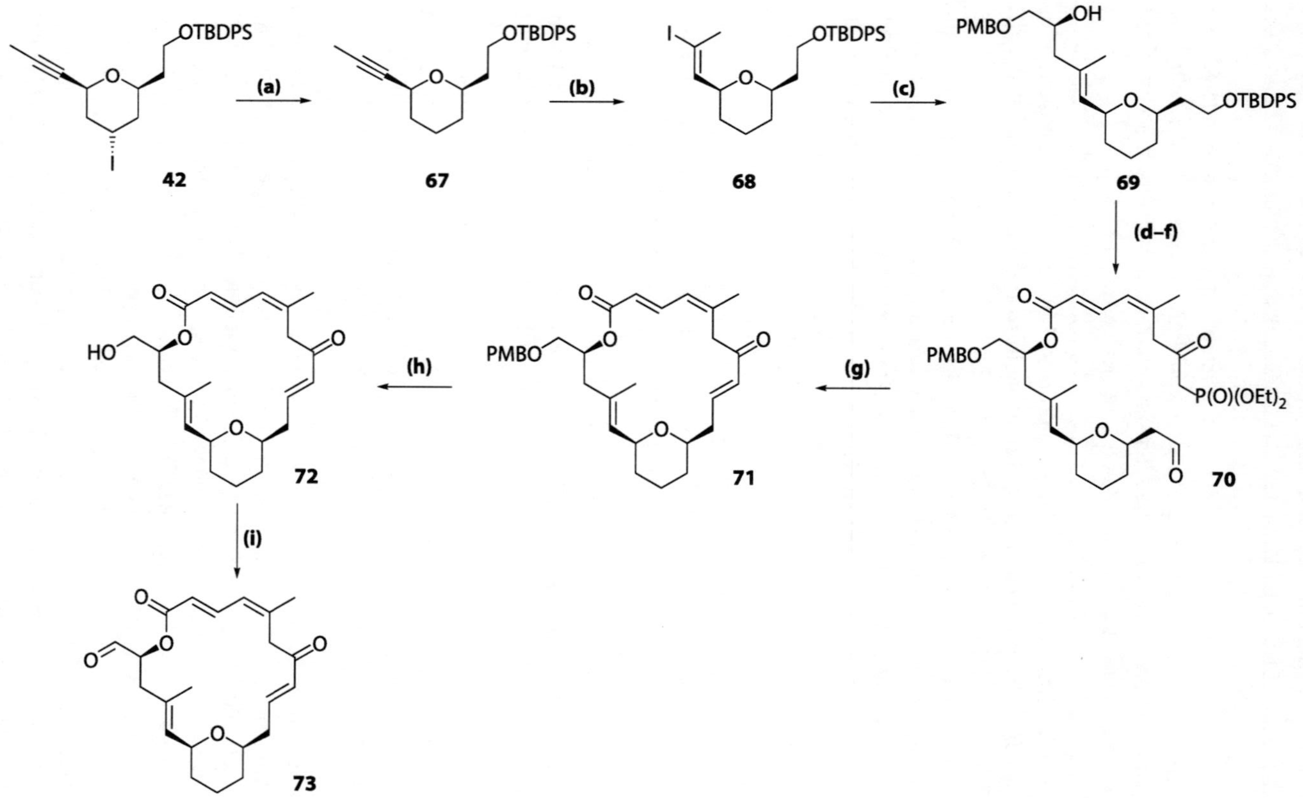

SCHEME 15.15 Reagents and conditions: **(a)** Bu_3SnH, AIBN (cat.), toluene, 60°C, 88%; **(b)** i. Bu_3SnH, *n*-BuLi, CuCN, THF, MeOH, −78°C → −10°C, ii. NIS, THF, −78°C, 73%; **(c)** *t*-BuLi, **35**, $BF_3 \cdot OEt_2$, toluene, −78°C, 31%; **(d)** 2,4,6-trichlorobenzoyl chloride, NEt_3, DMAP, **33**, toluene, RT, 74%; **(e)** HF·py, THF, 0°C → RT, 80%; **(f)** DMP, CH_2Cl_2, RT, 72%; **(g)** NaHMDS, THF, −78°C → RT, 2 d, 49%; **(h)** DDQ, CH_2Cl_2/H_2O (5/1), RT, 72%; **(i)** DMP, CH_2Cl_2, 77%.

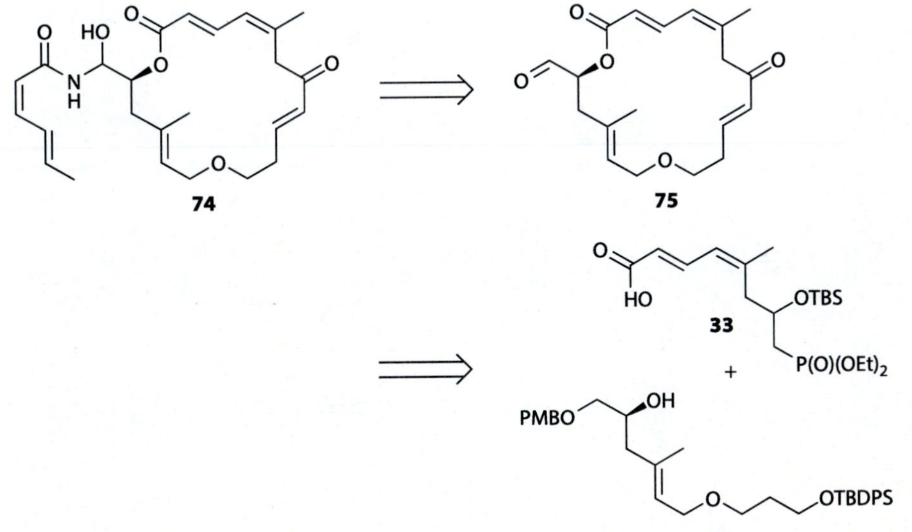

74 75 33 76

SCHEME 15.16 Retrosynthesis of desTHP-zampanolide **74**.

Reaction of **77** with allyl bromide in the presence of NaH gave allyl ether **78**; hydroboration and TBDPS protection of the resulting terminal hydroxy group then furnished **79** in 21% overall yield from **78**. While this sequence is not very step efficient, all attempts at the direct alkylation of **77** with TBDPS-protected 3-bromo-1-propanol proved to be unsuccessful. The latter could be used to alkylate **55** in 45% yield, but the resulting propargylic ether could not be converted into the required vinyl iodide **79** under the conditions previously established for the conversion of **36** into **34** and of **67** into **68**, respectively.

Reaction of lithiated **79** with epoxide **35** in toluene in the presence of $BF_3 \cdot OEt_2$ afforded secondary alcohol **76** in 61% yield; that is, the reaction proceeded with equal efficiency as for the conversion of **36** into **34** (Scheme 15.11). The subsequent coupling of **76** with acid **33** under Yamaguchi conditions, as in all other cases discussed here, was uneventful and gave the desired ester in excellent yield. The latter was then transformed into aldehyde **80**, which set the stage for the critical macrocyclization reaction. It was unclear at this point if the lack of the THP ring in this modified cyclization precursor would lead to a loss of preorganization for macrocyclization; however, this concern proved to be unfounded. In the event, the $Ba(OH)_2$-promoted intramolecular HWE reaction proceeded smoothly and the desired macrolactone **81** was obtained in 85% yield. PMB removal from **81** under oxidative conditions (DDQ) followed by DMP oxidation then gave (−)-dactylolide analog **75** in 56% overall yield. Finally, aza-aldol reaction between **75** and amide **4** gave desTHP-zampanolide (**74**) in 28% yield as a *ca.* 1.6/1 mixture of isomers at the hemiaminal center after HPLC purification. Compared to **1** and *epi*-**1**, isomer separation was significantly more difficult and initial biological testing, thus, was performed with the mixture of isomers **74**. In the meantime, both isomers have also been obtained in pure form (M. Jordi, unpublished results).

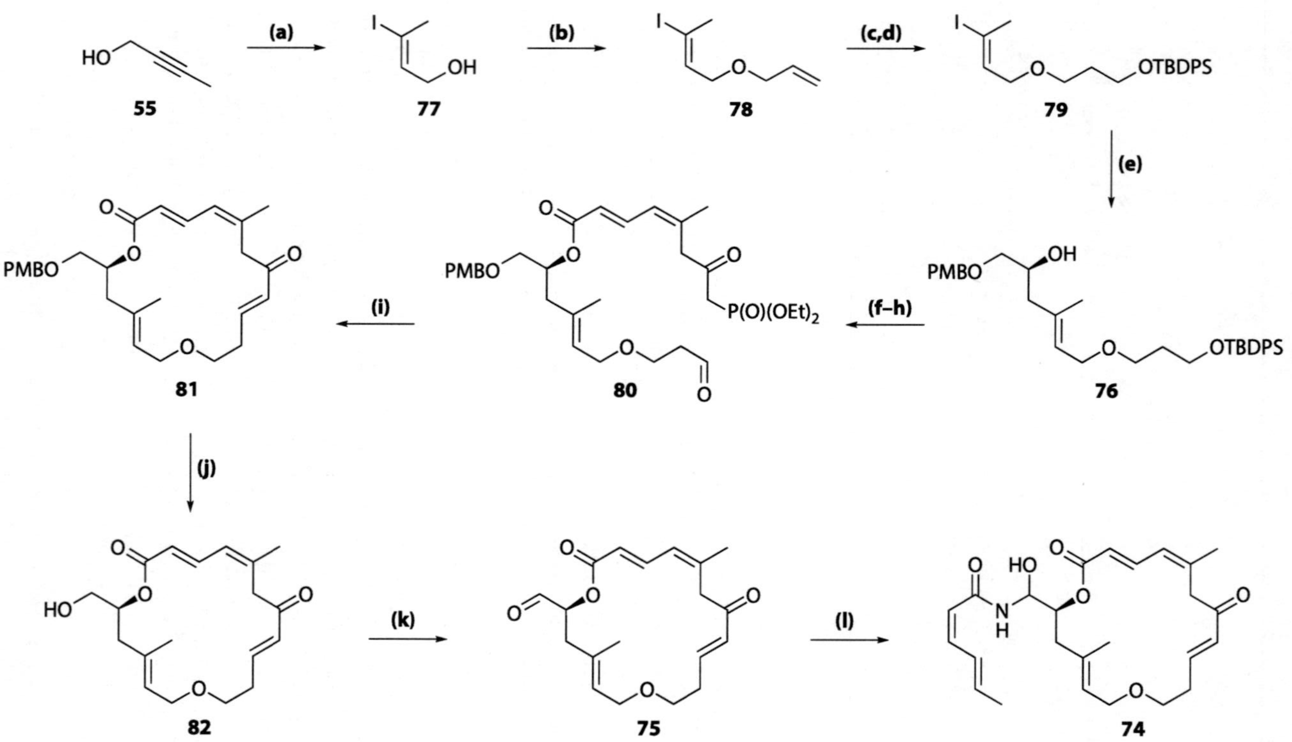

SCHEME 15.17 Reagents and conditions: **(a)** i. Bu_3SnH, *n*-BuLi, CuCN, THF, MeOH, −78°C → −15°C, 74% ii. I_2, THF, −17°C, 94%; **(b)** $CH_2{=}CHCH_2Br$, NaH, THF, 0°C, 42%; **(c)** $BH_3{\cdot}THF$, THF, 0°C, NaOH/H_2O_2, 0°C, 52%; **(d)** TBDPSCl, DMAP (cat.), NEt_3, CH_2Cl_2, RT, 94%; **(e)** *t*-BuLi, **35**, $BF_3 \cdot OEt_2$, toluene, −78°C, 61%; **(f)** 2,4,6-trichlorobenzoyl chloride, NEt_3, DMAP, **33**, toluene, RT, 81%; **(g)** HF·py, THF, 0°C → RT, 86%; **(h)** DMP, CH_2Cl_2, RT, 73%; **(i)** $Ba(OH)_2 \cdot 0.8H_2O$, THF/H_2O (40/1), 0°C → RT, 85%; **(j)** DDQ, CH_2Cl_2/H_2O (5/1), RT, 77%; **(k)** DMP, CH_2Cl_2, 75%; **(l)** **4**, DIBALH, then **75**, THF, RT, 72% for the mixture of epimers (*ca.* 1.6/1).

15.6 STRUCTURE–ACTIVITY RELATIONSHIPS

All of the dactylolide/zampanolide analogs discussed above were tested for their antiproliferative activity against a panel of four solid human tumor–derived cell lines (A549, MCF-7, HCT116, PC-3). Table 15.1 summarizes the IC_{50} values that were obtained against the colon tumor cell line HCT116, which are largely representative of the effects observed across the entire panel (for the whole dataset, see Reference 61). Importantly, our synthetic (−)-zampanolide (**1**) was found to inhibit human cancer cell proliferation with single-digit nanomolar IC_{50} values, which is in excellent agreement with the reported activity of natural[38,47] as well as synthetic[54] material against other cell lines. The C20*R* epimer of **1**, *epi*-**1**, was *ca.* one order of magnitude less active than the natural product **1** across all cell lines investigated; a similar activity difference had also been reported data by Uenishi and coworkers against SKM-1 and U937 cells.[54] The IC_{50} values obtained for (−)-dactylolide (*ent*-**2**) were even higher than those for *epi*-**1** and in the same range as those that had been reported for the compound by Ding and Jennings against a much broader panel of 19 cell lines. As indicated in Section 15.2, no biological data are available for **2** other than those reported together with the isolation work. It is, therefore, unclear, how the activities of **2** and *ent*-**2** really compare, even if data from Ding and Jennings suggest that the activity should be similar (based on the activity of synthetic *ent*-**2** against one of the cell lines also investigated in the isolation paper, *vide supra*).[53] Assuming this general conclusion to be justified (and also assuming that natural dactylolide is dextrorotatory), the hemiaminal-linked side chain of **1** is the major determinant of the activity difference between **1** and **2**, rather than the absolute configuration of the macrocycle.

TABLE 15.1
Antiproliferative Activity of 1, *epi*-1, *ent*-2, and Analogs of 1 and *ent*-2

Compound	IC_{50} (HCT116) (nM)[a]
1	7.2 ± 0.8
epi-**1**	88 ± 5
ent-**2**	210 ± 5
63	155 ± 2
64	1603 ± 122
65	12733 ± 379
66	1204 ± 63
72	74 ± 2
73	249 ± 28
82	1846 ± 92
75	2653 ± 68
74[b]	309 ± 47

[a] Cells were exposed to compounds for 72 h.
[b] Mixture of diastereoisomers at C18.

The IC_{50} value for alcohol **63** was comparable with that of *ent*-**2**, with a slight trend toward higher potency for the former. Independent of the true significance of this difference, the data clearly suggest that the aldehyde functionality in *ent*-**2** is not essential for its antiproliferative activity. However, conversion of **63** into its methyl ether **64** led to a clear (>10-fold) loss in antiproliferative potency. Oxidation of C20 to the carboxylic acid stage produced an even more substantial drop in activity, with carboxylic acid **65** being >50-fold less potent than *ent*-**2**. We have not investigated if this could perhaps be a consequence of poor cell penetration, due to the negatively charged carboxyl group. Interestingly, and encouragingly, the activity of amide **66** was *ca.* 10-fold enhanced over that of **65** and it seemed well conceivable that fine-tuning of the substituent moiety on the amide nitrogen could lead to improved potency. As a consequence, we have prepared a number of analogs of desTHP-zampanolide with amide-based side chains (M. Jordi, unpublished results) and the investigation of these compounds is currently ongoing.

As for the importance of the exocyclic methylene moiety on the THP ring, the data for 13-desmethylene *ent*-**2** (**73**) and the corresponding primary alcohol derivative **72** clearly indicate that this group is dispensable. We are currently investigating if the removal of the 13-methylene moiety is equally well tolerated for (−)-zampanolide (**1**). In agreement with our findings on *ent*-**2** and alcohol **63**, no meaningful activity difference was observed between aldehyde **73** and alcohol **72**. Like (−)-zampanolide, *ent*-**2** as well as analogs **63**, **72**, and **73**, all promote tubulin polymerization (41%–74% induction of tubulin polymerization relative to the effect of 25 μM of epothilone B [10 μM tubulin, 2 μM test compound] vs. 82% for epothilone A).[56] Details on the interactions of **63**, **72**, **73**, and different desTHP derivatives with the tubulin/microtubule system will be reported elsewhere. For (−)-zampanolide (**1**) and *ent*-**2**, their binding to dimeric tubulin and microtubules will be discussed in more detail below.

Arguably, the most intriguing finding that has emerged from our early SAR work on (−)-zampanolide (**1**) is the submicromolar activity of desTHP-zampanolide (**74**). Given the removal of two (out of four) chiral centers and of a rigidifying structural element, the significant extent to which **74** retained its antiproliferative activity relative to the parent natural product was truly surprising (even if the compound is 25- to 80-fold less active than **1**). As for **1** and *ent*-**2**, desTHP-(−)-dactylolide (**75**) is less active than desTHP-(−)-zampanolide (**74**). We also note that the data in Table 15.1 are for a 1.7/1 mixture of diastereoisomers at C18. In the meantime, we have also prepared pure C18*S*-**74** and C18*R*-**74** and found the former to be *ca.* sevenfold more potent against several leukemia cell lines than the latter (B. Pena and M. Jordi, unpublished data). We are currently investigating whether improvements in the activity of **74** are possible without any undue (re)increase in structural complexity.

15.7 MODE OF ACTION AND BINDING TO TUBULIN

As highlighted in Section 15.3 of this chapter, the potent antiproliferative activity of (−)-zampanolide (**1**) is a result of its stabilizing effect on cellular microtubules, as was first demonstrated in 2009 by Northcote, Miller, and coworkers.[47] Cellular microtubules are hollow filaments of *ca.* 240 Å outer diameter that are composed of α/β-tubulin heterodimers as the constituent subunits.[104] They form an essential part of the cytoskeleton and play a central role in cell division. Compounds that functionally interfere with the tubulin/microtubule system (*tubulin modulators*) are potent inhibitors of cancer cell proliferation and as such constitute an important group of chemotherapeutic agents for the treatment of cancer.[105] In general, tubulin modulators either prevent (noncovalent) tubulin assembly into microtubule polymers (as is the case for the clinically relevant vinca alkaloids vinblastine/vincristine or the halichondrin B derivative eribulin) or stabilize preexisting microtubules under otherwise nonstabilizing conditions.[106] Microtubule stabilizers also promote the assembly of tubulin heterodimers into microtubule polymers; that is, they shift the equilibrium between soluble and polymerized tubulin to the polymer side. Currently, four microtubule-stabilizing agents are employed in the clinical treatment of cancers, namely taxol (paclitaxel; Taxol®), the semisynthetic taxol analogs docetaxel (Taxotere®) and cabazitaxel (Jevtana®), and the semisynthetic epothilone derivative ixabepilone (Ixempra®).[107] In addition to taxol and epothilones, a number of other natural products have been described over the last 20 years to be microtubule-stabilizing agents, including several marine secondary metabolites.[108] The majority of these natural products bind to the taxol-binding site on β-tubulin, the exceptions being the marine macrolides peloruside A and laulimalide, which bind to an alternative site on β-tubulin.[109] Quite intriguingly, all potent microtubule-stabilizing agents identified to date, without exception, are natural products or derived from a natural product.

As for (−)-zampanolide (**1**), its exact mode of binding to tubulin or even its gross binding site was not determined at the time of discovery of its microtubule-stabilizing properties.[47] This was only the case in 2012, when Díaz and coworkers, in collaboration with the groups of Northcote/Miller and our own group and using natural as well as synthetic material, found the compound to bind to the taxol-binding site on β-tubulin.[62] Moreover, and of equal significance, they discovered that the compound bound to β-tubulin in an irreversible fashion, as preincubation of microtubules with **1** for a period as short as 30 min completely abolished its displacement by the fluorescent taxol derivative Flutax 2. Following the disappearance of available Flutax-binding sites, the reaction of **1** with the protein was determined to be very rapid (with an apparent first-order rate constant of 1 h^{-1}) and to occur essentially immediately after formation of the prereaction noncovalent complex (Figure 15.3).

Based on MS/MS analysis of tryptic digests of the zampanolide-modified tubulin, the reaction sites on the protein were identified as the Asn228 and His229 side chains, with the latter being the more important and acting as the major nucleophile. The site of electrophilic attack on the ligand could not be clearly elucidated at the time, although modeling suggested that nucleophilic attack was more likely to occur at C3 (i.e., on the enoate moiety) rather than at C9 (i.e., on the enone moiety). The reaction of **1** with unassembled tubulin heterodimers was also investigated and found to be complete within less than 4 h, producing the same signature in MS/MS experiments of tryptic protein digests as for the reaction of **1** with microtubules. These data strongly suggest that the binding of

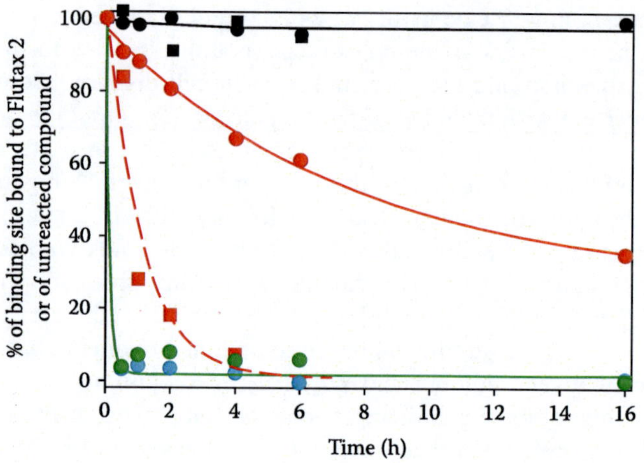

FIGURE 15.3 Kinetics of the reaction of **1** and *ent*-**2** with microtubules. Circles, remaining Flutax-2 binding sites when incubated with DMSO (black), natural **1** (cyan), synthetic **1** (green), and *ent*-**2** (red). Squares, unreacted *ent*-**2** when incubated with buffer (black) or stabilized binding sites in microtubules (red). (Reproduced from Field, J.J. et al., *Chem. Biol.*, 19, 686, 2012. With permission.)

(−)-zampanolide (**1**) to soluble tubulin heterodimers occurs at the same site and involves similar key interactions as for the polymerized state of the protein as part of a microtubule. Similar observations were made for (−)-dactylolide (*ent*-**2**), except that the rate of the reaction with the protein was significantly slower than for **1** (apparent first-order rate constant for the covalent modification of microtubules of 0.12 h^{-1}). Thus, in contrast to **1**, noncovalently bound *ent*-**2** is in fast exchange with the medium.

Employing different NMR techniques, the conformations of *ent*-**2** free in solution, bound to soluble tubulin heterodimers, or bound to microtubules were all found to be virtually identical (Figure 15.4).

No transfer NOE experiments were possible to determine the tubulin-bound conformation of (−)-zampanolide (**1**), as these require the bound ligand to be in reasonably rapid exchange with the surrounding solvent.

As part of their study on the interactions of **1** and *ent*-**2** with tubulin, Díaz and coworkers also confirmed that (−)-zampanolide (**1**) retains its antiproliferative activity against P-gp-overexpressing multidrug-resistant cells. They also suggest that the design of covalent tubulin binders should in fact be a viable strategy to overcome drug resistance.[62]

The question of the bioactive conformation of (−)-zampanolide (**1**) and its interactions with tubulin at the molecular level was resolved in a ground-breaking crystallographic study by Steinmetz and coworkers, again using material that had become available as a result of the synthetic work in our laboratory.[110] In this study, the structure of a complex incorporating $\alpha\beta$-tubulin (T), the stathmin-like protein RB3 (R), tubulin tyrosine ligase (TTL), and **1** was solved at 1.8 Å resolution, thus providing the very first experimentally based molecular picture of a microtubule-stabilizing

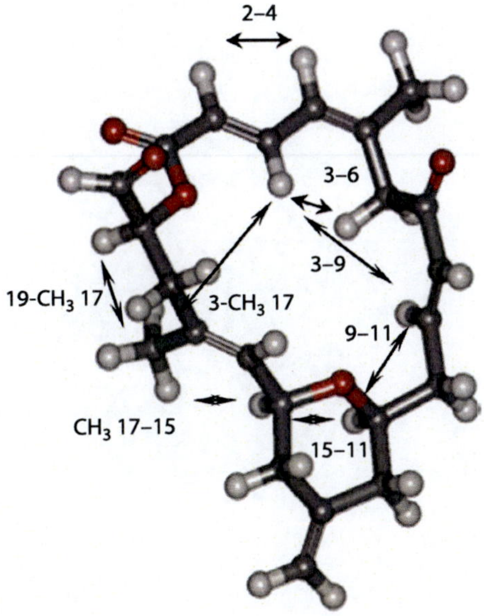

FIGURE 15.4 Microtubule-bound conformation of *ent*-**2**. Strong NOEs were observed for the following proton pairs: H2–H4, H3–H6, H11–H15, H15–C17 methyl, H19–C17 methyl. Medium-intensity NOEs were observed between H3 and H9 and between H9 and H11. A weak NOE was observed between H3 and the C17 methyl group. (Reproduced from Field, J.J. et al., *Chem. Biol.*, 19, 686, 2012. With permission.)

agent in complex with tubulin. In this structure, the C9 atom of the enone moiety of **1** was found to be covalently linked to the imidazole moiety of His229 (Figure 15.5); this is consistent with the biochemical data from Díaz and coworkers,[62] although they had suggested that attack on C3 of the enoate moiety would be the more likely reaction pathway.

One of the key features of the structure of the tubulin–zampanolide complex (and also of the tubulin–epothilone A complex that is described in the same publication) is the presence of a short helix involving residues Arg278 to Tyr 283 in the M-loop of β-tubulin (Figure 15.5). Importantly, this stretch of amino acids is largely disordered in the absence of (−)-zampanolide (**1**), while all other secondary structure elements of the binding pocket could be readily superimposed for the ligand-bound and the nonligated states of the protein. The induction of a helical conformation of the M-loop by the ligand is the result of different hydrophobic and polar interactions between the zampanolide side chain and M-loop residues. The helix is further stabilized by intramolecular hydrogen bonding interactions between residues of the M-loop and helix H9 of β-tubulin.

In the T_2R–TTL–zampanolide complex, the tubulin dimers are present in what is called a “curved” conformation, which is characteristic of unassembled, free tubulin,[111,112] as opposed to a “straight” conformation in microtubules.[113,114] However, a comparison of the overall architecture of the taxane (zampanolide)-binding site in the T_2R–TTL–zampanolide complex and of tubulin in a straight

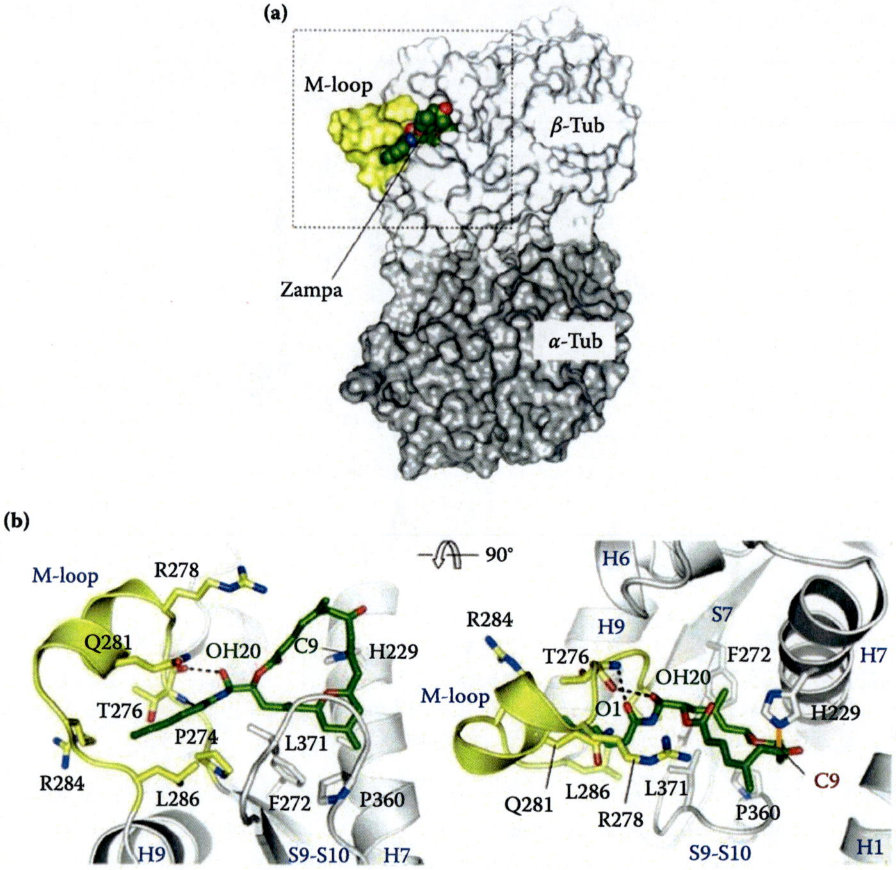

FIGURE 15.5 Structure of the tubulin-(−)-zampanolide (**1**) complex. **(a)** Overall view of the complex (tubulin: gray surface, M-loop in yellow; **1**: green spheres). The dashed box depicts the area shown in more detail in **(b)**. **(b)** Closeup views of the interaction network observed between **1** (green sticks) and *β*-tubulin (gray cartoon). Interacting residues of *β*-tubulin are shown in stick representation. Oxygen and nitrogen atoms are red and blue, respectively, and carbon atoms are green (**1**) or gray and yellow (*β*-tubulin). Hydrogen bonds are depicted as black dashed lines. The covalent bond between the C9 atom of **1** and the NE2 atom of His229 of *β*-tubulin is indicated by an orange stick. Single-letter abbreviations for the amino acid residues are as follows: A, Ala; C, Cys; D, Asp; E, Glu; F, Phe; G, Gly; H, His; I, Ile; K, Lys; L, Leu; M, Met; N, Asn; P, Pro; Q, Gln; R, Arg; S, Ser; T, Thr; V, Val; W, Trp; and Y, Tyr. (Reproduced from Prota, A.E. et al., *Science*, 339, 587, 2013. With permission.)

conformational state revealed only marginal differences.[11] This finding is in line with the fact that microtubule stabilizers, including (−)-zampanolide (**1**), can also bind to unassembled and/or oligomeric forms of tubulin, albeit with lower affinity than for microtubule polymers.[62] This affinity difference can be rationalized by the disordered nature of the M-loop in unassembled tubulin, which causes ligand binding to be energetically costly, due to conformational readjustments in the M-loop and rigidification. In contrast, the M-loop is structured in microtubules, even in the absence of a ligand, as this is linked to favorable lateral contacts between protofilaments, which results in microtubule stabilization. In this configuration, the energetic benefit of ligand binding is no longer counterbalanced by the need for intrinsically unfavorable structural changes. Within this landscape of structural changes and requirements, Prota et al.[110] have proposed that the helical structuring of the M-loop of β-tubulin by (−)-zampanolide (**1**) facilitates the curved-to-straight conformational change that occurs upon incorporation of tubulin into microtubules. In this model, the binding of a microtubule-stabilizing agent, such as (−)-zampanolide (**1**) or epothilone A, leads to tubulin preorganization according to the gross structural requirements of the assembly process, thus reducing the entropy loss associated with microtubule formation.

15.8 CONCLUSIONS

In this chapter, we have provided a detailed account of our work on the total synthesis of the marine macrolide (−)-zampanolide (**1**) and shown how this work has contributed to new discoveries in biology. The successful total synthesis of **1** has also provided a platform for the preparation and subsequent biological assessment of synthetic analogs that would not have been accessible by semisynthetic approaches. To put our own work into a broader context, we have also reviewed some of the basic aspects of the chemistry that has been developed by other groups in pursuit of the total synthesis of (−)-zampanolide (**1**) and (+)-dactylolide (**2**) and of the levorotatory, (presumably) nonnatural (−)-dactylolide (*ent*-**2**). These studies have led to methodological advances for the synthesis of polyunsaturated macrolides in general, and of those with an embedded THP ring in particular. A unique challenge in the total synthesis of **1** is the construction of the hemiaminal connector module that links the core macrocycle to the dienamide side chain; while reasonably practical solutions to this problem have been developed, no fully stereoselective, high-yielding approach toward this substructure has been elaborated to date.

Through highly productive collaborations with groups specializing in the biochemistry/biophysics/cell biology or the structural biology of tubulin and tubulin modulators, our synthetic work has enabled the elucidation of the covalent mode of action of (−)-zampanolide (**1**). Even more importantly, the availability of synthetic **1** was of major significance on the way to the first high-resolution crystal structure of a complex between tubulin and a microtubule-stabilizing agent. This structure has offered the first experimental insights into the *molecular* processes that underlie the stabilization of microtubules and the promotion of tubulin assembly by particular ligands. It has also enhanced our understanding of tubulin assembly and disassembly in general, as it constantly occurs in cells. We are of course aware that these studies, in principle, could have been conducted with natural (−)-zampanolide (**1**); however, the fact is that the larger part of the work described by Field et al.[62] and by Prota et al.[110] could not have been performed had it not been for the availability of synthetic material, simply because the available quantities of natural **1** were insufficient. To the best of our knowledge, there is no prospect for additional (−)-zampanolide (**1**) becoming available from a natural source any time soon. In contrast, total synthesis today represents a reliable means of material supply. Finally, our work on zampanolide analogs once again highlights the critical importance of *de novo* chemical synthesis to elucidate a natural product's SAR and to create simplified, yet potent analogs as new templates for optimization.

REFERENCES

1. Cragg, G. M.; Newman, D. J. Natural products: A continuing source of novel drug leads. *Biochim. Biophys. Acta Gen. Subj.* **2013**, *1830*, 3670–3695.
2. Newman, D. J.; Cragg, G. M. Natural products as sources of new drugs over the 30 years from 1981 to 2010. *J. Nat. Prod.* **2012**, *75*, 311–335.
3. Fusetani, N. **2009**. Marine toxins as research tools. In *Progress in Molecular and Subcellular Biology*, eds. N. Fusetani; W. Kem, pp. 1–44. Berlin, Germany: Springer.
4. Molinski, T. F.; Dalisay, D. S.; Lievens, S. L.; Saludes, J. P. Drug development from marine natural products. *Nat. Rev. Drug Discov.* **2009**, *8*, 69–85.
5. Montaser, R.; Luesch, H. Marine natural products: A new wave of drugs? *Future Med. Chem.* **2011**, *3*, 1475–1489.
6. Over, B.; Wetzel, S.; Grütter, C.; Nakai, Y.; Renner, S.; Rauh, D.; Waldmann, H. Natural-product-derived fragments for fragment-based ligand discovery. *Nat. Chem.* **2013**, *5*, 21–28.
7. Rizzo, S.; Waldmann, H. Development of a natural-product-derived chemical toolbox for modulation of protein function. *Chem. Rev.* **2014**, *114*, 4621–4639.
8. Paterson, I.; Anderson, E. A. The renaissance of natural products as drug candidates. *Science* **2005**, *310*, 451–452.
9. Vaske, Y. M.; Crews, P. **2014**. An update on the biomedical prospects of marine-derived small molecules with fascinating atom and stereochemical diversity. In *Bioactive Compounds from Marine Foods: Plant and Animal Sources*, eds. B. Herandez-Ledesma; M. Herrero, pp. 1–26. Chichester, U.K.: John Wiley & Sons, Ltd.
10. Coll, J. The chemistry and chemical ecology of octocorals (Coelenterata, Anthozoa, Octocorallia). *Chem. Rev.* **1992**, *92*, 613–631.
11. Flam, F. Chemical prospectors scour the seas for promising drugs. *Science* **1994**, *266*, 1324–1325.
12. Kelman, D.; Kashman, Y.; Hill, R. T.; Rosenberg, E.; Loya, Y. Chemical warfare in the sea: The search for antibiotics from Red Sea corals and sponges. *Pure Appl. Chem.* **2009**, *81*, 1113–1121.
13. Haefner, B. Drugs from the deep: Marine natural products as drug candidates. *Drug Discov. Today* **2003**, *8*, 536–544.
14. Blunt, J. W.; Copp, B. R.; Keyzers, R. A.; Munro, M. H. G.; Prinsep, M. R. Marine natural products. *Nat. Prod. Rep.* **2015**, *32*, 116–211.
15. Martin, G. E. Small-sample cryoprobe NMR applications. *eMagRes.* **2012**, *1*, 883–894.
16. Pean, E.; Klaar, S.; Berglund, E. G. et al. The European Medicines Agency review of eribulin for the treatment of patients with locally advanced or metastatic breast cancer: Summary of the scientific assessment of the committee for medicinal products for human use. *Clin. Cancer Res.* **2012**, *18*, 4491–4497.
17. Petitt, G. **1997**. The dolastatins. In *Fortschritte der Chemie organischer Naturstoffe [Progress in the Chemistry of Organic Natural Products]*, eds. W. Herz; G. W. Kirby; R. E. Moore; W. Steglich; Ch. Tamm, pp. 1–79. Wien, Austria: Springer.
18. Pham, A.; Chen, R. Brentuximab vedotin for the treatment of Hodgkin's lymphoma. *Expert Rev. Hematol.* **2015**, *8*, 403–412.
19. Schmidtko, A.; Loetsch, J.; Freynhagen, R.; Geisslinger, G. Ziconotide for treatment of severe chronic pain. *Lancet* **2010**, *375*, 1569–1577.

20. Butler, M. S.; Robertson, A. A. B.; Cooper, M. A. Natural product and natural product derived drugs in clinical trials. *Nat. Prod. Rep.* **2014**, *31*, 1612–1661.
21. Newman, D. J.; Cragg, G. M. Marine-sourced anti-cancer and cancer pain control agents in clinical and late preclinical development. *Mar. Drugs* **2014**, *12*, 255–278.
22. Rubiolo, J.; Alonso, E.; Cagide, E. **2014**. Marine compounds as a starting point to drugs. In *Seafood and Freshwater Toxins*, ed. L. M. Botana, pp. 1141–1178. Boca Raton, FL: CRC Press.
23. Newman, D. J.; Cragg, G. M. Drugs and drug candidates from marine sources: An assessment of the current "state of play". *Planta Med.* **2016**, *82*, 775–789.
24. Cuevas, C.; Francesch, A. Development of Yondelis (trabectedin, ET-743). A semisynthetic process solves the supply problem. *Nat. Prod. Rep.* **2009**, *26*, 322–337.
25. Yu, M. J.; Zheng, W.; Seletsky, B. M.; Littlefield, B. A.; Kishi, Y. Case history: Discovery of Eribulin (HALAVEN™), a Halichondrin B analogue that prolongs overall survival in patients with metastatic breast cancer. *Annu. Rep. Med. Chem.* **2011**, *46*, 227–241.
26. Yeung, K.-S.; Paterson, I. Advances in the total synthesis of biologically important marine macrolides. *Chem. Rev.* **2005**, *105*, 4237–4313.
27. Mickel, S. J. Total synthesis of the marine natural product (+)-discodermolide in multigram quantities. *Pure Appl. Chem.* **2007**, *79*, 685–700.
28. Klar, U.; Platzek, J. Asymmetric total synthesis of the epothilone sagopilone: From research to development. *Synlett* **2012**, *23*, 1291–1299.
29. Maier, M. E. Structural revisions of natural products by total synthesis. *Nat. Prod. Rep.* **2009**, *26*, 1105–1124.
30. Usami, Y. Recent synthetic studies leading to structural revisions of marine natural products. *Mar. Drugs* **2009**, *7*, 314–330.
31. Suyama, T. L.; Gerwick, W. H.; McPhail, K. L. Survey of marine natural product structure revisions: A synergy of spectroscopy and chemical synthesis. *Bioorg. Med. Chem.* **2011**, *19*, 6675–6701.
32. Yoo, H.-D.; Nam, S.-J.; Chin, Y.-W.; Kim, M.-S. Misassigned natural products and their revised structures. *Arch. Pharm. Res.* **2016**, *39*, 143–153.
33. Shaw, S. J. The structure activity relationship of discodermolide analogues. *Mini-Rev. Med. Chem.* **2008**, *8*, 276–284.
34. Feyen, F.; Cachoux, F.; Gertsch, J.; Wartmann, M.; Altmann, K.-H. Epothilones as lead structures for the synthesis-based discovery of new chemotypes for microtubule stabilization. *Acc. Chem. Res.* **2008**, *41*, 21–31.
35. Smith, A. B., III; Risatti, C. A.; Atasoylu, O.; Bennett, C. S.; Liu, J.; Cheng, H.; TenDyke, K.; Xu, Q. Design, synthesis, and biological evaluation of diminutive forms of (+)-spongistatin 1, lessons learned. *J. Am. Chem. Soc.* **2011**, *133*, 14042–14053.
36. Wach, J.-Y.; Gademann, K. Reduce to the maximum: Truncated natural products as powerful modulators of biological processes. *Synlett* **2012**, *23*, 163–170.
37. Ulanovskaya, O. A.; Janjic, J.; Suzuki, M.; Sabharwal, S. S.; Schumacker, P. T.; Kron, S. J.; Kozmin, S. A. Synthesis enables identification of the cellular target of leucascandrolide A and neopeltolide. *Nat. Chem. Biol.* **2008**, *4*, 418–424.
38. Tanaka, J.-I.; Higa, T. Zampanolide, a new cytotoxic macrolide from a marine sponge. *Tetrahedron Lett.* **1996**, *37*, 5535–5538.
39. Smith, A. B., III; Safonov, I. G.; Corbett, R. M. Total synthesis of (+)-zampanolide. *J. Am. Chem. Soc.* **2001**, *123*, 12426–12427.
40. Smith, A. B., III; Safonov, I. G.; Corbett, R. M. Total syntheses of (+)-zampanolide and (+)-dactylolide exploiting a unified strategy. *J. Am. Chem. Soc.* **2002**, *124*, 11102–11113.
41. Traber, R.; Keller-Juslén, C.; Loosli, H.-R.; Kuhn, M.; Von Wartburg, A. Cyclopeptid-Antibiotika aus Aspergillus-Arten. Struktur der Echinocandine C und D. *Helv. Chim. Acta* **1979**, *62*, 1252–1267.

42. Perry, N. B.; Blunt, J. W.; Munro, M. H. G.; Pannell, L. K. Mycalamide A, an antiviral compound from a New Zealand sponge of the genus *Mycale*. *J. Am. Chem. Soc.* **1988**, *110*, 4850–4851.
43. Perry, N. B.; Blunt, J. W.; Munro, M. H. G.; Thompson, A. M. Antiviral and antitumor agents from a New Zealand sponge, *Mycale* sp. 2. Structures and solution conformations of mycalamides A and B. *J. Org. Chem.* **1990**, *55*, 223–227.
44. Takeuchi, T.; Iinuma, H.; Kunimoto, S.; Masadu, T.; Ishizuka, M.; Takeuchi, M.; Hamada, M.; Naganawa, H.; Kondo, S.; Umezawa, H. A new antitumor antibiotic, spergualin: Isolation and antitumor activity. *J. Antibiot.* **1981**, *34*, 1619–1621.
45. Umezawa, U.; Kondo, S.; Iinuma, H.; Kunimoto, S.; Ikeda, Y.; Iwasawa, H.; Ikeda, D.; Takeuchi, T. Structure of an antitumor antibiotic, spergualin. *J. Antibiot.* **1981**, *34*, 1622–1624.
46. Jiménez, J. I.; Goetz, G.; Mau, C. M.; Yoshida, W. Y.; Scheuer, P. J.; Williamson, R. T.; Kelly, M. Upenamide: An unprecedented macrocyclic alkaloid from the Indonesian sponge *Echinochalina* sp. *J. Org. Chem.* **2000**, *65*, 8465–8469.
47. Field, J. J.; Singh, A. J.; Kanakkanthara, A.; Halafihi, T.; Northcote, P. T.; Miller, J. H. Microtubule-stabilizing activity of zampanolide, a potent macrolide isolated from the Tongan marine sponge *Cacospongia mycofijiensis*. *J. Med. Chem.* **2009**, *52*, 7328–7332.
48. Cutignano, A.; Bruno, I.; Bifulco, G.; Casapullo, A.; Debitus, C.; Gomez-Paloma, L.; Riccio, R. Dactylolide, a new cytotoxic macrolide from the Vanuatu sponge *Dactylospongia* sp. *Eur. J. Org. Chem.* **2001**, *2001*, 775–778.
49. Sanchez, C. C.; Keck, G. E. Total synthesis of (+)-dactylolide. *Org. Lett.* **2005**, *7*, 3053–3056.
50. Aubele, D. L.; Wan, S.; Floreancig, P. E. Total synthesis of (+)-dactylolide through an efficient sequential Peterson olefination and Prins cyclization reaction. *Angew. Chem. Int. Ed.* **2005**, *44*, 3485–3488.
51. Hoye, T. R.; Hu, M. Macrolactonization via Ti(IV)-mediated epoxy-acid coupling: A total synthesis of (–)-dactylolide [and zampanolide]. *J. Am. Chem. Soc.* **2003**, *125*, 9576–9577.
52. Louis, I.; Hungerford, N. L.; Humphries, E. J.; McLeod, M. D. Enantioselective total synthesis of (–)-dactylolide. *Org. Lett.* **2006**, *8*, 1117–1120.
53. Ding, F.; Jennings, M. P. Total synthesis of (–)-dactylolide and formal synthesis of (–)-zampanolide via target oriented β-C-glycoside formation. *J. Org. Chem.* **2008**, *73*, 5965–5976.
54. Uenishi, J.; Iwamoto, T.; Tanaka, J. Total synthesis of (–)-zampanolide and questionable existence of (–)-dactylolide as the elusive biosynthetic precursor of (–)-zampanolide in an okinawan sponge. *Org. Lett.* **2009**, *11*, 3262–3265.
55. Yun, S. Y.; Hansen, E. C.; Volchkov, I.; Cho, E. J.; Lo, W. Y.; Lee, D. Total synthesis of (–)-dactylolide. *Angew. Chem. Int. Ed.* **2010**, *49*, 4261–4263.
56. Zurwerra, D.; Gertsch, J.; Altmann, K.-H. Synthesis of (–)-dactylolide and 13-desmethylene-(–)-dactylolide and their effects on tubulin. *Org. Lett.* **2010**, *12*, 2302–2305.
57. Chen, Q.-H.; Kingston, D. G. I. Zampanolide and dactylolide: Cytotoxic tubulin assembly agents and promising anticancer leads. *Nat. Prod. Rep.* **2014**, *31*, 1202–1226.
58. Smith, A. B., III; Safonov, I. G. Total synthesis of (+)-dactylolide. *Org. Lett.* **2002**, *4*, 635–637.
59. Troast, D. M.; Porco, J. A., Jr. Studies toward the synthesis of (–)-zampanolide: Preparation of *N*-acyl hemiaminal model systems. *Org. Lett.* **2002**, *4*, 991–994.
60. Pera, B.; Nieves Calvo-Vidal, M.; Ambati, S.; Jordi, M.; Kahn, A.; Díaz, J. F.; Fang, W.; Altmann, K.-H.; Cerchietti, L.; Moore, M. A.. High affinity and covalent binding microtubule stabilizing agents show activity in chemotherapy-resistant acute myeloid leukemia cells. *Cancer Lett.* **2015**, *368*, 97–104.

61. Zurwerra, D.; Glaus, F.; Betschart, L.; Schuster, J.; Gertsch, J.; Ganci, W.; Altmann, K.-H.. Total synthesis of (–)-zampanolide and structure-activity relationship studies on (–)-dactylolide derivatives. *Chem. Eur. J.* **2012**, *18*, 16868–16883.
62. Field, J. J.; Pera, B.; Calvo, E. et al. Zampanolide, a potent new microtubule-stabilizing agent, covalently reacts with the taxane luminal site in tubulin α,β-heterodimers and microtubules. *Chem. Biol.* **2012**, *19*, 686–698.
63. Ghosh, A.; Cheng, X. Enantioselective total synthesis of (–)-zampanolide, a potent microtubule-stabilizing agent. *Org. Lett.* **2011**, *13*, 4108–4111.
64. Ghosh, A. K.; Cheng, X.; Bai, R.; Hamel, E. Total synthesis of the potent antitumor macrolide (–)-zampanolide: An oxidative intramolecular cyclization-based strategy. *Eur. J. Org. Chem.* **2012**, *2012*, 4130–4139.
65. Nicolaou, K. C.; Seitz, S. P.; Pavia, M. R. Carbohydrates in organic synthesis. Synthesis of 16-membered-ring macrolide antibiotics. 6. Total synthesis of *O*-mycinosyltylonolide: Coupling of key intermediates and macrocyclization. *J. Am. Chem. Soc.* **1982**, *104*, 2030–2031.
66. Kadota, I.; Hu, Y.; Packard, G. K.; Rychnovsky, S. D. A unified approach to polyene macrolides: Synthesis of candidin and nystatin polyols. *Proc. Natl. Acad. Sci. USA* **2004**, *101*, 11192–11195.
67. Berger, G. O.; Tius, M. O. Total synthesis of (±)-terpestacin and (±)-11-epi-terpestacin. *J. Org. Chem.* **2007**, *72*, 6473–6480.
68. Frick, J. A.; Klassen, J. B.; Bathe, A.; Abramson, J. M.; Rapoport, H. An efficient synthesis of enantiomerically pure (*R*)-(2-benzyloxyethyl)oxirane from (*S*)-aspartic acid. *Synthesis* **1992**, *92*, 621–623.
69. Clive, D. L. J.; Murthy, K. S. K.; Wee, A. G. H.; Prasad, J. H.; Da Silva, G. V. J.; Majewski, M.; Anderson, P. C.; Evans, C. F.; Haugen, R. D. Total synthesis of both (+)-compactin and (+)-mevinolin. A general strategy based on the use of a special titanium reagent for dicarbonyl coupling. *J. Am. Chem. Soc.* **1990**, *112*, 3018–3028.
70. Trost, B. M.; Rudd, M. T. Ruthenium-catalyzed cycloisomerizations of diynols. *J. Am. Chem. Soc.* **2005**, *127*, 4763–4776.
71. Jasti, R.; Vitale, J.; Rychnovsky, S. D. Axial-selective Prins cyclizations by solvolysis of α-bromo ethers. *J. Am. Chem. Soc.* **2004**, *126*, 9904–9905.
72. Liang, X.; Lohse, A.; Bols, M. Chemoenzymatic synthesis of isogalactofagomine. *J. Org. Chem.* **2000**, *65*, 7432–7437.
73. Kocovsky, P. J. A stereospecific, silver(I)-assisted solvolysis of cyclic halo ethers. Evidence for a push-pull mechanism involving neighboring group participation. *J. Org. Chem.* **1988**, *53*, 5816–5819.
74. Macdonald, T. L.; Narasimhan, N. Nucleophilic substitution of alkyl iodides via oxidative ligand transfer. *J. Org. Chem.* **1985**, *50*, 5000–5001.
75. Betzer, J. F.; Delaloge, F.; Muller, B.; Pancrazi, A.; Prunet, J. Radical hydrostannylation, Pd(0)-catalyzed hydrostannylation, stannylcupration of propargyl alcohols and enynols: Regio- and stereoselectivities. *J. Org. Chem.* **1997**, *62*, 7768–7780.
76. Betzer, J.-F.; Ardisson, J.; Lallemand, J.-Y.; Pancrazi, A. An efficient method in stannylcupration of a methyl substituted enyne or alkyne by kinetic control using methanol. *Tetrahedron Lett.* **1997**, *38*, 2279–2282.
77. Schwartz, J.; Labinger, J. A. Hydrozirconation: A new transition metal reagent for organic synthesis. *Angew. Chem. Int. Ed.* **1976**, *15*, 333–340.
78. Okukado, N.; Negishi, E.-I. One-step conversion of terminal acetylenes into terminally functionalized (*E*)-3-methyl-2-alkenes via zirconium-catalyzed carboalumination. A simple and selective route to terpenoids. *Tetrahedron Lett.* **1978**, *19*, 2357–2360.
79. Nicolaou, K. C.; Nold, A. L.; Milburn, R. R.; Schindler, C. S. Total synthesis of marinomycins A–C. *Angew. Chem. Int. Ed.* **2006**, *45*, 6527.

80. Saito, S.; Ishikawa, T.; Kuroda, A.; Koga, K.; Moriwake, T. A revised mechanism for chemoselective reduction of esters with borane-dimethyl sulfide complex and catalytic sodium tetrahydroborate directed by adjacent hydroxyl group. *Tetrahedron* **1992**, *48*, 4067–4086.
81. Tararov, V. I.; König, G.; Börner, A. Synthesis and highly stereoselective hydrogenation of the statin precursor ethyl (5*S*)-5,6-isopropylidenedioxy-3-oxohexanoate. *Adv. Synth. Catal.* **2006**, *348*, 2633–2644.
82. Corey, E. J.; Fuchs, P. L. A synthetic method for formyl → ethynyl conversion (RCHO → RC≡CH or RC≡CR′). *Tetrahedron Lett.* **1972**, *13*, 3769–3772.
83. Nicolaou, K. C.; Nold, A. L.; Milburn, R. R.; Schindler, C. S.; Cole, K. P.; Yamaguchi, J. Total synthesis of marinomycins A–C and of their monomeric counterparts monomarinomycin A and iso-monomarinomycin A. *J. Am. Chem. Soc.* **2007**, *129*, 1760–1768.
84. Ganem, B.; Dong, Y.; Zheng, Y. F.; Prestwich, G. D. Amidrazone and amidoxime inhibitors of squalene hopene cyclase. *J. Org. Chem.* **1999**, *64*, 5441–5446.
85. Racherla, U. S.; Brown, H. C. Chiral synthesis via organoboranes. 27. Remarkably rapid and exceptionally enantioselective (approaching 100% ee) allylboration of representative aldehydes at –100° under new, salt-free conditions. *J. Org. Chem.* **1991**, *56*, 401–404.
86. Brown, H. C.; Racherla, U. S.; Liao, Y.; Khanna, V. V. Chiral synthesis via organoboranes. 35. Simple procedures for the efficient recycling of the terpenyl chiral auxiliaries and convenient isolation of the homoallylic alcohols in asymmetric allyl- and crotylboration of aldehydes. *J. Org. Chem.* **1992**, *57*, 6608–6614.
87. Keck, G. E.; Tarbet, K. H.; Geraci, L. S. Catalytic asymmetric allylation of aldehydes. *J. Am. Chem. Soc.* **1993**, *115*, 8467–8468.
88. Keck, G. E.; Welch, D. S.; Vivian, P. K. Synthetic studies toward the bryostatins: A substrate-controlled approach to the A-ring. *Org. Lett.* **2006**, *8*, 3667–3670.
89. Hafner, A.; Duthaler, R. O.; Marti, R.; Rihs, G.; Rothe-Streit, P.; Schwarzenbach, F. Enantioselective syntheses with titanium carbohydrate complexes. Part 7. Enantioselective allyltitanation of aldehydes with cyclopentadienyldialkoxyallyltitanium complexes. *J. Am. Chem. Soc.* **1992**, *114*, 2321–2336.
90. Crosby, R.; Harding, J. R.; King, C. D.; Parker, G. D.; Willis, C. L. Oxonia-cope rearrangement and side-chain exchange in the Prins cyclization. *Org. Lett.* **2002**, *4*, 577–580.
91. Kende, A. S.; Liu, K.; Kaldor, I.; Dorey, G.; Koch, K. Total synthesis of the macrolide antitumor antibiotic Lankacidin C. *J. Am. Chem. Soc.* **1995**, *117*, 8258–8270.
92. Miyashita, K.; Ikejiri, M.; Kawasaki, H.; Maemura, S.; Imanishi, T. Total synthesis of an antitumor antibiotic, fostriecin (CI-920). *J. Am. Chem. Soc.* **2003**, *125*, 8238–8243.
93. Corey, E. J.; Katzenellenbogen, J. A.; Posner, G. H. New stereospecific synthesis of trisubstituted olefins. Stereospecific synthesis of farnesol. *J. Am. Chem. Soc.* **1967**, *89*, 4245–4247.
94. Chau, A.; Paquin, J.-F.; Lautens, M. Diastereoselective palladium-catalyzed formate reduction of allylic carbonates en route to polypropionate systems. *J. Org. Chem.* **2006**, *71*, 1924–1933.
95. Okano, K.; Tokuyama, H.; Fukuyama, T. Total synthesis of (+)-yatakemycin. *J. Am. Chem. Soc.* **2006**, *128*, 7136–7137.
96. Hsin, L.-W.; Dersch, C. M.; Baumann, M. H.; Stafford, D.; Glowa, J. R.; Rothman, R. B.; Jacobson, A. E.; Rice, K. C. Development of long-acting dopamine transporter ligands as potential cocaine-abuse therapeutic agents: Chiral hydroxyl-containing derivatives of 1-[2-[Bis(4-fluorophenyl)methoxy]ethyl]-4-(3-phenylpropyl)piperazine and 1-[2-(Diphenylmethoxy)ethyl]-4-(3-phenylpropyl)piperazine. *J. Med. Chem.* **2002**, *45*, 1321–1329.
97. Barlow, A. J.; Compton, B. J.; Weavers, R. T. Synthesis of the 3-methylene-2-vinyltetrahydropyran unit; the hallmark of the sesquiterpene, hodgsonox. *J. Org. Chem.* **2005**, *70*, 2470–2475.

98. Barrios, J.; Marinas, J. M.; Sinisterra, J. V. Ba(OH)2 as the catalyst in organic reactions. Part IV. Influence of catalyst structure and basicity in the alcohol condensation of acetone. *Bull. Soc. Chim. Belg.* **1986**, *95*, 107–117.
99. Paterson, I.; Yeung, K.-S.; Watson, C.; Ward, R. A.; Wallace, P. A. The total synthesis of scytophycin C. Part 1: Stereocontrolled synthesis of the C1–C32 protected seco acid. *Tetrahedron* **1998**, *54*, 11935–11954.
100. Bayer, A.; Maier, M. E. Synthesis of enamides from aldehydes and amides. *Tetrahedron* **2004**, *60*, 6665–6677.
101. Lindgren, B. O.; Nilsson, H. Preparation of carboxylic acids from aldehydes (including hydroxylated benzaldehydes) by oxidation with chlorite. *Acta Chem. Scand.* **1973**, *27*, 888–890.
102. Kraus, G. A.; Roth, B.; Synthetic studies toward verrucarol. 2. Synthesis of the AB ring system. *J. Org. Chem.* **1980**, *45*, 4825–4830.
103. Bal, B. S.; Childers, W. E.; Pinnick, H. W. Oxidation of α,β-unsaturated aldehydes. *Tetrahedron* **1981**, *37*, 2091–2096.
104. Amos, L. A.; Schlieper, D. Microtubules and MAPs. *Adv. Protein Chem.* **2005**, *71*, 257–298.
105. Dumontet, C.; Jordan, M. A. Microtubule-binding agents: A dynamic field of cancer therapeutics. *Nat. Rev. Drug Discov.* **2010**, *9*, 790–803.
106. Jordan, M. A.; Wilson, L. Microtubules as a target for anticancer drugs. *Nat. Rev. Cancer* **2004**, *4*, 253–265.
107. Chen, S.-M.; Meng, L.-H.; Ding, J. New microtubule-inhibiting anticancer agents. *Expert Opin. Investig. Drugs* **2010**, *19*, 329–343.
108. Altmann, K.-H.; Gertsch, J. Anticancer drugs from nature-natural products as a unique source of new microtubule-stabilizing agents. *Nat. Prod. Rep.* **2007**, *24*, 327–357.
109. Prota, A. E.; Bargsten, K.; Northcote, P. T.; Marsh, N.; Altmann, K.-H.; Miller, J. H.; Díaz, J. F.; Steinmetz, M. O.. Structural basis of microtubule stabilization by laulimalide and peloruside A. *Angew. Chem. Int. Ed.* **2014**, *53*, 1621–1625.
110. Prota, A. E.; Bargsten, K.; Zurwerra, D.; Field, J. J.; Díaz, J. F.; Altmann, K.-H.; Steinmetz, M. O.. Molecular mechanism of action of microtubule-stabilizing anticancer agents. *Science* **2013**, *339*, 587–590.
111. Ayaz, P.; Ye, X.; Huddleston, P.; Brautigam, C. A.; Rice, L. M. A TOG:$\alpha\beta$-tubulin complex structure reveals conformation-based mechanisms for a microtubule polymerase. *Science* **2012**, *337*, 857–860.
112. Pecqueur, L.; Duellberg, C.; Dreier, B. A designed ankyrin repeat protein selected to bind to tubulin caps the microtubule plus end. *Proc. Natl. Acad. Sci. USA* **2012**, *109*, 12011–12016.
113. Löwe, J.; Li, H.; Downing, K. H.; Nogales, E. Refined structure of $\alpha\beta$-tubulin at 3.5 Å resolution. *J. Mol. Biol.* **2001**, *313*, 1045–1057.
114. Nogales, E.; Wolf, S. G.; Downing, K. H. Structure of the $\alpha\beta$ tubulin dimer by electron crystallography. *Nature* **1998**, *391*, 199–203.

Index

A

C

D

F

G

H

I

N

U

V

W

X

Z